# FOOD ENGINEERING HANDBOOK

# Contemporary Food Engineering

**Series Editor**

Professor Da-Wen Sun, Director
*Food Refrigeration & Computerized Food Technology*
*National University of Ireland, Dublin*
*(University College Dublin)*
*Dublin, Ireland*
*http://www.ucd.ie/sun/*

Handbook of Food Processing and Engineering, Volume II: Food Process Engineering, *edited by Theodoros Varzakas and Constantina Tzia* (2014)
Handbook of Food Processing and Engineering, Volume I: Food Engineering Fundamentals, *edited by Theodoros Varzakas and Constantina Tzia* (2014)
Juice Processing: Quality, Safety and Value-Added Opportunities, *edited by Víctor Falguera and Albert Ibarz* (2014)
Engineering Aspects of Food Biotechnology, *edited by José A. Teixeira and António A. Vicente* (2013)
Engineering Aspects of Cereal and Cereal-Based Products, *edited by Raquel de Pinho Ferreira Guiné and Paula Maria dos Reis Correia* (2013)
Fermentation Processes Engineering in the Food Industry, *edited by Carlos Ricardo Soccol, Ashok Pandey, and Christian Larroche* (2013)
Modified Atmosphere and Active Packaging Technologies, *edited by Ioannis Arvanitoyannis* (2012)
Advances in Fruit Processing Technologies, *edited by Sueli Rodrigues and Fabiano Andre Narciso Fernandes* (2012)
Biopolymer Engineering in Food Processing, *edited by Vânia Regina Nicoletti Telis* (2012)
Operations in Food Refrigeration, *edited by Rodolfo H. Mascheroni* (2012)
Thermal Food Processing: New Technologies and Quality Issues, Second Edition, *edited by Da-Wen Sun* (2012)
Physical Properties of Foods: Novel Measurement Techniques and Applications, *edited by Ignacio Arana* (2012)
Handbook of Frozen Food Processing and Packaging, Second Edition, *edited by Da-Wen Sun* (2011)
Advances in Food Extrusion Technology, *edited by Medeni Maskan and Aylin Altan* (2011)
Enhancing Extraction Processes in the Food Industry, *edited by Nikolai Lebovka, Eugene Vorobiev, and Farid Chemat* (2011)
Emerging Technologies for Food Quality and Food Safety Evaluation, *edited by Yong-Jin Cho and Sukwon Kang* (2011)
Food Process Engineering Operations, *edited by George D. Saravacos and Zacharias B. Maroulis* (2011)
Biosensors in Food Processing, Safety, and Quality Control, *edited by Mehmet Mutlu* (2011)

Physicochemical Aspects of Food Engineering and Processing, *edited by Sakamon Devahastin* (2010)

Infrared Heating for Food and Agricultural Processing, *edited by Zhongli Pan and Griffiths Gregory Atungulu* (2010)

Mathematical Modeling of Food Processing, *edited by Mohammed M. Farid* (2009)

Engineering Aspects of Milk and Dairy Products, *edited by Jane Sélia dos Reis Coimbra and José A. Teixeira* (2009)

Innovation in Food Engineering: New Techniques and Products, *edited by Maria Laura Passos and Claudio P. Ribeiro* (2009)

Processing Effects on Safety and Quality of Foods, *edited by Enrique Ortega-Rivas* (2009)

Engineering Aspects of Thermal Food Processing, *edited by Ricardo Simpson* (2009)

Ultraviolet Light in Food Technology: Principles and Applications, *Tatiana N. Koutchma, Larry J. Forney, and Carmen I. Moraru* (2009)

Advances in Deep-Fat Frying of Foods, *edited by Serpil Sahin and Servet Gülüm Sumnu* (2009)

Extracting Bioactive Compounds for Food Products: Theory and Applications, *edited by M. Angela A. Meireles* (2009)

Advances in Food Dehydration, *edited by Cristina Ratti* (2009)

Optimization in Food Engineering, *edited by Ferruh Erdoğdu* (2009)

Optical Monitoring of Fresh and Processed Agricultural Crops, *edited by Manuela Zude* (2009)

Food Engineering Aspects of Baking Sweet Goods, *edited by Servet Gülüm Sumnu and Serpil Sahin* (2008)

Computational Fluid Dynamics in Food Processing, *edited by Da-Wen Sun* (2007)

# FOOD ENGINEERING HANDBOOK

## FOOD PROCESS ENGINEERING

EDITED BY
**Theodoros Varzakas**
**Constantina Tzia**

CRC Press
Taylor & Francis Group
Boca Raton London New York

CRC Press is an imprint of the
Taylor & Francis Group, an **informa** business

CRC Press
Taylor & Francis Group
6000 Broken Sound Parkway NW, Suite 300
Boca Raton, FL 33487-2742



Printed on acid-free paper
Version Date: 20141015

International Standard Book Number-13: 978-1-4822-6166-0 (Hardback)



**Visit the Taylor & Francis Web site at**
**http://www.taylorandfrancis.com**

**and the CRC Press Web site at**
**http://www.crcpress.com**

*Dedicated to my wife Elia and my daughter Fotini for their endless support and love. To my mother for her love and understanding and to the memory of my father.*

**Theodoros Varzakas**

*Dedicated to the memory of my parents.*

**Constantina Tzia**

# Contents

# Series Preface

## CONTEMPORARY FOOD ENGINEERING

Food engineering is the multidisciplinary field of applied physical sciences combined with the knowledge of product properties. Food engineers provide the technological knowledge transfer essential to the cost-effective production and commercialization of food products and services. In particular, food engineers develop and design processes and equipment to convert raw agricultural materials and ingredients into safe, convenient, and nutritious consumer food products. However, food engineering topics are continuously undergoing changes to meet diverse consumer demands, and the subject is being rapidly developed to reflect market needs.

In the development of food engineering, one of the many challenges is to employ modern tools and knowledge, such as computational materials science and nanotechnology, to develop new products and processes. Simultaneously, improving food quality, safety, and security continues to be a critical issue in food engineering studies. New packaging materials and techniques are being developed to provide more protection to foods, and novel preservation technologies are emerging to enhance food security and defense. Additionally, process control and automation regularly appear among the top priorities identified in food engineering. Advanced monitoring and control systems are developed to facilitate automation and flexible food manufacturing. Furthermore, energy saving and minimization of environmental problems continue to be important food engineering issues, and significant progress is being made in waste management, efficient utilization of energy, and reduction of effluents and emissions in food production.

The Contemporary Food Engineering Series, consisting of edited books, attempts to address some of the recent developments in food engineering. The series covers advances in classical unit operations in engineering applied to food manufacturing as well as topics such as progress in the transport and storage of liquid and solid foods; heating, chilling, and freezing of foods; mass transfer in foods; chemical and biochemical aspects of food engineering and the use of kinetic analysis; dehydration, thermal processing, nonthermal processing, extrusion, liquid food concentration, membrane processes, and applications of membranes in food processing; shelf-life and electronic indicators in inventory management; sustainable technologies in food processing; and packaging, cleaning, and sanitation. These books are aimed at professional food scientists, academics researching food engineering problems, and graduate-level students.

The editors of these books are leading engineers and scientists from different parts of the world. All the editors were asked to present their books to address the market's needs and pinpoint cutting-edge technologies in food engineering.

All contributions are written by internationally renowned experts who have both academic and professional credentials. All authors have attempted to provide critical, comprehensive, and readily accessible information on the art and science of a relevant topic in each chapter, with reference lists for further information. Therefore, each book can serve as an essential reference source to students and researchers in universities and research institutions.

**Da-Wen Sun**
*Series Editor*

# Preface

This book provides a stimulating and up-to-date review of food process engineering phenomena. It addresses the basic and applied principles of food process engineering methods used in food-processing operations around the world. While it does cover the theory, it combines this with a practical hands-on approach. The book explores the basic and applied aspects of food process engineering phenomena from chilling/freezing, dehydration, and mixing-emulsions to new emerging food-engineering technologies, encapsulation, and modeling of innovative and traditional food-processing technologies.

# Acknowledgments

The contributions by well-known, reputable, and distinguished researchers from around the world are much appreciated. We would like to acknowledge their valuable support and professionalism.

# Series Editor

Born in Southern China, Professor Da-Wen Sun is a world authority in food engineering research and education; he is a member of the Royal Irish Academy (RIA), which is the highest academic honor in Ireland; he is also a member of Academia Europaea (The Academy of Europe) and a fellow of the International Academy of Food Science and Technology. His main research activities include cooling, drying, and refrigeration processes and systems, quality and safety of food products, bioprocess simulation and optimization, and computer vision technology. Especially, his many scholarly works have become standard reference materials for researchers in the areas of computer vision, computational fluid dynamics modeling, vacuum cooling, and so on. Results of his work have been published in over 600 papers including about 300 peer-reviewed journal papers (Web of Science h-index = 45; Google Scholar h-index = 52). He has also edited 13 authoritative books. According to Thomson Scientific's *Essential Science Indicators*$^{SM}$, based on data derived over a period of 10 years from the ISI Web of Science, there are about 2500 scientists who are among the top 1% of the most cited scientists in the category of agriculture sciences. For many years, Professor Sun has consistently been ranked among the top 100 scientists in the world (he was at 31st position in 2010).

He received a first class BSc Honors and MSc in mechanical engineering, and a PhD in chemical engineering in China before working in various universities in Europe. He became the first Chinese national to be permanently employed in an Irish university when he was appointed college lecturer at the National University of Ireland, Dublin (University College Dublin [UCD]), in 1995, and was then continuously promoted in the shortest possible time to senior lecturer, associate professor, and full professor. Dr. Sun is now a professor of Food and Biosystems Engineering and the director of the Food Refrigeration and Computerised Food Technology Research Group at the UCD.

As a leading educator in food engineering, Professor Sun has significantly contributed to the field of food engineering. He has trained many PhD students, who have made their own contributions to the industry and academia. He has also delivered lectures on advances in food engineering on a regular basis in academic institutions internationally and delivered keynote speeches at international conferences. As a recognized authority in food engineering, he has been conferred adjunct/visiting/consulting professorships from 10 top universities in China, including Zhejiang University, Shanghai Jiaotong University, Harbin Institute of Technology, China Agricultural University, South China University of Technology, and Jiangnan University. In recognition of his significant contribution to food engineering worldwide and for his outstanding leadership in the field, the International Commission of Agricultural and Biosystems Engineering (CIGR) awarded him the "CIGR Merit Award" in 2000, and again in 2006, the Institution of Mechanical Engineers based in the United Kingdom named him "Food Engineer of the Year 2004." In 2008, he was awarded the "CIGR Recognition Award" in honor of his distinguished achievements as the top 1% of agricultural engineering scientists in the world. In 2007, he was presented with the only "AFST(I) Fellow Award" in that year by the Association of Food Scientists and Technologists (India), and in 2010, he was presented with the "CIGR Fellow Award"; the title of Fellow is the highest honor in CIGR and is conferred to individuals who have made sustained, outstanding contributions worldwide. In March 2013, he was presented with the "You Bring Charm to the World" Award by Hong Kong-based Phoenix Satellite Television. In July 2013, he received the "Frozen Food Foundation Freezing Research Award" from the International Association for Food Protection (IAFP) for his significant contributions to enhancing the field of food freezing technologies. This is the first time that this prestigious award was presented to a scientist outside the United States.

He is a fellow of the Institution of Agricultural Engineers and a fellow of Engineers Ireland (the Institution of Engineers of Ireland). He has also received numerous awards for teaching and research excellence, including the President's Research Fellowship, and has twice received the President's Research Award of the UCD. He is the editor-in-chief of *Food and Bioprocess Technology—An International Journal* (2012 Impact Factor = 4.115), former editor of *Journal of Food Engineering* (Elsevier), and editorial board member for a number of international journals, including the *Journal of Food Process Engineering, Journal of Food Measurement and Characterization,* and *Polish Journal of Food and Nutritional Sciences.* He is also a chartered engineer.

On May 28, 2010, he was awarded membership in the RIA, which is the highest honor that can be attained by scholars and scientists working in Ireland; at the 51st CIGR General Assembly held during the CIGR World Congress in Quebec City, Canada, on June 13–17, 2010, he was elected incoming president of CIGR and became CIGR president in 2013–2014—the term of his CIGR presidency is six years, two years each for serving as incoming president, president, and past president. On September 20, 2011, he was elected to Academia Europaea (The Academy of Europe), which is functioning as the European Academy of Humanities, Letters and Sciences, and is one of the most prestigious academies in the world; election to the Academia Europaea represents the highest academic distinction.

# Editors

**Theodoros Varzakas** earned a bachelor's (honors) in microbiology and biochemistry (1992), a PhD in food biotechnology, and an MBA in food from Reading University, United Kingdom (1998). Dr. Varzakas was a postdoctoral research staff member at the same university. He has worked in large pharmaceutical and multinational food companies in Greece for 5 years and has at least 14 years experience in the public sector. Since 2005, he has served as an assistant and associate professor in the Department of Food Technology, Technological Educational Institute of Peloponnese (ex Kalamata), Greece, specializing in issues of food technology, food processing, food quality, and safety. Dr. Varzakas has been a reviewer in many international journals such as the *International Journal of Food Science and Technology*, *Journal of Food Engineering*, *Waste Management*, *Critical Reviews in Food Science and Nutrition*, *Italian Journal of Food Science*, *Journal of Food Processing and Preservation*, *Journal of Culinary Science and Technology*, *Journal of Agricultural and Food Chemistry*, *Journal of Food Quality*, *Food Chemistry*, and *Journal of Food Science*. He has written more than 90 research papers and reviews and has presented more than 90 papers and posters at national and international conferences. He has written two books in Greek, one on genetically modified food and the other on quality control in food. He has edited a book on sweeteners, published in 2012 by CRC Press and another book on biosensors published by CRC Press in 2013.

Dr. Varzakas has participated in many European and national research programs as a coordinator or scientific member.

He is a fellow of the Institute of Food Science & Technology (2007).

**Constantina Tzia** earned a diploma (1977) and PhD (1987) in chemical engineering from the National Technical University of Athens (NTUA), Greece, and is a professor in the School of Chemical Engineering, NTUA, Greece.

Professor Tzia's main research fields are food science, food engineering, food quality, sensory evaluation of foods, food hygiene, food safety and HACCP (hazard analysis and critical control points), food processing (freezing, edible films/coatings, extraction, encapsulation), food technology (fats and oils, olives and olive oil, bakery products, dairy products), and utilization of food by-products (protein isolates).

Dr. Tzia has more than 70 articles in international scientific (SCI) journals, 2 books in the Greek language, 1 book (editor), and 6 chapters in international scientific books. She has more than 120 presentations at international conferences and more than 100 at Greek conferences. She has participated in European research projects and has coordinated many national research projects and projects in cooperation with Greek food industries. She has been a reviewer in many international journals.

# Contributors

**Alfredo Cassano**
Institute on Membrane Technology, ITM-CNR
University of Calabria
Rende, Italy

**Sofia Chanioti**
School of Chemical Engineering
National Technical University of Athens
Athens, Greece

**Charikleia Chranioti**
School of Chemical Engineering
National Technical University of Athens
Athens, Greece

**Paul Christakopoulos**
School of Chemical Engineering
National Technical University of Athens
Athens, Greece

**Enrico Drioli**
Institute on Membrane Technology, ITM-CNR
University of Calabria
Rende, Italy

**Robert H. Driscoll**
School of Chemical Engineering
UNSW Australia
Sydney, Australia

**Renè Ruby Figueroa**
Institute on Membrane Technology, ITM-CNR
Rende, Italy

**M. Giannakourou**
Department of Food Technology
Technological Educational Institute of Athens
Athens, Greece

**Virginia Giannou**
School of Chemical Engineering
National Technical University of Athens
Athens, Greece

**George I. Katsaros**
School of Chemical Engineering
National Technical University of Athens
Athens, Greece

**Constantinos Katsimpouras**
School of Chemical Engineering
National Technical University of Athens
Athens, Greece

**Kai Knoerzer**
CSIRO Animal Food and Health Sciences
Melbourne, Australia

**Magdalini K. Krokida**
School of Chemical Engineering
National Technical University of Athens
Athens, Greece

**George Liadakis**
Hellenic Army Chemical Laboratory
Keratsini, Greece

**Panagiotis A. Michailidis**
School of Chemical Engineering
National Technical University of Athens
Athens, Greece

**V. Polychniatou**
School of Chemical Engineering
National Technical University of Athens
Athens, Greece

**Henry Sabarez**
CSIRO Animal, Food and Health Sciences
Melbourne, Australia

**George Srzednicki**
School of Chemical Engineering
University of New South Wales Australia
Sydney, Australia

**Pawan S. Takhar**
Food Science and Human Nutrition
University of Illinois
Urbana-Champaign, Illinois

**Petros S. Taoukis**
School of Chemical Engineering
National Technical University of Athens
Athens, Greece

**Evangelos Topakas**
School of Chemical Engineering
National Technical University of Athens
Athens, Greece

**Constantina Tzia**
School of Chemical Engineering
National Technical University of Athens
Athens, Greece

**Theodoros Varzakas**
Department of Food Technology
Technological Educational Institute of Peloponnese
School of Agricultural Technology, Food Technology and Nutrition
Kalamata, Greece

**Epaminondas Voutsas**
School of Chemical Engineering
National Technical University of Athens
Athens, Greece

# 1 Membrane Separation

*Alfredo Cassano, Renè Ruby Figueroa, and Enrico Drioli*

## CONTENTS



## 1.1 INTRODUCTION

The first commercial production of microporous membranes on a small scale started in 1930 together with the first practical application of ion-exchange membranes and the development of the theory on ionic transport through charged membranes (Nunes and Peinemann 2001). However, until the late 1960s, membranes were used in a few laboratory and analytical applications, but not for industrial applications because they were too slow, too expensive, and too unselective. The seminal discovery that transformed membrane separation from a laboratory to an industrial process was the development of defect-free ultrathin cellulose acetate membranes by Loeb–Sourirajan process in the 1960s (Loeb and Sourirajan 1962; Mota et al. 2002).

The situation today is different because membranes are more robust, modules and equipment are designed better, and costs have come down significantly, partly because of the maturing technology and partly because of competition from an increasing number of membrane suppliers and the original equipment manufacturers

(Cheryan 1998). Membrane technology is currently an established part of several industrial processes such as producing drinking water from the sea, cleaning industrial effluents and recovering valuable constituents, separating gases and vapors, and to concentrate, purify, or fractionate macromolecular mixtures in the food and drug industries (Strathmann et al. 2006). Membrane separation technologies have attracted much attention in the food industries over recent decades as low-energy processes, providing a gentle treatment of the product at low–moderate temperatures and covering steps such as separation, fractionation, concentration, purification, and clarification of various streams.

Even though membrane processes are guided by four major driving forces—temperature, activity, electrical potential, and hydrostatic pressure gradients—this chapter will mainly focus on the use of pressure-driven membrane processes, such as microfiltration (MF), ultrafiltration (UF), nanofiltration (NF), and reverse osmosis (RO) in the food industry by referring to both well-established and potential applications.

These processes are based on the use of a permselective barrier through which fluids and solutes are selectively transported when a hydrostatic pressure is applied across it. As a result, the feed solution is converted into a *permeate* containing all the components that have permeated the membrane and a *retentate* containing all the compounds rejected by the membrane.

Section 1.1 of this chapter provides an overview on materials and structures of synthetic membranes. Section 1.2 is then focused on membrane design and evaluation of membrane performance. In Section 1.7, a summary of established and potential applications of this technology in the food industry will be provided.

## 1.2 MEMBRANE MATERIALS

A large variety of synthetic membranes have been reported in scientific and patent literature. The differences may be caused by material partitioning during membrane formation or by some selected surface postformation treatments. Membrane chemistry determines the important properties such as hydrophilicity or hydrophobicity, presence or absence of ionic charges, chemical and thermal resistance, binding affinity for solutes or particles, biocompatibility, and so on (Cheryan 1998; Strathmann et al. 2006).

Membrane materials must be chemically resistant to both feed and cleaning solutions, mechanically and thermally stable, and characterized by high selectivity and permeability.

The materials used for the preparation of membranes can be synthetic polymers, cellulose derivatives, ceramics, inorganics, and metals that may be neutral or carry electrical charges. Although more than 130 materials have been used to manufacture membranes, only a few have achieved commercial status, and fewer materials still have to obtain regulatory approval for use in food.

A brief summary of the typical materials suitable for pressure-driven membrane process is shown in Table 1.1.

**TABLE 1.1**
**Materials Used for the Manufacture of Membranes**

| Material | Processes |
|---|---|
| Alumina | MF |
| Carbon–carbon composites | MF |
| Cellulose esters (mixed) | MF |
| Cellulose nitrate | MF |
| Polyamide, aliphatic (e.g., nylon) | MF |
| Polycarbonate (track etch) | MF |
| Polyester (track etch) | MF |
| Polypropylene | MF |
| Polytetrafluoroethylene (PTFE) | MF |
| PCV | MF |
| Polyvinylidene fluoride (PVDF) | MF |
| Sintered stainless steel | MF |
| Cellulose (regenerated) | MF, UF |
| Ceramic composites (zirconia on alumina) | MF, UF |
| Polyacrylonitrile (PAN) | MF, UF |
| Polyvinyl alcohol (PVA) | MF, UF |
| Polysulfone (PS) | MF, UF |
| Polyethersulfone (PES) | MF, UF, and NF |
| Cellulose acetate (CA) | MF, UF, and RO |
| Cellulose triacetate (CTA) | MF, UF, and RO |
| Polyamide, aromatic (PA) | MF, UF, NF, and RO |
| Polyamide (PI) | UF, RO |
| CA/CTA blends | RO |
| Composites (e.g., polyacrylic acid on zirconia or stainless steel) | RO |
| Composites, polymeric thin film (e.g., PA or polyetherurea on polysulfone) | RO |
| Polybenzimidazole (PBI) | RO |
| Polyetherimide (PEI) | RO |

## 1.3 MEMBRANE STRUCTURE

Synthetic membranes can be classified on the basis of their structure as porous membranes, homogeneous membranes, solid membranes carrying electrical charges, and liquid or solid films containing selective carriers (Strathmann et al. 2006).

Porous membranes consist of a solid matrix with defined pores with diameters ranging from <1 nm to more than 10 μm. Porous membranes can be classified as macroporous, with average pore diameters larger than 50 nm (i.e., MF and UF membranes); mesoporous, with average pore diameters in the range between 2 and 50 nm (i.e., NF membranes); and microporous, if the average pore diameters are between 0.1 and 2 nm. Dense membranes, such as those used in RO, have no individual permanent pores but the separation occurs through fluctuating free volumes.

Furthermore, the structure of membranes may be symmetric, if pore diameters do not vary over the membrane cross section, or asymmetric, with pore diameters increasing from one side of the membrane to the other.

Symmetric porous membranes can be prepared by using different techniques such as sintering, track etching, and leaching (Strathmann et al. 2006). Specific details related to these techniques are extensively reported in literature (Bhave 1991; Cadotte 1985; Frommer et al. 1970; Hiatt et al. 1985; Kamide 1985; Kesting 1985; Smid et al. 1996; Sourirajan 1977; Strathmann 1985).

On the other hand, asymmetric membranes consist of a thin (0.1–1 μm) selective skin layer supported by a highly porous (100–200 μm) thick substructure. The skin represents the active layer of the membrane and the separation characteristics will depend on the nature of the material or the size of pores in the skin layer. These membranes are generally used for UF, NF, and RO applications. Asymmetric membranes can be prepared through two different techniques: (i) the phase inversion process, which leads to an integral structure with the skin and the support structure made from the same material in a single process (integral asymmetric membranes) (Kesting 1971); (ii) the deposition of an extremely thin polymer film on a preformed porous substructure in a two-step process, leading to the so-called composite membrane (Cadotte and Petersen 1981). Generally, the barrier and support are made from different materials.

Figure 1.1 shows the cross section of asymmetric membranes in flat sheet and hollow fiber configurations.

The filtration capability of pressure-driven membrane processes is shown in Figure 1.2. MF is used to separate particles with diameters of 0.1–10 μm from a solvent or other low-molecular-weight compounds. These particles are generally larger than those separated by UF and RO. Consequently, the osmotic pressure for MF is negligible and hydrostatic pressure differences used in MF are relatively small (in the range of 0.5–4 bar). UF is based on the use of asymmetric membranes with pore sizes in the skin layer of 2–10 nm. Typically dissolved molecules or small particles not larger than 0.1 μm in diameter are retained. An UF membrane is typically

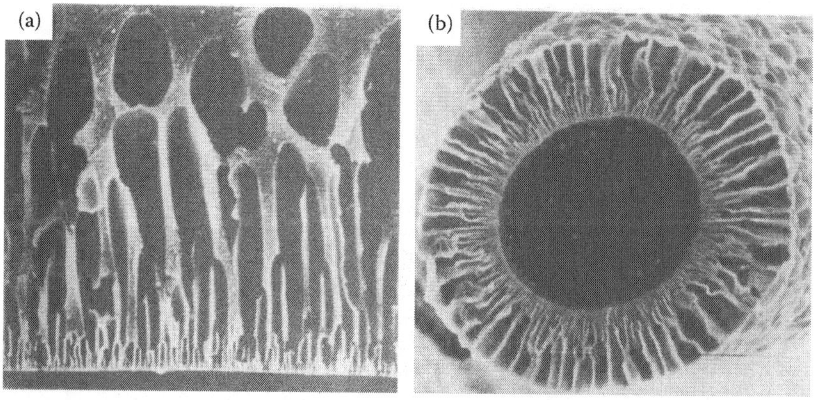


**FIGURE 1.1** A cross section of (a) cellulose acetate asymmetric flat sheet membrane; and (b) polyamide hollow fiber asymmetric membrane.

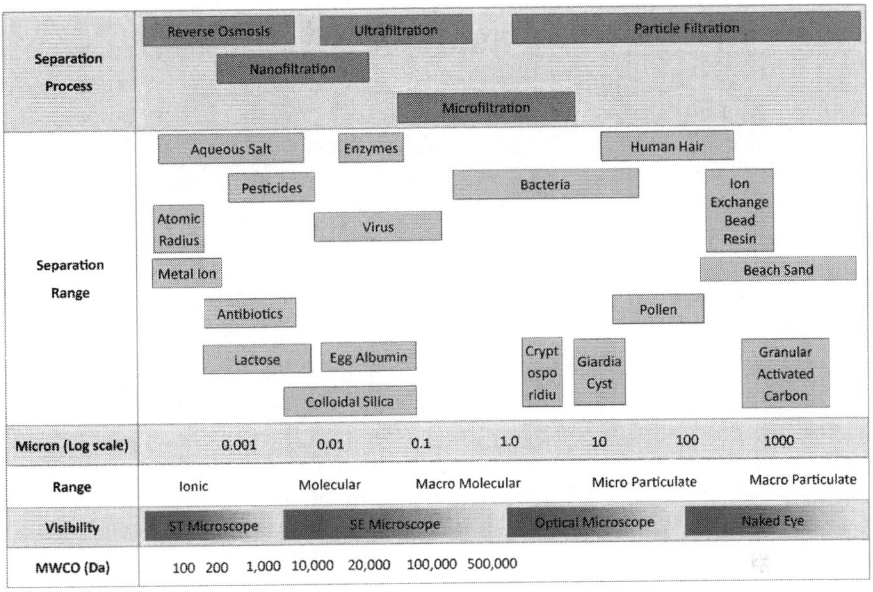


**FIGURE 1.2** Filtration capability of pressure-driven membrane operations.

characterized by its molecular weight cutoff (MWCO), defined as the equivalent molecular weight of the smallest species that exhibit 90% rejection. The MWCO of UF membranes is between $10^3$ and $10^6$ Da. Hydrostatic pressures of 2–10 bar are typically used.

In the NF process, components of a fluid are fractionated mainly according to their size and charge. Multivalent ions and uncharged organic molecules with molecular weight between 100 and 1000 Da are typically separated. NF membranes are characterized by a charged surface with pore diameters in the range 1–3 nm. They operate at lower pressures (generally in the range 3–30 bar) than RO membranes.

RO membranes are typically used to separate low-molecular-weight compounds from a solvent, usually water. The particle size range for applications of RO is 0.1–1 nm and solutes with molecular weight >300 Da are separated. The hydrostatic pressures applied as a driving force in RO are in the order of 10–100 bar.

## 1.4 PROCESS DESIGN AND OPERATION

The selection of the most effective membrane for a specific application plays an important role in determining the desired level of separation to be obtained. However, for an efficient utilization of membranes, the process design is equally important.

MF, UF, NF, and RO are filtration processes in which a hydrostatic pressure gradient is utilized to transport specific solutes through a membrane characterized by different permeability for different compounds. In these processes, the feed solution is converted into a permeate stream containing all the components that have permeated the membrane and a retentate containing all the compounds rejected by the membrane.

The performance of a membrane in a pressure-driven separation process depends on the filtration rate (the permeate flux) and on the membrane separation properties. These properties are a function of the membrane permeability of different compounds in the treated solution, the applied pressure, and the process design.

The process design is defined by different aspects concerning the membrane configuration, filtration methods (dead-end and cross-flow configuration), and the process configuration. It is also of importance for the control of concentration polarization and fouling phenomena, determining to a large extent the useful membrane life for a specific separation.

### 1.4.1 Membrane Modules

The membrane module concept denotes the device where the membrane must be installed to perform the separation process. On a large industrial scale, membrane modules are available in six basic designs: cartridge, hollow fiber, spiral wound, tubular, plate and frame, and capillary. They are quite different in their design, mode of operation, production costs, and energy requirement for pumping the feed solution through the module.

Pleated cartridge modules are mainly used in dead-end MF; they consist of a pleated membrane cartridge installed in a pressurized housing. These systems are operated at relatively low pressures of 1–2 bars. The main applications are related with the sterile filtration of water and beverages such as wine, beer, and fruit juices, as well as pharmaceutical solutions. At an industrial scale, they are used as prefilters in RO water desalination plants (Strathmann et al. 2006).

Membrane modules for cross-flow applications are illustrated in the following. The plate-and-frame configuration (Figure 1.3a) is mainly used for small-scale applications (production of pharmaceuticals, bioproducts, or fine chemicals). Membranes, feed flow spacers, and porous permeate support plates are layered together between two end plates and placed in a housing. The sheets are in the form of circular disks, elliptical sheets, or rectangular plates. The feed mixture is pressurized in the housing and forced across the membrane surface. A portion passes through the membrane, enters the permeate channel, and makes its way to a central permeate collection manifold.

Plate-and-frame modules are quite expensive and the membrane replacement is labor intensive. They are used in a limited number of UF applications with highly fouling feeds. The feed channels are often <1 mm and although more sensitive to fouling, they are easier to clean as no mesh support is used.

The spiral-wound configuration, widely used in UF, NF, and RO, is basically a variation of the plate-and-frame geometry. In this configuration, the feed flow channel spacer, the membrane, and the porous membrane support form an envelope that is rolled around a perforated central collection tube (Figure 1.3b). The module is placed inside a tubular pressure housing made from stainless steel or polyvinyl chloride (PVC). The feed solution passes axially through the feed channel across the membrane surface. The permeate stream is moved along the permeate channel and is collected in the collection tube.

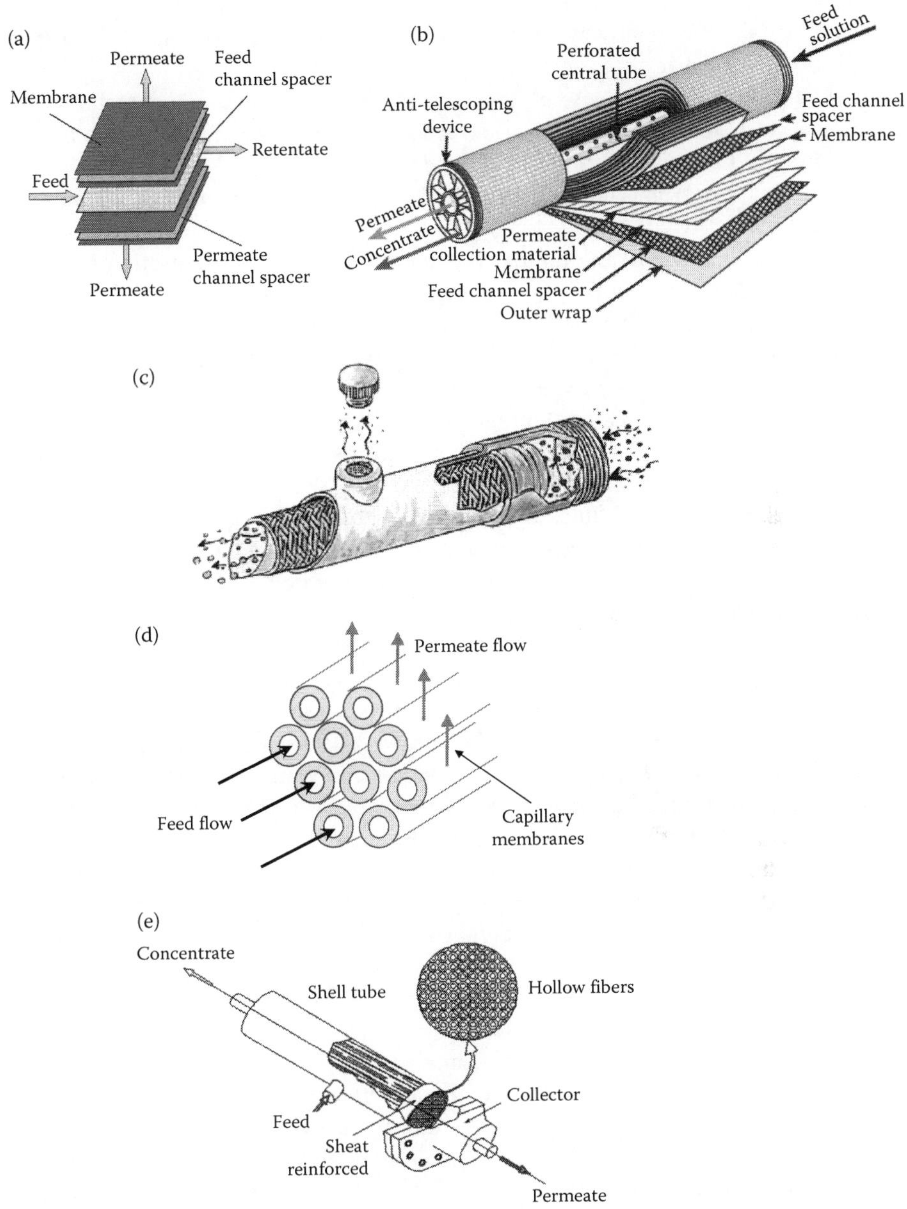


**FIGURE 1.3** A schematic representation of (a) plate and frame, (b) spiral wound, (c) tubular, (d) capillary, and (e) hollow fiber membrane module.

The spiral-wound configuration is compact providing a relatively large area per unit volume; so, it is relatively inexpensive, but prone to particulate fouling (thus prefiltration is needed).

Tubular membrane modules (Figure 1.3c) consist of membrane tubes placed into porous stainless steel or fiberglass-reinforced plastic pipes. The feed solution is pressurized internally along the tubes and the permeate is collected on the outer side of the porous support. Tube diameters are in the range of 1–2.5 cm and a number of tubes are placed in one pressure housing to increase the module productivity.

The main advantage of the tubular configuration is that concentration polarization effects and membrane fouling phenomena can be easily controlled; however, low surface areas can be installed in a given unit volume resulting in high production costs. Basically, tubular membrane modules are used to treat feed solutions characterized by high viscosity and a high content of solids.

Capillary and hollow fiber membrane modules have the same basic spinning process for the preparation. The principal differences are related with the inner diameter and also with the position of the selective layer.

Capillary membrane modules are constituted by a large number of capillaries with an inner diameter of 0.5–3 mm arranged in parallel as a bundle in a shell tube. The feed is pumped in the lumen of the membranes while the permeate is collected in the shell side (Figure 1.3d).

This configuration is characterized by a high membrane area per module volume and low production costs; another advantage is the possibility to control concentration polarization and membrane fouling through a physical process called back flushing in which the permeate flow is reversed allowing to dislodge the fouling material from the membrane surface. The required low operating pressure (4–6 bar as maximum values), due to the limited stability of the membranes, is the main drawback.

Hollow fiber membrane modules consist of a bundle of several membrane fibers with the free ends potted with an epoxy resin into a cylindrical housing. Typically, the outer diameter of fibers is between 50 and 200 μm; in this case, the selective layer is on the outside of the fibers where the feed fluid is applied, while the permeate is removed down the fiber bore (Figure 1.3e).

Hollow fiber membrane modules consisting of fibers with diameter between 200 and 500 μm are also available. In these systems, the solution is fed into the lumen of the fibers and the permeate is removed from the shell side.

Hollow fiber membrane modules are characterized by the highest packing density if compared with other configurations and their production is very cost-effective. Drawbacks are related to the difficult control of concentration polarization and membrane fouling. Consequently, extensive pretreatments of the solution are needed to remove particles, macromolecules, or other materials that can precipitate at the membrane surface.

A brief summary of membrane module characteristics is shown in Table 1.2.

Dynamic filtration modules, not discussed in this chapter, are used in very specific applications (such as in the pharmaceutical industry) when shear-sensitive compounds have to be processed. In these modules, such as rotating and vibrating membrane modules, the shear rate at the membrane surface is created by a relative motion between the membrane and the housing (Al Akoum et al. 2002, 2004; Ding

**TABLE 1.2**
**Characteristics of Different Membrane Modules**

| Module Type | Area of Standard Module ($m^2$) | Characteristics |
|---|---|---|
| Hollow fine fiber | 300–600 | • Low cost per square meter of membrane but modules easily fouled.<br>• Only suitable for clean fluids. |
| Capillary fiber | 50–150 | • Limited to low-pressure applications <200 psi; good fouling resistance can be backflushed.<br>• Important in UF and MF applications. |
| Spiral wound | 20–40 | • The most common RO module.<br>• Increasingly used in UF and gas-separation applications. |
| Tubular | 5–10 | • High cost limits applications. |
| Plate and frame | 5–10 | • High cost limits applications. |

et al. 2002). In this approach, transmembrane pressures and shear rates are independent; so, high shear with low pressures can be achieved.

### 1.4.2 Filtration Methods

Membrane filtration can operate at either dead-end or in cross-flow configurations. In the dead-end mode (Figure 1.4a), the feed flow is perpendicular to the membrane surface. It is forced through the membrane, which causes the retained particles to accumulate and form a type of cake layer at the membrane surface. The thickness of the cake layer increases with the filtration time. Therefore, the permeation rate decreases by increasing the cake layer thickness. Dead-end filtration is often used as a method to estimate the specific cake resistance for cross-flow filtration and usually gives reasonable data for spherical and ellipsoidal-shaped cells (Tanaka et al. 1994).

In cross-flow filtration (Figure 1.4b), the fluid to be filtered flows according to a parallel direction to the membrane surface and permeates through the membrane

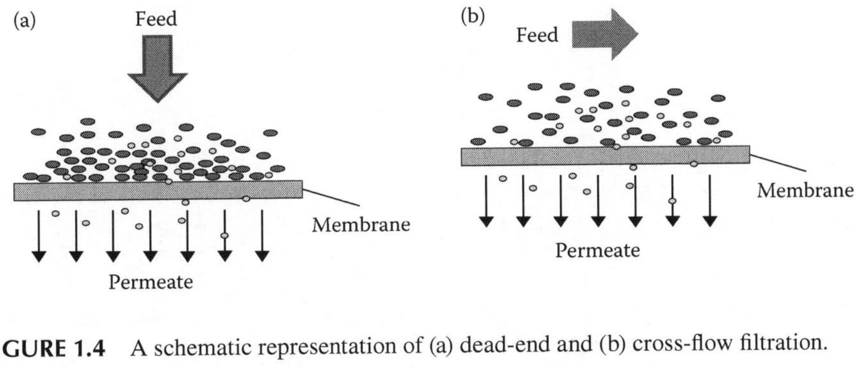


**FIGURE 1.4** A schematic representation of (a) dead-end and (b) cross-flow filtration.

due to the imposed transmembrane pressure difference. Unlike dead-end filtration, rejected particles form a cake layer on the membrane surface that does not build up indefinitely; so, the cake formed is of a limited thickness.

Under the action of pressure drop and frictional drag on the cake particles, the compressible cake phenomena can be enumerated as follows: (i) successive particles rearrangement inside the cake under stress; (ii) matrix compression in gel-like cakes; and (iii) complex cases where both the aforementioned phenomena interplay, such as in the case of a biofilm (Tiller and Cooper 1962; Tiller and Yeh 1985). The cake structure will be affected by different phenomena such as the collapse of the pore structure, pore compression, and pore distortion. This set of phenomena will affect, to a different extent, the porosity, the pore size, and the pore tortuosity of the filtration cake (Mota et al. 2002).

### 1.4.3 Process Configuration

The most common process configurations used in practical applications of pressure-driven membrane processes are illustrated in Figure 1.5a through d.

In the total recycle configuration (Figure 1.5a), permeate and retentate streams are recycled back to the feed reservoir so that a steady state is attained in fixed concentration of the feed. This configuration is mainly used to study the effect of different

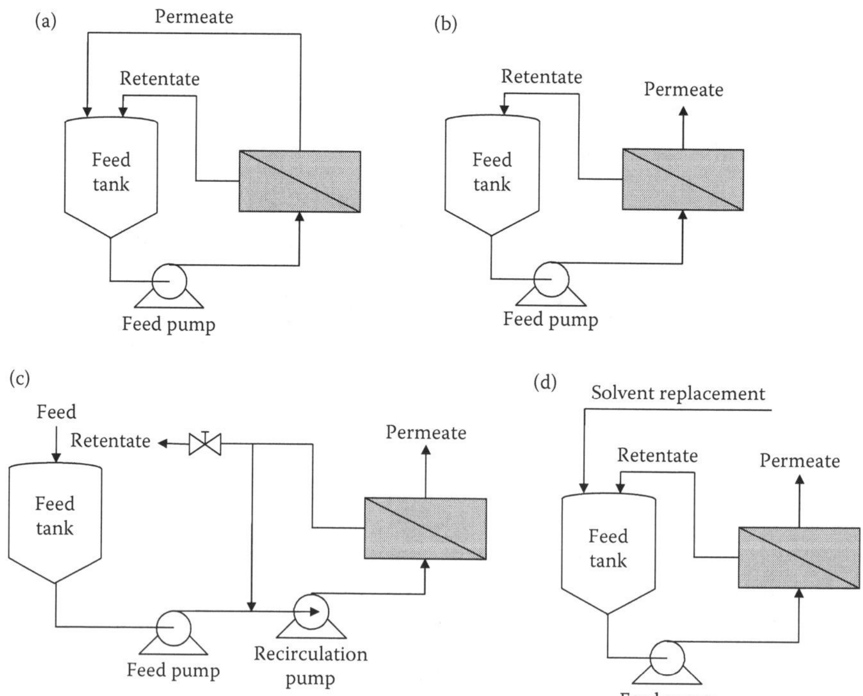


**FIGURE 1.5** A schematic diagram of (a) total recycle, (b) batch concentration, (c) feed-and-bleed, and (d) diafiltration configuration.

operating parameters (feed concentration, cross-flow velocity, transmembrane pressure, and temperature) on steady-state permeate flux and permeate quality (Rai et al. 2006).

In the batch concentration configuration, the retentate is returned to the feed reservoir and the permeate is collected separately (Figure 1.5b). This approach requires the least membrane area to achieve a given separation/unit time. Batch operations are used when the permeate is the product of interest, such as in fruit juice clarification or treatment of effluents in which the retentate has to be discharged.

The feed-and-bleed configuration is commonly used for continuous operation when a higher recovery rate must be obtained (e.g., in the food industries). In this case, the permeate is removed from the system together with a small part of the retentate (Figure 1.5c). Most of the retentate is recycled and mixed with the feed solution to maintain a high tangential velocity in the membrane module. A feed pump and a recirculation pump are required to provide the transmembrane pressure and the cross-flow, respectively.

Diafiltration (Figure 1.5d) involves the addition of water to the retentate to overcome low permeate fluxes at high concentrations or to improve the separation of permeable compounds. It can be carried out in sequential form (by replacing the permeate with an equal volume of pure solvent) or continuously (by replacing the permeate with the same flow rate of added water). This configuration is often used when a more complete separation of micro- and macrosolutes is required.

## 1.5 MEMBRANE PERFORMANCE

The membrane performance in pressure-driven membrane systems is determined by the filtration rate and membrane separation properties generally evaluated in terms of membrane rejection.

Several mathematical models available in literature attempt to describe the mechanism of transport through membranes. Although the operating techniques of MF, UF, NF, and RO are similar, some considerations should be taken into account. In MF and UF, the convection of a bulk solution is the dominant form of transport, while diffusion is generally insignificant. In NF and RO, matter is transported through the membrane mainly by diffusion of individual molecules through a more or less homogeneous membrane matrix, but convection can become significant with high flux membranes (Strathmann et al. 2006).

In this section, a brief description of separation capabilities of membranes, transport mechanisms, and concentration polarization and fouling phenomena in pressure-driven membrane systems is reported.

### 1.5.1 Membrane Rejection and Volume Reduction Factor

The separation capability of MF, UF, NF, and RO membranes can be expressed in terms of membrane rejection according to the following equation:

$$R = \left(1 - \frac{C_p}{C_f}\right) \tag{1.1}$$

where $R$ is the membrane rejection for a given component in defined conditions of hydrostatic pressure and feed solution concentration, while $C_p$ and $C_f$ are the concentrations of the component in the permeate stream and feed solution, respectively.

However, the concentrations in the retentate and permeate streams depend not only on the membrane rejection but also on the recovery rate (Δ) that is given by

$$\Delta = \frac{V_p}{V_0} \tag{1.2}$$

where $V_p = V_p(t)$ and $V_0$ are the permeate volume and the initial feed volume, respectively, and $t$ is the time. The recovery rate ranges between 0 and 1. Sometimes, data are also presented as volume concentration factor (VRF):

$$\text{VRF} = \frac{V_o}{V_r(t)} = \frac{V_o}{V_o - V_p(t)} \tag{1.3}$$

where $V_o$ is the initial volume (mL) of the feed, $V_r(t)$ is the final volume (mL) of the retentate at a particular time ($t$), and $V_p(t)$ is the volume collected on the permeate side at a particular time (Singh et al. 2013).

### 1.5.2 Transport Mechanisms

The transport mechanisms in membrane operations are generally described by the phenomenological equations such as Darcy's law, Fick's law, Haugen–Poiseuille's law, and Ohm's law (Strathmann et al. 2006). In MF and UF processes, components that permeate through the membrane are transported by a convective flow through the membrane pores under a pressure gradient as the driving force and the separation occurs through a size-exclusion mechanism. Darcy's law (Mexis and Kontominas 2010; Cheryan 1998) describes this type of transport:

$$J_w = \frac{V_p}{A_m dt} = \frac{\Delta P}{\mu R_T} \tag{1.4}$$

where $J_w$ is the volume flux (m/s), $A_m$ is the membrane cross- sectional area (m$^2$), $V_p$ is the filtrate volume (mL) collected on the permeate side at a particular time interval $dt$ (s), $\Delta P$ is the applied transmembrane pressure (kPa), μ is the viscosity (Pa · s) of the permeate sample, and $R_T$ is the total membrane resistance (m$^{-1}$). $R_T$ is given by

$$R_T = R_C + R_F + R_M \tag{1.5}$$

where $R_C$ is the cake layer resistance (m$^{-1}$) due to the concentration polarization and the deposition of solids on the membrane surface, $R_F$ is the fouling layer resistance (m$^{-1}$) due to the internal fouling inside the pores, and $R_M$ is the intrinsic membrane resistance.

Experimentally, the resistances defined in Equation 1.5 can be determined from the values of the hydraulic permeability measured before and after the cleaning procedures. In particular, $R_M$ can be calculated by measuring the hydraulic permeability of the new or clean membrane as

$$R_M = \frac{\Delta P}{\mu_w L_p^0} \tag{1.6}$$

where $\mu_w$ is the viscosity of pure water (Das et al. 2006; Youn et al. 2004) and $L_p{}^0 = J_w/\Delta P$ is the hydraulic permeability (m/s · Pa) of the new membrane.

$R_T$ can be calculated as

$$R_T = \frac{1}{\mu_w L_p^1} \tag{1.7}$$

in which $L_p^1$ is the hydraulic permeability of the membrane after the treatment with a specific solution.

$R_C$ is removed by cleaning the membrane with water. The hydraulic permeability measured after such cleaning is $L_p^2$, therefore

$$R_M + R_F = \frac{1}{\mu_w L_p^2} \tag{1.8}$$

Finally, the resistance caused by the cake layer formation can be estimated as

$$R_C = [R_T - (R_M + R_F)] \tag{1.9}$$

In an ideal situation (pores uniformly distributed, no fouling, and negligible concentration polarization), the fluid flow through a porous membrane can be described by the Haugen–Poiseuille law:

$$J_w = \frac{\varepsilon d_p^2 P_T}{32 \Delta x \mu} \tag{1.10}$$

where $\varepsilon$ is the surface porosity of the membrane, $d_p$ is the mean pore diameter, $P_T$ is the applied transmembrane pressure, $\Delta x$ is the length of the channel, and $\mu$ is the viscosity of the fluid permeating the membrane.

The net driving force $\Delta P$ for an ideal membrane process should be $(P_T - \Delta\Pi)$ where

$$P_T = P_F - P_P \tag{1.11}$$

$$\Delta\pi = \pi_F - \pi_P \tag{1.12}$$

where $P_F$ and $P_P$ are the pressures on the feed and permeate side of the membrane, respectively; similarly, $\pi_F$ and $\pi_P$ are the osmotic pressures on the feed and permeate side.

The cross-flow mode gives rise to a pressure drop from the inlet to the outlet of the membrane module, so that the feed side pressure is expressed as

$$P_F = \frac{P_i + P_o}{2} \tag{1.13}$$

where $P_i$ is the inlet pressure of the membrane module and $P_o$ is the outlet pressure.

The Haugen–Poiseuille law assumes that the flow through the pores is laminar and independent of time, the density is constant, the fluid is Newtonian, and end effects are negligible.

According to Equation 1.10, the flux is directly proportional to the applied pressure and inversely proportional to the viscosity. Basically, viscosity is controlled by solids concentration and temperature; for non-Newtonian liquids, it is also affected by shear rate and velocity.

The model is largely true under certain conditions such as low pressure, low feed concentration, and high feed velocity; when the process deviates from any of these conditions, flux becomes independent of pressure. In these conditions, mass transfer-limited models, such as the film theory, can effectively describe the process.

### 1.5.3 Concentration Polarization and Membrane Fouling

Concentration polarization is a natural consequence of the selectivity of a membrane. When a feed solution containing a solvent and a solute or suspended solids is filtered through a porous membrane, some components permeate the membrane under a given driving force while others are retained. This leads to an accumulation of particles or solutes in a mass transfer boundary layer adjacent to the membrane surface that can affect the flux. A concentration gradient between the solution at the membrane surface and the bulk is established that leads to a back transport of the material accumulated at the membrane surface by diffusion and, eventually, other means. This phenomenon, schematically represented in Figure 1.6, is referred to as *concentration polarization* (Aimar et al. 1989; Mulder 1991). It is not to be confused with the membrane-fouling phenomenon that is essentially due to a deposition of retained particles onto the membrane surface or in the membrane pores.

In NF and RO, mainly, low-molecular-weight materials are separated from a solvent such as water producing an increase in the osmotic pressure that is directly proportional to the solute concentration at the membrane surface and thus a decrease in the membrane flux at constant applied hydrostatic pressures (Belfort 1989). On the other hand, in MF and UF, the mechanism is different due to the fact that macromolecules and particles are retained by the membrane; thus, osmotic pressure is generally quite low. The retained components often are precipitated and form a solid layer at the membrane surface. This layer, which often exhibits membrane properties itself, can affect the membrane separation characteristics significantly by reducing

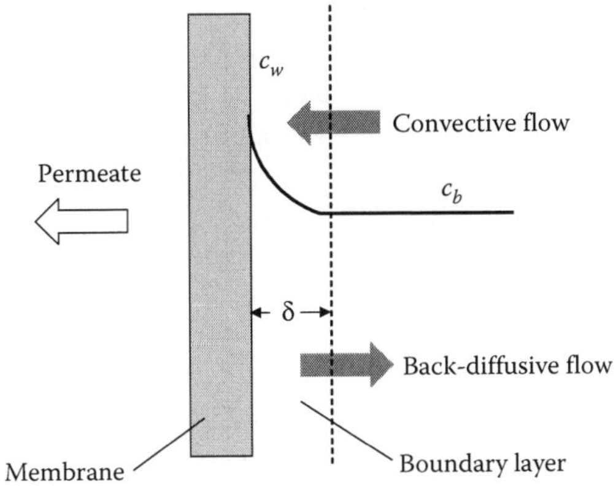


**FIGURE 1.6** Schematic representation of concentration polarization ($c_w$ = gel concentration; $c_b$ = bulk concentration).

the membrane flux and by changing the rejection of lower-molecular-weight components (Strathmann et al. 2006).

The term *fouling* is referred to a long-term flux decline caused by the deposition of retained particles (colloids, suspended particles, macromolecules, etc.) onto the membrane surface and/or within the pores of the membrane. The fouling behavior is strongly affected by the physicochemical nature of the membrane, the solutes, and the fluid dynamic system design.

According to Bacchin et al. (2006), the buildup of material may take different forms such as (1) adsorption, when attractive interactions between the membrane and particles exist; (2) pore blockage, with a reduction in flux due to the closure (or partial closure) of pores (this phenomenon is predominant in porous membranes such as MF and UF membranes); (3) a deposit of particles that can grow layer by layer at the membrane surface, leading to an additional hydraulic resistance; and (4) a gel formation due to concentration polarization of macromolecules such as pectins.

Although polarization concentration and fouling phenomena determine a reduction of the permeate flux, they can have an opposite effect on the observed rejection. The concentration polarization determines a reduction of the rejection; in case of fouling, if the buildup of solids on the membrane is significant enough, it may act as a secondary membrane and change the effective sieving and transport properties of the system: consequently, the rejection can be increased or maintained constant. In addition, while the concentration polarization is a reversible process based on diffusion taking place over a few seconds, fouling is generally irreversible and the flux decline is a long-term process. Finally, the concentration polarization can be minimized by hydrodynamic means, such as the feed flow velocity and the membrane module design; on the contrary, the control of membrane fouling is more difficult.

Methods generally accepted to control and minimize fouling phenomena include feed pretreatment, modification of membrane properties, modification of operating conditions, flow manipulation, and membrane cleaning with proper chemical agents.

A mathematical model generally accepted to describe concentration polarization phenomena is the well-known *film theory*. As previously reported, this model is more useful in describing transport phenomena through membranes when a polarized layer is reached and the flux becomes independent of pressure.

The film model assumes that the solute is brought to the membrane surface by convective transport at a rate $J_s$ defined as

$$J_s = JC_B \tag{1.14}$$

where $J$ is the permeate flux and $C_B$ is the bulk concentration of the rejected solute. The resulting gradient causes a back transport of the solute into the bulk of the solution due to diffusional effect, which can be described as

$$J_s = D\frac{dC}{dx} \tag{1.15}$$

where $D$ is the diffusion coefficient and $dC/dx$ is the concentration gradient over a differential element in the boundary layer. Once the steady state is reached, the two mechanisms will balance each other, and Equations 1.14 and 1.15 can be equated and integrated over the boundary layer, leading to the following equation:

$$J = \frac{D}{\delta}\ln\frac{C_G}{C_B} = k\ln\frac{C_G}{C_B} \tag{1.16}$$

where $C_G$ is the gel concentration (solute concentration at the membrane surface), and $k$ is the mass transfer coefficient given by

$$k = \frac{D}{\delta} \tag{1.17}$$

where $\delta$ is the thickness of the boundary layer over which the concentration gradient exists.

Several relationships to correlate the mass transfer coefficients with physical properties, flow channel dimensions, and operating parameters have been reported in literature (Gekas and Hallstrom 1987; Sherwood et al. 1975; Treybal 1981). These relationships that are generally adapted to specific applications are not universally applicable.

An estimation of the mass transfer coefficient can be approached through a dimensional analysis. By using the $\pi$ theorem, the following equation can be derived:

$$S_h = A(R_e)^{\alpha}(S_c)^{\beta} \tag{1.18}$$

where $S_h$, $R_e$, and $S_c$ are Sherwood, Reynolds, and Schmidt numbers, respectively. The dimensionless numbers can be obtained through Equations 1.19, 1.21, and 1.22.

$$S_h = \frac{kd_h}{D} \tag{1.19}$$

Here, $d_h$ is the hydraulic diameter that can be calculated as

$$d_h = 4\frac{\textit{Cross section available for flow}}{\textit{Wetted perimeter of the channel}} \tag{1.20}$$

The Sherwood number is a measure of the ratio of convective mass transfer to molecular mass transfer; the Reynolds number is a measure of the ratio of inertia effect to viscous effect and the state of turbulence in a system can be obtained as

$$R_e = \frac{d_h V \rho}{\mu} \tag{1.21}$$

In general, $R_e$ values <1800 are considered to be laminar flow, and Re values >4000 are considered to be turbulent flow (Cheryan 1998).

The Schmidt number represents the ratio of momentum transfer to mass transfer; it is given by

$$S_c = \frac{\mu}{\rho D} \tag{1.22}$$

The exponents α and β in Equation 1.18 are constants determined by the state of development of the velocity and concentration profiles along the channel.

On the other hand, in processes such as NF and RO where membranes are nonporous, the separation occurs by means of a different solubility and diffusivity of components across the membrane due to a concentration or chemical potential gradient. The molecules diffusion through homogeneous dense membranes occurs through the free-volume elements, or empty spaces between polymer chains caused by thermal motion of the polymer molecules, which fluctuate in the position and volume on the same timescale as the molecule permeates. This type of transport is described by Fick's law:

$$J_i = -D_i \frac{dC_i}{dZ} \tag{1.23}$$

where $C_i$ is the concentration and $D_i$ is the diffusion coefficient of a component $i$.

## 1.6 SELECTED APPLICATIONS

Membrane operations are well-established technologies in food and beverage processing. They offer several advantages when compared to conventional separation operations (decantation, filtration, centrifugation, chromatography, etc.) in terms of flexibility in equipment design and operations; easy scale-up; separation without phase change; minimum space requirements for plant operation; reduction of mechanical and thermal stress, hence retaining chemical properties of the feed components; possibility to separate compounds in a wide range of molecular weights; low operating costs; low-energy requirement; and improvement of the final product quality.

In addition, the possibility to realize integrated membrane systems in which all the steps of the food production are based on molecular membrane separations allows better performances in terms of product quality, plant compactness, environmental impact, and energetic aspects.

A general overview related to the application of pressure-driven membrane operations in some areas of the agro-food production (milk and dairy, fruit and vegetable juices, sugar, and brewing industries) will be provided in the following sections.

### 1.6.1 Milk and Dairy Industry

In milk processing, consolidate applications of pressure-driven membrane operations concern the removal of bacteria, milk standardization, the production of concentrated milk, and caseins fractionation (Atra et al. 2005; Daufin et al. 2001).

MF is suitable for cold sterilization to produce fresh milk with medium-term conservation; MF membranes with pores of 0.2 μm should be able to separate fat and bacteria from the rest of milk components, preserving its organoleptic properties due to the low temperature used in the process. Cycles of MF and pasteurization can be integrated to produce milk for long-term conservation (Strathmann et al. 2006).

UF membranes retain proteins, fats, and insoluble and bound salts allowing the permeation of lactose and soluble salts. The final content of proteins, lactose, and minerals in the retentate depends on the duration of the process and operating conditions (pH and temperature).

The UF process is generally carried out by using hydrophilic membranes due to their low fouling properties, capability to operate at temperatures around 50°C, resistance to chloride, and extreme pH-values used for sanitization and cleaning processes.

The UF process can be used to concentrate milk with the aim of reducing refrigeration and transportation costs. Renowned cheeses such as cheddar, camembert, roquefort, ricotta, and cream cheese are typical examples of cheese produced by using preconcentrated milk.

The use of preconcentrated milk offers different advantages such as an increase of yield, reduction in rennet enzyme, reduction in volume of milk to handle, a low production of whey, and the possibility to use continuous processes.

During the manufacturing of cheese, a subproduct named whey is produced. The whey has a considerable nutrition value and therefore a high biochemical oxygen demand (BOD) (30,000–50,000 ppm), which prevents a direct discharge in the environment. Membrane processes such as UF and RO are generally used to produce whey protein concentrates (WPCs), lactose, and waste streams with reduced BOD.

UF is largely used for the production of WPCs: the initial protein content of 10–12% (dry bases) can be increased up to 80% with a simultaneous decrease in lactose and some salts. Diafiltration can be used after the concentration step to purify the WPC above 50% protein (dry bases). WPC can be further fractionated into β-lactoglobulin and α-lactoglobulin fractions or can be used for the manufacture of macropeptides with pharmacotherapeutic value (Zydney 1998).

NF is a relatively new but competitive process that is applied for the demineralization of whey, achieving a desalination degree of up to 40%. Larger ions of $Ca^{++}$ or phosphates are retained by the membrane while univalent ions such as $Na^+$, $Cl^-$, and $K^+$ cross through the NF membrane. The advantages are associated with lower plant cost compared to conventional desalination and in the production of a desalted whey useful for special products with reduced content of ash components.

In Figure 1.7, a schematic representation of pressure-driven membrane operations in the milk and dairy industry is reported.

### 1.6.2 Fruit and Vegetable Juices

Juice clarification, stabilization, depectinization, and concentration are typical steps where pressure-driven membrane processes have been successfully used (Fukumoto et al. 1998; Koseoglu et al. 1990). In particular, MF and UF represent a valid alternative to the use of traditional fining agents such as gelatine, bentonite, and silica sol that cause problems of environmental impact due to their disposal (Eykamp 1995). Additional advantages of UF and MF processes over conventional methods include the continuous production of clarified juice, a simplification of the clarification process, the reduction of the quantity of pectinase used, the reduction of clarification times, and the reduction of production costs.

MF and UF membranes separate the juice into a fibrous concentrated pulp and a clarified fraction free of spoilage microorganisms. These membranes retain high-molecular-weight compounds (pectin or proteins) and allow low-molecular-weight solutes (sucrose, acids, salts, aroma, and flavor compounds) to permeate through the membrane.

Fruit juices clarified by membrane operations can be commercialized or submitted to a concentration process to obtain a product suitable for the preparation of juices and beverages.

Fruit juices are usually concentrated to reduce storage, package, and shipping costs (Luh et al. 1986). The concentration of fruit juices is usually obtained by multistage vacuum evaporation; however, this process results in a loss of fresh juice flavors, color degradation, and a "cooked" taste, recognized as off-flavors, due to thermal effects. Alternative techniques, such as freeze concentration systems (cryoconcentration), in which water is removed as ice rather than as vapor, permit to preserve the aroma compounds but they are characterized by high-energy consumptions. Besides, the achievable concentration (about 50°Brix) is lower than the values obtained in thermal evaporation (60–65°Brix).

The concentration of fruit juices by RO has been of interest in the fruit-processing industry for about 30 years. The advantages of RO over conventional concentration techniques are in terms of low thermal damage of the product, reduction of energy consumption, and lower capital investments (Medina and Garcia 1988) as the process is carried out at low temperatures and it does not involve phase change for water removal.

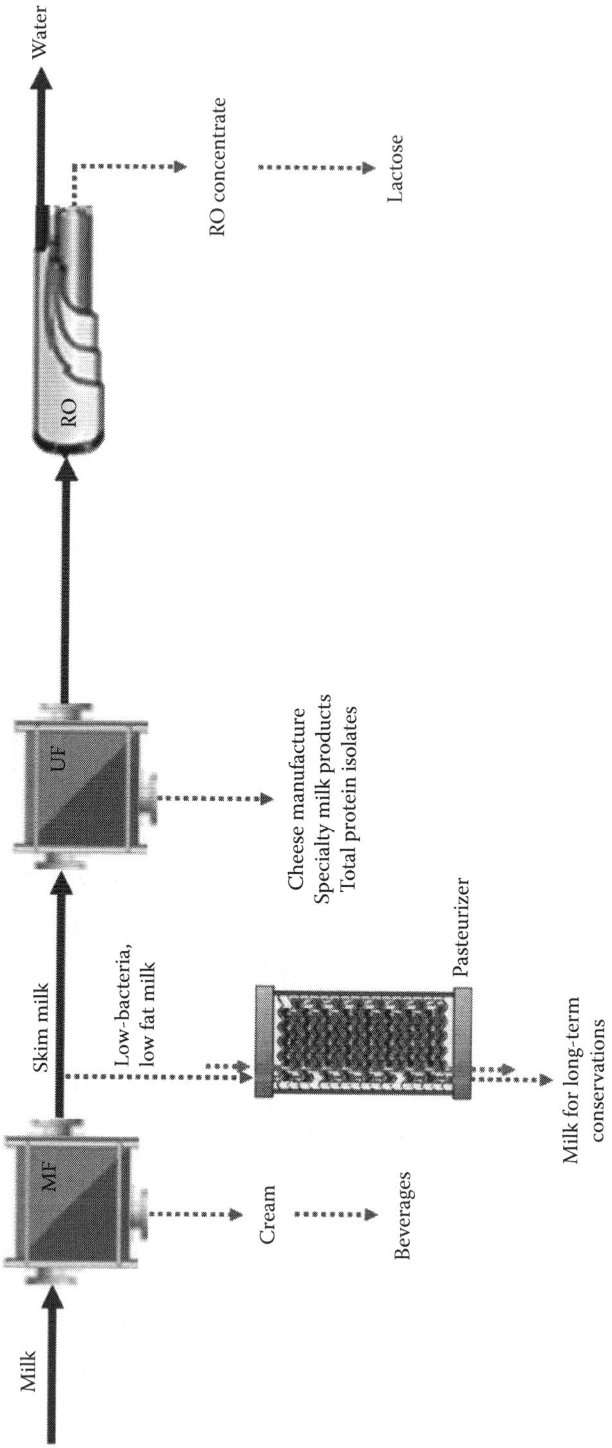


**FIGURE 1.7** Pressure-driven membrane operations in the milk and dairy industry.

However, the osmotic pressure and the viscosity of the juice increase rapidly with the increasing of the sugar concentration. Both factors influence the RO process; so, the final concentration cannot be higher than 20%: otherwise, the process is not convenient from an economical point of view. For these limitations, RO can be considered as an advantageous technique as a preconcentration step. To achieve concentrations higher than 60–70°Brix, the RO needs to be integrated with other processes, that is, evaporation or osmotic distillation (Cassano et al. 2003).

A pilot-scale system investigating an integrated process for the clarification (by UF), concentration (by thermal evaporation), and aroma compounds recovery (by pervaporation, PV) from apple juice has been exploited by Alvarez et al. (2000) (Figure 1.8).

The separation and concentration of polyphenolic compounds from apple juice by using 1 and 0.25 kDa MWCO spiral-wound NF membranes has been reported by Saleh et al. (2006). The concentration of apple and pear juices by NF at low pressures (between 8 and 12 bar) has also been investigated (Warczok et al. 2004).

### 1.6.3 Sugar Industry

Membrane technology can play an important role in the future of the sugar industry by supporting the development of sustainable processes. The initial research effort in the development of membranes and membrane processes for the sugar industry can be dated back to the beginning of the 1970s, when R. F. Madsen *(DDS—De Danske Sukkerfabrikker, nowadays Nordic Sugar, part of Nordzucker)* proposed the use of membrane processes in the beet sugar industry (Madsen 1971).

The use of NF and RO in the treatment of sugar beet press water and the recovery of salt from waste brine in sugar decolorization plants, as well as MF and UF of raw sugar juice as an alternative to chemical purification process, are typical applications of the membrane technology in this area (Hinkova et al. 2002).

The integration of NF and RO has been proposed to improve the efficiency of the conventional sugar extraction. By using NF or RO, it is possible to concentrate the most undesirable components, including sugar from the press water, and submit this stream to low-grade crystallization. The NF/RO permeate stream, which is almost pure water, can be recycled to the extraction unit. Hatziantoniou and Howell (2002) studied the isolation of the sugar beet pectin, which can be used as a thickener, fat replacer, or fat mimic in the food industry by applying UF with diafiltration.

One of the main challenges in the sugar production process is the separation of nonsucrose components. Attridge et al. (2001) have reported the use of UF spiral-wound and tubular membrane modules on a commercial scale to purify and concentrate a clarified and prefiltered juice up to a VRF of 6. Tubular modules can process clarified juices up to a VRF of 20. Similar results were reported by Tyndall (1999) for clarified juice up to a VRF of 5 using tubular membranes (ITT-PCI Membrane Systems, UK) on a pilot scale by using 18 tube housings in single-stage operation.

Other works concentrated their efforts on the removal of color impurities from raw sugar by UF. Mak (1991) reported the use of an Alfa Laval filtration unit

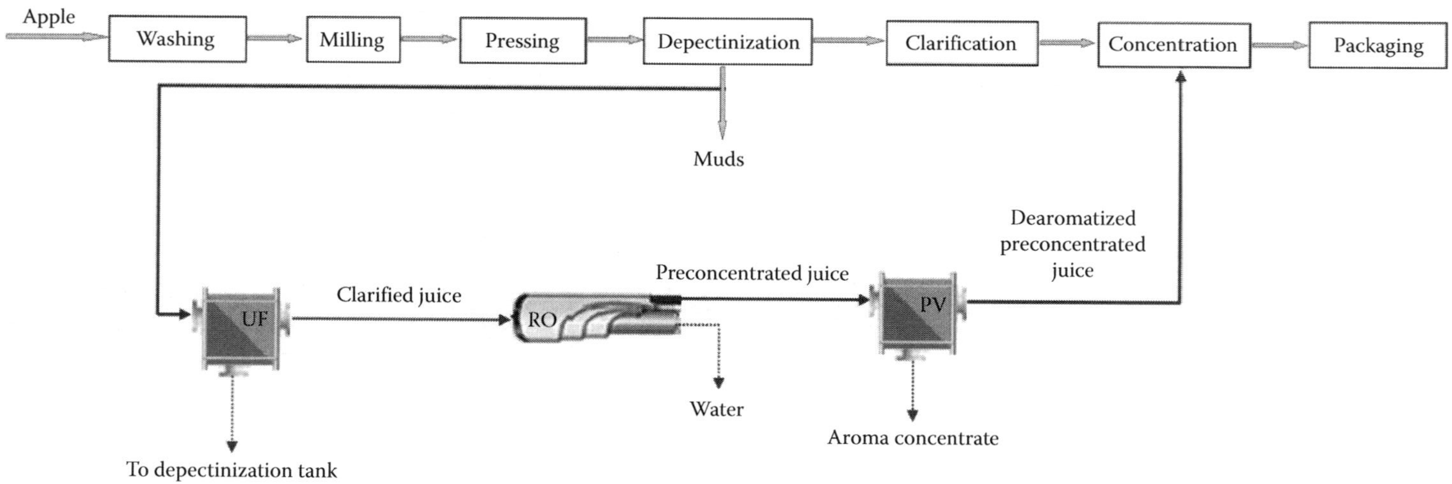


**FIGURE 1.8** Integrated membrane process for the production of apple juice concentrate. (Adapted from Alvarez, S. et al. 2000. *J Food Eng* 46:109–25.)

(Sweden) with PM10 hollow fiber modules that are 1.1 m long and contain a membrane area of 2.46 $m^2$ per module. Experimental results indicated the removal of proteins, starches, gums, and colloids other than color impurities.

After the demineralization process, the thin juice enters a multieffect evaporator system in which the sugar is concentrated from 15% up to 60% (Madaeni and Zereshki 2008). The use of membrane processes, such as NF or RO, represents an interesting alternative. Recently, Gul and Harasek (2012) presented a new multistage pressure-driven membrane separation process for the preconcentration of thin juice. This new membrane concept can concentrate thin sugar juice from an initial feed concentration of 15–50% by the use of membranes at moderate transmembrane pressure of 32 bar and at a temperature of 80°C. Hence, 82% of the water is removed without any phase change. After this, the preconcentrated juice can be further concentrated up to 65–70%. The process can save more than 80% energy in the concentration step and allows about 70% reduction in heat and heat-transfer area for preheating before evaporation.

The use of pressure-driven membrane operations in the sugar industry is summarized in Figure 1.9.

### 1.6.4 Brewing Industry

Beer is the fifth most consumed beverage in the world besides tea, carbonates, milk, and coffee and it continues to be a popular drink with an average consumption of 9.6 L per capita by population aged above 15 (OECD 2005). During production, beer alternately goes through four chemical and biochemical reactions (mashing, boiling, fermentation, and maturation) and three solid–liquid separations (wort separation, wort clarification, and rough beer clarification).

Beer clarification and pasteurization represent the most important operations in the brewing process. In this context, beer filtration with membrane technology has been proposed to overcome the problems associated with the traditional process using kieselguhr filtration. In 2003, the companies Alfa Laval (Sweden) together with Sartorius (Germany) developed a system incorporating filtration membranes from Sartorius that were specifically designed for beer filtration. This technology was created for full-scale breweries with a production from 100 to 500 hL/h or more. The overall advantages of the process include beer quality identical to kieselguhr filtration, simple operation and maintenance, continuous, and automated operation (Atkinson 2005).

At the industrial scale, vacuum distillation (rectification) and vacuum evaporators (single or multistage) have been implemented successfully for reducing the alcohol content in beer. Unfortunately, these processes lead to great loss of beer flavor and liveliness (Branyik et al. 2012). In this context, membrane technology appears as a great alternative because it can be operated automatically at lower temperatures, which is essential when sensitive aroma compounds have to be separated. Typically, thin-film composite RO membranes in a spiral-wound configuration can be used for alcohol reduction, achieving a low-alcohol beer with similar quality to standard beer—low-energy consumption compared to evaporation, no beer heating, no oxygen pickup, and long membrane lifetime. During this step, a permeate stream containing mainly water and alcohol is separated from a concentrated beer stream, which contains most of the flavors.

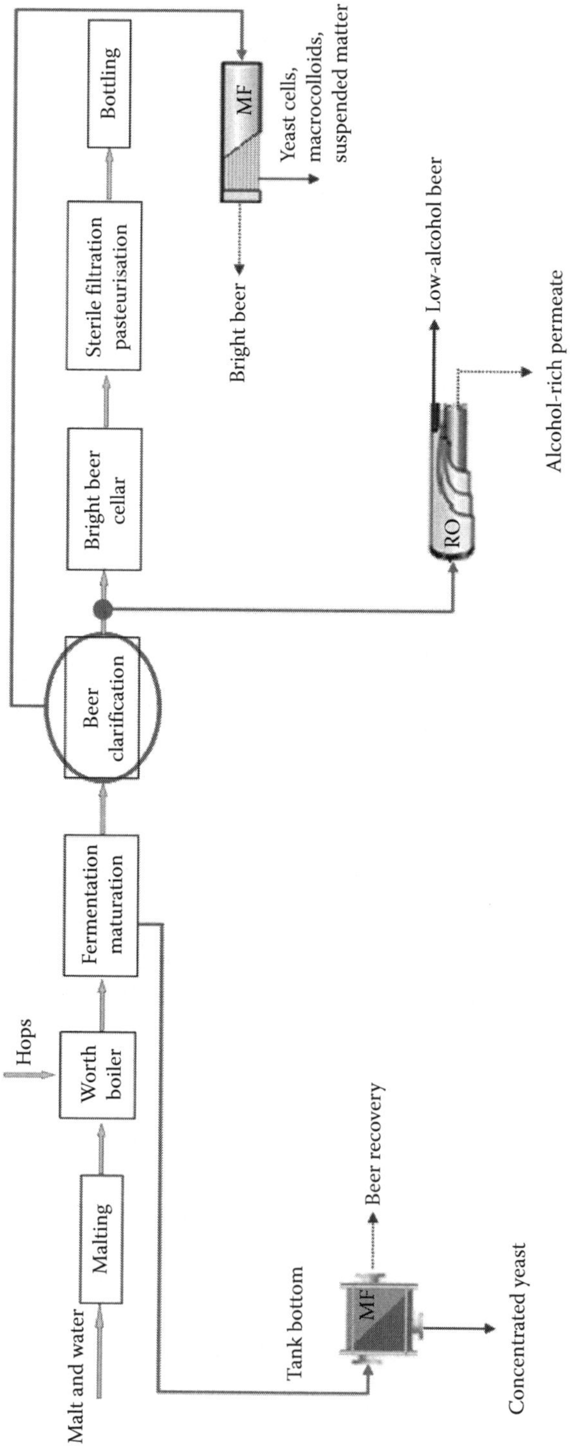


**FIGURE 1.9** Schematic representation of pressure-driven membrane operations in the sugar industry.

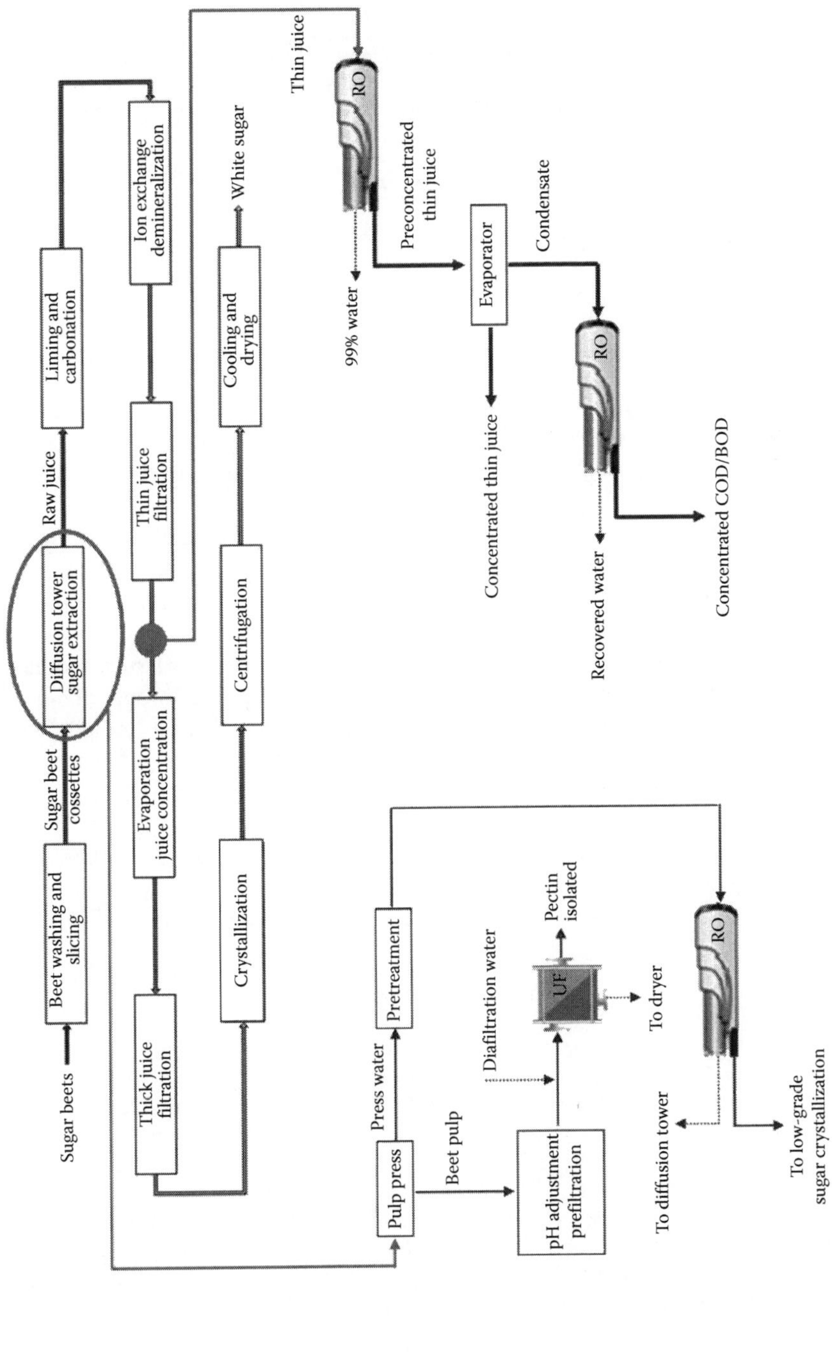


**FIGURE 1.10** Schematic representation of pressure-driven membrane operations in the brewing industry.

After fermentation, yeast settles at the bottom of the fermentation vessels. This precipitate represents 1.5–2% of the total beer volume and contains—apart from the yeast—a high proportion of beer, which is lost if not recovered. Membrane technology appears to be a novel and attractive application to recover the beer and concentrate the yeast up to 20% DM. In 2002, Pall (USA/France) presented the Keraflux system, a full batch process using ceramic membranes, Membralox membranes (Pall Exekia, France). The key advantage of this process includes a recovered beer of high quality (no oxygen pickup, virtually sterile beer). Other benefits are low membrane replacement costs and continuous operation that involves less yeast autolysis and a reduction in the number of tanks.

Pressure-driven membrane operations in the brewing industry are schematically represented in Figure 1.10.

## 1.7 CONCLUDING OUTLOOK

The success of membrane technology in different areas of the agro-food industry is today clearly established, thanks to its key advantages (high selectivity, gentle product treatment, low-energy consumption, and compact design) over conventional separation methodologies.

This chapter provides an overview of pressure-driven membrane operations, the largest and most developed membrane market in the food industry. Well-established applications in some agro-food productions such as milk and dairy, fruit and vegetable juices, beer, and sugar industries have been highlighted.

Pressure-driven membrane operations play an important role also in the treatment of process water and wastewaters of the agro-food industry. An additional driver for these processes is the continuous growth in the market of green and natural products, a segment in which membrane operations have a high potential. Specifically, this is exemplified by a range of development work on extracting plant compounds for food, cosmetics, and well-being products.

Finally, according to the process intensification strategy, technological inputs and economical benefits can be achieved through the integration of different unit membrane operations between themselves or with other conventional technologies.

## REFERENCES

Aimar, P., Howell, J.A. and M. Turner. 1989. Effects of concentration boundary layer development on the flux limitations of ultrafiltration. *Chem Eng Res Des* 67:255–61.

Al Akoum, O., Jaffrin, M.Y., Ding, L.H. and M. Frappart. 2004. Treatment of dairy process waters using a vibrating filtration system and NF and RO membranes. *J Membr Sci* 235:111–22.

Al Akoum, O., Jaffrin, M.Y., Ding, L.H., Paullier, P. and C. Vanhoutte. 2002. A hydrodynamic investigation of microfiltration and ultrafiltration in a vibrating membrane module. *J Membr Sci* 197:37–52.

Álvarez, S., Riera, F.A., Álvarez, R., Coca, J., Cuperus, F.P., Bouwer, S., Boswinkel, G. et al. 2000. A new integrated membrane process for producing clarified apple juice and apple juice aroma concentrate. *J Food Eng* 46:109–25.

Atkinson, S. 2005. Beer clarification. *Membr Technol* 7:8–9.

Atra, R., Vatai, G., Bekassy-Molnar, E. and A. Balint. 2005. Investigation of ultra and nanofiltration for utilization of whey protein and lactose. *J Food Eng* 67:325–32.

Attridge, L., Eringis, A. and I. Jaferey. 2001. Ultrafiltration of beet diffusion juice using spiral and tubular polymeric membranes. In *Biennial Meeting 2001 of the American Society of Sugar Beet Technologists (ASSBT)*, Vancouver, Canada, February 28–March 3, 2001.

Bacchin, P., Aimar, P. and R.W. Field. 2006. Critical and sustainable fluxes: Theory, experiments and applications. *J Membr Sci* 281:42–69.

Belfort, G. 1989. Fluid mechanics in membrane filtration. *J Membr Sci* 40:123–47.

Bhave, R.R. 1991. *Inorganic Membranes: Synthesis, Characterization and Application*. New York: Van Nostrand Reinhold.

Branyik, T., Silva, D.P., Baszczynski, M., Lehnert, R. and J.B.A.E. Silva. 2012. A review of methods of low alcohol and alcohol-free beer production. *J Food Eng* 108:493–506.

Cadotte, J.E. 1985. Evolution of composite reverse osmosis membranes. In *Material Science of Synthetic Membranes*, ed. D.R. Lloyd, ACS Symposium Series 269, 273–94. Washington: American Chemical Society.

Cadotte, J.E. and R.I. Petersen. 1981. Thin film reverse osmosis membranes: Origin, development, and recent advances. In *Synthetic Membranes*, ed. A.F. Turbak, ACS Symposium Series 153, 305–25. Washington: American Chemical Society.

Cassano, A., Drioli, E., Galaverna, G., Marchelli, R., Di Silvestro, G. and P. Cagnasso. 2003. Clarification and concentration of citrus and carrots juice by integrated membrane process. *J Food Eng* 57:153–63.

Cheryan, M. 1998. *Ultrafiltration and Microfiltration Handbook*. Lancaster: Technomic Publishing Company.

Das, C., Patel, P., De, S. and S. Das Gupta. 2006. Treatment of tanning effluent using nanofiltration followed by reverse osmosis. *Sep Purif Technol* 50:291–9.

Daufin, G., Escudier, J.E., Carrère, B.S., Fillaudeau, L. and M. Decloux. 2001. Recent and emerging applications of membrane processes in the food and dairy industry. *Trans Inst Chem Eng* 79:89–102.

Ding, L., Al Akoum, O., Abraham, A. and M.Y. Jaffrin. 2002. Dynamic microfiltration of yeast suspensions using rotating disks equipped with vanes. *J Membr Sci* 197:269–82.

Eykamp, W. 1995. Microfiltration and ultrafiltration. In *Membrane Separation Technology—Principles and Applications*, eds. R.D. Noble, and S.A. Stern, 1–43. Amsterdam: Elsevier.

Frommer, M.A., Feiner, I., Kedem, O. and R. Block. 1970. Mechanism for formation of skinned membranes. 2. Equilibrium properties and osmotic flows determining membrane structure. *Desalination* 7:393–402.

Fukumoto, L.R., Delaquis, P. and B. Girard. 1998. Microfiltration and ultrafiltration ceramic membranes for apple juice clarification. *J Food Sci* 63:845–50.

Gekas, V. and B. Hallstrom. 1987. Mass transfer in the membrane concentration polarization layer under turbulent dross flow: I. Critical literature review and adaptation of existing Sherwood correlation to membrane operations. *J Membr Sci* 30:153–70.

Gul, S. and M. Harasek. 2012. Energy saving in sugar manufacturing through the integration of environmental friendly new membrane processes for thin juice pre-concentration. *Appl Therm Eng* 43:128–33.

Hatziantoniou, D. and J.A. Howell. 2002. Influence of the properties and characteristics of sugar-beet pulp extract on its fouling and rejection behaviour during membrane filtration. *Desalination* 148:67–72.

Hiatt, W.C., Vitzthum, G.H., Wagner, K.B., Gerlach, K. and C. Josefiak. 1985. Microporous membranes via upper critical temperature phase separation. In *Material Science of Synthetic Membranes*, ed. D.R. Lloyd, ACS Symposium Series 269, 229–44. Washington: American Chemical Society.

Hinkova, A., Bubnik, Z., Kadlec, P. and J. Pridal. 2002. Potentials of separation membranes in the sugar industry. *Sep Purif Technol* 26:101–110.

Kamide, K. and S. Manabe. 1985. Role of microphase separation phenomena in the formation of porous polymeric membranes. In *Material Science of Synthetic Membranes*, ed. D.R. Lloyd, ACS Symposium Series 269, 197–228. Washington: American Chemical Society.

Kesting, R.E. 1971. *Synthetic Polymeric Membranes*. New York: McGraw-Hill.

Kesting, R.E. 1985. Phase inversion membranes. In *Material Science of Synthetic Membranes*, ed. D.R. Lloyd, ACS Symposium Series 269, 131–64. Washington: American Chemical Society.

Koseoglu, S.S., Lawhon, J.T. and E.W. Lusas. 1990. Use of membranes in citrus juice processing. *Food Technol* 44:90–7.

Loeb, S. and S. Sourirajan. 1962. Seawater demineralization by means of a semipermeable membrane. In *Advances in Chemistry*, ed. R. Gould, 117–32. Washington: American Chemical Society.

Luh, B.S., Feinberg, B., Chung, J.I. and J.G. Woodroof. 1986. Freezing fruits. In *Commercial Fruit Processing*, eds. J.G. Woodroof and B.S. Luh, 263–351. Westport: AVI Publishing Co.

Madaeni, S.S. and S. Zereshki. 2008. Reverse osmosis alternative: Energy implication for sugar industry. *Chem Eng Process* 47:1075–80.

Madsen, R.F. 1971. Process for the purification and clarification of sugar juice. British Patent No. 1.361.674.

Mak, F.K. 1991. Removal of colour impurities in raw sugar by ultrafiltration. *Int Sugar J* 93: 263–5.

Medina, B.G. and A. Garcia. 1988. Concentration of orange juice by reverse osmosis. *J Food Process Eng* 10:217–30.

Mexis, S.F. and M.G. Kontominas. 2010. Effect of oxygen absorber, nitrogen flushing, packaging material oxygen transmission rate and storage conditions on quality retention of raw whole unpeeled almond kernels (*Prunus dulcis*). *LWT—Food Sci Technol* 43:1–11.

Mota, M., Teixeira, J.A. and A. Yelshin. 2002. Influence of cell-shape on the cake resistance in dead-end and cross-flow filtrations. *Sep Purif Technol* 27:137–44.

Mulder, M. 1991. *Basic Principles of Membrane Technology*. Dordrecht: Kluwer Academic Publishers.

Nunes, S.P. and K.V. Peinemann. 2001. *Membrane Technology in the Chemical Industry*. Weinheim: Wiley-VCH Verlag GmbH.

OECD (Organization for Economic Co-operation and Development). 2005. Annual report. http://www.oecd.org/newsroom/34711139.pdf.

Rai, P., Majumdar, G.C., Sharma, G., DasGupta, S. and S. De. 2006. Effect of various cutoff membranes on permeate flux and quality during filtration of mosambi (*Citrus sinensis* L.) Osbeck juice. *Food Bioprod Process* 84:213–9.

Saleh, Z.S., Stanley, R. and R. Wibisono. 2006. Separation and concentration of health compounds by membrane filtration. *Int J Food Eng* 2:1–14.

Sherwood, R.K., Pigford, R.L. and C.R. Wilke. 1975. *Mass Transfer*. New York: McGraw-Hill.

Singh, V., Jain, P.K. and C. Das. 2013. Performance of spiral wound ultrafiltration membrane module for with and without permeate recycle: Experimental and theoretical consideration. *Desalination* 322:94–103.

Smid, J., Avci, C., Guenay, V., Terpstra, R.A. and J.P.G.M. Van Eijk. 1996. Preparation and characterization of microporous ceramic hollow fiber membranes. *J Membr Sci* 112:85–90.

Sourirajan, S. and B. Kunst. 1977. Cellulose acetate and other cellulose ester membranes. In *Reverse Osmosis and Synthetic Membranes*, ed. S. Sourirajan, 129–152. Ottawa: National Research Council, Canada.

Strathmann, H. 1985. Production of microporous media by phase inversion processes. In *Material Science of Synthetic Membranes*, ed. D.R. Lloyd, ACS Symposium Series 269, 165–195. Washington: American Chemical Society.

Strathmann, H., Giorno, L. and E. Drioli. 2006. *An Introduction to Membrane Science and Technology*. Roma: Consiglio Nazionale delle Ricerche.

Tanaka, T., Abe, K.I., Asakawa, H., Yoshida, H. and K. Nakanishi. 1994. Filtration characteristics and structure of cake in cross-flow filtration of bacterial suspension. *J Ferment Bioeng* 78:455–61.

Tiller, F.M. and C.S. Yeh. 1985. The role of porosity in filtration. Part X. Deposition of compressible cakes on external radial surface. *Am Inst Chem Eng J* 31:1241–8.

Tiller, F.M. and H.R. Cooper. 1962. The role of porosity in filtration. Part V. Porosity variation in filter cakes. *Am Inst Chem Eng J* 8:445–9.

Treybal, R.E. 1981. *Mass Transfer Operations*. New York: McGraw-Hill.

Tyndall, T.J. 1999. Recent developments in sugar clarification with tubular polymeric membranes. In *Symposium on Advanced Technology for Raw Sugar and Cane and Beet Sugar Refined Sugar Production*, New Orleans, USA, September 8–10, 1999.

Warczok, J., Ferrando, M., López, F. and C. Güell. 2004. Concentration of apple and pear juices by nanofiltration at low pressures. *J Food Eng* 63:63–70.

Youn, K.S., Hong, J.H., Bae, D.H., Kim, S.J. and S.D. Kim. 2004. Effective clarifying process of reconstituted apple juice using membrane filtration with filter-aid pretreatment. *J Membr Sci* 228:179–86.

Zydney, A.L. 1998. Protein separation using membrane filtration: New opportunities for whey fractionation. *Int Dairy J* 8:243–50.

# 2 Size Reduction

*Constantina Tzia and Virginia Giannou*

## CONTENTS



## 2.1 INTRODUCTION

Size reduction is the unit operation in which the average size of solid or liquid food material is reduced by the application of shearing (grinding, compression, or impact) forces. Depending on the material, the size reduction operation is called "grinding or cutting" in the case of solid food particles, or "emulsification, homogenization or atomization" in the case of immiscible liquid globules (oil globules in water).

Size reduction operations are commonly used in food-processing technology, as a preliminary treatment of raw materials required before a process (i.e., extraction, drying, etc.), as a process itself intended for a certain food product (i.e., wheat milling for flour production, emulsification for mayonnaise production, etc.), and for the production of particles within a specified size range (granular or powder products). Separation by screening may additionally be used in solid particles.

Various types of equipment specialized for grinding or cutting of solid food materials or for emulsification have been designed and are available for industrial purposes. They are categorized depending on the principle in which their function for size reduction is based; in solid foods, the type of force applied (shear, compression, or impact) and in emulsification, the type of energy for shearing (mechanical, hydro-shearing, or ultrasonic).

The energy requirements of the size reduction process are considerable and are generally described by theoretical models (Kick's, Rittinger's, and Bond's laws). The estimation of the energy of a given process is useful for industrial operations.

The extent or the efficiency of the process is expressed by the *reduction ratio* (RR) (the average size of the feed's particles to the average size of the product's particles). The control of the process is achieved through the measurement of the particle size. The maintenance of globule size and the stabilization of an emulsion are of interest as well.

After application of a size reduction process, the particle size is first reduced, but, at the same time, the surface area of the particle is increased. Therefore, the process concurrently affects the nutritional, rheological, and sensorial properties and the shelf life of foods.

## 2.2 CLASSIFICATION OF SIZE REDUCTION METHODS

The objective of size reduction methods is to produce from larger particles smaller ones with superior characteristics regarding the size of their surface area, or their shape, size, and number. The methods are classified according to the nature of food materials (solid or liquid) that are subjected to size reduction and analyzed in the following paragraphs (Fellows, 2000).

1. Solid foods
   a. Cutting (chopping, crushing, slicing, and dicing)
   b. Grinding (milling or comminution) to powders or pastes of increasing fineness
2. Liquid
   Emulsification and homogenization

Size reduction of solid materials is referred to as the operation that creates smaller particles of the same material by breaking or tearing through mechanical action. There are two types of operation: cutting and grinding. Depending on the food material, the terms are specified as crushing, grinding, milling, mincing, and dicing (i.e., milling of cereals, mincing of beef, dicing of tubers, and grinding of spices). In all types of operation, the solid material is subjected to stress by application of force.

Three different types of force are used to reduce the size of foods (Barbosa-Canovas et al., 2005a; Brennan, 2006; Fellows, 2000):

1. Compression forces
2. Impact forces
3. Shearing (or attrition) forces

Cutting is applied to break down large pieces of food into smaller pieces of a desired size and even of a defined shape. The size of the resulted particles may be adjusted to be suitable for further processing or specified for a final product. Both of them are often applied in food processing, for example, the cutting of meat for retail sales and the dicing of processed meat for ready-to-eat meal preparation.

Grinding is applied to reduce the material's size by fracturing. In the mechanism of the process, the material properties, the equipment used, the time stress applied, and the energy required are involved. Using a certain mill, the material is stressed by mechanical action and fracture will occur along the lines of weakness, when the local strain energy exceeds a critical level. The energy absorbed internally by the material is consumed for its fracture, while a significant part of energy is released as heat. The fracture depends on the mechanical properties (i.e., hardness, friability) of the material and on the magnitude of the force and time of its action—in general, fracture may happen at lower stress levels if force is applied for longer times. The common grinding processes in the food industry are the milling of grains for flour production, of corn for cornstarch production, of sugar, and of various dried foods.

All three types of forces can be used in solid-size reduction processes, depending on the mechanical properties of the material and the desired particle size of the final product, by using the suitable equipment. For hard, friable, or crystalline foods, compressive forces are generally used. The same forces can also be applied to break down externally hard grains and facilitate the separation of the internal part, as in the separation of the endosperm from the bran in wheat. For semisoft materials, impact forces are used and for soft materials, shear forces are used, respectively, while for fibrous materials, combined impact and shearing forces are necessary. For coarse crushing (to a size of about 3 mm), compressive forces are used, while for coarse, medium, and fine grinding of various food materials, impact forces can generally be used. Finally, for fine grinding (pulverization) and even ultrafine grinding that represent a product's size in the micrometer or submicron range, respectively, shear (or attrition) forces are applied. In size reduction equipment, more than one type of force may take part; however, in most mills, one force dominates their function (Barbosa-Canovas et al., 2005b; Brennan, 2006; Fellows, 2000).

Therefore, the type of mill most suitable for a certain food application can be selected according to products' characteristics (i.e., roller mills use compression, hammer mills apply impact forces, and disk attrition mills utilize shear forces while a combination of forces is necessary for the production of powdered sugar, flour, mustard, and cocoa).

In liquid materials, usually oil or fat globules in a water phase, the size reduction operation is related to the creation of smaller globules by shearing forces and is called homogenization or emulsification (Walstra et al., 2006). The process is applied to natural food systems or involved in new product designs. In both cases, emulsifying agents, present or added, contribute to the production and maintenance of stable emulsions. Suitable equipment has been developed to provide the energy needed for emulsification.

## 2.3 PURPOSE FOR USE/APPLICATION OF A SIZE REDUCTION PROCESS

Size reduction operations are often necessary in food-processing technology for different purposes; they may be used either for pretreatment of raw materials or as a basic or complementary processing step in food production. The most common purposes for the application of size reduction in food processing, either for solid (grinding or cutting) or for liquid (emulsification or homogenization) foods are presented below.

- *Preliminary treatment of raw materials*: If raw materials occur in too large sizes that cannot be easily treated, then they must be reduced in size. Apart from facilitating handling and treatment of small-sized materials, for certain processes size reduction and the consequent surface-area-to-volume ratio increase may be required to improve the efficiency or shorten the process time. For example, cutting vegetables or fruits increases the rate of drying and shortens the cooling and heat treatments (blanching, cooking); also grinding of oilseeds before expression or extraction results in increased oil recovery (Xu and Diosady, 2003). Furthermore, complete mixing of solid ingredients is achieved when they have a similar range of particle size (ingredients for cake mixes).
- *Individual processes intended for a certain food product*: There are technological methods based on the size reduction process that have been developed especially for certain products. Thus, the process itself intends to produce well-established commercial products. Milling of wheat combined with sieving for white flour production and emulsification for mayonnaise production are representative examples for solid and liquid foods, respectively. The raw milk homogenization is an operation conventionally included in dairy technology and allows the production of stable dairy products standardized in fats (Walstra et al., 2006; Wilbey, 2011).
- *Production of particles within a specified size range (granular or powder products)*: The particle size is important for the correct functional or processing properties of some products, such as granular or powdered products (i.e., icing sugar, spices, and cornstarch). Thus, after the size reduction process, screening follows to achieve the predetermined range of particle size.

## 2.4 SIZE REDUCTION OF SOLID FOODS

### 2.4.1 Principles of a Size Reduction Process

The behavior (mechanical resistance) of solid food materials subjected to a force is usually described by stress–strain diagrams (Barbosa-Canovas et al., 2005b;

Brennan, 2006; Fellows, 2000). A representative diagram is shown in Figure 2.1 in which the relation of stress versus strain is presented for foods with different mechanical properties as follows: (1) = hard, strong, and brittle material; (2) = hard, strong, and ductile material; (3) = soft, weak, and ductile material; and (4) = soft, weak, and brittle material.

The application of stress to a food causes the deformation of its tissues. If the tissues return to their original shape when the stress is removed, the food exhibits elastic behavior. This represents the linear relation between stress and strain, named as elastic region (O–E) (Figure 2.1). This is continued until the stress reaches a limit, named as *elastic stress limit (E)*. Energy is internally absorbed for tissues deformation without breaking the food and is released as heat.

When the applied stress exceeds the elastic stress limit, the food undergoes permanent deformation until the stress reaches the *yield point (Y)*. Beyond the yield point, the food will further deform or flow along the region of ductility (Y–B) (Figure 2.1) until it reaches the *breaking point (B)*. Finally, when the applied stress exceeds the breaking point, the food will fracture along a line of weakness. During the breaking part of the absorbed energy, it is released as sound and heat.

The particular behavior of each food depends on its structure and mechanical properties. So, as shown in Figure 2.1, hard food materials exhibit increased modulus of elasticity *E* (ratio of stress to strain) than soft food materials; strong food materials possess higher elastic stress limits than weak food materials; brittle food

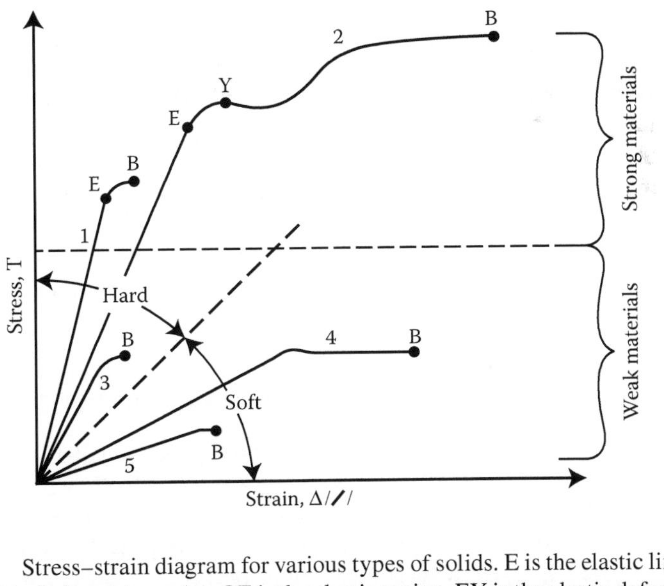


**FIGURE 2.1** Stress–strain diagram for various types of solids. E is the elastic limit, Y is the yield point, B is the breaking point, OE is the elastic region, EY is the elastic deformation, and YB is the region of ductility. Different curves are represented for different types of material depending on their mechanical behavior: material (1): hard, strong, and brittle; material (2): hard, strong, and ductile; material (3): hard, weak, and brittle; material (4): soft, weak, and ductile; and material (5): soft, weak, and brittle. (From Loncin M., Merson R.L. 1979. *Food Engineering*. 246–264. New York: Academic Press.)

materials break soon after the stress exceeds the yield stress, while ductile food materials can further deform without breaking; and tough food materials have the ability to resist the propagation of cracks, while the friable food materials rupture along the existing fissures and then along new-formed fissures or crack tips.

On the basis of the above-mentioned properties, the mechanical and structural properties of foods such as hardness and friability (tendency to crack) considerably determine the energy needed for their fracture; hard foods and/or those with only few lines of weakness available require higher energy input. In addition, the moisture and heat sensitivity of foods influence the energy input; therefore, in certain foods, the moisture content or the temperature is conditioned before milling (i.e., wet milling of maize, cryogenic grinding of spices) to achieve complete disintegration or preserve the heat-sensitive components, respectively.

Grinding proceeds more easily in large particles than the small particles; large particles contain more fissures and have more lines of weakness available, while in the small particles, new crack tips will be created during grinding. That means that the breaking strength of small particles is higher than the large particles, thus requiring additional energy for their fracture. Moreover, during grinding, fewer lines of weakness remain available, thereby increasing the energy requirements. Furthermore, to achieve very fine particles, intermolecular forces must be overcome. There is no such need in the usual food applications; however, control of the size distribution of a given product prevents overgrinding while saving time and energy (Barbosa-Canovas et al., 2005b; Brennan, 2006; Fellows, 2000).

Regarding the total energy consumed, the major part is employed in elastic and inelastic deformation of the particles, elastic distortion of the equipment, friction between particles or between particles and the equipment, friction losses in the equipment, and the heat, noise, and vibration generated by the equipment. However, the actual energy used in size reduction and generation of new surfaces constitutes only about 1% of the applied energy. Concluding, in the industrial processes of size reduction, the amount of energy that must be consumed (depending on the size of forces and the time of application), the extent of size reduction, and the new surface that can be achieved, as well as the losses from heat generated are all related. Theoretical approaches for the estimation of energy and new surface formed during grinding are therefore available in the literature.

### 2.4.2 Particle Properties and New Surface Formed by Grinding

As previously referred to, during the crushing of a material, its particle size is progressively changing; so, starting from sizes varying in range from relative coarse to fine or even powder, it inclines to a certain predominant size. As a result of the particles' size reduction, the surface area of the fine particulate material is increased, being available for reactions, thus influencing its properties (Earle, 1983). For example, the first crushing of wheat results in the coarse flour with a wide range of particle sizes. However, after further grinding, the predominant fraction becomes the one passing through a 250 mm sieve and being retained on a 125 mm sieve. Also, regarding the change of properties by grinding, wheat grains can be kept relatively stable under dry storage, while fine flour becomes liable to oxidation.

In addition to the size and the new surface formed by grinding of particles, their geometric characteristics may be important for the final product's evaluation. The actual processes yield products that normally consist of a mixture of particles varying in sizes and even shapes. The shape of particles obtained after grinding resembles a polyhedron with several nearly plane faces and as their size becomes smaller, they can be considered to be spherical.

The terms generally used to describe the shape of a particle and its typical dimensions are "sphericity" and "diameter" respectively. The typical dimension ($x_p$) is related to the surface area ($s_p$), volume ($V_p$), and sphericity ($\Phi_s$) of the particles as

$$s_p = \alpha_s x_p^2 \tag{2.1}$$

$$V_p = \alpha_v x_p^3 \tag{2.2}$$

and

$$\Phi_S = \frac{6V_P}{x_P s_P} \tag{2.3}$$

where $s_p$ is the area of the particle, $V_p$ is the volume of the particle's surface, $x_p$ is the equivalent diameter of the particle, and $\alpha_s$, $\alpha_v$ are the surface and volume factors, respectively (Barbosa-Canovas et al., 2005a).

The numerical values of $\alpha_s$, $\alpha_v$ are factors that connect the particles' geometries and are dependent on the particles' shapes. For spherical particles, the value of sphericity is 1, while for crushed materials, it ranges between 0.6 and 0.7. The area or the mass of particles can be combined with the results of sieve analysis to estimate the total surface area of a powder.

### 2.4.3 Process Efficiency: Evaluation of the Size Reduction Result

In size reduction operations, the aim is to produce, from a starting material, particles within a specified size range. In such a process, the efficiency, expressed as the breakdown extent in relation to the energy consumed, as well as the average size and/or the new formed surface area of the products' particles are important.

*RR* is usually used for evaluation/measurement of the breakdown extent of a food material expressed as the ratio of the average size of the feed's particles to the average size of the product's particles. Although the RR depends on the specific type of equipment used, it appears that the coarser the reduction, in general, the smaller the ratio. Size RRs may vary depending on the properties of the starting material and the target particle size in the final product (i.e., from <8:1 in coarse crushing to more than 100:1 in fine grinding). The RR can also be used as an estimate of the performance of a size reduction operation (Brennan, 2006; Fellows, 2000).

Otherwise, the extent of grinding is influenced by both the magnitude of force applied (compression, impact, or shear) and the time of its application. Therefore, an efficient grinding must exhibit the lowest possible energy consumption. Theoretical estimates of the energy required for a specific grinding are provided in literature. However, in practice, excess energy is needed and a significant amount of it is lost as heat.

The *average size* of particles obtained by cutting or grinding is of interest whether further processing is required or when it is the final product. Moreover, the product's specification may require to contain particles of a specified size.

In the food industry, separation by sieving or classification by screening usually follows after the grinding process for the product's recovery or for obtaining ground fractions, respectively. According to common industrial practices, a multistage grinding can be applied in which each stage may employ different types of mills and may be further combined with sieving (i.e., crushing rolls in series of decreasing diameters in the milling of wheat for flour production).

Apart from the average size of particles formed by grinding, the amount of new surface formed is of great interest. The resultant surface area of the particles is important for their further processing (extraction, freezing, drying, etc.) or for their stable storage due to the large surface area available for reactions (oxidations in fine particulate materials).

It must be noted that the determination of the average particle size depends on the method of measurement. Screen analysis is widely used for determining the particle size distribution in granular materials and powders. In practice, particle size is referred to as screen aperture size. Other analytical techniques currently available for particle size measurement are microscopy techniques, sedimentation, light-scattering techniques (DLS), stream scanning, and online measurement techniques (Bancarz et al., 2008).

### 2.4.4 Energy Used in Grinding

Grinding is considered possibly the most inefficient unit operation in the food-processing industry. For the estimation of the amount of energy required for grinding processes, mathematical models and laws based on theoretical assumptions are available. Such estimations are useful for processors allowing the best energy management, that is, the minimum energy consumption with minimum losses as heat for the efficient operation of a specific size reduction process (Brennan et al., 1990; Leniger and Beverloo, 1975; Meuser, 2003).

The common base in these theories is the following assumption: The energy, d$E$, required to produce a change, d$x$, in the typical size dimension $x$ (i.e., length, diameter) of a material's particle is a simple power function of the size of the material, thus expressed as

$$\frac{\mathrm{d}E}{\mathrm{d}x} = \frac{K}{x^n} \tag{2.4}$$

where d$E$ is the differential energy required, d$x$ is the change in a typical dimension, $x$ is the magnitude of a typical size dimension, and $K$ and $n$ are constants.

The three following laws and equations are used for the estimation of the energy required to reduce the size of solid foods.

**Kick's law** states that the energy required to reduce the particle size of a material is directly proportional to the size RR, namely, the ratio of the initial size of the typical dimension in the feed's particles to the final size of this dimension in the product's particles. That gives $n = 1$ in Equation 2.4, thus obtaining

$$E = K_K \ln\left(\frac{x_1}{x_2}\right) \tag{2.5}$$

where $E$ (J) is the energy required per mass of feed, $K_K$ is Kick's constant, and $x_1/x_2$ is the *size RR* with $x_1$ (m): the average initial size of the feed's particles and $x_2$ (m): the average size of the product's particles.

**Rittinger's law** states that the energy required to reduce the particle size of a material is proportional to the new surface area produced. That gives $n = 2$ in Equation 2.4, thus obtaining

$$E = K_R\left(\frac{1}{x_2} - \frac{1}{x_1}\right) \tag{2.6}$$

where $E$ (J) is the energy per mass of feed required to produce the new surface area, $x_1$ (m) is the average initial size of the feed's particles, $x_2$ (m): is the average size of the product's particles, and $K_R$: is Rittinger's constant.

**Bond's law** considers that the energy required for the reduction is inversely proportional to the square root of the particle size produced. So, putting $n = 3/2$ in Equation 2.4, it results in

$$E = 2K\left(\frac{1}{\sqrt{x_2}} - \frac{1}{\sqrt{x_1}}\right) \tag{2.7}$$

where $E$ (kWh ton$^{-1}$) is the energy required for size reduction, $x_1$ and $x_2$ (μm) are the average size of the feed's and product's particles, respectively, and $K = 5E_i$, where $E_i$ (J kg$^{-1}$) is the Bond Work Index defined as the energy required to reduce a unit mass of material from an infinite particle size to a size such that 80% passes a 100-μm sieve. Values of Bond Work Index obtained experimentally range from 40.000 to 80.000 J kg$^{-1}$ for hard foods such as sugar or grains.

Kick's law is best applied providing reasonable results to coarse crushing, where most of the energy is used in causing fracture along the existing cracks and the increase in surface area per unit mass is relatively small (for *RR*s below 8:1). Rittinger's law applies better in fine grinding that results in a large increase in surface area (*RR*s exceeding 100:1). Bond's law can be applied reasonably well to a variety of materials undergoing coarse, intermediate, and fine grinding. It can be concluded that the above equations can be used in comparative estimations of the energy requirements between processes with various degrees of size reduction (Barbosa-Canovas et al., 2005b; Brennan, 2006; Fellows, 2000; Meuser, 2003).

## 2.4.5 Equipment

### 2.4.5.1 Food Factors Affecting the Selection of the Size Reduction Process: Criteria for the Selection of Equipment

Food processors have to design a size reduction process and decide on the appropriate equipment. When selecting a size reduction machine, the following factors of the food material feed should be considered: (a) mechanical and structural properties, (b) moisture content, (c) temperature sensitivity, and (d) the size distribution of the feed as well as of the product (Barbosa-Canovas et al., 2005b; Brennan, 2006).

On the basis of these criteria, the most appropriate equipment can be selected in each case. Additionally, the size and other features, such as the capacity and rotational velocity of the machine, should be evaluated and its operating variables should be normally adjusted. Finally, good practices regarding loading, operation, or stopping, which do not exceed the critical speed, cleaning, maintenance, and so on, should be followed to exploit the complete grinding capacity of the mill.

#### *2.4.5.1.1 Mechanical and Structural Properties of the Feed*

The mechanical and structural properties of the feed material (i.e., hardness and friability) are of major importance in selecting the suitable equipment for size reduction. The knowledge of these parameters is useful to determine the type of force needed in performing the size reduction. Moreover, the design of a size reduction operation for a certain food material may require experimental determination of the processing parameters. Accordingly, requirements for the wearing material, the working surfaces, the easy replacement of moving parts, and the entire construction of mills should be met.

Friable and crystalline materials may fracture easily, with larger particles breaking down more easily than smaller particles, starting from the current cleavage planes and then creating new crack tips. So, crushing using compressive forces is initially recommended followed by impact and shear forces. Roller mills are usually employed for such materials.

Hard food materials may be either brittle, fracturing rapidly above the elastic limit, or ductile, deforming extensively before breakdown. As already mentioned, the harder the material, the more difficult it is to break down and more energy is required. This should be taken into account in the case of very hard materials, where larger mills or extended milling may be required. Since many dry food materials are brittle and fragile, compressive forces are recommended. Finally, for most solid foods, ball mills, hammer mills, roller mills, and attrition mills are commonly used in the food industry.

Foods with fibrous structure exhibit increased toughness and resistance to breakdown, requiring cutting than crushing (i.e., cutting of roasted whole beans to ground coffee with particles of suitable size and shape). Disk mills, pin-disk mills, or cutting devices are usually employed to break down such materials.

#### *2.4.5.1.2 Moisture Content of the Feed*

The moisture content and distribution significantly affect both the degree of size reduction and the mechanism of breakdown in some foods. Each food material must have the optimum moisture content for milling to achieve the expedient performance;

if it is above or below that limit, problems may arise in the product's quality or the mill's safety. Therefore, before milling, the food materials are subjected to conditioning or soaking to obtain the optimum moisture content; such examples are the dry milling of wheat after conditioning and the wet milling of maize after soaking to attain complete disintegration of the starchy materials. When exceeding the optimum moisture content upward or downward, agglomeration of the product's particles, which generates difficulty in the product's free flowing and blocking of the mill, or problems due to dust formation associated to human health or inflammation and explosion may arise, respectively. Anyway, in both cases, these have an adverse impact on the efficiency of the process. Thus, in the case of wheat, excessive moisture content may deform rather than break, without allowing the release of the endosperm, while, in the case of dry milling, fine particles of bran are created that cannot be separated by the screens, thereby contaminating the white flour.

##### *2.4.5.1.3 Temperature Sensitivity*

Considerable amounts of heat, generated during the mill's operation, lead to a considerable temperature increase that can adversely affect the quality of food materials subjected to grinding. To protect the heat-sensitive and volatile components of foods, milling should be performed under reduced temperatures. For this reason, mills equipped with cooling systems or cryogenic milling (by mixing the feed with solid carbon dioxide or liquid nitrogen before milling) are recommended. Such examples are the milling of fibrous materials such as meat or the fine grinding of roasted coffee that is facilitated under cryogenic conditions and the cool milling of spices to prevent the undesirable heating effects on sensitive volatile components. Also, the increase of surface temperature during grinding of sugar may lead to overstepping, locally, the glass transition temperature. This results in the generation of an amorphous surface layer that is hygroscopic and may induce recrystallization and consequently agglomeration of the sugar particles.

##### *2.4.5.1.4 Size Distribution of Feed and Product*

Each type of crusher or grinder is intended for a certain particle size of the feed and product. Recommendations for selecting the appropriate equipment in relation to the food material and the reduction range, as shown in Table 2.1, should be taken into account. Size control and/or screening of the feed material allow the proper operation of the equipment, thereby protecting from jamming, and so on, while monitoring of the product's size and distribution is necessary, in particular for the quality control of the final product.

#### 2.4.5.2 Mills

##### *2.4.5.2.1 Classification*

There are various types and sizes of equipment available for use in size reduction or comminution operations in the food- processing industry. They are commonly classified according to their basic function into one of three classes:

(a) Crushers that use compressive forces and are mainly used for coarse reduction;

**TABLE 2.1**

**Size Reduction Equipment: Types, Characteristics, and Applications in Food Processing**

| Generic Name of Equipment | Type of Equipment | Operating Force | | | | | Fineness Range | | | | Typical Products |
|---|---|---|---|---|---|---|---|---|---|---|---|
| | | Compression | Impact | Shearing | Attrition | Cutting | Coarse lumps[1] | Coarse grits[2] | Interme-diate[3] | Fine–Ultra fine[4] | |
| Crushers | Crushing rolls | * | | | | | * | | | | Cane sugar, wheat |
| Grinders—mills | *(i) Roller mills (crushing rolls)* | | | | | | | | | | |
| | Roller mills | * | | * | | | | * | * | * | Wheat, roasted coffee, and chocolate refining |
| | *(ii) Impact (percussion) mills* | | | | | | | | | | |
| | Hammer mill | | * | | | | | * | * | | Spices, sugar, dried milk, cocoa press cake, dry fruits, and vegetables |
| | Beater bar mill | | * | * | | | | * | * | | Dried vegetables, spices |
| | Comminuting mill | | * | | | * | | | | | Fruits, vegetables |
| | Pin (pin-disk) mill | | * | * | | | | * | * | | Sugar, starch, roasted nuts, cocoa powder, and spices |

| | | | | | | | | | | |
|---|---|---|---|---|---|---|---|---|---|---|
| | *(iii) Attrition mills* | | | | | | | | | |
| | Single-disk attrition mill | | | * | | | | * | * | Wheat, corn, cocoa kernels, sugar, spices, and nuts |
| | Double-disk attrition mill | | | * | | | | * | * | |
| | Buhrstone mill | | | * | | | | * | * | |
| | Colloid mill | | | * | | | | | | Pastes and purees |
| | *(iv) Tumbling mills* | | | | | | | | | |
| | Ball mills | * | * | | | | | | * | Fine grinding, cocoa mass |
| | Rod mills | * | * | | | | | | * | Sticky products |
| Cutters | *(i) Slicing and flaking equipment* | | | | * | * | | | | Meat, fruits and vegetables, and dry particulate foods |
| | *(ii) Dicing equipment* | | | | * | * | | | | |
| | *(iii) Shredding equipment* | | | | * | * | * | | | |
| | *(iv) Pulping equipment* | | | | * | | * | * | | |
| | Bowl chopper | | | | | | | | | |

[1]:1–0.1 m, [2]:10–1 cm, [3]:10–1 mm, and [4]:100–10 μm.

(b) Grinders or mills that combine shear and impact with compressive forces. They can be additionally divided into force grinders, employed in intermediate and fine reduction, or ultrafine grinders; and
(c) Cutting machines used for exact reduction.

Another classification of the size reduction equipment can be made according to the feed material into: (a) an equipment for size reduction of dry foods (including categories A and B of the previous classification) and (b) equipment for size reduction of fibrous foods (including category C of the previous classification).

There are several types of machines included in each category of the above classification (Barbosa-Canovas et al., 2005b; Brennan et al., 1990; Brennan, 2006; Fellows, 2000; Leniger and Beverloo, 1975; Meuser, 2003). The appropriate equipment is selected according to its operation principles and individual characteristics as well as the desired reduction range for its applications in specific food materials as shown in Table 2.1. Nowadays, to address claims, such as reduced noise, operating temperatures, and dust generation, or to satisfy the requirements of modern food industries (i.e., controlled comminution, high speed of production, micro-fine powders specification), new, improved types of size reduction equipment (i.e., ultrasonic cutters) are commercially available.

###### *2.4.5.2.1.1 Crushers*

*Jaw and gyratory crushers*: The jaw crusher consists of two heavy jaws, one fixed and the other reciprocating (placed sideways). The feed material is introduced between the jaws and as it proceeds downward, into the narrowest space of the machine, it is forced to crush. The gyratory crusher consists of a truncated conical fixed casing and inside this, a crushing head, shaped as an inverted cone, which is rotated eccentrically. The material is fed between the outer and the inner gyrating cones and crushed as it is forced into a narrowest space of the machine. These crushers are no longer widely used in the food industry (Earle, 1983).

*Crushing rolls*: This type of crusher consists of two large horizontal, smooth-surfaced cylinders, placed close together in parallel, which are driven in opposite directions at the same or different speeds. The material is fed into the space between the rotating rolls, and as it passes through it is nipped and crushed. Such crushers usually achieve an RR of 4 or lower. An example of their use is with the cane sugar industry.

###### *2.4.5.2.1.2 Grinders or Mills*

*Roller mills (crushing rolls):* A *roller mill* contains two (or more) smooth or finely fluted, corrugated, and grooved steel horizontal rolls, placed close together in parallel and rotating at different speeds. The food particles are fed from above in the space between the rollers (nip), then nipped and pulled through (Figure 2.2). The main force contributing to the breakdown is compression; if the rollers are rotated at different speeds, or if the rollers are fluted, then there is the addition of shear force. The nip is adjustable,

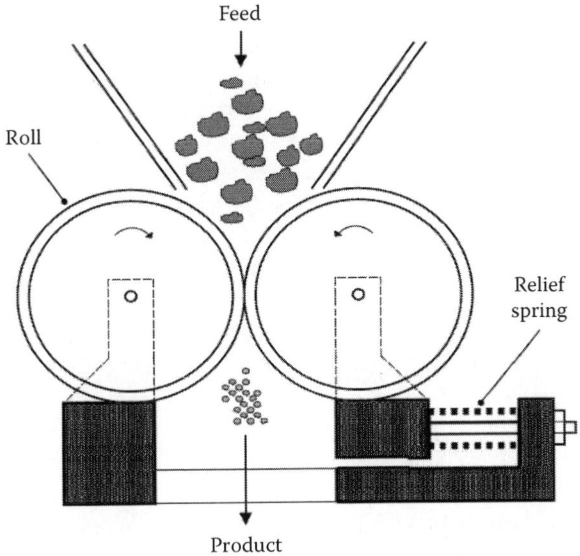


**FIGURE 2.2** A roller mill (crushing rolls).

depending on the particle size of the feed, and their overloading must be avoided, protecting the rollers against damage. If necessary, the surface of the rolls may be cooled or heated (Barbosa-Canovas et al., 2005b; Brennan, 2006; Fellows, 2000; Meuser, 2003). There are alternative roller mills that carry breaker bars or teeth on the roller surfaces, increasing the friction and facilitating the trapping of particles, or, as in the case of *edge mills*, they consist of one roll operating against a flat or curved stationary plate (Brennan, 2006).

The surface of the rolls are selected and their operating speeds are adjusted according to the feed and product material (i.e., fluted, corrugated, or grooved surfaces for high RR whereas smooth surfaces for fine milling). For a certain roller mill system, the maximum permissible particle size of the feed and the theoretical mass flow rate of the product can be estimated. In practice, the available roller mills have diameters >500 mm and rotating speeds in the range of 50–300 rpm; smaller rolls may operate at higher speeds. Roller mills are widely applied in the milling of wheat, in roasted coffee grinding, and in the refining of chocolate. Usually, high RRs are achieved using two or more pairs of rolls in sequence with increasing smoothness and decreasing nip size from one pair to the next, with each milling step to be accompanied by sieving. Wheat grains are milled by this process.

*Impact (percussion) mills:* A *hammer mill* contains a high-speed rotor mounted on a horizontal shaft inside a cylinder casing; a toughened plate (breaker plate) is also fitted inside the casing. The rotor carries swinging hammerheads around its periphery that pass within a small clearance of the casing (Figure 2.3). As the hammers drive, the fed material is forced against the breaker plate. The ground particles may pass through a screen mounted in the bottom of the casing; so, a product of a certain diameter is obtained (choke feeding). Fracture of particles is mainly by impact forces, although shear forces or even attrition forces under choke feeding can also evolve. Hammer mills are usually heavy devices of large capacity, with rotors rotating at 1.000–1.200 rpm. The operation of mills with fibrous or sticky materials may be associated with screen blockage; therefore, mills must be equipped with removable screens and/or with cooling systems. Also, a high proportion of fine particles in the product may result in the grinding of friable materials (Barbosa-Canovas et al., 2005b; Brennan, 2006; Fellows, 2000; Meuser, 2003).

A wide variety of materials can be handled in hammer mills, from hard and friable to fibrous and sticky; for fibrous materials, knives instead of hammers may be used to give a cutting action. They are widely used in the food industry for various materials including spices, sugar, dried milk, cocoa press cake, dry fruits, and vegetables.

The *beater bar mill* carries bars instead of hammers, the tips of which pass within a small clearance of the casing.

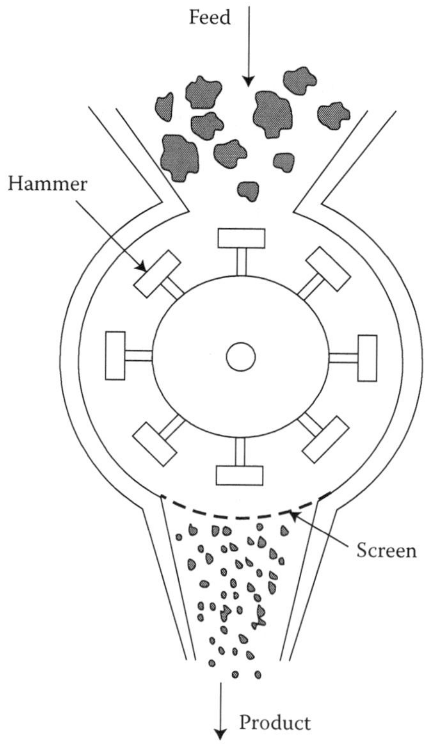


**FIGURE 2.3** A hammer mill.

The *comminuting mill* carries knives instead of bars or hammers that can be either hinged to the shaft and swing out as it rotates or be rigidly fixed to it. They are used in relatively soft materials, such as fruits and vegetables. Moreover, the knives may be sharp on one edge and blunt on the other; when they rotate in the first direction, they perform a cutting action while on the opposite direction, they act as beater bars (Earle, 1983).

A *pin (pin-disk) mill* contains two disks, a stationary and a rotating, located with a small clearance between them. Both disks have pins, pegs, or teeth attached on the surface so that the rows of one disk fit alternately into the rows of the other disk. The material is subjected to impact and shear forces between the stationary and rotating pins (Figure 2.4). Disk speeds may be up to 10.000 rpm. In other types of mills, it is possible to operate in a choke-feed mode (with the product's screen attached), to rotate the two disks in the same direction at different speeds, or in opposite directions, to adjust the clearance between the disks, to feed the material by air stream, or to control the temperature. Improved effectiveness is attained in pin-and-disk mills with intermeshing pins fixed either to the single disk and casing or to double disks. Such mills are used for fine grinding of friable materials and for breaking down fibrous substances (i.e., sugar, starch, roasted nuts,

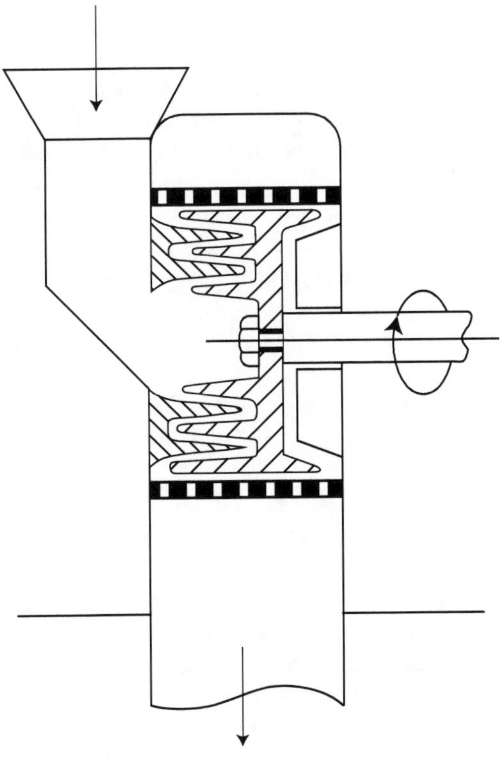

**FIGURE 2.4** A pin mill.

cocoa powder, and spices) (Barbosa-Canovas et al., 2005b; Brennan, 2006; Fellows, 2000; Meuser, 2003).

In *fluid energy (jet) mill,* the feed material is suspended in a high-velocity gas stream into a chamber where the breakdown occurs through the impact between individual particles and the chamber wall, while the product is recovered by an air–solids separation cyclone system.

*Attrition mills*: There are several designs of attrition mills (disk or plate mills). They use shear forces for fine grinding or shearing and impact forces for coarser grinding. The attrition mills are employed in the dry milling of wheat and wet milling of corn. Other applications include grinding of cocoa kernels, sugar, spices, and nuts (Barbosa-Canovas et al., 2005b; Brennan, 2006; Fellows, 2000; Meuser, 2003).

A *single-disk attrition mill* has a high-speed rotating grooved disk close to a stationary disk with matching grooves. The material is fed at the center of the stationary disk, passing through the disks and breaks due to intense shearing between the disks and pressure subjected (Figure 2.5a). The gap between the two disks is adjustable, and a product screen may be attached (Brennan, 2006; Fellows, 2000).

A *double-disk attrition mill* contains two counter-rotating disks with matching grooves, located close to each other. The breakdown of the material is caused by shear forces and the product is discharged, passing through a screen; however, a greater degree of shear is achieved compared to the single-disk mill (Figure 2.5b). Both types of disk attrition mills are used for corn and rice milling (Brennan, 2006; Fellows, 2000).

The *Buhrstone mill* consists of two circular stones mounted on a vertical axis of which the upper stone is fixed, while the bottom stone rotates. The feed material moving between the grooved stones is gradually ground by a scissor action and is discharged over the edge of the lower stone (Figure 2.5c). In some Buhrstone mills, both stones rotate, in opposite directions. It is the oldest attrition mill, used for wheat milling for centuries, and is still used today for wet milling of corn and whole-grain flour production.

The *Foos mill* is a modified pin mill with studs instead of grooves, with similar applications as the other disk mills.

The *colloid mill* is also an attrition mill used for emulsification as well as in pastes and purees.

*Tumbling mills:* Tumbling mills consist of a horizontal slow-speed rotating cylinder, sometimes with conical ends, half-filled with a solid-grinding medium, most commonly balls or rods. As the cylinder shell rotates, the grinding medium units are lifted up the sides of the shell and then dropped onto the material being milled, which fills the void spaces between the grinding medium units. In this method, effective size reduction is achieved by exerting impact and shearing action on the feed material (Figure 2.6). A common problem in tumbling milling is the product's contamination because of

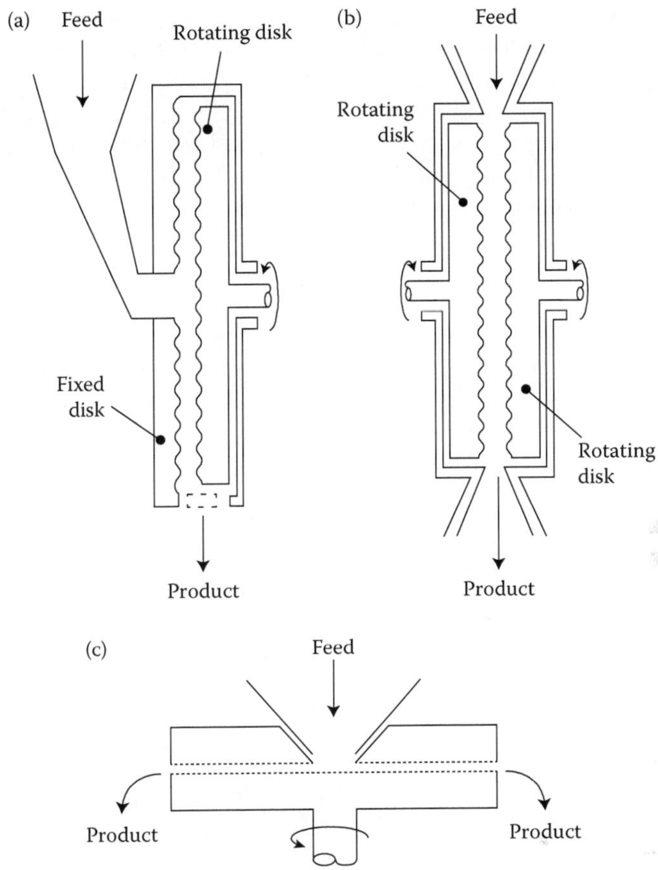


**FIGURE 2.5** Disk attrition mills: (a) single-disk mill; (b) double-disk mill; and (c) Buhr mill.

the wear of the grinding medium. A usual application of tumbling mills is in fine grinding (Barbosa-Canovas et al., 2005b; Brennan, 2006; Fellows, 2000; Meuser, 2003).

In *ball mills,* the cylinder shell is commonly made of steel lined with carbon-steel plate, porcelain, silica rock, or rubber while the grinding medium are steel (or porcelain or zircon) balls (25–150 mm in diameter), occupying 50% of the shell volume. *Pebble mills* use flint pebbles as grinding materials. At low rotation speeds or if the balls used are small, shearing forces predominate. At higher rotation speeds or with the use of larger balls, impact forces are generated, contributing to the size reduction of the feed material (Brennan, 2006; Fellows, 2000). Ball mills operate under batch or continuous processes. They are used for fine grinding, both under dry or wet conditions (i.e., fluid cocoa mass).

Another type is the *trunnion overflow mill* in which the feed's material passes through a hollow trunnion at one end, and the ground product overflows at the opposite end (Figure 2.6a). In *compartment mills*, grinding can be performed over sequential compartments of the mill using different grinding medium (large balls, small balls, and pebbles), thus increasing the efficiency of the milling operation (Figure 2.6b). Finally, the *conical ball mills* (Figure 2.6c) consist of a single cone-shaped section that contains balls of different sizes. The material is fed through the widest part of the mill while the product leaves through the narrowest part of the cone. As the shell rotates, the large balls move toward the feed point while the small balls migrate toward the discharge outlet. The initial breaking of the feed material is thereby performed by the largest balls dropping the greatest distance, whereas the final reduction of small particles is done by small balls dropping a smaller distance (Barbosa-Canovas et al., 2005b; Brennan, 2006).

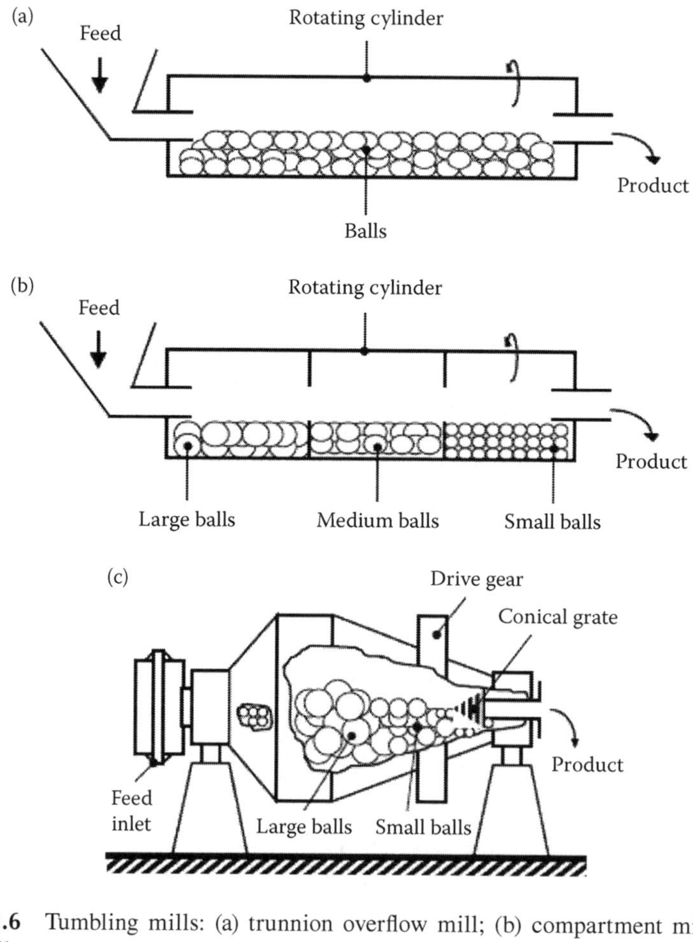


**FIGURE 2.6** Tumbling mills: (a) trunnion overflow mill; (b) compartment mill; and (c) conical mill.

In vibration-type ball mills, impact forces predominate and very fine grinding is attainable. In the *attritor mill*, grinding is achieved by slow stirring of the balls together with the feed particles, which generates shear and probably impact forces. It is commonly used in wet milling (i.e., chocolate manufacture).

In *rod mills,* instead of balls, rods (of circular, square, or hexagonal cross section) are used. These are made of high carbon steel with 25–125 mm diameter and occupy 35% of the shell volume. In these mills, size reduction is attained mainly by attrition forces, while impact forces may also contribute. Rod mills arc more suitable for sticky products than ball mills (Brennan, 2006; Fellows, 2000).

*2.4.5.2.1.3 Cutters* Cutting equipment is generally simple, consisting of rotating knives in various arrangements. As a general requirement in all types, the knives blades must be kept sharp, to minimize the force needed for cutting and also to reduce the products' damage. Selected equipment is used to reduce the size of fibrous foods (i.e., meats, fruits, and vegetables) or pulps as well as of dry particulate foods into powders. The main types of size reduction equipment are presented below, classified in order of decreasing particle size (Fellows, 2000).

*Slicing and flaking equipment: High-speed slicers* (Figure 2.7a) are employed for the precise cutting of foods (i.e., bread, cheese, meat, vegetables, etc.). Meat cutting is usually done by a circular rotary action and is more effective when its mass is frozen. In fruits and vegetables, slicing and simultaneous decoring is obtained in ambient or chilled temperatures by stationary knives fitted inside a tube with or without the use of water (*hydrocutter*). In some types, the food can be held against the blades by centrifugal force so that the slices fall away easily. Finally, food-grade lubricants may be required in slicing of sticky products. Nowadays, computer-controlled and programmed cutters and slicers, are able to cut products into special shapes and controlled size, (Fellows, 2000).

*Ultrasonic cutters* achieve an improved cutting capability and have found applications particularly in difficult-to-cut products such as bakery goods, ice cream, fresh meats, fish, and vegetables.

Flaking equipment is similar to slicing equipment; by adjusting the blades' type and spacing, they are capable of producing flakes (i.e., flaked fish, nuts, and meat).

*Dicing equipment:* Dicing is a two-step process. The first step involves slicing and cutting of food into strips by rotating blades. It is then followed by cutting of strips into cubes of the desired dimensions by a second set of rotating knives, which operate at right angles to the first set of knives (Figure 2.7b). Such applications are used in vegetables, fruits, and meats (Fellows, 2000).

*Shredding equipment:* Typical shredding equipment resembles hammer mills, though the equipment uses knives instead of hammers. Another type,

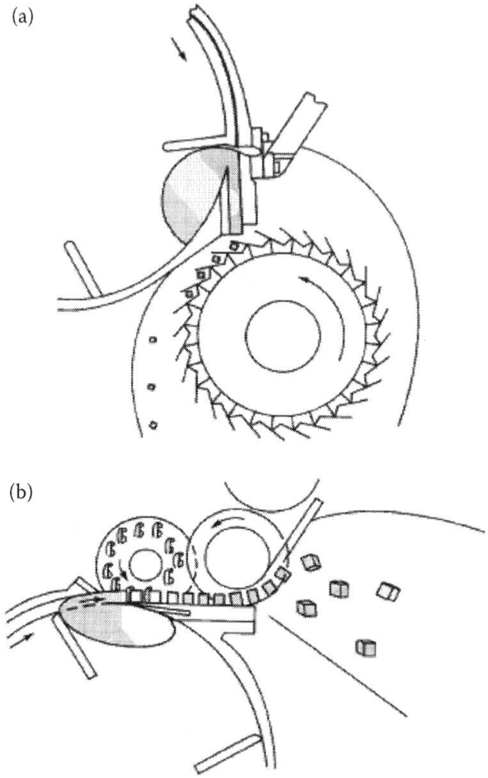


**FIGURE 2.7** Slicing equipment (a) and dicing equipment (b).

known as *squirrel-cage disintegrator*, is composed of two concentric cylindrical, cages fitted with knife blades along their length, inside a casing. Shredding is performed as the food passes between the two cages, which move in opposite directions, by shearing and cutting forces (Fellows, 2000).

*Pulping equipment:* A *bowl chopper* consists of a horizontal bowl, rotating at slow speeds, forcing feed material beneath a set of high-speed rotating blades. The extent of size reduction depends on the number of times the food passes beneath the knives. It is used for chopping meat and reducing harder fruits and vegetables into a pulp (Fellows, 2000).

### 2.4.6 Effect on Foods

Size reduction is widely used in food processing as it facilitates and improves the performance of various processes, such as mixing and heat transfer, while also contributing to the control of important quality characteristics of foods (i.e., the texture in bakery products, the rheological properties in fruit juices, the color in flour, and the flavor/aroma in herbs/spices). However, the size reduction induces considerable alterations in the foods' properties that significantly affect their shelf life (Fellows, 2000; Taoukis et al., 1997).

The increase in surface area by size reduction contributes to the promotion of oxidative deterioration and the enhancement of microbiological and enzymic activity. All these will adversely affect the sensorial and nutritional properties of foods and consequently their keepability.

In particular, oxidation of fats incurred due to the increase of temperature during milling, generates rancid flavor and taste; oxidation with similar effects also continues during storage of ground products, depending on the moisture content of food and the storage conditions (temperature, humidity, and oxygen concentration in the atmosphere). Owing to the same reasons, oxidation and losses of volatile constituents occur during grinding and storage of spices and nuts. In addition, oxidation of certain substances, such as carotenes, provokes bleaching in flour, but mainly reduction of the nutritional value of relative foods. Substantial losses of nutritional value due to vitamin C and thiamin oxidation are observed in some products, such as chopped or sliced fruits and vegetables. Loss of nutritive components can also arise in dry foods during the separations that follow milling (i.e., loss of vitamins in milled rice, loss of fiber in wheat flour). As far as the microbiological and enzymic activity is concerned, they are accelerated more in moist foods, such as fruits and vegetables, deteriorating their texture (due to the release of hydrolytic enzymes) and resulting in the development of undesirable color and off-flavors (due to microbiological growth). The above changes must be controlled prior to subsequent preservation operations.

## 2.5 SIZE REDUCTION IN LIQUID FOODS

### 2.5.1 Emulsification and Homogenization

Emulsification and homogenization are relative size reduction processes for liquid or semiliquid food systems. They are determined as follows:

Emulsification is the formation of a stable emulsion by the intimate mixing of two or more immiscible liquids (usually oil and water), so that one (dispersed phase) is formed into very small droplets within the second (continuous phase) (Friberg, 2004; McClements, 2005). Homogenization is the reduction in size (up to 0.5–30 μm), and hence the increase in number, of solid or liquid particles in the dispersed phase by the application of intense shearing forces (Walstra et al., 2006).

There are two types of liquid–liquid emulsion: oil-in-water (o/w) and water-in-oil (w/o). To maintain the droplet structure, the oil–water interface is coated with a necessary layer of emulsifier (Friberg, 2004; McClements, 2005; Tadros, 2009). In o/w emulsions, the stability is highly dependent on the emulsifier. However, in some o/w systems (i.e., butter), the continuous phase is partially solid, which enhances stability and provides food with a firm texture. Several examples of these simple food systems can be cited, such as milk, a natural o/w emulsion, and margarine, a formulated w/o emulsion respectively, while various food products (i.e., ice cream, sausages, etc.) are considered as more complex emulsions (Fellows, 2000).

Homogenization is, in general, a more severe operation than emulsification. In both processes, the stability of the emulsions produced is required, as determined mainly by the size of globules in the dispersed phase (McClements, 2007). In practice, the emulsions are considered stable when the globules' size is between 1 and 10 μm, the densities of the two phases are reasonably close, and the viscosity of the continuous phase is high.

Homogenizers reduce the size of droplets in the dispersed phase and emulsifying agents form micelles around the droplets reducing the interfacial tension between the phases, thus preventing the droplets from coalescing. Emulsifiers are naturally present (i.e., phospholipids) in foods or added (glycerol esters or sorbitan esters of fatty acids) to food systems. Polar-emulsifying agents are suitable for o/w emulsions and non-polar-emulsifying agents are suitable for w/o emulsions, respectively. They are characterized by the hydrophile–lipophile balance (HLB) value; agents with low HLB values (below 9) (lipophilic) are used in w/o emulsions while those with high HLB values (11–20) (hydrophilic) are used in o/w emulsions, respectively. Polar-emulsifying agents can either be ionic, presenting different surface activities over the pH range, or nonionic, with activities independent of the pH value. Stabilizers are polysaccharide hydrocolloids that dissolve in water to form viscous solutions or gels. In o/w emulsions, they increase viscosity and form a three-dimensional network that stabilizes the emulsion and prevents coalescence. Concluding, as emulsifying agents lower the energy input required to form an emulsion, their proper selection is important to create the desired emulsion in a given food system (Fellows, 2000; Whitehurst, 2004).

### 2.5.2 Equipment

The main types of homogenizer and the principles of their operation are described in a separate chapter. The operation of most types is based on shearing, turbulence, and cavitation actions. *High-speed mixers* are used to premix emulsions of low-viscosity liquids. *Pressure homogenizers* applying pressures of 10–50 MPa reduce the globule size (Walstra et al., 2006; Wilbey, 2011). *Colloid mills* are more effective than pressure homogenizers for high-viscosity liquids. *Ultrasonic homogenizers* via high-frequency sound waves (18–30 kHz) form emulsions with droplet sizes of 1–2 μm and are used for salad creams, ice cream, and essential oil emulsions or for dispersing powders in liquids (Ashokkumar et al., 2010). The *hydroshear* produces droplets in the dispersed phase within a range of 2–8 μm while the *microfluidizer* produces droplets of <1 μm in diameter.

### 2.5.3 Effect on Foods

Size reduction and emulsification are used to prepare raw materials before processing, to produce stable products and improve the properties and the eating quality of foods (i.e., homogenization results in stable milk and dairy products) as well as to develop new products. However, as in dry foods, the changes induced due to the increase in the exposed surfaces may promote oxidative, hydrolytic, enzymic, and microbiological degradation activity. Changes in food properties, nutritional value, and the shelf life of products provoked by homogenization and emulsification processes are also of interest (Fellows, 2000).

Homogenization and emulsification may affect the physicochemical and sensory properties of several foods (i.e., appearance, color, texture, aroma, and flavor). For example, the homogenized milk, which contains fat globules ranging in size from 4 to <1 μm, displays the following characteristics: homogeneous appearance without fat separation in the top of the milk's volume, lighter white color, because of the greater reflectance and scattering of light from the larger number of globules, increased viscosity, due to the higher number of globules and adsorption of casein onto the globule surface, creamer texture, improved flavor and aroma, as volatile components are better dispersed throughout the milk, and improved digestibility. It must be noted that the control of homogenization conditions is required (severe conditions may induce burnt taste and flavor). In emulsified products, the nutritional value as well as other properties are determined by the formulation used (i.e., vitamins or color, aroma materials, etc. are incorporated in margarine and low-fat spreads). It must be noted that the role of the emulsifier and stabilizer in the stability of emulsified products is very important. The type and concentration of the emulsifier determines the structure and thickness of the interfacial layer being crucial for the resulting oxidative stability of the emulsion. In addition, some of these agents can furthermore exhibit antioxidative properties or attribute to the desired mouthfeel properties. Finally, in food emulsions, the degradative changes in natural or added components, such as hydrolysis or oxidation of pigments, aroma compounds and vitamins, and microbial growth, should be controlled through the enforcement of adequate processing, hygienic packaging, and proper storage conditions (Frisenfeldt Horn et al., 2012).

## 2.6 EXAMPLES OF SIZE REDUCTION APPLICATIONS IN FOOD PROCESSING

Size reduction is normally applied to various food materials with different mechanical properties and a variety of grinding characteristics. Representative examples of size reduction applications in food processing that require different technological approaches are presented below.

### 2.6.1 Wheat

Wheat consists of the outer layer or bran, the white starchy endosperm, and the embryo or germ. If necessary, grains are conditioned to obtain the optimum moisture content and then subjected to milling. Milling breaks the bran, separates the endosperm from the embryo and bran and grinds it into white flour. The *percentage extraction rate* determines the number of parts of flour by weight produced per 100 parts of wheat. In practice, the extraction rate ranges between 70% and 80%.

The milling system includes break (4–5 pairs of rolls) and reduction (up to 16 pairs of rolls) sections, each composing of fluted rolls with decreasing distance between them and successively shorter flutes, which rotate in the same direction but with different speeds. The grain is fed and crushed through the first pair of break rolls resulting in large fragments of endosperm (semolina), a small amount of small particles (flour), and fragments of bran containing a small part of the endosperm; the fractions are separated by sieving. The bran passes through the next pair of crush rolls so that more semolina and flour is released. Through this procedure, more flour is successively produced and semolina is refined in the reduction section passing through smooth rolls rotating at slow speeds; thus, by breaking of the larger fragments, additional flour is released (by sieving) while the germ is separated.

The processing efficiency of flour mills has significantly improved over the past few years due to the recent developments in the optimization of both the equipment's design and capacity (more compact flour mills), and the milling process through advanced control applications (DCS, SCADA systems) (Owens, 2001).

### 2.6.2 Chocolate

Size reduction processes are applied in several stages of the chocolate manufacture. Initially, raw cocoa beans are predried and undergo breaking of their shells using an impact mill to release the nibs. The beans are fed onto plate or disk mills, rotating at high speeds and flung against breaker plates by centrifugal force; a second rotating disk with the same function can be potentially used. Separation of nibs is achieved by a combination of sieving and air classification. Through the above process, the particle size is reduced to the range of 1.5–7 mm (Brennan, 2006).

Separated cocoa nibs, containing about 55% butter within cells (20–30 μm in size), are ground to cocoa liquor for cocoa butter release. Grinding of tough fibrous nibs can be achieved in two steps using mills working in series. In the first step, pregrinding is carried out by a hammer, disk or pin mill, or more frequently by ball mills with balls of decreasing size in the direction of the liquor flow (from 15 to 2 mm). In the second step, grinding is carried out using an agitated ball or a three single-disk attrition mill to attain a particle size ranging between 15 and 50 μm. The refining step is important in providing a sufficient particle size of the dispersed phase (i.e., sugar or milk powder) that does not impart gritty texture to the chocolate. For this purpose, a five-roll system is used, and the gap between them decreases from bottom to top while their rotation speed increases in that direction; their temperature is controlled. The well-mixed cocoa liquor is fed between the two bottom rolls and moves to the faster roll upward where the product is received. The desired RR is 5–10. Finally, in cocoa powder production, the press cake of partially defatted ground nibs, after blending, is milled in a pin mill to a powder that is further cooled and stabilized by solidification of the remaining fat (Ziegler and Hogg, 1999).

### 2.6.3 Coffee Beans

Coffee beans, after roasting, are cooled and then ground to the desired particle size according to the characteristics and intended use of the final coffee product. Therefore, the actual particle sizes may vary from larger than 850 μm (very coarse)—for instant coffee production, 850 μm (coarse)—to be percolated, to 450 μm (medium)—to be filtered, and to 50 μm (finely)—for Turkish and espresso coffee production.

In large-scale grinding of coffee beans, the available systems mainly consist of three or four pairs of rolls (for coarse–medium and fine grinding, respectively) located one above the other with the rolls in each pair rotating at different speeds.

The fast roll, in each pair, features U-shaped corrugations running lengthwise, while the other has peripheral corrugations. The clearance between successive pairs of rolls decreases while the roll speed increases as the beans are passing down. For smaller-scale grinding of roasted coffee beans, attrition and hammer mills are employed. The original method for fresh grinding of coffee beans used Buhrstone-type mills and even now, similar mills, using serrated, metallic disks instead of stones, are in use. Nowadays, cryogenic grinding of coffee beans (using solid carbon dioxide or liquid nitrogen) is considered as the most appropriate method to maximize the flavor volatiles retention; however, care should be taken for carbon dioxide release during grinding (Brennan, 2006; Clarke, 1987).

### 2.6.4 Oilseeds and Nuts

Oilseeds prior to oil recovery, either by mechanical (pressing or centrifugation in olives) or solvent extraction (i.e., hexane extraction in soybeans) are usually ground and/or flaked. In such cases, hammer and attrition mills are commonly used for the preliminary breakdown of large oilseeds, such as palm or soybean, which are then flaked. Another system frequently used in similar materials includes five rolls located one above the other, from which, the top roll is grooved with the rest being smooth. The seeds are fed in between the two top rolls and then pass back and forth between the rolls down to the bottom of the stack. Therefore, seeds are subjected to increasing pressure, as each roll supports the weight of the rolls above it. Such systems are employed for cottonseed, flaxseed, and peanuts milling (Kemper, 2005; Xu and Diosady, 2003).

### 2.6.5 Sugarcane

Sugarcanes are subjected to breaking and tearing before the sugar extraction step. The pretreatment includes crushing the cane under high pressures, using large, two- or three-roll crushers with a grooved surface followed by breaking using shredding equipment (large-capacity hammer mills) to tear open the cane cells and release the juice. The hammers in the mill are pivoted to disks or plates and pass close to a plate made up of rectangular bars. This allows the shredded cane to obtain a fibrous structure that holds it together, facilitating the extraction of the juice. The juice extraction is achieved by pressing the shredded cane with added water through five triple-roll mills with grooved surfaces (Brennan, 2006; Schiweck et al., 2007).

### 2.6.6 Other Applications in Solid Foods

Mustard seeds are milled by a procedure similar to that of wheat. Breaking is performed using grooved-surfaced rolls while reduction is performed using smooth-surfaced rolls. In spices or similar materials, although impact and attrition mills can be used, the heat generated reduces their sensorial properties leading to the development of undesirable aromas and flavors. Thus, cryogenic milling or cooling of mills' surfaces are recommended to avoid volatile components' losses and attain milled products of high quality. Other common applications in the food industry are the grinding of sugar crystals using impact mills to produce icing sugar and the size reduction of milk powder, lactose, and dry whey using impact mills (Brennan, 2006).

### 2.6.7 Milk Homogenization

Milk homogenization is a standard process commonly applied in the dairy industry. Milk is a natural o/w emulsion with the milk fat globules acting as the dispersed phase. In raw milk, the fat globules have diameters of 2–10 μm and tend to collide and aggregate, thus rising to the surface of milk's volume forming two phases. The aim of homogenization is to prevent the unsolicited fat separation occurring in milk. During homogenization, the fat globules are subjected to severe processing conditions; thereby, high shear stress and temperature gradient develop and cavitation occurs, decreasing their diameter to 1–0.1 μm. The two-step pressure homogenization method is conventionally used in the dairy industry, with applied pressures ranging between 10 and 20 MPa (Walstra et al., 2006; Wilbey, 2011). The homogenization may further affect the quality of certain dairy products, for example, the texture of yogurt. Ultrasonication is considered as an alternative method for milk homogenization with promising results (Ashokkumar et al., 2010).

## REFERENCES

Ashokkumar M., Bhaskaracharya R., Kentish S., Lee J., Palmer M., Zisu B. 2010. The ultrasonic processing of dairy products—An overview. *Dairy Science and Technology*, 90(2–3): 147–168.

Bancarz D., Huck D., Kaszuba M., Pugh D., Ward-Smith S. 2008. Particle sizing in the food and beverage industry (Chapter 7). In: *Nondestructive Testing Quality,* eds., Irudayara J., Reh C. Ames: Blackwell Publishing and Institute of Food Technologists.

Barbosa-Canovas G.V., Ortega-Rivas E., Juliano P., Yan H. 2005a. Particle properties (Chapter 2). In: *Food Powders, Physical Properties, Processing, and Functionality.* 19–53. New York: Kluwer Academic/Plenum Publishers.

Barbosa-Canovas G.V., Ortega-Rivas E., Juliano P., Yan H. 2005b. Size reduction (Chapter 6). In: *Food Powders, Physical Properties, Processing, and Functionality*. 117–173. New York: Kluwer Academic/Plenum Publishers.

Brennan J.G. 2006. Mixing emulsification and size reduction (Chapter 15). In: *Food Processing Handbook*. 537–558. Weinheim: Wiley-VCH Verlag GmbH & Co.

Brennan J.G., Butters J.R., Cowell N.D., Lilly A.E.V. 1990. Size reduction and screening of solids (Chapter 4). In: *Food Engineering Operations* (3rd edn.). 121–122. London: Elsevier Applied Science.

Clarke R.J. 1987. Roasting and grinding. In: *Coffee Technology*, eds., Clarke R., Macrae R., Vol. 2, 273–107. London: Elsevier Applied Science.

Earle R.L. 1983. Size reduction (Chapter 11). In: *Unit Operations in Food Processing*. Oxford: Pergamon Press.

Fellows P. 2000. Size reduction (Chapter 4). In: *Food Processing Technology: Principles and Practice* (2nd edn.). 98–117. Boca Raton: Woodhead Publishing Limited and CRC Press, LLC.

Friberg S.E. 2004. *Food Emulsions*. New York: Marcel Dekker.

Frisenfeldt Horn A., Skall Nielsen N., Sogaard Jensen L., Horsewell A., Jacobsen C. 2012. The choice of homogenisation equipment affects lipid oxidation in emulsions. *Food Chemistry*, 134: 803–810.

Kemper T. 2005. Oil extraction. In: *Bailey's Industrial Oil and Fat Products*, ed. Shahidi F., Vol. 5, 57–97. Hoboken: John Wiley & Sons, Inc.

Leniger H.A., Beverloo W.A. 1975. Mechanical operations: Size reduction and size enlargement. In: *Food Process Engineering*. 169–188. Dordrecht: Reidel Publishing.

Loncin M., Merson R.L. 1979. *Food Engineering*. 246–264. New York: Academic Press.

McClements D.J. 2005. *Food Emulsions Principles, Practices and Techniques* (2nd edn.). New York: CRC Press.

McClements D.J. 2007. Critical review of techniques and methodologies for characterization of emulsion stability. *Critical Review of Food Science Nutrition*, 47: 611–649.

Meuser F. 2003. Types of mill and their uses. In: *Encyclopedia of Food Science and Nutrition* (2nd edn.), eds. Caballero B., Trugo L.C., Finglas P.M., 3987–3997. London: Academic Press.

Owens W.G. 2001. Wheat, corn and coarse grains milling (Chapter 3). In: *Cereals Processing Technology*. 27–52. Boca Raton: Woodhead Publishing Limited and CRC Press, LLC.

Schiweck H., Clarke M., Pollach G. 2007. Sugar. In: *Ullmann's Encyclopedia of Industrial Chemistry*. Weinheim: Wiley-VCH Verlag GmbH & Co. DOI: 10.1002/14356007.a25_345.pub2.

Tadros T.F. 2009. Emulsion science and technology: A general introduction. In: *Emulsion Science and Technology*. 1–56. Weinheim: Wiley-VCH Verlag GmbH and Co.

Taoukis P.S., Labuza T.P., Saguy I. 1997. Kinetics of food deterioration and shelf-life prediction (Chapter 10). In: *The Handbook of Food Engineering Practice*. eds., Valentas K.J., Rotstein E., Singh R.P., 361–403. Boca Raton: CRC Press, LLC.

Walstra P., Wouters J.T.M., Geurts T.J. 2006. Homogenization (Chapter 9). In: *Dairy Science and Technology*. 279–296. Boca Raton: Taylor & Francis Group, LLC.

Whitehurst R.J. 2004. *Emulsifiers in Food Technology*. London: Blackwell Publishing.

Wilbey R.A. 2011. Homogenization of milk, principles and mechanism of homogenization, effects and assessment of efficiency: Valve homogenizers. In: *Encyclopedia of Dairy Sciences* (2nd edn.), ed. Fuquay J.W., 750–754. Cambridge: Elsevier Ltd.

Xu L., Diosady L.L. 2003. Fats and oils from plant materials (Chapter 6). In: *Extraction Optimization in Food Engineering*, eds., Tzia C., Liadakis G., 181–215. New York: Marcel Dekker.

Ziegler G., Hogg R. 1999. Particle size reduction. In: *Industrial Chocolate Manufacture and Use* (3rd edn.), ed. Beckett S.T., 115–136. Oxford: Blackwell Science.

# 3 Centrifugation–Filtration

*Theodoros Varzakas*

## CONTENTS

## 3.1 INTRODUCTION

Centrifuges are used to clarify nonsettling solutions containing finely divided solids; to break emulsions of two immiscible liquids; to obtain a clear solution from a liquid–solid mixture; or to separate two immiscible liquids.

Separation operations involve properties such as particle size, density, and solubility, and have to do with the separation of solids from solids, solids from liquids, liquids from liquids, and solids from gas.

Examples of separation of solids from solids with regard to particle size include size grading of fruits and vegetables, and screening of wheat, tea, sugar, and cocoa beans.

On the basis of particle size and density, we have the phenomenon of settlement, that is, suspension of solid in fluid, liquid, or gas and typically, the finer solids (such as kaolin) of smaller size and/or density in the feed slurry are separated in the centrate stream as product (e.g., 90 percent of particles less than 1 mm, etc.), whereas the larger and/or denser solids are captured in cake as reject. Furthermore, separation can be in the form of thickening, where solids settle under centrifugal force to form a stream with concentrated solids (Perry, 1999). The larger and denser particles will settle faster than the smaller and less dense ones. Classification implies the sorting of particulate material into size ranges. Use can be made of the different rates of movement of particles of different sizes and densities suspended in a fluid and differentially affected by imposed forces such as gravity and centrifugal fields. Separation can also be in the form of classification and degritting at which separation is effected by means of particle size and density (Perry, 1999).

Regarding solubility, examples are the extraction of sugar from sugar beet or coffee from ground-roasted beans where one of the components is soluble.

Regarding the separation of solid from liquid on the basis of particle size and filtration, a solid suspension is forced in the liquid through a filter medium, cloth, or screen by the development of a pressure gradient across the filter medium. This is carried out in a filter press. Examples are filtration of sugar solutions during refining, and filtration of canning brines and syrups.

Regarding the particle size and density and settlement, centrifugation subjects the solid suspension in a liquid to a cyclic motion in a bowl with a perforated wall. Centrifugal force forces the liquid out through the perforations of the bowl wall. The size of perforations determines the portion of solid retained in the bowl. Examples include dewatering of sugar crystals and milk clarification.

Examples of separating liquid from liquid includes the separation of cream from milk and dewatering oils. Settlement allows the more dense liquid to be rejected. Centrifugation subjects the mixture to a cyclic motion in a solid bowl removing the more dense liquid through an outlet near the circumference of the bowl and the less dense liquid through an outlet near the center of the bowl's rotation.

Regarding solubility examples are the crystallization of sugar from solution and extraction where one of the liquid components dissolves out as it occurs in solvent extraction of oils and flavouring materials (Butters, 1993; Loncin and Merson, 1979).

Finally, an example of the separation of a solid from a gas includes the bag filtration method of removing dried powder from the air of a spray drier on the basis of particle size and filtration.

Regarding particle size and density, cyclone separation of dried powder from the spray drier subjects the solid suspension in gas to a cyclic motion, hence separating out the more dense solids. On the basis of solubility, wet scrubbing of the powder (dissolving out the solid) occurs out of the exhaust air of a spray drier using the solvent as a feed material.

Diffusion coefficients for glucose and malic acid have been measured by the lag-time model using cell-free and cell-occupied calcium alginate membranes placed in a plate diaphragm diffusion cell (Teixeira et al., 1994). The results indicated that neither carbon dioxide release nor solute concentration or cocurrent diffusion affected the diffusion coefficients of either solute. The presence of cells caused a decrease in solute transport across alginate membranes and a reduction in the diffusion coefficients. The mass transfer mechanism in cell-occupied membranes was best characterized by the random pore model, a model describing diffusion through gel membranes. This model described the influence of cell concentration on the effective diffusion coefficient for the experimental results tested. The value obtained for the diffusion coefficient of glucose was $6.6 \times 10^{-10}$ $m^2$ $s^{-1}$ and was of the same order of magnitude as the molecular diffusion coefficient of glucose in water. However, $D_{eff}$ for malic acid was found to be equal to $4.36 \times 10^{-10}$ $m^2$ $s^{-1}$ and was significantly lower than that in aqueous solution ($8 \times 10^{-10}$ $m^2$ $s^{-1}$). The lag-time analysis method has also been used to evaluate diffusion coefficients of lactose and lactic acid through a 3% agarose gel membrane (Bassi et al., 1987). Average diffusion coefficients were estimated to be $3.97 \times 10^{-10}$ and $2.9 \times 10^{-10}$ $m^2$ $s^{-1}$ for lactose and lactic acid, respectively. The effective diffusion coefficient was affected by several factors, including the shape, density, length, and diameter of the pores within and on the surface of the matrix.

A diaphragm diffusion cell has been used to evaluate the effective diffusion coefficient of ethanol in a 4% (w/v) agarose gel at 25°C. A mean value of $9.5 \times 10^{-10}$ $m^2$ $s^{-1}$ was reported by Westrin and Axelsson (1991). This value agreed with the value of the molecular diffusion coefficient of ethanol in water. Djelveh et al. (1989) also used an improved cell diffusion method based on a diaphragm cell to determine diffusion constants in gels or foods. A two-compartment diffusion cell has been used by Coca et al. (1986) to study the permeation of potassium chromate solutions through rat epidermis at 37°C. They reported a change in diffusivities with solute concentration. Partition coefficients were also concentration dependent.

## 3.2 POLYSULFONE MEMBRANE CHARACTERISTICS: DEFINITIONS, USES, STRUCTURE, AND FUNCTIONS

Membrane polymers are often chemically inert. Polysulfone (PS) is one material suited to meet this goal and possesses very good chemical, thermal, and biological stability combined with excellent mechanical strength and flexibility (Rodemann and Staude, 1995). PS membranes are microfiltration (MF) or ultrafiltration (UF), synthetic polymeric, hydrophilic membranes. They are used because of their ability to cope with wide temperature limits, wide pH tolerance (1–13), good chlorine resistance, wide range of pore sizes (10–200 Å), and molecular weight cut-off (MWCO) ranging between 1000 and 500,000. Their only limitation is the restriction to low pressure drops. They are characterized by having diphenylene sulfone repeating units in their structure. The $-SO_2$ group in the polymeric sulfone is very stable because of the electronic attraction of resonating electrons between adjacent aromatic groups. The oxygen molecules projecting from this group each have two pairs of unshared electrons to donate to strong hydrogen bonding with solute or solvent molecules. Repeating phenylene rings create both steric hindrance to rotation within the molecule and electronic attraction of resonating electron systems between adjacent molecules; both contribute to a high degree of molecular immobility, producing high rigidity, strength, creep resistance, dimensional stability, and heat deflection temperature. Phenyl sulfone groups have long-term high-temperature stability (Mulder, 1991).

An Intersep Nadir PS membrane (Intersep Filtration Systems, Wokingham, Berkshire) was used in these experiments. This membrane was polypropylene backed for improved handling and repeated use with a skin modified by polyvinyldipuridone to be hydrophilic. The industrial production process employed in the manufacture of this membrane offers controlled retention characteristics, water permeability, and solute transport. Such asymmetric membranes consist of a thin top layer supported by a porous sublayer with the resistance to mass transfer being almost completely determined by the top layer. Therefore, L, the effective thickness of the membrane, has been taken to be approximately equal to 0.05 mm + 0.25 μm ≅ 0.05 mm = 50 μm. The porous sublayer thickness was not taken into account for the calculation of the effective thickness of the membrane since it was very small.

Other important properties show that it is inert, noncytotoxic, and does not denature biological materials; also, it has broad chemical resistance and wide pH range compatibility. Its nominal MWCO is 100 kDa, implying that solutes with a MW greater than 100,000 are rejected more than 90%. According to Intersep's membrane catalog, the PS membrane shows less than 10% rejection for bovine serum albumin (BSA), which has a MW of 67 kDa. It has been assumed that the rejection for lysozyme and cellulase is approximately zero. Typically, the membranes may be reused 10 times in laboratory stirred cells and withstand over a year of constant use in the cartridge configuration.

The membrane cleaning procedure followed was the following:

- Flushing with warm water
- Cleaning with 0.5% P3-ultrasil 10
- Flushing with water
- Cleaning with 0.5% P3-ultrasil 75
- Flushing with water
- Cleaning with 1% P3-ultrasil 10
- Flushing with water
- Sanitizing with 1% P3-ultrasil 25

This hydrophilic PS membrane was washed with deionized water and equilibrated with the buffer solution for 1 h before being placed in the diffusion cell. Then it was left in the buffer for another hour at stirring conditions.

### 3.2.1 Polysulfone Membrane

According to the membrane catalog provided by Intersep, the permeability water flux for the PS membrane is 450 LMH (L $m^{-2}$ $h^{-1}$) tested at 3 bar, 700 RPM, 20°C, stirred cell.

The pore geometry is also very important in the permeability method and needs to be determined so that the experimental results could be interpreted. Therefore, assuming cylindrical, perpendicular pores, which imply tortuosity equal to unity (T = 1) and assuming that all pores have the same radius, using the Hagen–Poiseuille equation as described in Mulder (1991), the radius of the pores could be estimated as follows:

$$J = (\varepsilon r^2/8\mu T) \cdot (\Delta P/L) \tag{3.1}$$

where J is the volume flux through the pores equal to $450 \times 10^{-3}$ $m^3/3600$ $m^2$ s; $\varepsilon$ is the surface porosity, which is equal to the ratio of the effective area (S) to the area of the membrane (A), and hence $\varepsilon = (2 \times 10^{-5} m^2)((\pi / 4)(4 \times 10^{-2})^2 m^2) = 0.0159$ and usually has a low value ranging from 0.1% to 1% for UF membranes; $\mu$ is the viscosity and is equal to $10^{-3}$ N s $m^{-2}$; T is the tortuosity equal to unity; $\Delta P$ is the pressure difference across the membrane and is equal to $3 \times 10^5$ N $m^{-2}$; and L is the thickness equal to $50 \times 10^{-6}$ m. This equation indicates that the solvent flux is proportional to the driving force, that is, the pressure difference across a membrane of thickness L and inversely proportional to the viscosity.

By substituting these data in Equation 3.1, it is evident that the average value of the radius of the pores is 102 nm. In UF membranes, the average pore diameter (i.e., diameter of pore in the dense skin) is generally in the range of 2–100 nm. The Hagen–Poiseuille equation gives a good description of transport through membranes consisting of a number of parallel pores. However, very few membranes have such a structure in practice.

Insulin transport phenomena across a series of porous charged membranes were studied at two pH conditions (pH 3.3 and pH 7.4) by Zhang et al. (2009). The membranes were prepared by pore-surface modification of porous poly(acrylonitrile) (PAN) membranes by grafting with weak acidic and basic functional groups. The insulin partition coefficient *K* between the membrane and the solution was estimated from the equilibrium adsorption amount in the batch adsorption experiment. The insulin-effective diffusion coefficient *D* inside the membrane was determined as a fitting parameter by matching the diffusion model with the experimental data of the diffusion measurement. Both *K* and *D* correlated well with the charge properties of the insulin and membrane. When the insulin and membrane carried opposite net charge, the partition coefficient showed relatively larger values, while the effective diffusion coefficient was reduced. The insulin permeability coefficient *P* obtained from the experimental results agreed with that estimated from the partition coefficient and effective diffusion coefficient. These results suggested that the combined effects of the solubility and diffusivity on the permeability coefficient complicated the relationship between the permeability and the charged properties of the insulin and membrane. Moreover, insulin permeability was reduced by the boundary layer between the membrane and solution (Varzakas, 1998).

## 3.3 CHARACTERISTICS OF MEMBRANES: MASS TRANSFER IN MEMBRANE TECHNOLOGY

For the separation of macromolecules, pressure could be applied depending on the membranes used or the difference in electric potential. The last case refers to electrodialysis (ED), which is an electrochemical process in which a direct electric current causes ions to move from a less concentrated solution to a more concentrated solution. Anion (negative) and cation (positive) ions transfer through selective membranes (Figure 3.1).

In conventional ED, flat-sheet membranes are stacked in alternating layers of cation-exchange and anion-exchange membranes. With the application of electrical current, cations migrate through the cation-exchange membranes toward the cathode, but the cations are stopped by anion-exchange membranes. Similarly, anions migrate toward the anode through anion-exchange membranes but are stopped by cation-exchange membranes. An extension of conventional ED uses bipolar membranes, which splits water into its component $H^+$ and $OH^-$ ions. When used along with conventional cation- and anion-exchange membranes, it allows a salt stream to be converted into an acid and a base stream. This is particularly useful in downstream processing of organic acids such as citric, lactic, acetic, and gluconic acids, in that it produces the more desirable acid form of the compound while regenerating the alkali, which is used in the fermentation vessel (Cheryan, 2007).

Like other membrane processes, ED is affected by concentration polarization and membrane fouling. In addition, ED also exhibits simultaneous water transport,

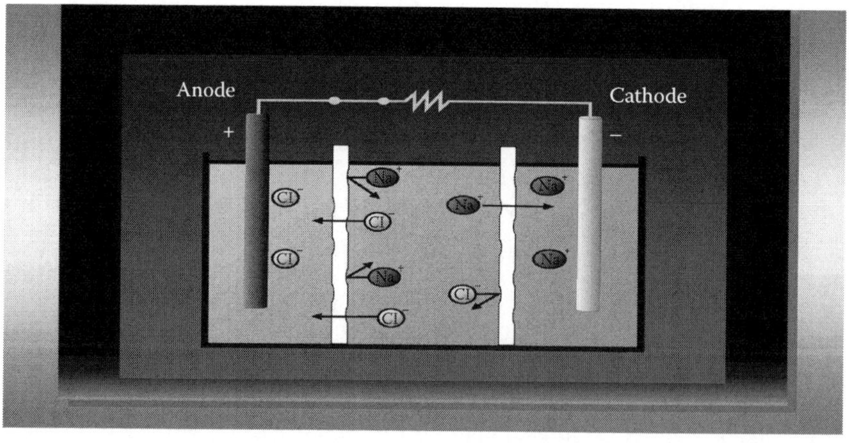


**FIGURE 3.1** Example of conventional electrodialysis. (Adapted from Coca Cola Manufacturing Manual, 2003. TCCQS auditors toolkit. The Coca-Cola quality system. Coca-Cola Europe, Eurasia and Middle East. Confidential report; Cheryan, M. 2007. In: Heldman, D. R. and Lund, D. (eds.), *Handbook of Food Engineering*, Chapter 9, 2nd edn., CRC Press, Boca Raton, FL, USA, pp. 554–596.)

which limits the maximum concentration that can be attained for permeable species. With bipolar membranes, fouling by divalent cations is quite severe, since they can precipitate with the hydroxyls encountered in the cationic membranes. ED is generally more expensive than pressure-driven membrane processes, primarily due to the electrical energy requirements. It is justified in some applications, such as water desalting, tartarate removal from wine, demineralization of protein solutions, and separation and concentration of organic acids (Bailly et al., 2001).

Pervaporation is also a pressure-driven process except that unlike all the others discussed so far, the permeate is a vapor and not a liquid. The solutes permeating the membrane are relatively more volatile than the solvent.

The driving force for transport is a chemical potential gradient that arises due to a decrease in the activity of the permeating components. The activity decrease can be accomplished by pressure reduction, for example, a vacuum can be applied on the permeate side. The vapor is then condensed and the noncondensables are removed by the vacuum pump (Rajagopalan et al., 1994; Rajagopalan and Cheryan, 1995).

Process fluid passing through the membrane is named permeate, whereas fluid that does not pass through is called (concentrate-retentate) (Walstra et al., 1999). The schematic diagram of flow through a membrane system is shown in Figure 3.2.

The same phenomenon occurs during the operation of cross-flow filtration. In a cross-flow ultrafilter used for the concentration or separation of macromolecules, process fluid is passed over a filtration membrane. The pressure being higher in the feed side, there is a flow of permeate through the membrane and the retentate leaves the membrane system since it cannot pass through the membrane being more concentrated in species.

If one considers what happens to a protein being unable to pass through the membrane and assumes that the species concentration in the bulk stream is Cb and that the flow rate across the membrane is high then the bulk concentration remains constant in the membrane (Pyle, 1992). We also assume that the stream flowing across the membrane comprises a well-mixed turbulent core and a thin boundary layer. Then the flux J is Q/A, where Q is the flow rate of the permeate and A is the membrane area. The flux of permeate is expected to follow Darcy's law and hence

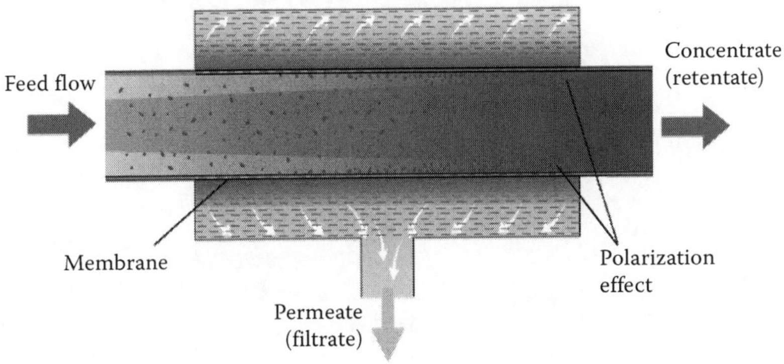


**FIGURE 3.2** Schematic diagram of flow through a membrane system. (Adapted from Bylund, G. 1995. *Dairy Processing Handbook.* Tetra Pak Processing Systems, Lund, Sweden.)

J = ΔP/Wm, where ΔP is the pressure drop and Wm the hydraulic resistance of the membrane depending on its pore size distribution, voidage fraction, and viscosity of the permeate.

If the protein cannot pass through the membrane, it will accumulate in the boundary film. Hence, concentration increases toward the membrane surface, and protein, due to Fick's law, will diffuse back toward the bulk. If a steady state is established within the boundary layer and no net transfer toward the membrane occurs, then the bulk flux and the back-diffusional flux must be equal (Pyle, 1996). This concentration increase toward the membrane surface is called concentration polarization. Owing to the transport through the membrane of molecules/particles below a certain size, molecules/particles above this size are concentrated on the feed side of the membrane surface (Gekas and Hallström, 1987; Hallström et al., 2007). It should also be noted that the mean driving force over the boundary layer is the log-mean concentration difference.

Finally, a negative feature of polarized membranes is when permeate flux becomes independent of the applied pressure drop.

Depending on the type of membrane, the size of the membrane pores, the applied pressure, and the type of particles retented (Figure 3.3), we have the following filtration operations: reverse osmosis (RO), NF, UF, MF, and ED. All these have applications in the food and drink industries.

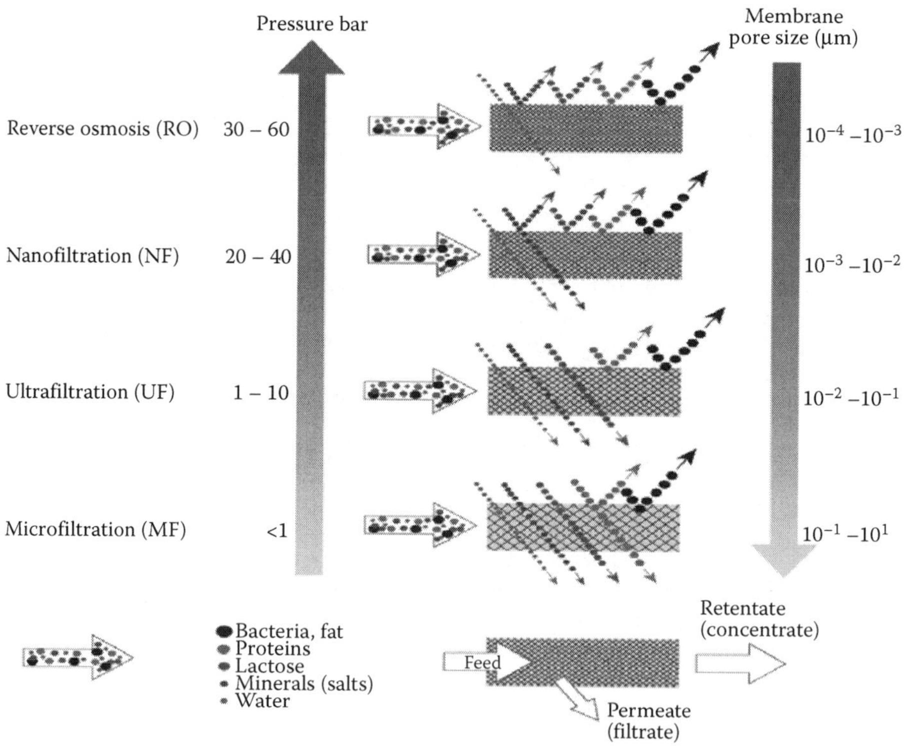


**FIGURE 3.3** Membrane types. (Adapted from Bylund, G. 1995. *Dairy Processing Handbook*. Tetra Pak Processing Systems, Lund, Sweden.)

### 3.3.1 Membrane Technology and Its Uses

Membrane technology is useful for the selective enrichment of some ingredients. For the manufacture of yogurt, different technologies can be employed such as RO, NF, UF, and MF in order to increase the solids by removing water. Their use is limited in the evaporation of skimmed milk, which will be the raw material for the manufacture of different types of yogurt. Some of these methods remove part of lactose and inorganic salts of milk with the result of increasing the protein content. These technologies can be used to evaporate skimmed milk containing a maximum of 9–12% soluble solids. In the retentate, enough lactose remains to facilitate fermentations (Kilara, 2006).

Membrane applications in conventional food processing technologies are well described (Cuperus, 1998; Cuperus and Nijhuis, 1993), involving dairy (Daufin et al., 2001; Rosenberg, 1995; Zydney, 1998), fruit juice and beverages (Girard and Fukumoto, 2000; Jiao et al., 2004), fats and oils (Bhosle and Subramanian, 2005; Koseoglu and Engelgau, 1990; Koseoglu 1991; Manjula and Subramanian, 2006; Snape and Nakajima, 1996), and starch (Rausch, 2002; Singh and Cheryan, 1998).

Membrane applications of lipid-, carbohydrate-, and protein-based nutraceuticals and some minor bioactive components have been critically evaluated by Akin et al. (2012). Both nonporous and porous membranes were employed for lipid-based nutraceutical separation. The use of nonporous membranes together with nonaqueous solvents brought about the impact of solution-diffusion theory on transport through membranes. Both organic and inorganic membranes gave encouraging results for the recovery of lipid components with single- and/or multistage membrane processing. Two-stage UF–NF systems with polymeric membranes provided an efficient approach for the removal of high- and low-MW unwanted components, resulting in higher-purity oligosaccharides in the NF retentate. The charged nature of protein-based nutraceutical components had a major effect on their separation. Operating at optimal pH levels was critical for fractionation, especially for low-MW peptide hydrolysates. Processing of minor components such as polyphenols utilized all types of porous membranes from prefiltration to concentration stages. The coupling of membrane separation and supercritical fluid technologies would combine unique advantages of each process, resulting in a novel separation technology offering great potential for the nutraceutical and functional food industry.

#### 3.3.1.1 Ultrafiltration

UF can be used to concentrate molecules such as peptides, proteins, or other particles. Membranes have been constructed in such a way as to allow molecules to pass until a certain MW and a pore size of approximately 0.1 μm (Rosenberg, 1995). The pressure applied in ultrafiltration ranges between 1 and 6 bar resulting in concentration of molecules of MW 1000–50000 Da. Moreover, the supply rate in UF is high with the aim to prevent the obstruction of the membrane pores. The construction material in these membranes is cellulose with derivatives such as cellulose acetate (CA) and some thermotolerant polymers such as polysulphone and polyethersulphone.

UF separates effectively macromolecules (proteins) and particles (casein micelles, fat cells, somatic cells, and bacteria) of milk. The main aim of ultrafiltration is the increase in protein concentration and is applied in skimmed milk and whey. Initially, it could cause significant changes in the composition of dairy products manufactured with ultrafiltrated milk and for this reason, it allows the production of novel products (Walstra et al., 1999).

UF is applied in skimmed milk and whey for the manufacture of fresh and feta cheeses, since it leads to an increase in the yield, decrease in the quantity of added rennet, and a better quality of the produced cheese. It is used in the preconcentration of milk, fractionation of whey, and micellar casein enrichment for cheese production, because of the low cost, energy efficiency, and consistent product quality.

UF can also be used in the manufacture of yogurt. Following UF, milk presents increased solids due to the concentration of its macromolecules (lipids and proteins) (Kilara, 2006). These yogurts have a better texture, higher cohesion, gentle aroma, and pleasant taste. They have anincreased concentration of proteins—up to 50%—and low concentration in lactose—approximately 50%—compared to commercial-type yogurts, whereas they have increased calcium and iron concentration, which give specific characteristics and high added value to the product (Rinaldoni et al., 2009).

Cross-flow filtration in the final steps provides important advantages to designed processes since membrane polarization is reduced when compared to plug-flow filtration. Transmembrane pressure is relatively constant and high-throughput operations can be implemented with proper industrial equipment (Ghosh and Cui, 2000).

One of the most attractive segments in the food and cosmetic industries is that of natural pigments. Natural pigments obtained through biotechnological processes represent an attractive alternative. Ruiz-Ruiz et al. (2013) has previously worked on the development of an aqueous two-phase system (ATPS)-based prototype process for the recovery of B-phycoerythrin (BPE), a natural high-value pigment obtained from *Porphyridium cruentum*.

Detailed studies describing the scaling up of ATPS processes from bench scale to pilot plant facilities are not common. They described experiences derived from the scale-up of a previously developed process for production and recovery of highly purified (purity defined as the absorbance ratio $A_{545}/A_{280} > 4$) BPE, where a scale-up factor of 850 was implemented. Characterization of cell disruption with a pilot-scale bead mill allowed efficient BPE release at 2900 rpm, 10% (w/v) sample load, 60% (v/v) bead load, and 0.5 mm glass beads and 22 min of residence time with a yield of 1.35 mg BPE $g^{-1}$ of wet biomass.

BPE was recovered and purified using a strategy comprising isoelectric precipitation, aqueous two-phase fractionation and UF. A 54% global BPE recovery yield, with final purity of 4.1, was achieved under optimal process conditions. Considering total costs for raw materials and energy expenditures for one batch, it was determined that the production cost of BPE was \$1.17 USD $mg^{-1}$, which is less than the commercial price of a BPE standard (>\$30 USD $mg^{-1}$).

Liquid–liquid extraction systems such as ATPS allow rapid recuperation and purification of biological products, integration and intensification of bioprocesses, development of a biocompatible medium for biologic compounds and scale-up feasibility (Rito-Palomares, 2004; Benavides and Rito-Palomares, 2008).

Yokota et al. (2012) has developed a novel analytical method for the quantification of bromate in fresh foods using high-performance liquid chromatography (HPLC) with a postcolumn reaction. The fresh food sample solutions were pretreated with homogenization, centrifugal UF, and subsequent solid-phase extraction using a strong anion-exchange resin. After separation on a strong anion-exchange chromatography column using a highly concentrated NaCl solution (0.3 M) as the eluent, the bromate was quantified by detection using a postcolumn reaction with a noncarcinogenic reagent (tetramethylbenzidine). The developed HPLC technique made it possible to quantify bromate in salt-rich fresh foods. The recoveries from fresh foods spiked with bromate at low levels (2 or 10 ng $g^{-1}$) satisfactorily ranged from 75.3% to 90.7%.

The lowest quantification limit in fresh foods was estimated to be 0.6 ng/g as bromic acid. The method should be helpful for the quantification of bromate in fresh foods disinfected with hypochlorite solutions.

Solid–liquid separation is a very important unit operation related to chemical, engineering, and environmental processes. A number of conventional separation methods and hardware, such as centrifugation, filtration, or sedimentation, have been in use for a long time (Rossignol et al., 2000; Sharma, 1994). However, nowadays, we are looking for techniques and processes with less energy and capital.

Biomimetics is a technique by which phenomena in nature are used as the basis to modify existing technologies or design new ones (Bulger et al., 2008). Hung et al. (2012) developed a new particle separator based on the mechanism of cross-flow filtration employed by suspension-feeding fish. When water enters a suspension-feeding fish's oral cavity, it brings in food particles suspended in large quantities of water. A high proportion of the particles that enter are retained inside the fish's oral cavity and eventually swallowed while most of the water exits via branchial slits.

To construct the model of the bioinspired particle separator, computational fluid dynamics techniques are used, and parameters related to separator shape, fluid flow, and particle properties that might affect the performance in removing particles from the flow are varied and tested. The goal is to induce a flow rotation, which enhances the separation of particles from the flow, reduce the particle-laden flow that exits via a collection zone at the lower/posterior end of the separator, while at the same time increase the concentration of particles in that flow. Based on preliminary particle removal efficiency tests, an exiting flow through the collection zone of about 8% of the influent flow rate is selected for all the performance tests of the separator, including trials with particles carried by air flow instead of water. Under this condition, the simulation results yield similar particle removal efficiencies in water and air, but with different particle properties. Particle removal efficiencies (percentage of influent particles that exit through the collection zone) were determined for particles ranging in size from 1 to 1500 μm with a density between 1000 and 1150 kg $m^{-3}$ in water and 2 and 19 mm and 68 and 2150 kg $m^{-3}$ in air.

In UF of milk, nonprotein nitrogen and soluble components such as lactose, salts, and some vitamins pass through the membrane, whereas milk fat, proteins, and insoluble salts are retained by the membrane (Mehaia, 1997). The growing use of UF in the dairy industry, especially in the area of value addition, promises to dramatically change the technology of concentrated and dried milk products. Its main advantages are the higher dry matter and milk protein contents and the increased ratio of protein to dry matter compared to native milk.

Dairy whitener is widely used as a substitute for fresh milk, cream, or evaporated milk in tea, coffee, cocoa, or drinking chocolate and is also suitable for adding to foods like soups, sauces, puddings, and cereal dishes. The main advantages of using dairy whitener are ease of handling, improved shelf life, which may be specific requirements for use in restaurants, railways, airways, and waterways (Khatkar et al., 2012a).

Dairy whiteners should also have the ability to withstand the high temperature (80–90°C) and low pH (4.6–5.2) of coffee solution (Khatkar et al., 2012a).

The whitening effect is produced in coffee as a result of light scattered from the surface of finely emulsified particles. The whitening powder comes mainly from a well-emulsified and finely dispersed fat and protein in a colloidal state (Khatkar et al., 2012b).

Khatkar et al. (2013) studied the physicochemical and functional quality attributes of dairy whitener prepared from a UF process. Developed dairy whitener had significantly ($P < 0.01$) greater protein (40.07 ± 0.66%) and calcium (1.42 ± 0.05%) contents compared to market samples and also had good solubility index (0.25 mL) and significantly ($P < 0.01$) higher dispersibility (92.08%) and $L^*$ value (93.87). Dairy whitener, even at lower solids level, had an edge over the market samples in terms of sensory and instrumental color characteristics in both tea and coffee without leaving any undissolved suspended particles. Dairy processors would be able to prepare value added dairy whitener commercially that would increase their profitability.

Recovery of high value-added products such as proteins, aromas, and flavors can be done using UF processes (Vandajon et al., 2002). The application of membrane technology as the main method of separation, concentration, and purification (Murado et al., 2010) of valuable compounds from industrial waste materials has been applied to diverse sources, including fish meal (Afonso et al., 2004), palm oil mill effluents (Wu et al., 2007), and solid by-products of the brewing industry (Tang et al., 2009).

Rodriguez-Amado et al. (2013) focused on the production of antihypertensive and antioxidant activities using enzymatic hydrolysis of protein concentrates recovered by UF of different wastewaters from the industrial processing of cuttlefish (*Illex argentinus*). The effluents were produced in the processes of thawing (E1), softening (E2), boiling (E3), and gelation (E4). Their results showed that membranes with cut-off at 100, 30, and 10 kDa were an effective resource to protein concentration of E2 and E3 but limited for E1 and E4. In addition, E2 and E3 retentates led to remarkable antihypertensive and antioxidant activities, further improved by enzymatic hydrolysis. Also, sequential UF revealed the enrichment of these protein concentrates in peptides with high angiotensin-converting enzyme (ACE)-inhibitory activity.

Thereby, UF-fractionation followed by proteolysis of protein concentrates from cuttlefish wastewaters offers new opportunities for the development of bioactive hydrolysates with application in the food industry. In addition, this approach contributes to an improved depuration of industrial wastewaters, reducing the treatment costs and leading to a decrease in its contaminating effect.

Different cascades of UF–DF were employed using waste effluents. For this purpose, Prep/Scale-TFF cartridges (Millipore Corporation, Bedford, MA, USA) of 100, 30, 10, and 1 kDa MWCO were used by Rodriguez-Amado et al. (2013). According to the manufacturer, cartridges were made of polyethersulfone (PES), except for 1 kDa, which was from regenerated cellulose.

The operation mode was the following: an initial phase of UF at 40°C with total recirculation of retentate was performed, immediately followed by a DF step. During UF, the inlet pressure remained constant to determine the drops of flow rate due to the increased concentration of the retentate and possible adhesions to the membrane.

Castel et al. (2012) compared protein yield, protein concentration, and physicochemical characteristics of *Amaranth mantegazzianus* protein concentrates (APCs) obtained at pilot scale by a conventional process (CP) (alkaline extraction and isoelectric precipitation) and two alternative processes (AP): (1) acid pretreatment process combined with isoelectric precipitation and (2) acid pretreatment process combined with UF. Although AP resulted in higher protein concentration, protein yield was lower than in CP. SDS–PAGE and size-exclusion chromatography showed high-molecular-weight fractions only for isoelectric precipitation concentrates (obtained by CP and AP). The amino acids concentration, especially phenylalanine, isoleucine, and methionine, increased in all protein concentrates with respect to the amaranth flour. Particularly, the product obtained by UF was rich in phenylalanine and lysine, and presented no limiting amino acid with respect to the recommendation of the Food and Agriculture Organization of the United Nations (FAO). In conclusion, process (2) improved protein concentration and nutritional quality (balanced amino acid composition) of APCs with respect to CP and process (1), suggesting that the UF process is a viable alternative to CP and a promising method for obtaining protein concentrates.

Simultaneously, several biomass-based processes were developed over the past decade suggesting scenarios from a classic biofuel plant to a new biorefinery concept, which produces, for instance, polymers that were previous fossil resource based. The growth of bioresource-based chemicals, functional monomers, as well as fuels, leads to an increased demand for new separation processes. This review by Abels et al. (2013) highlights the role of membrane separations within current and future biofuel and biorefinery scenarios. Membrane processes reviewed are, for instance, pervaporation for alcohol recovery and UF of canola oil, as well as new developments such as the UF/NF of lignin in a solvent-based lignocellulose conversion process, or the recovery of amino acids via ED. The membrane processes are classically categorized as concentration-driven, pressure-driven, electrical-driven, and prospective. It follows the transition of a classic biofuel production plant to a new sophisticated biorefinery. The review closes with a reflection of membrane-based downstream processes required in a biorefinery, transforming cellulose into itaconic acid.

Parés et al. (2012) tested the use of serum from porcine blood as functional ingredient in frankfurter production. Three pilot productions of sausages were carried out to compare serum containing frankfurters and sausages, based on a standard commercial formula that included caseinate and polyphosphate. Both products were very similar for proximate composition, water-holding capacity, cooking and purge losses, instrumental texture, and microstructure. The sensory descriptive profile and the overall acceptance were also comparatively evaluated. Although significantly higher values for the animal taste and odor attributes of sausages with serum compared to control ones were obtained, the differences were lower than those reported in a previous study using whole plasma. Thus, UF could be useful to reduce animal off-flavor in blood-based protein ingredients. Moreover, overall acceptance did not significantly differ between the two types of products, being 6.7 and 6.5, for control and test sausages, respectively.

Kawa-Rygielska et al. (2013) investigated the feasibility of the concentrate obtained after membrane UF of sugar beet thin juice for ethanol production and selection of fermentation conditions (yeast strain and media supplementation). The resulting concentrate was subjected to batch ethanol fermentation using two strains of *Saccharomyces cerevisiae* (Ethanol Red and Safdistill C-70). The effect of different forms of media supplementation (mineral salts: $(NH_4)_2SO_4$, $K_2HPO_4$, $MgCl_2$; urea + $Mg_3(PO_4)_2$, and yeast extract) on the fermentation course was also studied. It was stated that sugar beet juice concentrate is suitable for ethanol production yielding, depending on the yeast strain, ca. 85–87 g $L^{-1}$ ethanol with ca. 82% practical yield and more than 95% of sugar consumption after 72 h of fermentation. Nutrient enrichment further increased ethanol yield. The best results were obtained for media supplemented with urea + $Mg_3(PO_4)_2$ yielding 91.16–92.06 g $L^{-1}$ ethanol with practical yield ranging 84.78–85.62% and full sugar consumption.

A novel UF membrane with controllable selectivity for protein separation was obtained by Li et al. (2013) by altering the structure of the thick sieving layer on the membrane surface and subsurface. Poly(vinyl pyrrolidone) (PVP) was first cross-linked on/in the poly(vinylidenefluoride) (PVDF) hollow fiber MF membrane to attract more sulfobetaine (SB) monomer adjacent to membrane for subsequent grafting polymerization and the formed thick sulfobetaine polymer (PSB) layer on the membrane surface and subsurface acted as the sieving layer with environment sensitivity. After immersed in 20 mmol $L^{-1}$ NaCl solution at 60°C, the novel UF membrane with the sieving layer of 4.8 μm showed high permeate capacity with a water flux of 590 L $m^{-2}$ $h^{-1}$ and a good selective behavior with MWCO of 95–110 kDa. The protein mixture (BSA and Lys) could be separated through the novel UF membrane efficiently by isoelectric focusing of one component with a larger size (BSA). By means of the simple immersion in pure water, the membrane permeated mostly proteins and the degree of flux decline reduced obviously. After the membrane is swelled in NaCl solution again, the membrane restored the selectivity and the protein separation efficiency. Such smart UF membranes are attractive candidates for the batch separation of protein mixtures, expanding the membrane application in the fields of agrofood, biomedicine, and other biofiltration.

The antioxidant properties of barley glutelin hydrolysates were evaluated by Xia et al. (2012) based on their radical-scavenging capacity (DPPH/$O_2^-$/$OH^-$), $Fe^{2+}$-chelating effect, and reducing power. Alcalase hydrolysates (AH) demonstrated significantly higher antioxidant capacity than those treated by flavorzyme in most of the assays. The AH was separated using ultrafiltration and reversed-phase chromatography, and assessment of the fractions indicated that the large-sized peptides (Mw > 10 kDa) possessed stronger DPPH-scavenging activity and reducing power, whereas small-sized peptides (Mw < 1 kDa) were more effective in $Fe^{2+}$-chelating and $OH^-$-scavenging effect. The hydrophobic fraction contributed more to $Fe^{2+}$-chelating and $OH^-$-scavenging activity. Four peptides contributing to antioxidant activities were identified using LC–MS/MS: Gln–Lys–Pro–Phe–Pro–Gln–Gln–Pro–Pro–Phe, Pro–Gln–Ile–Pro–Glu–Gln–Phe, Leu–Arg–Thr–Leu–Pro–Met, and Ser–Val–Asn–Val–Pro–Leu. Compared to the positive controls, AH exhibited excellent $Fe^{2+}$-chelating activity and strong DPPH/$OH^-$-scavenging effect. Thus, hydrolyzed barley glutelin is a potential source of antioxidant peptides for food and nutraceutical applications.

In the food processing industry, shrimp shells (*Parapenaeus longorostris*) have great commercial value because they are rich in chitin (24 wt%), protein (40 wt%), lipids, pigments, and flavor compounds. Benhabiles et al. (2013) examined protein recovery by UF during isolation of chitin from shrimp shell. Up to 96 wt% of the proteins could be removed (i.e., deproteinization) from the shrimp shells by incubating them in NaOH (2 N) over 2 h, at T = 45°C, and solid to solvent ratio of 1:2 (w/v). A solute rejection coefficient ($R_0$) of 97% was obtained in the UF process to recover proteins from deproteinized shell waste water. The protein concentration process that was carried out beyond the critical flux of 380 L $h^{-1}$ $m^{-2}$, at a transmembrane pressure of 3 bars, and a tangential velocity of 5 m $s^{-1}$ was found to reduce the hydrolysate volume by a factor of 2.4. Owing to a reduction in organic matter in the effluent, the chemical oxygen demand (COD) of the permeate was reduced by 87%.

The extracellular α-L-rhamnosidase has been purified by growing a new fungal strain, *Aspergillus awamori* MTCC-2879, in the liquid culture growth medium containing orange peel (Yadav et al., 2013). The purification procedure involved UF using PM-10 membrane and anion-exchange chromatography on diethyl amino ethyl cellulose. The purified enzyme gave a single protein band in SDS–PAGE analysis corresponding to molecular mass 75.0 kDa. The native PAGE analysis of the purified enzyme also gave a single protein band, confirming the purity of the enzyme. The $K_m$ and $V_{max}$ values of the enzyme for *p*-nitrophenyl-α-L-rhamnopyranoside were 0.62 mM and 27.06 μmol $min^{-1}$ $mg^{-1}$, respectively, yielding $k_{cat}$ and $k_{cat}/k_m$ values 39.90 $s^{-1}$ and 54.70 $mM^{-1}$ $s^{-1}$, respectively. The enzyme had an optimum pH of 7.0 and an optimum temperature of 60°C. The activation energy for the thermal denaturation of the enzyme was 35.65 $kJ^{-1}$ $mol^{-1}$ $K^{-1}$. The purified enzyme can be used for specifically cleaving terminal α-L-rhamnose from the natural glycosides, thereby contributing to the preparation of pharmaceutically important compounds like prunin and L-rhamnose.

UF experiments of polysaccharide macromolecule have been performed in a batch, a stirred as well as unstirred membrane cell using a fully retentive membrane over a wide range of operating conditions as described by Sarkar (2013).

A model based on Hermia's approach for constant pressure dead-end filtration laws is proposed to analyze the flux decline behavior during UF in a batch cell. Two model parameters, namely, complete pore blocking coefficient and cake filtration coefficient are obtained by minimizing the error involved between calculated and experimental flux data. These parameters along with known operating conditions, membrane permeability, and physical properties of feed enable one to predict the transient permeate flux decline. The effect of various operating conditions, such as feed solute concentration, stirrer speed, and transmembrane pressure on the flux decline is studied. Experimental results show that operating conditions have significant effect on the onset of cake formation as well as on the flux decline behavior. The model results are successfully compared with the experimental data.

### 3.3.1.2 Osmotic Distillation

Osmotic distillation (OD), also called osmotic evaporation or isothermal membrane distillation, can be used to remove water selectively from aqueous solutions under atmospheric pressure and room temperature, avoiding thermal degradation (Courel et al., 2000; Hogan et al., 1998; Kunz et al., 1996). It involves the use of a microporous hydrophobic membrane to separate two circulating aqueous solutions at different solute concentrations: a dilute solution and a hypertonic salt solution. If the operating pressure is kept below the capillary penetration pressure of liquid into the pores, the membrane cannot be wetted by the solutions.

The difference in solute concentrations, and consequently in water activity of both solutions, generates, at the vapor–liquid interface, a vapor pressure difference causing a vapor transfer from the dilute solution toward the stripping solution.

Cross-flow UF and OD were implemented on a laboratory scale to obtain formulations of interest for food and/or the pharmaceutical industry starting from the blood orange juice produced in the Calabria region (Destani et al., 2013). Freshly squeezed juice, after a depectinization step, was submitted to an UF process in order to recover natural antioxidants, such as hydroxycinnamic acids, hydroxybenzoic acids, flavanones, flavan-3-ols, and anthocyanins. The UF permeate, with an initial total soluble solids (TSS) content of 10.5°Brix, was concentrated by OD up to a final concentration of 61.4°Brix.

The performance of both processes was analyzed in terms of productivity (permeate fluxes in UF and evaporation fluxes in OD) and quality of clarified and concentrated samples through the identification and quantization of phenolic compounds.

The UF membrane showed a rejection toward the identified phenolic compounds in the range 0.4–6.9% and a little decrease of the TAA (8.2%) was observed in the UF permeate in comparison with fresh juice. Phenolic compounds were also well preserved in the retentate of the OD process as demonstrated by the constant value of the ratio between the concentration of phenolic compounds in the OD retentate and the concentration of these compounds in the UF permeate stream (in the range 5.54–6.39).

#### *3.3.1.2.1 Membrane Fouling and Cleaning*

The common practice of membrane cleaning often involves a combination of hydraulic and chemical cleaning. The selection of chemical cleaning agents is largely limited by the compatibility of membrane media and other filter components to cleaning chemicals, and the avoidance of potential product contamination. According to Kuzmenko et al. (2005) and Rabiller-Baudry et al. (2006), membrane cleaning is typically performed daily for 2–3 h after 6–8 h of filtration in the dairy and water treatment industries.

The cleaning agents commonly used for membrane plants are alkalis, acids, enzymes, surface-active agents, formulated cleaning agents, combined cleaning and disinfecting agents, and disinfectants (Ghosh, 2003; Kazemimoghadam and Mohammadi, 2007).

Cleaning membranes fouled with protein solutions is a common part of the normal operational procedure in membrane applications for the dairy and other food industries. Chemical cleaning with simple caustic and acid cleaners is still a common practice, although the response of different protein components in a foulant layer to cleaning agents is not well understood. Norazman et al. (2013) examined the efficiency of chemical cleaning of PES membranes fouled during UF of whey protein isolate solution and the protein residuals remaining on the membranes after cleaning. Using a combination of the Lowry assay and gel electrophoresis for the estimation of protein amount and composition on the membrane before and after cleaning, it was observed that the high-molecular-weight components in the foulant layer could more easily be removed than the smaller components, and the proteins trapped in the membrane pores were the most difficult to clean. In repeated cycles of filtration and cleaning with NaOH followed by HCl, the flux recovery due to NaOH cleaning remained constant after the second cycle, while the flux recovery due to HCl cleaning increased with the repeated cycle, indicating the importance of HCl cleaning stage in the removal of residual protein components.

The resistance-in-series model was used by Baklouti et al. (2013) to analyze flux behavior, which involved the resistances of the membrane itself, and the fouling and solute concentration polarization. Response surface methodology was used to establish the relationships between operating parameters and UF efficiency and, thus, determined the optimal conditions. Experiments were performed according to Box–Behnken design by changing the levels of three parameters, namely, transmembrane pressure, feed flow rate, and temperature. The fitted mathematical models were employed to plot isoresponse curves. It was shown that the resistance due to solute concentration polarization ($R_{cp}$) dominated the flux decline (40–74%). The fouling resistance ($R_f$) varied from 12% to 46%. To simultaneously optimize the three responses studied ($R_f$, $R_{cp}$, and permeate limit flux), the desirability function approach was applied, which determined the best acceptable compromise.

The selected UF conditions of the compromise were as follows: 3 bars, 0.95 L $min^{-1}$, and 30°C. Optimal values of $R_f$, $R_{cp}$, and permeate limit flux were equal to 18%, 72%, and 19 L $h^{-1}$ $m^{-2}$, respectively.

Many techniques have been implemented to reduce membrane fouling such as hydrodynamic factors considering feed pretreatment, working at critical flux, backwashing, increase in shear at the membrane surface, and use of effective chemical cleaning agents. However, the critical flux approach has opened up interesting perspectives, particularly subcritical flux operations or close to them (Bacchin et al., 2006; Le Clech et al., 2003). The critical flux is a method to minimize fouling and extend the industrial process as much as possible before cleaning steps are necessary. Working at main process parameters (transmembrane pressure and linear flow velocity, largely) near to the critical point, fouling phenomena are drastically reduced and, as a consequence, the productivity and membrane life are significantly increased (Bacchin et al., 1995, 2006; Badan et al., 2008; Field et al. 1995; Iaquinta et al., 2009; Lipnizki, 2008; Metsamuuronen and Nystrom, 2005; Mizubuti et al., 2000; Pollice et al., 2005; Ribeira et al., 2002; Sogi et al., 2003; Wani et al., 2008).

Two UF membranes with different geometries (spiral polymeric and tubular ceramic) but similar cut-offs were used by Muro et al. (2013) to treat wastewater from the food industry. Hydrodynamic conditions were optimized by statistical methods as a strategy to get more accurate values of the critical parameters thereby producing and minimizing membrane fouling. The validation of the optimization method was obtained by determining experimental critical flux at critical parameters. Membrane fluxes revealed significant differences during filtration.

The polymeric membrane showed an optimal flux of 45.60 L $h^{-1}$ $m^{-2}$ at 3.21 bar while operating at a stable time of 11.61 h, whereas optimal flux of the ceramic membrane was 32.43 L $h^{-1}$ $m^{-2}$ at 3.98 bar for 16.03 h. Experimental critical flux values were only slightly lower than optimal fluxes for both membranes, showing the validity of the statistics models applied. Negligible osmotic pressure was found on two membranes at critical flux parameters, indicating irreversible fouling for both cases. The polymeric membrane revealed strong fouling behavior and the ceramic membrane showed a weak form; the flux decline occurred first in the polymeric membrane, whereas the ceramic membrane exhibited high stability during the filtration operations. A high degree of purification of wastewater was obtained by this membrane at critical flux conditions.

Cleaning-in-place (CIP) protocols are often automated, requiring that the temperature and chemical composition of the cleaning solution be tailored to the specific foulant. With growing pressure to reduce both the water footprint and environmental impact of a process, CIP procedures must be optimized to minimize water and chemical use, an exercise that will help to reduce costs and cleaning outages. MF and UF are filtration technologies that can separate micron-sized bacteria or even large macromolecules from process streams using a selectively permeable membrane, often without any required heating or energy-intensive mechanical action. Biofilms are associations of microorganisms in an aquatic environment, bound together by an extracellular polymer matrix, and attached as a layer to a substrate such as a pipe or wall. Biofilms are readily deformable soft deposits, which make the use of probes for measurement unsuitable. Researchers have been able to observe their removal from a surface by fluid shear and estimate their initial thickness (Lewis et al., 2012).

Stevioside is one of the naturally occurring sweeteners that can be widely applied in food, drinks, medicine, and daily chemicals. Membrane separation has a potential application in the clarification of stevioside from pretreated stevia extract by UF. Mondal et al. (2013) have used 5-, 10-, 30-, and 100-kDa MWCO membranes. Quantification of membrane fouling during UF is essential for improving the efficiency of such filtration systems. A systematic analysis was carried out to identify the prevailing mechanism of membrane fouling using a batch unstirred filtration cell. It was observed that the flux decline phenomenon was governed by cake filtration in almost all the membranes. For 100 kDa membrane, both internal pore blocking and cake filtration are equally important. Resistance in series analysis shows that the cake resistance is several orders of magnitude higher than the membrane resistance. The cake resistance is almost independent of transmembrane pressure drop, which indicates the incompressible nature of the cake. A response surface analysis was carried out to quantify the development of cake resistance with time during UF of various membranes. Quality parameters show that the 30-kDa membrane is better suited for clarification purposes. Identification of the fouling mechanism would aid in the process of design and scaling up of such clarification setup in future.

#### 3.3.1.2.2 Reverse Osmosis

Osmotic pressure is a critically important property in RO based upon Gibbs free energy, which, on a molar basis, is called the thermodynamic or chemical potential (μ). It is an intensive quantity, that is, dependent on its nature and concentration, but independent of the size of the system.

It is a driving force that describes changes in free energy (*G*) when 1 mol of a component is added to or removed from the system.

This difference in the chemical potential of the water is the driving force for permeation of water from the high-potential side to the low-potential side, a phenomenon called osmosis (Figure 3.4).

The most important structural properties of an RO membrane are its chemical nature, its pore statistics (pore size, pore size distribution and density, and void volume) and its degree of asymmetry. From a functional point of view, the most important are its permeability (measure of the rate at which a given molecule permeates) and its permselectivity (measure of the rate of permeation of one molecule relative to another). These characteristics are more commonly termed as "flux" and "rejection" (Cheryan, 1998) (Figure 3.5).

Selenium-enriched mushroom aqueous enzymatic extracts (MAEE) were obtained from the white button mushroom (*Agaricus bisporus*) by a procedure based on enzyme and membrane technology (Cremades et al., 2012).

The mushroom hydrolysate (MH) was concentrated by three different procedures: vacuum evaporation, RO, and NF at 1 kDa, according to standard procedures.

RO concentration was performed using an RO tubular module (Paterson Candy International, England), which consisted of five perforated stainless-steel tubes connected in series. Each tube was lined with a membrane element, 1.2 m in length and 12.5 mm in diameter (total area of 0.25 $m^2$). The module contained an AFC 80 polyamide (PA) tubular membrane.

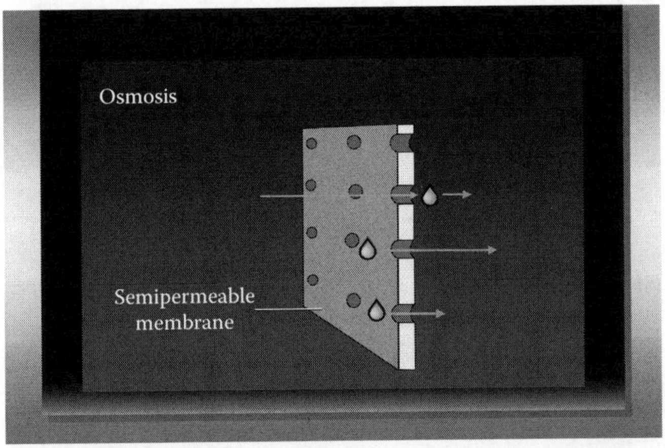


**FIGURE 3.4** Osmosis. (Adapted from Coca Cola Manufacturing Manual, 2003. TCCQS auditors toolkit. The Coca-Cola quality system. Coca-Cola Europe, Eurasia and Middle East. Confidential report.)

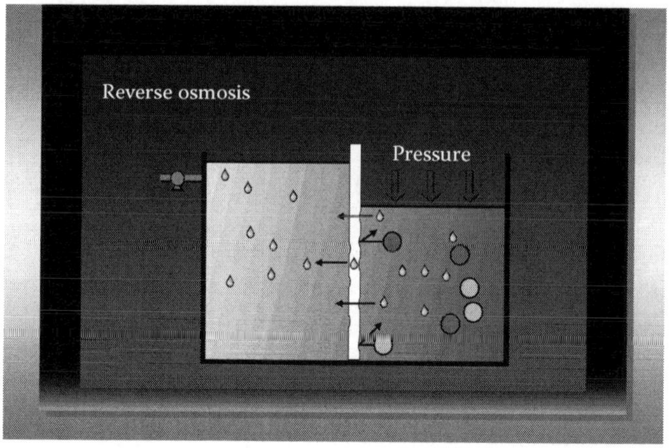


**FIGURE 3.5** Reverse osmosis. (Adapted from Coca Cola Manufacturing Manual, 2003. TCCQS auditors toolkit. The Coca-Cola quality system. Coca-Cola Europe, Eurasia and Middle East. Confidential report.)

NF experiments were performed out with tubular membranes (SeIRO MPT-34, Koch Membrane Systems Inc., Stafford, UK) with a 0.5 kDa cut-off. The filtration area was 0.366 m$^2$. Experiments were performed in batch mode, using a laboratory-scale plant.

Concentrations of hydrolysate were of 23.5 ×, 18.2 ×, and 17.3 ×, respectively. Selenium recoveries measured were 71.7%, 65.3%, and 60.1%, respectively. The lower yields observed in the concentration by RO and nF could be attributed to the loss of Se products during the membrane treatments.

MAEE, with a selenium concentration of 51.8 ± 5.58 μg g$^{-1}$, is a product suitable for achieving the recommended daily dose (RDD) of 55 μg with a small amount of around 1 g of product, which can be incorporated into any type of solid or liquid food without modifying its organoleptic properties. Chemical characterization and selenium speciation are also reported; more than 86% of the selenium-containing products are organic in nature. The utilization of this product would help in the treatment and/or prevention of diseases associated with low selenium concentrations, such as aging and neurodegenerative, cardiovascular, and immunological diseases, while avoiding the risk of reaching high plasma selenium concentrations, which has recently been associated with deleterious effects.

RO is a separation method of water from its solutions with the use of membranes having very small diameter pores. It is an alternative to concentration for water removal, but with lower energy consumption. However, it is expensive to buy and maintain, whereas its efficiency depends on the operation conditions (Walstra et al., 1999).

RO aims to concentrate skimmed milk, whey, and liquid waste pollutants with a high microbial load with the final aim their concentration. Operating at low temperatures and the hold-up of volatile substances are some of its advantages whereas the disadvantages are that milk cannot be concentrated at high levels and the filtrate is pure water (Walstra et al., 1999). RO membranes retain particles of MW until 100 Da and values of applied pressure are 5–10 times higher than those of ultrafiltration

(Rosenberg, 1995). RO of milk and whey removes only water and is similar to thermal evaporation (Mistry and Maubois, 1992). RO is used in whey during powder preparation. Concentration of whey occurs until solids reach 25% followed by evaporation until 50% and then drying. Moreover, RO is applied in whey concentration before ED and in the filtrate arising from UF (Tamime and Robinson, 1999).

RO can be used in milk concentration as well as yogurt preparation. However, it cannot be used in the preparation of strained yogurt since it causes problems due to the increased percentage of lactose and salts in the final product (Tamime and Robinson, 1999).

#### 3.3.1.2.3 Cellulose Acetate Membranes

CA polymers are produced by acetylation of cellulose with acetic anhydride, acetic acid, and a catalyst such as sulfuric acid. Acetylation is carried out until modification of the three hydroxyl groups of each unit occurs (esterified by acetyl groups).

The hydrophilic properties of the membrane are a function of the number of hydroxyl groups remaining after acetylation. Thus, the degree of substitution (DS) must be carefully controlled. The hydrophobic acetyl groups act as cross-links due to dipole–dipole interaction and restrict the swelling resulting in the increased permselectivity of the membrane, that is, water and not hydrated ions are allowed to pass the polymer micellar matrix (Kesting, 1985).

CA membranes are fairly easy to manufacture, and a wide variety of pore sizes are available, from RO to MF. In addition, cellulose (the raw material) is an abundant and renewable resource. There are, however, several limitations to CA that restricts its use, especially in food and biotechnology applications where standards and requirements are quite rigorous. The factors to account for include temperature and pH of operation, use of acidic or basic cleaners, chlorine and other oxidizing agents used for sanitation, microbiological activity, and other mechanical influences such as pressure and high shear in the system.

CA is very temperature sensitive, which limits the maximum operating temperature to 30°C, with some blends of CA and cellulose triacetate (CTA) tolerating 35°C. This low temperature poses several problems such as low flux, since viscosity, diffusivity, and solubility are aided by high temperatures; microbial growth may be a problem at these temperatures (which RO and NF will only make worse since all solutes are concentrated during the process, unless the process is operated at refrigeration temperatures), and cleaning is also more difficult since cleaning agents work best at higher temperatures.

CA is also pH sensitive; recommended pH limits are pH 2–8, preferably pH 3–6. CA will hydrolyze in water at either high or low pH. The rate of hydrolysis, which is lowest at pH 4.5–5.0, is temperature dependent with the recommended pH range decreasing with increasing temperature (Cheryan, 2007).

Thus, the use of MF in the first step of separation can be more effective in subsequent downstream purification by ultra- or NF membrane.

It has been well established that CA polymer can impart high hydrophilicity to a membrane, thus reducing the fouling potential (Kim and Lee, 1998; Muthusamy et al., 2006; Shin et al., 2005; Sossna et al., 2007; Wang et al., 2005, 2006) while the presence of PS can substantially improve mechanical strength. PVP and polyethylene glycol (PEG) are best-known examples (Wienk et al., 1996) of hydrophilic polymer ingredients that can enhance pore formation on a polymer membrane.

Sikder et al. (2009) focused on synthesis and characterization of a polymer blend MF membrane for separation of microbial cells from lactic acid fermentation broth in a continuous process. The membranes were prepared by blending hydrophilic cellulose diacetate (CA) polymer with hydrophobic polysulfone (PSF) polymer in the wet phase inversion method. Immersion precipitation is the most important and best-studied method in the preparation of phase inversion polymer membranes.

Polymers were blended in *N*-methyl-2-pyrrolidone (NMP) solvent (70 wt.%), where PEG was added as a pore former. The membranes were characterized in terms of morphology, porosity, flux, and microbial separation capability. The best prepared membrane with PSF/CA weight ratio of 25/75 yielded a pure water flux of 1830 LMH (L $m^{-2}$ $h^{-1}$) and a fermentation broth flux of 1430 LMH at around 1.5 bar TMP (transmembrane pressure). The membrane was successful in the complete retention of microbial cells from the broth in a continuous cross-flow membrane module integrated with the fermentor.

#### 3.3.1.2.4 Thin-Film Composites

Hydrophilic polymers are superior to CA. Membranes made of PAs, polyamidehydrazides, polybenz-imidazole, and others are very promising examples (Cheryan, 1998). The mechanical, chemical, and biological properties of PA membranes are generally superior to CA. These membranes are not as susceptible to hydrolysis or to microbial attack and they can tolerate alkaline conditions and temperatures up to 50°C, but are extremely sensitive to chlorine. In the latter respect, they are much worse than CA, tolerating a maximum exposure of 0.1 ppm active chlorine.

The first commercially successful composite membrane is the Dow-FilmTec FT-30 membrane. It is composed of a PS support cross-linked by interfacial polycondensation with the PA polymer. The skin layer may itself be composed of several layers, which enhances its strength, flexibility, and abrasion resistance. The hydrophilic carboxyl groups are responsible for its relatively high water permeability, similar to the hydroxyl groups in CA membranes.

Thin-film composite membranes are available from almost every company that manufactures RO membranes.

For the food industry, composites are available as spirals, as flat sheets in plate systems, and in tubular form. It has been used in seawater desalination and has been field tested in food processing facilities for several years (Cheryan, 2007).

Composite hydrophilic pervaporation membranes were prepared from chitosan (CS) blended with hydroxyethylcellulose (HEC) using CA as a porous support by Jiraratananon (2002). The membrane with a CS/HEC blend ratio of 3/1 exhibited the highest pervaporation separation index (PSI) and was selected to be cast on a porous CA support.

#### 3.3.1.2.5 Nanofiltration

NF is effective in separating the mixtures of small organic solutes such as oligosaccharides, low- MW peptides, inorganic salts, amino acids, and other low-MW materials.

NF, which is widely used in many industrial sectors, offers several advantages such as low operating pressure, high flux, and high retention of multivalent anion salts and organic molecules (Eriksson, 1988). In addition to the capability of reducing the ionic strength of the solution, NF membranes can remove hardness ions, organics, and particulate contaminants. NF has been employed in water treatment (Bowler et al., 1998; Frenzel et al., 2006), the food industry (Samhaber, 2005), pretreatment for desalination (Hassan et al., 1998), and other areas.

The use of a porous microresonator placed in a microelectrofluidic system for integrated functions of NF and sensing of small biomolecules and chemical analytes in extremely dilute solution was proposed and investigated by Huang and Guo (2012). As an example, aminoglycosides in drug residues in food and livestock products were considered as the trace chemical analyte. The filtration process of the charged analyte in aqueous solution driven by an applied electrical field and the accompanying optical whispering-gallery modes in the resonator are modeled.

The dynamic process of adsorption and desorption of the analyte onto the porous matrix is studied. Deposition of the analyte inside the porous structure will alter the material refractive index of the resonator, and thus induce an optical resonance frequency shift. By measuring the optical frequency shift, the analyte concentration as well as the absorption/desorption process can be analyzed. Through an intensive numerical study, a correlation between the frequency shift, the analyte concentration, and the applied electrical voltage gradient was obtained.

This reveals a linear relationship between the resonance frequency shift and the analyte concentration. The applied electrical voltage substantially enhances the filtration capability and the magnitude of the optical frequency shift, pushing the porous resonator-based sensor to function at the extremely dilute picomolar concentration level for small bio/chemical molecules down to the subnanometer scale. Moreover, the use of the second-order whispering-gallery mode is found to provide better sensitivity compared with the first-order mode.

NF separates mixtures of proteins and peptides of MW from 300 to 3000 Da. NF can concentrate organic compounds with the removal of monovalent ions such as sodium and chlorine with the result the desalination (Kilara, 2006). Some NF membranes can be used for desalination when high pressures are applied and this is an alternative separation method for ED (Walstra et al., 1999).

NF membranes are membranes where pressure is applied targeting on the small retention of monovalent ions (Kelly et al., 1992). Separation characteristics of NF fall between RO and UF (Jelen, 1992). Applied pressures range between 6 and 40 bar, whereas in RO, pressures are applied until 40 bar, and in UF, they fall below 10 bar (Jeantet et al., 2000). The mechanism of separation of NF membranes depends on the simultaneous effect of electrical separation and separation based on the size of particles. NF membranes do not have visible pores, but have free pores with different structure and opening (Nyström et al., 1995). Most commercial NF membranes are fine synthetic membranes bringing an active layer consisting of aromatic PAs and are negatively affected at neutral pH values (Yaroshchuk et al., 2000).

NF has applications in whey concentration with the retention of high lactose quantities. By using NF, whey is concentrated approximately 15–25% and total inorganic compounds are reduced by 40–50% whereas losses in lactose range between 1% and

5% (Van der Horst et al., 1995, Vasilievic and Jelen, 2000). Moreover, NF applies in the production of fresh whey cheese (Zambrini et al., 1990) and in the recovery of amino acids and peptides from the hydrolysis of β-lactoglobulin (Wijers et al., 1998). NF may be used by the yogurt industry to remove part of lactose from the filtrate of milk UF, which will then be used in yogurt manufacturing, because the product will have a lower percentage of lactose making it more digestible (Rinaldoni et al., 2009).

A dynamic mathematical model was developed by Dey et al. (2012) for a flat sheet cross-flow NF membrane module for the separation of lactic acid from fermentation broth. The model developed with extended Nernst–Planck approach included pH effects during Donnan exclusion of lactate ions through the Henderson–Hasselbalch equation.

The most relevant structural characteristics (pore radius, porosity-to-thickness ratio, and membrane charge density) of the membranes were completely determined by comparison and convergence of the model-predicted and experimental data on flux and rejection using standard solute–solvent systems as well as actual fermentation broth. Thus, the pore sizes of the investigated three NF membranes NF2, NF3, and NF20 were determined as 0.57, 0.55, and 0.54 nm, respectively. Through the modeling and experimental investigation, the best membrane (NF2) that could retain and recycle more than 90% sugars while allowing more than 70% lactic acid to permeate could also be selected. Finally, the refined model could very well predict transport of lactic acid through NF membranes operating in flat-sheet cross-flow module as evident in a very small relative error (<0.1) and the value of overall correlation coefficient ($R^2$) of greater than 0.980.

Economic evaluation of a membrane-integrated bioreactor system for lactic-acid production from sugarcane juice was performed by Sikder et al. (2012). The production process consisted of sterilization, fermentation, MF, NF, and final concentration by vacuum evaporation. Membrane recycle fermentor operating at a cell concentration of 22 g $L^{-1}$ resulted in a productivity of 53 g $L^{-1}$ $h^{-1}$ with a lactic acid concentration of 106 g $L^{-1}$ and a yield of 0.96. The membrane units (cross-flow MF and NF) and pump contribute about 2% to the total fixed capital cost whereas fermentation unit along with holding tank contribute about 36% to the total fixed capital cost. The two largest cost components were raw material and yeast extract costs contributing about 6% and 87%, respectively, to the total operating cost. Total product cost stood at 3.15 US$ $kg^{-1}$ of 80% (w/w) concentrated and 95% pure lactic acid. The study revealed that the operating cost could be reduced further by using a cheaper nitrogen source such as silkworm larvae or yeast autolysate and installing the lactic acid plant in the sugarcane-growing areas or by optimizing the recycle of NF retentate to the fermentor.

Olive milling produces huge amounts of wastewater (OMWW) characterized by an extremely high organic load. Its polyphenols content is a hindrance to conventional biological treatment and as a growing medium for common microbial biomasses. The practice to dump it on soil is in conflict with the latest EU directives on waste management.

OMWW can be effectively and efficiently treated by means of membrane technology to a fraction of the initial volume, but membrane-processing concentrates still require treatment. Reversing the overall cost balance of membrane processing and subsequent treatment requires valorizing the concentrates through their reuse, as well as ensuring long-term service of the membrane system through effective wastewater pretreatment and sustainable, fouling-controlling, membrane operation conduits.

Cicci et al. (2013) reused and valorized the ultra- and NF-membrane concentrates as media for biomass production of microalgae and cyanobacteria. *Scenedesmus dimorphus* and *Arthrospira platensis*, usable as a food, feed, nutraceutical component, or feedstock for biofuels, were selected for this investigation. The performances on microalgal growth obtained by using the different membrane concentrates and the pretreated olive mill wastewater (OMWW) feedstock were experimentally determined and related to the composition of the culture media thus obtained. They were also related to the irradiance distribution within the photobioreactor volume to decouple light limitation and medium chemical composition effects in the production of microalgal and cyanobacterial biomass.

Climate changes are inducing increased sugar levels of must, which produces negative effects on wine quality as unbalanced wines with high degrees of alcohol. So, effective strategies to control the increase of sugar levels in must have been studied. The most recently applied technologies are based on the use of membrane processes that retain the sugars of musts (Bonnet and De Vilmorin, 2004; Calvin, 2001; García-Martín et al., 2009; Gresch, 1996).

One of them is the use of a membrane process, and this is applied in the work by Mihnea et al. (2012). Single NF treatments to the initial must were applied. Three different types of membranes, HL (HL2540FM), DL (DL1812C-28D), and DK (DK1812C-28D), all provided by GE Water & Process Technologies (Barcelona, Spain), were used. According to the commercial company, DL is characterized by 96% cut-off retention for magnesium sulfate and 2.27 water permeability, and DK by 98% cut-off retention for magnesium sulfate and 1.98 water permeability. A combination of UF and NF was also studied.

The sugar level of white must from Verdejo (*Vitis vinifera* variety) was reduced using diverse membrane processes, and the effect of this on the volatile composition of the corresponding wines is studied. The study was carried out during three consecutive vintages. An important impact of the reduction of sugar levels of must on the volatile composition of the obtained wines was detected, which was due to some retention phenomena of aromatic and precursor compounds. To minimize the volatile composition modifications, an appropriate selection of the NF membrane must be done.

During desalination of feed with highly concentrated salt by NF, predictive modeling was difficult due to the effect of salt on the retention of organic solutes. Consequently, a better understanding of the salt effect on membrane and organic solutes was required. Luo and Wan (2011) used four well-known commercially available NF polymeric membranes, NF270, NF-, Desal-5 DL, and Nanomax50, and analyzed them by a model based on an extended Nernst–Planck equation, using highly concentrated glucose and sodium chloride (NaCl) solutions. The results showed that with increasing salt concentration, the solute-to-pore size ratio ($\lambda i$) decreased while the ratio of effective membrane thickness to porosity ($\Delta x/\varepsilon$) increased, indicating that the effect of salt may include decreasing solutes size, increasing membrane pore size, and increasing effective membrane thickness. Moreover, such salt effect appeared to be independent of membrane and solute types, and the correction model could well predict the retention of charged solutes at high salt concentration because electrostatic repulsion effect between charged solutes and membranes was completely screened by the salt ions. Meanwhile, several hypotheses such as membrane swelling, hydration layer thinning, and particle collision were provided to explain the change of model parameters by highly concentrated salt.

Bouchoux et al. (2005) and Bargeman et al. (2005) proposed that membrane swelling (i.e., an increase of average pore size) occurred due to the stronger repulsive interaction between ions with same charge inside the pores when salt was added, implying that electrostatic effect was enhanced by increasing salt concentration. However, such explanation contradicted charge screening by salt. Nilsson et al. (2008) argued that membrane swelling due to a salting-in effect (also called lyotropic effect (Piculell and Nilsson, 1990)) was more suitable to explain their experimental results.

It is well known that pH and salt conditions havc significant influence on NF performance. In order to manipulate NF process and optimize its efficiency, it is very important to get insights into the effects of pH and salt on NF. Luo and Wan (2013) reviewed the reports on NF performance at different pH and salt conditions, focusing on the mechanisms behind various phenomena induced by pH and salt. The effects of pH and salt on NF are mainly reflected in the variations of membrane flux/permeability, solute rejection, and fouling behavior, which also depend on both solute type and solution composition. In order to explain these effects, the changes of membrane properties are evaluated by physical, chemical, and mathematical characterization methods. Eight mechanisms for pH and salt effects are summarized and several practical advices for NF operation are provided. Besides, some interesting opinions such as dominant ions, co-ions competition, salting-out induced pore swelling, and charge-induced concentration polarization are reviewed. This review also intends to provide a guide to optimize NF separation and maintenance.

#### *3.3.1.2.6 Diafiltration*

DF is a useful membrane filtration technique to separate macrosolutes from microsolutes based on their molecular size difference (Paulen et al., 2011). In view of its advantage in the maximization of the purity of desired substances such as proteins, polysaccharides, and so on, the DF process combined with a UF or NF membrane was widely studied. For instance, Gonzalez-Munoz and Paraja (2010) operated the aqueous eucalyptus wood DF through a ceramic NF membrane, and developed a mathematical model to predict the time-course of flux (Jv) and the concentrations of the different compounds. The retention was also introduced, which was considered as a constant. Sun et al. (2011) demonstrated the potential application of UF for the recovery of the polysaccharides from rapeseed, as a key step in the large-scale development of rapeseed meal. The MW distribution of the polysaccharides in bunk solution or the solutions rejected by the membranes was analyzed by size exclusion chromatography. However, the MW distribution of the polysaccharides in the permeate was not determined, and the retention of each MW fraction of polysaccharides was not analyzed during DF.

The polysaccharides were considered as total carbohydrates, and the recovery and purity were computed by a calibration curve of absorbance at 530 nm.

To investigate the membrane separation performance of multicomponent solution throughout continuous volumetric diafiltration (CVD), sucrose and glucose were implemented as model solute (Zhao et al., 2013). Two commercial NF membranes (QY-NF-1-SW with MWCO of 150 Da and QY-NF-3-D with MWCO of 250 Da) were operated in batch recycling mode. Flux (Jv), transmembrane pressure (ΔP), concentrations in feed ($C_f$), and permeate ($C_p$) were collected, and the retentions (R) were calculated. Based on the Spiegler and Kedem (S–K) equation, a theoretical R model was established. The model was employed to establish the mass balance equations in a feed tank during CVD. A novel revised retention equation was established to simulate CVD of sucrose and glucose mixture solution. Further, the model was popularly applied in UF (UF) of the multicomponent solution–soybean molasses by a commercial membrane (QY-UF-1-G with MWCO of 1 kDa). It was predicted in theory that the total retention of multicomponents increased, since the concentration of low-retention solute declined with DF process time when the flux was insignificantly changed, which was then demonstrated by the experiment. The results showed that the retention curve predicted by the model agreed well with the experiment, and the revised retention equation was well designed for industrial food fluids.

###### *3.3.1.2.7 Microfiltration*

Carrere et al. (2002) dealt with the first unit operation of the downstream process for the production of lactic acid: the clarification of fermentation broths by cross-flow MF. MF experiments conducted under constant transmembrane pressure and under constant permeate fluxes (higher and lower than the critical flux) were represented by the resistance in series model in which the membrane resistance, the adsorption resistance, the bacteria cake resistance, and the soluble compounds concentration polarization resistance were taken into account. The different operating modes were compared in terms of two industrial interest criteria: the productivity and fouling rates. Higher productivities were obtained during constant transmembrane pressure runs whereas the lowest fouling rate was observed during the run conducted with a constant permeate flux lower than the critical flux. However, this fouling was mainly due to adsorption and solute components concentration polarization.

Concentration polarization refers to the reversible accumulation of solute within a thin boundary layer adjacent to the membrane surface (Blatt et al., 1970).

The process for the manufacture of milk protein concentrate (MPC) powders involves the UF and DF, and optional evaporation of the retentate prior to drying. MPC powders may be made as low- or high-heat products (Getler et al. 1997; Huffman and Harper, 1999). MPC powders have been used in cheese and a range of food applications, including meat, bakery and dairy products, high protein drinks, and desserts (Zwijgers, 1992).

The solubility of MPCs powders was influenced by the method used for preparing the concentrate, drying conditions, and the type of dryer used (Augustin et al., 2012). Increasing total solids of the ultrafiltered concentrates (23% total solids, TS) by DF to 25% TS or evaporation to 31% TS decreased the solubility of MPC powders (80–83% protein, w/w dry basis), when UF was followed by evaporation a higher total solids was obtained and this had the greater detrimental effect on solubility. High shear treatment (homogenization at 350/100 bar, microfluidization at 800 bar or ultrasonication at 24 kHz, 600 W) of ultrafiltered and diafiltered MPCs prior to spray drying increased the nitrogen solubility of MPC powders (82% protein, w/w dry basis). Of the treatments applied, microfluidization was the most effective for increasing nitrogen solubility of MPC powders after manufacture and during storage. Manufacture of MPC powders (91% protein, w/w dry basis) prepared on two different pilot-scale dryers (single stage or two stage) from MPCs (20% TS) resulted in powders with different nitrogen solubility and an altered response to the effects of microfluidization. Microfluidization (400, 800, and 1200 bar) of the concentrate prior to drying resulted in increased long-term solubility of MPC powders that were prepared on a single-stage dryer but not those produced on a two-stage spray dryer. This work by Augustin et al. (2012) demonstrates that microfluidization can be used as a physical intervention for improving MPC powder solubility. Interactions between the method of preparation and treatment of concentrate prior to drying, the drying conditions, and dryer type all influence MPC solubility characteristics.

A novel method for the extraction of (–) cpigallocatechin gallate of high purity from green tea leaves is proposed by Kumar et al. (2012). The method comprised a two-stage water-based extraction followed by successive use of MF and UF. MF was used as a pretreatment to UF. The best process conditions of each unit operation were estimated by performing well-planned experiments. The clarified liquor was dried to powder by freeze drying. Chemical analyses revealed that the tea powder contained about 90% of polyphenols. The purity of (–) epigallocatechin gallate was found to be about 80%, while its average yield was 1.22 g $L^{-1}$. The method outlined in this study may have remarkable importance for the bulk production of high-purity (–) epigallocatechin gallate with potential applications in pharmaceutical, cosmetic, and food processing industries.

Besides being a green process, this method can be easily scaled up for the commercial production of (–) epigallocatechin gallate.

MF removes small-size particles such as bacteria, yeast cells, and colloid particles. The pores of the MF membranes have a size ranging between 0.1 and 10 μm and retain and separate the above-mentioned particles (Huisman, 2000). Pressures applied are lower than those applied in UF membranes, ranging between 0.1 and 8 bar, with a higher flow (Rosenberg, 1995).

Hemicelluloses with a high molecular mass are needed for the manufacture of value-added products such as food packaging barrier films. Krawczyk et al. (2013) recovered such molecules from chemithermomechanical pulp (CTMP) process water using an innovative three-stage process comprising membrane separation and enzymatic treatment with laccase. MF followed by UF was found to be a suitable combination in the first stage, providing a concentrated and purified hemicellulose fraction suitable for enzymatic treatment.

The MF membrane was a tubular ceramic membrane manufactured by Atech Innovations GmbH (Gladbeck, Germany). This membrane had seven parallel feed channels, each with an inner diameter of 6 mm. The total length of the membrane tube was 1 m and the outer diameter was 25 mm. Spiral-wound elements (Alfa Laval Nordic A/S, Soborg, Denmark) equipped with a 48-mil spacer were used in both UF stages. The UF membranes were made of the same material, but differed in their nominal MWCO.

In both membrane processes, a high average flux (260 and 115 L $m^{-2}$ $h^{-1}$) and a low fouling tendency were observed. A marked increase in the average molecular mass of hemicelluloses with bound lignin moieties was achieved by laccase treatment in the second stage. The enzymatically cross-linked hemicelluloses were finally recovered in the third stage using UF. In the final high-molecular-mass solution, the hemicellulose concentration was 54 g $L^{-1}$, the contribution of hemicelluloses to the total solids content was 43%, and the viscosity of the solution was 27 mPa s. The results demonstrate that a hemicellulose fraction of high quality can be produced from CTMP process water, and that this could constitute a suitable feedstock for the production of, for example, barrier films for renewable packaging.

MF was initially developed in Germany in 1929 from Sartorius-Werke. During the Second World War it was developed for bacteriological examination of water supplies. Until 1963, the manufacturing material of MF membranes was nitrocellulose or a mixture of cellulose esters (Merin and Daufin, 1990). Nowadays, these membranes are manufactured by glass, ceramic materials such as alumina, titanium dioxide, and zirconium oxide (Figure 3.6), and metals such as silver and stainless steel. The advantage of these inorganic compounds is their stability over extreme conditions during food processing, such as high temperature values, extreme pH values, and the contact with solutions different in composition than water. According to Espina et al. (2010), ceramic membranes in MF have a better yield compared to organic membranes during casein micellar separation from whey proteins in skimmed milk. Most metallic and some ceramic membranes are manufactured with concentration of their materials without melting, whereas the remaining ceramic membranes are manufactured with the fusion of a liquid collidal system or with anode oxidation. Some new membranes are prepared with lithographic techniques (Espina, 2010).

In 2002, MF market showed profits of approximately 400 million dollars and the annual growth rate was 6.6% (Huisman, 2000).

MF can take place in two different ways: (a) dead-end filtration, in-line filtration during which the direction of flow is perpendicular to the membrane and (b) crossflow or tangential MF, where the direction of flow is tangential to the membrane.

**FIGURE 3.6** Ceramic microfiltration membranes.

During dead-end MF, particles continuously move toward the direction of membrane and are deposited on its surface or within the pores of the membrane. This deposition leads to an increasing resistance to flow with the result of the continuous reduction in the rate of filtrate flow. The reduction in the deposited precipitate is accomplished by the application of tangential MF. The tangential flow to the membrane removes particles from the membrane surface and minimizes the particle deposition (Huisman, 2000).

MF is used for the manufacture of milk, cheeses, and products with a high shelf life such as milk powder and milk proteins because it reduces the bacteria population and increases their shelf life. In yogurt manufacturing, MF is used in conjunction with UF. When it is used before UF, it contributes to the maintenance of the membrane pores of UF, that is, keeping them clean and minimizing the deposition of particles on the membrane, keeping the flow at high rates. Moreover, it can be used in whey powder manufacturing since it can retain liposomes of small diameter that have escaped centrifugation (Rinaldoni et al., 2009) and can contribute to the increase of purity.

*3.3.1.2.7.1 Application of Microfiltration in the Dairy Industry* The diameter of the pores of MF membranes range between 10 and 0.1 μm and can be applied in specific particle separations dispersed in liquids. Milk particles can be separated according to their size as follows: somatic cells (15–6 μm), liposomes (6–0.2 μm), bacteria (15–0.2 μm), and casein micelles (0.3–0.03 μm) (Pierre et al., 1998).

Milk MF should be carried out with special attention to avoid the quick blocking of the membrane pores. Initially, the device should be cleaned with hot water (52°C) and with evaporation valves open in order to remove air bubbles. Then, the heating of the milk takes place at 50°C for 20 min with the aim to assure the physicochemical balance of milk (Saboya and Maubois, 2000).

An alternative technique to that of MF of milk is bactofugation. With this technique, specially designed separators centrifuge milk at 73°C, so bacteria being a bit more heavier than milk are removed with a small quantity of skimmed milk. With bactofugation, we also have removal of spores in the milk (Walstra et al., 1999). This technique requires high amounts of energy and the reduction in the total number of spores ranges between 90% and 95% (Guerra et al., 1998).

Microfiltrated milk contains approximately 3.5 log microbial load lower than conventional milk. Moreover, sporogenic bacteria surviving pasteurization can be better retained with MF due to the large cellular volume. The number of spores using MF decreases by 4.5 log. Regarding the effect of MF on pathogenic bacteria such as *Listeria monocytogenes*, *Brucella abortus*, *Salmonella typhimurium*, and *Mycobacterium tuberculosis*, their number diminishes by 3.4, 4, 3.5, and 3.7 log, respectively. Hence, microfiltrated skimmed milk contains pathogenic bacteria at a concentration of less than 1 cfu $L^{-1}$ (Madec et al., 1992). MF also removes somatic cells fully resulting in microfiltrated milk not containing their enzymes resistant to pasteurization (Law and Goodenough 1995).

*3.3.1.2.7.2 Microfiltration Applications in the Food and Drink Industry* An important application of MF is in fruit juices. Several researchers have reported on the effect of microfiltration on orange and apple juice as well as tropical fruits. One of the basic organoleptic characteristics of fruit juices is the aroma that depends on the volatile compounds, which in turn depend on thermal processing. Similarly, vitamins are sensitive during thermal processing and the presence of oxygen. Pasteurization at 90°C assures the microbiological safety of the juice; however, it negatively affects its organoleptic characteristics. MF is another alternative of juice processing (Matta et al., 2004). During thermal processing of apple juice, natural volatile aromatic compounds diminish, hence affecting its nutritional value (Perédi et al., 1981). However, using MF, concentrations of aromatic compounds increased except that of *iso*-butyl-acetic acid, whereas pasteurization of microfiltrated juice reduced their concentrations (Su and Wiley, 1998). Moreover, using MF membranes, yeasts and molds were removed fully in apple juice and its clarity was accomplished with removal of solids (Matta et al., 2004).

MF is also applied in brewing. Clarification of produced beer is important for product stability and is achieved with removal of active yeast cells and molecules formed with the interaction of phenolic and protein particles at low temperatures (colloid stability). Moreover, it is very important to retain aromatic compounds and reduce bound oxygen in beer to keep the aroma constant (Gana et al., 2001). Kieselguhr filtration is the well-known technology of beer filtration. However, the high cost leads to the use of MF membranes. Ceramic membranes with pores of 0.2–1.3 μm can easily remove solid substances and yeasts, reduce losses in beer production, and reduce energy required to manufacture a qualitative product (Kiefer, 1991).

In vinegar and wine production, MF is used as a pretreatment to remove solids and sterilize the product. In wineries, MF (a) stabilizes wine by removal of colloid particles, (b) avoids secondary fermentations with bacteria and yeasts removal and (c) clarifies wine (Van der Horst and Hanemaaijer 1990).

MF also removes ovomucin from egg white. It is a glycoprotein of high MW affecting the viscosity of egg white (Powrie and Nakai, 1986). According to Ferreira et al. (1999), MF membranes with pores of 1.4 μm reduced the number of microorganisms in egg white and separated the glycoprotein without blocking the membrane pores.

Finally, MF is used in the processing of liquid waste. Separation of water from oil emulsions is a very important process. Classical methods include chemical demulsification and precipitation. However, high energy consumption is required at a high cost. Hence, MF can be used alternatively (Hu and Scott, 2007). Many researchers have used MF to reduce pollutants of water-in-oil emulsions. Anderson et al. (1987) effectively separated oil emulsions with the use of MF. Moreover, MF was used as a pretreatment for the further processing of liquid wastes.

Li et al. (2013) used MF membranes to remove solid particles from the liquid waste of a fish processing industry followed by UF.

*3.3.1.2.7.3 Cleaning of Microfiltration* Water should not contain any colloid particles or microorganisms that could block the membrane pores. Washing should be done with hot water (50°C) followed by washing with an acid and then with a base. Finally, washing with deionized water takes place until the pH of water is neutral (Beolchini et al., 2005). The efficiency of washing is controlled with water flow and the bacteriological condition of the last washed water, which should not contain any microorganism (Saboya and Maubois, 2000).

## 3.4 CERAMIC MEMBRANE ELEMENTS AND MEMBRANE FILTRATION SYSTEMS

GEA Westfalia Separator Group (2013b) develops, manufactures, and installs ceramic membranes and complete membrane filtration systems. Thanks to cross-flow filtration with ceramic elements, a solution is now available from a single source for many application areas requiring the parallel employment of separators, decanters, and membrane filtration.

Benefits to the user: Intelligent technologies using centrifugation and membrane filtration, matched to each other, providing a significant improvement in the process line efficiency, for example, in the following applications:

Food industry

- Production and processing of dairy products
- Clarification of fruit juices, wine, and beer

Clarification, fractionation, and concentration of fruit juices and plant extracts have been described by Todisco et al. (1998), Jiao et al. (2004), Rai et al. (2006a, b), and Sarkar et al. (2008a) employing the use of UF, NF, and RO. However, the disadvantage of RO is related to its inability to reach high concentration levels because of limitations imposed by high osmotic pressures.

Single-stage RO systems permits to reach final concentrations of fruit juices of about 30°Brix (corresponding to osmotic pressures of 50 bar) quite below the value of 45–65°Brix for standard products obtained by evaporation (Rodrigues et al., 2004).

- Water treatment

Biotechnology

- Continuous contamination-free separation of biotechnologically recovered products from fermenters

Chemical industry

- Catalyst recovery
- Purification and recovery of raw materials
- Recycling of solvents

Recycling technology

- Water treatment
- Recycling of cleaning baths
- Oil–water emulsion separation

Different membrane configurations are shown in Figure 3.7.

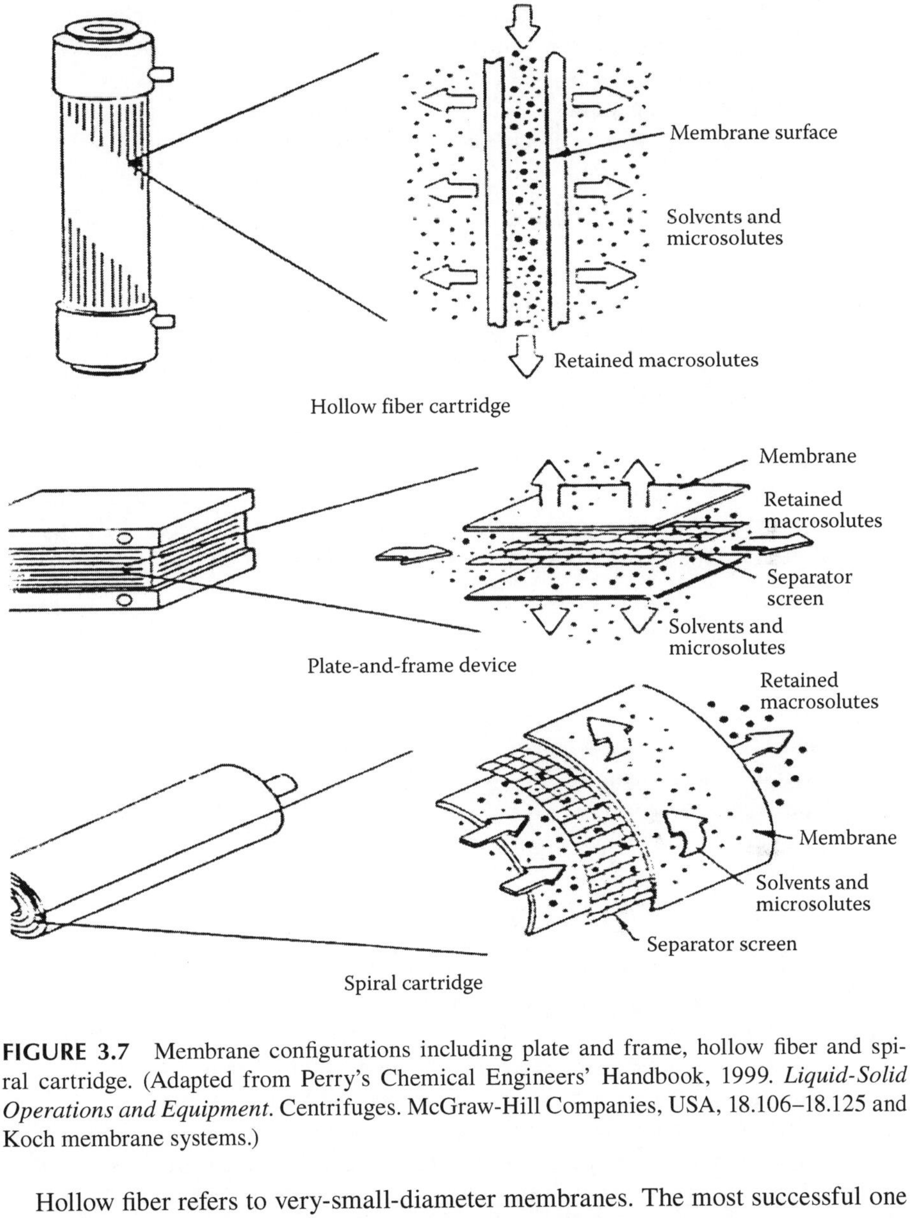


**FIGURE 3.7** Membrane configurations including plate and frame, hollow fiber and spiral cartridge. (Adapted from Perry's Chemical Engineers' Handbook, 1999. *Liquid-Solid Operations and Equipment.* Centrifuges. McGraw-Hill Companies, USA, 18.106–18.125 and Koch membrane systems.)

Hollow fiber refers to very-small-diameter membranes. The most successful one has an outer diameter of only 93 mm and is used for RO.

Low-pressure and high-pressure hollow-fiber membrane modules exist. For low-pressure applications such as air, they may run with tube-side feed. Gas membranes operating at high pressure (above 1.5 MPa) are almost always run with shell-side feed. The outer diameter for gas membranes may be as high as 500 mm (Perry and Greene, 1984; Perry's, 1999).

The spiral-wound module is used in many membrane processes. Permeate collection material is wound on a perforated permeate pipe. A membrane "sandwich" is constructed over the permeate carrier using glue seams as seals. Membrane "sandwiches" are separated by feed-channel spacers, through which the feed stream is passed (Figure 3.7).

The plate and frame module is very much like a filter press. Once found in RO, UF, and MF, it is still the only module commonly used in ED.

### 3.4.1 Technology and Function of the Ceramic Membrane Elements

Dynamic filtration with ceramic membrane elements involve coupling with the lowest possible flow resistance using a support made of pure $\alpha$-$Al_2O_3$ with a macroporous structure. The membrane is applied to this support material and consists of at least one, but usually several layers of highly porous ceramic, with a precisely defined texture. The layer with the finest porosity determines the filtration characteristics.

GEA Westfalia Separator Group offers an unrivaled broad, application-specific spectrum of membranes and element geometrics. Membranes with rated pore sizes of 1–1400 nm are available. Ceramic membranes are therefore ideal for use in MF and UF.

The superb properties of such membrane elements are used successfully in filter systems from GEA Westfalia Separator Group worldwide and have the following advantages:

- Inert material
- Acid and lye resistant
- Solvent resistant
- Regenerative
- Backflush no problem
- Wear resistant
- Long service life
- Reliable
- Heat resistant, steam sterilizable

Centrifugal technology such as the Westfalia Separator's FRUPEX process is commonly used since 2001 by juice manufacturers because it gently separates the liquid from the solids and can be tightly controlled. The company's ESE 500 clarifier was specially designed for this application and can process up to 50,000 L $h^{-1}$.

The demand for beer and fruit juice in countries such as China and Russia is growing and offers new opportunities to the beverage industry. The need for high-performance production machines parallels this. Recently, Westfalia Separator's FRUPEX process, comprising a decanter in the first stage, a decanter in the second stage, and a clarifier, was installed for the first time in the main apple-growing region in China.

Westfalia Separator has introduced a new process that uses centrifugal technology to obtain the "wine must," that is, juice from the grapes.

Known as the VINEX process, it produces a rapid, gentle, and hygienic preclarification of the wine must. The red wine mash is fed into the rotating decanter bowls, where pressure forces the grapes against the wall of the bowl and the centrifugal force separates the must from the solid components. The system continuously discharges the must, which is taken away for fermentation. The red wine mash is heated

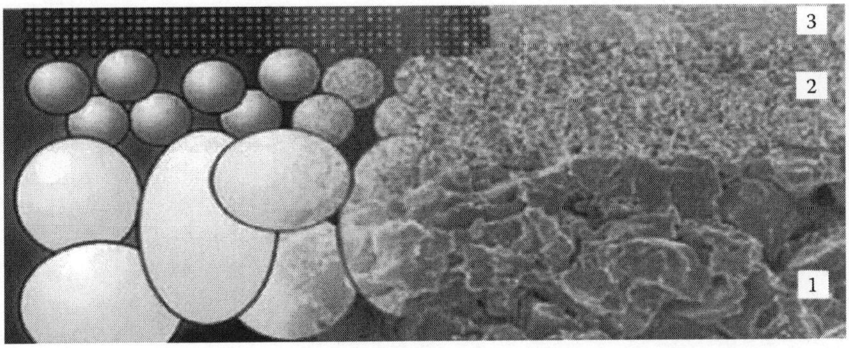


**FIGURE 3.8** The configuration of a ceramic membrane filter. http://www.westfaliaseparator/no_cache/products/cross-flow-filtration.html. 1. Coarse–porous (10 μm) carrier body of $-Al_2O_3$ in the form of a tube with a wall with a wall thickness of 1–2 mm or in the form of a multi channel element. 2. One or several intermediate layers of $Al_2O_3$ with a pore size of 0.2–2 μm and a layer thickness around 10–20 μm. 3. Separating layers of $Al_2O_3$ and $TiO_2$ with pore sizes of 0.1–1.4 μm (microfiltration) and 5–100 nm (ultrafiltration) respectively with layer thicknesses between 0.5 and 10 μm.

to a temperature of 80°C and then pumped continuously into decanters for dye recovery. The remaining pomace, consisting of grape pips and skin, is conveyed by a scroll out of the bowl.

The bitter grape pips are not damaged in the process and can, for example, be pressed to obtain grape pip oil. The remainder of the pomace can be turned into compost (Westfalia, 2001).

### 3.4.2 Ceramic Multichannel Element

The ceramic multichannel element, consists of highly porous ceramic containing several rounds of channels running parallel to its longitudinal axis on the surface of which the membrane is mounted (Figure 3.8).

The suspension to be clarified flows into the channels along the membrane, whereby a partial stream passes through the membrane as filtrate and is discharged by the carrier material. Owing to the very high permeability, the pressure loss on passing through the carrier is so low that it is negligible compared to the pressure drop when passing through the extremely thin membrane.

## 3.5 CENTRIFUGAL SEPARATIONS

In a centrifuge, great forces can be obtained by the centrifugal action. Gravity still acts and the net force is a combination of the centrifugal force with gravity as in the cyclone; however, gravity can be easily neglected since its forces are very weak.

Centrifugal force is generated when materials are rotated around an axis. The size of the force depends on the radius and speed of rotation and the density of the centrifuged material (Fellows, 2000). The force generated through the rotation acts in an outward direction. Depending on the speed of the rotating body, it increases or drops on the circular path. Mechanical separation technology makes use of this property when light substances, heavy substances, or substances of different densities have to be separated from each other.

Separation by means of centrifugal force is, however, faster when the vessel has an insert. The heavier particles deposit faster due to the insert. The settling path is shortened by the insert and hence a higher throughput capacity is attained. This means that larger volumes of liquid mixtures can be clarified or separated in the same period of time. The more inerts there are, the shorter the settling paths and the higher the throughput capacities.

The *centrifugal force* on a particle that is constrained to rotate in a circular path is given by

$$F_c = mr\omega^2 \tag{3.2}$$

where $F_c$ is the centrifugal force acting on the particle to maintain it in the circular path, $r$ is the radius of the path, $m$ is the mass of the particle, and $\omega$ the *angular velocity of the particle* = 2π N/60 (Earle, 1983; Heldman and Hartel 1997a, b).

And since $\omega = v/r$, where $v$ is the *tangential velocity of the particle*

$$F_c = (mv^2)/r \tag{3.3}$$

The centrifugal force depends upon the radius and speed of rotation and upon the mass of the particle, and the heavier the particle, the greater the centrifugal force per unit volume.

A boundary region between the liquids at a given centrifuge speed forms at a radius $rn$, where the hydrostatic pressure of the two layers is equal. This is termed the neutral zone and is important in equipment design to determine the position of feed and discharge pipes.

If we have a dense and light liquid, then

$$rn = (\rho_d r^2{}_d - \rho_l \mathrm{r}^2{}_1)/\rho_d - \rho_l$$

where $d$ and $l$ represent the dense and light liquid layers, respectively, ρ is the density in kg m$^{-3}$, and $r$ is the radius in $m$.

If liquid flow is streamlined, the rate of movement and volumetric flow rate is determined by the densities of the particles and liquid, the viscosity of the liquid, and the speed of rotation (Equation 3.2).

$$Q = \frac{D^2\omega^2\left(\rho_p - \rho_l\right)V}{18\,\mu\,\ln\left(r2/r1\right)} \tag{3.4}$$

where $Q$ is the volumetric flow rate (m$^3$ s$^{-1}$), $V$ is the volume of the centrifuge (m$^3$), $D$ is the diameter of the particle (m), $\rho_p$, $\rho_l$ are the density of particles and liquid, respectively, in kg $m^{-3}$, μ is the viscosity of the liquid (N s $m^{-2}$), and *r*2, *r*1 are the radius of centrifuge bowl and radius of liquid, respectively, in *m*, and *N* is the speed of rotation in rev $s^{-1}$.

The residence time *t* is = *V*/*Q*, that is, the time taken for a particle to travel through the liquid to the centrifuge wall (Fellows, 2000). Derivations and additional details of these equations are given by Brennan et al. (1990) and Earle (1983).

Secondary flows result from curvature as nonuniform centrifugal force-induced Dean vortices (Dean, 1927). As a result, cross-sectional mixing occurs, decreasing residence time distributions (Saxena and Nigam, 1984) while substantially increasing the heat transfer (Shah and Joshi, 1987). Density differences between the bulk and the wall induce Morton vortices (Morton, 1959), which promote cross-sectional mixing and increase heat transfer (Morcos and Bergles, 1975). When curvature and buoyancy forces are both significant, a mixed flow regime can be observed.

The heating, holding, and cooling section of the high-temperature region ($T > 100°C$) of coiled UHT sterilizers were simulated three-dimensionally by Kelder et al. (2002) to assess the impact of centrifugal and buoyant forces on viscous power-law food products. For a wide range of curvature (De = 0–500), heat transfer, lethality development, and thiamine concentration were analyzed.

Curvature and pseudoplasticity substantially increase heat transfer and render the axial velocity profile more uniform. As a result, processing times in the heater, holder, and cooler can be shorter, thus greatly improving thiamine retention. Significant lethality accumulated in the heating section, but the cooling section contributed very little to product sterility. Buoyant forces were shown to be of minor importance in coiled sterilizer flow at the current process conditions.

Decanoic acid reverse micelle-based coacervates were proposed by Garcia-Prieto et al. (2008) for the extraction of bisphenol A (BPA) from canned vegetables and fruits prior to its determination by liquid chromatography and fluorescence detection at $\lambda_{exc} = 276$ nm and $\lambda_{em} = 306$ nm.

Coacervates are water-immiscible liquids that separate from colloidal solutions under the action of a desolvating agent, affecting the temperature or the pH of the colloidal solution (Gander et al., 2002). The procedure involved the extraction of minute quantities (300–700 mg) of homogenized food sample with an aqueous solution containing 10% of THF and 0.5% of decanoic acid, conditions under which the coacervate (around 340 μL) formed *in situ* and instantaneously. The overall sample treatment, which included extraction and centrifugation, took about 25–30 min, and several samples could be simultaneously treated using conventional lab equipment. No clean-up or solvent evaporation were required. Extraction efficiencies mainly depended on the decanoic acid and THF concentration in the aqueous solution and were not affected by the pH or the temperature in the ranges studied (1–4 and 20–60°C, respectively). Recoveries in samples ranged between about 81% and 96%. The precision of the method, expressed as relative standard deviation, was about 3% and the quantitation limit was around 9 ng $g^{-1}$, which was far below the current specific migration limit (SML) set for BPA by the EU Commission (600 ng $g^{-1}$). The method was successfully applied to determine the BPA in the solid content of canned fruit salad, peaches in syrup, mango slices, red peppers, sweetcorn, green beans, and peas. BPA was present at concentrations in the range from 7.8 to 24.4 ng $g^{-1}$ in canned fruits and from 55 to 103 ng $g^{-1}$ in canned vegetables.

High-speed countercurrent chromatography (HSCCC) has been applied extensively to the separation of synthetic dyes (Weisz and Ito, 2011). HSCCC is a liquid–liquid partition technique that does not involve the use of a solid support. One of the two immiscible liquid phases is retained in an Ito multilayer-coil column by centrifugal force while the other liquid phase is pumped through the rotating column.

The performance of three types of HSCCC instruments was assessed for their use by Weisz and Ito (2011) in separating components in hydrophilic and hydrophobic dye mixtures. The HSCCC instruments compared were

(i) A J-type coil planet centrifuge (CPC) system with a conventional multilayer-coil column, where the column holder revolves around the central axis of the centrifuge while rotating about its own axis at the same angular velocity in the same direction and separates a broad range of hydrophobic and hydrophilic compounds, except for extremely polar compounds such as proteins and polysaccharides.

The system consisted of a column (three multilayer coils connected in series and made of 1.6 mm i.d. polytetrafluoroethylene (PTFE) tubing with a total capacity of ~320 mL) mounted on a rotating frame (centrifuge), a speed controller, and a Model 300 LC pump (Scientific Systems, State College, PA, USA). To this system they added a right-angle flow-switching valve to conveniently introduce into the column the stationary phase, a UV detector, a chart recorder, and a fraction collector.

(ii) A J-type CPC system with a spiral-tube assembly-coil column fitted with three spiral-tube assembly coils (CC Biotech LLC, Rockville, MD, USA) (Ito et al., 2008) for use as the column separating polar peptides, proteins, nucleic acids, and polysaccharides.

(iii) A cross-axis CPC system with a multilayer-coil column. It has an XL-type planetary motion where the column revolves around the vertical centrifuge axis while rotating about its horizontal axis at the same angular velocity.

The column consisted of four multilayer coils connected in series and made of 1.6 mm i.d. PTFE tubing with a total capacity of ~250 mL. The relevant parameters (Menet and Thiebaut, 1999) of the instrument are as follows: radius, $r$, of the multilayer-coil holder, ~4.5 cm; distance between the two axes, $R$, ~5 cm; and measure of the lateral shift of the multilayered-coil holder along its axis, $L$, ~7.6 cm.

The hydrophilic dye mixture consisted of a sample of FD&C Blue No. 2 that contained mainly two isomeric components, 5,5′- and 5,7′-disulfonated indigo, in the ratio of ~7:1. The hydrophobic dye mixture consisted of a sample of D&C Red No. 17 (mainly Sudan III) and Sudan II in the ratio of ~4:1. The two-phase solvent systems used for these separations were 1-butanol/1.3 M HCl and hexane/acetonitrile. Each of the three instruments was used in two experiments for the hydrophilic dye mixture and two for the hydrophobic dye mixture, for a total of 12 experiments.

In one set of experiments, the lower phase was used as the mobile phase, and in the second set of experiments, the upper phase was used as the mobile phase. The results suggest that (a) the use of a J-type instrument with either a multilayer-coil column or a spiral-tube assembly column, applying the lower phase as the mobile phase, is preferable for separating the hydrophilic components of FD&C Blue No. 2, and (b) the use of a J-type instrument with multilayer-coil column, while applying either the upper phase or the lower phase as the mobile phase, is preferable for separating the hydrophobic dye mixture of D&C Red No. 17 and Sudan II.

### 3.5.1 Protein and Oil Recovery by Centrifugation from Various Plants

The processes involved in the separation and recovery of protein and oil include a demulsification step, followed by the separation of the aqueous and oil phases by centrifugation. Demulsification has been carried out in a number of different ways. The aim of demulsification is to promote or accelerate (in a thermodynamic sense) the mechanisms responsible for separation of the components.

The tendency of emulsified components to separate into distinctive phases can be understood from the second law of thermodynamics, as explained by Tadros (1989). The free energy for the formation of an emulsion is given by

$$\Delta G^{form} = \gamma_{ow}\Delta a - \theta\Delta S^{Conf} \tag{3.5}$$

where $\gamma_{ow}$ is the interfacial tension and $\Delta a$ is the increase in interfacial area. During emulsification, the interfacial area (a) between the oil and water phases is greatly increased as a result of the oil droplet subdividing into much smaller units; this is accompanied by increased interfacial energy $\gamma_{ow}\Delta a$.

Liquid–liquid separation, similar to emulsions, occurs in the dairy industry where milk is fed into the bowl of a vertical continuous liquid centrifuge and separated into skimmed milk and cream.

Virgin olive oil is extracted in olive oil mills with pressure, centrifugation, and percolation systems by using different apparatus driven by physical forces, which, when correctly exerted on olive paste, enables the separation of the different phases of olives: liquid and solid. In the past, the percolation system was coupled with pressures; at present it is coupled with the centrifugal decanter (Mascolo 1978; Martinez-Suarez et al., 1974).

Centrifugation is a worldwide system based on the centrifugal force applied on olive paste, diluted with lukewarm water. The dilution increases the difference between the specific weights of the immiscible liquids (oil and vegetable water) and the solid matter.

Di Giovacchino et al. (2002) reported that the separation of oil from the solid and liquid phases of olive paste is performed by using pressure, percolation, or centrifugation. All systems may provide good-quality oil if olive fruits are sound and at the correct ripeness, but the centrifugation system helps to avoid or reduce the risk of an organoleptic contamination. The new centrifugal decanters operating without adding water (or only a minimal amount of water) to olive paste save heat energy, and the oils obtained are more fruity and have a higher content of natural phenolic antioxidants.

## 3.5.2 Centrifuge Equipment

The simplest form of centrifuge consists of a bowl spinning about a vertical axis. Liquids, or liquids and solids, are introduced into this, and under centrifugal force, the heavier liquid or particles pass to the outermost regions of the bowl, while the lighter components move toward the center.

The main categories are liquid–liquid centrifuges, centrifugal clarifiers for removal of small amounts of solids, and desludging or dewatering centrifuges.

### 3.5.2.1 Liquid–Liquid Centrifuges

The simplest type of equipment is the tubular bowl centrifuge. It consists of a vertical cylinder (or bowl), typically 0.1 m in diameter and 0.75 m long, which rotates inside a stationary casing at between 15,000 and 50,000 rev min$^{-1}$ depending on the diameter.

If the feed is all liquid, then collection pipes can be fitted accordingly to separate the heavier and the lighter components. Various arrangements are used for this. Centrifuge action is analogous to gravity settling, with the various weirs and overflows acting in the same way as in a settling tank even though the centrifugal forces are much greater than gravity. In liquid/liquid separation centrifuges, conical plates are arranged, giving smoother flow and better separation.

Better separation is obtained by thinner layers of liquid formed in the disk bowl centrifuge. Here, a cylindrical bowl, 0.2–1.2 m in diameter, contains a stack of inverted metal cones that have a fixed clearance of 0.5–1.27 mm and rotate at 2000–7000 rev min$^{-1}$.

Whereas liquid phases can easily be removed from a centrifuge, solids present much more of a problem.

### 3.5.2.2 Centrifugal Clarifiers

The simplest solid–liquid centrifuge is a *solid bowl clarifier*, which is a rotating cylindrical bowl, 0.6–1.0 m in diameter. Liquor, with a maximum of 3% w/w solids, is fed into the bowl and the solids form a cake on the bowl wall (Fellows, 2000).

The mixture to be clarified enters the rotor through a centrally arranged feed tube. The distributor accelerates the product and conveys it into the separating space to the disk stack where the actual separation of solids and liquid takes place. In a centrifugal separator, the separated phases are discharged under pressure by centripetal pump.

### 3.5.2.3 Disk Separators with Solid-Wall Bowl

By fitting a large number of conical disks separated by spacers between 0.3 and 2 mm, a substantially larger equivalent clarification area is achieved with the same bowl volume compared to chamber bowl separators. The equivalent clarification area is the product of the sum of the geometric surfaces of the disks multiplied with the acceleration due to gravity.

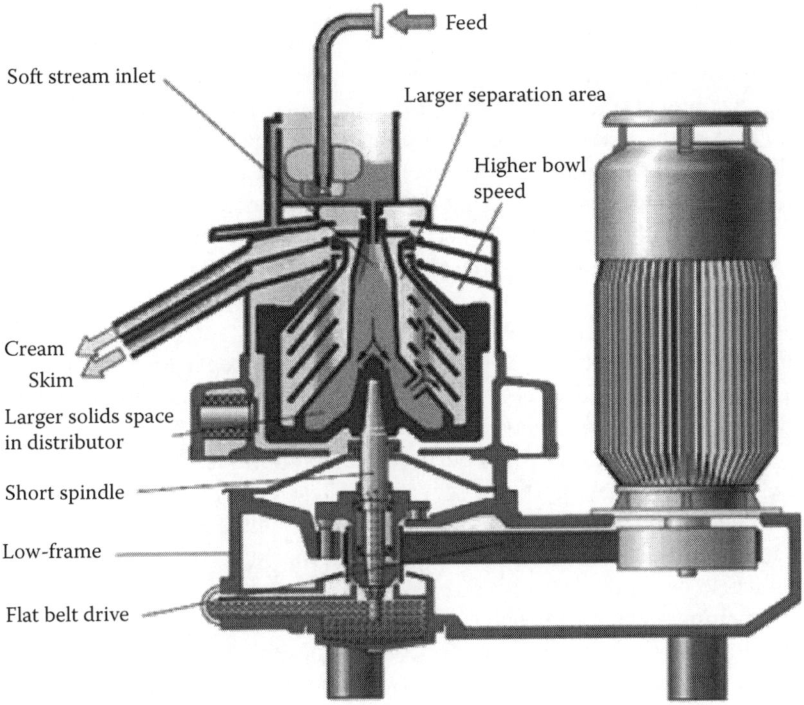


**FIGURE 3.9** Solid-wall disk separator. (Adapted from Westfalia, 2001. Centrifugal technology benefits the global beverage industry. *Filtration-Separation* 30–31.)

Solid-wall separators are available in two versions:

- Separators of two liquids dissolved in one another
- Clarifiers for the separation of solids from liquids

Solid-wall separators are used almost exclusively for the separation of liquid mixtures with little or no solids content since the separated solids can only be removed manually. Example of a disk wall separator is given in Figure 3.9.

#### 3.5.2.4 Chamber Bowl Separators

Chamber bowl separators are solid bowl centrifuges with annular inserts (chambers). They are used for clarification, that is, the separation of solids from suspensions. The suspension flows through the individual chambers from the inside to the outside, whereby the solids deposit in the chambers. The clarified liquid discharges under gravity over an overflow weir (regulating ring) or is discharged under pressure by means of a centripetal pump.

The machines must be shut down to empty chamber bowl separators; the separated solids are removed manually.

Application criteria for chamber bowl separators are

- They are ideal for polishing suspensions, that is, for separating smallest solid contents since they feature excellent flow characteristics.
- Owing to the long retention time in the centrifugal field, the separated solids are recovered in a very compact state.
- Chamber bowl separators cannot be used when processing with centrifuges equipped with an automatic discharge system due to insufficient shear stability or excessive erosive effect.

#### 3.5.2.5 Self-Cleaning Disk Separators

The disk separators with self-cleaning bowls are equipped with a stack of conical disks to create a large equivalent clarification area within a relatively small bowl volume. Separators with self-cleaning bowls are able to periodically discharge the separated solids at full speed. For this purpose, several ports are spaced evenly around the bowl periphery. These ports are opened and closed by means of a movable sliding piston located in the bowl bottom. The opening mechanism is actuated hydraulically. Water is normally used as a control medium; in special cases, low-viscous organic fluids can also be used.

This opening mechanism enables both partial ejections and total ejections. Total ejections involve discharging the entire contents of the bowl with closed feed valve. In the case of a partial ejection, by contrast, only part of the bowl content is ejected with an open feed valve.

#### 3.5.2.6 Westfalia Separator® Hydrostop

Westfalia Separator hydrostop is a special system developed by GEA Westfalia Separator (2013a), which can be adjusted to specific requirements in terms of solid concentration exactly and reproducibly. This patented ejection system makes it possible to optimize the ejection cycle to the shortest possible time.

The Westfalia Separator hydrostop system reduces the actual ejection time to less than a tenth of a second and permits partial ejections to be performed in a 30 s rhythm. This assures that even small volumes of 1.5 to 2 L are discharged reproducibly with a margin of error of less than 10%. This innovative technology enables precise, fast ejections and, hence, significantly higher and qualitatively better yields (http://www.westfalia-separator.com/products/separators/self-cleaning-disk-separators.html).

Feeds that contain a higher solid content are separated using *nozzle centrifuges* or *valve discharge centrifuges*. These are similar to disk bowl types, but the bowls have a biconical shape.

In the nozzle type, solids are continuously discharged through small holes at the periphery of the bowl and are collected in a containing vessel. In the valve type, the holes are fitted with valves that periodically open for a fraction of a second to discharge the accumulated solids.

Nozzle separators are continuously operating disk centrifuges. They are built as centrifugal clarifiers and separators and are more solids orientated than self-cleaning separators. When they are configured to function as clarifiers, they are called concentrators. They are used to thicken solids from suspensions.

Nozzle separators are absolute specialists in the processing of fermentation products in biotechnology and the pharmaceutical and food industries. The separated solids are discharged continuously through nozzles fitted at the bowl periphery. The nozzle separators are equipped with a hydraulically operated ejection system, which enables both partial and total ejections to be triggered during product separation.

*The advantage*: When processing difficult products, the separating time to the next CIP cycle can be extended. The time-controlled or process-dependent ejections additionally optimize cleaning of nozzles, bowl, and disks.

In addition to models equipped with standard nozzles, GEA Westfalia Separator also offers separators featuring patented special Westfalia Separator Viscon® nozzles (viscosity-controlled nozzles). In contrast to standard nozzles, they guarantee a virtually constant biomass concentration by means of an automatic, internal concentration regulation, even in the case of fluctuating feed volumes and concentrations. Volume streams of over 300 $m^3\ h^{-1}$ can be processed with the largest models in this series.

Centrifugal clarifiers are used to treat oils, juices, beer, and starches and to recover yeast cells. They have capacities up to 300,000 $L\ h^{-1}$. Centrifugal clarifiers can also be used in the manufacturing of whey and whey cream.

Whey is defined as the liquid that forms in cheese-making after the casein and fat have been separated when the milk clots. Continued processing of whey can lead to the production of whey protein concentrate, whey concentrate, whey powder, and lactose.

Separators can also be used to recover valuable constituents such as cheese dust or lactalbumin. Lactalbumin is a mixture of various whey proteins. It is traditionally recovered by heat denaturation of wheys from various origins, primarily from cheese wheys. Once the lactalbumin has become insoluble, it can be centrifuged by separators and dried in a spray-drying tower to form a powder. Lactalbumin has a high nutritional value.

The obtaining and further processing of unwanted products from whey can also be demonstrated by the example of cheese dust. Cheese dust is defined as very fine cheese particles of a few μm to 1 mm in size which are formed during the cutting, agitating, and pumping of curd. These protein particles get into the whey during the separation of curd and whey or when draining the cheese molds and pressing them. Many large dairies separate the cheese dust from the whey with clarifiers from GEA Westfalia Separator and then use decanters to remove more moisture from the concentrate thus obtained to form a free-flowing cheese mass with a dry matter of around 40%. This mass can then be made into processed cheese.

Westfalia separators can also be used in warm or cold milk skimming. Mechanical skimming by centrifugal force is necessary to separate the raw milk into cream and skimmed milk. After separation, both are mixed together again in a certain ratio (depending on whether low-fat milk, full-fat milk, or cream is to be produced) until the desired fat content is set.

Most dairies work with warm milk skimming; the raw milk is first heated and then skimmed warm. Because of the higher temperature, there is a significant difference in density between cream and skimmed milk and this has a positive impact on the skimming precision of a separator.

However, cold milk skimming is on the rise, especially in the United States, Mexico, Australia, and New Zealand. Here, the operating temperature is between 4°C and 20°C. This means lower energy consumption and thus reduced production costs for the dairy when compared with the previous skimming temperatures of 52–55°C.

GEA Westfalia's procool separator, the first cold milk centrifuge to use a belt drive and therefore consume less energy, supports cold milk skimming.

In liquid/solid separation, stationary ploughs cannot be used as these create too much disturbance of the flow pattern on which the centrifuge depends for its separation. One method of handling solids is to provide nozzles on the circumference of the centrifuge bowl as illustrated in Figure 3.10. These nozzles may be opened at intervals to discharge accumulated solids together with some of the heavy liquid. Alternatively, the nozzles may be open continuously relying on their size and position to discharge the solids with as little as possible of the heavier liquid. These machines thus separate the feed into three streams—light liquid, heavy liquid, and solids—the solids carrying with them some of the heavy liquid as well. Similar to the operation of disk nozzles, there are disks with intermittent discharge.

Another method of handling solids from continuous feed is to employ telescoping action in the bowl: sections of the bowl moving over one another and conveying the solids that have accumulated toward the outlet (Earle, 1983).

The horizontal bowl with scroll discharge, centrifuge, as illustrated in Figure 3.11, can discharge continuously. In this machine, the horizontal collection scroll (or screw) rotates inside the conical-ended bowl of the machine and conveys the solids with it, while the liquid discharges over an overflow toward the center of the machine and at the opposite end to the solid discharge. The speed of the scroll, relative to the bowl, must not be great. For example, if the bowl speed is 2000 rev min$^{-1}$, a suitable speed for the scroll might be 25 rev min$^{-1}$ relative to the bowl, which would mean a scroll speed of 2025 or 1975 rev min$^{-1}$ (Coulson et al., 1978; Foust et al., 1980; Heldman and Hartel, 1997a,b; McCabe et al., 1985; Trowbridge 1962).

Specific applications of centrifuges are described by Hemfort (1984) for the fermentation industries and by Hemfort (1983) for the food industry.

Finally, tubular bowls exist as shown in Figure 3.12.

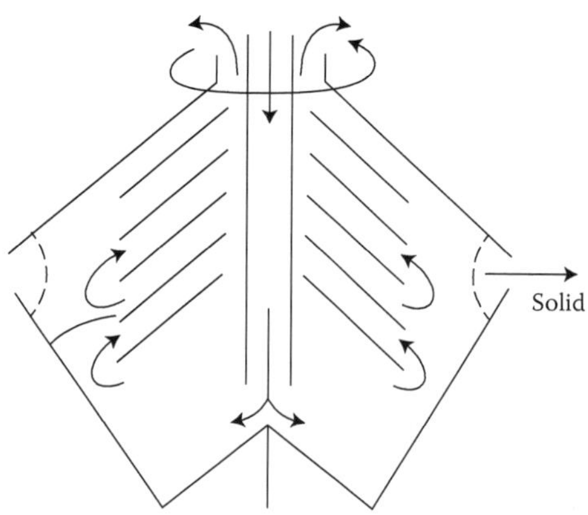


**FIGURE 3.10** Disk nozzle.

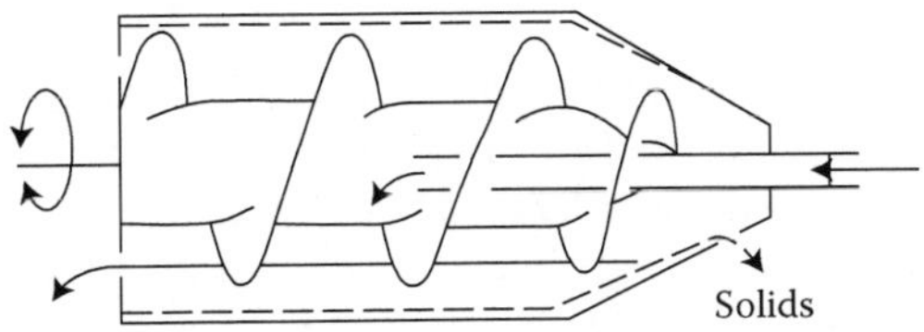


**FIGURE 3.11** Liquid/solid centrifuges, horizontal bowl, scroll discharge.

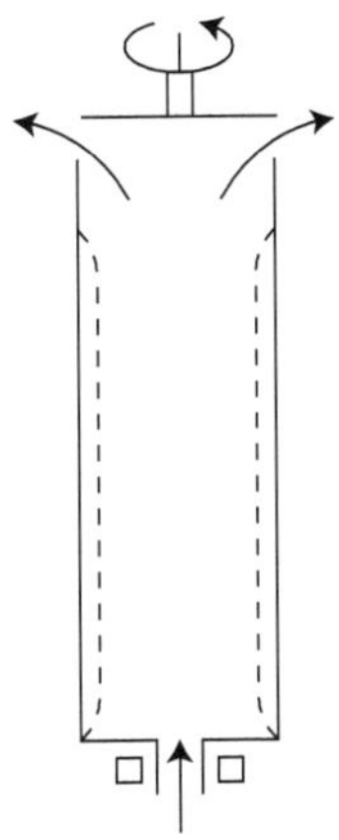

**FIGURE 3.12** Simplified diagram of a tubular bowl centrifuge.

#### 3.5.2.7 Desludging, Decanting, or Dewatering Centrifuges

Feeds with high solid contents are separated using desludging centrifuges, including conveyor bowls, screen conveyors, baskets, and reciprocating conveyor centrifuges. In the *conveyor bowl centrifuge*, the solid bowl rotates up to 25 rev min$^{-1}$ faster than the screw conveyor. This causes the solids to be conveyed to one end of the centrifuge, whereas the liquid fraction moves to the other larger-diameter end (Fellows, 2000).

The solids are relatively dry compared with other types of equipment. The *screen conveyor centrifuge* has a similar design but the bowl is perforated to remove the liquid fraction. The *reciprocating conveyor centrifuge* is used to separate fragile solids (e.g., crystals from liquor).

## 3.6 FILTRATION

Filtration is the phenomenon when a fluid is subject to a force that moves it past the retained particles. The particles suspended in the fluid, which will not pass through the apertures, are retained and build up to form a cake of increasing thickness, the so-called filter cake. The fine apertures necessary for filtration are provided by fabric filter cloths, meshes and screens of plastics or metals, or by beds of solid particles.

Rate of filtration = driving force/resistance

Resistance arises from the filter cloth, mesh, or bed, and to this is added the *resistance of the filter cake* as it accumulates. The filter cake resistance is obtained by multiplying the specific resistance of the filter cake, that is, the resistance per unit thickness, by the thickness of the cake. The resistances of the filter material and pre-coat are combined into a single resistance called the filter resistance.

Filtration can be classified by the following:

1. Driving force
   The filtrate is induced to flow through the filter medium by hydrostatic head (gravity), pressure applied upstream of the filter medium, vacuum or reduced pressure applied downstream of the filter medium, or centrifugal force across the medium.
2. Filtration mechanism
   When solids are trapped within the pores or body of the medium, it is termed *depth*, *filter-medium*, or *clarifying filtration*.
3. Objective
   The process goal of filtration may be dry solids, clarified liquid (filtrate), or both. Good solid recovery is best obtained by cake filtration, while clarification of the liquid is accomplished by either depth or cake filtration.
4. Operating cycle
   Filtration may be intermittent (batch) or continuous. Batch filters may be operated with constant-pressure driving force, at constant rate, or in cycles that are variable with respect to both pressure and rate. Batch cycle can vary greatly, depending on filter area and solid loading.
5. Nature of the solids
   Cake filtration may involve an accumulation of solids that is compressible or substantially incompressible, corresponding roughly in filter-medium filtration to particles that are deformable and to those that are rigid. The particle or particle-aggregate size may be of the same order of magnitude as the minimum pore size of most filter media (1–10 μm and greater), or may be smaller (1 μm down to the dimension of bacteria and even large molecules) (Perry, 1999).

It is convenient to express the filter resistance in terms of a fictitious thickness of filter cake. This thickness is multiplied by the specific resistance of the filter cake to give the filter resistance. Thus, the overall equation giving the volumetric *rate of flow* $dV/dt$ is

$$dV/dt = (A \cdot \Delta P)/R \tag{3.6}$$

As the total resistance is proportional to the viscosity of the fluid, we can write

$$R = \mu r(L_c + L) \tag{3.7}$$

where $R$ is the resistance to flow through the filter, μ is the viscosity of the fluid, $r$ is the specific resistance of the filter cake, $L_c$ is the thickness of the filter cake, and L is the fictitious equivalent thickness of the filter cloth and precoat, $A$ is the filter area, and $\Delta P$ is the pressure drop across the filter.

If the rate of flow of the liquid and its solid content are known and assuming that all solids are retained on the filter, the thickness of the filter cake can be expressed by

$$L_c = wV/A \tag{3.8}$$

where $w$ is the fractional solid content per unit volume of liquid, $V$ is the volume of fluid that has passed through the filter, and $A$ is the area of filter surface on which the cake forms.

The resistance can then be written as

$$R = \mu r[(wV/A) + L] \tag{3.9}$$

and the equation for flow through the filter, under the driving force of the pressure drop is then (Earle, 1983)

$$\mathrm{d}V/\mathrm{d}t = A\Delta P/\mu r[w(V/A) + L] \tag{3.10}$$

Moreover, Perry (1999) reported that in cake or surface filtration, there are two areas: continuous filtration, in which the resistance of the filter cake (deposited process solids) is very large with respect to that of the filter media and filtrate drainage, and batch pressure filtration, in which the resistance of the filter cake is not very large. Batch pressure filters are generally fitted with heavy, tight filter cloths plus a layer of precoat, and continuous filters, except for precoats, use relatively open cloths that offer little resistance compared to that of the filter cake.

Simplified theory for both batch and continuous filtration is based on Hagen–Poiseuille equation:

$$\mathrm{d}V/\mathrm{d}t = A\Delta P/\mu(awV/A + r) \tag{3.11}$$

where $V$ is the volume of filtrate collected, $t$ is the filtration time, $A$ is the filter area, $P$ is the total pressure across the system, $w$ is the weight of cake solids/unit volume of filtrate, μ is the filtrate viscosity, a is the cake-specific resistance, and $r$ is the resistance of the filter cloth plus the drainage system.

In the filtration process, the components that are rejected accumulate near the membrane surface. This phenomenon is defined as concentration polarization and has significant consequences, especially in UF. The other phenomenon that can occur is the formation of the so-called gel layer, which acts as an extra hydraulic resistance. This also results in a lowering of the flux.

Fouling can become more important if the components in the gel layer react with each other and form a dense cross-linked layer on the top of the membrane (de Bruijn et al., 2003).

### 3.6.1 Constant-Rate Filtration

In the early stages of a filtration cycle, it frequently happens that the filter resistance is large relative to the resistance of the filter cake because the cake is thin. Under these circumstances, the resistance offered to the flow is virtually constant and so filtration proceeds at a more or less constant rate.

### 3.6.2 Constant-Pressure Filtration

Once the initial cake has been built up, and this is true of the greater part of many practical filtration operations, flow occurs under a constant-pressure differential. Under these conditions, the term $\Delta P$ in Equation 3.10 is constant and so

$$\mu r[w(V/A)+L]\mathrm{d}V = A\Delta P\mathrm{d}t \tag{3.12}$$

and integrating from $V = 0$ at $t = 0$, to $V = V$ at $t = t$
$\mu r[w(V^2/2A) + LV] = A\Delta Pt$ and rewriting this

$$tA/V = \mu rw/2\Delta P] \times (V/A) + \mu rL/\Delta P \tag{3.13}$$

$$t/(V/A) = (\mu rw/2\Delta P) \times (V/A) + \mu rL/\Delta P \tag{3.14}$$

Equation 3.14 is useful because it covers a situation that is frequently found in a practical filtration plant. It can be used to predict the performance of filtration plant on the basis of experimental results. If a test is carried out using constant pressure, collecting and measuring the filtrate at measured time intervals, a filtration graph can be plotted of $t/(V/A)$ against $(V/A)$ and from the statement of Equation 3.14, it can be seen that this graph should be a straight line. The slope of this line will correspond to $\mu rw/2\Delta P$ and the intercept on the $t/(V/A)$ axis will give the value of $\mu rL/\Delta P$. Since, in general, $\mu$, $w$, $\Delta P$, and $A$ are known or can be measured, the values of the slope and intercept on this graph enable $L$ and $r$ to be calculated (Earle, 1983; Perry, 1999).

In constant filtration, we assume that the resistance of the filter cloth plus filtrate drainage is negligible compared to the resistance of the filter cake and also that both pressure drop and specific cake resistance remain constant throughout the filter cycle.

$$W = \sqrt{\frac{2wPt}{\mu\left((awV/A) + r\right)}} \tag{3.15}$$

and

$$V_f = \sqrt{\frac{2Pt}{\mu aw}} \tag{3.16}$$

where $W$ is the weight of dry filter cake solids/unit area, $V_f$ is the volume of cake formation filtrate/unit area, $V_w$ is the volume of cake wash filtrate/unit area, $t$ is the cake formation time, and $N$ is the wash ratio, the volume of cake wash/volume of liquid in the discharged cake.

### 3.6.3 Filter Cake Compressibility

With some filter cakes, the specific resistance varies with the pressure drop across it. This is because the cake becomes denser under higher pressure and so provides fewer and smaller passages for flow. The effect is known as the compressibility of the cake. Soft and flocculent materials provide highly compressible filter cakes, whereas hard granular materials, such as sugar and salt crystals, are little affected by pressure. To allow for cake compressibility, the following empirical relationship has been proposed:

$$r = r'\Delta P^{s} \tag{3.17}$$

where $r$ is the average specific cake resistance under pressure $P$, $\Delta P$ is the pressure drop across the filter, $r'$ is the specific resistance of the cake under a pressure drop of 1 atm and is a constant determined largely by the size of the particles forming the cake, and $s$ is a constant for the material, called cake compressibility, varying from 0 for rigid, incompressible cakes, such as fine sand and diatomite, to 1.0 for very highly compressible cakes. For most industrial slurries, $s$ lies between 0.1 and 0.8.

This expression for $r$ can be inserted into the filtration equations, such as Equation 3.14, and values for $r'$ and $s$ can be determined by carrying out experimental runs under various pressures (Perry, 1999; Earle, 1983).

### 3.6.4 Filter Media

The selection of filter media depends on the optimization of the following factors (Perry, 1999):

1. Ability to bridge solids across its pores quickly after the feed is started (i.e., minimum propensity to bleed)
2. Low rate of entrapment of solids within its interstices (i.e., minimum propensity to blind)
3. Minimum resistance to filtrate flow (i.e., high production rate)
4. Resistance to chemical attack
5. Sufficient strength to support the filtering pressure
6. Acceptable resistance to mechanical wear
7. Ability to discharge cake easily and cleanly
8. Ability to conform mechanically to the filter with which it will be used
9. Minimum cost

### 3.6.5 Enhanced Filtration

Enhanced filtration is used by the Coca Cola company to treat 100% of raw water by using one of the following processes:

- Coagulation/flocculation followed by granular media filtration (conventional chemical treatment) Fe or Al coagulant

Coagulation/flocculation is the combination of two processes working together to destabilize stable particulate suspensions in water and combine these destabilized particles into particles large enough for sedimentation and filtration. Another name for stable particulate suspensions is "colloidal suspension." Coagulation is the process of particle destabilization. Flocculation is the process of combining these destabilized particles. A suspension is a system in which very small particles (solid, semisolid, or liquid) are more or less uniformly dispersed in a liquid or gaseous medium. If the particles are small enough to pass through a filter membrane, between 1 nm and 1 μm, the system is called a "colloidal suspension."

Contributing to the stability of the suspension is that, in general, in natural waters the particles in the suspension carry a negative surface charge. These negative surface charges interact to push (repulse) the particles away from each other (Coca Cola Company, 2003).

Particulate suspensions commonly removed by coagulation/flocculation include.

Clay, silt-based turbidity, natural organic matter (including disinfection by-product precursors), microbial matter (including bacteria, yeast, molds, virus, protozoa), metals, synthetic organic chemicals, iron, manganese, off-taste causing compounds, odor-causing compounds.

There are two major groups of flocculants:

Inorganic flocculants
Organic flocculants (commonly referred to as coagulant aids)
There are two main types of inorganic flocculants: cationic iron flocculants and cationic aluminum flocculants.
The two common forms of cationic iron flocculants are ferrous ($Fe^{+2}$), such as ferrous sulfate, and ferric ($Fe^{+3}$), such as ferric sulfate or ferric chloride.
The two common forms of cationic aluminum flocculants are aluminum sulfate and polyaluminum chloride, both $Al^{+3}$.

There are three steps in coagulation/flocculation:

1. Coagulant formation
   The flocculant is added to the water that is to be treated and immediately dissolves and hydrolyzes.
2. Coagulation (also called charge neutralization or particle destabilization)
   The coagulant cation's positive charge neutralizes the negative surface charges typically carried by the suspended particulates. This neutralization step takes place in fractions of a second and allows the suspended particles to come close enough together to begin building a floc.
3. Flocculation (also called particle aggression) (Figure 3.13)
   Particle charge is now no longer important. A gelatinous floc begins to build and becomes a "sticky" trap for material suspended in the water.

Iron-based flocculants:

Ferrous sulfate {$FeSO_4 \times 7H_2O$}
Ferrous ($Fe^{+2}$) must be oxidized to the ($Fe^{+3}$) valence state

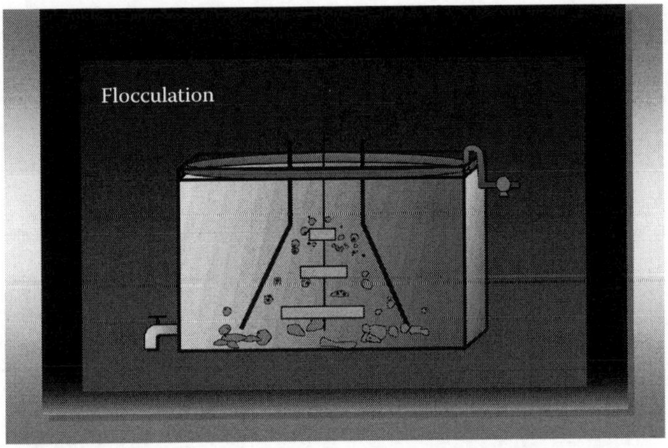


**FIGURE 3.13** Flocculation. (Adapted from Coca Cola Manufacturing Manual, 2003. TCCQS auditors toolkit. The Coca-Cola quality system. Coca-Cola Europe, Eurasia and Middle East. Confidential report.)

Ferric chloride {$FeCl_3$}
Ferric sulfate {$Fe_2(SO_4)_3$}
Ferric hydroxide {$Fe_2(OH)_3$} coagulant

Aluminum-based flocculants:

Aluminum sulfate {$Al_2(SO_4)_3 \times 14H_2O)$}
Polyaluminum chloride (a partially hydrolyzed salt)
Aluminum hydroxide $Al_2(OH)_3$ coagulant

Ferric iron coagulants are recommended since they are insoluble over a much wider range of pH, and as a result, offer significant operational advantages.

Coagulant aids are very sensitive to tiny changes in the water's quality, so they can unbalance the treatment process. If coagulant aids remain in the water after treatment, they also can affect the final product. Many countries have legal limits on doses of coagulant aids because they may be toxic.

Granular media filtration is the last step in enhanced filtration using coagulation/filtration.

Water filtration separates suspended matter from water by passing it through porous media. Coagulation destabilizes particles and flocculation makes them cluster into "macrofloc" particles. These flocs are large enough to be retained by the porous media as the water flows through it.

There are four types of media for enhanced filtration: anthracite, silica sand, garnet, and gravel.

Manufacturers build granular media filters to operate in two main conditions: downflow or upflow under pressure and under gravity feed. Filters that operate under pressure in upflow or downflow usually are of lined steel or stainless steel, so they can withstand the pressure and have typical cylindrical shape. Filters operating under

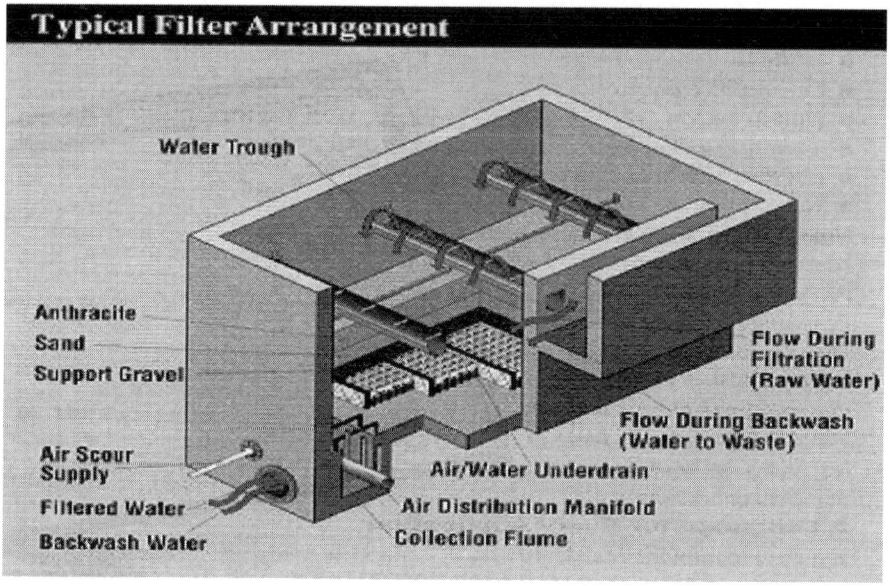


**FIGURE 3.14** Typical gravity-feed filter using granular media. (Adapted from Coca Cola Manufacturing Manual, 2003. TCCQS auditors toolkit. The Coca-Cola quality system. Coca-Cola Europe, Eurasia and Middle East. Confidential report.)

gravity feed usually are of lined steel and often are rectangular. They can use single, dual, or multiple media depending on the source water's characteristics (Figure 3.14).

Other filtration processes include direct filtration (injection of flocculant in-line followed by static mixing and deep bed multimedia filtration), which is only used if raw water meets the following recommended quality criteria:

| | |
|---|---|
| Turbidity: | <5 NTU |
| $KMnO_4$: | <5 ppm |
| Iron: | <0.3 ppm |
| Manganese: | <0.05 ppm |
| TOC: | <3 ppm (total organic carbon) |

UF (<20,000 MWCO) can also be used with ion exchange.

NF is another filtration method and RO using CA membrane or PA (Figure 3.15).

*Alkalinity reduction* If necessary (if alkalinity > 85 mg $L^{-1}$), a system can be designed to reduce enough alkalinity in raw water so that it meets the specifications for treated water.

One of the following processes can be employed to reduce alkalinity: hydrated lime treatment with lime $Ca(OH)_2$, ion exchange, NF, RO, or ED.

*Disinfection (for control of microorganisms)* Sodium hypochlorite disinfects to control microorganisms remaining in source water after it passes through enhanced filtration. Sodium hypochlorite (NaOCl) comes as a liquid, at a pH of about 11 for stability.

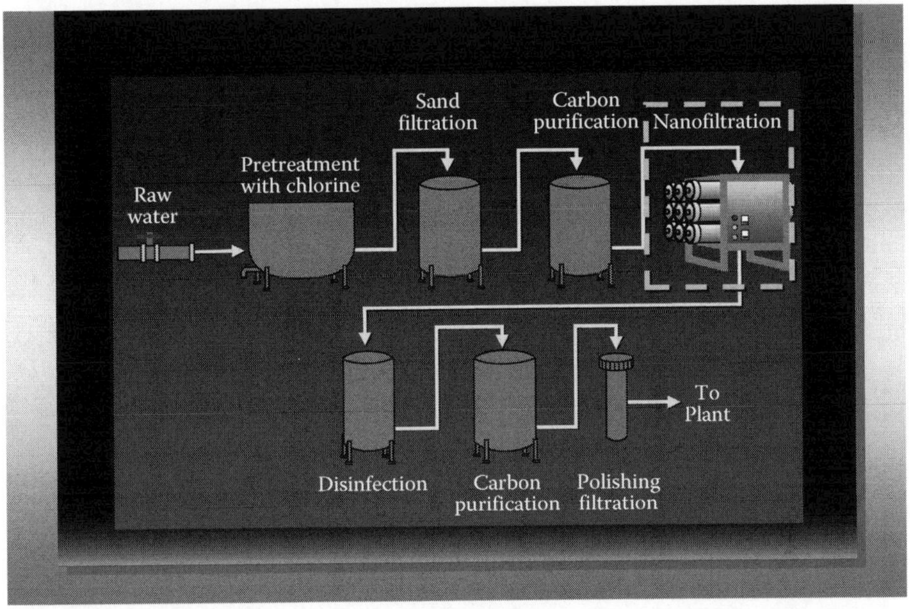


**FIGURE 3.15** Example of incorporation of nanofiltration after purification. Ultrafiltration or reverse osmosis are placed on the same position before disinfection. (From Coca Cola Manufacturing Manual, 2003. TCCQS auditors toolkit. The Coca-Cola quality system. Coca-Cola Europe, Eurasia and Middle East. Confidential report.)

When sodium hypochlorite dissolves in water, calcium carbonate and other insolubles precipitate. Coagulation, flocculation, and filtration then remove the insolubles from the water.

As sodium hypochlorite reacts with water, it forms hypochlorous acid (HOCl)—the best source of free chlorine for disinfection.

Chlorine disinfection controls microorganisms remaining in the source water after the enhanced filtration stage.

Hypochlorous acid (HOCl) and hypochlorite ion ($OCl^-$) both are known as free chlorine, but HOCl more effectively kills microorganisms.

Chlorine can also be used to disinfect 100% of raw water assuring the water contacts the chlorine for at least 30 min. Ultraviolet light (UV) can only be used for supplemental disinfection after enhanced filtration systems that use membranes. The UV lamp should produce at least 40,000 μWs $cm^{-2}$ on the 254 nm wavelength at the end of life. Residual chlorine of at least 1 ppm provides the primary disinfection.

*Carbon purifiers (to control taste-odor)* Carbon purifiers can be employed to treat 100% of raw water by using the following specifications: bed depth of at least 1.5 m, 50% freeboard, stainless-steel type 316, and a flow rate of 140 L $min^{-1}$ $m^{-3}$.

Through adsorption, granular activated carbon reduces or removes certain chemicals and materials remaining after other treatment.

Organic materials that are regulated are materials that cause off-tastes and odors, chlorine and ozone.

Bituminous coal works best for water treatment. It filters various materials in water because it is very hard, contains little fines or dust, and has a good pore size distribution.

*Polishing filtration (to control appearance)* It is used by removing small particles of carbon in treated water.

Treated water passes through filters with pore size of 5–20 microns absolute. (This pore size results in at least 3-log removal of 30 μm particles or larger with a beta factor of 1000.) Filtering media compatible with steam or chlorine sanitation are used. Filter elements are usually single-ended and sealed by double O-rings (Coca Cola, 2003).

Total trihalomethane (TTHM) needs to be covered by strict regulations. TTHM is the sum of chloroform ($CHCl_3$), bromoform ($CHBr_3$), dichlorobromomethane ($CHCl_2Br$), and dibromochloromethane ($CHClBr_2$). In Europe, the following countries have a tighter spec on TTHM: Germany: 10 ppb, Austria: 30 ppb, Italy: 30 ppb, Sweden: 50 ppb.

### 3.6.6 Filtration Equipment

The basic requirements for filtration equipment are: mechanical support for the filter medium, flow accesses to and from the filter medium, and provision for removing excess filter cake.

In some instances, washing of the filter cake to remove traces of the solution may be necessary. Pressure can be provided on the upstream side of the filter, or a vacuum can be drawn downstream, or both can be used to drive the wash fluid through.

#### 3.6.6.1 String Discharge Filter

A system of endless strings or wires spaced about 13 mm apart pass around the filter drum but are separated tangentially from the drum at the point of cake discharge, lifting the cake off as they leave contact with the drum. The strings return to the drum surface guided by two rollers, the cake separating from the strings as they pass over the rollers (Figure 3.16).

#### 3.6.6.2 Plate-and-Frame Filter Press: Recessed Chamber Filter Press and/or Diaphragm (Membrane) Filter Press

In the plate-and-frame filter press, a cloth or mesh is spread out over plates that support the cloth along ridges, but at the same time leave a free large area below the cloth for flow of the filtrate. The plates with their filter cloths may be horizontal or vertical with a number of plates operated in parallel to give sufficient area.

Perry (1999) defined this press as an alternate assembly of plates covered on both sides with a filter medium, usually a cloth, and hollow frames that provide space for cake accumulation during filtration. The frames have feed and wash manifold ports, while the plates have filtrate drainage ports.

Filter cake builds up on the upstream side of the cloth that is the side away from the plate. In the early stages of the filtration cycle, the pressure drop across the cloth is small and filtration proceeds at a more or less constant rate. As the cake increases, the process becomes more and more a constant pressure, and this is the

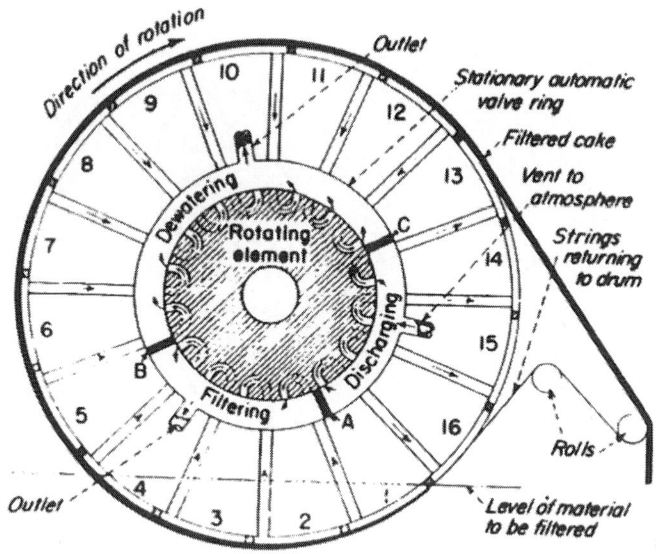


**FIGURE 3.16** Diagram of a string filter. (Courtesy of Ametek. With permission.)

case throughout most of the cycle. When the available space between successive frames is filled with cake, the press has to be dismantled and the cake scraped off and cleaned, after which a further cycle can be initiated.

Plates, which may be rectangular or circular, are supported on a central hollow shaft for the filtrate and the whole assembly enclosed in a pressure tank containing the slurry (Figure 3.17). Filtration can be done under pressure or vacuum. The advantage of vacuum filtration is that the pressure drop can be maintained while the cake is still under atmospheric pressure and so can be removed easily. The disadvantages are the greater costs of maintaining a given pressure drop by applying a vacuum and the limitation on the vacuum to about 80 kPa maximum. In pressure filtration, the pressure driving force is limited only by the cost and the mechanical strength of the equipment.

The above plate-and-frame filter press can be used to treat wastewater with high solids. These presses provide excellent solids retention, producing 35–50% solids in the filter cake. These presses routinely outperform rotary vacuum, horizontal vacuum, and belt filter presses.

Plate-and-frame filter presses are dewatering machines that utilize pressure (60–80 psi, typically) to remove the liquid from a liquid–solid slurry. They are particularly suited for low solids (<2% solids), or solids composed of fines (–200 mesh); however, they will essentially dewater many combinations of particle size distribution and percent solid slurries.

The basic operation of a plate-and-frame filter press in Figure 3.17 shows the feed entering the press at the bottom of the plate, using a pump suitable for pumping up to 80–90 psi. Then, the feed travels the path of least resistance (up between the filter plates), which has filter media inserted between the plates, and the void between the plates is filled with the slurry, as the liquid passes through

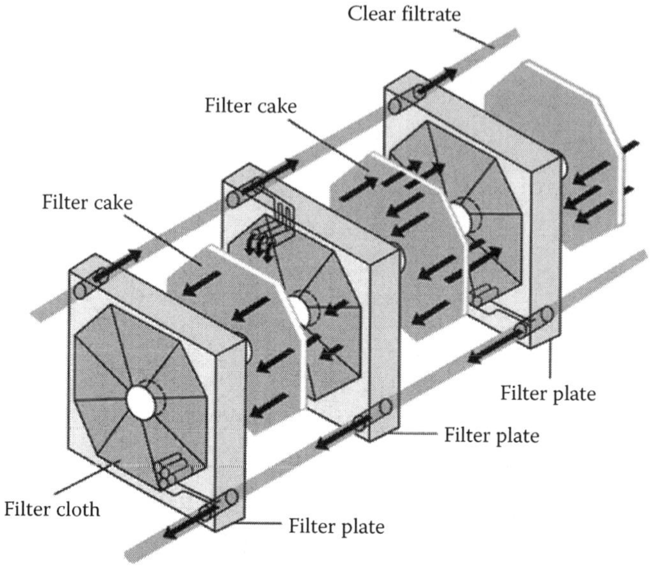


**FIGURE 3.17** Plate-and-frame filter press http://www.beckart.com/wastewater_treatment/filter_presses.php.

the filter media, and travels up to the outlet port at the top of the plate. This liquid is referred to as the "filtrate," and is discharged from the press. The solids remain in the void between the plates, until the plates discharge the filtered solids (http://www.beckart.com/wastewater_treatment/filter_presses.php).

Large plate-and-frame filter presses have mechanical "plate shifters," to move the plates, allowing the rapid discharge of the solids stuck in between them. Also, they have the capability of blowing compressed air into the plate, to dry the cake, and to aid in its discharge. Typical capacities for a plate-and-frame filter will depend upon the solids being dewatered.

The filter press may also be used as a "polishing" filter to remove minute quantities of solids from an influent stream. In these applications, the press is not sized for the quantity of solid-holding capacity but for maximum filtration area and hydraulic throughput. When used as a polishing filter, generally a dry filter cake is not developed. Rather, when throughput flow rates drop to an unacceptable level, the cycle is ended.

The last most common usage of the filter press is as a "variable volume" filter through the use of a filter plate known as a diaphragm or membrane plate. This type of plate has a flexible drain field, which when sealed around the edges forms an integral bladder or diaphragm that may be inflated to physically press additional liquid from the filter cake. This process can significantly reduce the typical elapsed time for a press cycle and produce a dryer cake product, or more uniform cake dryness from cycle to cycle (http://www.water.siemens.com/en/products/sludge_biosolids_processing/filter_press/Pages/dewatering_systems_generic_what_is_a_filter_press.aspx).

The filter press is made up of two principal components: the skeleton and the filter pack.

The skeleton holds the filter pack together against the pressures developed internally during the filtration process and includes a stationary and follower head, hydraulics, and manifold.

The filter pack is where the actual liquid/solid separation process takes place. The pack consists of a series of filter elements that form a series of chambers when held together in the press skeleton. Each chamber wall has a series of raised cylinders or "pips," which are then covered with a porous cloth medium. These pips form a flow path for the liquid draining from the press.

The prime function of the filter media is to provide a porous support structure for the filter cake as it develops and builds. Initially, some solids may pass through the cloth media, causing a slight turbidity in the filtrate, but gradually the larger particles within the slurry begin to bridge the openings in the media, reducing the effective opening size. This allows smaller particles to bridge these reduced openings initiating the cake filtration process. Once a layer of solid particles achieves 1–2 mm in thickness, this "precoat" layer serves to separate out finer and finer particles as the cake builds in thickness, yielding a filtrate that is very low in turbidity.

The pressure behind the slurry (typically 100 psi, but up to 900 psi [7–60 bar]) is provided by a feed pump—sometimes a positive displacement or centrifugal pump. It is the existence of this pressure differential (between the feed pressure and the gravity discharge), not just the feed pump pressure, that causes the filtering action to occur. Solids within the slurry will flow to the area of cake development with the lowest pressure differential, resulting in a filter cake that builds uniformly over the drain field on either side of the chamber walls.

Two wash techniques are used in plate-and-frame filter presses. In simple washing, the wash liquor follows the same path as the filtrate. If the cake is not extremely uniform and highly permeable, this type of washing is ineffective in a well-filled press. A better technique is thorough washing, in which the wash is introduced to the faces of alternate plates (with their discharge channels valved off). The wash passes through the entire cake and exits through the faces of the other plates (Perry, 1999).

#### *3.6.6.2.1 Rotary Filters*

In rotary filters, the flow passes through a rotating cylindrical cloth from which the filter cake can be continuously scraped. Either pressure or vacuum can provide the driving force, but a particularly useful form is the rotary vacuum filter. In this, the cloth is supported on the periphery of a horizontal cylindrical drum that dips into a bath of the slurry. Vacuum is drawn in those segments of the drum surface on which the cake is building up. A suitable bearing applies the vacuum at the stage where the actual filtration commences and breaks the vacuum at the stage where the cake is being scraped off after filtration. Filtrate is removed through trunnion bearings. Rotary vacuum filters are expensive, but they do provide a considerable degree of mechanization and convenience. A rotary vacuum filter is illustrated diagrammatically in Figure 3.18.

Vacuum filters can be divided into bottom feed (precoat and drum filters), side feed (disk filters), and top feed (belt, tray, table, and tilting pan filters).

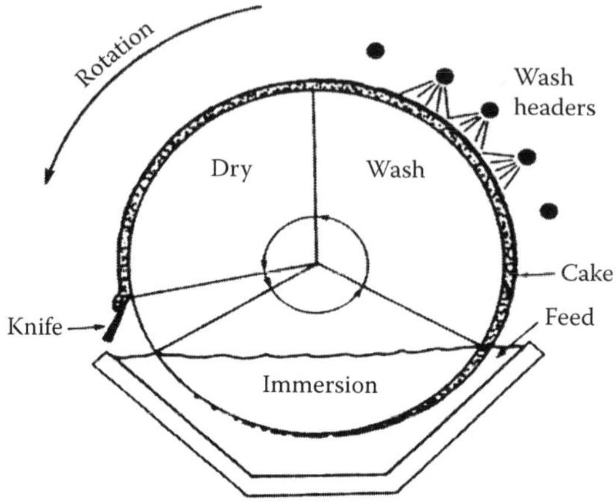


**FIGURE 3.18** Rotary vacuum filter.

There are two concepts in the layout design of vacuum filters: barometric leg layouts and floor-mounted layouts. In barometric leg layouts, the filter is mounted 8–9 m above ground level to ensure that the liquid level in the leg will not flood the filtrate receiver. This applies to liquids with a specific gravity of 1; however, for heavier liquids, the filter elevation with regard to the ground level may be lowered. Barometric leg layouts require, therefore, a structure for the filter, but the advantage is that the filtrate pump operates under a positive suction head. On the other hand, floor-mounted layouts require a receiver-mounted filtrate pump designed to effectively deliver the filtrate under full vacuum on the suction side.

The DORR-OLIVER Horizontal Pan Filter is a continuous vacuum filter designed for economic filtration of granular, fibrous and coarse fast-settling solids, including cake washing and drying, if required. The DORR-OLIVER Horizontal Pan Filter provides a circular filtration surface ("pan") rotating in a horizontal plane. Multiple filter designs, filter sizes, and process components are available, suitable for a wide range of industrial applications and process needs.

The unique filter design allows for cake washing in countercurrent mode thereby achieving highest cake washing efficiency at reduced wash media consumption. The cake moisture is effectively reduced by means of the applied vacuum and additional methods if required before it is removed by a wear-resistant cake discharge scroll. The filtration surface can be dressed with a wide range of textile and metal type of filter media, thereby using most advanced cloth fixing methods, including cassette designs. (http://www.flsmidth.com/en-US/Products/Product + Index/All + Products/Vacuum + Filtration/Horizontal + Pan + Filter/DORR-OLIVER + Horizontal + Pan + Filter).

All drum filters (except the single-compartment filter) utilize a rotary-valve arrangement in the drum-axis support trunnion to facilitate removal of filtrate and wash liquid and to allow introduction of air or gas for cake blowback if needed. The valve controls the relative duration of each cycle as well as providing "dead" portions of the cycle through the use of bridge blocks (Perry, 1999).

#### 3.6.6.3 Centrifugal Filters

Centrifugal force is used to provide the driving force in some filters. These machines are centrifuges fitted with a perforated bowl that may also have filter cloth on it. Liquid is fed into the interior of the bowl and under the centrifugal forces, it passes out through the filter material.

During filtration, the filter vessel is fed under pressure; the filtrate passes through the plates and out through the shaft. The filter cake forms on the upper side of the filter elements. After filtration, the remaining feed in the vessel is either drained or filtered via the scavenge system. The cake may then be washed or dried by an appropriate heated gas. Spinning the entire stack at moderate speeds generates a centrifugal force that discharges the cake. The cake can be discharged in slurry or dry form. This is illustrated in Figure 3.19.

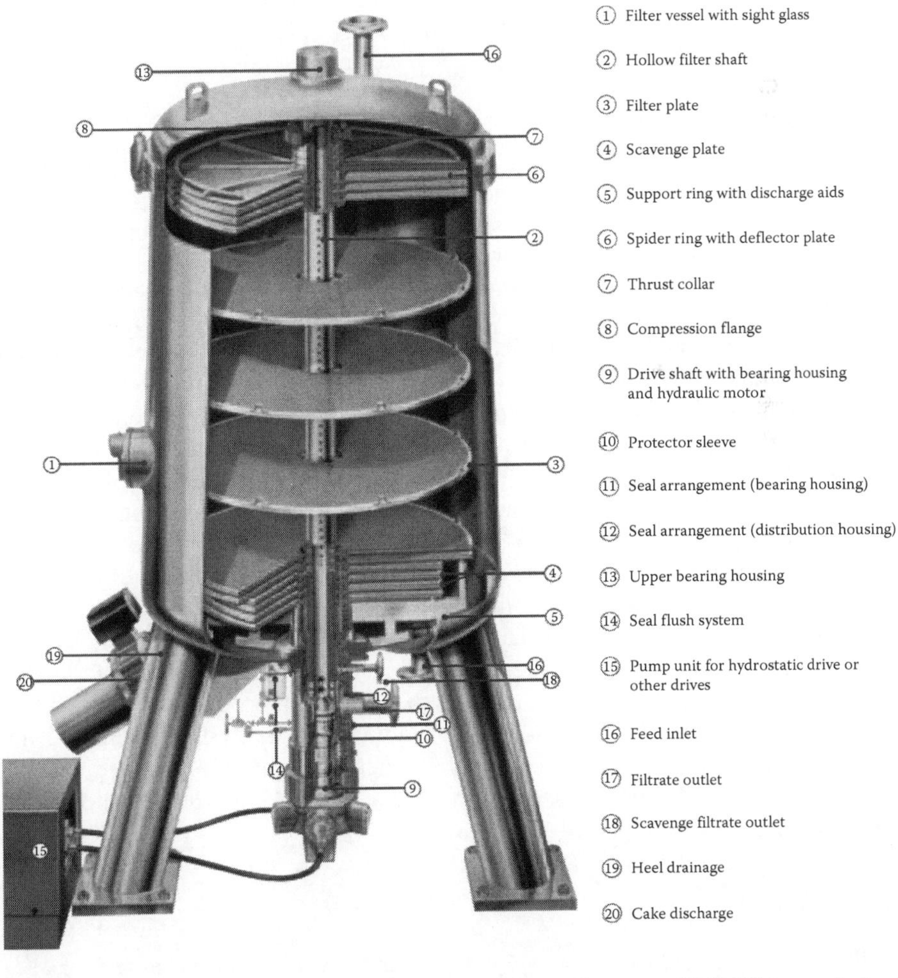


**FIGURE 3.19** SeitzSchenk Centrifugal Discharge Filter. (Courtesy of Pall. With permission.)

The operating advantages of the SeitzSchenk Centrifugal Discharge Filter is the use of horizontal filter elements and the ability of automatic cake discharge without having to open the filter as described by Pall.

The Pall SeitzSchenk Centrifugal Discharge Filter Type ZHF consists mainly of a pressure vessel with a hollow center shaft around which series of round filter elements are vertically stacked at specific, but variable, spacing. The filter stack, consisting of both the hollow shaft and the elements, is installed in the vessel, so that it can freely rotate. To clean the filter, the whole stack is spun by means of a drive system. The hollow shaft that serves as a filtrate discharge manifold is connected to an external drive motor, permitting the removal of cake by centrifugal action.

*Funda filter* (marketed in the United States by Steri Technologies Horizontal) is a centrifugal discharge filter where top-surface filter plates may be mounted on a hollow motor-connected shaft that serves both as a filtrate discharge manifold and as a drive shaft to permit centrifugal removal of the cake.

The Funda filter is driven from the top, leaving the bottom unobstructed for inlet and drainage lines; however, the Schenk filter employs a bottom drive, providing a lower center of mass and ground-level access to the drive system (Perry, 1999).

#### 3.6.6.4 Air Filters: Tube Filters or Liquid Bag Filters

Air filters are used quite extensively to remove suspended dust or particles from air streams. The air or gas moves through a fabric and the dust is left behind. These filters are particularly useful for the removal of fine particles. One type of bag filter consists of a number of vertical cylindrical cloth bags 15–30 cm in diameter, the air passing through the bags in parallel. Air bearing the dust enters the bags, usually at the bottom and the air passes out through the cloth. A well-known example is a bag filter for dust found in the vacuum cleaner. Some designs of bag filters provide for the mechanical removal of the accumulated dust. For removal of particles less than 5 μm diameter in modern air sterilization units, paper filters and packed tubular filters are used. These cover the range of sizes of bacterial cells and spores.

The advantages of the tubular filter are that it uses an easily replaced filter medium, its filtration cycle can be interrupted and the shell can be emptied of prefilt at any time without loss of the cake, the cake is readily recoverable in dry form, and the inside of the filter is conveniently accessible.

A cylindrical filter bag sealed at one end is inserted into the perforated tube. The open end of the filter bag generally has a flange or special seal ring to prevent leakage.

Slurry under pressure is admitted to the chamber between the head of the shell and the tube sheet, and then enters and fills the tubes. Filtration occurs as the filtrate passes radially outward through the filter medium and the wall of each tube into the shell and on out the filtrate discharge line, depositing cake on the medium. The filtration cycle is ended when the tubes have filled with cake or when the media have become plugged (Perry, 1999).

## REFERENCES

Abels, C., F. Carstensen, and M. Wessling, 2013. Membrane processes in biorefinery applications *J. Membr. Sci.* 444: 285–317.

Afonso, M. D., J. Ferrer, and R. Bórquez, 2004. An economic assessment of proteins recovery from fish meal effluents by ultrafiltration. *Trends Food Sci. Tech.* 15: 506–512.

Akin, O., F. Temelli, and S. Koseoglu, 2012. Membrane applications in functional foods and nutraceuticals. *Crit. Rev. Food Sci. Nutr.* 52: 347–371.

Anderson, M., B. E. Brooker, and E. C. Needs, 1987. The role of proteins in the stabilization/destabilization of dairy foams. In: Dickinson, E. (ed.) *Food Emulsions and Foams*. Royal Society of Chemistry, London, pp. 100.

Augustin, M. A., P. Sanguansri, R. Williams, and H. Andrews, 2012. High shear treatment of concentrates and drying conditions influence the solubility of milk protein concentrate powders. *J. Dairy Res.* 79: 459–468.

Bacchin, P., P. Aimar, and R. W. Field, 2006. Critical and sustainable fluxes: Theory, experiments and applications. *J. Membr. Sci.* 281: 42–69.

Bacchin, P., P. Aimar, and V. Sanchez, 1995. Model for colloidal fouling of membranes. *AIChE J.* 41: 368–377.

Badan, A. P., N. Bei, A. Lireny, G. Goncalves, J. Cunha, and A. Viotto, 2008. The optimization of soybean oil degumming on a pilot plant scale using a ceramic membrane. *J. Food Eng.* 87: 514–521.

Bailly, M., H., Roux-de Balmann, P. Aimar, F. Lutin, and M. Cheryan, 2001. Production processes of fermented organic acids targeted around membrane operations: Design of the concentration step by conventional electrodialysis. *J. Membr. Sci.* 191: 129–142.

Baklouti, S., A. Kamoun, R. Ellouze-Ghorbel, and S. Chaabouni, 2013. Optimising operating conditions in ultrafiltration fouling of pomegranate juice by response surface methodology. *Int. J. Food Sci. Technol.* 48: 1519–1525.

Bargeman, G., J. M. Vollenbroek, J. Straatsma, C. G. P. H. Schroen, and R. M. Boom, 2005. Nanofiltration of multi-component feeds. Interactions between neutral and charged components and their effect on retention. *J. Membr. Sci.* 247: 11–20.

Bassi, A. S., S. Rohani, and D. G. Macdonald, 1987. Measurement of effective diffusivities of lactose and lactic acid in 3% agarose gel membrane. *Biotechnol. Bioeng.* 30: 794–797.

Benavides, J. and M. Rito-Palomares, 2008. Practical experiences from the development of aqueous two-phase processes for the recovery of high value biological products. *J. Chem Technol. Biotechnol.* 83: 133–142.

Benhabiles, M. S., N. Abdi, N. Drouiche, H. Lounici, A. Pauss, M. F. A. Goosen, and N. Mameri, 2013. Protein recovery by ultrafiltration during isolation of chitin from shrimp shells *Parapenaeus longirostris*. *Food Hydrocolloids* 32: 28–34.

Beolchini, F., S., Cimini, L., Mosca, F., Veglio, and D. Barba, 2005. Microfiltration of bovine and ovine milk for the reduction of microbial content: Effect of some operating conditions on permeate flux and microbial reduction. *Sep. Sci. Technol.* 40: 757–772.

Bhosle, B. M. and R. Subramanian, 2005. New approaches in deacidification of edible oils—A review. *J. Food Eng.* 69: 481–494.

Blatt, W. F., A. Dravid, A. S. Michaels, L. Nelson, 1970. Solute polarization and cake formation in membrane ultrafiltration: Causes, consequences and control techniques. In: Flinn, J. E. (ed.), *Membrane Science and Technology*, Plenum Press, New York.

Bonnet, J. and H. De Vilmorin, 2004. Procedure for the controlled reduction of the sugar level in fruit juices and device to accomplish this procedure. EU Patent 04360028.7.

Bouchoux, A., H. Roux-de Balmann, and F. Lutin, 2005. Nanofiltration of glucose and sodium lactate solutions—Variations of retention between single- and mixed solute solutions. *J. Membr. Sci.* 258: 123–132.

Bowler, D. R., M. J. Murphy, W. J. Bowman, P. Brandhuber, and G. Amy, 1998. Alternative methods for membrane filtration of arsenic from drinking water. *Desalination* 117: 1.

Brennan, J. G., J. R. Butters, N. D. Cowell, and A. E. V. Lilley, 1990. *Food Engineering Operations*, 3rd edn., Elsevier Applied Science, London.

Bulger, P. G., S. K. Bagal, and R. Marquez, 2008. Recent advances in biomimetic natural product synthesis *Nat. Prod. Rep.* 25: 254–297.

Butters, J. R. 1993. *Notes on Separation Operations*. Department of Food Science and Technology, University of Reading, UK.

Bylund, G. 1995. *Dairy Processing Handbook*. Tetra Pak Processing Systems, Lund, Sweden.

Calvin, R. W. Process for making a low-alcohol wine. US Patent 6,203,826, 2001.

Carrere, H., F. Blaszkowa, and H. R. de Balmann, 2002. Modelling the microfiltration of lactic acid fermentation broths and comparison of operating modes. *Desalination* 145: 201–206.

Castel, V., O. Andrich, F. M. Netto, L. G. Santiago, and C. R. Carrara, 2012. Comparison between isoelectric precipitation and ultrafiltration processes to obtain *Amaranth mantegazzianus* protein concentrates at pilot plant scale. *J. Food Eng.* 112: 288–295.

Cheryan, M. 1998. *Ultrafiltration and Microfiltration Handbook*, CRC Press, Boca Raton, FL.

Cheryan, M. 2007. Membrane concentration of liquid foods. In: Heldman, D. R. and Lund, D. (eds.), *Handbook of Food Engineering*, Chapter 9, 2nd edn., CRC Press, Boca Raton, FL, USA, pp. 554–596.

Cicci, A., M. Stoller, and M. Bravi, 2013. Microalgal biomass production by using ultra- and nanofiltration membrane fractions of olive mill wastewater. *Water Res.* 47: 4710–4718.

Coca Cola Manufacturing Manual, 2003. TCCQS auditors toolkit. The Coca-Cola quality system. Coca-Cola Europe, Eurasia and Middle East. Confidential report.

Coca, J., Alvarez, R., Alvarez, M. T., and Stefania, H. 1986. Skin permeability: Potassium chromate solutions through rat epidermis. Biomedical Engineering, *World Congress of Chemical Engineering*, Tokyo, pp. 1081–1084.

Coulson, J. M., J. F. Richardson, J. R. Backhurst, and J. H. Harker, 1978. *Chemical Engineering*, Vol. 2, 3rd edn., Pergamon Press, Oxford, UK.

Courel, M., M. Dornier, J. M. Herry, G. M. Rios, and M. Reynes, 2000. Effect of operating conditions on water transport during the concentration of sucrose solutions by osmotic distillation. *J. Membr. Sci.* 170: 281–289.

Cremades, O., M. M. Diaz-Herrero, P. Carbonero-Aguilar, J. F. Gutierrez-Gil, E. Fontiveros, B. Rodriguez-Morgado, J. Parrado, and J. Bautista, 2012. Preparation and characterisation of selenium-enriched mushroom aqueous enzymatic extracts (MAEE) obtained from the white button mushroom (*Agaricus bisporus*). *Food Chem.* 133: 1538–1543.

Cuperus, F. P. 1998. Membrane processes in agri-food, state-of-the-art and new opportunities. *Sep. Purif. Technol.* 14: 233–239.

Cuperus, F. P. and H. H. Nijhuis, 1993. Applications of membrane technology to food processing. *Trends Food Sci. Technol.* 4: 277–282.

Daufin, G., J. P. Escudier, H. Carrere, S. Berot, L. Fillaudeau, and M. Decloux, 2001. Review—Recent and emerging applications of membrane process in food and dairy industry. *Trans IChemE.* 79: 89–102.

De Bruijn, J. P. F., A. Venegas, J. A. Martinez, and R. Borquez, 2003. Ultrafiltration performance of Carbosep membranes for the clarification of apple juice. *Lebensmittel-Wissenschaft und Technologie* 36: 397–406.

Dean, W. 1927. Note on the motion of a fluid in a curved pipe. *Philos. Mag. Series* 7(4): 208–223.

Destani, F., A. Cassano, A. Fazio, J-P. Vincken, and B. Gabriele, 2013. Recovery and concentration of phenolic compounds in blood orange juice by membrane operations. *J. Food Eng.* 117: 263–271.

Dey, P., L. Linnanen, and P. Pal, 2012. Separation of lactic acid from fermentation broth by cross flow nanofiltration: Membrane characterization and transport modelling. *Desalination* 288: 47–57.

Di Giovacchino, L., S. Sestili, and D. Di Vincenzo, 2002. Influence of olive processing on virgin olive oil quality. *Eur. J. Lipid Sci. Technol.* 104: 587–601.

Djelveh, G., J. B. Gros, and B. Bories, 1989. An improvement of the cell diffusion method for the rapid determination of diffusion constants in gels or foods. *J. Food Sci.* 54(1): 166–169.

Earle, R. L. 1983. *Unit Operations in Food Processing*, 2nd edn., Pergamon Prcss, Oxford, UK, pp. 143–158.

Eriksson, P. 1988. Nanofiltration extends the range of membrane filtration. *Environ. Prog.*, 7: 58.

Espina, V., M. Y. Jaffrin, M. Frappart, and L. H. Ding, 2010. Separation of casein from whey proteins by dynamic filtration. *Desalination* 250: 1109–1112.

Fellows P. J. 2000. *Food Processing Technology-Principles and Practice.* 2nd ed. CRC Press, Woodhead Publishing, Cambridge, UK.

Ferreira, M., A. R. Fernada, and R. Jost, 1999. Application of microfiltration to egg white depleted in ovomucin. *Int. J. Food Sci. Technol.* 34: 27–32.

Field, R. W., D. Wu, J. A. Howell, and B. B. Gupta, 1995. Critical flux concept for microfiltration fouling. *J. Membr. Sci.* 100: 259–272.

Foust, A. S., L. A. Wenzel, C. V. Clump, L. Maus, and L. Bryce Anderson, 1980. *Principles of Unit Operations*, 2nd edn., John Wiley and Sons, Chichester, UK.

Frenzel, I., D. F. Stamatialis, and M. Wessling, 2006. Water recycling from mixed chromic acid waste effluents by membrane technology. *Sep. Purif. Technol.*, 49: 76.

Gana, Q., J. A. Howell, R. W. Field, R. England, M. R. Bird, C. L. O'Shaughnessy, and M. T. MeKechinie, 2001. Beer clarification by microfiltration—Product quality control and fractionation of particles and macromolecules. *J. Membr. Sci.,* 194: 185–196.

Gander, B., M. J. Blanco-Prieto, C. Thomasin, D. Wandrey, and D. Hunkeler, 2002. Coacervation/phase separation. In: Swarbrick, J. and Boylan, J. C. (eds.), *Encyclopedia of Pharmaceutical Technology*, Marcel Dekker, New York.

García-Martín, N., L. Palacio, P. Prádanos, A. Hernández, M. Ortega-Heras, S. Pérez-Magariño, and D. C. Gonza´lez-Huerta, 2009. Evaluation of several ultra- and nanofiltration membranes for sugar control in winemaking. *Desalination* 245, 554–558.

Garcia-Prieto, A., L. Lunar, S. Rubio, and D. Perez-Bendito, 2008. Decanoic acid reverse micelle-based coacervates for the microextraction of bisphenol A from canned vegetables and fruits. *Anal. Chim. Acta* 617: 51–58.

GEA Westfalia Separator Group, GEA Mechanical equipment, 2013a. Westfalia Separator hydrostop system (http://www.westfalia-separator.com/products/separators/self-cleaning-disk-separators.html (accessed 30 July 2013).

GEA Westfalia Separator Group, GEA Mechanical equipment, 2013b. Cross-Flow Filtration—Ceramic Membrane Elements and Membrane Filtration Systems. The configuration of a ceramic membrane filter. http://www.westfaliaseparator/no_cache/products/cross-flow-filtration.html (accessed 30 July 2013).

Gekas, V. and B. Hallström, 1987. Mass transfer in the membrane concentration polarization layer under turbulent cross flow. *J. Membr. Sci.* 30: 153–170.

Getler, J., A. Nielse, and J. Sprogø, 1997. Functional process for MPC. *Dairy Ind. Int.* 62(3): 25, 27.

Ghosh, R. and Z. F. Cui, 2000. Protein purification by ultrafiltration with pre-treated membrane. *J. Membr. Sci.* 167(1): 47–53.

Ghosh, R. 2003. *Protein Bioseparation Using Ultrafiltration: Theory, Applications and New Developments*. Imperial College Press, London.

Girard, B. and L. R. Fukumoto, 2000. Membrane processing of fruit juices and beverages. *Crit. Rev. Food Sci. Nutr.* 40(2): 91–157.

Gonzalez-Munoz, M. J. and J. C. Paraja, 2010. Diafiltration of Eucalyptus wood autohydrolysis liquors: Mathematical modeling. *J. Membr. Sci.* 346: 98–104.

Gresch, W. 1996. Process for the production of a low-sugar, Alcohol-free beverage. US Patent 5,496,577.

Guerra, A., G. Jonsson, A. Rasmussen, E. W. Nielsen, and D. Edelsten, 1998. Low cross flow velocity microfiltration of skim milk for removal of bacterial spores. *Int. Dairy J.* 7: 849–861.

Hallström, B., V. Gekas, I. Sjöholm, and A. M. Romulus, 2007. Mass transfer in foods. In: Heldman, D. R. and Lund, D. (eds.), *Handbook of Food Engineering*, Chapter 7, 2nd edn., CRC Press, Boca Raton, FL, USA, pp. 471–495.

Hassan, A. M., M. A. K. Al-Sofi, A. S. Al-Amoudi, A. T. M. Jamaluddin, A. M. Farooque, A. Rowaili, A. G. I. Dalvi, N. M. Kither, G. M. Mustafa, and I. A. R. Al-Tisan, 1998. A new approach to membrane and thermal seawater desalination processes using nanofiltration membranes (Part 1). *Desalination* 118: 35.

Heldman, D. R. and R. W. Hartel, 1997a. Liquid concentration. In: Heldman, D. R. and Hartel, R. W. (eds.), *Principles of Food Processing*, Chapman and Hall, New York, pp. 138–176.

Heldman, D. R. and R. W. Hartel, 1997b. Other separation processes. In: Heldman, D. R. and Hartel, R. W. (eds.) *Principles of Food Processing*, Chapman and Hall, New York, pp. 219–252.

Hemfort, H. 1983. *Centrifugal Separators for the Food Industry*, Westfalia Separator AG, 4740 Oelde 1, Germany.

Hemfort, H. 1984. *Centrifugal Clarifiers and Decanters for Biotechnology*. Westfalia Separator AG, 4740 Oelde 1, Germany.

Hogan, P. A., R. P. Canning, P. A. Peterson, R. A. Johnson, and A. S. Michaels, 1998. A new option: Osmotic distillation. *Chem. Eng. Progr.* July: 49–61.

Hu, B. and K. Scott, 2007. Influence of membrane material and corrugation and process conditions on emulsion microfiltration, *J. Membr. Sci.*, 294 (1–2): 30–39.

Huang, L. and Z. Guo, 2012. Nanofiltration and sensing of picomolar chemical residues in aqueous solution using an optical porous resonator in a microelectrofluidic channel. *Nanotechnology* 23: 065,502 (10pp) doi:10.1088/0957-4484/23/6/065502.

Huisman, H. I. 2000. *Membrane Separations. II Microfiltration*. Academic Press, New York, pp. 1764–1777.

Huffman, L. M. and W. J. Harper, 1999. Maximizing the value of milk through separation technologies. *J. Dairy Sci.* 82: 2238–2244.

Iaquinta, M., M. Stoller, and C. Merli, 2009. Optimization of a nanofiltration membrane process for tomato industry wastewater effluent treatment. *Desalination* 245: 314–320.

Ito, Y., R. Clary, J. Powell, M. Knight, T. M. Finn, and J. Liq. 2008. *Chromatogr. Rel. Technol.* 31: 1346.

Jeantet, R., J. Rodriguez, and A. Garen, 2000. Nanofiltration of sweet whey by spiral wound organic membranes. Impact of hydrodynamics. *Lait* 80: 155–163.

Jelen, P. 1992. Pressure-driven membrane processes: Principles and definitions. In: *New Applications of Membrane Processes*, Special Issue 9201, International Dairy Federation, Brussels, Belgium, pp. 7–14.

Jiao, B., A. Cassano, and E. Drioli, 2004. Recent advances on membrane processes for concentration of fruit juices: A review. *J. Food Eng.* 63: 303–324.

Jiraratananon, R., A. Chanachai, R. Y. M. Huang, and D. Uttapap, 2002. Pervaporation dehydration of ethanol–water mixtures with chitosan/hydroxyethylcellulose (CS/HEC) composite membranes I. Effect of operating conditions. *J. Membr. Sci.* 195: 143–151.

Kawa-Rygielska, J., W. Pietrzak, P. Regiec, and P. Stencel, 2013. Utilization of concentrate after membrane filtration of sugar beet thin juice for ethanol production. *Biores. Technol.* 133: 134–141.

Kazemimoghadam, M. and T. Mohammadi, 2007. Chemical cleaning of ultrafiltration membranes in the milk industry. *Desalination*, 204 (1–3 Spec. Iss.): 213–218.

Kelder, J. D. H., K. J. Ptasinski, and P. J. A. M. Kerkhof, 2002. Power-law foods in continuous coiled sterilisers. *Chem. Eng. Sci.* 57: 4605–4615.

Kelly, P. M., B. S. Horton, H. Burling, 1992. Partial demineralization of whey by nanofiltration. *Int. Dairy Fed. Special Issue* 9201: 130–140.

Kesting, R. E. 1985. *Synthetic Polymeric Membranes: A Structural Perspective*, John Wiley & Sons, NY.

Khatkar, S. K., V. K. Gupta, and A. B. Khatkar, 2012b. Studies on quality attributes of liquid dairy whitener prepared from ultrafiltration process in tea and coffee. *Indian J. Dairy Sci.* 65(4): 285–292.

Khatkar, S. K., V. K. Gupta, and A. B. Khatkar, 2013. Physicochemical and functional quality attributes of dairy whitener prepared from ultrafiltration process. *J. Food Process. Pres.* In Press.

Khatkar, S. K., V. K. Gupta, and S. Kumar, 2012a. Effect of homogenization, stabilizing and flavoring salts on the quality of liquid dairy whitener from buffalo milk employing ultrafiltration process. *Indian J. Dairy Sci.* 65(2): 115–121.

Kiefer, J. 1991. Cross-flow filtration of beer. In: *Proceedings of the European Brewery Convention Congress*, Lisbon, pp. 657–664.

Kilara, A. 2006. Basic dairy processing principles. In: Chandan, R. H., White, C. H., Kilara, A., and Hui, Y. H. (eds.), *Manufacturing Yoghurt and Fermented Milks*, Blackwell Publishing, Ltd, Oxford, UK, pp. 73–87.

Kim, J.-H. and K-H. Lee, 1998. Effect of PEG additive membrane formation by phase inversion. *J. Membr. Sci.* 138: 153–163.

Koseoglu, S. S. 1991. Membrane technology for edible oil refining. *Oils Fats Int.* 5: 16–21.

Koseoglu, S. S. and D. E. Engelgau, 1990. Membrane applications and research in the edible oil industry: An assessment. *J. Am. Oil Chem. Soc.* 67: 239–249.

Krawczyk, H., P. Oinonen, and A. -S. Jonsson, 2013. Combined membrane filtration and enzymatic treatment for recovery of high molecular mass hemicelluloses from chemithermomechanical pulp process water. *Chem. Eng. J.* 225: 292–299.

Kumar, A., B. Kumar Thakur and S. De, 2012. Selective extraction of (–) epigallocatechin gallate from green tea leaves using two-stage infusion coupled with membrane separation. *Food Bioprocess. Technol.* 5: 2568–2577.

Kunz, W., A. Benabiles, and R. Ben-Aim, 1996. Osmotic evaporation through macroporous hydrophobic membranes: A survey of current research and applications. *J. Membr. Sci.* 121: 25–36.

Kuzmenko, D., E. Arkhangelsky, S. Belfer, V. Freger, and V. Gitis, 2005. Chemical cleaning of UF membranes fouled by BSA. *Desalination* 179(1–3): 323–333.

Law, B. A. and P. W. Goodenough, 1995. Enzymes in milk and cheese production. In: Tucker, G. A. and Woods, L. F. J. (eds.), *Enzymes in Food Processing*, Blackie Acad. & Professional, Glasgow, UK, pp. 114–143.

Le Clech, P., B. Jefferson, I. S. Chang, and S. J. Judd, 2003. Critical flux determination by the flux-step method in a submerged membrane bioreactor. *J. Membr. Sci.* 227: 81–93.

Lewis, W. J. T., O. P. W. Peck, A. C. Muir, Y. M. Chew, and M. R. Bird, 2012. The fouling and cleaning of surfaces in the food sector. *Food Sci. Technol.* 26(4): 30–32.

Li, Q., Q.-Y., Bi, H.-H. Lin, L.-X. Bian, and X.-L. Wang, 2013. A novel ultrafiltration (UF) membrane with controllable selectivity for protein separation. *J. Membr. Sci.* 427: 155–167.

Lipnizki, J. 2008. Optimization of membrane processes in white biotechnology. *Desalination* 224: 105–110.

Loncin, M. and L. M. Merson, 1979. *Food Engineering—Principles and Selected Applications*, Academic Press, New York.

Luo, J. and Y. Wan, 2011. Effect of highly concentrated salt on retention of organic solutes by nanofiltration polymeric membranes. *J. Membr. Sci.* 372: 145–153.

Luo, J. and Y. Wan, 2013. Effects of pH and salt on nanofiltration—A critical review. *J. Membr. Sci.* 438: 18–28.

Madec, M. N., S. Méjean, and J. -L. Maubois, 1992. Retention of *Listeria* and *Salmonella* cells contaminating skim milk by tangential membrane microfiltration (Bactocach process). *Lait* 72: 327–332.

Manjula, S. and R. Subramanian, 2006. Membrane technology in degumming, dewaxing, deacidifying and decolorizing edible oils. *Crit. Rev. Food Sci. Nutr.* 46: 569–592.

Martinez Suarez, J. M., E. Munoz Aranda, J. Alba Mendoza, and A. Lanzon Rey, 1974. Elaboracion del aceite por centrifugacion en continuo. *Grasas Aceites* 25: 148–159.

Mascolo, A., A. Cucurachi, L. Di Giovacchino, M. Solinas, and L. Seghetti, 1978–1980. Confronto tra i rendimenti in olio conseguibili con i sistemi della pressione e della centrifugazione nella lavorazione delle olive. *Ann. Ist. Sper. Elaiotecnica* VIII: 165–175.

Matta, V. M., R. H. Moretti, and L. M. C. Cabral, 2004. Microfiltration and reverse osmosis for clarification and concentration of acerola juice. *J. Food Eng.* 61: 477–482.

McCabe, W. L., J. C. Smith, and P. Harriott, 1985. *Unit Operations of Chemical Engineering*, 4th edn., McGraw Hill, London, UK.

Mehaia, M. A. 1997. Studies on the rennet coagulation of skim camel milk concentrated by ultrafiltration. *J. King Saud Univ.* 9(1): 11–123.

Merin, U. and G. Daufin, 1990. Cross flow microfiltration in the dairy industry: State-of-the art. *Le Lait* 70: 281–291.

Menet, J.-M. and D. Thiebaut (eds.) 1999. *Countercurrent Chromatography* (Chromatographic Science Series Vol 82). Marcel Dekker, Inc., New York.

Metsamuuronen, S. and M. Nystrom, 2005. Critical flux in cross-flow ultrafiltration of protein solutions. *Desalination* 175: 37–47.

Mihnea, M., M. L. González-SanJosé, M. Ortega-Heras, S. Pérez-Magariño, N. García-Martin, L. Palacio, P. Prádanos, and A. Hernández, 2012. Impact of must sugar reduction by membrane applications on volatile composition of verdejo wines. *J. Agric. Food Chem.* 60: 7050–7063. doi: org/10.1021/jf301433j.

Mistry, V. V. and J. L. Maubois, 2004. Application of membrane separation technology to cheese production. In: P. F. Fox, P. L. H. McSweeney, T. M. Cogan and T. P. Guinee (eds.) *Cheese: Chemistry, Physics and Microbiology*, 3rd edn. vol. 2, Elsevier Academic Press, Amsterdam, pp. 261–285.

Mizubuti, I. Y., O. B. Junior, L. W. Souza, R. S. Silva, and E. L. Ida, 2000. Response surface methodology for extraction optimization of pigeon pea protein. *Food Chem.* 70: 259–265.

Mondal, S., C. Rai, and S. De, 2013. Identification of fouling mechanism during ultrafiltration of stevia extract. *Food Bioprocess. Technol.* 6(4): 931–940.

Morcos, S. and A. Bergles, 1975. Experimental investigation of combined forced and free laminar convection in horizontal tubes. *J. Heat Transfer—Trans. ASME* 97(C2): 212–219.

Morton, B. 1959. Laminar convection in uniformly heated horizontal pipes at low Rayleigh numbers. *Quart. J. Mech. Appl. Math.* 12: 410–420.

Mulder, M. 1991. *Basic Principles of Membrane Technology*. Kluwer Academic Publishers, Dordrecht. Chapters II and V. pp. 46, 174.

Murado, M. A., J. Fraguas, M. I. Montemayor, J. A. Vázquez, and P. González, 2010. Preparation of highly purified chondroitin sulphate from skate (*Raja clavata*) cartilage by-products. Process optimization including a new procedure of alkaline hydroalcoholic hydrolysis. *Biochem. Eng. J.* 49: 126–132.

Muro Urista, C., J. Escobar Jimenez, M. D. C. Diaz Nava, R. E. Zavala Arce, B. Garcia Gaitan, and F. Riera Rodriguez, 2013. Critical flux determination in ultrafiltration of wastewater from a food industry by optimization method. *Chem. Eng. Comm.* 200: 163–177.

Muthusamy, S., R. M. Doraisamy, and R. Ramamoorthy, 2006. Studies on cellulose acetate-polysulfone ultrafiltration membranes II. Effect of additive concentration. *J. Membr. Sci.* 268: 208–219.

Nilsson, M., G. Tragardh, and K. Ostergren, 2008. The influence of pH, salt and temperature on nanofiltration performance. *J. Membr. Sci.* 312: 97–106.

Norazman, N., W. Wu, H. Li, V. Wasinger, H. Zhang, and V. Chen. 2013. Evaluation of chemical cleaning of UF membranes fouled with whey protein isolates via analysis of residual protein components on membranes surface. *Sep. Purif. Technol.* 103: 241–250.

Nyström, M., L. Kaipia, and S. Luque, 1995. Fouling and retention of nanofiltration membranes. *J. Membr. Sci.* 98: 249–262.

Parés, D., E. Saguer, N. Pap, M. Toldrà, and C. Carretero, 2012. Low-salt porcine serum concentrate as functional ingredient in frankfurters. *Meat Sci.* 92: 151–156.

Paulen, R., M. Fikar, Z. Kovacs, and P. Czermak, 2011. Process optimization of diafiltration with time-dependent water adding for albumin production. *Chem. Eng. Process.* 50: 815–821.

Perédi, K., L. Vamos-Vigyazo, and N. Kiss-Kutz, 1981. Flavor losses in apple juice manufacture. *Nahrung. Food.* 25: 573–582.

Perry, R. H. and D. Greene, 1984. *Perry's Chemical Engineer's Handbook.* McGraw-Hill, New York.

Perry's Chemical Engineers' Handbook, 1999. *Liquid-Solid Operations and Equipment.* Centrifuges. McGraw-Hill Companies, USA, pp.18.106–18.125.

Piculell, L. and S. Nilsson, 1990. Effects of salts on association and conformational equilibria of macromolecules in solution. *Prog. Colloid Polym. Sci.* 82: 198–210.

Pierre, A., H. Goudédranche, A. Garem, G. Daufin, 1998. Industrie Laitière. In: Daufin, D., René, F., Aimar, P. (Coords.), *Les Séparations par Membrane Dans les Procédés de L'industrie Alimentaire.* Tec. Doc., Lavoisier, Paris, pp. 282–371.

Pollice, A., A. Brookes, B. Jefferson, and S. Judd, 2005. Sub-critical flux fouling in membrane bioreactors—A review of recent literature. *Desalination* 174: 221–230.

Powrie, W. and S. Nakai, 1986. The chemistry of eggs and egg products. In: Stadelman, W. J. and Cotterill, O. J. (eds.), *Egg Science and Technology* Chapter 6, Haworth Press, Binghampton, New York, p. 97.

Pyle, D. L. 1992. Chapter 9. *Mass Transfer in Food and Bioprocesses.* Chemical Engineering for the food industry. Cambridge programme for the food industry. Department of Chemical Engineering, University of Cambridge.

Pyle, D. L. 1996. Mass transfer in food and bioprocesses. In: Fryer, P. J., Pyle, D. L., Rielly, C. D. (eds) *Chemical Engineering for the Food Industry.* Blackie Academic and Professional, UK.

Rabiller-Baudry, M., L. Paugam, L. Begoin, D. Delaunay, M. Fernandez-Cruz, C. Phina-Ziebin, C. L. G. de Guadiana, and B. Chaufer, 2006. Alkaline cleaning of PES membranes used in skimmed milk ultrafiltration: From reactor to spiralwound module via a plate-and-frame module. *Desalination* 191(1–3): 334–343.

Rai, P., Majumdar, G. C., DasGupta, S., and De, S., 2006a. Modeling of Sucrose permeation through pectin gel during ultrafiltration of depectinized mosambi (*Citrus sinensis* (L.) Osbech) juice. *J. Food Sci.* 71(2): E87–E94.

Rai, P., C., Rai, G. C., Majumdar, S., DasGupta, and S., De, 2006b. Resistance in series model for ultrafiltration of mosambi (*Citrus sinensis* (L.) Osbeck) juice in a stirred continuous mode. *J. Membr. Sci.* 283: 116–122.

Rajagopalan, N. and M. Cheryan, 1995. Pervaporation of grape juice aroma. *J. Membr. Sci.* 104: 243–250.

Rajagopalan, N., M. Cheryan, and T. Matsuura, 1994. Recovery of diacetyl by pervaporation. *Biotechnol. Tech.* 8: 869–872.

Rausch, K. D. 2002. Front end to backpipe: Membrane technology in the starch processing industry. *Starch/Starke* 54: 273–284.

Ribeira, R., R. Bergamasco, and M. Gimenes, 2002. Membranes synthesis study for colour removal of a textile effluent. *Desalination* 145: 136–143.

Rinaldoni, A. N., M. Campderros, C. J. Menendez, and A. Prez Padilla, 2009. Fractionation of skim milk by an integrated membrane process for yoghurt elaboration and lactose recuperation. *Int. J. Food Eng.* 5(3): art. no. 1.

Rito-Palomares, M. 2004. Practical application of aqueous two-phase partition to process development for the recovery of biological products. *J. Chromatogr. B.* 807: 3–11.

Rodemann, K. and E. Staude, 1995. Polysulfone affinity membranes for the treatment of amino acid mixtures. *Biotechnol. Bioeng.* 46: 503–509.

Rodrigues, R. B., H. C. Menezes, L. M. C. Cabral, M. Dornier, G. M. Rios, and M. Reynes, 2004. Evaluation of reverse osmosis and osmotic evaporation to concentrate camu–camu juice (*Myrciaria dubia*). *J. Food Eng.* 63: 97–102.

Rodriguez-Amado, I., J. A. Vazquez, M. P. Gonzalez, and M. A. Murado, 2013. Production of antihypertensive and antioxidant activities by enzymatic hydrolysis of protein concentrates recovered by ultrafiltration from cuttlefish processing wastewaters. *Biochem. Eng. J.* 76: 43–54.

Rosenberg, M. 1995. Current and future application for membrane processes in the dairy industry. *Trends Food Sci. Technol.* 6: 12–19.

Rossignol, N., P. Jaoucn, J. M. Robert, and F. Quemeneur, 2000. Production of exocellular pigment by the marine diatom *Haslea ostrearia* Simonsen in a photobioreactor equipped with immersed ultrafiltration membranes *Bioresour. Technol.* 73: 197–200.

Ruiz-Ruiz, F., J., Benavides, and M. Rito-Palomares, 2013. Scaling-up of a B-phycoerythrin production and purification bioprocess involving aqueous two-phase systems: Practical experiences. *Process. Biochem.* 48: 738–745.

Saboya, L. V. and J. L. Maubois, 2000. Current developments of microfiltration technology in the dairy industry. *Lait* 80: 541–553.

Samhaber, W. M. 2005. Uses and problems of nanofiltration in the food industry. *Chem. Ing. Tech.* 77: 583.

Sarkar, B. 2013. A combined complete pore blocking and cake filtration model during ultrafiltration of polysaccharide in a batch cell. *J. Food Eng.* 116: 333–343.

Sarkar, B., S. DasGupta, and S. De, 2008a. Cross-flow electro-ultrafiltration of mosambi (*Citrus sinensis* (L.) Osbeck) juice. *J. Food Eng.* 89: 241–245.

Saxena, A. and K., Nigam, 1984. Coiled configurations for flow inversion and its effect on residence time distribution. *AIChE J.* 30: 363–368.

Shah, R. and S., Joshi, 1987. Convective heat transfer in curved ducts. In: Kakac, S., Shah, R. K., and Aung, W. (eds.), *Handbook of Single-Phase Convective Heat Transfer*, Wiley, New York.

Sharma, B. P. 1994. In: Lydersen, B. K., D'elia, N. A., and Nelson, K. L. (eds.), *Cell Separation Systems Bioprocess Engineering*, Wiley, New York, pp. 87–112.

Shin, S.-J., J.-P. Kim, J.-H. Jeon, and B.-R. Min, 2005. Preparation and characterization of polysulfone microfiltration membrane by a 2-methoxyethanol additive. *Desalination* 186: 1–10.

Sikder, J., C. Pereira, S. Palchoudhury, K. Vohra, D. Basumatary, and P. Pal, 2009. Synthesis and characterization of cellulose acetate-polysulfone blend microfiltration membrane for separation of microbial cells from lactic acid fermentation broth. *Desalination* 249: 802–808.

Sikder, J., M. Roy, P. Dey, and P. Pal, 2012. Techno-economic analysis of a membrane-integrated bioreactor system for production of lactic acid from sugarcane juice. *Biochem. Eng. J.* 63: 81–87.

Singh, N. and M. Cheryan, 1998. Membrane technology in corn refining and bioproduct-processing. *Starch/Starke* 50: 16–23.

Snape, J. B. and M. Nakajima, 1996. Processing of agricultural fats and oils using membrane technology. *J. Food Eng.* 30: 1–41.

Sogi, D. S., M. S. Arora, S. K. Garg, and A. S. Bawa, 2003. Response surface methodology for the optimization of tomato seed protein. *J. Food Sci.* 40: 267–271.

Sossna, M., M. Hollas, J. Schaper, and T. Scheper, 2007. Structural development of asymmetric cellulose acetate microfiltration membranes prepared by single-layer dry-casting method. *J. Membr. Sci.* 289: 7–14.

Su, S. K. and R. C. Wiley, 1998. Changes in apple juice flavor compounds during processing. *J. Food Sci.* 63: 688–691.

Sun, H., D. Qi, J. Xu, S. Juan, and C. Zhe, 2011. Fractionation of polysaccharides from rapeseed by ultrafiltration: effect of molecular pore size and operation conditions on the membrane performance. *Sep. Purif. Technol.* 80: 670–676.

Tadros, T. F. 1989. Colloidal aspects of pesticidal and pharmaceutical formulation—An overview. *Pestic. Sci.* 26: 51–77.

Tamime, A. Y. and R. K. Robinson, 1999. *Yoghurt, Science and Technology*, 2nd edn., Woodhead Publishing Limited & CRC Press LLC, London, UK, pp. 1–128, 432–485.

Tang, D.-S., G.-M. Yen, Y.-Z. He, S.-Q. Hu, B. Li, L. Li, and H.-L. Liang, 2009. Recovery of protein from brewer's spent grain by ultrafiltration. *Biochem. Eng. J.* 48: 1–5.

Teixeira, J. A., M. Mota, and A. Venancio, 1994. Model identification and diffusion coefficients determination of glucose and malic acid in calcium alginate membranes. *The Chem. Eng. J.* 56: B9–B14.

Trowbridge, M. E. 1962. O'K.: Problems in the scaling-up of centrifugal separation equipment. *Chem. Eng. Lond.* 162: A73.

Hung, T-C., R. H. Piedrahita, and A. Cheer, 2012. Bio-inspired particle separator design based on the food retention mechanism by suspension-feeding fish. *Bioinspir. Biomim.* 7: 046003 (12pp) doi:10.1088/1748-3182/7/4/046003.

Todisco, S., P. Tallarico, and E. Drioli, 1998. Modelling and analyses of the effects of ultrafiltration on the quality of freshly squeezed orange juice. *Italian Food Beverage Technol.* XII (May): 3–8.

Van der Horst, H. C. and J. H. Hanemaaijer, 1990. Cross flow microfiltration in the food industry. State of the art. *Desalination*, 77: 235–258.

Van der Horst, H. C., J. M. K. Timmer, T. Robbertsen, and J. Leenders, 1995. Use of nanofiltration for concentration and demineralization in the dairy industry: Model for mass transport. *J. Membr. Sci.* 104: 205–218.

Vandajon, L., S. Cros, P. Jaouen, F. Quemeneur, and P. Bourseau, 2002. Recovery by nanofiltration and reverse osmosis of marine flavours from seafood cooking waters. *Desalination* 144: 379–385.

Varzakas, 1998. Ph.D. Thesis, University of Reading, UK.

Vasiljevic, T. and P. Jelen, 2000. Comparison of nanofiltration and high pressure ultrafiltration of cottage cheese whey and whey permeate. *Milchwissenschaft* 55: 145–149.

Walstra, P., T. J. Geurts, A. Noomen, A. Jellema, and M. A. J. S. Van Boeckel, 1999. *Dairy Technology—Principles of Milk Properties and Processes*. Marcel Dekker Inc., New York, pp. 517–537.

Wang, M., L.-G. Wu, J.-X. Mo, and C.-J. Gao, 2006. The preparation and characterization of novel charged polyacrylonitrile/PES-C blend membranes used for ultrafiltration. *J. Membr. Sci.* 274: 200–208.

Wang, Y. Q., Y. L. Wang, F. B. Su, H. Wu, and Z. Y. Zion, 2005. Remarkable reduction of irreversible fouling and improvement of the permeation properties of poly(ether sulfone) ultrafiltration membranes by blending with Pluronic F127. *Langmuir* 21: 11856–11862.

Wani, A. A., D. Kaur, I. Ahmed, and D. S. Sogi, 2008. Extraction optimization of watermelon seed protein using response surface methodology. *Food Sci. Technol.* 41: 1514–1520.

Weisz, A. and Y. Ito, 2011. Performance comparison of three types of high-speed counter-current chromatographs for the separation of components of hydrophilic and hydrophobic color additives. *J. Chromat. A* 1218: 6156–6164.

Westfalia, 2001. Centrifugal technology benefits the global beverage industry. *Filtration-Separation* 30–31.

Westrin, B. A. and A. Axelsson 1991. A diaphragm diffusion cell applied to ethanol diffusion in agarose gel: A reproducibility study. *Biotechnol. Tech.* 5(4): 303–306.

Wienk, I. M., R. M. Boom, M. A. M. Beerlage, A. M. W. Bulte, C. A. Smolders, and H. Strathmann, 1996. Recent advances in the formation of phase inversion membranes made from amorphous or semi-crystalline polymers. *J. Membr. Sci.* 113: 361–371.

Wijers, M. C., Y. Pouliot, S. F. Gauthier, M. Pouliot, and L. Nadeau, 1998. Use of nanofiltration membranes for the desalting of peptide fractions from whey protein enzymatic hydrolysates. *Lait* 78: 621–632.

Wu, T. Y., A. W. Mohammad, J. M. d. Jahim, and N. Anuar, 2007. Palm oil mill effluent (POME) treatment and bioresources recovery using ultrafiltration membrane: Effect of pressure on membrane fouling. *Biochem. Eng. J.* 35: 309–317.

Xia, Y., F. Bamdad, M. Ganzle, and L. Chen, 2012. Fractionation and characterization of antioxidant peptides derived from barley glutelin by enzymatic hydrolysis. *Food Chem.* 134: 1509–1518.

Yadav, S., R. S. S. Yadav, and K. D. S. Yadav, 2013. An α-L-rhamnosidase from *Aspergillus awamori* MTCC-2879 and its role in debittering of orange juice. *Int. J. Food Sci. Technol.* 48: 927–933.

Yaroshchuk, A. E., A. L. Makovetskiy, Y. P. Boiko, and E. W. Galinker, 2000. Non-steady-state membrane potential: theory and measurements by a novel technique to determine the ion transport numbers in active layers of nanofiltration membranes, *J. Membr Sci.* 172: 203–221.

Yokota, A., H. Kubota, S. Komiya, K. Sato, H. Akiyama, and I. Koshiishi, 2012. Sensitive and simple determination of bromate in foods disinfected with hypochlorite reagents using high performance liquid chromatography with post-column derivatization. *J. Chromat. A* 1262: 219–222.

Zambrini A. V., A. Giroletti, G. Sudati, and A. P. Mistretta, 1990. Separation of lactate from acid whey permeates by nanofiltration. *XXIII International Dairy Congress Brief Communications and Abstract of Posters*, Montreal, Canada, Vol. II, Subject K, p. 382.

Zhang, S., H. Matsumoto, K. Saito, M. Minagawa, and A. Tanioka, 2009. Insulin transport across porous charged membranes: Effect of the electrostatic interaction. *Biotechnol. Progr. AIChE* 25(5): 1379–1386.

Zhao, L., H. Zhao, P. Nguyen, A. Li, L. Jiang, Q. Xia, Y. Rong, Y. Qiu, and J. Zhou, 2013. Separation performance of multi-components solution by membrane technology in continual diafiltration mode. *Desalination* 322: 113–120.

Zwijgers, A. 1992. Outline of milk protein concentrate. *Int. Food Ingredients* 3: 18–23.

Zydney, A. L., 1998. Protein separations using membrane filtration: New opportunities for whey fractionation. *Int. Dairy J.* 8: 243–250.

# 4 Crystallization

*Theodoros Varzakas*

## CONTENTS

The theory of gas–liquid equilibria, solid–liquid equilibria, equilibrium–concentration relationships, operating conditions, the calculation of separation in contact/equilibrium processes are discussed in this chapter.

The chapters deals with crystallization equilibrium (solubility and saturation, nucleation, metastable region, seed crystals), rate of crystal growth, stage-equilibrium crystallization, crystallization equipment (scraped heat exchangers, evaporative crystallizers), problem solving, and includes the following:

1. Whey Crystallization
2. Margarine Crystallization
3. Sugar Crystallization
4. Other Crystallization Processes in the Food Industry
5. Crystallization Equipment

## 4.1 INTRODUCTION

Crystallization is a separation process in which mass is transferred from a liquid solution, whose composition is generally mixed, to a pure solid crystal. Soluble components are removed from solution by adjusting the conditions so that the solution becomes supersaturated and excess solute crystallizes out in a pure form. This is generally accomplished by lowering the temperature, or by concentrating the solution, to form a supersaturated solution from which crystallization can occur (Earle and Earle, 2004). The equilibrium is established between the crystals and the surrounding solution, the mother liquor. The manufacture of sucrose, from sugar cane or sugar beet, is an important example of crystallization in food technology. Crystallization is also used in the manufacture of other sugars, such as glucose and lactose, in the manufacture of food additives, such as salt, and in the processing of foodstuffs, such as ice cream. In the manufacture of sucrose from cane, water is added and the sugar is pressed out from the residual cane as a solution. This solution is purified and then concentrated to allow the sucrose to crystallize out from the solution.

Crystallization is also used in the dairy industry (where lactose is produced from cheese whey or casein whey). Crystallization (fractionation) is also used in the edible oil industry, to modify the properties of edible oils and fats.

## 4.2 CRYSTALLIZATION EQUILIBRIUM

Once crystallization is concluded, equilibrium is set up between the crystals of pure solute and the residual mother liquor, the balance being determined by the solubility (concentration) and the temperature. The driving force making the crystals grow is the concentration excess (supersaturation) of the solution above the equilibrium (saturation) level. The resistances to growth are the resistance to mass transfer within the solution and the energy needed at the crystal surface for incoming molecules to orient themselves to the crystal lattice (Earle and Earle, 2004).

Solubility is a function of temperature. For most food materials an increase in temperature increases the solubility of the solute as shown for sucrose in Figure 4.1. Pressure has very little effect on solubility.

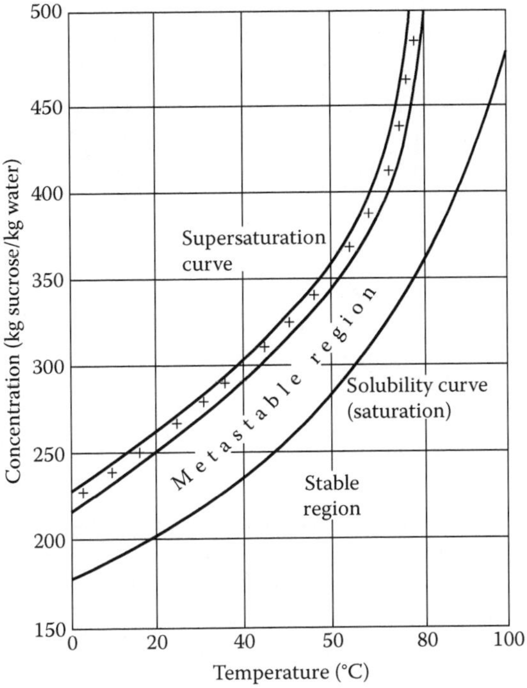


**FIGURE 4.1** Solubility and saturation curves for sucrose in water. (Ref: http:/www.nzifst.org.nz/unitoperations/contegseparation10.html.)

Solubility is usually measured as how many grams of solvent can be dissolved in 100 g of solute.

A saturated solution is one which contains as much solute as the solvent can hold. A supersaturated solution contains more dissolved solute than a saturated solution, that is, more dissolved solute then can ordinarily be accommodated at that temperature.

Two forms of supersaturation exist. Metastable (just beyond saturation) and labile (very supersaturated). Crystallization is normally operated in the metastable region. Crystallization is possible and spontaneous in the supersaturated or labile zone, however crystallization is possible but not spontaneous in the metastable zone.

In general, crystallization is achieved by cooling a solution (if supersaturation is a function of temperature), removal of the solvent by evaporation (where supersaturation is independent of temperature, e.g., common salt) and addition of another solvent to reduce solubility (when solubility is high and the above methods are not desirable, or in combination with the above methods, or the new solvent is called the antisolvent and is chosen such that the solubility is less in this new solution than it was before).

During crystallization, the crystals are grown from solutions with concentrations higher than the saturation level in the solubility curves. Above the supersaturation line, crystals form spontaneously and rapidly, without external initiating action. This is called spontaneous nucleation (Figure 4.2). In the area of concentrations between the saturation and the supersaturation curves, the *metastable* region, the rate of

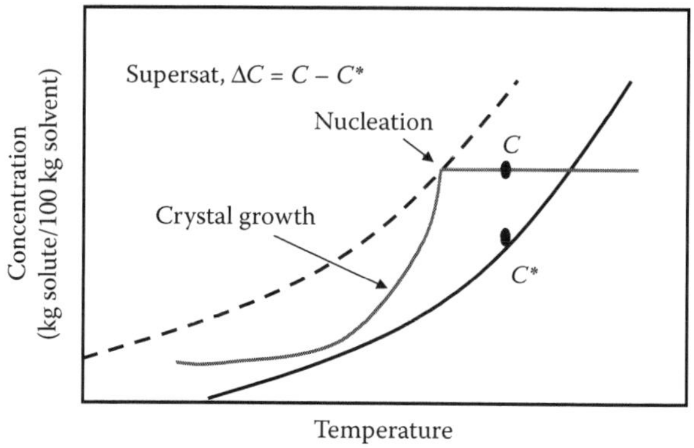


**FIGURE 4.2** Concentration versus temperature showing regions of supersaturation, nucleation, and crystal growth.

initiation of crystallization is slow; aggregates of molecules form but then disperse again and they will not grow unless seed crystals are added. *Seed* crystals are small crystals, generally from the solute, which then grow, by deposition on them, further solute from the solution. This growth continues until the solution concentration falls to the saturation line. Below the saturation curve there is no crystal growth; instead, crystals dissolve.

The degree of supersaturation of a solution is usually defined as $S = C_B/C_S^*$, where $C_B$ is the total solute concentration and $C_S^*$ is the saturation concentration of the solute in the solvent. The degree of supersaturation that can be achieved before any crystals will form depends on storage temperature, the nature of the solute and solvent, the presence of any contaminating materials, and the application of external forces (Kashchiev and van Rosmalen 2003; Lindfors et al., 2008).

A key variable during batch crystallization processes is the solution supersaturation which significantly determines the development of nucleation and growth phenomena (Srisa-nga et al., 2006) and, consequently, the final crystal yield and size. It is well established that the rate of cooling directly affects both nucleation and growth kinetics.

Kubota et al. (2001) showed that seeding plays a key role in crystallization to control crystal size distribution (CSD).

Supersaturation is of great importance in sugar crystallization. It has a profound effect on product quality and on the cost of production; parameters which determine the chances of survival of plants all over the world. It is also well known that seeding is a very critical step of crystallization and, therefore, this step should be carried out in a reproducible, reliable way, that is automatically based on the on-line monitoring of supersaturation. The SeedMaster optional software developed by Rozsa (2006), which can be run directly in the PR-01-S-type process refractometer of the K-PATENTS OY Company (Finland), addressed these needs and has already proved its worth in mills in the United States, Peru, Colombia, and Iran. Owing to the hardware limitations, it was designed to display and transmit supersaturation data only (4–20 mA standard current output), and to implement fully automatic seeding of crystallizers based on supersaturation or density. SeedMaster 2 is a further step in establishing a new class of instruments for crystallization control. It has a wide range of advanced features implemented in dedicated hardware and software, capable to serve two crystallizers simultaneously. It uses liquid concentration data provided by refractometers from the K-PATENTS family of products, and data from any one of the existing sensors measuring massecuite solids content, density, or power/current consumption of the stirrer motor. Based on on-line calculations SeedMaster 2 provides data on six massecuite parameters (per pan), including supersaturation and crystal content for transmission to external control equipment. It can be used to implement automatic seeding on its own as well.

## 4.3 HEAT OF CRYSTALLIZATION

When a solution is cooled to produce a supersaturated solution and hence to cause crystallization, the heat that must be removed is the sum of the sensible heat necessary to cool the solution and the heat of crystallization. When using evaporation to achieve the supersaturation, the heat of vaporization must also be taken into account. Because few heats of crystallization are available, it is usual to take the heat of crystallization as equal to the heat of solution to form a saturated solution. Theoretically, it is equal to the heat of solution plus the heat of dilution, but the latter is small and can be ignored. For most food materials, the heat of crystallization is positive, that is, heat is given out during crystallization. Note that heat of crystallization is the opposite of heat of solution. If a material takes in heat, that is, has a negative solution heat, then the heat of crystallization is positive. Heat balances can be calculated for crystallization (Earle and Earle, 2004).

## 4.4 RATE OF CRYSTAL GROWTH

Once nuclei are formed, either spontaneously or by seeding, the crystals will continue to grow so long as supersaturation persists. The factors controlling the rates of both nucleation and of crystal growth are temperature, degree of supersaturation, and interfacial tension between the solute and the solvent. If supersaturation is maintained at a low level, nucleus formation is not encouraged but the available nuclei will continue to grow and large crystals will result. If supersaturation is high, there may be further nucleation and so the growth of existing crystals will not be so great. Slow cooling produces large crystals since a low level of supersaturation is maintained whereas fast cooling produces small crystals (Earle and Earle, 2004; Hartel, 2001; Walstra, 2003).

The overall crystal growth rate depends on a number of factors, including mass transfer of the liquid molecules to the solid–liquid interface, mass transfer of noncrystallizing species away from the interface, incorporation of the liquid molecules into the crystal lattice, or removal of the heat generated by the crystallization process from the interface. Any of these processes can be rate limiting depending on the molecular characteristics of the system and the prevailing environmental conditions, for example, temperature profile and mechanical agitation. Consequently, a general theoretical model of crystal growth is difficult to construct. In crystallizing lipid systems, the incorporation of a molecule at the crystal surface is often rate limiting at high temperatures, whereas the diffusion of a molecule to the solid–liquid interface is often rate limiting at low temperatures (Hartel, 2001; McClements, 2012).

Once crystallization is complete, there may still be changes in crystal size and shape during storage due to postcrystallization processes, such as crystal aggregation or Ostwald ripening. Crystal aggregation occurs when two or more crystals come together and form a larger crystal, whereas Ostwald ripening occurs when oil molecules migrate from smaller crystals to large crystals through the intervening medium (Hartel, 2001; Walstra, 2003).

Nucleation rate is also increased by agitation. For example, in the preparation of fondant for cake decoration, the solution is cooled and stirred energetically. This causes fast formation of nuclei and a large crop of small crystals, which give the smooth texture and the opaque appearance desired by the cake decorator.

Once nuclei have been formed, the important fact in crystallization is the rate at which the crystals will grow. This rate is controlled by the diffusion of the solute through the solvent to the surface of the crystal and by the rate of the reaction at the crystal face when the solute molecules rearrange themselves into the crystal lattice.

These rates of crystal growth can be represented by the following equations

$$\mathrm{d}w/\mathrm{d}t = K_d A\,(c - c_i) \tag{4.1}$$

$$\mathrm{d}w/\mathrm{d}t = K_s(c_i - c_s) \tag{4.2}$$

where d$w$ is the increase in weight of crystals in time d$t$, $A$ is the surface area of the crystals, $c$ is the solute concentration of the bulk solution, $c_i$ is the solute concentration at the crystal/solution interface, $c_s$ is the concentration of the saturated solution, $K_d$ is the mass-transfer coefficient to the interface, and $K_s$ is the rate constant for the surface reaction (Earle and Earle, 2004). Similar equations have been quoted by Perry (1999) who employed a mathematical model which can correlate the nucleation rate to the level of supersaturation and/or the growth rate. Where $s$, the supersaturation, is defined as $(c - c_s)$, $c$ being the concentration of the solute and $c_s$ its saturation concentration.

$$\text{and since } \mathrm{d}w = A\rho_s \mathrm{d}L \text{ and } 1/K = 1/K_d + 1/K_s$$

where d$L$/d$t$ is the rate of growth of the side of the crystal and $\rho_s$ is the density of the crystal.

$$\mathrm{d}L/\mathrm{d}t = K(c - c_s)/\rho_s \tag{4.3}$$

It has been shown that at low temperatures diffusion through the solution to the crystal surface requires only a small part of the total energy needed for crystal growth and, hence has relatively little effect on the growth rate. However, at higher temperatures, diffusion is more important since diffusion energies are of the same order as growth energies. Experimental results have shown that for sucrose the limiting temperature is about 45°C, above which diffusion becomes the controlling factor. Finally, impurities in the solution retard crystal growth.

## 4.5 STAGE-EQUILIBRIUM CRYSTALLIZATION

When the first crystals have been separated, the mother liquor can have its temperature and concentration changed to establish a new equilibrium and so a new harvest of crystals. The limit to successive crystallizations is the build-up of impurities in the mother liquor that makes both crystallization and crystal separation slow and difficult. This is also the reason why multiple crystallizations are used, with the purest and best crystals coming from the early stages.

For example, in the manufacture of sugar, the concentration of the solution is increased and then seed crystals are added. The temperature is controlled until the crystal nuclei added have grown to the desired size, then the crystals are separated from the residual liquor by centrifuging. The liquor is next returned to a crystallizing evaporator, concentrated again to produce further supersaturation, seeded and a further crop of crystals of the desired size grown. By this method, the crystal size of the sugar can be controlled. The final mother liquor, called molasses, can be held indefinitely without producing any crystallization of sugar (Earle and Earle, 2004).

## 4.6 CRYSTALLIZATION OF WHEY CONCENTRATE

For the production of lactose from whey, whey is normally evaporated to a supersaturated solution (total solids content 60–73%). In cooling down the solution, crystallization starts and the crystals begin to grow. The crystals are removed from the liquid phase by centrifugation. Depending on the required grade, further purification (refining) can take place by washing the crystals, or redissolving them and recrystallizing them, followed by treatment with active carbon for the removal of any impurities.

Pre-crystallization of the concentrate shifts the sticking temperature upwards, as the crystallization yields a concentrate with much less amorphous lactose to be dried. Hence, it is possible to use considerably higher feed concentrations and inlet temperatures.

The three main steps in lactose production are concentration, crystallization, and separation.

The concentration process involves the evaporation of water in whey permeate to increase lactose concentration. Concentrated whey permeate has about 65–70% total solids, with about 80% of the total solids as lactose. The mixture is then cooled during the crystallization process where the lactose is separated as $\alpha$-lactose monohydrate crystal. Crude/food grade lactose is obtained after two stages of centrifugation, and final drying.

Crystallization is the most important separation step, but the crystallization process in the dairy industry is far from being optimized. The filling of the tank takes about 6 h, followed by gradual cooling and crystallization that lasts for 14–18 h, which means the crystallization process lasts for 20–24 h (Shi et al., 2006). The process is usually carried out in a large stirred tank cooling crystallizer. As the solution is gradually cooled, the supersaturation increases and lactose crystals are formed. The growth of crystals is typically accompanied by secondary nucleation, so the final product has a lot of small crystals.

Crystallization process could be greatly improved by operating at conditions that promote growth and minimize secondary nucleation, leading to the production of a narrow distribution of larger crystals. The ideal operating conditions can be selected by examining the lactose "supersolubility" diagram (Wong et al., 2011).

The metastable zone width (MSZW) is defined as the region between the solubility and supersolubility lines. The region between forced crystallization and supersolubility lines was reported by Hunziker (1926) as the optimum region for mass crystallization in condensed milk.

A strategy to minimize fines production was developed by Wong et al. (2012). On lab scale units, lactose crystals were produced from three crystallizers (draft-tube baffled, anchor, and paddle) operated with three cooling profiles (at different regions inside metastable zone [MSZ]). Computational fluid dynamics was used to simulate the flow profile. Among all combinations investigated, anchor crystallizer (lowest shear) operated at slow cooling rate (in upper MSZ) produces the largest crystals with minimal fines. Then, the design strategy was applied in industrial scale crystallizer. The 13 h cooling profile created for operation in the medium MSZ region successfully produced crystals with 28% less fines than the typical process. Therefore, depending on the crystallizer design and operational region (in MSZ), production of fines can be minimized.

To understand the theoretic background for the crystallization process we should look at the physico-chemical properties of the lactose which form about 3/4 of the whey solids.

Lactose is a disaccharide quite different in its behavior from other common sugars. A distinctive feature of the lactose is its appearance in different modifications with physico-chemical interrelations determined by the temperature.

In an aqueous solution the lactose molecule is present in an α and a β form.

The α and β forms are in a reversible equilibrium, that is, there is a continuous transformation of the α form into the β form and vice versa, called mutarotation. The proportion of the α form to the β form is determined by the temperature.

In whey powder produced according to the process described above with a pneumatic conveying system, the lactose is present in an amorphous or glassy state. This form of lactose is extremely hygroscopic absorbing water from the air forming α-lactose monohydrate.

The α-lactose monohydrate, which is not hygroscopic, is also formed in liquid whey concentrates by crystallization of the lactose from supersaturated solutions. As the α form is less soluble than the β form at a given temperature, the α form reaches the point of supersaturation first and forms crystals of α-lactose monohydrate. The removal of α-lactose from the solution due to the crystallization means that the proportion between α- and β-lactose changes, so that the solution contains more β-lactose than corresponding to the equilibrium.

Owing to the mutarotation, the solution of α-lactose becomes again supersaturated, so that the crystallization continues. This process will continue as long as the solution is supersaturated and will not stop until the saturation point is reached.

Lactose can be supersaturated by:

Increasing the content of lactose in relation to the water content. This is done by evaporation.

Cooling the solution as lactose becomes less soluble in water at lower temperatures.

A rich crystallization is therefore achieved by concentrating the whey to a solids content as high as possible and by cooling the concentrate.

The rate of mutarotation is directly influenced by the temperature of the solution, and it proceeds relatively fast at high temperatures, whereas it goes very slow at temperatures near the freezing point. It is thus clear that the conditions promoting mutarotation and crystallization are conflicting.

This means that by cooling the concentrate from too fast to too low a temperature the proportion of lactose which will crystallize will be low, even though the crystallization proceeds quickly. This is due to the slow mutarotation, resulting in only a small amount of β-lactose being transformed into α-lactose.

A temperature compromise is therefore necessary to obtain an optimal degree of crystallization.

The actual obtainable solids content depends on the type of whey, degree of denaturation of the whey proteins, pH of the whey, and the ability of the evaporator.

Although the crystallization process is not fully understood, it is a well-known fact that the crystallization takes place on the surface of already existing crystals. In order to promote the crystallization, lactose crystals or well-crystallized whey powder is seeded to the supersaturated solution. As the surface area of the seeding material is an important factor, a sufficient amount of small crystals should be added to the concentrate. If lactose crystals (α-lactose monohydrate) are used, approximately 0.1% w/w should be added to the concentrate.

Cooling of the concentrate is done instantaneously to 30°C in a flash cooler (see page 59) connected to the evaporator after which it is cooled slowly (1–3°C/h) to about 15°C in specially designed crystallization tanks, see Figure 4.3. This may allow the mutarotation to proceed at a reasonable speed resulting in a rich crystallization of about 80% of the lactose. During the whole crystallization time it is of significant importance that the content of the crystallization tank be vigorously agitated continuously. This is done in order to transport supersaturated solution to the surface of the crystals, simultaneously replacing the saturated solution. The agitation also prevents the viscosity of the thixotropic suspension from getting too high, and furthermore sedimentation of the lactose crystals does not take place.

The viscosity of the crystallized whey concentrate is mainly influenced by:

> Heat treatment before the evaporation, solids content of the concentrate, and size of the lactose crystals.

Lactose has a number of unique properties that can be used in many food and pharmaceutical products. The recovery of lactose from whey is accomplished by concentration and crystallization. Preparation of various grades of lactose generally includes methods of removing whey protein and ash prior to crystallization. The most important pretreatments considered by Akbari et al. (2012), have been carried out by using lime, $Al_2O_3$, $FeCl_3$, and $AlCl_3$ along with heating and also by using the acidic–alkaline method. Whey after pretreatments and lactose crystals were analyzed for lactose, ash, protein, and mineral composition. Results showed that the best way for removal of protein is accomplished with adding HCl to decrease pH from 6 to 4, boiling for 20 min, and filtering protein aggregates. Moreover, partial demineralization in whey

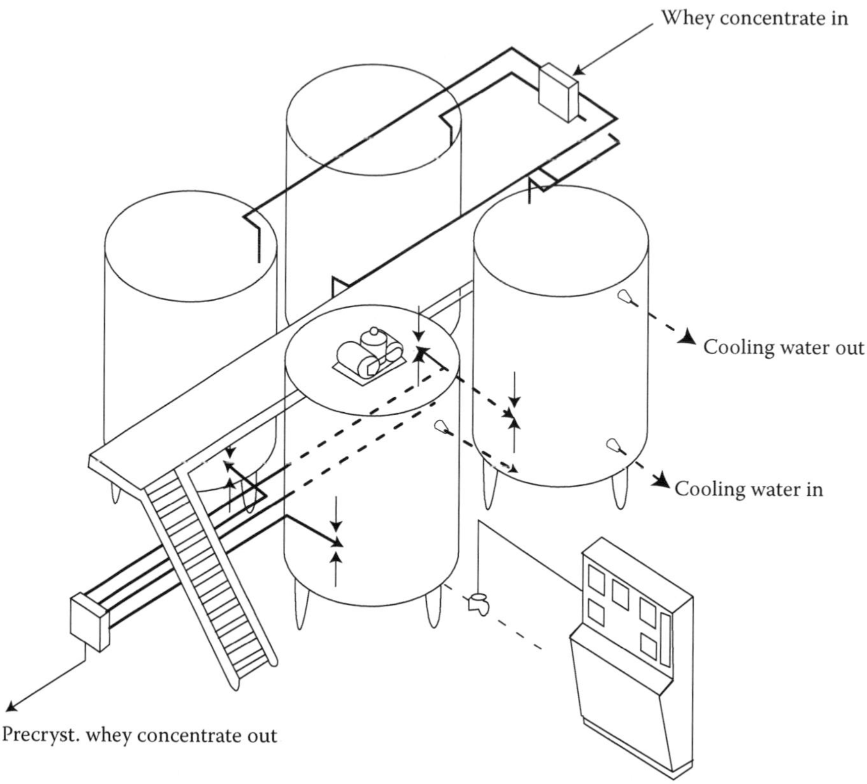


**FIGURE 4.3** Whey crystallization tanks. (From Westergaard, V. 2010. Milk powder technology-evaporation and spray drying. GEA-NIRO. http://www.niro.com/.)

solution for reducing ash is accomplished by adding NaOH to increase pH to 7.2 and heating solution for 20 min. Whey concentration was carried out in evaporator under vacuum at 62°C. For lactose crystallization, a crystallizer equipped with temperature controller and stirrer is used. Results showed that removal of protein along with partial demineralization would improve lactose purity up to 99%.

### 4.6.1 Heat Treatment before the Evaporation

The degree of denaturation of the whey proteins has a great influence on the viscosity of the concentrate as well as on the rate of crystallization, the drying properties, and the final powder properties. The properties of the proteins may also vary in different types of whey, and consequently they may require different heat treatment.

As a guideline, the preheating temperature should be in a narrow range of 80°C. Higher preheating temperature normally results in higher viscosity, and problems with pumping of the crystallized concentrate may occur. Lower temperature than 8°C will decrease the viscosity, but at the same time, problems with deposits in the drying chamber may occur due to an increased thermo-plasticity calling for higher particle/outlet temperature with an inferior product as a result.

It is thus a general rule that the preheating temperature is kept at such a level that the viscosity is about 2000 cP.

### 4.6.2 Solids Content of the Concentrate

It goes without saying that the higher the solids content the higher the viscosity in the concentrate, but also during the crystallization the viscosity undergoes deep changes, partly not only due to the decrease in temperature, but also due to the crystallization itself.

### 4.6.3 Size of the Lactose Crystals

A typical development of the viscosity during the crystallization is shown in Figure 4.4 (GEA-NIRO, 2010). The maximum viscosity at the time to max. for a given amount of concentrate is reached after a short time (1/2-1 h) and may go as high as several thousand cP. The reason is that the size of the lactose crystals is very small (big specific surface) and their mother liquor solution still has a relatively high solids content, which means that the friction between the crystals and the mother liquor is big. As time passes the lactose crystals grow, and the solids content of the mother liquor solution decreases resulting in a decrease of the viscosity.

The viscosity at $t_0$ max. may be so high that agitation is no longer possible. However, at the time $t_i$ noncrystallized concentrate with low viscosity is pumped into the crystallization tank, and thereby the content of the tank is "diluted" in the way of viscosity (Figure 4.4).

The curve $\Sigma_{t_{0-n}}$ is therefore the interesting one, and under normal conditions the viscosity does not present any problems, provided the heat treatment is controlled. The crystal size aimed at in the concentrate is 20–30 μ and the biggest crystal should not exceed 50 μ. This is in order to ensure that all spray particles contain at least one lactose crystal enabling a further crystallization to proceed during the final drying. This is possible as the mother liquor solution again becomes supersaturated due to the evaporation.

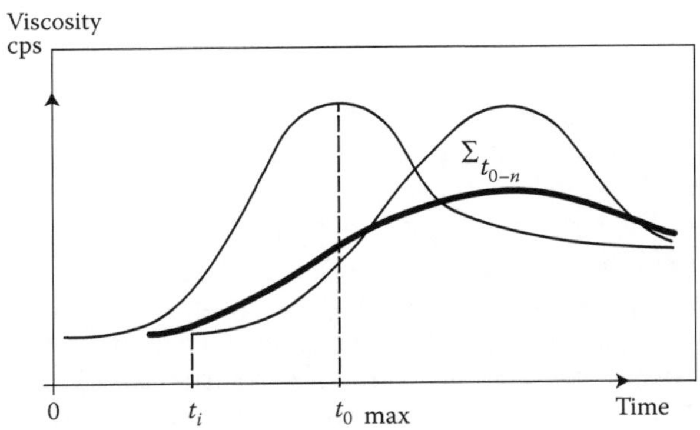


**FIGURE 4.4** Viscosity development during crystallization of whey concentrate.

Under normal conditions it is possible to obtain 80–90% crystallization of the lactose. This may be determined in a very simple way by using an ordinary hand refractometer, which gives the refractive index or direct% of sugar in solution.

The refractive index or direct% of sugar is measured at short intervals of 15–30 min direct from the evaporator, and the average arithmetical value is calculated = $S_1$. In an average sample of the crystallized whey concentrate a second refractometric reading is made = $S_2$. The degree of crystallization with a good approximation will be as follows.

$$\%\text{Crystallization} = \frac{(S_1 - S_2) \times 9500 \times 100}{L \times TS \times (95 - S_2)}$$

where $S_1$ = % sugar (ref. index) of the concentrate direct from the evaporator, $S_2$ = % sugar (ref. index) of the crystallized concentrate, $L$ = % lactose (74% may be used), and $TS$ = total solids content in%.

## 4.7 SEMI-CRYSTALLIZATION OF WHEY CONCENTRATE

In a continuous semi-crystallization process, the majority of the lactose is crystallized after 3–4 h. The resulting powder exhibits good physical properties.

The whey is precondensed to 50–52% solids, after which it is flash-cooled to 30°C. The concentrate is pumped to an ordinary single-walled tank equipped with an agitator. The volume of the tank corresponds to 2–3 h production. The concentrate is seeded as described above in order to start the crystallization.

When the tank is full the product is discharged at the same speed as new product is pumped in. The semi-crystallized concentrate is pumped to flash cooler No. 2 where it is cooled to 15°C and pumped to a new tank with a volume corresponding to 2–3 h in order to finish the crystallization. Because of the heat created by the crystallization, the temperature will increase by 2–3°C. When the tank is full the spray drying is started. (GEA-NIRO, 2010).

Scotta is a by-product from the Ricotta cheese production, which has been studied insufficiently, and that may be the reason for scotta not being widely used for industrial purposes. Pisponen et al. (2013) studied scotta as a raw material for the lactose production and the process of crystallization at different pH levels and concentration factors. Crystallization was carried out at six different pH levels and five concentration factors. The optimum acidity for lactose crystallization was determined to be close to pH 4; at higher or lower pH levels the growth of crystals was inhibited. The relationship between the dimensions of crystals and the concentration factor corresponded to data provided in the literature for the conventional cheese whey. The qualitative properties of the crystals obtained under the examined and recommended conditions were similar to the qualitative properties of crystals reported in the related literature about the cheese whey. It was demonstrated that scotta has a great potential for lactose production, and the results can be used for optimizing the industrial process.

Combined crystallization and drying of lactose solutions was performed in a pilot-scale spray dryer over a wide range of operating conditions by Das and

Langrish (2012). The effect of different parameters, including temperature, moisture content, atomizing air flow rate, liquid feed rate, main drying air flow rate, and particle size, on the degree of crystallinity of the spray-dried powders was analyzed. Water-induced crystallization (WIC) and modulated differential scanning calorimetry (MDSC) were used to assess the effect of these parameters on the degree of crystallinity of the spray-dried powders. The particles were characterized in terms of the final moisture content using WIC. Distinctive differences in the peak heights, which are indicative of the particle crystallinity, were found for spray-dried particles using different drying conditions, supporting the results from MDSC. MDSC showed that decreasing the inlet air temperature by 40°C increased the degree of crystallinity in the particles threefold from 22% to 72%. A decrease in the inlet air temperature may decrease the particle temperature, resulting in wetter particles, and a lower temperature meant a longer particle drying time and allowed the particles to rearrange themselves into a more crystalline form. Up to 72% crystallinity is achievable in a pilot-scale spray dryer by suitable adjustment of the operating conditions. The results suggest differences in the rate of crystallization and particle size between small and pilot-scale spray dryers.

Whey, a liquid by-product of the cheese manufacturing process, is normally converted for practical use to a powder within a spray dryer. The lactose portion of whey powder is usually in an amorphous state, due to rapid drying and solidification within the spray dryer. It is possible to reduce the hygroscopic nature of lactose in whey powder by converting a portion of the lactose to the more stable crystalline form in a pre-crystallisation step before spray drying. This process involves seeding a supersaturated whey solution with fine lactose crystals to nucleate the crystallization process, and allows the lactose to crystallize over a period between 4 and 24 h (Fox and McSweeney, 1998). According to Hynd (1980), up to 85% of the lactose in whey can be crystallized in a carefully controlled pre-crystallization step.

The fluidization of partially crystallized whey powder above the glass-transition temperature of lactose has been investigated by Nijdam et al. (2008), with the intent of crystallizing the amorphous-lactose fraction in order to reduce the propensity of the powder to cake during storage. Partially crystallized whey powder can be fluidized in a vibrated fluidized bed at temperatures of 25–40°C above the glass-transition point of lactose, depending on the relative humidity of the air, before the powder becomes too sticky to fluidize. This temperature difference can be increased up to 80°C by fluidizing the powder with fine, relatively nonsticky, fully crystallized whey powder in order to coat and protect the sticky partially crystallized whey particulates during fluidization. Despite this temperature-difference increase, the time required to crystallize the amorphous-lactose fraction in partially crystallized whey powder is not reduced sufficiently for this process to be viable in industry. An amorphous whey powder crystallization process is likely to be more feasible, because the reduced salt and protein concentrations in this powder would ensure that lactose crystallization is faster. Finally, they highlighted the potential of using the phenomenon of lactose plasticization above the glass-transition temperature and fines coating to improve the instant properties of milk-based powders.

## 4.8 CRYSTALLIZATION OF ALPHA-LACTOSE MONOHYDRATE

Three consecutive steps of lactose crystallization exist:, mutarotation, nucleation and crystal growth rate.

Two anomeric forms of lactose (alpha and beta) exist simultaneously in solution, and a reversible reaction called mutarotation occurs between them (Hartel, 2001; Hartel and Shastry, 1991). A slight temperature dependence of the β:α ratio at equilibrium ($K_m$) exists as calculated by Roetman and Buma (1974). Mutarotation does not occur instantaneously and its kinetics have been investigated by several authors since the 1930s.

During a batch process, as the alpha-form crystallizes from supersaturated solution, the concentration of alpha-lactose present in solution decreases so that the numerical value of the ratio between the two forms increases. The molecules of beta anomer then mutarotate into the alpha-form to regain the equilibrium (Hartel, 2001).

Nucleation is the second step. Only few studies on nucleation kinetics of alpha-lactose monohydrate are available. Griffiths et al. (1982) Shi et al. (1990), Liang et al. (1991) and, earlier, calculated parameters of nucleation kinetics for continuous lactose crystallization using mixed suspension, mixed product removal (MSMPR) population balance technique.

A continuous melt suspension crystallization process has been presented for the purification of the phosphoric acid by Chen et al. (2013), which is conducted in the cascade of a continuous mixed suspension and mixed product removal (MSMPR) crystallization and subsequent countercurrent solid–liquid contact in a gravity wash column. With magma withdrawn from the MSMPR crystallizer being fed into the top of the gravity wash column, crystals settle down toward the bottom of the column and the reflux melt rises upward, whereas the sweating of crystals and counter-current contact between crystal particles and the melt lead to high separation capability of impurities. The gravity wash column used in this study is a vertical agitated cylindrical column, whose configuration is different from the inclined column crystallizer. The feasibility of suspension crystallization equipment for purification of phosphoric acid was studied. Distribution of impurities in the wash column was also investigated as a function of the feed concentration, crystal bed height, reflux ratio, and stirring speed. Reflux ratio has been found to play an important role in the process. Food grade phosphoric acid can be achieved when the process is operated with feed concentrations of 84–86 wt.%, reflux ratios of 3.4–6.2, stirring speed of 10 rpm and crystal bed height of 150–250 mm in a gravity wash column.

Kinetics of mutarotation, nucleation, and crystal growth are therefore interrelated phenomena.

Nucleation rate depends on crystal mass and hence on the rate of crystal growth. Conversely, mass growth rate depends on nucleation rate since total crystal surface area is a function of crystals number. Additionally, both rates of crystal growth and nucleation steps depend on the concentration of alpha-form and hence on mutarotation rate.

A kinetic model combining first-order differential equations of the three consecutive steps of lactose crystallization—mutarotation, nucleation, and crystal growth rate—was developed by Mimouni et al. (2009). Numerical solutions successfully fitted the variations of crystal mass growth rate as a function of lactose concentration during unseeded isothermal batch crystallization, at different initial lactose concentrations and temperatures.

The model allowed the induction phase and the influence of seeding to be simulated by taking into account the dependency of crystal growth rate on total crystal surface area and, consequently, on nucleation rate. The proposed model highlights the large influence of the mutarotation step, even in unseeded crystallization kinetics. The effects of high lactose supersaturations that prevail at industrial scale during whey powder manufacturing and the effects of alkaline pH (9.5) on lactose crystallization kinetic were successfully predicted.

Attrition is the general term that describes the breakdown of crystals due to collisions with each other or some other body, or by a shearing fluid. These interactions produce new crystals from the parent crystals and are likely to be a major source of secondary nucleation in industrial crystallizers. However, these collision interactions are difficult to devolve because they occur simultaneously.

Agrawal et al. (2011) studied lactose monohydrate attrition within the small sample presentation unit of a laser diffraction particle sizer (Malvern Mastersizer, 2000, UK). Stirrer speed, particle concentration, and initial crystal size were varied and particle size distributions (PSDs) were recorded every 2 min for 1 h. From these, three basic attrition mechanisms were recognizable: shattering, chipping, and abrasion. The degree of attrition was quantified using indices: the shattering index (SI) and attrition index (AI), which are both typical of literature indices; and the differential attrition index (DAI) proposed here. All indices show similar trends although the DAI is more sensitive to subtle changes in PSDs.

Of the three variables studied, particle concentration least affects attrition. Experiments with low initial crystal size and low impeller speed produce fines in the <10-μm range, indicating that abrasion is the dominant attrition mechanism. In contrast, when the initial crystal size is large and impeller speed is high, crystal fragments are produced across all size ranges indicating that all three attrition mechanisms occur. This variation across the results indicates that the three mechanisms are differentiated by the collision energy intensity, which is a function of both crystal size and impact velocity. These experimental observations were found to closely follow theoretical predictions for crystal–impeller impacts (Mersmann et al. 1988; Synowiec et al. 1993). When applied to industrial crystallizers, this work demonstrates that careful selection of pumps and impellers is needed to control secondary nucleation.

A microfluidic process for producing crystals of controlled size by confining the crystallization within drops is demonstrated by Dombrowski et al. (2007). The process consists of a drop producing stage that segments the mother liquor into monodisperse drops followed by crystallization in a temperature-controlled tubular crystallizer. The process is implemented to produce lactose crystals with significantly narrower crystal size distribution (CSD) as compared with crystals produced in a stirred bulk crystallizer. By controlling the drop size and initial supersaturation the mean crystal size can be controlled. The distribution of the number of crystals per drop and the CSD are measured as a function of supersaturation at temperatures between 20°C and 40°C for 150 and 300 μm drops. In the case where drops contain only one crystal, a very narrow CSD is obtained with a coefficient of variation of crystal size as low as 7%. The crystallization of lactose in a microfluidic tubular crystallizer is modeled by treating nucleation in drops as a Poisson process with a nucleation rate based on classical nucleation theory. Experimental results are in good agreement with predictions from a Poisson process model over the range of temperatures, supersaturations, and drop sizes tested.

## 4.9 SKIMMED MILK POWDER AND CRYSTALLIZATION

Milk powders are widely used in food processing industries or for reconstituting and often must be stored for months. Spray-dried milk powders, in amorphous states, are very unstable. They have a tendency to sorb moisture and form caked powders that are not free flowing.

Physicochemical qualities and properties decline during long-term storage due to the hygroscopicity of amorphous lactose.

Deteriorative changes start by sorbing moisture from the environment, leading to a decrease in the glass-transition temperature of amorphous components to below ambient conditions, which initiates crystallization and an increase in the stickiness of the powders. The expelled water from crystallization further increases the water activity and enhances the changes.

In addition, crystallization of lactose from the matrix rearranges the protein–lactose bonds and de-stabilizes the proteins, meaning that less amorphous lactose is available around the proteins to be attached to the H-bond active sites of the proteins (Buera et al., 2005).

There are different methods that are commonly used to prevent the powders from sorbing moisture, such as keeping them in cool and dry conditions and using sealed impermeable packaging.

Crystallization of lactose in milk powders in pre-crystallization or post-crystallization facilities has been suggested (Hynd, 1980; Yazdanpanah and Langrish, 2011b) to improve the powder stability against moisture sorption and enhancing the physical properties. Yazdanpanah and Langrish (2011a,b) showed the decrease in moisture sorption for the crystallized lactose and milk powders that were crystallized in a fluidized-bed dryer/crystallizer. An example of a fluidized-bed drier is shown in Figure 4.5.

Surface modification of particles by crystallization of the outer layer, giving the so called "egg-shell" particles, has shown significant improvements in the physical properties of the powder in their fresh state.

Yazdanpanah and Langrish (2013) investigated the effect of aging on raw powder and the processed powder (with modified surface) stored at around 33% relative humidity and 25–30°C for 30 weeks. Agglomeration, large lactose crystal formation on the surface, surface composition changes, and protein modifications were studied. The changes between the raw powder and process powder were compared after aging. The nonhygroscopic crystalline surface layer showed significant benefits in maintaining the physicochemical qualities of the powders over long storage times. The aged raw powder showed a 24% change in the crystallinity, a 3% change in the lactose/protein ratio on the surface and 6% protein denaturation compared with the aged processed powder with a 4% change in the crystallinity, a 1.5% change in the lactose/protein ratio on the surface and 2% protein denaturation, respectively, after 30 weeks of storage.

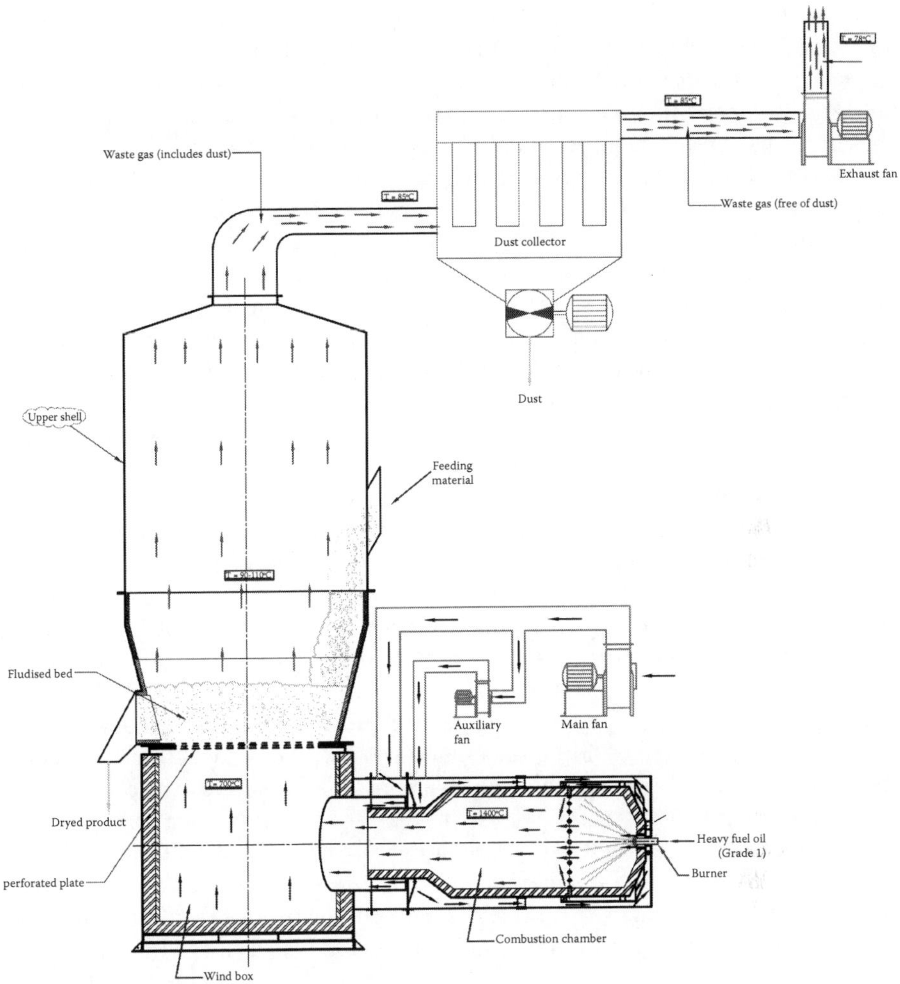


**FIGURE 4.5** Fluidized-bed drier.

Processed powder with egg-shell structure particles has been made by a fluidized-bed crystallization technique described before (Yazdanpanah and Langrish, 2011c). The processing condition was 60°C, 40% RH, and 20 min processing time. After processing, the powder was transferred to a vacuum dryer and left under vacuum at room temperature for 4 h to slow down further crystallization.

The potentially negative effects of low molecular weight disaccharides, especially lactose, on spray-drying efficiency and storage stability of dairy powders are often counterbalanced by the presence of intact milk proteins. Hydrolysis of proteins, however, may impair such protective effects and contribute to a loss in production performance. Hydrolyzed or nonhydrolyzed whey protein/lactose (WP/L) dispersions were spray dried, in order to examine the effects of protein hydrolysis on relaxation behavior and stickiness of model powders (Hogan and O'Callaghan, 2013).

Whey proteins included a nonhydrolyzed, whey protein isolate control and three hydrolyzed whey protein powders (WPH), with DH values of 8, 11, and 32, where DH = degree of hydrolysis. Hydrolysis of whey proteins increased moisture sorption in WP/L powders. Moisture sorption was higher in powders containing hydrolyzed proteins. Whey proteins delayed the time-dependent onset of lactose crystallization, and this effect was greatest in powders containing WPH32. Glass–rubber transition (T gr) temperatures of WP/L powders were not affected by protein hydrolysis but were dominated by the lactose fraction. Powders containing hydrolyzed whey proteins were more susceptible to sticking compared to intact proteins. Surface coverage by proteins or peptides was lower in powders containing hydrolyzed WP, and this would have contributed to the greater susceptibility of these powders to sticking. Results suggest that hydrolysis of WP affected the relaxation behavior of WP/L powders and altered the rate at which lactose underwent viscous flow behavior.

Spray-dried whole milk powders have been produced to a high bulk-to-surface-free-fat ratio due to the susceptibility to oxidation of surface free fat. Processing spray-dried whole milk powders in a fluidized-bed crystallizer by hot and high humidity air can cause the release of bulk fat and increase both inner- and surface-free fat. Yazdanpanah and Langrish (2012) investigated the fat release and its migration to particle surfaces by two different mechanisms: (a) fat melting due to high temperature processing in a fluidized-bed crystallizer so that liquid fat can diffuse outwards to the particle surfaces; (b) fat release from the matrix due to lactose crystallization, causing component segregation and releasing molecular bonds. More free fat is preferable in some dairy-based processing, such as chocolate making, where cocoa butter could be substituted by milk fat to adjust the viscosity of chocolate paste and formulate good chocolate taste. They showed the possibility of releasing more than 90% of the bulk fat to increase free fat (inner and surface) in 15 min processing time. Different temperatures and humidities (45°C and 78% RH) have been found to provide suitable conditions to release the bulk fat by crystallizing lactose in higher amount, as a form of inner free fat which is ready for further use, such as in chocolate industries. Meanwhile the released fat, as inner-free fat, has less contact with oxygen than surface-free fat and maintains the powder quality better. The changes in particle structure and fat release processes have also been studied during fluidized-bed drying of spray-dried whole milk powders that could improve the control and design of the drying process for milk powders.

## 4.10 SUGAR CANE JUICE PRODUCTION

The production of sucrose from sugar cane juice includes lime and sulfur dioxide addition to clarify the raw juice stemming from sugar cane in the mill (double-sulfitation process). The raw juice is a complex mixture, which is initially clarified by addition of sulfur dioxide and calcium hydroxide (lime). As a result, the suspended colloids of the raw juice coagulate and precipitate along with calcium sulfite. The clarified juice is then concentrated by evaporation (steam heating) and bleached by a second sulfitation. Finally, the product stream is crystallized to obtain white sugar crystals.

The double-sulfitation process suffers from inefficient removal of substances from the raw juice, like gum, ashes, silica, colorants, and reversible colloids, and an inefficient removal of sulfite from the product stream. As a result, the quality of plantation white sugar is noticeably inferior to refined sugar. Hence, ultrafiltration membranes are investigated to replace the second sulfitation step for the purification of clarified sugarcane juice (Abels et al., 2013).

In sugarcane processing, the juice extracted across tandem mills or a diffuser is subsequently purified in the clarification unit process. The clarified juice (13–18 Brix or% dissolved solids) is then concentrated through a series of multiple-effect evaporators to yield the final evaporator syrup (FES) of ~65 Brix. FES is further concentrated under vacuum to ~90–95 Brix (at lower temperatures than in evaporation to minimize the chemical degradation of sucrose) and crystallized. The vacuum pans are seeded with finely ground sucrose to allow larger sucrose crystals to form. The mixtures of sucrose crystals and mother liquor (molasses) discharged from a vacuum pan are massecuites, which are subsequently separated in centrifuges; the mother liquor is further re-concentrated and recrystallized to give two more crops ("B and C crops") of crystals. The final liquor is the by-product molasses (also known as blackstrap or final molasses).

Louisiana hard-to-boil (HTB) massecuites (mixtures of sucrose crystals in molasses) with markedly low heat-transfer properties are a sporadic problem in sugarcane factories, which cause raw sugar and molasses production to decrease and increase, respectively. This usually occurs after severely deteriorated sugarcane has been processed, but the specific cause is unknown and only limited correction has been achievable. At the end of the 2006 sugarcane-processing season, HTB and normal massecuites and molasses were collected from four Louisiana factories by Eggleston et al. (2011). Compared to normal samples, the HTB samples had 9.1–33.2% lower heat conductivity and 10.0–49.2% higher heat resistivity. The more HTB a sample is, the greater the increase in heat resistivity compared to the corresponding decrease in heat conductivity.

Excess lime addition to neutralize acids during juice clarification is not the direct cause of hard boiling. Oscillatory deformation rheology applied at 20°C to normal molasses samples gave typical mechanical spectra of concentrated solutions. In contrast, a highly viscous, intermolecular (gel) network was present in the HTB molasses, which would explain the difficulty of removing entrapped water on boiling. Polysaccharides in the samples were characterized. GFC, TLC, and methylation analyses suggested the presence of an arabinogalactan and endo-dextranase-resistant dextran structures. The HTB phenomenon may have different causes and mannitol is a contributing factor.

There are three types of postharvest deterioration of sugarcane leading to the degradation of sucrose and formation of unwanted degradation products or impurities: (1) chemical, (2) enzymatic, and (3) microbial. However, the latter is mostly responsible for the deterioration of sugarcane. The major (but not sole) contributor to microbial deterioration of sugarcane, particularly in areas where humid and warm conditions prevail, is infection by *Leuconostoc mesenteroides* hetero-fermentative, lactic acid bacteria.

Mannitol is a major deterioration product of *Leuconostoc mesenteroides* bacterial metabolism of sucrose and fructose from both sugarcane and sugar beet. The effect of crystallization conditions on the mannitol partition coefficient ($K_{eff}$) between impure sucrose syrup and crystal has been investigated by Eggleston et al. (2012) in a batch laboratory crystallizer and a batch pilot plant-scale vacuum pan. Laboratory crystallization was operated at 65.5°C, 60.0°C, and 51.7°C with a 78.0 Brix (% refractometric dissolved solids) pure sucrose syrup containing 0%, 0.1%, 0.2%, 1.0%, 2.0%, 3.0%, and 10% (at 65.5°C only) mannitol on a Brix basis. Produced mother liquor and crystals were separated by centrifugation and their mannitol contents measured by ion chromatography with integrated pulsed amperometric detection (IC-IPAD). The extent of mannitol partitioning into the crystals depended strongly on the mannitol concentration in the feed syrup and, to a lesser extent, the crystallization temperature. At 65.5°C and 60.0°C, the $K_{eff}$ varied from 0.4% to 3.0% with 0.2% to 3.0% mannitol in the feed syrup, respectively. The mannitol $K_{eff}$ was lower than that reported for dextran (9–10% $K_{eff}$), another product of *Leuconostoc* deterioration, under similar sucrose crystal growth conditions. At 10% mannitol concentration in the syrup at 65.5°C, co-crystallization of mannitol with sucrose occurred and the crystal growth rate was greatly impeded. In both laboratory and pilot plant crystallizations (95.7% purity; 78.0 Brix; 65.5°C), mannitol tended to cause conglomerates to form, which became progressively worse with increased mannitol syrup concentration.

At the 3% mannitol concentration, crystallization at both the laboratory and pilot plant scales was more difficult. Mannitol incorporation into the sucrose crystal results mostly from liquid syrup inclusions, but adsorption onto the crystal surface may play a minor role at lower mannitol concentrations.

## 4.11 SUGAR CRYSTALLIZATION

An evaporator is used at the stage of concentration to remove some of the water prior to crystallization. The resulting syrup, which is known as evaporated liquor, is about 74% solids.

The crystallization process takes place in vacuum pans, which boil the juice at lower temperatures (about 80°C) under vacuum. When the juice reaches a predetermined concentration it is "seeded" with tiny sugar crystals that provide the nucleus for larger crystals to form and grow.

When the crystals reach the desired size the process is stopped and the resultant mixture of crystal sugar and syrup—known as massecuite—is spun in centrifuges to separate the sugar from the "mother liquor." The sugar crystals are washed and after drying and cooling, are conveyed to storage silos.

The growth of the sucrose crystal only involves sucrose and water. The nonsugars contained in the sugar juice are not incorporated into the crystal structure, instead most of them remain in the liquid phase while some are released to the vapor phase. The sugar crystals are removed from the liquid phase by centrifugation.

Some sugar remains in the separated liquid so it is boiled again in a further set of vacuum pans to produce raw sugar. This process is repeated a third time resulting in final product sugar and molasses. Raw and final product sugars are redissolved into the thick juice http://www.britishsugar.co.uk. The sugar manufacturing process is shown in Figure 4.6.

The British Sugar factory at Bury St. Edmunds manufactures ~2500 te/day granulated sugar from sugar beet. Sugar is extracted from the beet into aqueous solution in

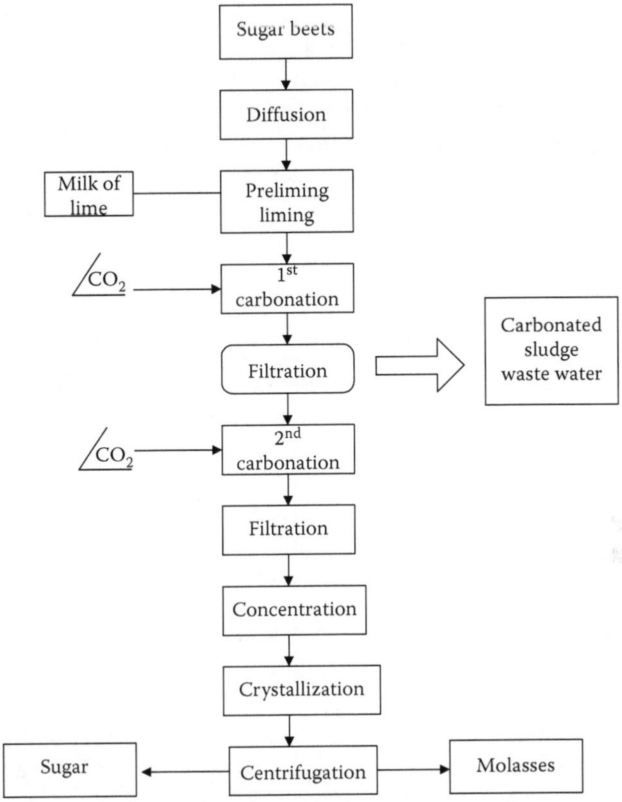


**FIGURE 4.6** Sugar manufacturing process.

"diffuser" units and the resultant "thin juice" is concentrated from a 16 wt% sugar solution to c. 67 wt% by evaporation in two parallel series of forward-feed falling/rising film evaporators. The concentrated solution then passes to crystallizers where further evaporation occurs in the course of crystallization.

Energy efficiency is maximized by preheating the incoming thin juice from 29°C to c. 90°C with a mixture of crystallizer and evaporator vapors and condensates as described by Smaili et al. (1999).

Heavy fouling occurs in several of the heat exchangers [HEX] with the result that a higher temperature vapor stream (5$^{th}$ effect, at 97°C, cf. 6th effect, at 92°C) has to be used in both HEX 10 and 11 in order to achieve the raw juice target temperature. The use of this higher temperature utility reduces the overall thermal efficiency of the plant and introduces control and productivity issues.

For beet-sugar manufacturing a general requirement is that the concentration of juice in the evaporator outlet (thick juice) be as high as possible, reaching 72–75% in most advanced factories. It is then possible to apply crystallization technologies that ensure a high sugar quality while also facilitating low-energy consumption. For the widely applied three-stage crystallization, the relative energy consumption as a function of juice concentration is shown by Urbaniec (2004).

The temperature range of evaporation is determined by the constraints of juice temperature. To avoid the deterioration of juice quality, especially its pH and color, the juice temperature at inlet should not exceed 128°C. To avoid uncontrolled crystal formation in thick juice (at outlet), its temperature should not be lower than about 90°C.

Three types of thin film evaporator units are in use:

Falling film, heat transfer in tubes; falling film, heat transfer in plates; climbing film, heat transfer in plates.

Falling film units can be operated at temperature differences as small as 3–4 K. When using such units in an evaporator station comprising five or six stages, a correct operation is possible at initial juice temperature about 121–122°C. This facilitates high quality of thick juice making it possible to carry out sugar crystallization very efficiently.

Crystallization can be a major determinant of quality in sugar-based products. In products such as hard candies, the formation of sugar crystals is inhibited during formation of the glassy state, whereas in products such as fondants, the presence of crystals is necessary for the desired texture.

Crystallization of an amorphous sugar matrix is affected by numerous factors, including water content, ingredients or additives, and environmental conditions of relative humidity and temperature. Both additives and water content can affect the glass transition temperature (Tg) of a product. Tg is a determining factor in the stability and, thus, crystallization of a sugar glass. Tg is the temperature at which a sugar glass softens from solid-like to liquid-like characteristics and begins to flow. Below Tg, a sugar glass is stable to many physical and chemical reactions including crystallization. A storage temperature above Tg can be related to various relaxation properties in the amorphous material, which are often represented with the Williams–Landel–Ferry (Williams et al., 1955).

The level and type of additives can affect the stability of an amorphous sugar matrix. The addition of corn syrup to sugar confections both inhibits sucrose nucleation and increases the induction time before onset of nuclei formation. In hard candies, corn syrup prevents unwanted crystallization, or graining. The effect of corn syrup on sucrose nucleation may either be through its effect on Tg or by some specific interaction.

Gabarra and Hartel (1998) found that the amount of sucrose that crystallized in a sucrose–corn syrup mixture decreased with increasing percentage of corn syrup added. Tjuradi and Hartel (1995) found that higher dextrose equivalent (DE) corn syrups inhibited the rate of crystallization most, whereas Bhandari et al. (1997a,b); Bhandari and Hartel (2002) showed that lower DE corn syrups inhibited crystallization most.

The nucleation rate (number of crystals/mL min) in highly viscous fluids is often described by

$$
\begin{aligned}
J &= A\exp\left\{-\frac{\Delta G_c}{kT} + \frac{\Delta G_v'}{kT}\right\} \\
&= A\exp\left\{-\frac{16\pi\sigma_s^3 v^2}{3k^2T^2\ln^2(S)} + \frac{\Delta G_v'}{kT}\right\}
\end{aligned}
$$

where k is Boltzman's constant ($J/K$), $T$ is absolute temperature ($K$), $\Delta G_c'$ is the critical free energy for nuclei formation, $\Delta G_v'$ is the free energy related to diffusion or mobility effects, $S$ is supersaturation ratio, rs is interfacial tension ($J/m^2$), and v is molecular volume ($m^3$) (Hartel, 2001; Mullin, 1993).

Induction times for onset of nucleation and nucleation rates were determined by Levenson and Hartel (2005) for amorphous sugar matrices made with sucrose and corn syrup held at temperatures between 60°C and 110°C. Induction times decreased and nucleation rates increased exponentially with increasing temperature. Both the classical nucleation rate equation and the Williams–Landel–Ferry (WLF) equation fitted the data. Within the measurement variability, the type of corn syrup used did not cause a significant difference in either the induction times or nucleation rates. The kinetic barrier to nucleation was determined to be −76.7 to −127.2 kJ/mol, which was 4–6 times higher than the activation energy for translational diffusivity of sucrose, suggesting that rotational diffusivity of sucrose and the interaction between corn syrup and sucrose also must play an important role in nucleation.

Crystalline structures of sugars, particularly that of sucrose, depend on crystallization conditions and the presence of impurities. Sugar crystals show melting that often occurs at low temperatures with time- and temperature-dependent characteristics. Melting at low temperatures can be accounted for by the presence of impurities and defects. Sugar crystals also contain noncrystalline regions that may undergo decomposition and subsequent dissolution at the decomposition interface and acceleration of decomposition reactions. Such processes with melting establish a supersaturated condition for the remaining crystals, leading to a time-dependent melting point depression and subsequent melting of the remaining crystals (Roos et al., 2013). Decomposition of sugars, as well as dissolution and melting of sugar crystals, are separate phenomena, although they are commonly found to coincide. Decomposition of sugars requires the presence and mobility of molecules for reactions outside the crystal lattice; that is, the molecular mobility of amorphous or melted regions is a prerequisite for decomposition, whereas melting of sugar crystals occurs as a separate thermodynamic process with no chemical change of the molecules.

Amorphous matrices made up of sugar molecules, are frequently used in food and pharmaceutical industries. A drawback to their use is that they are susceptible to collapse, as a result of water uptake and an increase in temperature and subsequently crystallize. The crystallization characteristics of amorphous sugar (sucrose and α-lactose) preparations were analyzed by Imamura et al. (2012), with the purpose of obtaining knowledge that could lead to the prediction of how long the amorphous state is retained under various conditions. The impact of compression, physical aging, and freezing rate on the induction period ($t_{ind}$) for crystallization were examined. Freeze-dried sugar samples were compressed at 74 or 443 MPa (5 min) and then rehumidified at specified RHs. Some freeze-dried sucrose samples were physically aged, and alternatively freeze-drying was conducted under different conditions. The isothermal crystallization of the prepared samples at different temperatures ($T$), the glass transition and the crystallization temperature ($T_{cry}$) were measured, using differential scanning calorimetry. The compression markedly decreased the $t_{ind}$, while significantly lowered the hygroscopicity. Physical aging and slower freezing also shortened the $t_{ind}$. The $t_{ind}$ was found to be correlated exclusively with ($T_{cry}$–$T$), regardless of rehumidification, compression, sugar type, physical aging, and freezing rate in the freeze-drying process.

The crystallization behavior of lactose/sucrose mixtures during water-induced crystallization was studied by Sormoli et al. (2013) to gain more insight about their crystallization during storage. Solutions with different ratios of lactose and sucrose, 75:25 and 50:50, were spray dried to produce amorphous powders. The powders were kept at a controlled temperature and humidity to study their sorption–desorption behavior. X-ray diffraction and light microscopy analysis were performed to study their crystallization behavior. Two-step desorption was observed after sieving the powders as sample preparation. Sieving decreased the crystallization time for lactose/sucrose mixture 75:25 from 22 days to 2.5 days. Based on the x-ray diffraction analysis during this two-step process of water desorption, it was concluded that lactose crystallizes first and more quickly than sucrose. The degree of crystallization for the lactose crystals increases by 89% (relative to their final level of crystallinity), whereas sucrose crystals increase their level of crystallinity by only 28% during the first step of crystallization in the lactose/sucrose (75:25) mixtures. The light microscopy images also suggested that the crystallization of amorphous lactose/sucrose powders during water-induced crystallization may occur as a solution rather than in the solid phase.

In sugar manufacturing, dextrose monohydrate is produced by crystallization of high dextrose equivalent (DE) syrup. At the start of crystallization, the syrup is seeded with massecuite from a previous batch. The massecuite is a slurry containing grown crystals suspended in the syrup at the end of crystallization. During production, the mixing ratio of massecuite to syrup is often varied and hence the concentration of dextrose at the start of crystallization process changes. The increase or decrease in initial dextrose concentration may influence the crystallization rate and yield of the process as well as the final crystal size distribution resulting from nucleation and growth phenomena. In relation to the dextrose crystallization, the amount of crystals produced is very sensitive to variations in the initial dextrose concentration and the temperature profile (Parisi et al., 2007).

A batch-seeded cooling crystallizer was used to study dextrose monohydrate crystallization. Experiments were conducted by Markande et al. (2012) to investigate how a 2% increase in the initial dextrose concentration (from 65.5% to 67.5%) would influence final crystal yield and size. The crystallizations were performed for three different seed masses and cooling profiles, consequently the influence of these parameters was also investigated. The parameters were varied in accordance with an industrial scale process. An in-line focused beam reflectance measurement probe and an in-line process refractometer were used to continuously monitor the crystallizations. The experimental results showed that the 2% increase in initial dextrose concentration had a major influence on the rate of crystallization and yield over a 24 h crystallization period, and only a minor influence on the median crystal size.

Two in-line monitoring probes were inserted by Markande et al. (2013) into a 3.2 L batch cooling crystallizer to monitor the progression of a dextrose monohydrate crystallization and how it was influenced by process variables including impurities, seed size, and initial dextrose concentration. The probes used were a K-patents refractometer and a Lasentec FBRM probe. These probes were applied to continuously monitor dextrose concentration in solution and crystal chord length distribution throughout the duration of the crystallization.

The influence of process variables on crystallizer performance, as determined by final crystal yield and chord size, may not be intuitive and some unexpected results occurred. The continuous in-line measurement data proved very useful for interpreting how variations in the process variables influenced the crystallizer performance. Supersaturation is a key variable in the control of the crystallization process as it influences both nucleation and crystal growth. In-line monitoring of solution dextrose concentration can be applied to continuously evaluate supersaturation and this can then be applied to control supersaturation so as to achieve a desired crystallizer performance.

Paz Suarez et al. (2011) developed feasible, advanced control strategies for the operation of industrial fed-batch multistage sugar crystallization processes. The operation of such processes poses very challenging problems mainly those inherent to its batch nature and also those due to the difficulties in measuring key process variables. Inadequate control policies lead to out-of-spec batches, with consequent losses resulting from the need of product recycling.

To address these problems, a modification of the general nonlinear model predictive control (NMPC) is proposed by Paz Suarez et al. (2011), where the NMPC is executed only when the tracking error is outside a prespecified bound $\alpha$. Once the error converges toward the $\alpha$-strip, the NMPC is switched off and the control action is kept constant. In order to further reduce the complexity of the control system, the proposed modification, termed error tolerant MPC (ETMPC), is provided with a recurrent neural network (RNN) predictive model. The ETMPC + RNN control scheme was extensively tested on a crystallizer dynamic simulator, tuned with data from two industrial units, and compared with the classical NMPC and PI strategy. The results demonstrate that both NMPC and ETMPC controllers lead to improved end-point process specifications, when compared with the PI controller. The explicit introduction of the error tolerance in the optimization relaxes the computational burden and can complement several other suggestions in the literature for feasible industrial real-time control.

Growth rates of sucrose crystallization from pure solutions of initial relative supersaturation levels between 0.094 and 0.181 were studied by Khaddour et al. (2010) in agitated crystallizer at 313.13 K. Birth and spread model was applicable for the obtained growth rate data in this range of supersaturation and was used to estimate the principal growth parameters. The estimated interfacial free energy varied inversely with supersaturation from 0.00842 to 0.00461 J/m$^2$, respectively. The obtained kinetic coefficient changed with the initial supersaturation from $9.45 \times 10^{-5}$ to $2.79 \times 10^{-7}$ m/s. The corresponding radius of the 2D (two dimensional) critical nucleus varied from $7.47 \times 10^{-9}$ to $1.46 \times 10^{-9}$ m. Predominance of surface integration or volume diffusion mechanism during the growth process was assessed using the calculated activation free energies of the 2D nucleation process. An acceptable confirmation of the calculated radius of the critical 2D nucleus was found using atomic force microscopy (AFM) technique. The calculated interfacial free energy between the saturated sucrose solution and the crystal surface was found to be 0.02325 J/m$^2$.

## 4.12 CRYSTALLIZATION UNIT

Thick juice arising after concentration and filtration is led to crystallization for receiving of saccharose.

This is carried out in two stages:

1. Crystallization and mixing equipment where the crystals are formed and
2. Centrifugers where separation of sugar from mother liquor is carried out.

The aim of crystallization is to:

Obtain a high yield in crystals with exhausting the mother liquor.
Create sugar mass that can be easily centrifuged and make sugar of high quality.

Factors affecting crystallization are:

Degree of saturation of mother liquor, viscosity, nonsugars percentage, crystallization speed, and difference in temperature.

## 4.13 CRYSTALLIZATION–FRACTIONATION OF EDIBLE OILS AND FATS

Fractionation is based on the principle that the solubilities of the higher melting components in the liquid phase change at different temperatures. This difference can be extended by using an organic solvent that has the effect of decreasing the viscosity and leading to better washing of the crystals.

The equipment includes tanks for pre-heating, stirred and cooled tanks for crystallization, band or membrane filters for the separation of the crystals from the liquor, and distillation vessels for solvent recovery. The oil is heated to 10°C above the melting point of the highest tricylclycerol present, to give a fully liquid starting material (e.g., the heating point is typically 75°C for palm oil [PO]). The molten oil is then cooled and stirred to form crystal nuclei, and the temperature is maintained at a lower temperature to induce crystal growth (typically 12 h at 28–30°C for PO). If a solvent is used, it is added to the molten oil prior to cooling. The mixtures containing the crystallized solids and the dissolved liquids are separated by filters. If a solvent is used, it is removed from the fractions by distillation (Hyfoma, 2013).

Research in bulk lipids has shown that cooling rate affects crystal formation, which can affect the smoothness or graininess of margarine, the snap and gloss of chocolate, and the spreadability of butter and margarine (Campos et al., 2002; Martini et al., 2002).

Crystallization of lipids in oil-in-water emulsions has been well reviewed by Coupland (2002) and Rousseau (2000). They both stated the importance of lipid crystals in emulsions and how they affect emulsion stability. Lopez et al. (2002) studied the effect of cooling rate on milk fat and cream and found that though crystallization of the lipid state did not change significantly; there was a difference in the melting profiles. These authors showed that the slower the cooling rate the less fractionated the milk-fat fractions.

The effect of cooling rate on the destabilization of oil-in-water (o/w) emulsions was studied as a function of oil content (20% and 40% o/w), homogenization conditions, and crystallization temperatures (10°C, 5°C, 0°C, –5°C, and –10°C) by Tippetts and Martini (2009). The lipid phase was a mixture of anhydrous milk fat (AMF) and soybean oil, and whey protein was used as the emulsifier. Differential scanning calorimetry was used to analyze the crystallization and melting behaviors; while a vertical scan macroscopic analyzer measured the physicochemical stability.

The slow cooling rate increased the stability of emulsions with 20% oil. In addition, slow cooling promoted the onset of crystallization and delayed crystal growth. These effects were more significant in emulsions formulated with 20% oil and formulated under processing conditions that resulted in bigger droplet sizes (~0.9 μm).

Upon crystallization, fats such as cocoa butter form a three-dimensional crystal network with distinct levels of structure. When triglycerides present in the sample are supercooled, they undergo a liquid–solid transformation to form primary crystals with characteristic polymorphism (Chapman, 1962). These crystals then aggregate via a heat- and mass-transfer-limited process into larger crystal aggregates. The aggregation process continues until a continuous three-dimensional network is formed (Heertje, 1993; Marangoni, 2002). The solid fat content (SFC) will also strongly influence the mechanical behavior of a fat (deMan, 1976). Furthermore, the particular processing conditions utilized during crystallization can influence all these levels of structure, and this, in turn, will affect macroscopic mechanical properties and product quality (Herrera and Hartel, 2000).

The mechanical properties and microstructure of cocoa butter crystallized statically at 5°C, 15°C, 20°C, 22°C, and 24°C were characterized in time up to 35 days of storage. In general, after an initial increase, the storage modulus (G′), yield force (Fy), elastic constant (K) and solid fat content (SFC) remained constant at all temperatures, except at times corresponding to polymorphic transformations. At 22°C, however, G′ increased throughout the 35-day storage period. Brunello et al. (2003) suggested that SFC cannot be used as the sole indicator of mechanical strength in cocoa butter, and that polymorphism and microstructure have a profound effect. This effect can be characterized using fractal microstructural analysis.

Milk fat can crystallize in different phases, characterized by a particular composition and molecular arrangement or polymorphic form. The TAG molecules form lamellar structures by stacking in the longitudinal direction, usually of two (2 L) or three (3 L) fatty acid lengths. Typically, the crystals form plate-like structures with molecules lying perpendicular to the flat surface. The more common polymorphic forms, analogous to those formed by pure triacylglycerols, are usually termed $\alpha$, $\beta'$, and $\beta$ in order of increasing melting point, packing density and thermodynamic stability. This polymorphism is a consequence of the variety of arrangements of lateral packing of the CH2 groups of the triacylglycerol molecules in a fat crystal (Sato, 2001).

The study of crystallization of fats under shear flow using synchrotron radiation is very recent (Macmillan et al., 2002; Mazzanti et al., 2006, 2007), though shear flow has been used to crystallize fats for a long time in the industry.

A detailed synchrotron x-ray diffraction study on the kinetics of crystallization of AMF and milk fat triacylglycerols (MFT) was done in a Couette cell at 17°C, 17.5°C, and 20°C under shear rates between 0 and 2880 $s^{-1}$ (Mazzanti et al., 2009).

They observed shear-induced acceleration of the transition from phase α to β′ and the presence of crystalline orientation, but no effect of shear on the onset time of phase a was observed. A two-stage regime was observed for the growth of phase β′. The first stage follows a series–parallel system of differential equations describing the conversion between liquid and crystalline phases.

The second stage follows a diffusion-controlled regime. These mechanisms are consistent with the crystalline orientation, the growth of the crystalline domains and the observed displacement of the diffraction peak positions. The absence of the polar lipids explains the faster kinetics of MFT.

The size and shape of the crystals present within an emulsion-based delivery system will influence its physical stability (e.g., particle aggregation and gravitational separation) (Boode et al., 1993; Boode and Walstra, 1993; Rousseau, 2000), its physicochemical properties (e.g., rheology, appearance, and mouthfeel) (Helgason et al., 2008), and its biological activity (e.g., aggregation state, dissolution and absorption rates in the GIT) (Golding et al., 2011; Kesisoglou et al., 2007).

Blends of AMF with vegetable oils can lower the costs relative to butter while having the preferred taste of butter.

The macroscopic properties and functionalities of fat-based products are closely related to their microscopic properties. Macroscopic properties depend not only on the solid fat content of the system, polymorphism and crystal network are also key factors (Braipson-Danthine and Deroanne, 2004; Marangoni and Narine, 2002).

Kaylegian and Lindsay (1995) presented a review of milk fat fractionation technology. The most common technique is dry fractionation, which involves melting of the fat, controlled cooling, and crystallization, followed by a separation of the crystals from the remaining liquid phase.

Dry fractionation of AMF can produce a range of products with different physical and chemical characteristics, which are interesting for use as food ingredients and suitable for a wide range of applications (De et al., 2007).

The physicochemical properties of fat model systems made of commercial samples of AMF or a fraction blended with PO were studied. Physical properties such as solid fat content, melting curves by differential scanning calorimetry, textural properties, and polymorphism were investigated.

Danthine (2012) systematically mapped interactions (compatibility/incompatibility) that occur in models of compound fat systems with respect to butyric-based shortening or butter-like spreads formulations.

For that purpose, iso-solid diagrams have been constructed from p-NMR data. Molecular interactions have been highlighted for all the blends, especially at low temperatures. Compositions at which molecular interactions were detected depend on the TAGs composition of the fractions involved in the blends. For example, under dynamic conditions, a minimum was observed (eutectic interaction) for all the blends. This minimum was shifted to higher PO content for blends made of AMF fractions with lower iodine value (IV) and to lower PO content for AMF with higher IV. After static crystallization followed by a tempering at 15°C, interactions also existed for all the blends. It was shown that the deviations found in hardness after this tempering procedure can be explained by intersolubility, polymorphic and microstructure arguments.

Results reported here concern physical characteristics of several compound fat blends (butyric–vegetal). A comprehensive analysis of binary fat blends made of AMF, or its fractions, blended with a vegetal fat was conducted. This better understanding of crystallization phenomena occurring is required to further enhance the use of AMF in compound-margarines and shortenings. Indeed, phase properties of fat blends have a significant influence on the sensorial characteristics of the final products (hardness, brittleness, grainy, or smooth texture.

A novel continuous laminar shear structuring crystallizer with a suitable cooling system was designed and built by Maleky and Marangoni (2008). This is a new method to continuously crystallize edible fat in the desirable polymorphic form from the melt while being uniformly sheared.

The machine consists of four main sections: Feed unit, shearing mechanism, cooling system, and power unit. In each of these sections specific design considerations are taken into account which makes the process controllable and continuous. The shearing unit is made of two concentric cylinders. The internal cylinder is stationary and has a cooling system inside for temperature control. The outer cylinder rotates to produce a uniform shear in the sample fluid placed in the 1.5 mm gap between the cylinders.

The sample's feed rate is controlled while it is pumped to the gap. The cooling system has three segments and provides an individual temperature gradient for each region.

Cocoa butter and a binary mixture of cocoa butter and milk fat were crystallized from the melt under shear and different cooling regimes. A major modification in the samples' phase transition behavior was observed; laminar shear induces acceleration of phase transition from less stable polymorphic form to the more stable form, βV. Moreover, x-ray diffraction patterns clearly showed crystalline orientation of the samples. This machine may open up new avenues for the processing and manufacturing of chocolate, shortenings, and margarine.

## 4.14 CRYSTALLIZATION IN THE FOOD INDUSTRY

Scraped surface heat exchangers (SSHEs) are widely used in the food industry for crystallization applications in several food processes, such as the crystallization of margarine (Shahidi, 2005), the tempering of chocolate (Dhonsi and Stapley, 2006), the freeze concentration of milk (Sanchez et al., 2011), and the freezing of sorbet or ice cream (Cook and Hartel, 2010). To a certain degree, all of these food products behave as shear thinning fluids.

The final quality of these food products is mainly governed by sensory properties related to crystal size distribution and apparent viscosity. The crystallization phenomenon in these food processes is closely related to the residence time distribution (RTD) of the product. During the crystallization process, the phase transition involved leads to an increase in the crystal volume fraction, and to an increase in the apparent viscosity of the product. This effect modifies the fluid flow behavior, the RTD of the product, and the temperature profile inside the equipment (Arellano et al., 2013).

However, there is not much information available on the RTD of food products when crystallization occurs under cooling conditions (Belhamri et al., 2009; Russell et al., 1997).

RTD analysis and flow visualization were used to characterize the flow behavior of both a model shear thinning fluid (Carbopol) and a food stream (ice cream) in a commercial scale scraped surface heat exchanger (SSHE). Different flow behavior was observed in the isothermal model system compared to that for ice cream, due to a radial gradient of temperature and therefore viscosity in the ice cream system, which produced more axial dispersion. Nevertheless, the model solution showed similar relative changes in flow behavior with changing operating conditions, thus representing a viable technique for SSHE optimization. Experiments with the model system revealed that the degree of shear thinning is a more important factor in determining the RTD than the magnitude of the apparent viscosity. This is due to the axial velocity profile across the annular gap becoming flatter as the flow behavior index decreases, causing less axial flow dispersion. It was also found that gaps between blades created disturbances in the flow pattern, which contributed to axial dispersion. This effect increased with increasing rotor speed (Russell et al., 1997).

At cooling conditions, during the crystallization of water in ice cream, Belhamri et al. (2009) found that an increase in product flow rate and in rotational speed led to a narrowing of the RTD.

During the freezing of sorbet, the increase in the ice volume fraction leads to an increase in the apparent viscosity of the product. This effect modifies the fluid flow behavior, the RTD and the temperature profile inside the equipment. Arellano et al. (2013) aimed at studying the influence of the operating conditions on the RTD and the axial temperature profile of the product in a SSHE, so as to characterize the product flow behavior.

RTD experiments were carried out in a continuous laboratory pilot-scale SSHE by means of a colorimetric method. Experiments showed that high product flow rates led to a narrowing of the RTD and thus to less axial dispersion, due to the enhancement of the radial mixing with the decrease in the apparent viscosity of the product. Spreading of the RTD was obtained for lower refrigerant fluid temperatures, due to a higher radial temperature gradient between the wall and the center of the exchanger, leading to a higher gradient of the apparent viscosity. This effect increased the difference in axial flow velocities and thus the axial dispersion. These results can be useful for the optimization and modeling of crystallization processes in SSHEs.

The ice crystal size in the product exiting the SSHE is affected by a variety of operating and formulation parameters. Operating variables, such as draw temperature, dasher speed, and throughput rate, affect the mechanisms by which ice crystals form and ripen into the disc-shaped crystals observed exiting the freezer (Marshall et al., 2003).

During processing, nuclei formation must be promoted and ice crystal growth minimized to create many small ice crystals. A small mean ice crystal size, perhaps between 10 and 20 μm, is desired in order to give a creamy mouthfeel, which can increase consumer acceptance of ice cream, especially the reduced fat variety. When ice crystals are larger than about 50 µm, they can be detected in the mouth and an excess of these larger crystals can result in a coarse product (Miller-Livney and Hartel, 1997).

In the SSHE, there is a large subcooling between the refrigerant, at approximately –30°C, and the ice cream mix, initially at 1–2°C (Hartel, 2001). When the mix enters the freezing chamber, the liquid refrigerant removes enough heat to bring the mix below its freezing point.

Nucleation then begins at the barrel wall at a rate determined by the local subcooling (Fennema et al., 1973). Schwartzberg (1990) proposed that dendrites grow out from the barrel wall into the bulk solution due to the large temperature gradient present at the wall surface. Currently, it is thought that the layer at the cylinder wall is probably not fully crystallized, but exists as a slush layer containing small, dendritic crystals (Marshall et al., 2003).

As ice crystals are formed and grow, latent heat is released into the coolant and the bulk ice cream. Latent heat generation is greater when larger ice phase volumes are formed, which requires lower coolant temperatures to maintain constant draw temperature.

During freezing, nuclei formation must be promoted and ice crystal growth and recrystallization minimized to create many small ice crystals. The effects of sweetener type, draw temperature, dasher speed, and throughput rate on ice crystal size distributions during freezing of ice cream were investigated by Drewett and Hartel (2007). These operating variables affect the mechanisms by which ice crystals form and ripen into the disc-shaped crystals observed exiting the freezer. Increasing the throughput rate and reducing the residence time in the freezer allows less time at warmer temperatures where recrystallization occurs rapidly, which leads to smaller ice crystals at the freezer exit. Smaller ice crystals were also found at lower draw temperatures, since coolant temperatures were reduced, increasing the driving force for nucleation. Increased ice crystal size at draw occurred when faster dasher speeds were utilized due to the addition of frictional heat into the system causing the melting of small crystals. Residence time was found to have the most pronounced effect on mean ice crystal size followed by draw temperature and dasher speed, whereas type of sweetener had no significant effect.

Crystallization during ice cream processing has been largely studied, not only in SSHE but in different type of equipments (Adapa et al., 2000; Habib and Farid, 2006; Hartel, 1996; Qin et al., 2006).

During ice cream processing, 3 or 4 phases have to be taken into account: one or two liquid phases (water and oil) a solid phase (ice crystals), and a gas phase. The final texture of the product results from a combination of crystal size and ratio, and bubble distribution due to gas injection. Crystal size and ratio are controlled by heat transfer and mixing. Foaming is not only controlled by gas injection ratio, but also by mixing which creates the foam. Regulation is done on inlet flow rates.

Scraped surface heat exchangers are used in the food industry to process highly viscous fluids, and ice cream in particular, but most of the time, the influence of operating conditions on product quality is poorly understood. Fayolle et al. (2013) developed simple tools to help industrials understand, and then optimize their fabrication process. RTD has been characterized in an industrial pilot system during real ice cream production, after the method had been validated in an experimental set-up with a simple mixture of water and sucrose. It has been shown that in its dimensionless form, RTD depends slightly on flow rate and scraper rotational speed. A simple model of flow pattern applicable to SSHE during crystallization was developed to reproduce the observed RTD. It distinguishes two zones: the volume of fluid near the cooling wall where ice is generated and swept by the blades and the volume of fluid near to the rotor. Therefore, the model considers two parallel plug flow reactors with axial dispersion that exchange fluid by radial mixing. After adjustment of the model parameters, a good agreement was obtained with the experiment results. The flow rate is lower in the zone near the cooling wall; this can be due to a higher ice concentration leading to higher viscosity.

This approach can contribute to better understand, optimize, and control SSHE used for ice cream production.

## 4.15 CHOCOLATE TEMPERING

The tempering of chocolate is an important stage in the manufacture of chocolate. The tempering process produces seed crystals of the cocoa butter polymorph Form V or b, which enables this polymorph to be the dominant form in the subsequent molding step (Talbot, 1994). This is achieved by subjecting the chocolate to a carefully defined shear and temperature history, whereby the chocolate is first melted, then cooled to initiate nucleation of b crystal form seed crystals, and finally warmed by a few degrees to melt out crystals of any lower melting polymorphs that may have formed. The application of shear during the cooling and holding stages appears to be a key factor in the success of tempering processes.

Tempering is a directed pre-crystallization that consists of shearing chocolate mass at controlled temperatures to promote cocoa butter crystallization in a thermodynamically stable polymorphic form.

Tempering has four key steps: melting to completion (at 50°C), cooling to the point of crystallization (at 32°C), crystallization (at 27°C), and conversion of any unstable crystals (at 29–31°C) (Talbot, 1999).

The effect of shear on chocolate or cocoa butter tempering has been studied in a number of different flow geometries; for example, scraped surface heat exchanger with cocoa butter and chocolate (Bolliger et al., 1999), Couette geometry with milk chocolate (Stapley et al., 1999) and cocoa butter (Mazzanti et al., 2003), cone and plate system with cocoa butter (MacMillan et al., 2002), parallel plate viscometer with milk chocolate (Briggs & Wang, 2004), and a helical ribbon device with cocoa butter (Toro-Vazquez et al., 2004).

The influence of shear rate and temperature on the tempering of different mixtures of cocoa butter, sugar, and lecithin has been studied by Dhonsi and Stapley (2006) using a concentric cylinder viscometer as a shearing device and using viscosity measurement to monitor crystallization. Shear rates ranging from 1 to 50 $s^{-1}$ were tested at four different isothermal temperatures (13°C, 17°C, 20°C, and 23°C). Three different material compositions were investigated: plain cocoa butter, a cocoa butter/sugar mixture (44 wt% sugar), and a cocoa butter/sugar/lecithin mixture (44 wt% sugar and 0.2% lecithin). In each case a gradual remelt was performed in the viscometer to obtain an indication of the sample onset melting point. The results for cocoa butter show that at lower temperatures induction times were much shorter, unaffected by shear and generally lead to a lower melting point sample. At 23°C, induction times were shear dependent, with higher shear rates producing higher melting samples, suggestive of higher melting polymorphs. The addition of sugar caused a universal and substantial decrease in induction times, and although tempering at 23°C was still shear dependent, the onset melting point was lower than for the sugar only sample. The addition of lecithin caused a slight delay in the onset of crystallization. The results tend to suggest that sugar crystals provide sites for heterogeneous nucleation, which is slightly weakened by lecithin that coats the sugar surfaces.

The central composite rotatable design (CCRD) for $K = 2$ was used by Afoakwa et al. (2008) to study the combined effects of multistage heat exchangers for stages 1 (14–30°C) and 2 (12–28°C) coolant temperatures at constant stage 3 coolant and holding temperatures during tempering of dark chocolates using a laboratory-scale mini-temperer.

Quantitative data on the chocolate temper index (slope) were obtained for products with varying particle size distribution (PSD) (D90 of 18, 25, 35, and 50 μm) and fat (30% and 35%) content. Regression models generated using stepwise regression analyses were used to plot response surface curves, to study the tempering behavior of products. The results showed that both stage 1 and stage 2 coolant temperatures had significant linear and quadratic effects on the crystallization behavior causing wide variations in chocolate temper index during tempering of products with variable PSD and fat content. Differences in fat content exerted the greatest variability in temperature settings of the different zones for attaining well-tempered products. At 35% fat content, changes in PSD caused only slight and insignificant effect on tempering behavior. No unique set of conditions was found to achieve good temper in dark chocolate with a specified tempering unit. Thus, different combinations of temperatures could be employed between the multistage heat exchangers to induce nucleation and growth of stable fat crystal polymorphs during tempering. Variations in tempering outcomes of the dark chocolates were dependent more on the fat content than PSD.

Tempering consists of shearing chocolate mass at controlled temperatures to promote cocoa butter crystallization in a stable polymorphic form. During industrial processing, multistage heat exchangers are used to control temperature adjustments to promote formation of appropriate stable polymorphic crystals to obtain products with good snap, color, contraction, gloss, and shelf-life characteristics. The process employs varying time–temperature throughputs of the multistage units making it difficult to obtain standard tempering conditions for products with variable particle sizes and fat content, thus prolonging equipment standardization periods with consequential effects on processing times and product quality characteristics. Modeling the tempering behavior of dark chocolates from varying PSD and fat content would enhance our knowledge and understanding on the optimal temperature conditions for obtaining good tempered products during industrial manufacture, with significance for reducing processing (tempering) times and assurances in quality and shelf characteristics.

## 4.16 RECRYSTALLIZATION

Fennema (1973) gives five mechanisms for recrystallization. Iso-mass recrystallization refers to any change in surface or internal structure of an individual crystal of fixed mass as it moves to a lower energy level. Surface iso-mass recrystallization occurs when a crystal of irregular shape and large surface-to-volume ratio assumes a more compact structure with a smaller surface-to-volume ratio and a lower surface energy. Migratory recrystallization (also referred as "grain growth"), is the tendency of large crystals in a polycrystal system to grow at the expense of small crystals, since small crystals have a slightly lower equilibrium melting point in melt crystallization systems, as predicted by the Kelvin equation. Fluctuating temperatures enhance migratory recrystallization. Accretive recrystallization is the process where two crystals that are in a point of contact, join together into one crystal. The point of contact of two equal-sized spheres has a high surface energy. This is a driving force for material to transfer to this area and a neck is formed. Pressure-induced and irruptive recrystallization conditions are not likely to occur in conventional storage of ice cream, so it will not be discussed here. Martino and Zaritzky (1987) give another mechanism, melt-refreeze recrystallization, which occurs when the temperature fluctuates. When the temperature increases, the size of large crystals decreases, but small crystals may melt completely and disappear.

The effect of storage conditions on ice recrystallization in vanilla ice cream was investigated using a simulation program by Ben-Yoseph and Hartel (1998). Changes in mean size and coefficient of variance of the ice-crystal size distribution were determined from data in the literature. Thermal diffusivities during the heating and cooling of ice cream were found experimentally. The unsteady-state heat-transfer equation for a finite cylinder was solved simultaneously with the recrystallization kinetics equation. A good agreement was found between experimental and simulation results. Simulation of typical storage conditions gave final ice-crystal mean size of ~46 μm and coefficient of variance of 0.58. The recrystallization rate was high during the initial stages of storage. Storage temperature fluctuations had a greater influence near the containers' surface, resulting in a higher mean size closer to the surface at the end of the storage. Very low storage temperatures (~−30°C) during initial steps of storage decreased the recrystallization rate significantly. High heat-transfer coefficient of the cooling medium inhibited recrystallization at storage stages and involved initial cooling and low-amplitude temperatures.

## 4.17 CRYSTALLIZATION EQUIPMENT

In a crystallizer, sizing is normally done on the basis of the volume required for crystallization.

The mechanical design of the crystallizer has a significant influence on the nucleation rate due to contact nucleation (caused by contact of the crystals with each other and with the pump impeller, or propeller when suspended in a supersaturated solution). This phenomenon yields varying rates of nucleation in scale up, and differences in the nucleation rates when the same equipment is used with different materials.

In continuous crystallizers, crystal size distribution deviates from the desired distribution due to the presence of external disturbances, such as changes in the concentration of solute in the feed stream, and due to the randomly occurring changes in the operating conditions. In batch processes, batch crystallizer suffers from a changing level of supersaturation during their operation (Markande et al., 2012).

## 4.18 JACKETED BOILING PANS

Jacketed pans have been used considerably in the food and bioindustries for crystallization, evaporation, and simple boiling or heating operations. They are simple, cheap, and easy to clean although they have low heat-transfer coefficients.

In the batch method the "pan" consists of a cylinder with a dished bottom and a domed top from which the vapor was carried off to the condenser through a large gooseneck. The lower part of the vacuum pan was dished for strength, and jacketed to provide the necessary heating surface (Armerding, 1960).

The pan can be closed hermetically by means of a flanged cover and can be operated under reduced pressure by a vacuum pump. This makes low-temperature processing possible.

The pan also has a mechanical stirrer. In the takeoff line from the pan an impingement baffle is fitted and any liquid carried past this point in the vapor is removed in the cyclone separator. Vapor is condensed on the coil condenser.

For more than 20 years, the BMA vertical continuous vacuum pan (VKT) has been in operation in the sugar industry. Equipped with mechanical agitators, the VKT system has the advantage that it can be operated at very low-temperature differences. The equipment for evaporation and crystallization can thus be designed to provide several options for reduced energy consumption (Hempelmann, 2006). One of these is double-effect evaporation: The vapor from one crystallizer or crystallization step is used as heating steam for another crystallizer. Especially with increasing capacities, applications in refineries have become a very interesting feature for modern energy concepts. Hempelmann (2006) described the currently used BMA system with vertical and horizontal installations including control concept and cleaning procedures. Various installations are presented with the main focus on modern energy concepts, including double-effect evaporation.

For decades, the impellers in evaporating crystallizers (vacuum pans) have been installed within the central draft tube of the calandria.

The term "calandria," unique to evaporating equipment, simply means a tube bundle or a tube chest with the product flowing through the tubes and the heating medium around the tubes. The tubes are relatively short and seldom longer than six feet. In the center of the vertical tube bundle, a large downtake pipe was installed, usually larger than necessary, but this was considered good practice. The pattern of the vertical tubes surrounding the down-pipe was dictated by manufacturing convenience. The entire tube nest, or calandria, was installed in the lower part of the vacuum pan (Armerding, 1960).

The axial flow impellers—so called "Kaplan type"—are nowadays the standard impellers in evaporating crystallizers. In comparison with this standard type the performance of radial flow impellers installed below the heating chambers of two evaporating crystallizers at the GosJrawice sugar factory was investigated during the 2004 campaign by Bruhns et al. (2007). The two evaporating crystallizers were operated continuously as the 3rd as well as the 2nd stage of B (raw sugar) and C (after product sugar) cascade crystallizers. They investigated the running of each impeller at three different values of the rotational speed and measured the power consumption of the agitator drive while determining the heat-transfer coefficient in the respective evaporating crystallizer. Moreover, the performance of standard Kaplan impellers installed in the central tubes of the heating chambers of both evaporating crystallizers was also investigated. In relation to Kaplan impellers, the radial flow impellers attained markedly higher heat-transfer coefficients at lower values of the rotational speed. By comparing their power consumption at similar levels of the heat-transfer intensity, a comparison between the two impeller types with respect to the circulating efficiency was also made. In B3 evaporating crystallizer, the circulating efficiency of the radial flow impeller was higher than that of the Kaplan impeller by a factor of 2.8. In C2 evaporating crystallizer, the corresponding factor was 2.6.

According to product suspension crystallizers are classified as follows.

## 4.19 MIXED-SUSPENSION, MIXED-PRODUCT-REMOVAL CRYSTALLIZERS

Crystallizers employing vaporization of solvents or refrigerants are usually preferred. The primary reason for this preference is that heat transferred through the critical supersaturating step is through a boiling-liquid–gas surface, avoiding the troublesome solid deposits that can form on a metal heat-transfer surface (Perry, 1999).

## 4.20 COOLING CRYSTALLIZERS: *NON-AGITATED VESSELS*

The simplest type of cooling crystallizer is the unstirred tank: a hot feedstock solution is charged to the open vessel where it is allowed to cool, often for several days, by natural convection. Metallic rods may be suspended in the solution so that large crystals can grow on them and reduce the amount of product that sinks to the bottom of the crystallizer. The product is removed by hand.

Because cooling is slow, large interlocked crystals are usually obtained and retention of the mother liquor is unavoidable. As a result, the dried crystals are generally impure. Because of the uncontrolled nature of the process, product crystals range from a fine dust to large agglomerates.

Labor costs are generally high, but the method is economical for small batches; however, productivity of this type of equipment is low and space requirements are high.

## 4.21 AGITATED VESSELS

Installation of an agitator in an open-tank crystallizer generally results in smaller, more uniform crystals and reduced batch time. The final product tends to have a higher purity because less mother liquor is retained by the crystals after filtration and more efficient washing is possible. Water jackets are usually preferred to coils for cooling because the latter often become encrusted with crystals and cease to operate efficiently. Where possible, the internal surfaces of the crystallizer should be smooth and flat to suppress encrustation.

An agitated cooler is more expensive to operate than a simple tank crystallizer, but it has a much higher productivity. The design of tank crystallizers varies from shallow pans to large cylindrical tanks.

The large agitated cooling crystallizer has an upper conical section, which slows down the upward velocity of liquor and prevents the crystalline product from being swept out with the spent liquor. An agitator located in the lower region of a draft tube circulates the crystal slurry (magma) through the growth zone of the crystallizer. If required, cooling surfaces may be provided inside the crystallizer.

Use of external circulation allows good mixing inside the crystallizer and high rates of heat transfer between the liquor and coolant. An internal agitator may be installed in the crystallization tank if needed (Mullin, 2003).

## 4.22 DIRECT-CONTACT COOLING

The use of a conventional heat exchanger and the problems caused by crystal encrustation can be avoided by employing direct-contact cooling (DCC) in which supersaturation is achieved by allowing the process liquor to come into contact with a cold heat-transfer medium. Other potential advantages of DCC over conventional indirect-contact methods include better heat transfer and smaller cooling load. However, problems are also associated with DCC crystallization; these include product contamination from the coolant and the cost of extra processing required to recover the coolant for further use.

A solid, liquid, or gaseous coolant can be used; heat exchange may occur via the transfer of sensible or latent heat. The coolant may or may not boil during the operation, and it can be miscible or immiscible with the process liquor. Thus, four basic types of DCC crystallization are possible:

1. Immiscible, boiling, solid, or liquid coolant: heat is removed mainly by transfer of latent heat of sublimation or vaporization.
2. Immiscible, nonboiling, solid, liquid, or gaseous coolant: mainly sensible heat transfer.
3. Miscible, boiling, liquid coolant: mainly latent heat transfer.
4. Miscible, nonboiling, liquid coolant: mainly sensible heat transfer.

Crystallization processes employing DCC have been used successfully in the dewaxing of lubricating oils, the desalination of water, and the production of inorganic salts from aqueous solution (Mullin, 2003).

## 4.23 FORCED-CIRCULATION EVAPORATOR-CRYSTALLIZER (SEE SWENSON EQUIPMENT)

Forced circulation is the most widely practiced method of crystallization.

For feeds where high rates of evaporation are required, where there are scaling compounds, where crystallization is achieved in inverted solubility solutions, or where the solution is of relatively high viscosity, the *forced-circulation* crystallizer is the best choice. Its most frequent use is in the continuous processing of such materials as sodium chloride, sodium sulfate, sodium carbonate monohydrate, citric acid, monosodium glutamate, urea, and other similar crystalline materials (Swenson Crystallization Equipment Handout, 2013).

Slurry leaving the body is pumped through a circulating pipe and through a tube-and-shell heat exchanger, where its temperature increases by about 2–6°C. Since this heating is done without vaporization, materials of normal solubility should produce no deposition on the tubes. The heated slurry, returned to the body by a recirculation line, mixes with the body slurry and raises its temperature locally near the point of entry, which causes boiling at the liquid surface.

The crystal magma is circulated from the lower conical section of the evaporator body through the vertical tubular heat exchanger and reintroduced tangentially into the evaporator below the liquor level to create a swirling action and prevent flashing (sudden evaporation) (Mullin, 2003).

During the consequent cooling and vaporization to achieve equilibrium between liquid and vapor, the supersaturation that is created causes deposits on the swirling body of suspended crystals until they again leave via the circulating pipe.

If the crystallizer is not of the evaporative type but relies only on *adiabatic evaporative cooling* to achieve the yield, the heating element is omitted (Perry, 1999).

The effect of the dynamic regulation of vacuum pressure was analyzed by Bolaños-Reynoso et al. (2008) in an adiabatic batch crystallizer using data and image acquisition, virtual instruments, and supervisor control to obtain a controlled vacuum pressure, confirming that the process follows programmed routes in the central computer to obtain both programmed vacuum pressure and cooling dynamic profiles inside the crystallizer (direct design approach). The dynamical regulation profile of vacuum pressure called "adiabatic natural cooling" originated an abrupt cooling effect by adiabatic evaporation inside of the crystallizer (e.g., natural cooling in batch crystallizers operated to atmospheric pressure), benefiting the seeded crystal growth from cane sugar until it reached a size of 732.7 μm (% volume). In turn, the increase in crystal growth, formed mass, and the available supersaturation depletion in the system, contributed to final molasses reduction. From a process dynamics viewpoint, the use of dynamic regulation profiles of vacuum pressure reduces batch time and energy consumption (steam) in the process operation.

For applications where the solution's boiling point elevation is low, such as for sodium carbonate, it is possible to significantly reduce the cost of the operation by vapor recompression.

In this technique the vapor, which is removed to precipitate the crystalline product, is compressed by a positive displacement or centrifugal compressor to a pressure high enough so that it may be used as steam in the heat exchanger.

By proper heat exchange, many mechanical recompression applications can operate efficiently without any external source of heat, except for start-up purposes. For other applications, a small quantity of steam is required at the heating element to maintain the system on a continuous basis.

The recompression crystallization technique can also be applied to the DTB or SWENSON OSLO (growth type) crystallizers.

## 4.24 FLUIDIZED-BED CRYSTALLIZERS

In an *Oslo fluidized-bed crystallizer,* a bed of crystals is suspended in the vessel by the upward flow of supersaturated liquor in the annular region surrounding a central downcomer. Fluidized-bed Oslo units are now frequently operated in a mixed-suspension mode to improve productivity, although this reduces product crystal size. A cooling-type Oslo crystallizer operates in the classifying mode as follows. The hot, concentrated feed solution is fed into the vessel at a point directly above the inlet to the circulation pipe. Saturated solution from the upper regions of the crystallizer, together with the small amount of feedstock, is circulated through the tubes of the heat exchanger, which is cooled by forced circulation of water or brine. On cooling, the solution becomes supersaturated, but not enough for spontaneous nucleation to occur; great care, in fact, is taken to prevent this. Product crystal magma is removed from the lower regions of the vessel (Mullin, 2003).

## 4.25 SURFACE-COOLED CRYSTALLIZER

The surface-cooled unit is a type of forced-circulation crystallizer consisting of a shell and tube heat exchanger through which is pumped the slurry of growing crystals, a crystallizer body to provide retention time, and a recirculation pump and piping. Within the crystallizer body is a baffle designed to keep excessively fine crystals separated from the growing magma for size and slurry density control purposes.

Surface-cooled crystallizers are used where the solution's boiling point elevation is extremely high, as in the case of caustic solutions, or when the temperature level is so low evaporation by vacuum is impossible.

## 4.26 DRAFT-TUBE-BAFFLE EVAPORATOR-CRYSTALLIZER

Because mechanical circulation greatly influences the level of nucleation within the crystallizer, a number of designs have been developed that use circulators located within the body of the crystallizer, thereby reducing the head against which the circulator must pump. This technique reduces the power input and circulator tip speed and therefore the rate of nucleation. A typical example is the draft-tube-baffle (DTB) evaporator-crystallizer (Swenson Process Equipment, Inc. <location of company; also please confirm company name.>).

Where excess nucleation makes it difficult to achieve a crystal size in the range of 10–30 mesh, the DTB crystallizer is preferred (U.S. Patents 3,873,275 and 3,961.904). This crystallizer is built in both the adiabatic cooling and evaporative type, and includes a baffle section surrounding a suspended magma of growing crystals from which a stream of mother liquor is removed containing excess fine crystals (Figure 4.7). These fines can be destroyed by adding heat (as in an evaporative crystallizer) or by adding water or unsaturated feed solution. The magma is suspended by means of a large slow-moving propeller circulator, which fluidizes the suspension

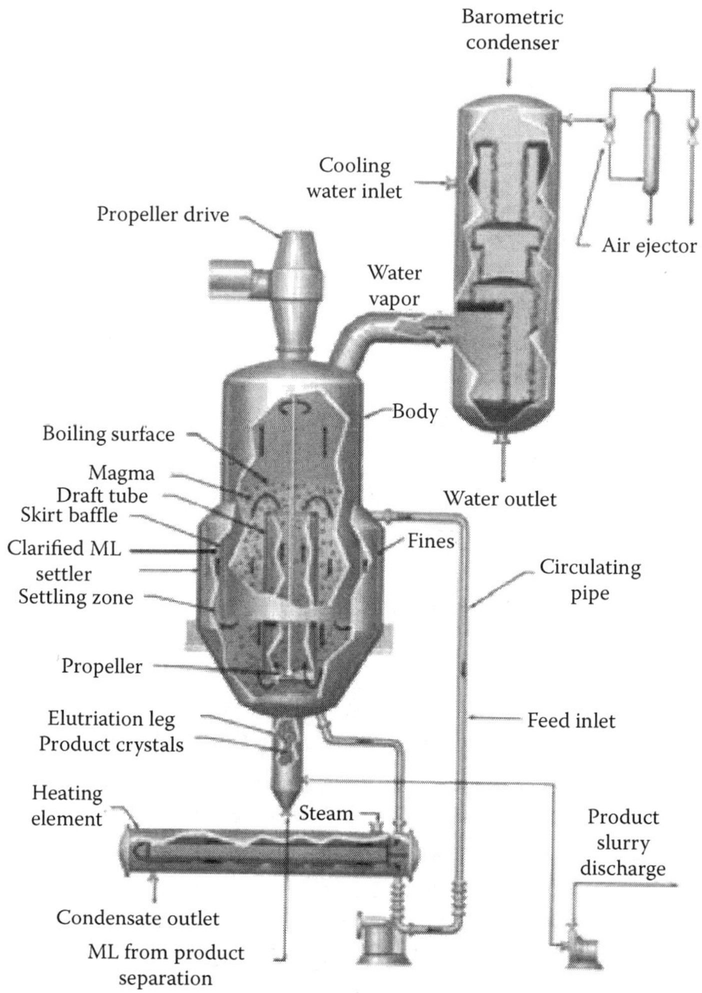


**FIGURE 4.7** Draft tube baffle crystallizer. (Courtesy of Swenson.)

and maintains relatively uniform growth zone conditions. This crystallizer design has proven very useful for producing such materials as ammonium sulfate, potassium chloride, diammonium phosphate, hypo, Epsom salts, potassium sulfate, monosodium glutamate, borax, sodium carbonate decahydrate, trisodium phosphate, urea, XP soda, and so on (Swenson Crystallization Equipment Handout, 2013).

The suspension of product crystals in maintained by a large, slow-moving propeller surrounded by a draft tube within the body. The propeller directs the slurry to the liquid surface so as to prevent solids from short-circuiting the zone of the most intense supersaturation.

The settling zone surrounds the crystallizer body, the circulating pump, and the heating element. The heating element supplies sufficient heat to meet the evaporation requirements and to raise the temperature of the solution removed from the settler so as to destroy any small crystalline particles withdrawn. Coarse crystals are separated from the fines in the settling zone by gravitational sedimentation.

This type of equipment can also be used for applications in which the only heat removed is that required for adiabatic cooling of the incoming feed solution (Perry, 1999).

When crystallization occurs at such a low temperature that it is impractical to use surface cooling or when the rapid crystallization of solids on the tube walls would foul a conventional surface-cooled crystallizer, a DTB crystallizer (or a forced-circulation unit) utilizing the direct contact refrigeration technique can be used. In this operation, a refrigerant is mixed with the circulating magma within the crystallizer body where it absorbs heat and is vaporized. Refrigerant vapor leaves the surface of the crystallizer similar to water vapor in a conventional evaporative crystallizer.

Reactive crystallization, where a solid phase crystalline material results from the reaction of two components, can often be performed more profitably in a crystallizer than in a separate reactor.

An example of reactive crystallization is the production of ammonium sulfate from liquid or gaseous ammonia and concentrated sulfuric acid.

The DTB crystallizer is particularly suited for reactive crystallization. The reactants are mixed in the draft tube of the DTB unit where a large volume of slurry is mixed continuously with the materials to minimize the driving force (supersaturation) created by the reaction. Removal of the heat produced by the reaction is accomplished by vaporizing water or other solvents as in a conventional evaporative-type crystallizer (Swenson, 2013).

The *Standard–Messo turbulence crystallizer* is another successful draft-tube vacuum unit. Two liquor flow circuits are created by concentric pipes: an outer ejector tube with a circumferential slot and an inner guide tube. Circulation is effected by a variable-speed agitator in the guide tube. The principle of the Oslo crystallizer is utilized in the growth zone; partial classification occurs in the lower regions, and fine crystals segregate in the upper regions. The primary circuit is created by a fast upward flow of liquor in the guide tube and a downward flow in the annulus; liquor is thus drawn through the slot between the ejector tube and the baffle, and a secondary flow circuit is formed in the lower region of the vessel. Feedstock is introduced into the guide tube and passes into the vaporizer section where flash evaporation takes place. Nucleation, therefore, occurs in this region, and the nuclei are swept into the primary circuit. Mother liquor can be drawn off via a control valve, thus providing a means of controlling crystal slurry density (Mullin, 2003).

The *Escher–Wyss Tsukishima double-propeller (DP) crystallizer* is essentially a draft-tube agitated crystallizer with some novel features. The DP unit contains an annular baffled zone and a double-propeller agitator, which maintains a steady upward flow inside the draft tube and a downward flow in the annular region, as well as a heat exchanger and an elutriating leg. Very stable suspension characteristics are claimed.

## 4.27 BATCH VACUUM CRYSTALLIZERS

For special cases requiring very low operating temperatures achieved only by very high vacuum, and for those applications involving relatively small amounts of material—or when the material being processed must be handled on less than a continuous basis—it is often both convenient and economical to use a Swenson batch vacuum crystallizer.

Cycles on batch equipment range from 2 to 8 h. At the conclusion of the cycle, the material is deposited in an agitated tank from which it is removed on either a batch or continuous basis for separation and drying. The entire cycle for such equipment may be automated.

Where the material is cooled through a very wide range and/or to a final temperature that requires a very high vacuum, a large ejector or booster is utilized to compress the vapor to a pressure high enough for condensation with available cooling water (Swenson, 2013).

## 4.28 REACTION-TYPE CRYSTALLIZERS

This type of equipment is used in chemical reactions in which the end product is a solid-phase material such as a crystal or an amorphous solid. By mixing the reactants in a large circulated stream of mother liquor containing suspended solids of the equilibrium phase, it is possible to minimize the driving force created during their reaction and remove the heat of reaction through the vaporization of a solvent, normally water.

## 4.29 CLASSIFIED-SUSPENSION CRYSTALLIZER

This equipment is also known as the growth or Oslo crystallizer and is characterized by the production of supersaturation in a circulating stream of liquor. Supersaturation is developed in one part of the system by evaporative cooling or by cooling in a heat exchanger, and it is relieved by passing the liquor through a fluidized bed of crystals. The fluidized bed may be contained in a simple tank or in a more sophisticated vessel.

In an Oslo surface-cooled crystallizer supersaturation develops in the circulated liquor by chilling in the cooler. This supersaturated liquor is contacted with the suspension of crystals in the suspension chamber. At the top of the suspension chamber a stream of mother liquor can be removed to be used for fines removal and destruction (Perry, 1999).

## 4.30 ULTRASOUNDS AND CRYSTALLIZATION

High-powered ultrasound can assist the crystallization process in several ways: influence the initiation of crystal nucleation, control the rate of crystal growth, ensure the formation of small and even-sized crystals, and prevent fouling of surfaces by the newly formed crystals (Luque de Castro and Priego-Capote, 2007; Virone et al., 2006).

Ultrasonic crystallization technology can be applied to foods where it can be used to control the size and rate of development of ice crystals in frozen foods (Chow et al., 2003). As food is frozen, small crystals form within the matrix. Freezing using ultrasonics ensures rapid and even nucleation, short dwell times, and the formation of small, evenly sized crystals, greatly reducing cellular damage and preserving product integrity, even on thawing (Zheng and Sun, 2006).

Owing to significant technical advances in the last 5–10 years, high-power ultrasonics has become an alternative to many conventional food processing steps, such as homogenization, milling, high shear mixing, pasteurization, and solid/liquid separation. Also, it has been shown to improve the efficiency of traditional processes such as filtration/screening, extraction, crystallization, and fermentation (i.e., as an add-on technology). The use of ultrasonics is often driven by economic benefits, yet in some cases a unique product functionality can be achieved. Patist and Bates (2008) presented several examples of commercial installations of this technology in the food industry and highlighted some of the challenges in scale up and development. High-power ultrasound has only recently (<10 years) become an efficient tool for large-scale commercial applications, such as emulsification, homogenization, extraction, crystallization, dewatering, low-temperature pasteurization, degassing, defoaming, activation and inactivation of enzymes, particle size reduction, and viscosity alteration. This can be attributed to improved equipment design and higher efficiencies of large-scale continuous flow-through systems. Like most innovative food processing technologies, high-power ultrasonics is not an off-the-shelf technology and, therefore, needs to be developed and scaled up for each application. The authors presented examples of ultrasonic applications that have made it to commercialization and to share some key learnings involving scale up of an innovative food technology in general.

## REFERENCES


Abels, C., Carstensen, F., and Wessling, M. 2013. Membrane processes in biorefinery applications. *Journal of Membrane Science* 444, 285–317.

Adapa, S., Schmidt, K. A., Jeon, I. J., Herald, T. J., Flores, R. A., 2000. Mechanisms of ice crystallization and recrystallisation in ice cream: A review. *Food Reviews International* 16, 259–271.

Afoakwa, E. O., Paterson, A., Fowler, M., and Vieira, J. 2008. Modelling tempering behaviour of dark chocolates from varying particle size distribution and fat content using response surface methodology. *Innovative Food Science and Emerging Technologies* 9, 527–533.

Agrawal, S. G., Balandier, A., Paterson, A. H. J., and Jones, J. R. 2011. Study on lactose attrition inside the mixing cell of a laser diffraction particle sizer using a novel attrition index. *Powder Technology* 208, 669–675.

Akbari, Z., Ashtiani, F. Z., and Ghomashchi, T. 2012. Effect of whey pretreatments on lactose recovery. *International Journal of Food Engineering* 8(3), 21.

Arellano, M., Benkhelifa, H., Alvarez, G., and Flick, D. 2013. Experimental study and modelling of the residence time distribution in a scraped surface heat exchanger during sorbet freezing. *Journal of Food Engineering* 117, 14–25.

Armerding, G. D. 1960. Evaporation methods as applied to the food industry. *Advances in Food Research* 15, 303–358.

Bhandari, B. R. and R. W. Hartel. 2002. Co-crystallization of Sucrose at High Concentration in the Presence of Glucose and Fructose. *Journal of Food Science* 67(5), 1797–1802.

Bhandari, B.R., Datta, N., Crooks, R., Howes, T., and Rigby, S., 1997a. A semiempirical approach to optimise the quantity required to spray dry sugar-rich foods. *Drying Technology* 15(10), 2509–2525.

Bhandari, B.R., Datta, N., and Howes, T. 1997b. Problems associated with spray drying of sugar-rich foods. *Drying Technology* 15(2), 671–684.

Belhamri, R., Fayolle, F., and Flick, D. 2009. Simplified flow pattern model in SSHE during crystallisation process. In: *8th World Congress of Chemical Engineering Montreal*, Quebec, Canada.

Ben-Yoseph, E. and R. W. Hartel. 1998. Computer simulation of ice recrystallization in ice cream during storage. *Journal of Food Engineering* 38, 309–329.

Bolaños-Reynoso, E., Xaca-Xaca, O., Alvarez-Ramirez, J., and Lopez-Zamora, L. 2008. Effect analysis from dynamic regulation of vacuum pressure in an adiabatic batch crystallizer using data and image acquisition. *Industrial and Engineering Chemistry Research* 47(23), 9426–9436.

Bolliger, S., Zeng, Y., and Windhab, E. J. 1999. In-line measurement of tempered cocoa butter and chocolate by means of near-infrared spectroscopy. *Journal of the American Oil Chemists Society* 76(6), 659–667.

Boode, K. and Walstra, P. 1993. Partial coalescence in oil-in-water emulsions.1. Nature of the aggregation. *Colloids and Surfaces A: Physiochemical and Engineering Aspects* 81, 121–137.

Boode, K., Walstra, P. and deGroot-Mostert, A. E. A. 1993. Partial coalescence in oil-in-water emulsions. 2. Influence of the properties of the fat. *Colloids and Surfaces A: Physicochemical and Engineering Aspects* 81, 139–151.

Braipson-Danthine, S. and Deroanne, C. 2004. Influence of SFC, microstructure and polymorphism on texture (hardness) of binary blends of fats involved in the preparation of industrial shortenings. *Food Research International* 37, 941–948.

Briggs, J. L. and Wang, T. 2004. Influence of shearing and time on the rheological properties of milk chocolate during tempering. *Journal of the American Oil Chemists Society* 81(2), 117–121.

Bruhns, M., Grabowski, M., Lindner, J. P., Pajonk, P., Urbaniec, K., Wegrzynowski, Z., Wolf, K. 2007. Improvement in the circulating efficiency by radial flow impellers in evaporating crystallizers operated in horizontal cascade crystallizers. *Zuckerindustrie* 132(4), 241–246.

Brunello, N., McGauley, S. E., and Marangoni, A. 2003. Mechanical properties of cocoa butter in relation to its crystallization behavior and microstructure. *Lebensm.-Wiss. u.-Technol.* 36, 525–532.

Buera, M. D. P., Schebor, C., and Elizalde, B. 2005. Effects of carbohydrate crystallization on stability of dehydrated foods and ingredient formulations. *Journal of Food Engineering* 67(1–2), 157–165.

Campos, R., Narine, S. S. and Marangoni, A. G. 2002. Effect of cooling rate on the structure and mechanical properties of milk fat and lard. *Food Research International* 35, 971–981.

Chapman, D. 1962. The polymorphism of glycerides. *Chemical Review* 62, 433–456.

Chen, A., Zhu, J., Chen, K., Wu, B., Ji, L., and Wu, Y. 2013. Melt suspension crystallization for purification of phosphoric acid. *Asia-Pacific Journal of Chemical Engineering* 8(3), 354–361.

Chow, R., Blindt, R., Chivers, R., and Povey, M. 2003. The sonocrystallization of ice in sucrose solutions: Primary and secondary nucleation. *Ultrasonics* 41, 595–604.

Cook, K. L. K. and Hartel, R. W. 2010. Mechanisms of ice crystallization in ice cream production. *Comprehensive Reviews in Food Science and Food Safety* 9(2), 213–222.

Coupland, J. N. 2002. Crystallization in emulsions. *Current Opinion in Colloid and Interface Science* 7, 445–450.

Danthine, S. 2012. Physicochemical and structural properties of compound dairy fat blends. *Food Research International* 48, 187–195.

Das, D. and Langrish, T. A. G. 2012. Combined crystallization and drying in a pilot-scale spray dryer. *Drying Technology* 30(9), 998–1007.

De, B. K., Hakimji, M., Patel, A., Sharma, D., Desai, H., and Kumar, T. 2007. Plastic fats and margarines through fractionation, blending and interesterification of milk fat. *European Journal of Lipid Science and Technology*, 109, 32–37.

DeMan, J. M. 1976. Texture of fats and fat products. In J. M. deMan, P. W. Voisey, V. F. Rasper, and D. W. Stanley (Eds.), *Rheology and Texture in Food Quality*. Westport, CT: AVI Press, pp. 355–381.

Dhonsi, D. and Stapley, A. G. F. 2006. The effect of shear rate, temperature, sugar and emulsifier on the tempering of cocoa butter. *Journal of Food Engineering* 77(4), 936–942.

Dombrowski, R. D., Litster, J. D., Wagner, N. J., and He, Y. 2007. Crystallization of alpha-lactose monohydrate in a drop-based microfluidic crystallizer. *Chemical Engineering Science* 62, 4802–4810.

Drewett, E. M. and Hartel, R. W. 2007. Ice crystallization in a scraped surface freezer. *Journal of Food Engineering* 78, 1060–1066.

Earle, R. L. and Earle, M. D. 2004. *Unit Operations in Food Processing*. Web Edition. New Zealand: The New Zealand Institute of Food Science & Technology (Inc.).

Eggleston, G., Cote, G., and Santee, C. 2011. New insights on the hard-to-boil massecuite phenomenon in raw sugar manufacture. *Food Chemistry* 126, 21–30.

Eggleston, G. J. W. Tiu Yen, Alexander, C., and Gober, J. 2012. Measurement and analysis of the mannitol partition coefficient in sucrose crystallization under simulated industrial conditions. *Carbohydrate Research* 355, 69–78.

Fayolle, F., Belhamri, R., and Flick, D. 2013. Residence time distribution measurements and simulation of the flow pattern in a scraped surface heat exchanger during crystallisation of ice cream. *Journal of Food Engineering* 116, 390–397.

Fennema, O. R. 1973. Nature of freezing process. In: O. R. Fennema, W. D. Powrie, E. H. Marth (Eds.), *Low-Temperature Preservation of Food and Living Matter*. New York: Marcel Dekker, Inc., pp. 202–222.

Fennema, O. R., Powrie, W. D., and Marth, E. H. 1973. *Low-Temperature Preservation of Foods and Living Matter*. New York: Marcel Dekker.

Fox, P. F. and McSweeney, P. L. H. 1998. *Dairy Chemistry and Biochemistry*. London: Blackie Academic & Professional.

Gabarra, P. and Hartel, R. W. 1998. Corn syrup solids and their saccharide fractions affect crystallization of amorphous sucrose. *Journal of Food Science* 63(3), 523–528.

Golding, M., Wooster, T. J., Day, L., Xu, M., Lundin, L., Keogh, J., Clifton, P. 2011. Impact of gastric structuring on the lipolysis of emulsified lipids. *Soft Matter* 7, 3513–3523.

Griffiths, R., Paramo, G., and Merson, R. 1982. Preliminary investigation of lactose crystallization using the population balance technique. *AIChE Symposium Series* 78, 118–128.

Habib, B. and Farid, M. 2006. Heat transfer and operating conditions for freeze concentration in a liquid–solid fluidized bed heat exchanger. *Chemical Engineering and Processing* 45, 698–710.

Hartel, R. W. 1996. Ice crystallization during the manufacture of ice cream. *Trends in Food Science and Technology*, 7, 315–321.

Hartel, R. W. 2001. *Crystallization in Foods*. Gaithersburg, Maryland, USA: Aspen Publishers.

Hartel, R.W. and Shastry, A.V. 1991. Sugar crystallization in food products. *Critical Reviews in Food Science and Nutrition* 30, 49–112.

Heertje, I. 1993. Microstructural studies in fat research. *Food Structure*, 12, 77–94.

Helgason, T., Awad, T. S., Kristbergsson, K., McClements, D. J., and Weiss, J. 2008. Influence of polymorphic transformations on gelation of tripalmitin solid lipid nanoparticle suspensions. *Journal of the American Oil Chemists Society* 85, 501–511.

Hempelmann, R. 2006. The VKT continuous vacuum pan—More than 20 years of experience. *International Sugar Journal* 108(1296), 704–710.

Herrera, M. L. and Hartel, R. W. 2000. Effect of processing conditions on physical properties of a milk fat model system: Rheology. *Journal of the American Oil Chemists Society*, 77, 1189–1196.

Hogan, S. A. and O'Callaghan, D. J. 2013. Moisture sorption and stickiness behaviour of hydrolysed whey protein/lactose powders. *Dairy Science and Technology* 93(4–5), 505–521.

Hunziker, O. F. 1926. *Condensed Milk and Milk Powder*, 4th edn. La Grange, Illinois.

Hyfoma. 2013. European network for hygienic manufacturing of foods. Education material on technology-separation techniques and crystallization. Fractionation of edible oils and fats. http://www.hyfoma.com/en/content/processing-technology/separation-techniques/crystallization/. Accessed 09.08.2013.

Hynd, J. 1980. Drying of whey. *International Journal of Dairy Technology* 33(2), 52–54.

Imamura, K., K. Kinugawa, R. Kagotani, M. Nomura, and K. Nakanishi. 2012. Impact of compression, physical aging, and freezing rate on the crystallization characteristics of an amorphous sugar matrix. *Journal of Food Engineering* 112, 313–318.

Kashchiev D. and van Rosmalen, G. M. 2003. Review: Nucleation in solutions revisited. *Cryst Res Technol* 38, 555–574.

Kaylegian, K. and Lindsay, R. C. 1995. In K. E. Kaylegian, and R. C. Lindsay (Eds.), *Handbook of Milk Fat Fractionation Technology*. Champaign, IL, USA: AOCS Press, pp. 39–508.

Kesisoglou, F., Panmai, S., and Wu, Y. H. 2007. Nanosizing—Oral formulation development and biopharmaceutical evaluation. *Adv Drug Deliv Rev* 59, 631–644.

Khaddour, I. A., Bento, L. S. M., Ferreira, A. M. A., and Rocha, F. A. N. 2010. Kinetics and thermodynamics of sucrose crystallization from pure solution at different initial supersaturations. *Surface Science* 604, 1208–1214.

Kubota, N., Doki, N., Yokota, M., and Sato, A. 2001. Sccding policy in batch cooling crystallization. *Powder Technology* 121(1), 31–38.

Levenson, D. A. and Hartel, R. W. 2005. Nucleation of amorphous sucrose–corn syrup mixtures. *Journal of Food Engineering* 69, 9–15.

Liang, B., Shi, Y., and Hartel, R. W. 1991. Growth-rate dispersion effects on lactose crystal size distributions from a continuous cooling crystallizer. *Journal of Food Science* 56, 848–854.

Lindfors, L., Forssen, S., Westergren, J., and Olsson, U. 2008. Nucleation and crystal growth in supersaturated solutions of a model drug. *Journal of Colloid Interface Science* 325, 404–13.

Lopez, C., Bourgaux, C., Lesieur, P., Bernadou, S., Keller, G., and Ollivon, M. 2002. Thermal and structural behavior of milk fat 3. Influence of cooling rate and droplet size on cream crystallization. *Journal of Colloid and Interface Science* 254, 64–78.

Luque de Castro, M. D., and Priego-Capote, F. 2007. Ultrasound assisted crystallization (sonocrystallization). *Ultrasonics Sonochemistry* 14, 717–724.

MacMillan, S. D., Roberts, K. J., Rossi, A., Wells, M. A., Polgreen, M. C., and Smith, I. H. 2002. *In situ* small angle X-ray scattering (SAXS) studies of polymorphism with the associated crystallization of cocoa butter fat using shearing conditions. *Crystal Growth and Design* 2(3), 221–226.

Maleky, F. and Marangoni, A. G. 2008. Process development for continuous crystallization of fat under laminar shear. *Journal of Food Engineering* 89, 399–407.

Marangoni, A. G. 2002. The nature of fractality in fat crystal networks. *Trends in Food Science and Technology* 13, 37–47.

Marangoni, A. G. and Narine, S. S. 2002. Identifying key structural indicators of mechanical strength in networks of fat crystals. *Food Research International*, 35, 957–969.

Markande, A., Fitzpatrick, J., Nezzal, A., Aertsc, L., and Redl, A. 2012. Effect of initial dextrose concentration, seeding and cooling profile on the crystallization of dextrose monohydrate. *Food and Bioproducts Processing* 90, 406–412.

Markande, A., Fitzpatrick, J., Nezzal, A., Aerts, L., and Redl, A. 2013. Application of in-line monitoring for aiding interpretation and control of dextrose monohydrate crystallization. *Journal of Food Engineering* 114, 8–13.

Marshall, R. T., Goff, H. D., and Hartel, R. W. 2003. *Ice Cream* (6th ed.). New York: Kluwer Academic/Plenum Publishers.

Martini, S., Herrera, M. L., and Hartel, R. W. 2002. Effect of cooling rate on crystallization behavior of milk fat fraction/sunflower oil blends. *Journal of American Oil Chemists Society* 79, 1055–1062.

Martino, M. and Zaritzky, N. 1987. Effect of temperature on recrystallization in polycrystalline ice. *Sci. Aliments* 7(l), 147–166.

Mazzanti, G., Guthrie, S. E., Marangoni, A. G., and Idziak, S. H. J. 2006. Synchrotron advances at the frontiers of food physics: Studies of edible fats such as chocolate under shear. *Physics in Canada*, 5(1), 313–320.

Mazzanti, G., Guthrie, S. E., Marangoni, A., and Idziak, S. H. J. 2007. A conceptual model for shear-induced phase behavior in crystallizing cocoa butter. *Crystal Growth and Design*, 7(7), 1230–1241.

Mazzanti, G., Guthrie, S. E., Sirota, E. B., Marangoni, A. G., and Idziak, S. H. J. 2003. Orientation and phase transitions of fat crystals under shear. *Crystal Growth and Design* 3(5), 721–725.

Mazzanti, G., Marangoni, A. G., and Idziak, S. H. J. 2009. Synchrotron study on crystallization kinetics of milk fat under shear flow. *Food Research International* 42, 682–694.

McClements, D. J. 2012. Crystals and crystallization in oil-in-water emulsions: Implications for emulsion-based delivery systems. *Advances in Colloid and Interface Science* 174, 1–30.

Mersmann, A., Sangl, R., Kind, M., Pohlisch, J. 1988. Attrition and secondary nucleation in crystallizers. *Chemical Engineering Technology* 11, 80–88.

Miller-Livney, T. and Hartel, R. W. 1997. Ice recrystallization in ice cream: Interactions between sweeteners and stabilizers. *Journal of Dairy Science* 80, 447–456.

Mimouni, A., Schuck, P., and Bouhallab, S. 2009. Isothermal batch crystallization of alpha-lactose: A kinetic model combining mutarotation, nucleation and growth steps. *International Dairy Journal* 19, 129–136.

Mullin, J. W. 1993. *Crystallization* (3rd ed.). London: Butterworth.

Mullin, J. W. 2003. *Crystallization and Precipitation.* In: *Ullman's Encyclopedia of Industrial Chemistry*. Wiley-VCH. Verlag GmbH, Weinheim, Germany. DOI:10.1002/14356007.b02_03.

Nijdam, J., Ibach, A., and Kind, M. 2008. Fluidisation of whey powders above the glass-transition temperature. *Powder Technology* 187, 53–61.

Parisi, M., Terranova, A., and Chianese, A. 2007. Pilot plant investigation on the kinetics of dextrose cooling crystallization. *Industrial & Engineering Chemistry Research* 46, 1277.

Patist, A. and Bates, D. 2008. Ultrasonic innovations in the food industry: From the laboratory to commercial production. *Innovative Food Science and Emerging Technologies* 9, 147–154.

Paz Suarez, L. A., Georgieva, P., and Feyo de Azevedo, S. 2011. Nonlinear MPC for fed-batch multiple stages sugar crystallization. *Chemical Engineering Research and Design* 89(6), 753–767.

Perry, 1999. *Crystallization from Solution. Liquid-Solid Operations and Equipment.* McGraw-Hill Companies, USA, pp. 18-35–18-54.

Pisponen, A., Pajumagi, S., Mootse, H., Karus, A., and Poikalainen, V. 2013. The lactose from Ricotta cheese whey: The effect of pH and concentration on size and morphology of lactose crystals. *Dairy Science and Technology* 93(4–5), 477–486.

Qin, F., Chen, X. D., Ramachandra, S., and Free, K. 2006. Heat transfer and power consumption in a scraped-surface heat exchanger while freezing aqueous solutions. *Separation and Purification Technology* 48, 150–158.

Roetman, K. and Buma, T. 1974. Temperature dependence of the equilibrium beta/alpha ratio of lactose in aqueous solution. *Netherlands Milk and Dairy Journal* 28, 155–165.

Roos, Y. H., Karel, M., Labuza, T. P., Levine, H., Mathlouthi, M., Reid, D., Shalaev, E., and Slade, L. 2013. Melting and crystallization of sugars in high-solids systems. *Journal of Agricultural and Food Chemistry* 61(13), 3167–3178.

Rousseau, D. 2000. Fat crystals and emulsion stability—A review. *Food Research International* 33, 3–14.

Rozsa, L. 2006. Seed Master 2: A universal crystallization transmitter and automatic seeding device. *International Sugar Journal* 108(1296), 683–695.

Russell, A. B., Burmester, S. S. H., and Winch, P. J. 1997. Characterization of shear thinning flow within a scraped surface heat exchanger. *Trans IChemE* 75 (C), 191–197.

Sanchez, J., Hernandez, E., Auleda, J. M., and Raventos, M. 2011. Review: Freeze concentration technology applied to dairy products. *Food Science and Technology International* 17(1), 5–13.

Sato, K. 2001. Crystallization behavior of fats and lipids—A review. *Chemical Engineering Science* 56, 2255–2265.

Schwartzberg, H. G. 1990. Food freeze concentration. In H. G. Schwartzberg and M. A. Rao (Eds.), *Biotechnology and Food Process Engineering*. New York: Marcel Dekker, Inc., pp. 127–202.

Shahidi, F. 2005. *Edible Oil and Fat Products: Processing Technologies*. Wiley-Interscience, John Wiley & Sons, Inc., Hoboken, New Jersey, USA and Canada.

Shi, Y., Liang, B., and Hartel, R. W. 1990. Crystallization kinetics of alpha-lactose monohydrate in a continuous cooling crystallizer. *Journal of Food Science* 55, 817–820.

Shi, Y., Liang, B., and Hartel, R. W. 2006. Crystal refining technologies by controlled crystallization. US 2006/0128953 A1.

Smaili, F., Angadi, D. K., Hatch, C. M., Herbert, O., Vassiliadis, V. S., and Wilson, D. I. 1999. Optimization of scheduling of cleaning in heat exchanger networks subject to fouling: Sugar industry case study. *Institution of Chemical Engineers Trans IChemE* 77(C), 159–164.

Sormoli, M. E., Das, D., and Langrish, T. A. G. 2013. Crystallization behavior of lactose/sucrose mixtures during water-induced crystallization. *Journal of Food Engineering* 116, 873–880.

Srisa-nga, S., Flood, A. E., and White, E. T. 2006. The secondary nucleation threshold and crystal growth of α-glucose monohydrate in aqueous solution. *Crystal Growth & Design* 6(3), 795.

Stapley, A. G. F., Tewkesbury, H., and Fryer, P. J. 1999. The effects of shear and temperature history on the crystallization of chocolate. *Journal of the American Oil Chemists Society* 76(6), 677–685.

Swenson Crystallization Equipment Handout, 2013, http://www.swensontechnology.com/equipment.php?id=2, Accessed August 2013.

Synowiec, P., Jones, A. G., Ayazi Shamlou, P. 1993. Crystal break-up in dilute turbulently agitated suspensions. *Chem. Eng. Sci.* 48, 3485–3495.

Talbot, G. 1994. Chocolate temper. In S. T. Beckett (Ed.), *Industrial Chocolate Manufacture and Use*. London: Blackie Academic and Professional, pp. 156–166.

Talbot, G. 1999. Chocolate temper. In S. T. Beckett (Ed.), *Industrial Chocolate Manufacture and Use* (3rd ed.). Oxford: Blackwell Science, pp. 218–230.

Tippetts, M. and Martini, S. 2009. Effect of cooling rate on lipid crystallization in oil-in-water emulsions. *Food Research International* 42, 847–855.

Tjuradi, P. and Hartel, R. W. 1995. Corn syrup oligosaccharide effects on sucrose crystallization. *Journal of Food Science* 60(6), 1353–1356.

Toro-Vazquez, J. F., Perez-Martinez, D., Dibildox-Alvarado, E., Charo-Alonso, M., and Reyes-Hernandez, J. 2004. Rheometry and polymorphism of cocoa butter during crystallization under static and stirring conditions. *Journal of the American Oil Chemists Society* 73(6), 195–202.

Urbaniec, K. 2004. The evolution of evaporator stations in the beet-sugar industry. *Journal of Food Engineering* 61, 505–508.

Virone, C., Kramer, H. J. M., van Rosmalen, G. M., Stoop, A. H., and Bakker, T. W. 2006. Primary nucleation induced by ultrasonic cavitation. *Journal of Crystal Growth* 1, 9–15.

Walstra, P. 2003. *Physical Chemistry of Foods*. New York, NY: Marcel Decker.

Westergaard, V. 2010. Milk powder technology-evaporation and spray drying. GEA-NIRO. http://www.niro.com/.

Williams, M. L., Landel, R. F., and Ferry, J. D. 1955. The temperature dependence of relaxation mechanisms in amorphous polymers and other glass-forming liquids. *Journal of the American Chemical Society* 20, 3701–3707.

Wong, S. Y., Bund, R. K., Connelly, R. K., and Hartel, R. W. 2011. Determination of the dynamic metastable limit for α-lactose monohydrate crystallization. *International Dairy Journal* 21(11), 839–847.

Wong, S. Y., Bund, R. K., Connelly, R. K., and Hartel, R. W. 2012. Designing a lactose crystallization process based on dynamic metastable limit. *Journal of Food Engineering* 111, 642–654.

Yazdanpanah, N. and Langrish, T. A. G. 2011a. Crystallization and drying of milk powder in a multiple-stage fluidized bed dryer. *Drying Technology* 29(9), 1046–1057.

Yazdanpanah, N. and Langrish, T. A. G. 2011b. Fast crystallization of lactose and milk powder in fluidized bed dryer/crystallizer. *Dairy Science and Technology* 91(3), 323–340.

Yazdanpanah, N. and Langrish, T. A. G. 2011c. Egg-shell like structure in dried milk powders. *Food Research International* 44(1), 39–45.

Yazdanpanah, N. and Langrish, T. A. G. 2012. Releasing fat in whole milk powder during fluidized bed drying. *Drying Technology* 30(10), 1081–1087.

Yazdanpanah, N. and Langrish, T. A. G. 2013. Comparative study of deteriorative changes in the ageing of milk powder. *Journal of Food Engineering* 114, 14–21.

Zheng, L. and Sun, D. W. 2006. Innovative applications of power ultrasound during food freezing processes—A review. *Trends in Food Science & Technology*, 17, 16–23.

# 5 Mixing-Emulsions

*Theodoros Varzakas, V. Polychniatou, and Constantina Tzia*

## CONTENTS

## 5.1 INTRODUCTION

Mixing occurs widely in the food and drink processing industries. Mixing brings about physico-chemical changes in the foods being processed. The rates of mass and heat transfer are improved by agitation. Agitation may be used to disperse multiphase or multi-component mixtures prior to further processing and packaging.

Mixing is an operation in which two or more components are interspersed in space and/or tend to remove nonuniformities in the properties of material in bulk, for example, color, temperature, composition. The objective of mixing is the achievement of a uniform distribution of components by means of flow.

The efficiency of a mixing process depends on the effective utilization of the energy used to generate the flow of components, the design of mechanical means providing the energy, the configuration of the containing vessel, and the physical properties of the components (Butters, 1993; EEUA, 1963; Harnby, 1985; Holland and Chapman, 1966; Loncin and Merson, 1979; Mutsakis et al., 1986; Oldshue, 1983).

The degree of uniformity in mixing varies between miscible and immiscible liquids. With the latter, as well as paste-like materials and dry powders, the degree of uniformity is less. With immiscible liquids or soluble solids in liquids very intimate mixing is possible. However, mixing becomes very difficult when viscous liquids appear with differing densities.

Dough making is quite different to the other mixing processes since it involves shear and extensional forces generated by the mixer and used to develop the flour and water mixture into a viscoelastic protein matrix capable of retaining the gas produced during proofing and baking.

Structured fluids produced by mixing include creams, butters, and margarines. The flow from the mixer disperses one liquid phase to another forming a stable emulsion.

The processes described include liquid-blending, gas–liquid dispersion, emulsification, solids suspension, and dissolution.

Assessment of the quality of any mixture depends on the scale of scrutiny. Danckwerts (1953) defined the scale of scrutiny of a mixture as the maximum size of regions of segregation, which would cause it to be regarded as unmixed. The scale of segregation and intensity of segregation are important parameters in the description of a mixture. The scale of segregation is a measure of the size of clumps of unmixed components in an imperfect mixture whereas the intensity of segregation is a measure of the difference in composition from the mean, averaged over all points in the mixture, and depends on the interdiffusion between components of the mixture. By increasing the length scale of segregation, the size of the dark regions increases. Increasing the intensity of segregation the mixture does not become very diffuse. At the perfectly mixed state both of these quantities tend to zero. Danckwerts (1952) defined the scale and scrutiny of segregation for a mixture of components in terms of their average and root mean square fluctuating concentrations and correlation functions. These quantities are difficult to measure, however, they determine the quality of a mixture and are useful in understanding the way in which fluid or solid mixtures approach homogeneity.

### 5.1.1 Mixing Equipment–Mixing Mechanisms

Many forms of mixers have been produced from time to time and can be classified according to whether they mix liquids, dry powders, or thick pastes.

Flow regimes for fluid mixing may be divided into laminar and turbulent regions depending on the impeller Reynolds number defined below. In both regions mixing takes place by convective transport of material throughout the mixer and by high shear dispersion in local regions of the mixer close to the moving blades. Mixing rates are enhanced by turbulent diffusion in the turbulent regime. Molecular diffusion operates in both regimes and is responsible for molecular state homogeneity although it is a very slow process and effective over very long times.

### 5.1.2 Laminar Mixing

The full laminar flow regime is restricted to impeller Reynolds numbers Re <10. This can only be achieved using liquids with viscosities greater than 10 Pa s or 10 kg/ms. These high viscosity fluids may also be non-Newtonian. Since the resistance to flow is high, these fluids will require to be mixed in vessels carrying impellers sweeping out the greater proportion of the mixer vessel volume. Impellers (anchors, helical ribbons, helical screws) will be of a large diameter and will have a low speed of rotation.

High shear rates are developed in the laminar region near the impellers and these deform and stretch the fluids leading to an increase in the surface area and a reduction in thickness. Suspended solids, droplets, or bubbles present in the fluid are reduced in size and dispersed (Butters, 1993; EEUA, 1963; Harnby, 1985; Holland and Chapman, 1966; Loncin and Merson, 1979; Mutsakis et al., 1986; Oldshue, 1983).

In low Reynold's numbers flow viscous effects dominate over inertial effects so the impeller sweeps through the vessel volume in order to facilitate adequate agitation. The regions close to the moving blade have large velocity gradients and high shear rates and associated extensional flows with stretching and elongation of fluids taking place.

The change in the interfacial area or striation thickness (thickness of the element) may be used as a measure of the degree of mixing. By increasing the interfacial area, the striation thickness decreases and the material becomes better mixed (Rielly, 1992).

These effects are more intense near the blades of the mixer. The distorted fluid elements are convected into the bulk flow where they are reoriented before passing once more through the region of high shear or accelerated flow.

The continued process of stretching and thinning followed by convective mixing and reorientation in the bulk gradually reduces the striation thickness of fluids and increases homogeneity of the tank contents.

In some laminar mixers (static mixers) mixer blades physically cut and twist the fluids and reorientate them in the flow.

Other examples of static mixers are the so-called in-line static mixers because they have no moving parts and do not require a mechanical drive. The mixer consists of a number of mixer elements inserted to a length of a straight pipe. Mixing is achieved using the energy available in the liquid and results in a pressure loss as the process fluid passes through the mixing elements. Pumping of the raw ingredients through a length of pipe containing many static mixer elements is carried out. As described before in the laminar regime, each static mixer element performs a cutting and twisting operation and after passing through a number of elements the scale of segregation is reduced to an acceptable level.

The number of mixer elements, $M$, required to reduce the striation thickness from $\delta_o$

$$\delta_o/\delta = 1/2^M \tag{5.1}$$

Doubling the volume of the static mixers squares the amount of mixing.

In-line continuous mixers may be dynamic or static. Static in-line mixers utilize the movement of materials flowing over specially contoured stationary mixing elements placed in a pipeline. They are used both in laminar and turbulent mixing and do not need the aid of any moving parts. The major variables are the flow rates and viscosities of the liquid components. The number of mixing elements required increases with the difficulty of mixing. The degree of homogeneity increases exponentially with increased mixer length. Fluids flowing under pump pressure are split and caused to flow along the length of the unit. These mixers require 10–100 times less energy compared with agitated vessels or extruders of high viscosity applications.

Mixing elements can take a variety of geometric forms (helices, vanes, or corrugated plates). Loncin and Merson (1979) discuss one type of static mixer in which the tubular housing contains elements generating helical flow paths a few centimeters in length, each juxtaposed 90° to the preceding element and each with the opposite spiral direction. Mixing arises with a combination of flow splitting, changes in flow direction, and through the development and breakup of boundary layers. Fluids can be miscible or immiscible and units can handle gas/liquid systems.

Static mixers can also be used in the processing of heat-sensitive liquids. Heat-transfer coefficients are increased as in the case of low-pressure drop mixers. Residence time distribution characteristics are also much more uniform than in equivalent empty pipes leading to better control over chemical reactions and plug flow can be approached.

In-line mixers can also be dynamic such as the Oakes continuous mixer. They are widely used for the breakdown and dispersion of particles and droplets in liquids and for the production of emulsions and foams. They have a high-speed rotor moving adjacent to a shaped close-clearance stator developing intense shearing in fluids flowing continuously through the action zone.

Under turbulent flow, the elements generate high turbulence intensities and promote rapid heat and mass transfer. The energy for mixing is provided by sizing the pump to give a larger pressure differential than would be required to overcome friction in an empty pipeline. Wilkinson and Cliff (1977) investigated the laminar flow pressure drop in a Kenics static mixer and proposed a modified form of the pipe flow pressure drop equation

$$\Delta\rho = K\, 4\, C_f\, \rho\, u^2\, L/2D \qquad (5.2)$$

where $u$ is the velocity in the empty pipe, $L$ is the length of the mixer, and $D$ is the internal diameter of the empty pipe. Factor $K$ is a modification factor depending on Reynolds number and the type of mixer elements.

### 5.1.3 Turbulent Mixing

At high impeller Reynolds numbers ($Re > 10^4$) the flow is turbulent. Inertial effects predominate over viscous effects and the fluid can be pumped by a small diameter impeller to all regions of the tank. Transport processes in turbulent flow are enhanced by turbulent eddy diffusion, that is, turbulent velocity fluctuations give much larger mass, heat, and momentum transfer rates than for molecular diffusion.

Turbulent mixing occurs when liquid viscosities are less than 1 kg/ms. The resistance to flow is less than in laminar mixing and currents initiated by a rotating impeller circulate through the vessel. Impellers (propellers, turbines, or simple paddles) are considered as pumps imparting energy to the fluid. Impeller diameters are smaller and have high speed of rotation compared to laminar flow.

Turbulence is not distributed uniformly throughout the vessel; regions close to the impeller have high turbulent energy dissipation rates and hence high turbulent eddy diffusivity values along the complete flow path. Rapid mixing takes place close to the impeller and less intense mixing occurs in the bulk flow. Mixing is more rapid compared to the laminar flow (Butters, 1993; EEUA, 1963; Harnby, 1985; Holland and Chapman, 1966; Loncin and Merson, 1979; Mutsakis et al., 1986; Oldshue, 1983).

The smallest turbulent motions in a stirred tank have length scales (typically O (10) μm) much larger than the molecular scale and molecular diffusion is necessary for homogeneity.

In both laminar and turbulent regimes an input of energy to the liquid that generates flow and mixing takes place. This energy is eventually dissipated as heat by the action of viscosity (Rielly, 1992).

### 5.1.4 Liquid Mixers

Food liquid mixtures could, in theory, be sampled and analyzed in the same way as solid mixtures, but little investigational work has been published on this or on the mixing performance of fluid mixers. Most of the information that is available concerns the power requirements for the most commonly used liquid mixer—some form of *paddle, impeller,* or *propeller stirrer.* In these mixers, the fluids to be mixed are placed in containers and the stirrer is rotated. Power introduced into a liquid mixing system by an agitator is determined by its speed of rotation, configuration of the mixer, and the physical properties of the liquid.

Measurements have been made in terms of dimensionless ratios involving all of the physical factors that influence *power consumption.* Dimensional analysis expresses the behavior of a physical system in terms of the minimum number of independent variables. The results have been correlated by Rushton et al. in an equation of the form using dimensional analysis.

$$(\mathrm{Po}) = \mathrm{K}\,(\mathrm{Re})^{n}(\mathrm{Fr})^{m} \tag{5.3}$$

$$\text{where } (\mathrm{Re}) = (D^2 N \rho/\mu), \tag{5.4}$$

is the dimensionless Reynolds number (relating the ratio of applied to viscous drag forces) then 3 becomes

$$(\mathrm{Po}) = (P/D^5 N^3 \rho) \tag{5.5}$$

is the *Power number* (relating drag forces to inertial forces), and

$$(\mathrm{Fr}) = (DN^2/g) \tag{5.6}$$

is the *Froude number* (relating inertial forces to gravitational ones); $D$ is the diameter of the propeller, N is the rotational speed/frequency of the propeller (rev/s), $\rho_L$ is the liquid density, $\mu_L$ is the viscosity of the liquid, $g$ is the gravitational acceleration and $P$ is the power consumed by the propeller, K is a constant, and n, m depend on the system.

If linear dimensions such as the height of the liquid in the tank, tank diameter, number, size and position of baffles are all in a definite geometrical ratio with the agitator diameter, then the power input to the agitator can be expressed as a function of $D, N, g, \rho, \mu$. This equation has been applied to liquid mixing with single impeller-type agitators in vertical cylindrical tanks by many researchers.

Notice that the Reynolds number in this instance uses the product $DN$ for the velocity, which differs by a factor of $p$ from the actual velocity at the tip of the propeller.

For the deliberate mixing of liquids, the propeller mixer is probably the most common and the most satisfactory. In using propeller mixers, it is important to avoid regular flow patterns such as an even swirl round a cylindrical tank, which may accomplish very little mixing. To break up these streamline patterns, baffles are often fitted, or the propeller may be mounted asymmetrically.

Various baffles can be used to prevent vortexing (a gravitational effect) and the placing of these can make very considerable differences to the mixing performances especially for low viscosity liquid mixing systems. It is tempting to relate the amount of power consumed by a mixer to the amount of mixing produced, but there is no necessary connection and very inefficient mixers can consume large amounts of power. Moreover, a power curve plotting the Power number versus the Reynolds number holds for a particular geometric configuration, but is independent of vessel size and is widely used in the scale-up of liquid mixers from pilot studies. In the viscous range (Reynolds number <10) the plot is linear.

### 5.1.5 Powder and Particle Mixers

The essential feature in these mixers is to displace parts of the mixture with respect to other parts. The ribbon blender, for example, as shown in Figure 5.1a consists of a trough in which rotates a shaft with two open helical screws attached to it, one screw being right-handed and the other left-handed. As the shaft rotates sections of the powder move in opposite directions and so particles are vigorously displaced relative to each other.

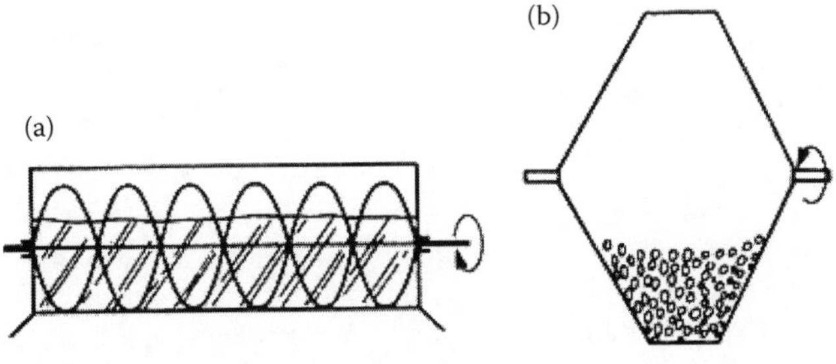


**FIGURE 5.1** Mixers. (a) ribbon blender and (b) double-cone mixer.

A commonly used blender for powders is the double-cone blender in which two cones are mounted with their open ends fastened together and they are rotated about an axis through their common base. This mixer is shown in Figure 5.1b.

### 5.1.6 Standard Geometry Stirred Tanks

The geometry of stirred tanks for low viscosity applications is shown in Figure 5.2.

Tanks may be baffled or unbaffled. More effective mixing is obtained by placing baffles on the tank wall which generate axial and radial velocities rather than a swirling flow. Full baffling may be achieved using four vertical baffles mounted radially, 90° apart. Baffles often have a small clearance from the base of the vessel. For fluid mixing with dispersed solid particles, baffles may be supported off the wall, leaving a gap of approx. T/14. This is to prevent build-up of particles in the crevice between baffles and the wall and facilitate cleaning.

In unbaffled tanks according to Nagata (1975) and Kay and Nedderman (1985), a theoretical model for the flow is described consisting of a central solid body (forced vortex) region with an outer free vortex. In the forced vortex region there is no relative movement of fluid elements and hence no mixing, whereas in the outer region mixing only takes place in the tangential direction.

In the food industry prismatic baffles with flush welds are used for the elimination of residues in corners and for ease of cleaning. In low viscosity fluids small diameter impellers (small D/T ratios) can generate flow in all parts of the tank at moderate power inputs. Impeller flow types include radial (flat paddle, Rushton disc turbine W = D/5), generating flow currents in a radial or tangential direction (transverse and

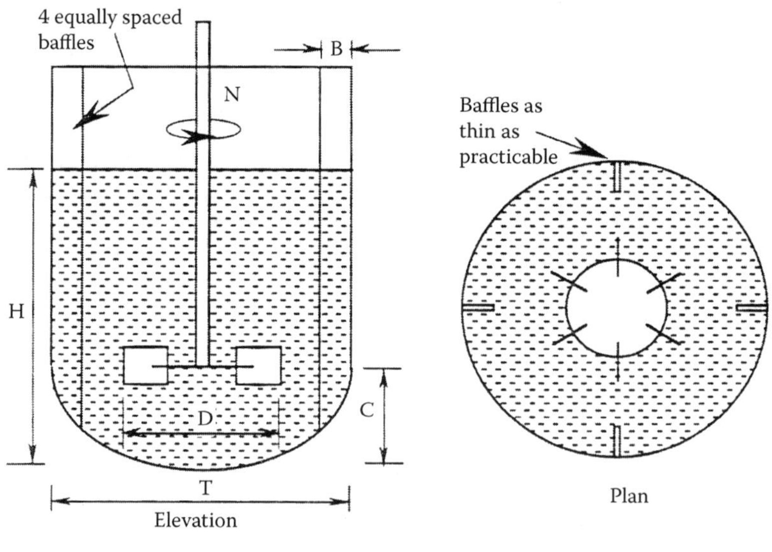


**FIGURE 5.2** Standard geometry for stirred tanks (low viscosity applications). (From Rielly, 1992. Chapter 8. Mixing operations in the food industry. Chemical Engineering for the food industry. Cambridge programme for the food industry. Dept. of Chemical Engineering, University of Cambridge. With permission.)

rotational motion), axial (marine propeller with pitch ratio of 1.5) generating flow currents parallel to the axis of the rotating shaft (longitudinal motion), and axial and radial mixed flow (pitched blade turbine, with W, the blade width, to be 1/5 of the impeller diameter, hydrofoil with various blade angles and pitches). Modern hydrofoil impellers produce strong liquid circulations with a low overall power input and a uniform dissipation of this energy throughout the tank volume.

These flows also show a pseudo-periodicity due to shedding of trailing vortices from the impeller blades, that is, showing unsteadiness in the flows (Yianneskis et al., 1987).

With aspect ratios (H/T) greater than 1.5 it is usual to have multiple impellers on the same shaft to give effective agitation throughout the tank volume. These impellers have standard geometry designs, for example, a typical width-to-diameter ratio as described before. Recent research has shown that hydrofoil or pitched bladed impellers are very advantageous for low-viscosity operations. Marine propellers (3-bladed) are also used since they operate at high rotational speeds with low D/T ratios/typical impeller tip speeds should be about 3 m/s.

Impeller discharge flow rates, Q can be calculated from velocity measurements in the vicinity of the blades (Nagata, 1975). Results are presented in a dimensionless form as a discharge coefficient, $N_Q$ versus the impeller Reynolds number.

$$N_Q = Q/ND^3 = f(\mathrm{Re}) \tag{5.7}$$

For low viscosity fluids the flow is turbulent and the discharge coefficient is constant, independent of impeller speed and diameter, hence the discharge flow rate is directly proportional to $ND^3$. Uhl and Gray (1966) presented a large number of discharge coefficients for various impellers in baffled and unbaffled vessels. Discharge flow is also linked to blend times or solids suspension criteria (Joshi et al., 1982).

The impeller agitators include paddles, propellers, and turbines.

The paddle agitators consist of a flat blade fixed to a rotating shaft. The shaft is mounted centrally in the mixing vessel and rotates at speeds in the range 20–150 rpm. Flow currents produced have high radial and rotational components but little perpendicular motion. Multivane paddles or gate agitators are used with viscous liquids. Anchor paddles are close-clearance paddles contoured to fit against the walls of the containing vessel. They are used to promote heat transfer and minimize wall deposition in jacketed vessels. With the exception of the close clearance variety, paddles usually measure 1/2 – 3/4 of the vessel diameter and the width of the blades 1/10–1/5 of their length.

Finally, counter-rotating paddles are another type of paddle agitator.

Propeller agitators are of small diameter and rotate at high speeds (500,000 rpm). They develop longitudinal flow currents, however, they can be more effective if mounted-off center and have their shafts inclined to the vertical. Side-entry propellers are commonly used with large diameter vessels. They can be employed in emulsification of liquids and dispersion of solids in liquids. Typical propeller impellers include open three bladed ones, saw-toothed, and guarded propellers.

Finally, turbine impellers are effective in mixing moderately viscous liquids and producing gas–liquid and liquid–liquid dispersions.

The impeller of this agitator consists of a number of blades mounted on a rotating boss carried on the shaft. Impeller diameters are in the range of 1/3–1/2 of the vessel diameter and rotational speed is less than that for propellers.

The shaft is usually centrally mounted. With turbines, strong radial and rotational currents are set and the baffles that are employed reduce swirling and vortexing. Vertical currents are also generated due to the deflection of radial currents at the vessel walls. Typical turbine impellers include straight blades, pitched, curved blades, and vaned discs (Butters, 1993; EEUA, 1963; Harnby, 1985; Holland and Chapman, 1966; Loncin and Merson, 1979; Mutsakis et al., 1986; Oldshue, 1983).

Small diameter impellers are only suitable for liquids with viscosities up to 2 Pa s for propellers and 50 Pa s for turbines (Edwards and Baker, 1985). At high viscosities small impellers only generate significant flows in the vicinity of the blades. Anchors and helical ribbon impellers (W = D/6) are used with close clearances between the blades and the wall. This prevents the formation of stagnant zones in the fluid. Anchors are often used to scrape sticky materials off the wall, but are not employed for liquid blending since they produce weak vertical circulations. Ribbon mixers producing strong flows can be used in mixing of miscible liquids.

For very high viscosity mixing ($\mu > 1000$ Pa s) ribbons are unsuitable and kneaders, Z or sigma blade mixers are used. Kneaders or Z-blade mixers are used in bread or biscuit doughs.

These mixers are mounted horizontally and have two counter-current blades. Clearances between the blades and trough are very close in order to eliminate stagnant regions and build-up of sticky material on the wall. Mixing is achieved by a combination of bulk movement and intense shearing and extensional flow as the material passes between the two blades or between the wall and a blade.

The Kneadermaster mixer (Patterson Industries, Inc.) is an adaptation of a sigma-blade mixer for continuous operation. Each two pairs of blades establish a mixing zone, the first pair pushing materials toward the discharge end of the trough and the second pair pushing them back. Forwarding to the next zone is by displacement with more feed material. Control of mixing intensity is by variation in rotor speed.

### 5.1.7 Dough and Paste Mixers

Dough and pastes are mixed in machines that have, of necessity, to be heavy and powerful. Because of the large power requirements, it is particularly desirable that these machines mix with reasonable efficiency, as the power is dissipated in the form of heat, which may cause substantial heating of the product. Such machines may require jacketing of the mixer to remove as much heat as possible with cooling water.

Perhaps, the most commonly used mixer for these very heavy materials is the kneader, which employs two counter-rotating arms of special shape that fold and shear the material across a cusp, or division, in the bottom of the mixer. The arms are of the so-called sigmoid shape as indicated in Figure 5.3.

They rotate at differential speeds, often in the ratio of nearly 3:2. Developments of this machine include types with multiple sigmoid blades along extended troughs, in which the blades are given a forward twist and the material makes its way continuously through the machine (American Institute of Chemical Engineers, 1979).

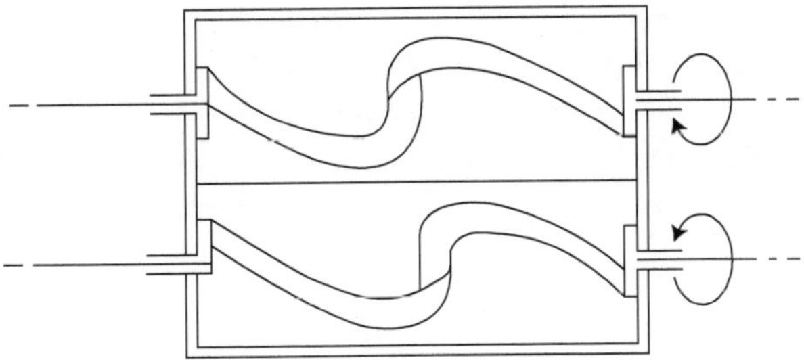

**FIGURE 5.3** Kneader.

Another type of machine employs very heavy contra-rotating paddles, whilst a modern continuous mixer consists of an interrupted screw, which oscillates with both rotary and reciprocating motion between pegs in an enclosing cylinder. The important principle in these machines is that the material has to be divided and folded and also displaced, so that fresh surfaces recombine as often as possible (Earle, 1983).

### 5.1.8 Double Spiral Mixer

The features of a double spiral mixer include: heavy steel structure, hydraulic head rise and descent, spirals' transmission by parallel axes, gearboxes, and belts, bowl transmission by means of a double set of friction wheels, rotating cylinder, bowl spiral tools in highly resistant stainless steel, stainless-steel dust tight bowl cover with inspection hole, automatic hydraulic hooking system of the bowl trolley (at the connection in the basement), automatic descent start of the head, and head descent completion by hold-to-run control. At the mixing end are an automatic bowl trolley release and head rise by hydraulic unit; the mixer is lifted from the ground (about 85 mm) by no. 4 stainless-steel feet.

Technical characteristics of blending tank for cream mixing are shown below.

### 5.1.9 Blending Tank for Cream Mixing

**Technical Characteristics**

Parts in contact with product made of AISI 316

Dome top with circular manhole 400 mm

Cone bottom 15°

Jacketed on wall and bottom for water or steam circulation, maximum operating pressure 4 bar.

Mineral wool insulation covered with AISI 304 sheet.

Anchor agitator with scraper blades, driven with one geared motor 5/3.1 kW; 23/46 rpm

Special impeller dia. 300 mm, housed in the bottom:

Driven by a geared motor of 18 kW, 1450 RPM, with 90° angle gear transmission.

Baffle

Surface finish: inside fine polished, outside 2B
No. 3 connection 50 DIN for ingredient feeding
CIP connection 40 DIN, n. 2 CIP spray balls
Bottom discharge 70 DIN
Vent 65/90 DIN
Adjustable sanitary feet
Power supply: 380 V, 50 Hz, 3 phases

### 5.1.10 Double-Arm Kneading Mixers

The universal mixing and kneading machine consists of two counter-rotating blades in a rectangular trough curved at the bottom to form two longitudinal half cylinders and a saddle section (Figure 5.4). The blades are driven by gearing at either or both ends. The oldest style empties through a bottom door or valve and is still in use when 100% discharge or thorough cleaning between batches is not an essential requirement. More commonly, however, double-arm mixers are tilted for discharge.

The tilting mechanism may be manual, mechanical, or hydraulic. A variety of blade shapes have evolved. The mixing action is a combination of bulk movement, smearing, stretching, folding, dividing, and recombining as the material is pulled and squeezed against blades, saddle, and sidewalls. The blades are pitched to achieve end-to-end circulation. Rotation is usually such that material is drawn down over the saddle. Clearances are as close as 1 mm (0.04 in).

The blades may be tangential or overlapping. Tangential blades are run at different speeds, with the advantages of faster mixing from constant change of relative

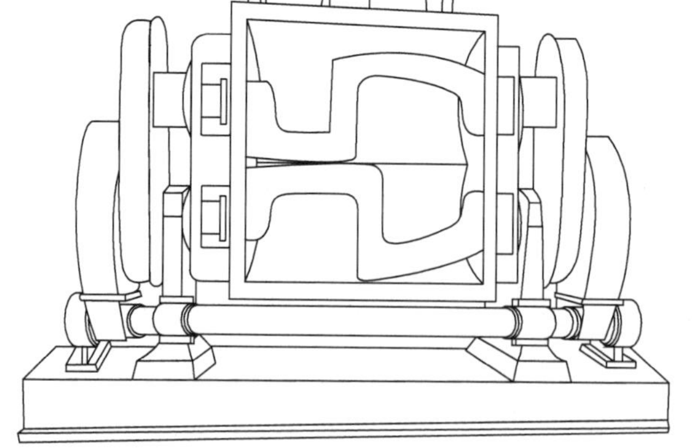


**FIGURE 5.4** Double-arm kneader mixer. *(APV Baker Perkins Inc.)*

position, greater wiped heat-transfer area per unit volume, and less riding of material above the blades. Overlapping blades can be designed to avoid build-up of sticky material on the blades.

The agitator design most widely used is the sigma blade. The sigma-blade mixer is capable of starting and operating with either liquids or solids or a combination of both. Modifications in blade-face design have been introduced to increase particular effects, such as shredding or wiping. The sigma blade has good mixing action, readily discharges materials which do not stick to the blades, and is relatively easy to clean when sticky materials are being processed (Perry and Green, 1999).

Other agitator blades include the dispersion blade, developed particularly to provide compressive shear higher than that achieved with standard sigma blades, multiwiping overlapping (MWOL) blades commonly used for mixtures which start tough and rubberlike, the single-curve blade developed for incorporating fiber reinforcement into plastics, the double-naben blade, a good blade for mixes which "ride," that is, form a lump which bridges across the sigma blade.

### 5.1.11 Gas–Liquid Mixing

#### 5.1.11.1 Surface Aeration in Stirred Tanks

In many food-processing applications entraining air during the mixing process should be avoided since this causes spoilage during product storage.

Surface aeration occurs above a minimum impeller speed denoted by $N_{SA}$ in the absence of gas sparging. Van Dierendonck et al. (1968) and Joshi et al. (1982) have correlated $N_{SA}$, the critical speed for the onset of aeration against physical properties and geometric parameters for a standard disc turbine in a baffled tank.

Greaves and Kobbacy (1981) described the aeration phenomena qualitatively from the free surface in baffled tanks at $N > N_{SA}$.

Strong eddies formed by the interaction of the discharge flow from the impeller with the baffles, induce other strong eddies which process slowly around the impeller shaft and form a hollow vortex at the surface.

Sparging gas bubbles into a stirred tank reduces power consumption of the impeller. In gas–liquid applications the gassed power input is fairly high (1–2 kW/m$^3$) since large energy dissipation rates produce small bubbles and large interfacial areas. Power consumption under aerated conditions is reported by Greaves and Barigou (1986), Michel and Miller (1962), Nagata (1975), and Uhl and Gray (1966).

Bruijn et al. (1974) and van't Riet and Smith (1973) plotted the gassed power ratio Pg/Po against the aeration number at constant impeller speed and explained their results in dimensionless terms.

The power input partly determines the rates of mass transfer in gas–liquid dispersions and the gassed power number should not drop off too rapidly as the aeration number increases. The decrease in gassed power consumption is explained as a consequence of the formation of stable gas cavities behind the blades.

The size and shape of these cavities depend on gas volumetric flow rate and impeller speed.

Smith (1986) showed flow regime maps showing cavity types as a function of the Froude number and aeration number (Figure 5.5).

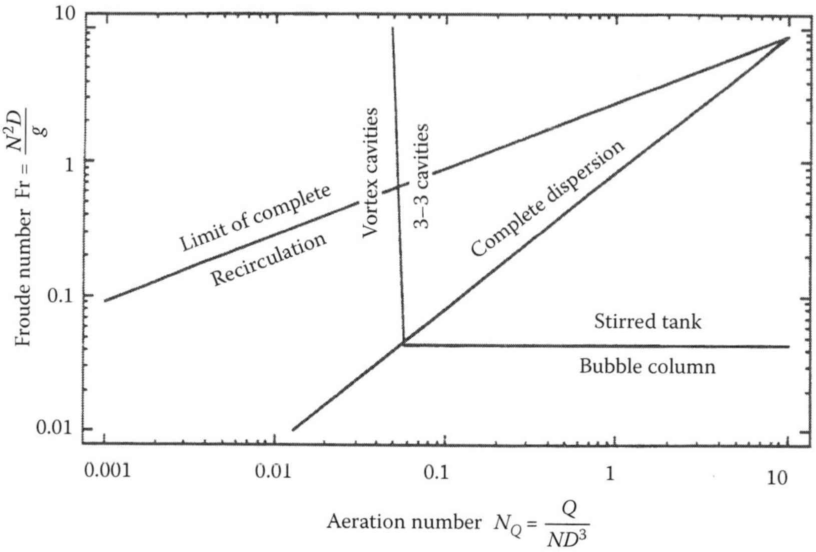


**FIGURE 5.5** Gas flow map for standard disc turbine with regions of different cavity formation. (Adapted from Rielly, 1992. Chapter 8. Mixing operations in the food industry. Chemical Engineering for the food industry. Cambridge programme for the food industry. Dept. of Chemical Engineering, University of Cambridge; Smith, J.M. 1986. *5th European Conference Mixing*, paper 13, BHRA Fluid Engng, Cranfield.)

These cavities alter the liquid streamlines around the blade and the separation point occurs further downstream from the leading edge of the blade. Power consumption is reduced in the presence of gas, which depends on the size and shape of the gas cavities.

Figure 5.6 shows the effect of gradually increasing the gas throughput or decreasing the impeller speed on gas flow patterns in a gas–liquid stirred tank (Nienow et al., 1978).

At low gas flow rates and high impeller speeds, bubbles are well dispersed above and below the impeller. Increasing gas flow leads to a much worse gas dispersion.

Nienow et al. (1978) defined a critical speed for complete dispersion, $N_{CD}$, at the change from condition c to d as shown in Figure 5.6.

For $H = T$, $C = T/4$, 6-bladed disc turbines ($T < 1.8$ m) the following equations are given for pipe spargers and ring spargers, respectively.

$$N_{CD} = 4\ Q^{0.5}\ T^{0.25}/D^2 \tag{5.8}$$

$$N_{CD} = 3\ Q^{0.5}\ T^{0.25}/D^2 \tag{5.9}$$

where $Q$ is the volumetric flow gas rate. These equations are valid for noncoalescing systems and turbines with more than six blades (Middleton, 1985).

Increasing the gas might lead to flooding. Then the impeller stops pumping in the radial direction and rising bubbles are formed as reported by Warmoeskerken and Smith (1984).

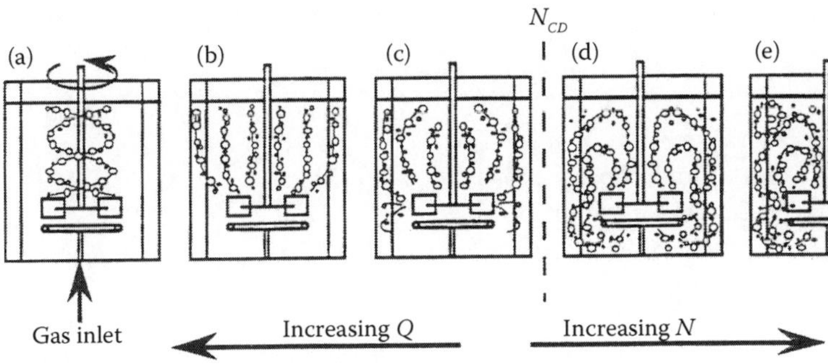


**FIGURE 5.6** Gas flow patterns as a function of impeller speed and gas flow rate. (From Rielly, 1992. Chapter 8. Mixing operations in the food industry. Chemical Engineering for the food industry. Cambridge programme for the food industry. Dept. of Chemical Engineering, University of Cambridge; Nienow, A.W., Wisdom, D.J., and Middleton, J.C. 1978. *2nd European Conference Mixing*, paper F1, BHRA Fluid Engng, Cranfield; Nienow, A.W., Chapman, C.M. and Middleton, J.C., 1979. *Chem. Eng. J.*, 17( 2): 111–118; Nienow, A.W. 1985. *Mixing in the process industries*, eds. Hornby, N., Edwards, M.F., and Nienow, A.W. Chapter 18, London, Butterworth.)

At the flooding point

$$Q/N_F D^3 = 1.2\, N_F^2\, D/g \tag{5.10}$$

where $N_F$ is the critical impeller speed for flooding at a given gas volumetric flow rate, $Q$.

#### 5.1.11.2 Gas Voidage Fraction in Stirred Tanks

The gas volume fraction held up in the liquid is important in the determination of gas–liquid reactors or fermenters. The mean gas voidage fraction ε is defined as

$$\varepsilon = V_G / V_G + V_L \tag{5.11}$$

where $V_G$, $V_L$ are the gas and liquid volumes in the stirred tank, respectively. Calderbank (1958) presented a semi-theoretical method to predict the mean gas voidage fraction. This is based on Kolmogoroff"s theory of local isotropic turbulence valid only at high Reynolds numbers. Analysis showed that the largest bubble size existing in a given turbulent flow depends on the power input per unit volume, the fluid density, ρ, and the surface tension, σ.

#### 5.1.11.3 Gas–Liquid Mass Transfer

Mixing might also involve the transfer of a solute gas into the liquid phase for subsequent aerobic fermentation. The purpose of this process is to cause gas–liquid interfacial area from the dispersion of sparged gas as small bubbles and hence increase mass-transfer rates. Large interfacial areas require small bubble sizes and large gas voidage fractions.

The equation for mass transfer between a liquid and a gas is shown below.

$$J = K_L\, a\, (C_L^* - C_L)\, V \tag{5.12}$$

$J$ is the molar transfer rate of species per unit volume
$K_L$ is the overall liquid-phase mass-transfer coefficient
$a$ is the interfacial area per unit volume
$C_L$ is the liquid-phase molar composition
$C_L^*$ is the equilibrium liquid-phase molar composition
and $V$ is the volume of the dispersion.

In some cases the liquid-phase resistance predominates and the overall mass-transfer coefficient $K_L$ is equal to the liquid film coefficient $k_L$ (Kay and Nedderman, 1985). Most researchers usually measure the product $k_L\, a$.

### 5.1.12 Solid–Liquid Mass Transfer

In solid–liquid stirred tank applications the objectives are listed below.

1. Formation and maintenance of homogeneous suspension of particles.
2. Suspension of particles to obtain flocculation. This requires gentle agitation to suspend solids and promote collisions between particles.
3. Suspension of particles that are resting on the bottom of the tank in order to expose the maximum solid-liquid surface area for mass transfer.

Typical solid–liquid operations include dissolution (sucrose syrups formation), fermentations (suspension of cell cultures and growth media), crystallization and continuous slurry draw offs from stirred tanks used as premixers or holding vessels (avoiding sedimentation and segregation).

Homogeneous suspension requires that solids are uniformly distributed. The impeller speed to obtain homogeneous suspension is considerably higher than for complete suspension and operation under these conditions is not economically feasible.

Sampling the mixture at various heights in the vessel is a way to determine experimentally the homogeneous condition. It is difficult to ensure representative, isokinetic sampling.

If samples are obtained isokinetically from all parts of the tank the homogeneous condition may be defined in terms of a mixing index, $M$.

$$M = \sigma/\bar{x} = \frac{\sqrt{\sum_{i=1}^{n} \frac{(x_i - \bar{x})^2}{(n-1)}}}{\bar{x}} \tag{5.13}$$

$x_i$ is the solids concentration (kg solids/kg mix.), $\sigma^2$ = variance of solids concentration $(\text{kg/kg})^2$, $n$ is the number of samples, and $\bar{x}$ is the mean solids concentration in tank.

For a perfectly homogeneous tank the mixing index would be zero (typically the tank is homogeneous when $M \leq 0.05$).

Zweitering (1958) used an equation covering the widest range of fluid properties, particle size, concentration and tank geometric parameters to measure the impeller speed in order to obtain the just suspended state (Perry and Green, 1999). This particular technique is suitable only for laboratory investigations using tanks that are transparent and well illuminated.

A dimensionless constant s in the equation depends on impeller type, clearance ratio, T/C, and diameter ratio T/D.

For dissolution of solid particles the rate of mass transfer (kmol/s) is given by

$$J = ks\,A\,(Cs - C\infty) \tag{5.14}$$

where $A$ is the exposed solid–liquid area and ks the mass-transfer coefficient.

$Cs$ is the saturation concentration at the surface of the particles. The latter equation applies to operations in which diffusion controls mass transfer from the particles to the bulk fluid at concentration $C\infty$.

Nienow and Miles (1978) showed that the diffusion controlled mass-transfer coefficient, $k_{JS}$, at the just suspended speed $N_{JS}$, was independent of the impeller and tank configuration and the impeller speed should be in the range $N_{JS} < N < N_{SA}$.

To achieve the just suspended condition large specific power inputs may be required.

Mass-transfer coefficient may be predicted from a correlation similar to the Froessling equation for particle–fluid systems (Rowe et al., 1965).

$$\mathrm{Sh} = 2 + 0.72\ \mathrm{Re}^{1/2}\ \mathrm{Sc}^{1/3} \tag{5.15}$$

$$\text{The Schmidt number is } \mathrm{Sc} = \mu/\rho\, D \tag{5.16}$$

$$\text{and the Sherwood number is } \mathrm{Sh} = k\, d_{32}/D \tag{5.17}$$

$d_{32}$ is the mean particle diameter, $D$ is the liquid diffusivity and the particle Reynolds number is

$$\mathrm{Rep} = \rho\, d_{32}\, u_s/\mu \tag{5.18}$$

$u_s$ is the slip velocity between the particle and the liquid which is difficult to calculate.

### 5.1.13 Mixing of Particulate Materials

Mixing of solids is not an irreversible process (mixtures of particles tend to segregate or unmix due to differences in size, density, shape, roughness). Segregation occurs due to the percolation of fine particles within the mixture (large particles rise to the top and fine particles fall into gaps between large particles slowly percolating to the bottom).

Examples of percolation segregation are the vibration of a solid mixture (as it happens with the case of nuts and raisins moving to the top of a pack of muesli), shearing of a solid mixture where small particles percolate to the lower layer, and during

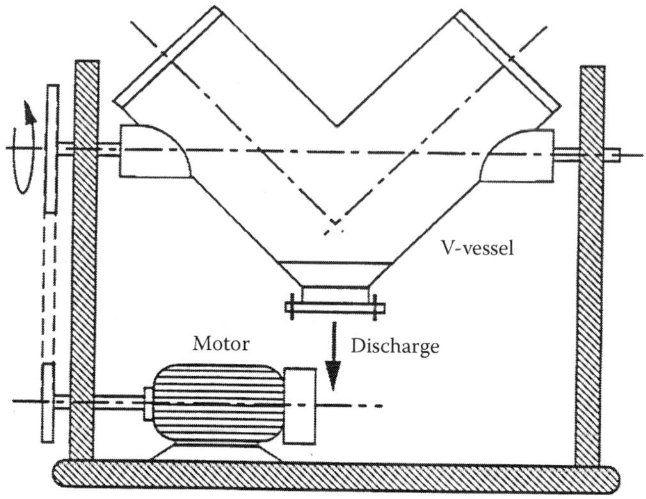


**FIGURE 5.7** V-shaped tumbler for solid-solid mixing. (From Rielly, 1992. Chapter 8. Mixing operations in the food industry. Chemical Engineering for the food industry. Cambridge programme for the food industry. Dept. of Chemical Engineering, University of Cambridge; Harnby, 1985.)

pouring or discharge from a vessel or hopper into a heap, which is a method used to separate coarser table sugar from castor sugar.

In solids mixing the motion of particles can only be achieved by input of energy to the system and the final equilibrium state is determined by the mixer design and the particle segregating properties (Streiff, 1979).

Harnby (1967, 1985) categorized solid mixers into tumbler, convective, and hopper mixers according to their mixing action.

The vessel of a tumbler mixer is designed as a roto-cube, double cone, Y- or V-shaped mixer (Figure 5.7). The vessel is filled to about half of its total capacity and is rotated on an axis between two bearings causing the solid particles to roll continuously or tumble over each other. Rotational speed is about half of the critical speed at which centrifugal forces throw the solids out to the extremities of the container and there is no relative motion of the particles.

In convective mixers, solids are mixed in a stationary vessel by a rotating impeller or screw, which convects particles within the mixture. Examples of convective mixers are the ribbon blender in which a helical or Z blade sweeps through a horizontal open trough or closed cylindrical vessel. The mixer speed is low, minimizing power input, heating of the mixture, and particle degradation.

Another example of a convective mixer is the Nautamix in which a small diameter Archimedian screw rotates within a vertical, conical hopper (Figure 5.8). Only particles near the screw are convected and there exists an epicyclic path within the hopper, which progressively sweeps the blade throughout the whole mixture. Reversing the direction of the screw rotation aids discharge of pasty materials.

Finally, hopper mixers achieve radial and axial mixing of different particulate components by recycling the discharged material to the top of the hopper. During the

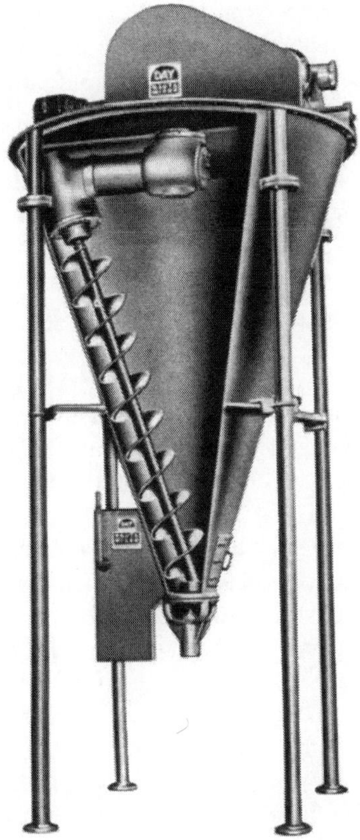

**FIGURE 5.8** Nauta mixer. (From Perry, R.H. and Green, D.W. *Perry's Chemical Engineers' Handbook*. McGraw Hill, USA, 1999.)

discharge of the granular material from the hopper significant relative motions of the particles are formed due to velocity gradients near the outlet orifice (Rielly, 1992).

## 5.1.14 Scaling Up Mixing Performance

### 5.1.14.1 Scale-Up of Batch Mixers

Scaling up of batch mixers depends on equal power per unit volume, although the most desirable practical criterion is equal blending per unit time. As size is increased, mechanical-design requirements may limit the larger mixer to lower agitator speeds.

If the power is high, the lower surface-to-volume ratio as size is increased may make temperature build-up a limiting factor. Since the impeller in a paste mixer generally comes close to the vessel wall, it is not possible to add cooling coils. In some instances, the impeller blades can be cored for additional heat-transfer area.

Experience with double-arm mixers indicates that power is proportional to the product of blade radius, blade-wing depth, trough length, and average of the speeds of the two blades (Irving and Saxton, 1967). The mixing time scales up inversely with blade speed. Mixing depends primarily on the number of revolutions that the blades have made. As indicated previously, the minimum possible mixing time may become dependent on heat-transfer rate.

Frequently, the physical properties of a paste vary considerably during the mixing cycle. As the equipment scale is increased, geometric similarity being at least approximated, there is a loss in surface-to-volume ratio. As size is increased, changing shear rate or length-to-diameter ratio may be required because of equipment-fabrication limitations.

### 5.1.15 Grinding

The critical step in the aqueous extraction process affecting the oil and protein yields is the grinding operation, which determines the oilseed particle size. Efficient grinding, which breaks down the walls of the oil-containing cell, is considered essential (Cater et al., 1974; Southwell and Harris, 1991). Smaller particle size allows not only easier diffusion of water-soluble components, thereby disintegrating the original structure and facilitating oil release, but also enhances enzyme diffusion rates, which can then act on the substrates more easily.

The grinding operation may be carried out either wet or dry depending on the oilseed. The choice between wet or dry grinding would be governed by several factors such as the initial moisture content, chemical composition, and physical structure of the oilseeds (Cater et al., 1974). In the case of materials with high moisture level, like coconut, wet grinding is more appropriate, thus avoiding an additional intensive drying step before grinding. On the other hand, in the case of materials with low initial moisture content, like peanuts, rapeseed, and soybean, dry grinding is considered more suitable. Wet grinding is sometimes considered better for rupturing cells, mainly because the water softens the cell walls. However, with some materials, like peanuts, wet grinding can produce a stable emulsion that has to be broken to recover oil (Cater et al., 1974).

Conflicting effects of wet grinding are thus evident: excessive grinding favors cell rupture and increases the efficiency of oil and protein extraction; however it also produces smaller oil globules which makes demulsification more difficult. Insufficient grinding, on the other hand, results in unacceptable losses of oil in the residue (Cater et al., 1974; Hagenmaier et al., 1972). In the case of coconut, wet-milling and grinding which conditions the coconut so that only less than 5% of the oil remains with the residue, results in the formation of oil globules of about 10 μm in diameter. This generally corresponds to the formation of a stable emulsion, similar to milk. It is evident that a stable emulsion is the price one must pay for efficient wet extraction of oil from fresh coconut (Hagenmaier et al., 1972).

### 5.1.16 Extraction

The extraction step basically consists of dispersing the ground seeds in water and then agitating the dispersion to enhance oil and protein extraction. Optimum extraction conditions vary according to the oilseed composition and structure (Lusas et al., 1982). Factors which influence the efficiency of extraction include: solid-to-water ratio, particle size, pH, time, temperature, degree of agitation (Cater et al., 1974; Lusas et al., 1982), and the number of extraction stages.

### 5.1.17 Power Dissipation Level

Simple stirring is sometimes sufficient to obtain high yields as in case of peanuts (Lawhon et al., 1981a) and sunflower (Hagenmaier, 1974). However, high power dissipation levels may be required occasionally and can be achieved by vigorously stirring the dispersion. High shear stirring also achieves further disintegration of the individual cells, and eases the release of oil from comminuted seeds. Increasing the blending time, in general, improves the oil extraction yield, but results in the formation of an emulsion with greater stability which can adversely affect the total yield after subsequent separation steps (Embong and Jelen, 1977). Hence, optimization of agitation speeds and power dissipation levels needs to be carried out, considering not only the extraction yield, but also the stability of the resulting dispersion. Moreover, the optimum solid-to-water ratio and extraction time may depend on the energy input level on the system, which in turn depends on the size and shape of the vessel (extractor).

To calculate the power consumed during agitation it is assumed that all the energy transmitted in the vessel filled with water and stirred at high speed is dissipated as heat, so that the measured power consumption is equivalent to the energy corresponding to the increase in the temperature of the correspondent amount of water, according to the expression:

$$P = mc_p \frac{dT}{dt} \tag{5.19}$$

where $P$ is the rate of transmitted energy or power dissipation rate, $m$ is the mass of water, $c_p$ is the mean specific heat of water (equal to 4.2 kJ/kg/K), and $dT/dt$ is the rate of change of temperature ($K$).

To determine the power input inside a baffled vessel with a fixed water agitation rate we use the equation applicable for nongassed Newtonian fluids, found to be independent of the Reynolds number in the turbulent regime ($Re > 10^4$) (Atkinson and Mavituna, 1991):

$$N_p = \frac{P}{N^3 D^5 \rho} \tag{5.20}$$

Here $P$ is the power supplied to the agitator, $N$ is the rotational speed of the impeller, $D$ is the diameter of the impeller, and $\rho$ is the density of the fluid.

### 5.1.18 Continuous Mixing

#### 5.1.18.1 Screw Extruders

Screw extruders are commonly used in polymer and plastics processing, however, they can also be used in the manufacture of reconstituted potato or corn curls snacks.

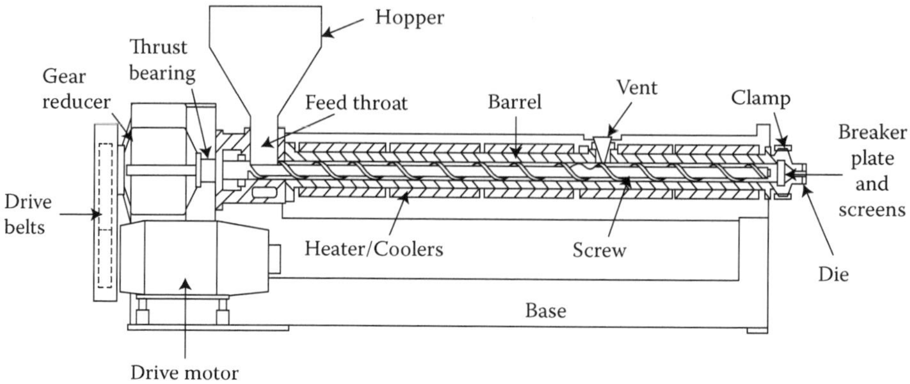


**FIGURE 5.9** Single-screw extruder. (Davis standard from Perry, R.H. and Green, D.W. *Perry's Chemical Engineers' Handbook*. McGraw Hill, USA, 1999.)

Raw ingredients such as corn grits or potato flour and rice flour are added through a feed hopper and conveyed by a rotating horizontal screw. High pressures and temperatures are generated within the barrel of the extruder such that the food material is rapidly cooked before being extruded through a die. Single or double counter-rotating screws are used although the latter give more intense mixing but with greater capital and running costs (Figure 5.9). Single-screw machines performance can be improved by adding extra heads or by using in-line static mixers.

The extruder combines the process functions of melting the base resin, mixing in the additives, and developing the pressure required for shaping the product into pellets, sheet, or profiles.

The dry ingredients, sometimes premixed in a batch blender, are fed into the feed throat where the channel depth is the deepest. As the root diameter of the screw is increased, the plastic is melted by a combination of friction and heat transfer from the barrel. Shear forces can be very high, especially in the melting zone, and the mixing is primarily a laminar shearing action.

Single-screw extruders can be built with a long length-to-diameter ratio to permit a sequence of process operations, such as staged addition of various ingredients. Capacity is determined by diameter, length, and power. Although the majority of extruders are in the 25- to 200-mm-diameter range, much larger units have been made for specific applications such as polyethylene homogenization.

Mixing enhancers (Maddock (straight); Maddock (tapered); pineapple; gear; pin) are utilized to provide both elongational reorienting and shearing action to provide for both dispersive and distributive mixing (Perry and Green, 1999).

#### 5.1.18.2 Rietz Extruder

This extruder has orifice plates and baffles along the vessel. The rotor carries multiple blades with a forward pitch, generating the head for extrusion through the orifice plates as well as battering the material to break up agglomerates between the baffles. Typical applications include wet granulation of pharmaceuticals, blending color in bar soap, and mixing and extruding cellulose materials.

### 5.1.18.3 Twin-Screw Extruders

With two screws in a Figure-eight barrel, the interaction between the screws as well as between the screw and the barrel is advantageous. Twin-screw machines are used in continuous melting, mixing, and homogenizing of different polymers with various additives, or to carry out the intimate mixing required for reactions in which at least one of the components is of high viscosity.

The screws can be tangential or intermeshing, and the latter either co-or counter-rotating. Tangential designs allow variability in channel depth and permit longer lengths.

Counter-rotating intermeshing screws provide a dispersive milling action between the screws and behave like a positive displacement device with the ability to generate pressure more efficiently than any other extruder.

The most common type of twin-screw mixing extruder is the corotating intermeshing variety (APV, Berstorff, Davis Standard, Leistritz, Werner and Pfleiderer). The two keyed or splined shafts are fitted with pairs of slip-on kneading or conveying elements, as shown in Figure 5.10. The arrays of these elements can be varied to provide a wide range of mixing effects.

Each pair of kneading paddles causes an alternating compression and expansion effect which massages the contents and provides a combination of shearing and elongational mixing actions (Perry and Green, 1999).

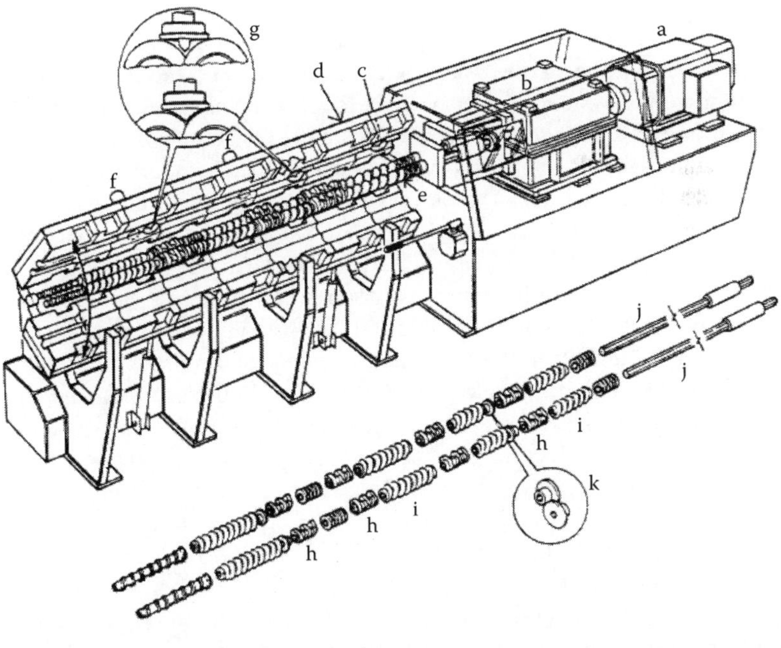


**FIGURE 5.10** Intermeshing corotating twin screw extruder: (a) drive motor; (b) gearbox; (c) feed port; (d) barrel; (e) assembled rotors; (f) vent; (g) barrel valve; (h) kneading paddles; (i) conveying screws; (j) splined shafts; and (k) blister rings. (*APV Chemical Machinery, Inc.*) (From Perry, R.H. and Green, D.W. *Perry's Chemical Engineers' Handbook*. McGraw Hill, USA, 1999.)

## 5.2 EMULSIONS

### 5.2.1 Introduction to Emulsions

Food emulsions are food forms familiar to consumers. Numerous natural and processed foods consist either partly or totally of emulsions, or processing of certain foods includes a stage in an emulsified state. Such examples are: milk, cream, fruit beverages, infant formula, soups, cake batters, salad dressings, mayonnaise, cream-liqueurs, sauces, deserts, salad cream, ice cream, coffee whitener, spreads, butter, and margarine (McClements, 2005).

Emulsions are relative unstable systems consisting mainly of water and oil. Thus, various ingredients are necessarily used to formulate and stabilize the emulsions and provide them specified properties. Food emulsions are required to remain stable over extended periods with a shelf life that usually reaches several months or longer. It is important that, as they must be edible, all ingredients used should be approved as safe for human consumption. Another essential requirement is related to their microbiological safety, that means the absence of pathogenic or spoilage organisms even after extended storage. The above requirements create restrictions and guidelines on the design and formulation of food emulsions. Recently, the food industry is emphasizing the preparation of natural, healthy, or low calorie as well as functional food products. The latter motivates research in order to prepare emulsions with natural ingredients such as proteins and phospholipids, with lower fat content and functional properties (Dalgleish, 2001).

Food emulsion behavior is basically defined by the three containing ingredients; the fat or oil (fat/oil phase), the water (aqueous phase), and the interfacial material used to stabilize the system. However, most food emulsions are more complex than the simple three-component systems. The aqueous phase may contain a variety of water-soluble ingredients, including sugars, salts, acids, bases, alcohols, surfactants, proteins, and polysaccharides. The oil phase usually contains a complex mixture of lipid-soluble components, such as triacylglycerols, diacylglycerols, monoacylglycerols, free fatty acids, sterols, and vitamins. The interfacial region may be composed of proteins or small emulsifiers such as monoglycerides, esters, or phospholipids, or mixtures of these components. In addition, these components may form various types of structural entities in the oil, water, or interfacial regions (such as fat crystals, ice crystals, biopolymer aggregates, air bubbles, liquid crystals, and surfactant micelles), which in turn may associate to form larger structures (such as biopolymer or particulate networks). Therefore, to understand the functional properties of the emulsions, it is necessary to understand the properties of these three compositional parts (Dalgleish, 2006).

Moreover it should be taken under consideration that food emulsions are usually subjected under severe treatment variations such as temperature, pressure, and mechanical agitation during their production, storage, and handling that can cause significant alterations in their properties. As a result there is a great need to control the factors affecting the emulsion morphology in order to be able to design and even to predict the properties and the shelf life of the produced emulsions.

Emulsions are stable suspensions of one liquid in another, the liquids being immiscible. Stability of the emulsion is obtained by dispersion of very fine droplets of one liquid, called the *disperse phase*, through the other liquid, which is called the *continuous phase*. The emulsion is stable when it can persist without change, for long periods of time, without the droplets of the disperse phase coalescing with each other, or rising or settling. The *stability of an emulsion* is controlled by

- Interfacial surface forces,
- Size of the disperse phase droplets,
- Viscous properties of the continuous phase and
- Density difference between the two phases.

The dispersed particles in the emulsion have a very large surface area, which is created in the process of emulsification. Surface effects depend upon the properties of the materials of the two phases, but very often a third component is added which is absorbed at the interface and which helps to prevent the droplets from coalescing.

These added materials are called **emulsifying agents** and examples are phosphates and glycerol monostearate (Earle, 1983).

The size of the disperse phase droplets is important and these are commonly of the order 1–10 μm diameter. Below 0.1 μm droplet diameter, the dispersion is often spoken of as colloidal. Coalescence of the disperse phase droplets is hindered by increased viscosity in the continuous liquid phase. The nearer the densities of the components are to each other, the less will be the separating effect of gravitational forces. Stokes' law gives a qualitative indication of the physical factors that influence the stability of an emulsion. This is because the relative flow of the particles under gravitational forces may break the emulsion, so stability is enhanced by small settling velocities.

$$v = D^2 g(\rho_p - \rho_f)/18\mu \tag{5.21}$$

The critical importance of particle size, occurring as a squared term, can be seen from Equation 5.21. Also it shows why emulsions are more stable when density differences are small and when the viscosity of the continuous phase is high.

The essential feature of an emulsion is the small size of the disperse phase droplets. This can be achieved by imposing very high shearing stresses upon the liquid that is to be dispersed and the shearing forces break the material into the multitude of fine particles.

*Shearing* is, generally attained by passing the liquid through a high-pressure pump, to bring it up to pressures of the order of $7 \times 10^3$ kPa, and then discharging this pressure suddenly by expansion of the liquid through a small gap or nozzle; the equipment is often called a homogenizer. In passing through the nozzle, very large shear forces are exerted on the liquid, disrupting cohesion and dispersing it into very small particles.

Centrifugal forces may also be used to obtain the shearing action. Discs spinning at high velocities give rise to high shearing forces in liquids flowing over them. Flow between contra-rotating discs, which may have pegs on the disc faces, can be used to produce emulsions. Designs in which small clearances are used between a stationary disc and a high-speed flat or conical rotating disc are called colloid mills.

Another source of energy for shearing is from ultrasonic vibrations induced in the liquid (Earle, 1983).

Existing emulsions can be given increased stability by decreasing the size of the droplets either by impact or shearing the emulsion still further; the process is called *homogenization*. Homogenizing results in smaller and more uniform droplet sizes and a practical example is the homogenizing of milk.

Examples of emulsions met with frequently in the food industry are—milk (fat dispersed in water), butter (water dispersed in fat), mayonnaise (oil in water), and ice cream (fat in water which is then frozen).

Milk is an emulsion of fat in water, which is not indefinitely stable as it separates on standing, into skim milk and cream. This is caused by the density differences between the fat and the water, the fat globules rising as predicted by Stokes' law and coalescing at the surface to form a layer of cream. After homogenizing, this separation does not occur as the globules are much reduced in size. Homogenizing is also used with ice cream mixes, which are dispersions of fat and air in sugar solutions, and in the manufacture of margarine.

Emulsion systems are found in milk (naturally occurring) and milk products, ice cream, butter, margarine, salad dressings, meat, bread, and cake products. Emulsifying agents could be natural (proteins, phospholipids, and sterols) or synthetic (glycerol esters, cellulose esters, and sorbitan esters of fatty acids).

Food emulsions vary in composition, for example, milk contains 3–5% milk fat whereas mayonnaise may contain up to 80% oil as the internal phase. Emulsions occur as oil in water, for example, salad cream or water in oil, for example, margarine.

In milk the fat is dispersed throughout the aqueous phase (serum) in the form of globules. Emulsion is stabilized by an adsorbed layer at the fat–serum interface also known as the milk-fat globule membrane. The main instability of milk emulsion is its tendency to cream on standing and, hence, concentrating fat globules at the surface. Creaming is reduced by pressure homogenization.

Ice cream is an oil-in-water emulsion containing about 10–12% fat dispersed in water containing proteins and inorganic salts in a colloidal state and carbohydrates and inorganic salts in solution.

It should be a very stable emulsion to resist breakdown during freezing.

Incorporation of air in the form of foam, formation of ice crystals, and concentration of solubles in the aqueous phase promote emulsion breakdown.

An ice cream mix contains milk or skimmed milk, water, cream, butter or butter oil as fat, condensed milk or skimmed milk powder as solid-nonfat, sweeteners, dextrose, or invert sugar as sugar and gelatine, alginates or gums as stabilizers. The latter stabilize the emulsions by increasing the viscosity and controlling ice crystal growth (Arbuckle, 1986; Marshall et al., 2003).

In butter making the starting material, milk, is an oil-in-water emulsion whereas the final product, butter, is an water-in-oil emulsion (Varnam and Sutherland, 2003).

In the manufacture of butter from fresh cream, pasteurized milk is separated by centrifugation to give cream with a fat content of 35–40%. The cream is then chilled and aged at low temperature for some hours to promote fat crystal growth. Then churning and working follows. Churning involves agitation of the cream in large hollow rotating vessels (churns). They may or may not contain baffles to assist in agitation of the cream. Wooden churns contain ribbed rollers to aid in working the butter. Air is incorporated and foam is produced. The fat globules concentrate and some are disrupted and release free fat, which binds the remaining globules in clumps (butter grains).

As churning continues these clumps grow in size, the foam breaks and the butter milk separates. Churning stops when the clumps reach an optimum size. The buttermilk is run off. Then the butter grains are washed and salted and working starts.

Butter grains are agitated slowly, kneaded, and folded. This results in the release of more free fat, dispersion of water and dispersion of added salt.

Working continues until butter has a fine, uniform texture. Then finally butter is chilled and stored under refrigeration or frozen conditions. A complete phase inversal does not occur; however, we have a water-in-oil emulsion.

At continuous manufacturing of butter in the United Kingdom accelerated churning and working occur at a high speed with beaters. The butter milk runs off and butter grains are worked continuously using twin screw-type devices (Brennan, 1993).

Another methodology involves the production of a concentrated cream containing about 80% fat in an oil-in-water form. This is achieved by a second-stage centrifugal separation. This involves agitation of the cream in jacketed cylinders to bring simultaneous cooling, leading to fat crystallization and phase inversion. Then product is extruded in the form of a ribbon.

In US production of a highly concentrated cream containing up to 90% fat takes place and the oil-in-water emulsion is broken down before, during, or after the second-stage separation. The pure butter oil produced forms the continuous phase of the water-in-oil emulsion. Centrifugal pumps, separators, and homogenizing valves break the concentrated oil-in-water emulsion and a scraped surface heat exchanger is used to cool and agitate the butter oil.

Margarine is a water-in-oil emulsion containing approx. 16–18% aqueous phase (skimmed milk or reconstituted skimmed milk powder inoculated with starter organisms such as *Streptococcus cremoris* and *citrovoris* and ripened until acidity reaches a suitable stage and chilled). The continuous phase consists of both solid and liquid fat. The fat phase is formed from a blend of vegetable and fish oils. The type and proportion of fats and oils used depend on melting point and texture.

The emulsifiers used are oil soluble and maybe natural or synthetic (monoglycerides, sorbitans). Other ingredients added are vitamins A and D, coloring agents, salt, and water. The method of preparation involves feeding fat and aqueous phases into a churn (high-speed paddle agitator in a jacketed vessel) or a series of scraped surface heat exchangers as in the Votator process, and simultaneous emulsification and cooling. Then chilling of the crude emulsion with ice-water or a refrigerated drum and tempering of the emulsion for more than 1 day for the development of correct solid/liquid ratio. Working is then carried out by rollers, working tables, and worms.

Mayonnaise is an oil-in-water emulsion containing 70–85% oil, 10–11% vinegar, 9% egg yolk, 2–3% sugar, salt, mustard, and pepper. Mixing is carried out at low temperatures and the emulsion may be further refined in a colloid mill.

Salad cream is also an oil-in-water emulsion containing some 30–40% oil dispersed in globules. It contains some 55–60% total solids made up of the oil, sugar, salt, egg, mustard, herbs and spices, colors, and stabilizers. Oil is groundnut or cottonseed and the stabilizer is tragacanth gum. Gum is dispersed in part of the vinegar and water and allowed to stand for 4 days after which it is beaten and sieved. Aqueous ingredients are premixed and heated, then cooled and sieved. The premix is agitated vigorously without aeration and the oil is added gradually to prepare an emulsion premix. Then emulsification takes place in a pressure homogenizer, colloid mill, or ultrasonic homogenizer and vacuum-fillled into glass jars.

Emulsions also occur in various meat products such as sausage meat and meat pastes. Emulsification prevents fat separation and affects product texture.

The production of meat emulsions-based products requires the use of different ingredients and additives in order to stabilize the products' structure, and to minimize the losses of fat and water during processing and cooking (Feiner, 2006). Proteins such as caseinate, and salts as polyphosphates, are widely used with this aim (Knipe, 2004).

These are oil-in-water two-phase system emulsions consisting of dispersions of solid fat in an aqueous phase. Emulsifiers are salt-soluble proteins. Emulsions are affected by quantity of soluble protein, temperature, rate of fat addition, mixing speed, types of fats (pork, beef), pH of mix and type of equipment.

Mincers (grinders), bowl choppers, roto-cut machines and emulsitators are used in the preparation of emulsions. Mincers are used for crude size reduction of ingredients. These are forced by a worm, through a perforated plate with knives rotating in contact with its surface. Shearing occurs assisting the emulsion.

The bowl chopper consists of a hemispherical bowl rotating slowly about a vertical axis and contains knives rotating about a horizontal axis. In addition to size reduction, soluble protein is released and emulsification occurs.

The roto-cut machine consists of a vertical rotating cylinder and a set of knives rotating near its edge. Baffles turn the material over just prior to the knives. The emulsitator consists of one or two sets of perforated plates and rotating knives horizontally positioned. Knives drag over the plates and create a vacuum drawing the feed in from the hopper mounted above the plates. Vigorous cutting and shearing occurs leading to a good emulsion.

### 5.2.2 Methods to Form an Emulsion

The formation of emulsion presumes that energy is imparted to the system to overcome the resistance to the creation of new surfaces arising from interfacial tension. Theoretically, the work required to produce an emulsion is equivalent to the product of the interfacial tension and the newly formed surface. Work is done on the phases by subjecting them to violent agitation of the type that subjects large droplets of the internal phase to shear (Brennan, 1993). A stable emulsion is formed if optimum temperature is used along with the correct type and quantity of emulsifier and adequate time. These factors influence the right adsorption of the protective film of emulsifier at the new interface.

### 5.2.3 Particle Size and Size Distribution

The size of an individual particle can be determined by a traveling microscope or a pair of calipers; however, neither of these methods is convenient for fine particles or for a material with a distribution of particle sizes.

Size distribution is determined by sieve analysis (particle size distribution by weight) or automated processes such as image analysis or a Coulter counter (number of particles in a given size range). Image analysis can also give information on particle shape.

Particle size distribution is normally expressed either as $p(D)$ where $p(D)\ dD$ is the fraction of the material with size between $D$ and $D + dD$ or as the cumulative distribution $P(D)$ where $P(D)$ is the fraction of the material with size less than D.

$$P\ (D)\ = \int_0^D p(D)dD \tag{5.22}$$

$D$ can be any measure of the particle size such as diameter $d$ or in the case of irregular particles the volume equivalent diameter de, defined by $V = \pi de^3/6$, where $V$ is the volume of the particle.

Two-particle size distribution is given by the Rosin-Rammler and the log normal distribution. Rosin-Rammler gives the cumulative distribution by weight.

$$P(z) = 1 - \exp\ (-kz) \tag{5.23}$$

where $z = D/D_R$ and $D_R$ and $k$ are parameters chosen to fit the experimental results. The log normal distribution is given by

$$p(x) = \frac{1}{\sigma\sqrt{2\pi}} \exp\left(-\frac{(x-\mu)^2}{2\sigma^2}\right) \tag{5.24}$$

$x$ is $\ln D$ and $\mu = \ln Dg$ where $Dg$ is the geometric mean diameter. If experimental results are on a straight line the distribution is log normal. Values of $\mu$, $\sigma$ can be obtained from the graph (Nedderman, 1992).

## 5.2.4 Emulsion Characteristics

### 5.2.4.1 Emulsion Type

An emulsion consists of two immiscible liquids (usually oil and water), with one of the liquids dispersed as small spherical droplets in the other. In order to maintain the droplet structure, the oil–water interfacial region is coated with necessary layer of emulsifier (Figure 5.11).

Emulsions can be basically classified according to the distribution of the oil and aqueous phases. Systems consisting of oil droplets dispersed in an aqueous continuous phase are called *oil-in-water* or O/W emulsions (e.g., milk, cream, dressings, mayonnaise, beverages, soups, and sauces). O/W emulsions are mainly fluid, although they may have partly crystalline oil phases. Stability of these emulsions may be maintained by adsorption of small-molecule emulsifiers, protein molecules, or aggregates of protein molecules (casein micelles, egg yolk granules), or by their mixtures.

On the other hand, systems consisting of water droplets dispersed in an oil continuous phase are called *water in-oil* or W/O emulsions (e.g., margarine and butter). Their stability depends more on the properties of the fat or oil and the surfactant

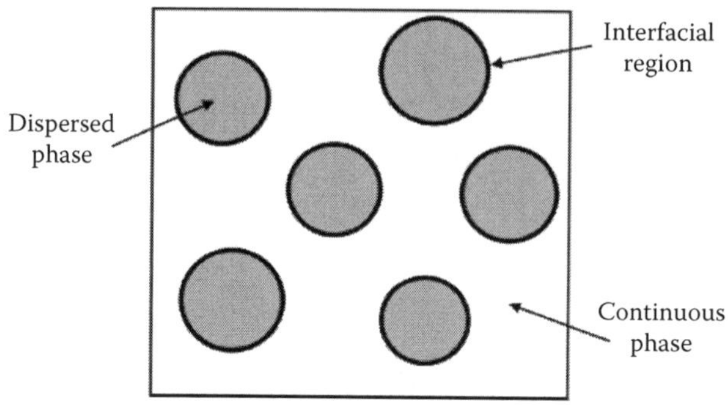


**FIGURE 5.11** Emulsion illustration. The dispersed phase, surrounded by the continuous phase is separated by a layer of surfactant. (Adapted from McClements D.J. 2005. *Food Emulsions: Principles, Practice and Techniques*, 2nd edn., Boca Raton, FL, CRC Press.)

used than on the properties of the aqueous phase, and thus fewer parameters can be varied for their control. Important factors in these emulsions are the crystallinity of the oil phase, the presence of rigid surfactants on the O/W interface, and the presence of agents, which increase the viscosity of the aqueous phase in the droplets. A degree of mechanical stabilization is, therefore, more important in W/O than in O/W emulsions (Krog and Sparso, 2004).

In oil-in-water (O/W) emulsions the stability is highly dependent on the emulsifier. However, in some water-in-oil (W/O) systems (such as butter) the continuous phase is partially solid, which enhances stability and gives the food a firm texture (Robins and Wilde, 2003). In most foods, the droplets' diameters usually range between 0.1 and 100 μm. It should be noted that although the terms colloid and emulsion are sometimes used interchangeably, emulsion should be used when both the dispersed and the continuous phase are liquids (Dickinson and Tanai, 1992).

In some cases, however, it is possible to make multiple emulsions such as water-in-oil-in-water (W/O/W) and oil-in-water-in-oil (O/W/O) emulsions by homogenizing a W/O or an O/W emulsion, respectively, in the presence of suitable surfactants and continuous phase. Multiple emulsions are multi-compartmentalized systems in which oil-in-water (O/W) and water-in-oil (W/O) coexist, where the globules of the dispersed phase themselves contain even smaller dispersed droplets (Dalgleish, 2001, 2006).

W/O/W emulsions have some advantages over conventional O/W emulsions such as they are better delivery systems for bioactive lipids and for encapsulation, protection and release of hydrophilic components (Jiménez-Colmenero, 2013; McClements et al., 2007). These emulsions have been used as a means of micro-encapsulation in pharmacology (carriers for anticancer agents, hormones, steroids, etc.), cosmetics (easy application of creams with encapsulated compounds) and other industrial uses because of their ability to trap and protect various substances and control their release from inside one phase to another (Benichou et al., 2004; Kukizaki and Goto, 2007). Double emulsions can also be used for the preparation of micro- and nano-capsulates (in solid or semi-solid form) containing hydrophilic and lipophilic compounds. Multiple emulsions may offer some advantages for food applications, since it has been found to be a potentially useful strategy for producing low calorie and reduced fat products, masking flavors, preventing oxidation, and improving sensory characteristics of foods, or controlling the release and protecting labile ingredients during eating and digestion (Douglas, 2004; Muschiolik, 2007).

#### 5.2.4.2 Micro- and Nano-Emulsions

Nano and microemulsions are two types of colloidal dispersions suitable for encapsulation and delivery of hydrophilic or lipophilic components within the food industry. However, there are some common misapprehensions in the literature regarding nano-emulsions that arise from their similarities to microemulsions.

Microemulsions are thermodynamically stable, transparent, and isotropic dispersions with particles sizes ranging from 10 to 100 nm (Lawrence and Rees, 2000). They are composed of two phases. Each phase is being dispersed in the other as small spherical droplets. Between the two phases exists an interfacial layer consisting of some necessary surfactant material (Flanagan et al., 2006). Microemulsions are of great interest in the food and pharmaceutical areas, since they are suitable as micro-reactors for the synthesis of food flavors, as a media for the solubilization of water or oil insoluble nutraceuticals, and as the delivery system for water-soluble nutrients, flavors, and colorants in foods. Moreover, microemulsion can provide a well-controlled medium for the incorporation of active ingredients and may protect the solubilized components from undesired degradation (Tadros, 2009). They are composed of two phases that are dispersed in each other as small spherical droplets with an interfacial layer consisting of some necessary surfactant substance (Flanagan and Singh, 2006).

Nano-emulsions on the other hand are emulsions with an extremely small droplet size, which can overlap those of microemulsions (Gutiérrez et al., 2008). Nanoemulsions are transparent or bluish, kinetically stable systems with a typical particle size range of 10–500 nm, as established by most researchers (Ghosh et al., 2013; Rashidi, 2011; Tang et al., 2012); 10–100 nm is suggested by some researchers as well (Rashidi, 2011). They are nonequilibrium systems with a spontaneous tendency to phase separation (substantial separation of the dispersed phase from the dispersant). Nevertheless, they may have a long kinetic stability and are reasonably resistant to creaming or sedimentation as well as to flocculation (Figure 5.12). Being suitable as media for the solubilization of nutraceuticals and as the delivery system for controlled release of water-soluble nutrients, flavors, and colorants in foods, they are also of great interest in various pharmaceuticals applications (Velikov and Pelan, 2008). They can be prepared from food-grade ingredients using relatively simple processing operations, such as mixing, shearing, and homogenization (Velikov and Pelan, 2008).

Therefore, micro- and nano-emulsion are fundamentally different. Microemulsions are equilibrium systems (i.e., thermodynamically stable), while nano-emulsions are nonequilibrium systems with a spontaneous tendency to separate into the constituent phases. Nevertheless, nano-emulsions may possess a relatively high kinetic stability, even for several years (Gutiérrez et al., 2008).

Moreover they differ notably in their behavior toward dilution and temperature fluctuations. The morphology of micro-emulsions is strongly affected and even

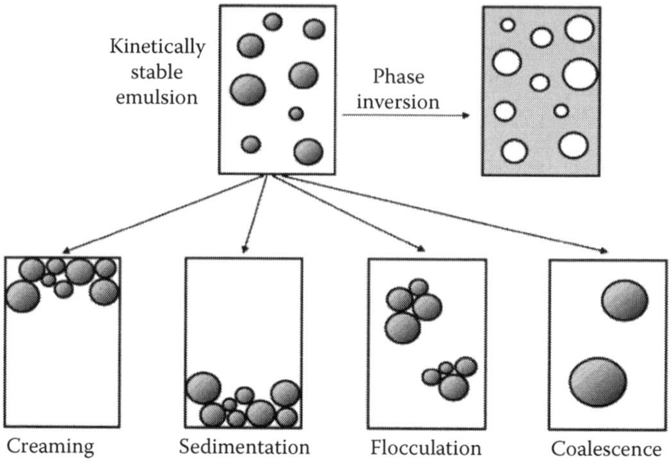


**FIGURE 5.12** Emulsion instability processes. (Adapted from McClements D.J. 2005. *Food Emulsions: Principles, Practice and Techniques*, 2nd edn., Boca Raton, FL, CRC Press.)

broken-up by temperature changes and/or dilutions, whereas nano-emulsion droplets will remain stable in such conditions of stress. This could have a significant effect at the final product where the emulsion is meant to be incorporated (Anton and Vandamme, 2011). Table 5.1 shows the properties of each type of emulsion discussed above in comparison with the conventional macroemulsion systems.

After deciding the most convenient type (micro- or nano-) of emulsion made, depending on the application, thought must be given concerning the ingredients to be used. The widespread application of nano- and micro-emulsions in many food products is currently limited because of a number of technical and practical reasons. First, there are only a limited number of food-grade surfactants currently available for preparing and stabilizing these systems (Kralova and Sjoblom, 2009). Also, some surfactants are not permitted in foods whereas others must be added only at low concentrations due to regulatory, financial, or sensory reasons. In addition, food

**TABLE 5.1**
**Different Emulsions**

| Property | Macroemulsion | Nanoemulsion | Microemulsion |
|---|---|---|---|
| Appearance | Formulation-dependent | Transparent to milky | Transparent |
| Preparation methods | Classic homogenization | High energy (pressure) | Low-energy emulsification |
| Surfactant load | Fairly low | Medium (<10%) | Fairly high (10–20%) |
| Droplet size | 0,5–100 μm | 100–1000 nm | 10–100 nm |
| Thermodynamic stability | Unstable; kinetically stable | Unstable; kinetically stable | Stable |

*Source:* Adapted from Jafari, S.M. et al. 2008. *Food Hydrocoll.*, 22: 1191–1202.

products are complex systems and there is currently a poor understanding of the influence of sample composition and environmental conditions on the formulation and stability of specific kinds of colloidal delivery systems (Salager et al., 2005). Finally, the edible oils have huge differences in their composition. For example, in the case of viscous oils it is difficult to produce low droplet size emulsions during high-pressure homogenization due to the relatively high oil-phase viscosity and interfacial tension (Salager et al., 2005).

## 5.2.5 Emulsion Components

### 5.2.5.1 Aqueous Phase

The water concentration plays an extremely crucial role since it determines the physicochemical and sensorial properties of food emulsions (solubility, conformation, and interactions of the other components present in aqueous solutions). It is therefore important to understand the contribution of water to the overall properties of food emulsions (McClements, 2005).

The role of the aqueous phase is mainly as a solvent to the water-soluble components (salts, emulsifiers, proteins, polysaccharides, etc.), either as the continuous phase in the case of O–W emulsions or as the dispersed phase in W–O emulsions. Water has great interfacial tension because of its molecular structure; strong orientation-dependent attractive hydrogen bonds, which make the breaking of the dispersed phase into fine droplets difficult. Thus, emulsifiers are needed in order to reduce the interfacial tension forces. The pH, ionic strength, and emulsifier concentration of the aqueous phase influence the formation and physical characteristics of the emulsion, by altering the size of droplets and the interactions between them (Robins and Wilde, 2003).

### 5.2.5.2 Oil Phase

Fats and oils are part of a group of compounds known as lipids. Lipids by definition are soluble compounds in organic solvents, but insoluble or only sparingly soluble in water (McClements, 2005). The oil phase in food emulsions, whether from animal or plant sources, is mainly in the form of triglycerides. The oil additionally contains small amounts of di- and monoglycerides, polar lipids, and free fatty acids. These lipids may be surface active, more water soluble, and are commonly used as food emulsifiers. Moreover, fat and lipids are essential nutritional compounds and provide a characteristic mouth feel to products such as cream, butter, and cheese. The oil phase is also used as a solvent for the more lipophilic emulsifiers, oil-soluble nutrients (e.g., oil-soluble vitamins and essential fatty acids) and flavor compounds (Robins and Wilde, 2003).

### 5.2.5.3 Emulsifiers

Emulsifiers are defined as substances that reduce surface tension between oil and water or air and water, thus enhancing emulsification and increasing emulsion stability (Dickinson and Tanai, 1992). The emulsifiers need to be surface active in order to form an interfacial layer around the emulsion droplets. This adsorbed layer lowers the interfacial tension, which leads to emulsification, and prevents the emulsion destabilization. Thus, emulsifiers must have amphiphilic properties. They can be either low-molecular-weight emulsifiers or macromolecular polymers (e.g., proteins).

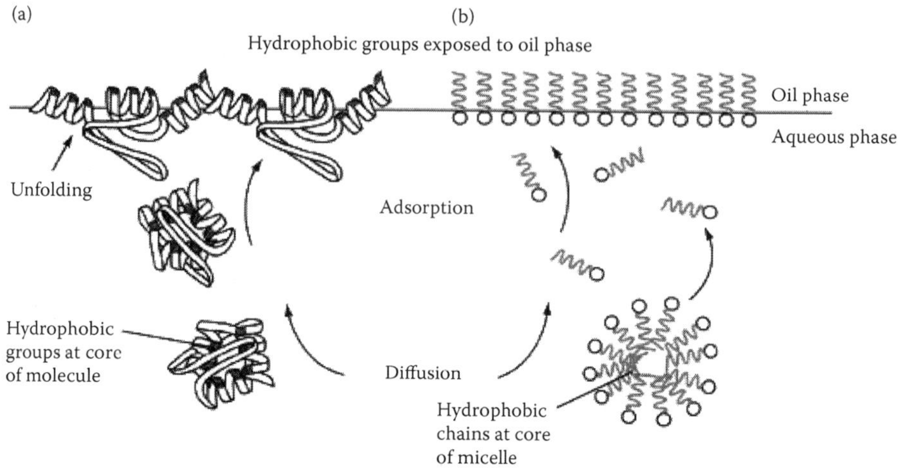


**FIGURE 5.13** Emulsifier structure and adsorption of (a) proteins/polymers and (b) low-molecular-weight emulsifiers/surfactants. Hydrophobic groups are shaded in gray. Proteins form a folded structure in solution, keeping the hydrophobic amino acids away from aqueous solution. Upon adsorption, they unfold to expose these groups to the oil phase. Similarly, surfactants and emulsifiers form micelles, holding the hydrophobic chains in the core of the structure, which then dissociate and adsorb to the interface. (Adapted from Robins, M. and Wilde, P. 2003. In: *Encyclopedia of Food Sciences and Nutrition*, 2nd edn., ed. Caballero, A., pp. 1517–1524, Oxford, UK, Academic Press.)

The low-molecular-weight emulsifiers have a small hydrophilic head group and one or more hydrocarbon chains; they include lecithins, mono- and diglycerides, acid esters of mono- and diglycerides, polyglycerol esters, sucrose esters, sorbitan esters and polysorbates and propylene glycol fatty acid esters. Figure 5.13 illustrates the emulsifier structure and adsorption of (a) proteins/polymers and (b) low-molecular-weight emulsifiers/surfactants (Robins and Wilde, 2003).

#### *5.2.5.3.1 Proteins*

Proteins act as emulsifiers because of their individual molecular structures and they provide many food characteristic properties to emulsions (McClements, 2005). The main function of proteins is either to increase stability toward coalescence or creaming/precipitation during long-term storage (e.g., recombined milk, coffee whiteners, salad dressings, etc.) or to induce destabilization and increase whippability of emulsions to be aerated (e.g., ice cream mix, cream, toppings, etc.) (Dickinson and Tanai, 1992).

Proteins are large biological molecules, or macromolecules, consisting of one or more chains of amino acids. All proteins in their native states possess specific three-dimensional structures, which are maintained in solution, unless they are subjected to extreme conditions such as intensive heating. They have hydrophilic characteristics because of the presence of carboxyl or amido groups in their structure. Consequently, when they are adsorbed to an oil–water interface, it is very likely that the side chains of the amino acids penetrate the interface. However, some proteins possess amino acids with hydrophobic side chains that will adsorb to the oil–water interface. When a protein is adsorbed, the structure of the protein itself will prevent close packing of the points of contact with the interface, and, as a result, the adsorption of a protein reduces the interfacial tension less than does the adsorption of small molecules (Dalgleish, 2001).

Although some proteins are excellent emulsifiers, not all proteins can be strongly absorbed to an O/W interface, because their side chains are strongly hydrophilic or because they possess rigid structures that do not allow the protein to attach to the interface. Gelatin acts in this way due to its hydrophilic character and its large and rather rigid molecule. Lysozyme, although can be absorbed to the droplet interfaces, is not a good emulsifier probably because of its relatively inflexible structure (Dalgleish, 2001).

#### *5.2.5.3.2 Small-Molecule Emulsifiers*

Various surfactants permitted for human consumption are used in the food industry for emulsion preparations. Small-molecule surfactants (monoglycerides and diglycerides, sorbitan esters of fatty acids, polyoxyethylene sorbitan esters of fatty acids, phospholipids, and many others) generally contain long-chain fatty acid residues that act as the hydrophobic group which is attached to the lipid phase of the oil–water interface and causes adsorption. The head groups of these compounds may vary according to its molecular structure, ranging from glycerol (in monoglycerides and diglycerides) and substituted phospho-glyceryl moieties (in phospholipids) to sorbitan highly substituted with polyoxyethylene chains. The classification of these emulsifiers is generally performed by the hydrophile–lipophile balance (HLB) system (Dalgleish, 2001).

#### *5.2.5.3.3 Hydrophile–Lipophile Balance*

The HLB system was developed in order to facilitate the choice of the optimum emulsifying agent depending on the use (Kruglyakov, 2000). The hydrophilic–lipophilic balance (hydrophile–lipophyle balance) system for the classification of emulsifiers was first introduced by Griffin for the Atlas Powder Company in America in the trade catalogue *Atlas of Surface Active Agents*. According to Griffin, the HLB theory signifies the balance between the hydrophilic and hydrophobic parts of the surfactant molecules with respect to their strength and efficiency, and may be defined as

$$\mathrm{HLB} = 20^{\frac{M_0}{M}} \tag{5.25}$$

where $M_0$ is the molecular weight of the hydrophilic part of the emulsifier molecule and $M$ is the total molecular weight. The scale is therefore from 0 to 20 (Griffin, 1949; Friberg, 1990).

Apart from Griffin, various authors have tried to offer a more accurate way for surfactant classification. A widely used method for the determination of HLB number is the following:

$$\mathrm{HLB} = 7 + \sum(\text{hydrophilic group numbers}) - \sum(\text{lipophilic group numbers}) \tag{5.26}$$

**TABLE 5.2**
**HLB Group Numbers**

| Hydrophylic | | Lipophylic | |
|---|---|---|---|
| Group | Number | Group | Number |
| $-SO_4Na$ | 39 | –CH | 0.5 |
| –COOH | 21 | $-CH_2$ | 0.5 |
| –COONa | 19 | $-CH_3$ | 0.5 |
| Sulfonate | 11 | $-CH_2CH_2O$ | 0.33 |
| Ester | 7 | | |
| –OH | 2 | | |

The sums of the group numbers of all the lipophilic groups and hydrophilic groups are substituted into the above equation and the HLB number is calculated. Although originally the above equation was considered semiempirical, it was proved that the sums of each group corresponds to the free energy changes in the hydrophilic and lipophilic parts of the molecule when the micelles are formed during emulsification (McClements, 2005). Table 5.2 summarizes the HLB group numbers of the major hydrophilic and lipophylic groups that exist in the surfactant molecules.

Moreover, it has been established that HLB is related to the distribution coefficient, $K_D$, of the emulsifier in the two separate phases of the emulsion. $K_D$ can be calculated as

$$K_D = \frac{C_{water}}{C_{oil}} \tag{5.27}$$

where $C_{water}$ and $C_{oil}$ are the equilibrium molar concentrations of the emulsifier in the water and oil phases, respectively. The HLB number can be thus calculated in the following way (Birdi, 2010):

$$\text{HLB} = 7 + 0.36\ln(K_D) \tag{5.28}$$

During the addition of an emulsifier into an emulsion, the emulsifier tends to get distributed both in the water and in the oil phase. The degree of distribution in each phase depends on its structure and thus in its HLB number. Emulsifiers with a highly hydrophilic nature have high HLB values, whereas those which are mostly lipophilic (low hydrophilic nature) have a low HLB value (Birdi, 2010).

Often more than one surfactant is used in an emulsion. In this case, it is assumed that the HLB number of the surfactant mixture is an additive function of the HLB numbers characteristic to the individual surfactants proportionally to the concentration of each surfactant (Kruglyakov, 2000).

The HLB value of an emulsifier may be used as a guide to its most appropriate application. The emulsifiers with a high HLB value stabilize oil-in-water emulsions and those with a low HLB value stabilize water-in-oil systems. Table 5.3 summarizes

**TABLE 5.3**
**HLB Values According to the Application and Water Solubility Behavior**

| Properties in Water | HLB-Value |
|---|---|
| Non-dispersible | 0–2 |
| Poor dispersibility | 2–6 |
| Milky dispersion | 6–10 |
| Stable milky dispersion | 8–12 |
| Transparent, clear dispersion | 12–15 |
| Clear colloidal solution | 15–20 |
| **Field of Application** | |
| Antifoaming agent | 1–3 |
| Water-in-oil emulsifier | 3–6 |
| Wetting agent | 7–9 |
| Oil-in-water emulsifier | 8–18 |
| Detergent | 13–15 |
| Solubilizer | 15–18 |

*Source:* Adapted from Friberg, S.E. 1990. *Emulsion Stability in Food Emulsions*, 3rd ed., eds Larsson, K., and Friberg, S.E., Marcel Dekker, New York.

the HLB values needed for different applications according to water solubility behavior.

However, the HLB system appears to have some disadvantages. It does not take into account the effect of temperature, the presence of additives and the concentration of emulsifying agents. Therefore, each surfactant must be tested to the applicable food product. The HLB system has been discussed thoroughly by Griffin and reviewed by Friberg (Griffin, 1949; Friberg, 1990).

#### *5.2.5.3.4 Lecithins*

Lecithins are among the most commonly used emulsifiers in the food industry. However, it is difficult to find organized information about the emulsification and application properties of lecithin (Bueschelberger, 2004). Lecithins are natural surface-active molecules that can be extracted from a variety of natural sources, including soybeans, rapeseed, and egg (McClements, 2005). Lecithin is not a distinct, standard compound but a natural mixture of a series of surface-active components whose overall emulsifying performance should be taken into account.

They are apparently complex systems of surface-active components, which through processes like modification, fractionation, de-oiling, compounding, and combinations thereof, can be further designed to their final application.

Concerning their dispersing behavior, lecithins rarely have an HLB value higher than 9 (this would be for hydrolyzed acetylated lecithins). Typically they are classified between 2 and 7 HLB values. This means that lecithins are actually not capable of forming oil-in-water emulsions. However, a standard fluid lecithin (HLB 2) can form such emulsion (Bueschelberger, 2004).

Lecithins are widely used in the food industry for the production of chocolates, chocolate products with fat-based fillings, soft and hard caramels, chewing gums, baked products, frozen doughs and instant products like whole milk powder, skim milk powder, infant formulations, protein drinks, cocoa and chocolate drinks, soups, sauces, and so on (Bueschelberger, 2004; Friberg, 1990).

#### *5.2.5.3.5 Mono- and Diglycerides*

Mono- and diglycerides are generally manufactured by interesterification (glycerolysis) of triglycerides with glycerol. They are commonly used to describe a series of surfactants produced by interesterification of fats or oils with glycerol. This procedure produces a complex mixture of monoacylglycerides, diacylglycerides, triacylglycerides, glycerol, and frcc fatty acids, which is usually referred to as monodiglycerides (McClements, 2005).

Monoglycerides possess a lipophilic character and are, therefore, assigned with a low HLB number (3–6). They dissolve in oil and in, stabilized water-in-oil (W/O) emulsions to form reversed micelles in oil (Moonen and Bas, 2004).

The main applications of mono- and diglycerides in food are typically in fat-based products, such as margarine, spreads and bakery fats (shortenings) and cake mixes. In dairy emulsions, mono- and diglycerides are used in ice cream and recombined milk, in combination with hydrocolloids. Distilled monoglycerides are also used in fat-free foodstuffs or products with very low fat content (Moonen and Bas, 2004).

#### *5.2.5.3.6 Acid Esters of Mono- and Diglycerides*

The free hydroxyl groups in monoglycerides can be esterified with organic acids such as acetic, lactic, diacetyltartaric, citric, and succinic acids and, thus, from more lipophilic or hydrophilic derivatives of monoglycerides (Dickinson and Tanai, 1992).

*5.2.5.3.6.1 ACETEM* ACETEM are also known as acetic acid esters of mono- and diglycerides, acetoglycerides, acetylated mono- and diglycerides, acetic- and fatty acid esters of glycerol and mono- and diglycerides of fatty acids esterified with acetic acid (Gaupp and Adams, 2004). ACETEM are non-ionic substances. During their manufacturing process one or both of the free hydroxyl groups of the mono- and diglycerides, which are responsible for the polar properties in the molecule, are esterified with acetic acid. Their HLB value will be lower compared to the mono- and diglycerides. The use of ACETEM as emulsifiers in water-in-oil emulsions is rather limited as they may reduce the surface tension of the phase barrier (Dickinson and Tanai, 1992).

Acetic acid esters of mono- and diglycerides have excellent aerating and foam stabilizing properties. They are also used as lubricants and release agents. Applications include topping powders, whipped topping concentrates, chewing gum base, coatings cakes (Gaupp and Adams, 2004; Dickinson and Tanai, 1992).

*5.2.5.3.6.2 LACTEM* LACTEM are also known as lactic acid esters of mono- and diglycerides, lactoglycerides, lactic acid and fatty acid esters of glycerol, mono- and diglycerides of fatty acids esterified with lactic acid, glycerol-lacto esters of fatty acids, lactated mono- and diglycerides and GLP (Dickinson and Tanai, 1992; Gaupp and Adams, 2004). LACTEM are nonionic substances. During the esterification with lactic acid, the free hydroxyl groups of the mono- and diglycerides will disappear which are responsible for their hydrophilic character. However, as a result of the esterification of lactic acid, hydroxyl groups will now be included into the molecule. When the esterification with lactic acid is complete, their number will remain same, as in the nonesterified mono- and diglyceride, but the total molecular weight of the resulting molecule is now much higher. Therefore, its HLB value will be lower compared to the mono- and diglycerides (Dickinson and Tanai, 1992; Gaupp and Adams, 2004; Wooster et al., 2008). LACTEM are able to reduce the surface tension of oil–water systems and can be classified as water-in-oil emulsifiers.

Applications of LACTEM include topping powders, nondairy creams, dairy and recombined creams, fine baked goods, shortening and chocolate compounds (Dickinson and Tanai, 1992; Gaupp and Adams, 2004).

*5.2.5.3.6.3 CITREM* CITREM are citric acid esters of mono- and diglycerides of edible fatty acids also known as mono- and diglycerides of fatty acids, esterified with citric acid, monoglycerol citric acid esters, monoglycerol citrates, citric acid monoglycerides, and so on. CITREM are ionic oil-in-water emulsifiers. Some of the free hydroxylic groups of the mono- and diglycerides disappear after the esterification, but they are partially replaced by the free hydroxylic groups of the citric acid. The hydrophilic part of the whole molecule is therefore formed by hydroxylic groups, deriving from the mono- and diglyceride part, the hydroxylic group of the citric acid, and their carboxylic group. During the partial or total neutralization of the carboxylic group, the hydrophilic part of the molecule will not increase substantially. But by its transformation into the salt, the hydrophilicity of the whole molecule will increase tremendously. This process allows CITREM to be manufactured depending on the application. The HLB value of CITREM will be much higher compared to the mono- and diglycerides they are derived from (Dickinson and Tanai, 1992; Wooster et al., 2008).

Citric acid esters of mono- and diglycerides are widely used within the food industry as emulsifiers, stabilizers, antispattering agents and as synergistic components with antioxidants. More specifically CITREM are used as fat stabilizers, in baked and confectionary products, in margarine, in mayonnaise, in meat products, in low-fat products, and as a flavor stabilizer and also as effective non-GMO lecithin substitute in margarine and chocolate (Dickinson and Tanai, 1992; Friberg, 1990; Gaupp and Adams, 2004).

*5.2.5.3.6.4 DATEM* DATEM are glycerol derivates esterified with edible fatty acids and mono- and diacetyl tartaric acids. DATEM are ionic oil-in-water emulsifiers. During the reaction of mono- and diglycerides with diacetylated tartaric acid anhydride, the free hydroxyl groups of the mono- and diglycerides, which are responsible for their hydrophilic properties, are esterified. At the same time, a huge hydrophilic and polar moiety with a free carboxylic group is transferred into the molecule via the esterified diacetylated tartaric acid. Due to the partial esterification of hydroxyl groups of the mono- and diglycerides with diacetylated tartaric acid anhydride, free hydroxyl groups, together with the diacetylated tartaric acid moiety, form the hydrophilic part into the DATEM molecules. DATEM are more hydrophilic compared to the mono- and diglycerides that they are made of. The hydrophilic part within these molecules is therefore bigger and hence they show a higher HLB value (Gaupp and Adams, 2004).

Mono- and diacetyl tartaric acid esters of mono- and diglycerides are used as dough conditioners for all baked products, particularly yeast-leavened products, white bread, rolls rusks, and in flour mixes for convenience food. Other applications include beverage whiteners, cream analogues, chewing gum, meat- and poultry products, emulsified sauces, canned coffee or tea and carriers of solvents for colors and food antioxidants (Gaupp and Adams, 2004).

*5.2.5.3.6.5 TATEM* Tartaric acid esters of mono- and diglycerides (TATEM) are permitted as food additives by the EU legislation. However, they are not preferred by the food manufacturers since their use can be well compensated for by other esterified mono- and diglycerides (Gaupp and Adams, 2004).

*5.2.5.3.6.6 MATEM* MATEM are mono- and diglycerides of fatty acids esterified with acetic- and diacetylated tartaric acid also known as diacetylatedtartaric, acetic and fatty acid esters of glycerol, mixed. They normally consist of all possible products obtained by the esterification of acetic-, tartaric-, diacetylated tartaric acid, and edible fatty acids with glycerol.

They are obtained either by reacting mono- and diglycerides of fatty acids with tartaric acid anhydride in the presence of acetic acid, or by esterification of mono- and diglycerides with tartaric acid and acetic acid in the presence of acetic acid anhydride (Dickinson and Tanai, 1992; Friberg, 1990; Gaupp and Adams, 2004; McClements and Rao, 2011).

Mono- and diacetyl tartaric acid esters of mono- and diglycerides of fatty acids are used as dough conditioners for all baked products, particularly yeast leavened products, white bread and rusks, and in ready mixed flours, particularly for use in the "all-in" method. In general MATEM have similar applications to DATEM (Gaupp and Adams, 2004).

#### *5.2.5.3.7 Polyglycerol Esters*

Polyglycerol esters are used extensively within the food industry because of their amphiphilic nature in various food products. The esters (fatty acid esters of polyglycerol) have characteristic interfacial properties due to the coexistence of hydrophilic and lipophilic moieties within the same molecule. The corresponding polyglycerol esters are produced from polyglycerol and fatty acids in a direct esterification of (or alternatively an inter-esterification between) triglycerides and polyglycerol.

Polyglycerol esters have an HLB value between 6 and 11 and thus can reduce the surface tension between oil and water, acting as an oil-in-water emulsifier (Dickinson and Tanai, 1992; Hasenhuettl, 2008; McClements, 2005; Norn, 2004; Wooster et al., 2008).

Polyglycerol esters are widely used in foods, cosmetic products. Such applications are margarine, low-fat cake batters, starch-based products, such as cakes and breads, creams and toppings, whipping cream substitutes, reduced fat-imitation cream with improved freeze- and acid-resistance properties, polyglycerol esters have also found applications in the pharmaceutical and cosmetic industries (Dickinson and Tanai, 1992; Norn, 2004; Orthoefer, 2008).

#### *5.2.5.3.8 Sucrose Esters*

Sucrose esters are nonionic compounds synthesized by esterification of fatty acids (or natural glycerides) with sucrose (Nelen and Cooper, 2004). Fully esterified sucrose fatty acid esters have been widely investigated as synthetic fat replacements and their synthesis have been reviewed. Partially esterified sucrose esters are versatile emulsifiers for food products. The distribution of mono-, di-, and triesters, and therefore their HLB value, depends on the ratio of fatty acid and sucrose in the reaction mixture. The degree of saturation and chain length of the fatty acid also influence their functional properties (Hasenhuettl, 2008).

Sucrose esters can have a wide range (1–18) of HLB value. Low ester substitution leads to high HLB, whereas higher substitutions lead to lower HLB values. Sucrose esters are commonly used in the food industry for the production of dressing and sauces, mayonnaise-like products and dressings, confectionery products, caramels and high boils (candies), chocolate, baked goods, bread, icings and fillings marshmallows, and other aerated products (Norn, 2004; Weyland and Hartel, 2008).

#### *5.2.5.3.9 Sorbitan Esters and Polysorbates*

Sorbitan esters of fatty acids are derived from a reaction of sorbitol and a commercial grade fatty acid. Sorbitan tristearate is the most common sorbitan ester used in the food industry. The sorbitan esters of stearic, lauric, oleic, and other fatty acids can all be further reacted with ethylene oxide to produce the polyoxyethylene sorbitan esters or polysorbates. Polysorbates are made by further modifying the sorbitan esters with ethylene oxide in specialized pressure vessels called autoclaves. Small amounts of a basic catalyst, such as potassium hydroxide, are used during the process (Cottrell and Van Peij, 2004; Dickinson and Tanai, 1992; Douglas, 2004).

Sorbitan esters of fatty acids are nonionic, low HLB lipophylic emulsifiers. The hydrophilic/lipophylic properties of the sorbitan ester depend on the degree and type of fatty acid esterified. The shorter the chain length of the fatty acid, the lower the HLB (Table 5.4). The polysorbate surfactants are amongst the most hydrophilic or water-soluble emulsifiers permitted for human consumption, because of their long polyoxyethylene chain. Their properties are controlled by the different fatty acid used in each product, since the ethylene oxide chain length, is stable for all products. All polysorbates, are considered hydrophilic and are truly water soluble (Table 5.5) (Cottrell and Van Peij, 2004).

Sorbitan esters are used for the production of baked products, cakes, bread, dried yeast, margarine, and spreads. Polysorbates are used for the production of beverages, chocolate, and other products. Both types of emulsifiers are limited because of their inherent flavor profile that lessens the sensory acceptance of the final product. They are, however, necessary for many applications food and beverage industry (Cottrell and Van Peij, 2004).

**TABLE 5.4**
**Nomenclature and Physical Characteristics of Sorbitan Esters**

| Generic Name | Common Name | Physical Form (25°C) | HLB* (±1) |
|---|---|---|---|
| Sorbitan monolaurate | Span 20 | Liquid | 8.6 |
| Sorbitan monopalmitate | Span 40 | Solid | 6.7 |
| Sorbitan monostearate | Span 60 | Solid | 4.7 |
| Sorbitan monooleate | Span 80 | Liquid | 4.3 |
| Sorbitan tristearate | Span 65 | Solid | 2.1 |
| Sorbitan trioleate | Span 85 | Liquid | 1.8 |

*Source:* Adapted from Cottrell, T. and Van Peij, J. 2004. In *Emulsifiers in Food Technology*, ed Whitehurst, R.J., UK, Blackwell Publishing.
*Note:* HLB* = Hydrophyle–lipophyle balance.

**TABLE 5.5**
**Nomenclature and Physical Characteristics of Ethoxylated Esters**

| Generic Name | Common Name | Other Names | Physical Form (25°C) | HLB (±1) |
|---|---|---|---|---|
| Polysorbate 20 | Tween 20 | Polyoxyethylene (20) sorbitan monolaurate | Liquid | 16.7 |
| Polysorbate 40 | Tween 40 | Polyoxyethylene (20) sorbitan monopalmitate | Liquid | 15.6 |
| Polysorbate 60 | Tween 60 | Polyoxyethylene (20) sorbitan monostearate | Gel | 14.9 |
| Polysorbate 80 | Tween 80 | Polyoxyethylene (20) sorbitan monooleate | Liquid | 15.0 |
| Polysorbate 65 | Tween 65 | Polyoxyethylene (20) sorbitan tristearate | Solid | 10.5 |
| Polysorbate 85 | Tween 85 | Polyoxyethylene (20) sorbitan trioleate | Liquid | 11.0 |
| Polyglycerate 60 | Ethoxylated mono- and di-glycerides | Polyoxyethylene (20) mono- and di-glycerides of fatty acids | Liquid | 13.0 |
| - | Myrj | Polyoxyethylene (8) stearate | Gel | 11.1 |

*Source:* Cottrell, T. and Van Peij, J. 2004. In *Emulsifiers in Food Technology*, ed Whitehurst, R.J., UK, Blackwell Publishing.

Moreover, sorbitan esters and polysorbates are often used as a mixture in various applications. A characteristic example is for the production of microemulsions. As has been discussed above microemulsions, similar in principle to solubilization, are thermodynamically stable suspensions that more closely resemble swollen micelles than conventional macroemulsions. There have been numerous studies proving the efficiency of polysorbate and sorbitan esters mixes to form microemulsions (Cottrell and Van Peij, 2004; Dickinson and Tanai, 1992; Flanagan et al., 2006; Flanagan and Singh, 2006; ).

#### *5.2.5.3.10 Propylene Glycol Fatty Acid Esters*

Propylene glycol fatty acid esters are produced by the esterification of propylene glycol with fatty acids, typically in the form of commercial stearic acid blends which are lipophilic, oil-soluble emulsifiers with specific crystalline properties.

The hydrophile–lipophile balance and mean molecular weight are controlled by the degree of glycerol polymerization and the fatty acid/polyglycerol ratio. These factors along with the nature of the fatty acid determine the product's melting point, whether it will be solid, liquid, or semisolid in ambient conditions (Friberg, 1990). This melting/crystallization behavior is related to the chemical composition of the compound regarding the fatty acid chain length, monoester content, and positional isomers (Sparso and Krog 2004). The degree of esterification and HLB depend on the ratio of fatty acid to polyglycerol in the reaction mixture (Sparso and Krog 2004).

Propylene glycol fatty acid esters are used in the food industry for the production of aerated bakery products and cake mixes, for sponge cakes, fat-free cakes, for dessert products, toppings and nondairy whipping creams (Orthoefer 2008; Sparso and Krog 2004).

### 5.2.5.4 Stabilizers

Although emulsifiers help stabilize emulsions, the term "stabilizer" in the context of emulsions refers to the polymers which are added to improve the stability of emulsions. Such polymers may be gums as xanthan, guar, carageenan, and so on. These ingredients can be divided into "thickening agents" and "gelling agents" (McClements, 2005). The addition of these polymers forms a network that can effectively encapsulate the emulsion droplets. This immobilization prevents flocculation of the droplets and reduces creaming. Lower concentrations of polymer can induce flocculation by the depletion mechanism. Under different conditions, flocculation by polymers can induce rapid creaming or sedimentation (Robins and Wilde, 2003).

### 5.2.5.5 Other Ingredients

Food emulsions are complex systems and may also contain salts, sugars, flavors, colors, and preservatives. These components may affect the colloidal structure of the emulsion. Salt, due to its ionic character, is perhaps the most important component as it can affect the solubility of functional ingredients such as proteins and polysaccharides. Salt also affects the electrostatic colloidal stability by screening the electrostatic repulsion forces between particles. Sugars can lower the dielectric constant of the continuous phase, which affects colloidal interactions, whereas certain sugars, such as trehalose, are known to protect proteins against denaturation during heating and drying. Flavors, colors, and preservatives are usually present at relatively low levels, and generally do not influence the physical characteristics. However, some components are volatile and may induce partitioning from one phase to the other (Robins and Wilde, 2003).

### 5.2.6 Properties of Emulsions

A wide variety of different analytical methods have been developed for the characterization of emulsion properties. Particle sizing characterization however, is of the most importance since it determines the type of the emulsion, and indicates the emulsion stability and other properties of emulsion (e.g., color, flavor, rheology).

#### 5.2.6.1 Emulsion Stability

Generally, emulsion stability is the ability of an emulsion to resist changes in its physicochemical properties over time (McClements, 2005). Emulsions are thermodynamically unstable and in sufficient time will eventually separate into two different phases of oil and water. However, temporary stability can be achieved providing food products with sufficient shelf-life. Thus, it is of high importance to identify the physical and chemical mechanisms responsible for the instability of emulsions (Dickinson and Tanai, 1992).

Food emulsions may become unstable due to a variety of different physicochemical mechanisms, including gravitational separation (creaming/sedimentation), flocculation, coalescence, partial coalescence, Ostwald ripening and phase inversion (Dickinson and Tanai, 1992; McClements et al., 2007).

*Gravitational separation* is the process during which the droplets of the dispersed phase move upward ("creaming") because of their lower density, or downwards ("sedimentation") because of their higher density in comparison with the surrounding liquid. Thus the disperse phase gets separated from an emulsion. The creaming rate (or settling rate for disperse phases more dense than the continuous phase) can be estimated from the Stokes' equation:

$$u = 2r^2(\rho - \rho_o)\frac{g}{9\eta} \tag{5.29}$$

where, $u$ is the creaming (settling) rate, $r$ is the droplet radius, $\rho$ is the density of the droplet, $\rho_o$ is the density of the dispersion medium, $\eta$ is the viscosity of the dispersion medium (continuous phase) and $g$ is the local acceleration due to gravity.

The density difference, $(\rho - \rho_o)$, is negative for creaming (an O/W emulsion), but positive for settling (a W/O emulsion). The Stokes' equation shows that creaming is inhibited by a small droplet radius, a highly viscous continuous phase and a low density difference between the oil and water phases (Dalgleish, 2001).

Some of the other breakdown processes are also related to the oil droplet size and the emulsion viscosity, like Ostwald ripening—which represents the growth of large droplets at the expense of smaller ones caused by excess (Laplace) pressure and coalescence.

*Flocculation* is the process where two or more droplets get attached to each other in order to form an aggregate in which each of the initial droplets retains its individual integrity and remain as totally separate entities. It is a resultant phenomenon due to the weak, net attractions between droplets which arise through various mechanisms (Dickinson and Stainsby, 1982; Fleer et al., 1984).

*Coalescence* is the process during which, two or more droplets merge together to form a single larger droplet. Partial coalescence is the process where two or more partly crystalline droplets merge together in order to form an irregularly shaped aggregate due to the penetration of solid crystals from one droplet into a fluid region of another droplet. Extensive droplet coalescence may eventually lead to the formation of a separate layer of oil on top of the emulsion, which is known as "oiling-off."

*Disproportionation* also known as Ostwald *ripening*, is a process that is dependent on the diffusion of disperse phase molecules from smaller to larger droplets through the continuous phase. During this procedure, the larger droplets grow at the expense of smaller droplets due to mass transport of dispersed phase material through the continuous phase (Strawbridge and Hallet, 1994).

Phase inversion is the process where an exchange happens between the disperse phase and the medium. O/W emulsion changes to a W/O and vice versa. In many cases, phase inversion passes through a transition state whereby multiple emulsions are produced (Velikov and Pelan, 2008).

The instability mechanisms that may occur during an emulsion disruption can be coacting, inducing each other. For example a droplet aggregation can lead to rapid creaming. As a result, in the certain case it is more important to prevent droplet aggregation, rather than droplet creaming.

The highest possible solid-to-water ratio is desirable in the extraction, so as to obtain less-stable emulsions and generate less effluent. However, to obtain the highest extraction rates and extraction yields it is usually necessary to use large quantities of water.

The time required to reach a desired extraction level depends on the oilseed, as well as the process variables mentioned. However, more stable emulsions are likely to be formed in a prolonged extraction.

The processes involved in the separation and recovery of protein and oil include a demulsification step, followed by the separation of the aqueous and oil phases by centrifugation. The aim of demulsification is to promote or accelerate (in a thermodynamic sense) the mechanisms responsible for separation of the components.

The tendency of emulsified components to separate into distinctive phases can be understood from the second law of thermodynamics, as explained by Tadros (1989). The free energy for the formation of an emulsion is given by

$$\Delta G^{form} = \gamma_{ow}\Delta a - \theta\Delta S^{Conf} \tag{5.30}$$

where $\gamma_{ow}$ is the interfacial tension and $\Delta a$ is the increase in interfacial area. During emulsification, the interfacial area (a) between the oil and water phases is greatly increased as a result of the oil droplet subdividing into much smaller units; this is accompanied by increased interfacial energy $\gamma_{ow}\Delta a$.

However, the formation of a large number of droplets is accompanied by an increase in the total entropy of the system. This increase in entropy facilitates emulsification although its value is relatively small compared to the change in the interfacial free energy. Hence, $\gamma_{ow}\Delta a > T\Delta S$ and therefore $\Delta G^{form}$ is large and positive. Thus, the process of emulsification is nonspontaneous so that, with time, the droplets tend to aggregate and/or coalesce to reduce the total energy of the system.

On the other hand, when surface-active agents (surfactants) are present in the system, they tend to get adsorbed at the interface, decreasing the interfacial tension and turning the emulsion into a much more stable state. This is what happens in the aqueous extraction process, where the proteins present in high concentration may act as surfactants. In the case of soybeans, the natural presence of other surfactants, such as lecithin and saponins (Snyder and Kwon, 1987), may strengthen the emulsion stability and worsen the problem of emulsion breakdown and component separation.

Demulsification operations like centrifugation can promote coalescence and creaming, but sometimes a more complete separation is obtained through phase inversion. Temperature changes can help make the emulsion less stable, therefore facilitating separation. Earlier studies have addressed demulsification in aqueous process. In coconut (Gunetileke and Laurentius, 1974), attempts to break the protein stabilized oil-in-water emulsion included (1) heating and centrifugation (Rajasekharan and Sreenivasan, 1967), (2) freezing and thawing (Roxas, 1963), (3) enzyme action followed by freezing and thawing, and (4) chilling and thawing the coconut cream obtained after centrifugation; the last method requiring significantly lower energy compared to when freezing alone was employed. In this last process, the emulsion was centrifuged before chilling and thawing, to obtain better packing of the coconut oil globules. The centrifugation step allowed the packing of cream oil globules that crystallized on lowering the temperature. On thawing, the oil globules lost their spherical shape and coalesced to form large droplets of varying sizes. Preliminary centrifugation, besides helping to pack the oil before breaking the emulsion, can also be used to remove undissolved solids after the extraction. The removal of solids present in high percentages in the dispersion of some oilseeds like coconuts and peanuts, is essential for efficient recovery of oil by centrifugation (Cater et al., 1974). The solids are mainly fibrous materials, carbohydrates and proteins, depending upon the pH employed in the extraction step. An alternative method for solid–liquid separation, besides centrifugation, is reported to be filtration through vibrating or pressing-type screens. Like demulsification, the selection of the most appropriate method of solid–liquid separation depends on the physical nature of the material in the dispersion, as well as the operating costs (Cater et al., 1974; Embong and Jelen, 1977).

Hagenmaier et al. (1972, 1973) used shearing to promote phase inversion and breakage of the coconut emulsion. Another method of achieving phase inversion, originally described by Sugarman (1956), involves the addition of clear oil to the emulsion. With the aid of high shear and temperature, almost all the oil can be freed from the emulsion (Cater et al., 1974). This is possible due to the reduction in the level of moisture below a threshold value, above which the emulsion cannot be broken (Cater et al., 1974).

Barrios et al. (1990) increased coconut oil yield from 69.3% to 79.4% by aqueous enzymatic extraction, simply by (1) adding extra water after the extraction to separate the cream through flotation followed by centrifugation; (2) recycling the water which carries the enzyme; and (3) draining and pressing the coconut meal after the extraction. This procedure considerably decreased the amount of water required for the process and the effluent to be treated and disposed; it also permitted the possibility of enzyme reuse which reduced the costs.

An alternative method proposed for increasing the amount of free oil, was by stirring the emulsion after extraction (Embong and Jelen, 1977). Appropriate levels of stirring promote coalescence of oil droplets, resulting in the formation of a greater amount of free oil after the separation step. These conclusions were reported for rapeseed (Embong and Jelen, 1977), as well as for peanut (Rhee et al., 1972, 1973) and sunflower oil (Hagenmaier, 1974).

After demulsification, the separation of the two, or three phases—in case solid removal was not complete—is carried out by centrifugation, which is one of the key steps in the aqueous process. Depending on the material and the extraction conditions, oil can either be recovered as free oil or as an oil-in-water emulsion (Cater et al., 1974).

#### 5.2.6.2 Disperse Phase Volume Fraction

Droplet concentration determines emulsion's cost, appearance, texture, flavor, and stability (McClements, 2005). The droplet concentration (2) is described as the disperse phase volume fraction ($\varphi$), which is equal to the volume of emulsion droplets ($V_D$) divided by the total volume of the emulsion ($V_E$):

$$\varphi = V_D/V_E. \tag{5.31}$$

The droplet concentration of an emulsion can be also expressed as the dispersed phase mass fraction ($\varphi_m$), which is equal to the mass of emulsion droplets ($m_D$) divided by the total mass of emulsion ($m_E$): $\varphi_m = m_D/m_E$. The relationship between $\varphi_m$ and $\varphi$ is given by the following equations:

$$\phi_m = \phi\left[\phi + (1-\phi)\frac{\rho 1}{\rho 2}\right]^{-1} \tag{5.32}$$

$$\phi = \phi_m\left[\phi_m + (1-\phi_m)\frac{\rho 1}{\rho 2}\right]^{-1} \tag{5.33}$$

where, $\rho 1$ and $\rho 2$ are the densities of the continuous and dispersed phases, respectively.

The mass fraction is equivalent to the volume fraction, when the densities of the two phases are equal. The droplet concentration may also be represented as either a dispersed phase volume percentage ($\varphi_{\%} = 100 \times \varphi$) or disperse phase mass percentage ($\varphi_m\% = 100 \times \varphi_m$). It is particularly important to convert the droplet concentration into the appropriate units when comparing experimental work with theoretical predictions (McClements, 2005).

The droplet concentration can also be calculated if the concentration of the ingredients used to prepare the emulsion is carefully controlled. Nevertheless, local variations in disperse phase volume fraction occur within emulsions as a result of the various instability phenomena occurring in the emulsion. Consequently, it is often important to have analytical techniques to measure disperse phase volume fraction (McClements et al., 2007).

#### 5.2.6.3 Droplet Size

After the homogenization of an emulsion it is critical to perform a classification according to its characteristics. The main characteristics of the emulsion are its concentration, size, charge, interfacial properties and interactions which determine its physicochemical properties (Gershanik et al., 1998).

The size of the dispersed particles or droplets determines the properties and stability of the dispersion. Even the color and the appearance of an emulsion are defined by the droplet size and distribution. Turbidity in an emulsion happens when the constituent particles scatter visible light, which means that their diameter is at least 0.4 μm. Moreover, the droplet size can affect the stability of an emulsion to gravitational separation or droplet aggregation. A decrease of droplet size can lead to stability prolongation since the velocity of sedimentation is proportional to the square of the droplet size (Dickinson et al., 1987).

The particles in colloidal foods are generally very polydisperse, either from the production of an emulsion or during its storage, making it very important to generate the optimum particle size distribution for each product, and to ensure that it does not change over time (McClements et al., 2007).

When all the droplets in an emulsion have the same size, the emulsion is known as "monodisperse," and can be characterized by a single value of droplet size, whether it is diameter or radius depending on the author. However, in actual food products, the dispersed droplets can be of different sizes and, therefore, the emulsion is referred to as being "polydisperse." A polydisperse emulsion is characterized by its "particle size distribution," which defines the concentration of droplets in different size classes (Figure 5.14) (McClements, 2005).

The droplet size distribution in the liquid phase after the extraction and before any centrifugation is usually profiled by a particle size analyser, such as that of Malvern. A sample of the suspension is taken at the end of the extraction and analyzed after removing the solid particles by filtration. The average diameters $d_{32}$ (or Sauter Mean diameter) and $d_{43}$, which are useful and sensitive parameters for describing coalescence stability (Dickinson and Galazka, 1991), are calculated respectively as

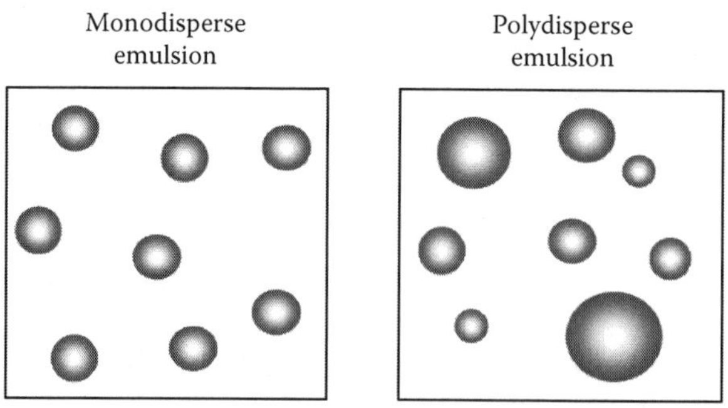


**FIGURE 5.14** Schematic representation of monodisperse and polydisperse emulsions. In a monodisperse emulsion all the droplets have the same size, but in a polydisperse emulsion they have a range of different sizes. (Adapted from McClements D.J. 2005. *Food Emulsions: Principles, Practice and Techniques*, 2nd edn., Boca Raton, FL, CRC Press.)

$$d_{32} = \frac{\sum_i n_i d_i^3}{\sum_i n_i d_i^2} \tag{5.34}$$

and

$$d_{43} = \frac{\sum_i n_i d_i^4}{\sum_i n_i d_i^3} \tag{5.35}$$

where $n_i$ is the number of droplets of diameter $d_i$.

The amount of cream (expressed either in volume or weight) formed during centrifugation can be used as an indicator of the emulsion stability, and also of the difficulty in breaking down the emulsion to separate oil from water and protein (Ogden and Rosenthal, 1997). He reported that high agitation rates not only produce high oil extraction yields, but also give rise to less stable emulsions which are more easily broken.

#### 5.2.6.4 Interfacial Properties

The emulsion droplets are surrounded by an interface consisting of a narrow region (≈1–50 nm thick) that contains a mixture of oil, water, and emulsifier molecules, as well as possibly other types of molecules and ions (e.g., mineral ions, hydrophilic polyelectrolytes, amphiphilic components) depending on the emulsion consistency. The interfacial region only makes up a significant fraction of the total volume of an emulsion, when the droplet size is less than about 1 μm) (McClements, 2005). Even so, it can have an appreciable impact on many of the most important bulk physicochemical and sensory properties of food emulsions, including their stability, rheology and flavor. The properties of the interfacial region are determined by the type, concentration and interactions of any surface-active species present, as well as by the events that occur before, during and after emulsion formation, for example, complexation, competitive adsorption, and layer-by-layer formation (McClements et al., 2007). The composition of the interfacial layer is governed mainly by what is present at the onset of emulsion formation (Dalgleish, 2001).

The electrical charge on the droplet interface influences its interaction with other charged molecules, as well as its stability to aggregation. The thickness and rheology of the interfacial region influences the stability of emulsions to gravitational separation, coalescence, and flocculation, and determines the rate at which molecules leave or enter the droplets (Dickinson and Tanai, 1992; McClements, 2005).

#### 5.2.6.5 Droplet Charge

In most food emulsions various molecules are absorbed on the surface of the droplets providing them with a certain electrical charge, for example, surfactants, proteins (McClements, 2005). The electrical charge of the droplet plays a crucial role in the emulsion properties and stability. It depends on the type and concentration of ionized charge species present at the surface, as well as the ionic composition and properties of the continuous phase. The electrical charge of the droplets determines the form of interaction between the droplets (McClements, 2005). The presence of repulsive electrical charges on the surfaces of emulsion droplets hinders droplet aggregation, thus leading to emulsion stability (Paraskevopoulou et al., 2009).

The electrical characteristics of emulsion droplets are mainly characterized in terms of ζ-potential (Hunter, 1986). ζ-potential is the electric potential in the interfacial double layer at the location of the slipping plane. This is defined as the distance away from the droplet surface below where the counterions remain strongly attached to the droplet when it moves in an electrical field, versus a point in the continuous phase away from the interface. ζ-potential is the potential difference between the dispersion medium and the stationary layer of fluid attached to the dispersed particle (Paraskevopoulou et al., 2009).

Various methods have been used for the determination of ζ-potential. A widespread method is the measurement of the migration velocity θ of an electrically charged droplet in an electric field ($E$). The emulsion droplets are usually negatively charged and migrate, with their electric double layer and diffuse ion cloud, toward the anode. The ζ-potential can be thus calculated from the following equation:

$$\zeta = \frac{4\pi\eta\theta}{\varepsilon E} \tag{5.36}$$

where ε is the dielectric constant and η the viscosity of the external phase (Heusch, 2000).

#### 5.2.6.6 Rheology

The application of a stress to a material causes it to deform or to flow (Macosko, 1994; Whorlow, 1992), the extent of which provides the emulsions with important functional properties, both during processing and in the final product where the texture is an important component of mouth feel (McClements and Weiss, 2004). The rheological properties are controlled primarily by the concentration and size of the droplets, and by the composition of the continuous phase. Emulsions of low droplet concentration, where the droplets are unflocculated and the continuous phase is a simple liquid, will be of low, constant viscosity at a given temperature (Newtonian behavior). As the droplet concentration increases and flocculation occurs, the emulsion becomes progressively more non-Newtonian. Its response to flow is now dependent on the magnitude of the applied stress, and the time scales of application (Robins and Wilde, 2003). Many of the sensory attributes of food emulsions are directly related to their rheological properties, for example, creaminess, thickness, smoothness, spreadability, pourabilty, flowability, brittleness, and hardness. A food manufacturer, therefore, must be able to design and produce a product that has the rheological properties expected by the consumer.

The shelf life of many food emulsions depend on the rheological characteristics of the component phases, for example, the creaming of oil droplets depends on the viscosity of the aqueous phase. Information about the rheology of food products is used by food engineers to design processing operations that depend on the way that a food behaves when it flows through a pipe, is stirred, or is packed into containers. Rheological measurements are also used by food scientists as an analytical tool to provide fundamental insights about the structural organization and interactions of the components within emulsions (McClements and Weiss, 2004; Tadros, 1994).

#### 5.2.6.7 Flavor

Flavor release is the process whereby flavor molecules move out of a food and into the surrounding saliva or vapor phase during mastication (McClements and Weiss, 2004). In many traditional foods, flavor components are present in both the oil and aqueous phases, and the full flavor is only released by droplet breakdown in the mouth. The droplet size and interfacial composition (emulsifiers) are also important factors (Robins and Wilde, 2003). The perception of a flavor depends on the precise location of the flavor molecules within an emulsion. The aroma is determined by the presence of volatile molecules in the vapor phase above an emulsion whereas most flavors are perceived more intensely when they are present in the aqueous phase, rather than in the oil phase (Kinsella, 1989; McNulty, 1987; Overbosch et al., 1991; Taylor and Linforth, 1996). Certain flavor molecules may associate with the interfacial region, which alters their concentration in the vapor and aqueous phases (Kinsella, 1989). It is therefore important to establish the factors that determine the partitioning of flavor molecules within an emulsion.

#### 5.2.6.8 Emulsion Appearance

The appearance of an emulsion is one of the most important parameters influencing its overall quality (McClements and Weiss, 2004). An understanding of the structural basis of the emulsion appearance depends on the physical processes occurring, when a light beam interacts with this emulsion and is subsequently detected by an eye or instrumental detector (McClements, 2002). The relative proportions of light transmitted and reflected at different wavelengths depend on the scattering and adsorption of the light wave by the emulsion. Light scattering and absorption depend on the size, concentration, refractive index, and spatial distribution of droplets (Wedzicha, 1988). Hence, the overall appearance of an emulsion is influenced by its structure and composition. Scattering is mostly responsible for the turbidity, opacity, or lightness of an emulsion, whereas absorption is largely responsible for chromaticity (blueness, greenness, redness, etc.). It should be stressed that the overall appearance of an emulsion also depends on the nature of the light source and detector used (Billmayer and Saltzman, 1981; Judd and Wyszecki, 1963; Wyszecki and Stiles, 1982). Finally, color of emulsion is also determined by the type of the lipids (number and position of double bonds) (McClements and Weiss, 2004).

#### 5.2.6.9 Lipid Oxidation in Food Emulsions

Since many common foods are emulsified materials (mayonnaise, coffee creamers, salad dressing, etc.), a better understanding of lipid oxidation mechanisms in emulsions is crucial for the formulation production and storage of these products (Kiokias et al., 2005; Ponginelbi et al., 1999). Moreover, apart from their technological importance, emulsion systems generally mimic the amphiphilic nature and the basic structural characteristics of important biological membranes (e.g., phospholipids), which are also prone to *in vivo* oxidative degradation when attacked by singlet oxygen and free radicals (Halliwell and Gutteridge, 1995; Kiokias et al., 2009). In that aspect, *in vitro* research on the oxidative stability and antioxidation of model emulsions could provide with useful information of nutritional interest and thereby serve as pilot studies for *in vivo* clinical trials (Kiokias and Gordon, 2003).

Emulsions are thermodynamically unstable systems because of the positive energy required to increase the surface area between the oil and water phases (Dickinson, 1997). Generally, the stability of food emulsions is complex because it covers a large number of phenomena, including flocculation, coalescence, creaming, and final phase separation (Friberg, 1996; Kiokias et al., 2004a,b,c). Oil-in-water emulsions consist of three different components: water (the continuous phase), oil (the dispersed phase), and surface-active agent (emulsifiers at the interface).

In such a system, the rate of oxidation is influenced by the emulsion composition (relative concentrations of substrate and emulsifier) and especially by the partition of the emulsifier between the interface and the water phase (Coupland and Mc Clements, 1996). Other factors influencing lipid oxidation in emulsions are particle size of the oil droplets, the ratio of oxidizable to nonoxidizable compounds in the emulsion droplets, and the packing properties of the surface-active molecules (Labuza, 1971; Kiokias et al., 2009). In addition, the amount and composition of the oil phase in an emulsion are important factors that influence oxidative stability, formation of volatiles, and partition of the decomposition products, between the oil and water phase (McClements and Decker, 2000).

A certain body of recent research has focused on the microstructural stability of protein stabilized oil-in-water emulsions, which are structurally similar to recently developed foodstuffs (e.g., dairy alternative or "fresh cheese type" products, etc.) (Dickinson, 2001; Kiokias and Bot, 2005, 2006). The image of such an emulsion has been visualized by use of confocal laser scanning microscopy (CSLM) as reported by Kiokias and Varzakas (2014). However, not much research has been done yet on the oxidative destabilization of these emulsion systems. A better understanding of the factors monitoring the oxidative deterioration of emulsions would offer antioxidant strategies to improve the organoleptic and nutritional value of the related products.

The replacement of synthetic antioxidants by "safer natural mixtures" is being increasingly advocated nowadays by food industry. This trend has been imposed by the worldwide preference of consumers for the use of natural antioxidants, some of which may exist inherently in foods or be added intentionally during their processing (Kiokias et al., 2009). The antioxidant potential of certain carotenoids and flavonoids has been summarized by a few review papers (Kiokias and Gordon, 2003; Kiokias et al., 2008). So far, a limited amount of research evidence has been reported in the literature concerning the antioxidant activity of both categories of compounds in multicomponent systems.

The antioxidant effects of flavonoids and β-carotene during the thermal auto-oxidation of food relevant oil-in-water emulsions were spectrophotometrically assessed by measuring the formation of primary oxidation products (conjugated dienes and lipid hydroperoxides). An oxidatively "sensitive" model emulsion was selected as substrate of this study in terms of processing and compositional factors. At a concentration of 1.5 mmol $kg^{-1}$, only quercetin among the tested compounds significantly reduced the oxidative deterioration of cottonseed oil-in-water emulsions. Structural characteristics (positioning of hydroxyl group) or partitioning behavior between the emulsion phases may modulate the flavonoid activity. The high oxygen pressure conditions of the experimental system may explain the lack of any antioxidant activity for β-carotene.

The antioxidant potential of quercetin increased with its concentration until a specific level. On the contrary, the antioxidant concentration within the same tested range (0.75–3 mmol $kgr^{-1}$) did not impact the activity of catechin and β-carotene. Mixtures of β-carotene with flavonoids did not exert a tendency for increasing the activity of each individual compound (Kiokias and Varzakas, 2014).

#### 5.2.6.10 TAG Crystallization in Emulsions

The same surface effects that govern liquid emulsions also apply to dispersions of solids in liquids and of liquids or solids in gases. Colloidal solutions of solids can be produced if the particle size is of the necessary order, below about 0.1 μm and, again, stability depends upon the surface properties of the materials. Aerosols, for example, fine mists in the atmosphere, can also be quite stable.

Many aspects of the texture of dairy spread (alternative) and fresh cheese products—such as firmness, melting behavior, temperature cycling stability—are determined in part by the crystallization behavior of the triacylglycerol (TAG) blend in the O/W emulsion (Bot and Blancher, 2001; Kiokias et al., 2004b; Reiffers-Magnani et al., 2002). However, TAG crystallization in O/W emulsions is still not understood as well as bulk TAG crystallization (Lopez et al., 2001; Coupland, 2002), and better control over composition and processing of the fat phase in the context of crystallization behavior could help to improve control over the product properties mentioned above.

The TAG crystallization mechanism in emulsions differs considerably from bulk crystallization. Crystallization in bulk occurs through heterogeneous nucleation involving trace impurities that provide the initial nucleation sites. Once crystallization starts, it rapidly spreads throughout the whole system by means of crystal growth and/or secondary nucleation (Berger, 1997). In emulsions, however, the lipid phase is finely dispersed in droplets and the number of impurities may be significantly less than the number of droplets, depending on the droplet size distribution in the emulsion. As a result, a considerable fraction of the fat droplets may be nucleus-free and remain in the metastable, supercooled (or alternatively, supersaturated) liquid state unless crystallization proceeds through an alternative mechanism. The final emulsion will consist of a mixture of droplets, part of which has crystallized to an "equilibrium" state and part of which remains liquid, giving rise to overall supercooling of the lipid phase in the emulsion.

One potential alternative mechanism to heterogeneous nucleation is homogenous nucleation in the supercooled liquid, the spontaneous formation of crystallites of sufficient size not to dissolve again (Davis et al., 2000; McClements et al., 1994; Rousseau, 2000). Homogeneous nucleation usually takes place well below the thermodynamic melting point (i.e., crystallization is thermodynamically favorable, but the required activation energy is high compared to the energy of the thermal fluctuations) (Dickinson et al., 1996). Deeper cooling increases the homogeneous nucleation rate and, therefore, ultimately reduces the degree of supersaturation in the lipid phase of the final emulsion because a larger fraction of supercooled droplets can crystallize through the homogeneous nucleation mechanism.

A second alternative to heterogeneous nucleation is surface-induced crystallization. O/W emulsions contain a high amount of interface and the presence of (low-molecular weight) emulsifiers may affect the crystallization kinetics by means of surface-induced heterogeneous nucleation as a result of the decreased surface tension of the nuclei (Vanapalli et al., 2002). Different types of emulsifiers are expected to affect crystallization in different ways. For example, SDS was found to reduce crystallization rates in the emulsion less than Tween20 (McClements et al., 1993).

TAG crystals occur in a number of different polymorphic forms (α, β′, β), that is, crystal structures with different states of molecular packing (Bot et al., 2003; de Bruijne and Bot, 1999). In crystallization of bulk animal or vegetable fats, the occurrence of large numbers of different types of TAGs makes a prediction of the preferred polymorph difficult. As a rule of thumb, however, asymmetric TAGs (such as UUS, USS, etc., form commonly found in palm oil and butterfat; U: unsaturated, S: saturated) tend to crystallize preferentially in β′ form, whereas symmetric TAGs (such as SUS, USU, SSS, etc., commonly found in lard and coconut) tend to crystallize preferentially in the more stable β form (Campbell et al., 2002; Coultate, 1996). The preferred polymorphic form of a complex crystallizing TAG mixture depends on the chemical composition of an individual TAG, the TAG solvent mixture, and the crystallization conditions (e.g., cooling rate, temperature, shear, presence of an interface) (see e.g. ten Grotenhuis et al., 1999). The sensitivity to crystallization conditions suggests that polymorphs formed in o/w emulsions could differ from those formed in bulk for a specific fat blend.

When the oil is cooled slowly, the degree of supercooling is less throughout the process, and the TAGs will crystallize in stages (Dickinson and McClements, 1995). Consequently, a smaller number of crystals will be formed with more pronounced differences in TAG composition between the crystals but a more homogeneous TAG composition per crystal than for fast cooling (i.e., some crystals are rich in high melting TAGs and other crystals are depleted in these TAGs) (Campos et al., 2002). Such systems are not very sensitive to recrystallization phenomena.

When oil is cooled rapidly all TAGs crystallize at approximately the same moment in time, resulting in a large number of small crystals with a relatively similar but complex TAGS composition. Because this process is governed more by kinetics than by thermodynamics, such crystals are prone to subsequent re-crystallization processes like polymorphic transitions, TAG segregation and Ostwald ripening. Transformations between polymorphs depend amongst others on the storage conditions, the mobility of the TAGs in the crystal and in the liquid phase, and the ease with which molecules fit within the crystal lattice. Whether these transitions during re-crystallization are the same in bulk and emulsion is currently an open question.

The presence of fat crystals in the dispersed phase of oil-in-water emulsion can reduce emulsion stability considerably through clumping or partial coalescence. The process is usually explained in terms of the formation of needle-shaped crystals protruding from the O/W interface (Boode et al., 1993; Goff, 1997), especially when Ostwald ripening facilitates the formation of larger crystals upon repeated temperature cycling (Mutoh et al., 2001). Partial coalescence upon temperature cycling has been linked to undesirable microstructural textural changes, such as increase of droplet size and firmness in protein-stabilized O/W emulsions upon temperature cycling (Boode et al., 1993; Kiokias et al., 2004b).

Kiokias et al. (2004c) reported that the presence of crystalline fat is a necessary but not sufficient condition for destabilization of the emulsion. Crystal size and shape are expected to be at least as important, and these factors are usually related to (changes in) the crystal polymorph. Some groups observe polymorphic transitions in vegetable-fat protein-stabilized model emulsions during temperature cycling (Mutoh et al., 2001) and other observe textural changes (Boode et al., 1991; Noda and Yamamoto, 1994). Sometimes these changes are found to be independent of the upper cycling temperature (Mutoh et al., 2001), whereas in other cases changes upon cycling are most pronounced if the upper cycling temperature is chosen such that a small percentage of solid fat is retained during cycling and the fat crystals do not completely melt (Boode et al., 1991; Noda and Yamamoto, 1994). Overall, however, few studies relate stability of emulsions to the polymorphic behavior of the lipid phase. Kiokias et al. (2004c) evaluated a series of potentially suited techniques to model emulsions that nevertheless contain all the essential elements of real dairy spread (alternative) and fresh cheese products including x-ray diffraction (XRD) and low-field time-domain proton NMR spectrometry.

## 5.2.7 Particle Sizing Characterization of Emulsions

### 5.2.7.1 Microscopy

Various microscopic techniques have been developed in order to observe and characterize the emulsions. Microscopic analysis can even identify whether the instability of an emulsion is due to reversible sedimentation or to irreversible coalescence (Heusch, 2000). The most widely used techniques of microscopy are optical microscopy, electron microscopy, and atomic force microscopy (Morris et al., 1999 and Murphy, 2001). They provide information about the structure of the droplets in the form of images (McClements et al., 2007). These techniques are based on different physicochemical principles and can be used to examine the structural organization of the emulsions (McClements et al., 2007).

The most representative image of the emulsion structure can be acquired by a relatively recent method, the laser scanning confocal microscopy (LSCM). This technique can provide three-dimensional images of emulsion by taking a series of two-dimensional slices in the vertical direction without the need to physically section the sample (McClements, 2002; Plucknett et al., 2001).

### 5.2.7.2 Particle Size Analysis

The particle size distribution of emulsions can be measured by various analytical instruments. They provide accurate measurements of the particle distribution of the emulsions and are based on different physical principles (scattering of light, velocity of particles in a field, scattering or absorption of ultrasonic waves, counting of particles, etc.) (McClements et al., 2007). However, these methods normally require low droplet concentration (<0.1 wt%) so as to avoid multiple scattering effects. As a result dilution of the sample may be necessary prior to the analysis.

#### 5.2.7.3 Static Light Scattering

Static light scattering instruments are based on the principle that the scattering pattern (intensity of scattered light versus scattering angle), which is produced when the laser beam is directed through the sample depending on the particle size distribution (McClements et al., 2007). For the interpretation of the intensity fluctuation data, specialized softwares are available that can predict the scattering pattern of an emulsion using a mathematical model, taking into account the characteristics of the particles that it contains (refractive index ratio, absorption coefficient, and diameter). Commercial static light scattering instruments are capable of determining particle diameters within the range of about 100 nm to 1000 μm (McClements et al., 2007).

#### 5.2.7.4 Dynamic Light Scattering

Dynamic light scattering techniques are based on the intensity fluctuations occurring when light is scattered by particles that change their relative spatial location due to Brownian motion. These intensity fluctuations depend on the speed of the particles movement and, therefore, indicate the size of the particles. Faster intensity fluctuations correspond to smaller particles that move more rapidly than the larger ones. By monitoring the intensity fluctuations of the scattered wave over time at a particular scattering angle, the particle size distribution of an emulsion can be derived (McClements et al., 2007). Once again specialized software is necessary in order to convert the intensity fluctuations into a particle size distribution as long as some information is given about the samples properties (refractive index, absorption, viscosity). Typical dynamic light scattering instruments can determine droplets from 3 nm to 5 μm (Berne and Pecora, 2000).

#### 5.2.7.5 Electrical Pulse Counting

Electrical pulse counting techniques are based on the changes of the electrical conductivity that happens across a small orifice when a dilute emulsion is pulled through it. The emulsion to be analyzed is placed in a beaker that has two electrodes dipping into it (Allen, 1990; Lines, 1996). The droplet concentration is determined by counting the number of pulses that pass through the hole per unit volume of emulsion. The particle size distribution is determined by measuring the height of each individual pulse, since the height of an electrical pulse is proportional to the volume of a particle. Such systems can measure droplets with 0.4–1200 μm diameter. However it is necessary to use glass tubes with different sized apertures in order to cover the whole range of droplet sizes (McClements, 2000, McClements et al., 2007). The main advantage of this technique is that no dilution of the sample is needed for the analysis (Hunter, 2001).

#### 5.2.7.6 Ultrasonic Spectrometry

Ultrasonic spectrometry is based on the change in ultrasonic attenuation and ultrasonic velocity as a function of frequency. Attenuation is determined by the energy losses in compressions and decompressions in ultrasonic waves, which include absorption and scattering contributions. Ultrasonic velocity is determined by the density and the elasticity of the medium (O'Driscoll et al., 2003). From the emulsion analysis derives an ultrasonic spectrum depending on the particle size distribution and concentration. The sample is placed in a chamber of stable temperature and then its ultrasonic spectrum is measured (typically from about 0.1 to 150 MHz) (McClements et al., 2007). Ultrasonic spectrometry can be used even in opaque samples since ultrasonic waves are able to propagate through turbid materials. The range of measurement is from 10 nm to 1000 μm droplet size (O'Driscoll et al., 2003)

#### 5.2.7.7 NMR Techniques

Particle sizing instruments using the technique of nuclear magnetic resonance (NMR) are based on the interactions between radio waves and the nuclei of hydrogen atoms to obtain information about the microstructure of emulsions (McClements et al., 2007). Within a liquid, the distance that a molecule can move in a certain time is limited by its translational diffusion coefficient. In the case of emulsion, its diffusion is limited from the interfacial matrix of the droplet. If the attenuation of the NMR signal is monitored through time it is possible to identify the diffusion restrictions and thus estimate the droplet size distribution using a suitable mathematical model (McClements et al., 2007). NMR techniques are used as standard measurements in the quality control of emulsion technology. It is a rapid, nondestructive, volumetric method, and also appropriate for multiple emulsions, in contrast to most techniques (Ambrosone et al., 2000). It is also a fitting method for high concentrated and multiple emulsions and does not require sample dilution. Moreover, it is unaffected by effects such as aggregation (Bernewitz et al., 2012).

#### 5.2.7.8 Other Techniques

Various other techniques have been developed for the particle characterization of emulsions. For example turbidity measurement is a method that provides indication for droplet size and dispersity. Other methods are also used such as neutron scattering, dielectric spectroscopy and electro-acoustics that may be useful for particular application (McClements et al., 2007).

### 5.2.8 Homogenization Methods

The process of converting two separate immiscible liquids into an emulsion, or of reducing the size of the droplets in a preexisting emulsion, is known as homogenization. In the food industry, this process is usually carried out using mechanical devices known as homogenizers, which usually subject the liquids to intense mechanical agitation (McClements, 2005).

Homogenization can be divided into two categories. The formation of an emulsion directly from two separate liquids is referred to as primary homogenization, whereas the reduction in size of the droplets in an existing emulsion is referred to as secondary homogenization. Both processes are very common in food industry. Often it is more efficient if the preparation of an emulsion is performed in two stages. Usually, the separate oil and water phases are mixed in order to be converted to a coarse emulsion that contains fairly large droplets using one type of homogenizer (e.g., a high-speed blender). Secondary homogenization is then performed using a high-energy homogenizing device (e.g., a high-pressure valve homogenizer). Various physical processes occur during primary and secondary homogenization, for example, mixing, droplet disruption, and droplet coalescence in the same time. Depending on the device used, different results are attained mainly concerning the size distribution of the emulsion droplets. The size of the emulsions droplets determines their stability, texture, appearance, and taste (McClements and Weiss, 2004). As a result the selection of the homogenization device and its condition is of high importance as it defines the properties of the resultant emulsion.

#### 5.2.8.1 Rotor–Stator Systems

The rotor–stator systems are widely used to emulsify liquids with medium to high viscosity (McClements, 2005). The rotor–stator assembly consists of a rotor housed concentrically inside the stator with two or more blades and a stator with either vertical or slant slots (Figure 5.15). As the rotor rotates, it generates a lower pressure to draw the liquid in and out of the assembly, thereby resulting in circulation and emulsification (Maa and Hsu, 1996). The liquid is fed into the colloid mill in the form of a coarse emulsion, or as separate phases, and flows through a narrow gap between a rotating disk (rotor) and a static disk (stator) (Jafari et al., 2008). In the case of discontinuous operation, agitators or gear-rim dispersion machines are usually used, whereas for continuous operation, colloid mills with smooth or toothed rotors and stators are preferred (Urban et al., 2006). The emulsion droplet size is thus reduced by the mechanical impingement against the wall due to high fluid acceleration and by the shear stress in the gap between rotor and stator, which is generated by the rapid rotation of the rotor. The intensity of the shear stress depends on the thickness of the gap, the rotation speed, and on the type of the used disks that have toothed surfaces or interlocking teeth (Becher, 2001; Jafari et al., 2008).

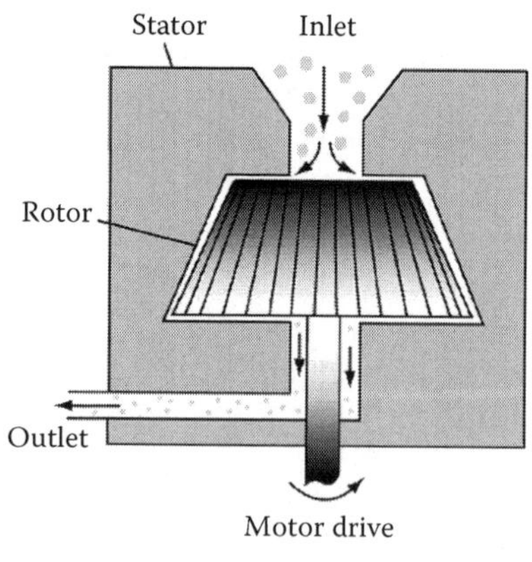


**FIGURE 5.15** Rotor–stator device or colloid mill. The rotor, incorporating grooves or blades, spins at high speed, generating high shear fields to break up droplets. (Adapted from Robins, M. and Wilde, P. 2003. In: *Encyclopedia of Food Sciences and Nutrition*, 2nd edn., ed. Caballero, A., pp. 1517–1524, Oxford, UK, Academic Press.)

### 5.2.8.2 High-Pressure Systems

High-pressure homogenization (microfluidization) is widely used for producing dairy and food emulsions. It is based on forcing the two fluids or a coarse premix to flow through an inlet valve, into a mixing chamber, under the effect of a very high pressure (Figure 5.16). The fluids undergo a combination of elongation and shear flows, impacts, and cavitations (Leal-Calderon, 2007; Soon et al., 2001). The crude emulsion is pumped through a central inlet bore, diverted by 90° and then passes through the radial gap between the valve seat and valve plug. As the coarse emulsion passes through the valve, it experiences a combination of intense shear, cavitation, and turbulent flow conditions, which cause the large droplets to be broken down into smaller ones (Jafari et al., 2008; Soon et al., 2001). Generally they are more effective at reducing the droplet size than at creating an emulsion directly from two separate phases (McClements, 2005). High-pressure homogenizers can be classified mainly by the nozzle geometry and design since the emulsifying nozzle is decisive for the efficiency of disruption using high-pressure devices. Moreover, they can be subdivided into radial diffusers (standard nozzles), jet dispersers, microfluidizers, and orifice valves. Standard nozzles are also called homogenizing valves and are the most common high-pressure systems used in industrial processes (Jafari et al., 2008; Phipps, 1985; Stang et al., 2001). The typical homogenizing pressures applied are between 5 and 50 MPa. However, microfluidizers and jet dispersers are able to produce emulsions at much higher pressures up to even 700 MPa (Jafari et al., 2008).

In the microfluidizer nozzle, two jets of crude emulsion from two opposite channels collide with one another (Jafari et al., 2008; Olson et al., 2004; Schultz et al., 2004). The process stream is delivered by a pneumatically powered pump, capable of pressurizing the in-house compressed air (150–650 kPa) up to about 150 MPa. Forcing the flow stream by high pressure through the microchannels toward an impingement area

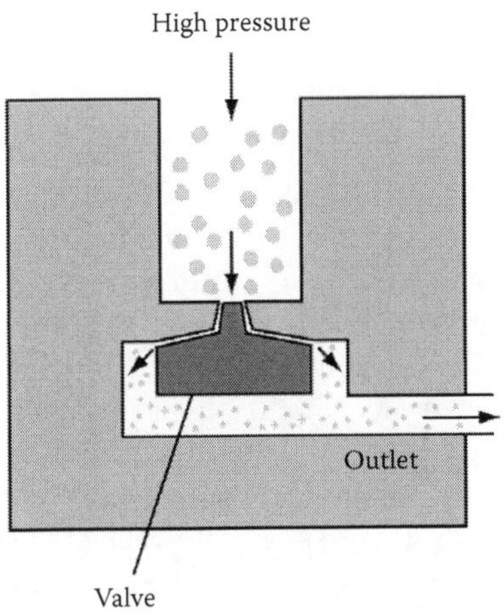


**FIGURE 5.16** High-pressure homogenizers force the droplets through narrow valves or orifices to break up droplets through high shear or elongational forces. (Adapted from Robins, M. and Wilde, P. 2003. In: *Encyclopedia of Food Sciences and Nutrition*, 2nd edn., ed. Caballero, A., pp. 1517–1524, Oxford, UK, Academic Press.)

a tremendous shearing action is created, which is able to provide emulsion with fine droplet sizes. In microfluidizers the inertial forces in turbulent flow along with cavitation phenomena are mainly responsible for droplet disruption. If the emulsion has high viscosity, disruption in laminar elongational flow may happen (Jafari et al., 2008).

In the "jet disperser," two or more jets of crude emulsion, each from opposing bores collide with one another but in a different way than microfluidizer (Urban et al., 2006). The diameters of the bores in jet dispersers are typically 0.3–0.5 mm. "Orifice plate" is the simplest construction form for a homogenizing nozzle. The diameter of the orifice bore is of the same order of magnitude as for the jet disperser, and the inlet head diameter of the orifice plate is typically 10–60 mm (Stang et al., 2001). In jet dispersers and orifice plates, droplets are disrupted predominantly due to laminar elongational flow ahead of the bores (Jafari et al., 2008). Unlike radial diffusers, the nozzle in microfluidizers, jet dispersers, and orifice plates contain no moving parts, thus they can be used at very high pressures, up to 300–400 MPa.

Hydroshear homogenizers have a double-cone-shaped chamber with a tangential feed pipe at the centre and outlet pipes at the end of each cone. The feed liquid enters the chamber at high velocity and is made to spin in increasingly smaller circles and increasing velocity until it reaches the centre and is discharged. The differences in velocity between adjacent layers of liquid causes high shearing forces, which together with cavitation and ultra-high-frequency vibration, break droplets in the dispersed phase to within a range of 2–8 μm (Jafari et al., 2001).

#### 5.2.8.3 Colloid Mills

In a colloid mill the crude emulsion is passed between the stationary surface (stator) and the rotating surface (rotor) separated by a small clearance. Liquids are subjected to turbulence and shear. Rotor speeds range from 3000 to 15,000 rpm. Rotors could also be mounted on vertical axes (paste mills). They handle high viscosity emulsions (>1000 cP).

#### 5.2.8.4 Ball Mills/Dispersion Mills

The dispersion of fine powders or emulsifications is carried out using ball mills or dispersion mills in order to break aggregates or form stable emulsions.

The ball mill consists of a horizontally mounted, cylindrical drum partially filled with high-density metal balls. The outer drum rotates slowly allowing the balls to roll over each other by gravity creating very large shearing forces in the nip between adjacent balls. The ball mill generally operates about half-full of balls at a rotational speed, which is about 60%, for the balls to be thrown out to the walls by centrifugal force. This gives a cascade angle of about 20–30°. These devices are not employed by the food industry since balls wear and may chip or shatter.

Dispersion mills use a very high-speed rotor blade which moves inside a slotted stator. Close clearances between the rotor and stator give high shear stresses, which may be used for droplet or aggregate breakup whereas flow through the slots drives gross circulations within the vessel.

These devices are suitable for the formation of emulsions-droplet breakup taking place in the high-energy dissipation regions close to the disperser head. Droplets are dispersed in the bulk liquid by convective flow coming from the head.

### 5.2.8.5 Pressure Homogenizers

Pressure homogenizers handle low viscosity materials (<200 cP). They are comprised of a valve and a high-pressure pump. Pressures of up to 10,000 psi are formed through the valve. The mechanism followed is when droplets of the internal phase are accelerated by entering the gap, then they shear against each other, become distorted and unstable and then disrupted. The break-up of the internal phase occurs by impingement at high speed on a hard surface. The overall disruptive effect is the result of the sudden drop in pressure on leaving the valve along with cavitational effects (Brennan, 1993).

### 5.2.8.6 Ultrasonic Devices

In a liquid irradiated with high-energy ultrasonic waves cavitation may occur, that is, each small region is alternatively under tension and compression. When the liquid is under tension bubbles will expand. In the other half of the cycle they will contract. The collapse of the bubbles can be very violent and energy is released when the pressure amplitude is high and bubbles small. If cavitation occurs at the interface between two immiscible liquids, one phase will be dispersed throughout the other with the immediate result being the formation of an emulsion.

Mechanical systems can generate ultrasonic waves and most specifically wedge resonators. This has a blade with wedge-shaped edges located in front of a rectangular slit and clamped in position at one or more nodal points. Liquid is pumped through the nozzle at pressures in the order of 50–200 psi, and the jet formed impinges on the blade causing its vibration at natural frequency. Then cavitation at the blade takes place bringing emulsification (Brennan, 1993).

In ultrasound emulsification, the energy input is provided through the so-called sonotrodes (sonicator probe) containing a piezoelectric quartz crystal that can expand and contract in response to alternating electrical voltage (Figure 5.17). As the tip of the sonicator probe contacts the liquid, it generates mechanical vibrations and therefore cavitation occurs (Abismai et al., 1999). Cavitation is the formation and the collapse of vapor cavities in a flowing liquid. The collapse of these cavities causes powerful shock waves to radiate throughout the solution in proximity to the radiating face of the tip,

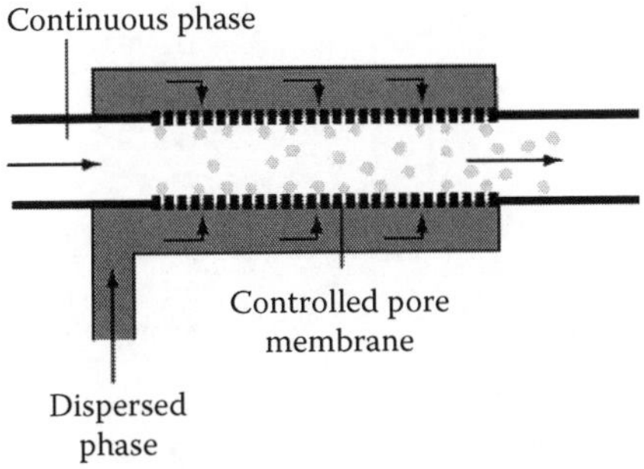


**FIGURE 5.17** Membrane emulsification. Porous membranes are used to extrude the dispersed phase into the flowing continuous phase. (Adapted from Robins, M. and Wilde, P. 2003. In: *Encyclopedia of Food Sciences and Nutrition*, 2nd edn., ed. Caballero, A., pp. 1517–1524, Oxford, UK, Academic Press.)

thereby breaking the dispersed droplets and homogenizing the liquid (Abismai et al., 1999). Only powerful ultrasound (16–100 kHz) frequencies are able to produce physical and chemical changes such as emulsification (Freitas et al., 2006). Ultrasound can be generated either mechanically (whistle, siren) or electrically (reverse piezoelectric effect or mangetostrictive transducers), which is the most widely used type of apparatus; a high-frequency oscillating electric field is converted into mechanical vibrations of the same frequency using a piezoelectric material such as quartz (Canselier et al., 2002). In batch emulsification, ultrasound is emitted by transducers fixed on the outside wall of the vessel or by cylindrical sonotrodes (Benichou et al., 2004) immersed in the liquid. In continuous operation, the circulating fluid may enter and leave a small reactor equipped with a probe; otherwise, one or several transducers are located inside the tubing assembly (Jafari et al., 2008). Starting from two separate phases, ultrasound can be used directly to produce emulsions, but since breaking an interface requires a large amount of energy, it is better to first prepare a coarse emulsion before applying acoustic power. In some cases, ultrasound can also act as a demulsifying technique (Canselier et al., 2002; Jafari et al., 2008).

### 5.2.8.7 Membrane Emulsification

Membrane emulsification is based on forcing the dispersed phase to permeate into the continuous phase through a membrane having a uniform pore with specific size distributions (Figure 5.18). The dispersed phase is pressed perpendicular to the membrane while the continuous phase is flowing tangential to the membrane (Leal-Calderon, 2007; Williams et al., 1998).

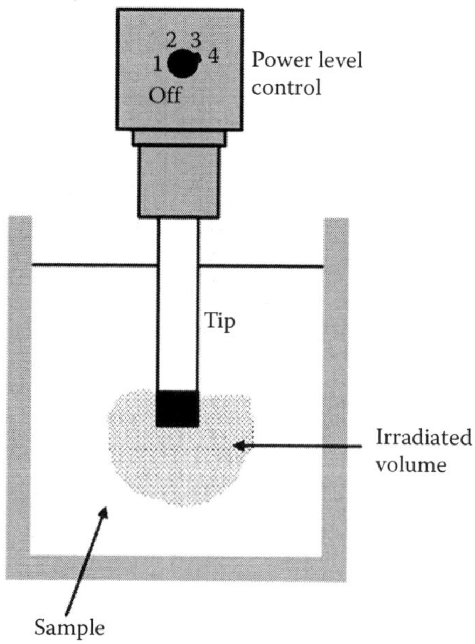


**FIGURE 5.18** Ultrasound emulsification. (Adapted from Dickinson, E. and Tanai, S. 1992. *Food Hydrocolloid.* 6: 163–71.)

Membrane emulsification depends mainly on the properties of the membrane (fluxes, and formulation, all influencing the emulsion size distribution). In order to obtain a monodisperse emulsion, the membrane pores must have a narrow size distribution (Williams et al., 1998). In most systems, the droplet size is proportional to the pore size. However, if the pore density is too large, coalescence of freshly formed drops is likely to occur, increasing polydispersity; in contrary, if the pore density is too low, the production rate is insufficient (Leal-Calderon, 2007). The emulsification ability of the membrane systems depends on phase velocity and interfacial tension. High continuous phase velocity and low interfacial tension will promote small droplets (Leal-Calderon, 2007). The necessary pressure that must be applied to the

**TABLE 5.6**
**Comparison of Different Types of Homogenization Systems**

| Emulsification System | Rotor–Stator Systems | High-Pressure Systems | Ultrasonic Systems | Membrane Systems |
|---|---|---|---|---|
| Examples | Mixers, agitators, colloid mills (Silverson, Ultra-Turrax) | Radial diffusers, valve homogenizers, jet dispersers, microfluidizer | Sonotrodes (sonication probes) | Glass/ceramic membranes |
| Droplet disruption mechanisms | Shear stress in laminar flow and/or shear and intertial stress in turbulent flow | Shear and inertial stress in turbulent flow; cavitation in laminar extension flow | Cavitation in microturbulent flows | Dispersed-phase flux |
| Throughput | Medium to high | High | Low | Low |
| Batch/continuous | Batch (mixers) or continuous (colloid mills) | Continuous | Batch or quasi-continuous | Continuous |
| Minimum droplet size (μm) | 1 | 0.1 | 0.1–0.2 | 0.2–0.5 |
| Optimal range of viscosity | Low to high (20–5000 mPa · s) | Low to medium | Low to medium | Low to medium |
| Application | Lab/industrial | Lab/industrial | Lab | Lab |
| Dominant flow regime | LV, TV | TI, TV (CL, LV) | CI | Injection |
| Energy density | Low-high | Medium-high | Medium-High | Low-medium |
| Change of energy input through | Rotation speed, exposure time, gap distance and disc design | Pressure, recirculation (exposure time) and nozzle design | Intensity and frequency of ultrasonic wave sonication time | Pore diameter |
| Residence time in dispersing zone, t | 0,1 < t < 1s | 0,1 < t < 3 ms | — | — |
| Required absorption rate of emulsifier | Low to high | High to very high | Middle to high | Middle to high |

*Source:* Adapted from Jafari, S.M. et al. 2008. *Food Hydrocoll.*, 22: 1191–1202.

dispersed phase depends on both the interfacial tension and the membrane pore size (Leal-Calderon, 2007; Schroder and Schubert, 1997). The most suitable operational conditions must be found in order to balance the pressure needed with the desired results in droplet size reduction.

#### 5.2.8.8 Comparison of Homogenization Methods

The selection of a homogenizer for a particular application depends on a number of factors, including the desired droplet size distribution, the volume to be homogenized, the viscosity of the emulsion and each fraction, the concentration and type of emulsifier, the energy consumption, the physicochemical properties of the component phases and the final product, the equipment available, and finally the initial and running costs (Jafari et al., 2008; McClements, 2005).

A comparative presentation of the various homogenization methods as previously discussed is shown in Table 5.6. After selecting the most suitable emulsification device, the operating conditions such as flow rate, pressure, gap distance, temperature, emulsification time, and rotation speed should be optimized to obtain the desired emulsion (Jafari et al., 2008).

## 5.3 CONCLUSIONS

The understanding of the colloidal basis that determines the properties of food emulsions has been of a major importance over the past few years. Various studies have been made in order to enlighten the contribution of each ingredient to the emulsion structure and properties. Technology has provided food scientists with powerful instruments capable to determine the colloidal characteristics of emulsions. Thus a wide variety of analytical, mathematical, and computational techniques have being developed suggesting the quantitative relation of the physiochemical and sensorial properties of food emulsions to the type, concentration, structure, and interactions of their constitutional components. The use of these new techniques will allow the researchers to determine, monitor and control the colloidal basis of emulsion properties providing consumers with high quality and safe emulsified food products.

## REFERENCES

Abismai, B., Canselier, J.P., Wilhelm, A.M., Delmas, H., and Gourdon, C. 1999. Emulsification by ultrasound: Drop size distribution and stability. *Ultrason. Sonochem.*, 6: 75–83.

Allen, T. 1990. *Particle Size Measurements*, 4th edition. London, Chapman & Hall.

Ambrosone, L., Ceglie, A., Colafemmina, G., and Palazzo, G. 2000. NMR studies of food emulsions: The dispersed-phase self-diffusion coefficient calculated by the least variance method. In *Trends in Colloid and Interface Science XIV*. Vol. 115, ed. Buckin, V., Germany, Verlag, Springer.

American Institute of Chemical Engineers 1979. *Dry Solids, Paste and Dough Mixing Equipment Testing Procedure*, New York, American Institute of Chemical Engineers.

Anton, N. and Vandamme, T.F. 2011. Nano-emulsions and micro-emulsions: Clarifications of the critical differences. *Pharmaceut. Res.*, 28(5): 978–985.

Arbuckle, W.S. 1986. *Ice Cream*, 6th edn., New York, AVI.

Atkinson, B. and Mavituna, F. 1991. Section 12.5. Power input. In: *Biochemical Engineering and Biotechnology Handbook*, eds. Atkinson, B. and Mavituna, F., London, McMillan Publishers Ltd, 714–720.

Barrios, V.A., Olmos, D.A., Noyola, R.A., and Lopez-Munguia, C.A. 1990. Optimization of an enzymatic process for coconut oil extraction. *Oleagineux*, 45(1), 35–42.

Becher, P. 2001. *Emulsions: Theory and Practice*. Oxford, Academic Press.

Benichou, A., Aserin, A., and Garti, N. 2004. Double emulsions stabilized with hybrids natural polymers for entrapment and slow release of active matters. *Adv Colloid Interfac.* 108–109:29–41.

Berger, K.G. Ice cream, In: *Food Emulsions*, 3rd ed. by S.E. Friberg, and K. Larsson, 1997, Marcel Dekker, New York.

Berne, B.J. and Pecora, R. 2000. Dynamic light scattering: With applications to chemistry. *Biology, and Physics*. New York, Courier Dover Publications.

Bernewitz, R., Guthausen, G., and Schuchmann, H.P. 2012. NMR on emulsions: Characterisation of liquid dispersed systems. *Magn. Reson. Chem.*, 49(1): 93–104

Billmayer, F.W. and Saltzman, M. 1981. *Principles of Color Technology*, New York, Wiley.

Birdi, K.S. 2010. *Surface and Colloid Chemistry, Principles and Applications*. Boca Raton, FL, CRC Press.

Boode, K., Walstra, P. and de Groot-Mostert, A.E.A. 1993. Partial coalescence in oil-in-water emulsions 2. Influence of the properties of the fat, *Colloids Surf. A Physicochem. Eng. Asp.*, 81: 139–151.

Bot, A. and Blancher, G. 2001. Modelling dairy spread alternatives as Matreshka composites: Firmness and syneresis, 2001, VD 01 0334.

Bot, A., Flöter, E., Lammers, J.G. and Pelan, E.G. 2003. Controlling the texture of spreads. In *Texture in Foods, Vol. 1: Semi-Solid Foods*, ed. B.M. McKenna, Cambridge, Woodhead, Chapter 14, p. 350–372. Also: VD 03 0120.

Brennan, J.G. 1993. Emulsions in food technology. Course notes Dept. of Food Science and Technology, University of Reading.

Bruijn, W., Van't Riet, K., and Smith, J.M. 1974. Power consumption with aerated Rushton turbines. *T. I. Chem. Eng.*, 52, 88–104.

Bueschelberger, H.G. 2004. Lecithins. In *Emulsifiers in Food Technology*, ed. Whitehurst, R.J., UK, Blackwell Publishing.

Butters, J.R. 1993. Notes on separation operations. Dept. of Food Science and Technology, University of Reading, UK.

Calderbank, P.H. 1958. Physical Rate Processes in Industrial Fermentation. Part 1. *T. I. Chem. Eng.*, 36: 443–463.

Campbell, S.D., Goff, H.D. and Rousseau, D. 2002. Comparison of crystallisation properties of a palm stearin/canola oil blend and lard in bulk and emulsion form. *Food Res. Int.*, 35: 935–944.

Campos, R., Narine, S.S., and Marangoni, A.G. 2002. Effect of cooling rate on the structure and mechanical properties of milk fat and lard. *Food Res. Int.*, 35: 971–981.

Canselier, J.R., Delmas, H., Wilhelm, A.M., and Abismail, B. 2002. Ultrasound emulsification—An overview. *J. Disper. Sci. Techol.* 23:333–349.

Cater, C.M., Rhee, K.C., Hagenmaier, R.D., and Mattil, K.F. 1974. Aqueous extraction—An alternative oilseed milling. *J. Am. Oil Chem. Soc.*, 51(4): 137–141.

Cottrell, T. and Van Peij, J. 2004. Sorbitan esters and polysorbates. In *Emulsifiers in Food Technology*, ed. Whitehurst, R.J., UK, Blackwell Publishing.

Coultate, T.P. 1996. *Food—The Chemistry of its Components*, 2nd edition. Cambridge, The Royal Society of Chemistry.

Coupland, J.N. 2002. Crystallisation in emulsions. *Curr. Op. Coll. Interface Sci.*, 7: 445–450.

Coupland, I.N. and Mc Clements, D.J. Lipid oxidation in food emulsions. *Trends Food. Sci. Technol.*, 7: 83, 1996.

Dalgleish, D.G. 2001. Food emulsions. In *Encyclopedic Handbook of Emulsion Technology*, ed. Sjoblom, J., New York, Marcel Dekker.
Dalgleish, D.G. 2006. Food emulsions—Their structures and structure-forming properties. *Food Hydrocoll.*, 20(4): 415–422.
Danckwerts, P.V. 1952. The definition and measurement of some characteristics of mixtures. *Appl. Sci. Research*, 3: 279.
Danckwerts, P.V. 1953. Theory of mixtures and mixing. *Research*, 6: 355.
Davis, E., Dickinson, E., and Bee, R. 2000. Shear stability of sodium caseinate emulsions containing monoglyceride and triglyceride crystals. *Food Hydrocoll.*, 14: 145–153.
de Bruijne, D.W. and A. Bot. 1999. Fabricated fat-based foods. In *Food Texture: Measurement and Perception*, ed. A.J. Rosenthal, Gaithersburg, Aspen, Chapter 7, p. 185–227; Rosenthal, A.J. and Ibarz Ribas, A. 2001. Alimentos elaborados basados en grasa. In *Textura de los alimentos: medida y percepción*, Acribia, Zaragoza, capitulo 7, p. 181–222; Also VD 98 0058.
Dickinson, E. 2001. Milk protein interfacial layers and the relationship to emulsion stability and rheology. *Colloids Surf. B Biointerfaces*, 20: 197–210. DOI:10.1016/S0927-7765(00)00204-6.
Dickinson, E. and Galazka, V.B. 1991. Emulsion stabilization by ionic and covalent complexes of β-lactoglobulin with polysaccharides. *Food Hydrocoll.*, 5: 281–296.
Dickinson E. and McClements, D.J. Fat crystallisation in oil-in-water emulsions. In *Advances in Food Colloids*, eds. E. Dickinson and D.J. McClements, 1995, Blackie Academic and Professional, Glasgow, UK.
Dickinson, E. 1997. *Introduction to Food Colloids.* Oxford University Press, Oxford.
Dickinson, E. and Stainsby, G. 1982. *Colloids in Food*, London, Applied Science Publishers.
Dickinson, E. and Tanai, S. 1992. Temperature dependence of the competitive displacement of protein from the emulsion droplet surface by surfactants. *Food Hydrocolloid.* 6: 163–71.
Dickinson, E., Whyman, R.H., and Dalgleish, D.G. 1987. Colloidal properties of model oil-in-water food emulsions stabilized separately by αs1-casein, β-casein and κ-casein. In *Food Emulsions and Foams*. Ed. Dickinson, E., Cambridge, Royal Society of Chemistry.
Dickinson, E., Ma, J., and Povey, M.J. 1996. Crystallisation kinetics in oil-in-water emulsions containing a mixture of solid and liquid droplets. *J. Chem. Soc. Faraday Trans.*, 92: 1213–1215.
Douglas, G.D. 2004. Food emulsions: Their structures and properties. In *Food Emulsions*, 4th edition, eds. Friberg, S.E., Larsson, K., and Sjoblom, J., New York, Marcel Dekker.
Earle, R.L. 1983. *Unit Operations in Food Processing*, 2nd edn., Pergamon Press, Oxford, UK, pp. 143–158.
Edwards, M.F. and Baker, M.R. 1985. A review of liquid mixing equipment. In *Mixing in the Process Industries*, eds. Harnby, N., Edwards, M.F., and Nienow, A.W., Chapter 7, London, Butterworths.
Embong, M.B. and Jelen, P. 1977. Technical feasibility of aqueous extraction of rapeseed oil—A laboratory study. *J. Inst. Can. Sci. Technol. Aliment.*, 10(4): 239–243.
Engineering Equipment Users Association (EEUA). 1963. *Agitator Selection and Design. Handbook no. 9 revised.* London, Constable and Co. Ltd.
Feiner, G. 2006. Meat products handbook. *Practical Science and Technology*. Cambridge: CRC Press and Woodhead Publishing Limited. *Fischer, Chem. Eng.* 69(3): 52, 1962.
Flanagan, J. and Singh, H. 2006. Microemulsions: A potential delivery system for bioactives in food. *Crit Rev Food Sci Nutr*. 46:221–237.
Flanagan, J., Kortegaard, K., Pinder, D.N., Rades, T., and Singh, H. 2006. Solubilization of soybean oil in microemulsions using various surfactants. *Food Hydrocoll.*, 20:253–260.
Fleer, G.J., Scheutjens, J.M.H.M., and Vincent, B. 1984. Polymer adsorption and dispersion stability. *ACS Symposium Series* 240, eds. Goddard, E.D., and Vincent, B., Washington, DC, American Chemical Society.

Freitas S., Hielscher G., Merkle H.P., and Gander B. 2006. Continuous contact- and contamination-free ultrasonic emulsification—A useful tool for pharmaceutical development and production. *Ultrason. Sonochem.*, 13:76–85.

Friberg, S.E. 1990. *Emulsion Stability in Food Emulsions*, 3rd ed., eds. Larsson, K., and Friberg, S.E., New York, Marcel Dekker.

Friberg, E. and Larsson K. 1996. *Food Emulsions*, 3rd ed. New York, Marcel Dekker..

Gaupp, R., and Adams, W. 2004. Acid esters of mono- and diglycerides. In *Emulsifiers in Food Technology*, ed. Whitehurst, R.J., UK, Blackwell Publishing.

Gershanik, T., Benzeno, S., and Benita, S. 1998. Interaction of a self-emulsifying lipid drug delivery system with the everted rat intestinal mucosa as a function of droplet size and surface charge. *Pharm Res.*, 15: 863–869.

Ghosh, V., Mukherjee, A., and Chandrasekaran, N. 2013. Ultrasonic emulsification of food-grade nanoemulsion formulation and evaluation of its bactericidal activity. *Ultrason. Sonochem.* 20:338–344.

Goff, H.D. 1997. Instability and partial coalescence in whippable dairy emulsions, *J. Dairy Sci.*, 80: 2620–2630.

Greaves, M. and Barigou, M. 1986. Estimation of gas hold up and impeller power in a stirred vessel. *Fluid mixing III, I. Chemical Engineering Symposium Series* 108, pp. 235–256.

Greaves, M. and Kobbacy, K.A.H. 1981. Power consumption and impeller dispersion efficiency in gas–liquid mixing. *Fluid mixing I, I. Chemical Engineering Symposium, Series* 64, paper L1.

Griffin, W.C., 1949. Classification of surface active agents by HLB. *J. Soc. Cosmet. Chem.* 1:311.

Gunetileke, K.G. and Laurentius, S.F. 1974. Conditions for the separation of oil and protein from coconut milk emulsion. *J. Food Sci.*, 39: 230–233.

Gutiérrez, J.M., González, C., Maestro, A., Solè, I., Pey, C.M., and Nolla, J. 2008. Nano-emulsions: New applications and optimization of their preparation. *Curr. Opin. Colloid Int.*, 13(4): 245–251.

Hagenmaier, R.D. 1974. Aqueous processing of full-fat sunflower seeds: Yields of oil and protein. *J. Am. Oil Chem. Soc.*, 51: 470–471.

Hagenmaier, R.D., Carter, C.M., and Mattil, K.F. 1972. Critical unit operations of the aqueous processing of fresh coconuts. *J. Am. Oil Chem. Soc.*, 49: 178–181.

Hagenmaier, R.D., Carter, C.M., and Mattil, K.F. 1973. Aqueous processing of fresh coconuts for recovery of oil and coconut skim milk. *J. Food Sci.*, 38: 516–518.

Harnby, N. 1967. A comparison of the performance of industrial solids mixers using segregating materials. *Powder Technol.*, 1(2): 94–102.

Harnby, N., Edwards, M.F., and Nienow, A.W. (Ed.). 1992. *Mixing in the Process Industries*. 2nd edn., Oxford, Butterworth-Heinemann Ltd.

Hasenhuettl, G.L. 2008. Synthesis and commercial preparation of food emulsifiers. In *Food Emulsifiers and Their Applications*. eds Hasenhuettl, G.L., and Hartel, R.W., FL, USA, Springer Science and Business Media, LLC.

Heusch, R. 2000. Emulsions. In *Ullmann's Encyclopedia of Industrial Chemistry*. Weinheim, Wiley-VCH.

Holland, F.A., and Chapman, F.S. 1966. Liquid mixing and processing in stirred tanks. New York,. Reinhold.

Halliwell, B., and Gutteridge, J. 1995. *Free Radicals in Biology and Medicine*. 2nd ed., Oxford, Clavendon Press.

Hunter, R.J. 1986. *Foundations of Colloid Science*. Oxford, Oxford University Press.

Hunter, R.J. 2001. Electroacoustic characterization of emulsions. In *Encyclopedic Handbook of Emulsion Technology*, ed. Sjoblom, J., New York, Marcel Dekker.

Irving and Saxton. 1967. Mixing of high viscosity materials. In Uhl and Gray, *Mixing Theory and Practice*, Vol. 2, New York, Academic, Chap. 8.

Jafari, S.M., Assadpoor, E., He, Y., and Bhandari, B. 2008. Re-coalescence of emulsion droplets during high-energy emulsification. *Food Hydrocoll.,* 22: 1191–1202.

Jiménez-Colmenero, F. 2013. Potential applications of multiple emulsions in the development of healthy and functional foods. *Food Res. Int.*, 52(1): 64–74.

Joshi, J.B., Pandit, A.B., and Sharma, M.M. 1982. Mechanically agitated gas–liquid reactors. *Chemical Engineering Science*, 37: 813–844.

Judd, D.B. and Wyszecki, G. 1963. *Color in Business, Science, and Industry*. New York, Wiley.

Kay, J.M. and Nedderman, R.M. 1985. *Fluid Mechanics and Heat Transfer*. Cambridge, UK, Cambridge University Press.

Kinsella, J.E. 1989. Flavor perception and binding to food components. In *Flavor Chemistry of Lipid Foods*, eds. Min, D.B. and Smouse, T.H., Champaign, IL, American Oil Chemists Society, p. 376.

Kiokias, S. and Gordon, M. 2003. Dietary supplementation with a natural carotenoid mixture decreases oxidative stress. *Eur. J. Clin. Nutr.*, 57: 1135.

Kiokias, S. and Bot, A. 2005. Effect of protein denaturation on temperature cycling stability of heat-treated acidified protein-stabilised o/w emulsions. *Food Hydrocolloids*, 19: 493.

Kiokias, S. and Bot, A. 2006. Temperature cycling stability of pre-heated acidified whey protein-stabilised o/w emulsion gels in relation to the internal surface area of the emulsion. *Food Hydrocoll.*, 20: 246.

Kiokias, S. and Varzakas, T. 2014. Activity of flavonoids and b-carotene during the auto-oxidative deterioration of model food oil-in water emulsions. *Food Chem.*, 150: 280–286.

Kiokias, S., Reszka, A.A., and Bot, A. 2004a. The use of static light scattering and pulsed-field gradient NMR to measure droplet sizes in heat-treated acidified protein stabilised oil-in-water emulsion gels. *Int. Dairy J.*, 14: 287–295.

Kiokias, S., Reiffers-Magnani, C.K. and Bot, A. 2004b. Stability of whey protein stabilized oil in water emulsions during chilled storage and temperature cycling. *J. Agric. Food Chem.*, 52(12): 3823–3830.

Kiokias, S., Rojers, E.C. and Bot, A. 2004c. Polymorphic transitions in TAGs during crystallisation and temperature cycling of o/w model emulsions. Research report, Unilever Vlaardingen.

Kiokias, S., Lampa, K., Tsimogiannis, D., and Oreopoulou, V. 2005. Inhibition of oxidative deterioration in food emulsions. In *Proceedings of INTRAFOOD- EFFoST Conference, Vol. 2, 1237, 2005*.

Kiokias, S., Varzakas, T., and Oreopoulou, V. 2008. *In vitro* activity of vitamins, flavonoids, and natural phenolic antioxidants against the oxidative deterioration of oil-based systems. *Crit. Rev. Food Sci. Nutr.* 48: 78–93.

Kiokias, S., Varzakas, T., Arvanitoyannis, I., and Labropoulos, A. 2009. Lipid oxidation and control of oxidation. In: *Advances in Food Biochemistry*, pp. 384–403. ed. Yildiz, New York, CRC Press.

Knipe, L. 2004. Use of phosphates in meat products. *Meat Industry Research Conference*, Ohio State University.

Kralova, I., and Sjoblom, J. 2009. Surfactants used in food industry: A review. *J. Disper. Sci. Techol.*, 30(9): 1363–1383.

Kruglyakov, P.M. 2000. *Hydrophile–Lipophile Balance of Surfactants and Solid Particles, Volume 9: Physicochemical Aspects and Applications (Studies in Interface Science)*, 1st edn., Amsterdam, the Netherlands, Elsevier Science.

Krog, N.J. and Sparso, F.V. 2004. Food emulsifiers: Their chemical and physical properties. In *Food Emulsions*, 4th edn., eds. Friberg, S., Larsson, K., and Sjoblom, J., Chapter 2, New York, Marcel Dekker.

Kukizaki, M. and Goto, M. 2007. Preparation and evaluation of uniformly sized solid lipid microcapsules using membrane emulsification. *Colloids Surf. A: Physicochemical and Engineering Aspects*, 293: 87–94.

Labuza, T.P., Kinetics of lipid oxidation in foods. *Crit. Rev. Food Techn.*, 2(355): 1971.

Lawhon, J.T., Manak, L.J., Rhee, K.C., and Lusas, E.W. 1981a. Production of oil and protein food products from raw peanuts by aqueous extraction and ultrafiltration. *J. Food Sci.*, 46: 391–395.

Lawrence, M.J. and Rees, G.D. 2000. Microemulsion-based media as novel drug delivery systems. *Adv. Drug Deliv. Rev.*, 45: 89–121.

Leal-Calderon, F. 2007. *Emulsion Science-Basic Principles*, 2nd edn. New York, Springer Verlag.

Lines, R.W. 1994. The electrical sensing zone method (the Coulter principle). In *Liquid and Surface Bourne Particle Measurement Handbook*, ed. Knapp, J.Z. New York, Marcel Dekker.

Loncin, M. and Merson, L.M. 1979. *Food Engineering—Principles and Selected Applications*. New York, London, San Francisco, Academic Press.

Lopez, C., Lesieur, P., Bourgaux, C., Keller, G., and Ollivion, M. 2001. Crystalline forms obtained by slow cooling of cream. *J. Colloid Interface Sci.*, 240: 150–161.

Lusas, E.W., Lawhon, J.T., and Rhee, K.C. 1982. Producing edible oil and protein from oilseeds by aqueous processing. *Oil Mill Gazetter*, 4: 28–34.

Maa, Y.F. and Hsu, C. 1996. Liquid–liquid emulsification by rotor/stator homogenization. *J. Control. Release*, 38, 219–228.

Macosko, C.W. 1994. *Rheology: Principles, Measurements and Applications*. New York, NY, VCH Publishers.

Marshall, R.T., Goff, H.D. and Hartel, R.W. 2003. Ice cream, 6th edition. New York, Kluwer Academic. ISBN 0-306-47700-9.

McClements, D.J. 2002. Theoretical prediction of emulsion color. *Adv. Colloid Interface Sci.*, 97: 63–89.

McClements D.J. 2005. *Food Emulsions: Principles, Practice and Techniques*, 2nd edn., Boca Raton, FL, CRC Press.

McClements, D.J. and Decker, E.A. 2000. Lipid oxidation in oil-in-water emulsions: Impact of molecular environment on chemical Reactions in heterogeneous food systems. *J. Food Sci.*, 65: 1270–1282.

McClements, D.J. and Weiss, J. 2004. Lipid emulsions. In *Baileys Industrial Fats and Oil Products*. Hoboken, NJ, Wiley Scientific.

McClements, D.J., Decker, E.A., and Weiss, J. 2007. Emulsion-based delivery systems for lipophilic bioactive components. *J. Food Sci.*, 72(8): 109–124.

McClements, D.J., Dungan, S.R., German, J.B., Simoneau, C., and Kinsella, J.E. 1994. Droplet size and emulsifier type affect crystallisation and melting of hydrocarbon-in-water emulsions. *J. Food Sci.*, 58: 1148–1151.

McClements, D.J., Han, S.W. and Dungan, S.R. 1993. Interdroplet heterogeneous nucleation of supercooled liquid droplets by solid droplets in oil-in-water emulsions. *J. Am. Oil. Chem. Soc.*, 71: 1385–1389.

McNulty, P.B. 1987. Flavor release–elusive and dynamic. In *Food Structure and Behaviour*, eds. Blanshard, J.M.V. and Lillford, P., London, UK, Academic Press.

McClements, D. J. and Rao, J. 2011. Food-grade nanoemulsions: Formulation, fabrication, properties, performance, biological fate, and potential toxicity. *Crit. Rev. Food Sci. Nutr.*, 51(4): 285–330.

Michel, B.J. and Miller, S.A. 1962. Power requirements of gas–liquid agitated systems. *AIChE J.*, 8: 262–266. doi: 10.1002/aic.690080226.

Middleton, J.C. 1985. In: Mixing in the process industries, eds. Harnby, N., Edwards, M.F., and Nienow, A.W., Chapter 17, London, UK, Butterworths.

Moonen, H., Bas, H. 2004. Mono- and diglycerides. In *Emulsifiers in Food Technology*, pp. 40–57, ed. Whitehurst, R.J. Oxford, Blackwell Publishing.

Morris, V.J., Gunning, A.P., Kirby, A.R. 1999. *Atomic Force Microscopy for Biologists,* London, UK, Imperial College Press.

Murphy, D.B. 2001. *Fundamentals of Light Microscopy and Electronic Imaging*, 1st ed., New York, NY, Wiley-Liss.

Muschiolik, G. 2007. Multiple emulsions for food use. *Curr. Op. Coll. Interface Sci.*, 12: 213–200.

Mutoh, T.A., Nakagawa, S., Noda, M., Shiinoki, Y., and Matsumura, Y. 2001. Relationship between characteristics of oil droplets and solidification of thermally treated creams. *J. Am. Oil Chem. Soc.*, 78: 177–182.

Mutsakis, M., Streiff, F.A., and Schneider, G. 1986. Advances in static mixing technology. *Chem. Eng. Prog.*, 82(7): 42–48.

Nagata, S. 1975. Mixing. *Principles and Applications*. Tokyo, Halstead Press.

Nelen, B.A.P. and Cooper, J.M. 2004. Sucrose esters. In *Emulsifiers in Food Technology*, pp. 131–158, ed. Whitehurst, R.J., Oxford, UK, Blackwell Publishing Ltd.

Nedderman, R.M. 1992. Granular materials. In: *Chemical Engineering for the Food Industry*. Cambridge programme for the food industry. Dept. of Chemical Engineering, University of Cambridge.

Nienow, A.W. 1985. The mixer as reactor: Liquid/solid systems. In *Mixing in the Process Industries*, eds. Harnby, N., Edwards, M.F., and Nienow, A.W. Chapter 18, London, Butterworths.

Nienow, A.W. and Miles, D. 1978. The effect of impeller/tank configuration on fluid-particle mass transfer. *Chem. Eng. J.*, 15: 13–24.

Nienow, A.W., Chapman, C.M. and Middleton, J.C., 1979. Gas recirculation rate through impeller cavities and surface aeration in sparged agitated vessels. *Chem. Eng. J.*, 17(2): 111–118.

Nienow, A.W., Wisdom, D.J., and Middleton, J.C. 1978. *2nd European Conference Mixing*, paper F1, BHRA Fluid Engng, Cranfield.

Noda, M. and H. Yamamoto, 1994. Effects of tempering on physical properties of whipped cream, *Nippon Shokuki Koyo Gakkaishi*, 41: 327–334.

O'Driscoll, B., Smyth, C., Alting, A.C., Visschers, R.W., and Buckin, V. 2003. Recent applications for high-resolution ultrasonic spectroscopy. *Am. Lab.*, 35: 54.

Ogden, L.G., and Rosenthal, A.J. 1997. Interaction between tristearin crystals and proteins at the oil–water interface. *J. Colloid Interface Sci.*, 190: 38–47.

Oldshue, J.Y. 1983. Fluid mixing technology and practice. *Chem. Eng.*, 90(12): 82–108.

Olson, D.W., White, C.H., and Richter, R.L. 2004. Effect of pressure and fat content on particle sizes in microfluidized milk. *J. Dairy Sci.* 87:3217–3223.

Orthoefer, F. 2008. Applications of emulsifiers in baked foods. In *Food Emulsifiers and Their Applications*. eds Hasenhuettl, G.L., and Hartel, R.W., FL, USA, Springer Science and Business Media, LLC.

Overbosch, P., Afterof, W.G.M., and Haring, P.G.M. 1991. Flavor release in the mouth. *Food Rev. Int.*, 7: 137.

Paraskevopoulou, A., Kiosseoglou, V., Alevisopoulos, S., and Kasapis, S. 1999. Influence of reduced cholesterol yolk on the viscoelastic behaviour of concentrated O/W emulsions. *Colloids Surf. B Biointerfaces*, 12: 107.

Perry, R.H. and Green, D.W. 1999. *Perry's Chemical Engineers' Handbook*. USA, McGraw Hill.

Phipps, L.W. 1985. *The High Pressure Dairy Homogenizer*. Reading, UK, The National Institute for Research in Dairying.

Plucknett, K.P., Pomfret, S.J., Normand, V., Ferdinando, D., Veerman, C., Frith, W.J., and Norton, I.T. 2001. Dynamic experimentation on the confocal laser scanning microscope: Application to soft-solid, composite food materials. *J. Micros. Oxford*, 201: 279.

Ponginelbi, I., Nawar, W.W., and Chinchoti, P., Oxidation of linoleic acid in emulsions. Effects of substrate, emulsifiers, and sugar concentration, *J. Am. Oil. Chem. Soc.*, 76(131): 1999.

Rajasekharan, N. and Sreenivasan, A. 1967. The use of coconut preparations as a protein supplement in child feedings. *J. Food Sci. Technol.* (India), 4, 59. Cited by Gunetileke and Laurentius, 1974.

Reiffers-Magnani, C.K., Lundin, L., Schnitker, M. and Tio, F. 2002. Microstructural model for dairy spread alternatives. 2. Emulsion instability during temperature abuse, VD 02 0123.

Rhee, K.C., Cater, C.M., and Mattil, K.F. 1972. Simultaneous recovery of protein and oil from raw peanuts in an aqueous system. *J. Food Sci.*, 37: 90–93.

Rhee, K.C., Cater, C.M., and Mattil, K.F. 1973. Effect of processing pH on the properties of peanut protein isolates and oil. *Cereal Chem.*, 50: 395–404.

Rielly, 1992. Chapter 8. Mixing operations in the food industry. Chemical Engineering for the food industry. Cambridge programme for the food industry. Dept. of Chemical Engineering, University of Cambridge.

Robins, M. and Wilde, P. 2003. Colloids and emulsion. In: *Encyclopedia of Food Sciences and Nutrition*, 2nd edn., ed. Caballero, A., pp. 1517–1524, Oxford, UK, Academic Press.

Rousseau, D. 2000. Fat crystals and emulsion stability. *Food Res. Int.*, 33: 3–14.

Rowe, P.N., Claxton, K.T., and Lewis, J.B. 1965. Heat and mass transfer from a single sphere in an extensive fluid. *T. I. Chem. Eng.*, 43: T14–T31.

Roxas, P.G. 1963. Recovering oils from oleaginous meats of nuts, beans and seeds. U.S. Patent 3083365.

Salager, J., Anton, R., Sabatini, D., Harwell, J., Acosta, E., and Tolosa, L. 2005. Enhancing solubilization in microemulsions—State of the art and current trends. *J. Surfactants Deterg.*, 8(1): 3–21.

Schroder, V. and Schubert, H. 1997. Emulsification using microporous ceramic membranes. In *Proceedings of the First European Congress on Chemical Engineering (ECCE 1)*, Florence, Italy, p. 2491 .

Schultz, S., Wagner, G., Urban, K., and Ulrich, J. 2004. High-pressure homogenization as a process for emulsion formation. *Chem. Eng. Technol.*, 27: 361–368.

Smith, J.M. 1986. *5th European Conference Mixing*, paper 13, BHRA Fluid Engng, Cranfield.

Snyder, H.E., and Kwon, T.W. 1987. *Soybean Utilization*. New York, Van Nostrand Reinhold Company Inc., pp 346.

Soon, S., Harbidge, J., Titchener-Hooker, N., and Shamlou, P., 2001. Prediction of drop breakage in an ultra high velocity jet homogenizer. *J. Chem. Eng. Jpn.*, 34(5): 640–646.

Southwell, K.H. and Harris, R.V. 1991. Extraction of oil from oilseeds using the hot water flotation method. *Tropical Sci.*, 32(3): 217–221.

Sparso, F.V. and Krog N. 2004. Propylene glycol fatty acid esters. In *Emulsifiers in Food Technology*, ed. Whitehurst, R.J., UK, Blackwell Publishing.

Stang, M., Schuchmann, H., and Schubert, H. 2001. Emulsification in high-pressure homogenizers. *Eng. Life Sci.*, 1:151–157.

Strawbridge, K.B. and Hallett, F.R. 1994. Size distributions obtained from the inversion of I(Q)using integrated light scattering spectroscopy. *Macromolecules*, 27: 2283–2287.

Streiff, F.A. 1979. Adapted motionless mixer design. *3rd European Conference in Mixing*, 171, BHRA Fluid Engng, Cranfield.

Sugarman, N. 1956. Processing for extraction of oil and protein simultaneously from oil bearing materials. U.S. Patent 2.762.820.

Tadros, T.F. 1989. Colloidal aspects of pesticidal and pharmaceutical formulation—An overview. *Pestic. Sci.*, 26: 51–77.

Tadros, T.F. 1994. Fundamental principles of emulsion rheology and their applications. *Colloid Sur. A*. 91: 39–55.

Tadros, T.F. 2009. Emulsion science and technology: A general introduction. In *Emulsion Science and Technology*, ed. Tadros, T.F., Weinheim, Wiley-VCH.

Tang, S.Y., Manickam, S., Wei, T.K., and Nashiru, B. 2012. Formulation development and optimization of a novel cremophore EL-based nanoemulsion using ultrasound cavitation. *Ultrason. Sonochem.* 19: 330–345.

Taylor, A.J. and Linforth, R.S.T. 1996. Flavour release in the mouth. *Trends Food Sci. Technol.* 7: 444–448.

ten Grotenhuis, E., G.A van Aken, K.F. van Malssen and H. Schenk. 1999. Polymorphism of milk fat studies by differential scanning calorimetry and real time X-ray powder diffraction. *J. Am. Oil. Chem. Soc.*, 76: 1031–1039.

Uhl, V.W. and Gray, J.B. 1966. *Mixing Theory and Practice*, 2nd edn., New York, Academic Press.

Urban, K., Wagner, G., Schaffner, D., Roglin, D., and Ulrich, J. 2006. Rotor–stator and disc systems for emulsification processes. *Chem. Eng. Technol.*, 29: 24–31.

Van Dierendonck, L.L., Fortuin, J.M.H., and Venderbos, D. 1968. *4th European Conference Chemical Reactor Engineering*, p. 205.

Van't Riet, K., Smith, J.M. 1973. The behaviour of gas–liquid mixtures near Rushton turbine blades. *Chem. Eng. Sci.*, 28: 1031–1037.

Vanapalli, S.A., Palanuwech, J. and Coupland, J.N. 2002. Stability of emulsions to dispersed phase crystallisation: Effect of oil type, dispersed phase volume fraction, and cooling rate, *Colloids Surf. A: Physicochem. Eng. Asp.*, 204: 227–237.

Varnam Alan, H. and Sutherland Jane, P. 2003. *Milk and Milk Products: Technology, Chemistry and Microbiology*, London, Chapman & Hall.

Velikov, K.P. and Pelan, E. 2008. Colloidal delivery systems for micronutrients and nutraceuticals. *Soft Matter.* 4: 1964–1980.

Warmoeskerken, M.M.C.G. and Smith, J.M. 1984. Fluid mixing II, *I. Chemical Engineering Symposium Series* 89, paper 5.

Wedzicha, B.L. 1988. Distribution of low-molecular-weight food additives. In *Advances in Food Emulsions*, eds Dickinson, E., and Stainsby, G. London, Elsevier Applied Science Publishers.

Weyland, M. and Hartel, R. 2008. Emulsifiers in confectionery. In *Food Emulsifiers and Their Applications*, eds Hasenhuettl, G.L., and Hartel, R.W., FL, USA, Springer Science and Business Media, LLC.

Whorlow, R.W. 1992. *Rheological Techniques*. New York, Ellis Horwood, Wiley.

Wilkinson, W.L. and Cliff, M.J. 1977. An investigation into the performance of a static in-line mixer. *Proceedings of 2nd European Conference in Mixing*, paper A2, BHRA Fluid Engng, Cranfield.

Williams, R.A., Peng, S.J., Wheeler, D.A., Morley, N.C., Taylor, D., Whalley, M., and Houldsworth, D.W. 1998. Controlled production of emulsions using crossflow membrane Part II. Industrial scale manufacture. *Chem. Eng. Res. Des.* 76(A): 902.

Wooster, T.J., Golding, M., and Sanguansri, P. 2008. Impact of oil type on nanoemulsion formation and Ostwald ripening stability. *Langmuir*, 24(22): 12758–12765.

Wyszecki, G. and Stiles, W.S. 1982. *Color Science: Concepts and Methods, Quantitative Data and Formulae*. Wiley, New York.

Yianneskis, M., Popiolek, Z., and Whitelaw, J.W. 1987. An experimental study of the steady and unsteady flow characteristics of stirred reactors. *J. Fluid Mechanics*, 175: 537–555.

Zweitering, T.N. 1958. *Chem. Eng. Sci.*, 8: 244.

# 6 Solid–Liquid Extraction

*Sofia Chanioti, George Liadakis, and Constantina Tzia*

## CONTENTS

## 6.1 INTRODUCTION

Extraction is a process where certain substances of a solid or a liquid mixture are dissolved, washed, or leached by the aid of a liquid solvent. The receiving or extract phase consists of the solvent and key substances transferred by the solvent from the feed or raffinate phase. Compared with other separation processes, regeneration of the solvent and purification of the valuable substances are required (Aguilera 2003, Sancey 2002, Sattler and Feidt 1995).

Solid–liquid extraction or leaching is a separation process where certain substances are dissolved out from a solid matrix by a liquid solvent forming a solution. Solid–liquid extraction is a unit operation with diverse applications in the food industry. Supercritical fluid extraction (SFE) involves the separating of a mixture, in solid or liquid state or in solid and liquid state, by contacting it with a fluid maintained under conditions of temperature and pressure above its critical point.

This chapter employs the solid–liquid extraction. The solvent used should be capable (having suitable properties) to dissolve the current substances from the insoluble permeable solid; thus its selection is important for the quantitative removal of them and the process yield. The removing soluble fraction is called solute and the resulting solution of the solute is called extract. The recovery of the solvent is commonly conducted and it is reused in the extraction process.

Extraction is a separation process widely used in the food- processing industry for various applications, while recently by SFE, it is possible to improve the quality of food products. Extraction may be a basic step in the processing of many food products, that is, sugar, edible oils, and so on, or it is used for the recovery of active constituents or for the removal of undesirable constituents from raw materials, while by extraction, important components may be separated from natural products with many applications to food industries (flavors, antioxidants, etc.).

The aim of an extraction process is therefore connected to the separation of usefulness and not of the solute substances. Thus, in such a process, the solute may be composed of nutritive components that can be separated from food sources and after purification or refining are further utilized for edible use, that is, the oilseeds are extracted from a recovering oil that is a staple food product. In particular, bioactive or functional components are contained in plant or animal sources that are separated by extraction and have found useful applications in enriched and nutraceutical/functional foods production, that is, extraction of various botanicals. Hence, the solute may consist of undesirable or poisoning substances that are removed by extraction, that is, extraction of toxic substances from oilseed meals. In this case, the solid material is of interest; thus, the extraction will result in a desirable or safe food. Leaching is industrially applied to extract lipids from oilseeds, oil fruits and press cake, proteins in oilseed meals, pharmaceutical substances from plants, and functional hydrocolloids from algae, or to remove undesirable contaminants and toxins in foods and feeds. In all these cases, the extraction occurs as a result of the effect of the solvent selectivity on the soluble solute (Aguilera 2003, Coulson et al. 1993, Sattler and Feidt 1995). Notwithstanding, in all the above cases, the design and operation of the extraction process that will be applied and the relative method with the involved equipment that is going to be selected are critical for a cost-effective extraction process.

## 6.2 EXTRACTION THEORY: PROCESS MECHANISM

### 6.2.1 Mechanisms of Extraction

If the solute is uniformly dispersed in the solid matrix, the material near the surface will be dissolved first, leaving a porous structure in the solid residue. To reach further solute, the solvent has to penetrate the outer layer, the operation becomes progressively more difficult, and the extraction rate falls. If the solute forms a large proportion of the solid, the porous structure may be destroyed giving a fine deposit of the insoluble residue and further solute may be easily accessed by the solvent (Aguilera 2003, Coulson et al. 1993, McCabe et al. 1993, Wakeman 1994).

Generally, a series of phenomenological steps can occur in solid–liquid extraction process:

1. Transfer of solvent from the bulk of the solution onto the surface of the oleaginous matrix
2. Penetration or diffusion of the solvent into the pores of the solid matrix
3. Dissolution of the solvent into the solute
4. Transport of the solute to the surface of the solid matrix
5. Migration of the extracted solute from the external surface of the solid material into the bulk solution
6. Movement of the extract with respect to the solid (i.e., extract displacement), and solid matrix

Any one of these phenomena may be responsible for limiting the extraction rate. The overall rate of the extraction process is determined by the step having the slowest rate or the rate-controlling step. In food extractions, the rate-controlling step is usually the transfer of solvent from the bulk solution to the solid surface and into the solid matrix (Aguilera 2003, Kemper 2005, Takeuchi et al. 2009).

### 6.2.2 Mass Transfer in Solid–Liquid Extraction Processes

Extraction rates from the porous residue are difficult to assess as it is difficult to define the shape of the pores through which mass transfer has to take place. The transfer of the solute inside the solid particle occurs because of the concentration gradient in the solid–liquid interface, and it could be mathematically described by the diffusion equation. The equation that describes this phenomenon is based on Fick's law and is given by (Coulson et al. 1993, Prabhudesai 1997, Takeuchi et al. 2009, Wakeman 1994)

$$\frac{N_C}{A_T} = -D\frac{dC_C}{dz} \tag{6.1}$$

where

$N_C$ is the rate of dissolution of the solute $C$ in the solution (kg/s),
$A_T$ is the area of the solid–liquid interface (m$^2$),

$C_C$ is the concentration of solute $C$ in the solution (kg/m³),
$z$ is the distance inside the porous of the solid matrix (m), and
$D$ is the diffusion coefficient (m²/s).

The minus sign gives a positive flux term because the gradient is negative (flow occurs down a concentration gradient, from high to low concentration).

Diffusion coefficient (D) is an important parameter in the diffusion model and its value is in the range $10^{-9}$–$10^{-10}$ m²/s for solids. In solid foods, the mass transport is strongly dependent on the size, shape, and porous presence. In these cases, the diffusion is expressed in terms of effective diffusivity $D_{\text{eff}}$, defined as follows (Aguilera 2003, Takeuchi et al. 2009):

$$D_{\text{eff}} = \frac{\varepsilon}{\tau} D \tag{6.2}$$

where

$\varepsilon$ is the void fraction space or porosity of the solid, and
$\tau$ is the tortuosity of the pores.

For solid materials, tortuosity varies between 2 and 6, and porosity varies between 0.3 and 0.8; thus $D_{\text{eff}}$ may be 6–15 times lower than $D$.

For a batch operation, where the total $V$ of solution is assumed to remain constant, onto the surface of the solid particle, the transfer of the solute occurs with simultaneous molecular and turbulent transport. Thus, the rate of mass transfer can be described by the following equation (Coulson et al. 1993, Prabhudesai 1997, Takeuchi et al. 2009, Wakeman 1994):

$$N_C = \frac{VdC_C}{dt} = A_T k_L (C_{CS} - C_C) \tag{6.3}$$

where

$k_L$ is the mass transfer coefficient in m/s,
$C_{CS}$ is the reference concentration of the solute $C$ in the solution in kg/m³,
$C_C$ is the concentration of the solute $C$ in the solution at time $t$ in kg/m³.

The time $t$ taken for the concentration of the solution to rise from its initial value $C_{C0}$ to a value $C_C$ is obtained by integration, assuming that $A$ remains constant, giving the following:

$$\int_{C_0}^{C} \frac{dC_C}{C_{CS} - C_C} = \frac{Ak_L}{V} \int_{t=0}^{t} dt \tag{6.4}$$

$$\frac{C_{CS} - C_C}{C_{CS} - C_{C0}} = e^{-\left(k_L A/V\right)t} \tag{6.5}$$

If pure solvent is used initially, $C_{C0} = 0$, and then:

$$1 - \frac{C_C}{C_{CS}} = e^{-\left(k_L A/V\right)t} \tag{6.6}$$

$$C_C = C_C\left(1 - e^{-\left(k_L A/V\right)t}\right) \tag{6.7}$$

## 6.3 EXTRACTION SYSTEMS OPERATION

### 6.3.1 Solvent Extractors Operating Methods

The types of flow used in leaching systems can occur in single or multiple stages and it can be countercurrent or crosscurrent.

The *single-stage leaching process* (Figure 6.1) involves only one stage of contacting the solid feed ($L_0$) and the fresh solvent ($V_2$) and, subsequently, separating the resulting insoluble solid by physical means. The extraction produces two outflows: the extract ($V_1$), which is constituted of a relatively large amount of solvent and the residue ($L_1$) containing the insoluble solid. Because of the low recovery of solute obtained and dilute solution produced, this type of flow is rarely used on an industrial scale (Couper et al. 2012, Prabhudesai 1997, Tandon and Rane 2008, Wakeman 1994, Xu and Diosady 2003).

The overall mass balance equation is

$$L_0 + V_2 = L_1 + V_1 = M \tag{6.8}$$

where $M$ is the mixture point in the single stage;
for the solute compound $A$:

$$L_0 y_{A0} + V_2 x_{A2} = L_1 y_{A1} + V_1 x_{A1} = M x_{Am} \tag{6.9}$$

where $y_{A0}$, $x_{A2}$, $y_{A1}$, and $x_{A1}$ are the mass fractions of compound $A$ in the feed, solvent, residue, and extract, respectively.

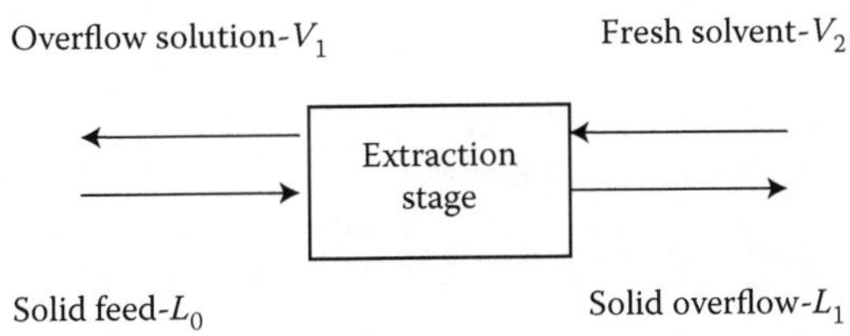


**FIGURE 6.1** Single-Stage solid-liquid extraction.

In *multistage crosscurrent extraction* (Figure 6.2), fresh batches of solvent and solids feed are continuously contacted. The above equations are applicable to the calculations for addition stages as the procedure for a single stage is repeated. Underflow from the first stage is sent to the second stage, where it is mixed with more fresh solvent. However, a large amount of solvent is required for this operation and at the last stages of extraction, dilute solutions of the solute are produced (Couper et al. 2012, Prabhudesai 1997, Takeuchi et al. 2009, Tandon and Rane 2008, Xu and Diosady 2003, ).

In the *continuous countercurrent multistage system* (Figure 6.3), the underflow and overflow streams flow countercurrently to each other. The system allows high recovery from the initial solid with a highly concentrated final extract utilizing the least amount of solvent. Thus, countercurrent extraction reduces solvent amounts and operating costs (Couper et al. 2012, Prabhudesai 1997, Takeuchi et al. 2009, Tandon and Rane 2008, Wakeman 1994, Xu and Diosady 2003).

### 6.3.2 Factors Influencing the Rate of Extraction

The maintenance of constant fluid flows, pressures and temperatures, and the provision of a sufficient contact time between the solvent and the solids are important in most leaching processes. The use of the control equipment and recording instruments provides a useful means for studying plant performance. Factors such as the particle size of the solid and the solvent employed have to be taken into account in leaching the operation design.

The nature of the compound to be extracted as well as the raw material that is going to be processed should be known in order to select the best extraction technology that will obtain a high recovery yield and also a high stability of the chemical

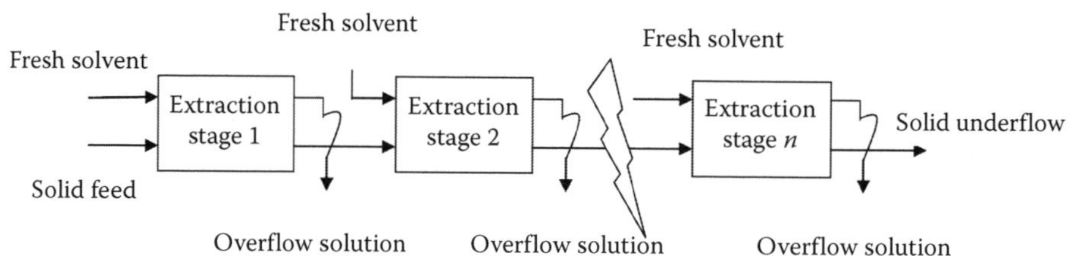


**FIGURE 6.2** Multistage crosscurrent solid-liquid extraction.

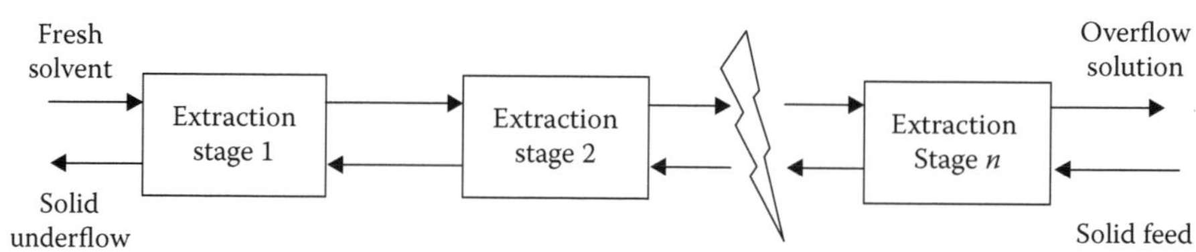


**FIGURE 6.3** Multistage counter-current solid-liquid extraction.

compounds of the solute. Thus, the selection of a proper extraction method of food sources/materials designates qualitative and quantitative studies of crude extract that lead to identification of its active components.

#### 6.3.2.1 Solvent

The following criteria are likely to be considered on solvent selection:

*Solubility of the specific substances in the solvent.* For example, vegetable oils consisting of triglycerides of fatty acids are normally extracted with hexane, whereas for free fatty acids extraction, more polar alcohols are used.

*Physical properties such as low interfacial tension and viscosity.* The solvent should be capable of wetting the solids and penetrating through pores and capillaries in the matrix. Also, its low viscosity assists diffusion rates in the solvent phase.

*Recovery, since the solvent will be reused in subsequent extractions.* If distillation or evaporation is used, the solvent should not form azeotropes and the latent heat of vaporization should be small.

*Hazards and cost.* Ideally, the solvent should be nontoxic, nonhazardous, nonreactive, nonflammable, harmless to the environment, and cheap. Avoidance of solvent losses may be obtainable through better process design.

As far as the laws on extraction solvents used in foods are concerned, they are primarily in accordance with human health requirements. In terms of human consumption, the presence of some solvents such as propane, butane, propyl acetate, ethyl acetate, ethanol, carbon dioxide, acetone, and nitrous oxide are acceptable in small residual percentages, according to good manufacturing practices (GMP). Also, substances such as hexane, methyl acetate, ethylmethylketone, and dichloromethane have been found acceptable by the Enzyme Commission (EC) when used under specific conditions and present limitations concerning pharmaceutical and food products because of their inherent toxicity. The PDEs (permissible daily exposures) of these solvents are given to the nearest 0.1 mg/d, and concentration limits vary from 50 to 3880 ppm, depending on the organic solvent used (Aguilera 2003, Coulson et al. 1993, Johnson and Lusas 1983, Prabhudesai 1997, Takeuchi et al. 2009, Wakeman 1994).

#### 6.3.2.2 Particle Size

In food materials, the cell structure is an important factor that needs to be considered. Although the solute can be on the surface of the cell, in most of the cases, it is stored in intracellular spaces, capillaries, or cell structures. Thus, the success of the solvent extraction strongly depends on the solid condition. One of the pretreatment steps that must be considered is the comminuting or grinding of the raw material. Grinding before solvent extraction promotes an increase of the contact area between the solvent and the solid matrix. The smaller the particle size, the higher is the rate of transfer of solute. The interfacial area between the solid and the liquid is greater and the intraparticle diffusion resistance becomes smaller because of the shorter diffusional path lengths. Hence, extraction efficiency increases with decreasing of the particle size. Thus, various oleaginous materials, such as soybeans, are formed into flakes reducing their thickness and the distance and the number of cell walls. However, smaller particle sizes make lower drainage rates from the solid residue and may create problems if circulation of the liquid is impended (Birch and Ian 2000, Coulson et al. 1993, Kemper 2005, Prabhudesai 1997, Takeuchi et al. 2009, Wakeman 1994).

#### 6.3.2.3 Solid Material Humidity

The water in the solid material can compete with the extraction solvent for the solute's dissolution, affecting the mass transfer. On the other hand, this humidity is necessary to permit the transport of the solute, as in coffee extraction. Nevertheless, in most of the cases, the material is dried under conditions that do not cause degradation of the compounds (Takeuchi et al. 2009).

#### 6.3.2.4 Temperature

The solubility of the material being extracted and its diffusivity normally increase with the rise in temperature and thus, the rate of extraction is improved. However, in the food industry, higher temperatures such as coffee, tea, and sugar beets may generate undesirable reactions such as the degradation of thermolabile compounds. For instance, in coffee processing elevated temperatures can cause hydrolysis. There is an upper limit for the temperature and it is determined by factors such as the avoidance of undesirable reactions. Additionally, the solvent must remain in a liquid state. For instance, as the boiling point of hexane is 64–69°C, the maximum temperature to prevent boiling is 63°C (Aguilera 2003, Coulson et al. 1993, Kemper 2005, Prabhudesai 1997, Takeuchi et al. 2009, Wakeman 1994).

#### 6.3.2.5 Agitation of the Fluid

Agitation of the solvent is important as it increases turbulent diffusion and the transfer rates of material from the surface of the particles to the bulk of the solution.

### 6.3.3 Process Design and Optimization

In an extraction process applied to recover or remove substances, the process product of interest may be the extracted material or the extract; the separation of solvent from the extracted solute is required by distillation or evaporation (Tiwari 1995). The design and optimization of an extraction process entails the knowledge and the role of the technological parameters of the process. The increased use of optimization on food manufacturers enables the efficiency, productivity, and quality to be increased and energy use, product loss, and environmental pollution to be reduced.

The most important parameters that significantly affect the extraction efficiency and are determinants for the process design and success, are the following (Tzia 2003):

(a) The preparation of solid material by size reduction (crushing, grinding, flaking, or cutting into pieces or cosettes); the solid size must be suitable (surface area per unit volume) to make the solute more accessible to the solvent and to favor the extraction, but not so fine as to to cause packing of solids and impeding free flow of the solvent.

(b) The selection of solvent for extraction based on a number of characteristics (capacity, selectivity, chemical inertness, thermophysical properties, flammability, toxicity, cost, and availability).

(c) The selection of operating temperature; the temperature must be high to give higher solubility of solute in solvents, but not so high as to to cause solvent losses, extraction of undesirable constituents, or damage of sensitive components.

(d) The equipment depending on the mode of operations (batch or continuous), the solids handled (fixed bed: percolation, full immersion, intermittent drainage, or dispersed/moving contact), or the performing arrangement (one stage or multistage).

The technical and economical feasibility of an extraction process is determined knowing the equipment size, operating conditions, solvent flow rates, and extraction yields. Extraction optimization involved both the design and determination of the optimal operating conditions for the equipment (separation allocation, differences in physical and/or chemical properties for separation, equipment type and sequence separators, fixation of separated phases, and entire process operating conditions).

A common problem in optimization of extraction processes is that operation time for the lowest cost is usually different from the time giving the best yield (effectiveness), so that it is impossible to have both optimum cost and yield simultaneously. In this case, the operation time becomes the independent parameter whose value determines the resulting value of the criteria: minimum cost and maximum yield. The relative importance of the two conflicting criteria in regard to the objective must be judged to obtain an optimum extraction time. So, if the lowering of the operation cost is of interest, the time of extraction will be optimized, while the yield will be the primary criterion in an isolation of desired constituents or removal of toxic substances for food sources. Suboptimum with respect to operation cost and yield, or optimum with respect to the combined criteria of cost and yield may be found. An overall criterion function can be established applying a more formalized optimization. Regional constraints on the range of possible times might be imposed; a very large time above a certain value is not cost-effective or the yield is not approved at a time below a certain value. After the optimum operation time has been determined, other parameters not included in the original problem (extraction medium, extraction system) may be used, thus decreasing the operation cost and increasing yield by formal optimization techniques. It must be noted that formal optimization theory should be used only to find optimum parameters in well-defined systems with quantitative-dependent and quantitative-independent parameters, criterion function, and functional and regional constraints (Tzia 2003).

Optimization studies on protein extraction process for protein isolates production have been studied from oilseeds (Liadakis et al. 1995).

## 6.4 SOLID–LIQUID EXTRACTORS

The solid–liquid extraction processes can be batch, continuous, or quasi-continuous. Continuous plants can be percolation, immersion, or direct extraction plants. In the food industry, continuous or quasi-continuous processes are commonly used (Berk 2009, Bernardini 1976, Birch and Ian 2000, Takeuchi et al. 2009, Tandon and Rane 2008).

### 6.4.1 Batch Extractors

A batch extraction system (Figure 6.4) consists of an agitated mixing vessel where the solids are treated with the solvent by percolation or immersion. A solid–liquid separation device follows. Batch extraction is carried out on certain cases such as the extraction of pigments from plants, the isolation of proteins from oilseeds, and so on. The advantage of batch extractors is that they are simple to operate. However, the limited capacity and the discontinuous output of the extract are its main drawbacks (Berk 2009, Takeuchi et al. 2009, Tandon and Rane 2008, Wakeman 1994).

### 6.4.2 Continuous Extractors

In continuous multistage extraction systems, the solids are moving from one stage to another continuously.

Moving-bed percolation extraction equipment is used for treatment of many types of vegetable seeds such as cottonseed, rice bran, soybeans, peanuts, and castor

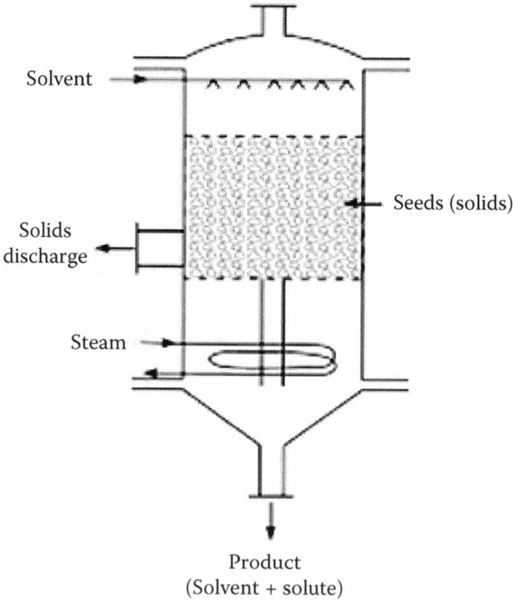


**FIGURE 6.4** Batch extractor. (Adapted from Rao D.G. 2010. In: *Fundamentals of Food Engineering*. PHI Learning Private Limited, pp. 365–371.)

beans. The pretreatment of the seeds is required and consists of dehulling, precooking the adjustment of water content and forming into flakes or rolls, and increasing the specific surface exposed to the solvent. The advantages of this extractor are that the miscella is free of solids since self-filtration takes place, the residue can be well drained when the equipment is controlled, and large amounts of solids can be treated continuously (Berk 2009, Pramparo et al. 2002, Prabhudesai 1997, Rao 2010, Tandon and Rane 2008, Wakeman 1994).

In a continuously immersion extraction process, the solid feed dips completely into the fresh solvent and is mixed with it. The disadvantage is the addition of a filtration step since no self-filtration of the solute extract takes place (Pramparo et al. 2002, Tandon and Rane 2008).

The *Bollman extractor* (Figure 6.5) is a continuous moving-bed perforated basket type of extractor. The solids arc loaded into baskets attached to a chain conveyor. The solid is dumped into the basket at the top of the descending side of the chain. The solid flakes are sprayed with half-miscella and the extraction process by percolation of the solvent through the flakes begins as the baskets move down. The final solvent, miscella, flows down through the remaining baskets concurrently. The miscella is collected at the bottom of the downward side. Control of flake size, thickness, and bulk density of solids is desirable (Badger and Banchero 1955, Bernardini 1976, Coulson et al. 1993, Eggers and Jaeger 2003, Prabhudesai 1997, Tandon and Rane 2008, Wakeman 1994, Xu and Diosady 2003).

The *Rotocel or Carousel extractor* (Figure 6.6) is a percolation system that obtains countercurrent extraction of solids through a sequence of discrete solid–liquid

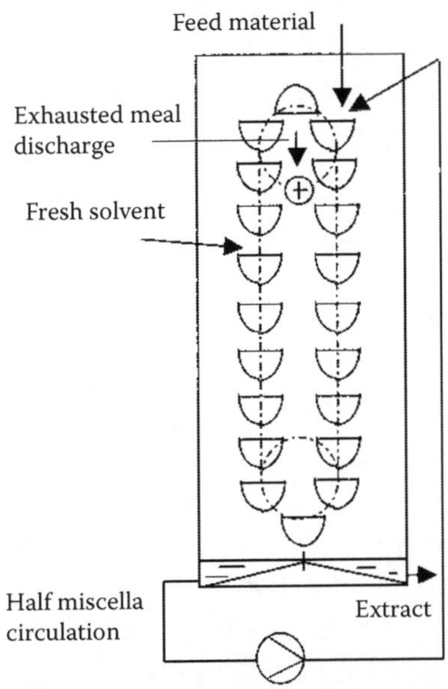


**FIGURE 6.5** Bollman extractor. (Adapted from Eggers R., Jaeger P. 2003. In: Tzia C, Liadakis G (eds.). *Extraction Optimization in Food Engineering*. Marcel Dekker, New York, pp. 1–42.)

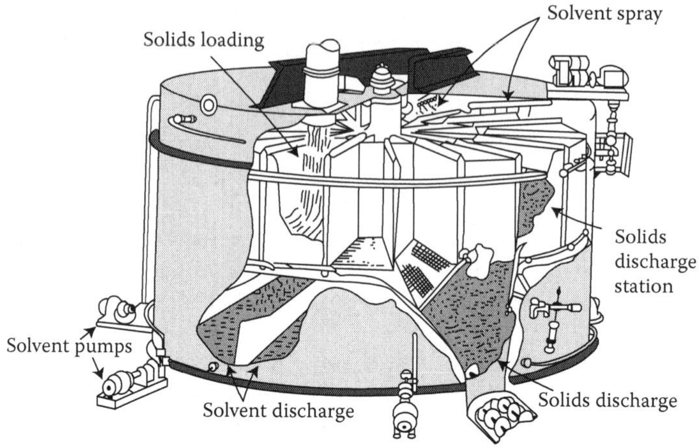


**FIGURE 6.6** Rotocel Extractor. (Adapted from Ramaswamy H., Marcotte M. 2006. In: *Food Processing: Principles and Application*. CRC Press, pp. 355–360.)

mixing. It consists of a rotor divided into sector-shaped cells. The rotor allows continuous introduction and discard of solids as it turns into a tight tank. Each cell passes under a special device for feeding the solids and under a series of solvent sprays, as the rotor rotates. Miscella goes from one cell to the other in countercurrent crossed flow in relation to the raw material flow, and is enriched of extracted oil. Therefore, in the beginning the miscella is weak and at the end of the process, it becomes concentrated. The extract solution is filtered in each cell (Badger and Banchero 1955, Coulson et al. 1993, Eggers and Jaeger 2003, Prabhudesai 1997, Ramaswamy and Marcotte 2006, Tandon and Rane 2008, Thomas et al. 2007, Wakeman 1994).

The *Hildebrandt total immersion extractor* (Figure 6.7) uses screw conveyors to transport the raw materials in three parts of a U-shaped extraction vessel. The solvent flows countercurrent to the solids through the extractor (Prabhudesai 1997, Suryanarayana 2002, Tandon and Rane 2008).

The *Bonotto extractor* (Figure 6.8) is an alternative tower design consisting of a tall cylindrical vessel with a series of slowly rotating horizontal plates and is used for countercurrent extraction according to the immersion method. Solids are fed to the top of the column continuously and are caused to fall through an opening onto each plate beneath in succession. The solvent is introduced at the bottom of the column and flows upward. The extract solution leaves the column at the top (Coulson et al. 1993, Prabhudesai 1997, Tandon and Rane 2008, Wakeman 1994).

The *belt extractor* (Figure 6.9) is a recently developed continuous system and is extensively used for extraction of oilseeds. The solid feed is continuously fed into a hopper and flows onto the slowly perforated belt of extractor, forming a constant thick mat. The bed height is kept constant by adjusting the feed rate. Fresh solvent is sprayed on the raw materials at the section nearest the discharge end of the extractor. The first extract is collected at the bottom of that section and pumped over the section, proceeding to the last. Discharge of the material into the outlet hopper is controlled by a rotary scraper (Berk 2009, Coulson et al. 1993, Spaninks and Brun 1979).

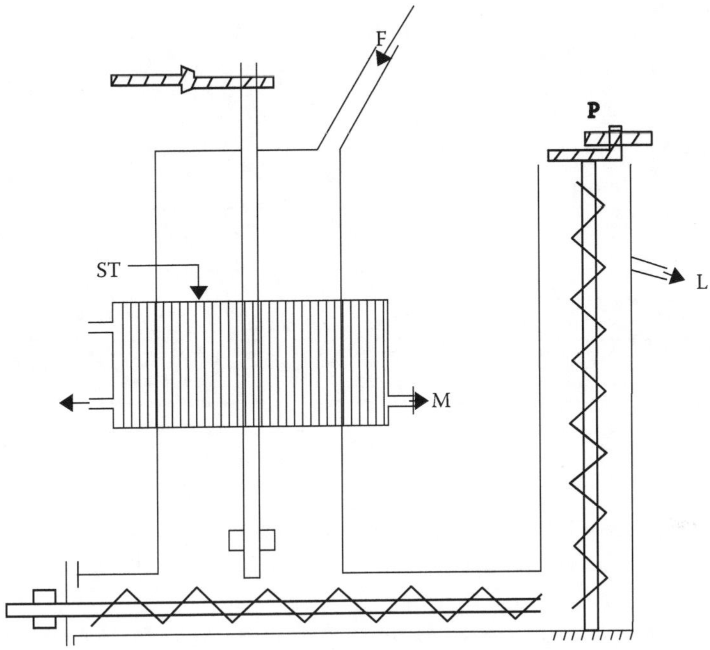


**FIGURE 6.7** Hildebrandt extractor. F = Solids to be leached, L = Leached solids, S = Solvent. (Adapted from Suryanarayana A. 2002. In: *Mass Transfer Operations*. New Age International (P) Limited Publishers, pp. 511–519.)

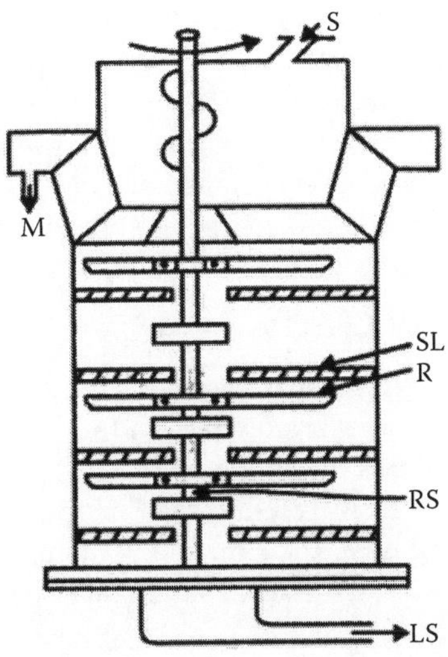


**FIGURE 6.8** Bonotto extractor. (Adapted from Suryanarayana A. 2002. In: *Mass Transfer Operations*. New Age International (P) Limited Publishers, pp. 511–519.)

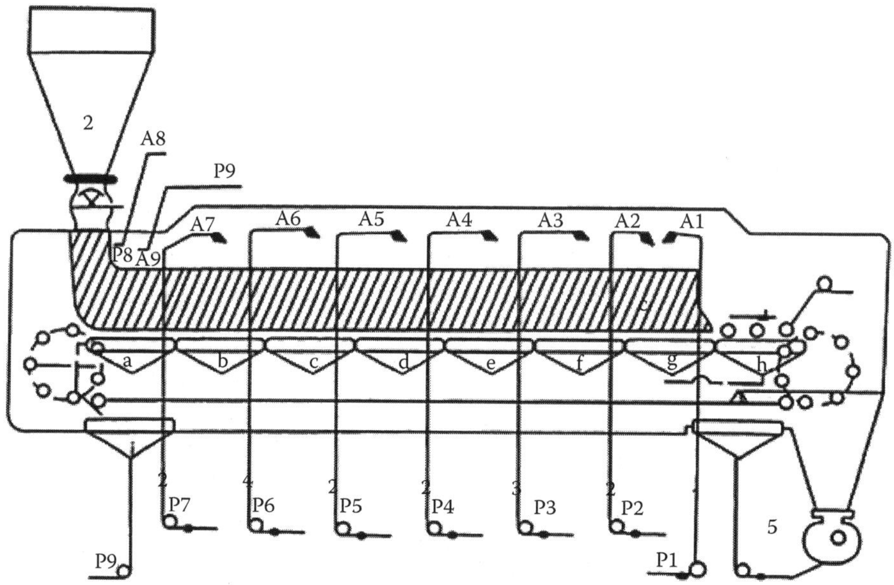


**FIGURE 6.9** Belt extractor. (Adapted from Chakrabarty M.M. 2003. In: *Chemistry and Technology of Oils and Fats.* Allied Publishers Pvt. Ltd., pp. 247.)

## 6.5 APPLICATION OF EXTRACTION IN FOOD PROCESSING

Specific extraction processes have been developed finding applications in the food industry. Depending on the nature of the solute, extraction processes may employ organic solvent or aqueous solutions as in the cases of oil recovery from oilseeds or of protein recovery from proteinaceous food materials (i.e., defatted oilseed meals), respectively. Moreover, in industrial practice, the steps required before and after the extraction process—raw materials pretreatment, refining and purification of extracts, as well as complementary procedures for cost minimizing, that is, solvent recovery and reuse—have been well designed and suggested. The following typical applications of extraction in food processing are presented.

### 6.5.1 Oil Extraction for Oilseeds

Many processes have been developed to recover oil from oilseeds, but the most common are hydraulic pressing, expeller pressing, and solvent extraction. The extraction of oil from the oilseeds requires the application of heat to disrupt oil-containing tissues and pressure to press the oil from the seed. Screw-press expellers are used for the mechanical extraction of oilseeds. However, it is impossible to achieve complete oil recovery by mechanical pressing since significant quantities of oil remain inside the matrix. Solvent extraction is the preferred operation process to recover the last 20% of the oil. The development of the solvent extraction process has increased oil recovery and also has made it economically attractive to process seeds that are relatively low in oil content such as soybeans (Hamm 2003, Xu and Diosady 2003).

Typically, the seeds are cleaned by passing through a magnetic separator since they carry foreign materials including sticks, leaves, and even metal shreds, and are then broken or formed into flakes to increase the area available for heat transfer. After preparation, the seed flakes or the cakes are conveyed to the solvent extractor. In the extractor, the solid material (flakes or cake) remains the required retention time, allowing large volumes of solvent to penetrate it. Industrially, hexane is the solvent used universally for the solvent extraction of oil from oilseed. After oil extraction, the solid matrix (meal) is conveyed in a closed transport system to the desolventizer–toaster (DT), where the meal is heated enough until the residual solvent is evaporated. To recover the crude oil from miscella, the solvent must be removed from the oil in distillation columns. The hot desolventized vapor is used to heat the vessel to evaporate some of the solvent. The last traces of solvent are removed by direct steam injection. The solvent vapors are then condensed in either water-cooled or air-cooled condensers. To minimize solvent losses, all the equipment is maintained under slightly negative pressure during the extraction process. The crude oil, that is, soybean, rapeseed (canola) oil, and so on, should be further treated by refining to remove undesirable compounds such as phospholipids or gums, free fatty acids, and coloring and odor substances, to be used as edible oil (Chakrabarty 2003, Hamm 2003, Xu and Diosady 2003).

### 6.5.2 Protein Extraction from Proteinaceous Materials

One of the major groups of food components that are found in vegetal and animal origin's organisms, are proteins (Minkiewicz et al. 2008). Many processes have been developed for the production of protein products by purifying or isolating the proteins from proteinaceous materials. Particularly, these processes include protein purification through nonprotein components, removal, or protein recovery through extraction and precipitation and the derived products are protein concentrates with 70% protein content or protein isolates with higher than 95% protein content, respectively (Oreopoulou and Tzia 2007). Such protein products are usually derived from oilseed meals, mainly soybeans, resulted as by-products of oilseeds processing for oil recovery by solvent extraction (Berk 1992). Then, the defatted flake desolventizing should be carried out either by steam or under vacuum (flash), avoiding protein denaturation.

#### 6.5.2.1 Protein Concentrate

The protein concentrate process is based on the extraction of carbohydrates, salts, and other low-molecular-weight substances from the flakes, leaving behind the insoluble proteins. There are three commonly extraction methods where the resulting products have different functional characteristics. In the first method, methanol, ethanol, or isopropanol aqueous solutions (60–70%, w/w) are used, since soluble carbohydrates are extracted without significantly affecting the dissolution of proteins. After flash desolventizing of the extracted materials, the protein concentrate is obtained and ground (Bernardini 1983, Oreopoulou and Tzia 2007). In the second method, defatted flakes or flour and a diluted acid solution at low pH are mixed and the separation of the insoluble material is carried out by centrifugation. The principle of the method is the fact that at the isoelectric point (pI), the proteins solubility is minimal, while the carbohydrates remain under solution. In the third method, the toasted flakes are extracted with hot water at 66–93°C, at an appropriately adjusted pH, that is, for soybean in the range of 5.0–7.5. The method is based on the decrease of proteins solubility after their denaturation while maximizing carbohydrate dissolution (Berk 1992, Oreopoulou and Tzia 2007).

### 6.5.2.2 Protein Isolate

The protein isolate process refers to the extraction, purification, and recovery of proteins from flakes. In contrast to protein concentration processes, the isolation approach is protein dissolution and most of the impurities remain in the solids (Berk 1992). Protein isolate production commonly comprises three main procedures: protein extraction, protein precipitation, and drying the protein precipitate (Liadakis et al. 1995, Oreopoulou and Tzia 2007).

#### *6.5.2.2.1 Protein Extraction*

The factors that influence protein extraction are the extracting medium, pH, temperature, ionic strength, agitation, liquid-to-solids ratio, extraction time, particle size, and formation of proteinaceous solid material. The extraction conditions should be taken into account so as to achieve the maximal protein dissolution (Oreopoulou and Tzia 2007, Zayas 1997).

As far as the extraction medium is concerned, in the past few years, many studies performed were aimed to compare different protein solubilization techniques using suitable solvents. The alcoholic extraction process after solid sample pretreatment (dehulling and deoiling) has a high efficiency of protein extraction yield. Aqueous alcohols such as ethanol, isopropyl alcohol, and butanol, are commonly used to remove phenolics, oligosaccharides, or inhibitors from defatted seeds. However, these alcoholic solvents cause protein coagulation and therefore degradation of their functional properties. Mechanical and thermal treatments are applied to avoid these problems (Barbin et al. 2011, Moure et al. 2006). Recently, aqueous extraction is gaining researchers' attention, because of the increasing environmental concerns over organic solvents use to protein recovery. Water is advantageous over alcohols as it is nonflammable and nontoxic. On an industrial scale, protein concentrates or isolates production consists of an aqueous solubilization of protein and carbohydrates at acid, neutral, or alkaline pH. Most of the oilseeds protein extraction is usually carried out with water by adjusting the pH at 8–10 (by NaOH, $Ca(OH)_2$, or $NH_3$) and at temperature of 50–55°C. Practically, protein solubility increases with temperature between 0°C and 40–50°C while at higher pH and temperatures, values or Maillard reactions with carbohydrate proteins may deteriorate (Moure et al. 2006, Oreopoulou and Tzia 2007, Zayas 1997). Aqueous enzymatic extraction is also alternatively suggested for protein recovery from proteinaceous solid materials (Tang et al. 2002).

For the operating conditions selection, the particle size of solid material should be suitable for extraction process facilitation and the protein solution separation. Agitation should be adequately adapted to enhance protein extraction without destroying the flakes formation and extraction time (45–60 min) should be adequate to achieve protein recovery maximizing. For the liquid-to-solids ratio selection, the water holding in the flakes and the water waste due to separating whey after extraction should be taken into account; the most common ratios used are 1:10–1:20.

##### 6.5.2.2.2 Protein Isolation by Precipitation

After protein extraction, the protein extract is separated by centrifugation or decanting and protein precipitation (Martínez-Maqueda et al. 2013, Berk, 1992). Isolation or enrichment/fractionation of proteins based on the molecular weight, size, and shape of each component can also be applied (Sharma et al. 2010). Precipitation is influenced by factors that cause protein solubility decrease. Various agents such as organic solvents, heat treatment, acid/bases, salts, nonionic polymers, and polyelectrolytes have been applied for protein precipitation (Oreopoulou and Tzia 2007). It is recognized that ammonium sulfate is the most widespread precipitant (Bodzon-Kulakowska et al. 2007). The addition of the appropriate amount of these precipitants provokes an increase of protein interactions followed by protein aggregation and, afterward, precipitation. Acid precipitation at the pI of proteins is a simple method commonly applied. Also, an alternative precipitation method known as salting-out process, is carried out by adding salt into a protein solution allowing selective protein separation (Martínez-Maqueda et al. 2013).

Concluding, in the whole protein isolation process, the extraction of proteins evaluated by protein extraction yield is determinant for the total protein yield (as protein percent in the isolated product to the protein percent in the raw material). Optimizing this process, the maximization of both protein extraction yield and total protein yield as well as for the protein quality of the isolated product (as protein content of the protein isolate) can be attained.

### 6.5.3 Other Applications

Sugar can be obtained from sugar beet using water as the extraction liquid. The extractors used are generally a diffusion battery, or Hildebrandt or Bonotto extractors. The washed beets are cut into thin strips called cossettes, to facilitate the extraction, and then immersed in hot water. The cossettes must be V-shaped and their uniformity regarding their size indicated by the mush content (which is the mass of cossettes <1 cm long in relation to the total cossettes mass) must not exceed 5%. The water is introduced under pressure liquid making the crossing of cossettes easy on the bed, while taking care not to damage their cellular structure and undesired components, besides sugar, to be extracted. The temperature should also be controlled to avoid undesired nonsugared compounds coextraction. Sugar is extracted from sugar beets in diffusers using a countercurrent of hot water, in continuous processes, thus obtaining through diffusion the raw juice with about 15% of dissolved solids. Raw juice is further purified by sedimentation and filtration and is then concentrated and the crystals are finally separated by centrifugation (Christodoulou 2003, Ibarz and Barbosa-Cánovas 2003, Vetter 1998).

The extraction process is one of the most important stages in the production of instant coffee and instant tea (Ibarz and Barbosa-Cánovas 2003). Roasted and crushed coffee beans are extracted in the equipment working under countercurrent, multistage systems, or fixed beds using water to obtain a final solution with 25–30% of solids. The temperature of extraction should be carefully selected to facilitate the solubilization of desirable hydrolyzed solids without transmitting undesirable aromas and flavors to the extract. Dehydrated tea leaves are extracted with hot water in fixed-bed extractors, initially at 70°C and then at 90°C under vacuum. The volatile aromatic products are eliminated from the extract by distillation, while the residual solution is concentrated by vacuum evaporation until 25–50% of soluble solids. The aromatic fraction is combined with the concentrated solution and instant tea is obtained by freeze- or spray-drying.

## 6.6 MODERN EXTRACTION TECHNIQUES FOR FOOD SAMPLES

An extraction process will obtain a high recovery yield and also a high stability of the solute substances. The common solid–liquid extraction techniques and the relative equipment conventionally used in the food industry as well as the main applications in certain food sectors, such as sugar, tea, coffee, vegetable oils, and functional compounds were already presented. However, conventional extraction methods require long extraction times, so that degradation of thermolabile active compounds may occur (Azmir et al. 2013, Chan et al. 2011, Eskilsson and Bjorklund 2000).

For this reason, over the past decade, there has been an increasing demand for modern sample extraction techniques with shortened extraction time, reduced organic solvent consumption, and minimized solute compounds degradation. Driven by this effort, modern methods of extraction are proposed including enzymic extraction, microwave-assisted extraction (MAE), ultrasonication-assisted extraction (UAE), SFE, pressurized solvent extraction (PSE), and even more for laboratory-scale extractions such as solid-phase microextraction (SPME), combination of Soxhlet with microwaves (Soxhwave). Enzymic extraction is applicable either in aqueous extracting systems or combined to solvent extraction. Solid oleaginous materials can be enzymically treated (i.e., by cellulose, protease) before oil extraction, or proteinaceous materials are extracted by aqueous enzyme solutions (i.e., of carbohydrase). Thus, the pretreatment of solid materials enhances oil extraction or the enzymic impact during aqueous extraction and facilitates protein solubilization, attaining increased extraction yields of solutes in both cases. Microwave or UAE are found as important applications that meet many of today's requirements in terms of environmental sustainability, speed operation, and automation. The fundamentals of the new extraction processes are different from those of conventional methods, since the extraction occurs because of changes in the cell structure caused by electromagnetic or sound waves. The basic principles of operation as well as method optimization are discussed in the following section (Azmir et al. 2013, Chan et al. 2011, Eskilsson and Bjorklund 2000, Harahsheh and Kingman 2004). Also, the selection of the suitable method should be based on the same criteria as in conventional methods (high-recovery yield and stability of key substances) while research on active compounds and raw material processed should be taken into account.

### 6.6.1 Microwave-Assisted Extraction

Microwave is nonionizing electromagnetic energy transmitted as waves with a frequency range from 0.3 to 300 GHz. It can penetrate into certain materials and interact with polar components inside the materials, such as water, to generate heat. MAE is a process that uses the effect of microwaves to extract biological materials; thus, it has drawn significant research attention in various fields, in particular medicinal plant research. MAE has been considered an important alternative to solid–liquid extraction because of its advantages: lower extraction time, lower solvent usage, selectivity, and volumetric heating and controllable heating process. In MAE, the process acceleration and high extraction yield may be the result of a synergistic combination of two transport phenomena: heat and mass gradients working in the same direction. On the other hand, in conventional extractions, although the heat transfer occurs from the outside to the inside of the substrate, the mass transfer occurs from inside to the outside. Moreover, despite the fact that in conventional process, the heat is transferred from the heating medium to the interior of the sample; in MAE, the heat is dissipated volumetrically inside the irradiated medium (Azmir et al. 2013, Chan et al. 2011, Destandau et al. 2013, Eskilsson and Bjorklund 2000, Harahsheh and Kingman 2004, Leonelli et al. 2013).

In the MAE process, a number of phenomenological steps occur during the interaction of solid matrix and solvent resulting in the separation, including (Takeuchi et al. 2009, Veggi et al. 2013):

1. Penetration of the solvent into the solid matrix.
2. Solubilization and/or breakdown of components.
3. Transport of the solute out of the solid matrix.
4. Migration of the extracted solute from the external surface of the solid into the bulk solution.
5. Movement of the extract with respect to the solid.
6. Separation and discharge of the extract and solid.

The physical principle of MAE is based on the direct effects of microwave energy on molecules of materials. Polar chemical compounds well absorb microwave energy; thus, targeted materials can be heated based on their dielectric constant. This absorbed energy is proportional to the medium dielectric constant, resulting in dipole rotation in an electric field and migration of ionic species. The ionic migration generates heat as a result of the resistance of the medium to the ion flow, causing collisions between molecules as the direction of ions changes as many times as the field changes the sign. Rotation movements of the polar molecules occur while these molecules are trying to line up with the electric field, with consequent multiple collisions that generate energy and increase the medium temperature (Azadmard-Damirchi et al. 2011, Chan et al. 2011, Destandau et al. 2013, Harahsheh and Kingman 2004, Takeuchi et al. 2009, Veggi et al. 2013).

#### 6.6.1.1 Mechanism of Microwave Heating

Microwaves cause molecular motion by migration of ions and rotation of dipoles, and by solvent heating improving its penetration. Ionic conduction is the electrophoretic migration of ions when an electromagnetic field is applied, while the resistance of the solution to this flow of ions results in friction heating the solution. Thus, the positively charged ions accelerate in the direction of the electric field in a microwave oven while the negatively charged ions move in the opposite direction. The direction of the motion of positive and negative ions changes, as the direction of electric field changes depending on frequency. Moving particles collide with the adjacent particles and result them in more agitation movement. Dipole rotation means rearrangement of dipoles with the applied field.

The main characteristic of microwave heating is energy transfer. Traditionally, in the conventional process, the energy is transferred to the material by convection, conduction, and radiation phenomena through the external material surface in the presence of thermal gradients, whereas in MAE, the microwave energy is delivered directly to materials through molecular interactions with the electromagnetic field via conversions of electromagnetic energy into thermal energy (Azadmard-Damirchi et al. 2011, Chan et al. 2011, Harahsheh and Kingman 2004, Sumnu and Sahin 2012, Veggi et al. 2013).

The efficiency of microwave heating strongly depends on the dielectric susceptibility of both the solvent and the solid matrix. The dielectric response of materials in an applied microwave field is expressed by the dielectric constant ($\varepsilon'$) and dielectric loss factor ($\varepsilon''$). The dielectric constant measures the ability of the material to absorb microwave energy (for vacuum, $\varepsilon' = 1$); that is, it quantifies the capacity of the material to be polarized. In contrast, the dielectric loss factor indicates the efficiency with which the microwave energy converted into heat. The two quantities are expressed in terms of the complex dielectric constant ($e^*$) and the loss tangent ($\tan\delta$) (Azmir et al. 2013, Destandau et al. 2013, Eskilsson and Bjorklund 2000, Harahsheh and Kingman 2004, Leonelli et al. 2013, Sumnu and Sahin 2012):

$$e^* = e' - je'' \tag{6.10}$$

$$\tan\delta = \frac{\varepsilon''}{\varepsilon'} \tag{6.11}$$

$$j = \sqrt{-1}$$

The loss tangent ($\tan\delta$ or dielectric loss) is the most important property in microwave processing; it provides an indication of the ability of how well the matrix can absorb microwave energy and dissipate heat to the surrounding molecules. Consequently, a material with high dielectric loss factor and $\tan\delta$ in combination with a moderate value of $\varepsilon'$ permits converting microwave energy into thermal energy (Destandau et al. 2013, Harahsheh and Kingman 2004, Leonelli et al. 2013, Veggi et al. 2013).

The first factor one must consider when selecting microwave physical constants is the solvent to be used. It is important to select a solvent with high extracting power and a strong interaction with the matrix and the solute. Polar molecules and ionic solutions (typically acids) strongly absorb microwave energy because of the permanent dipole moment. On the other hand, when exposed to microwaves, nonpolar solvents such as hexane will not heat up when exposed to microwaves. The degree of microwave absorption usually increases with the dielectric constant. In Table 6.1, the selected physical parameters, including dielectric constant and dissipation factors, are shown for commonly used solvents that are used in most of the applications (Chan et al. 2011, Destandau et al. 2013, Leonelli et al. 2013, Takeuchi et al. 2009, Veggi et al. 2013). A simple comparison between water and methanol shows that methanol has a lesser ability to obstruct the microwaves as they pass through but has a greater ability to dissipate the microwave energy into heat. The higher dielectric constant of water implies a significantly lower dissipation factor, which means that the system absorbs more microwave energy than it can dissipate. The larger the dipole moment of the solvent, the faster the solvent will heat under microwave irradiation. For example, hexane will not heat, whereas acetone with a dipole moment of 2.87 Debye will heat in a matter of seconds. Thus, a mixture of hexane and acetone (1:1) is an ideal solvent for applications (Chan et al. 2011, Leonelli et al. 2013, Lopez-Avila 2000, Veggi et al. 2013).

The second factor to be considered is the solid matrix. Its viscosity affects its ability to absorb microwave energy because it influences molecular rotation. When the molecular mobility is reduced, it is difficult for the molecules to align with the microwave field. Therefore, the heat produced by dipole rotation decreases, and the

**TABLE 6.1**
**Physical Constants and Dissipation Factors for Solvents Usually Used in MAE**

| Solvent | Dielectric Constant[a] | Dipole Moment[b] | Dissipater Factor tan δ (×10$^{-4}$) | Boiling Point[c] (°C) | Closed-Vessel Temperature[d] (°C) |
|---|---|---|---|---|---|
| Acetone | 20.7 | 2.69 | 5555 | 56 | 164 |
| Ethanol | 24.3 | 1.96 | 2500 | 78 | 164 |
| Hexane | 1.89 | 0.1 | | 69 | –[e] |
| Water | 78.3 | 2.3 | 1570 | 100 | |
| Methanol | 32.6 | 2.87 | 6400 | 65 | 151 |
| 2-Propanol | 19.9 | 1.66 | 6700 | 82 | 145 |
| Acetone: hexane = (1:1) | | | | 52 | 156 |

*Source:* All data adapted from Jassie L. et al. 1997. Microwave-assisted solvent extraction. In: Kingston HM, Haswell SJ (eds.). *Microwave-Enhanced Chemistry*. American Chemical Society, Washington, pp. 569.

[a] Determined at 20°C.

[b] Determined at 25°C.

[c] Determined at 101.4 kPa.

[d] Determined at 1207 kPa.

[e] Indicates no microwave heating.

higher the dissipation factor, the faster the heat will be transferred to the solvent (Chan et al. 2011, Leonelli et al. 2013, Veggi et al. 2013).

#### 6.6.1.2 Heat Transfer in MAE

The heat transfer in a material that receives microwave energy can be estimated by the general heat-transfer equation. Considering a transient heat transfer in an infinite slab, for one-dimensional flux, the corresponding equation is (Takeuchi et al. 2009, Veggi et al. 2013)

$$\frac{\partial^2 T}{\partial x^2} + \frac{q'''}{k} = \frac{1}{a}\frac{\partial T}{\partial t} \tag{6.12}$$

where

- $x$ is the heat flux direction,
- $q'''$ is the heat generation,
- $k$ is the thermal conductivity, and
- $\alpha$ is the thermal diffusivity.

Food materials have the ability to store and dissipate electric energy when subjected to an electromagnetic field—microwave energy in itself is not thermal energy. The heating is a result of the electromagnetic energy generated with the dielectric properties of the material combined with the electromagnetic field applied. The rate of conversion of electrical energy into thermal energy in the material is described by the following equation (Azadmard-Damirchi et al. 2011, Takeuchi et al. 2009, Veggi et al. 2013):

$$P_D = 2\pi E^2 f'\varepsilon'' \tag{6.13}$$

where

- $P_D$ is the power dissipation (W/cm$^3$),
- $E$ is electrical field strength (V/cm), and
- $f'$ is the applied frequency (Hz).

The geometry of the irradiated food–object and its dielectric properties affect the distribution of the electric field. The energy absorption inside the solid material causes an electric field that decreases with the distance from the material surface. The penetration depth of a wave (Dp) is the distance from the material surface where the absorbed electric field (e) falls to 1/e of the electric field at the surface, and is inversely proportional to the frequency and the dielectric properties of the material. For instance, for water, the greatest rate of heating occurs at 20 GHz where the dielectric loss factor is at its maximum value. Dp is given by (Harahsheh and Kingman 2004, Veggi et al. 2013)

$$Dp = \frac{c}{2\pi f'\sqrt{2e'}\left[\sqrt{1+\tan^2\delta}-1\right]^{1/2}} \tag{6.14}$$

where $c$ is the speed of light (m/s).

### 6.6.1.3 Variables Influencing MAE

The efficiency of MAE processes is related to a number of variables including the microwave power, extraction temperature and time, solvent used, and food matrix size used.

*Temperature:* Higher temperature promotes elevated extraction yields. At high temperatures, the solvent power increases because of a drop in viscosity and surface tension, facilitating the solvent to solubilize solutes, and improving matrix wetting and penetration. In addition, in closed-vessel microwave extraction, the temperature may reach well above the boiling point of the solvent, leading to improving extraction efficiency. However, the efficiency increases with the increase in temperature until an optimum temperature is reached and then the further increase in temperature causes efficiency decrease. This happens because of the direct connection of the stability and the extracted yield of the target compound with the selection of an ideal extraction temperature (Eskilsson and Bjorklund 2000, Lopez-Avila 2000, Takeuchi et al. 2009, Veggi et al. 2013).

*Pressure:* It is an important factor in MAE procedures in closed systems. The pressure of the system is related to the temperature applied, increasing the extraction efficiency and decreasing the exposure time (Lopez-Avila 2000, Takeuchi et al. 2009).

*Solvent:* The choice of the solvent is an important factor that affects MAE procedures. Solvent selection depends on the solubility of the compounds of interest, solvent penetration and its interaction with the sample matrix, its dielectric constant, and the mass transfer kinetics of the process. Another important aspect is the selection of the appropriate capacity of the solvent to absorb the microwave energy and consequently heat up. In general, when the solvent presents high dielectric constant and dielectric loss, its capacity to absorb microwave energy is high. Solvents that are transparent to microwaves do not heat when submitted to them such as hexane whereas ethanol is an excellent microwave-absorbing solvent. Both polar and nonpolar solvents can be used in MAE, and solvents such as ethanol, methanol, and water are sufficiently polar to be heated by microwave energy. The addition of salts to the mixture can also increase the heating rate, because besides dipole orientation, the ion conductivity is the main origin of polarization and corresponds to losses of heat in dielectric heating. Studies have shown that small amounts of water in the extracting solvent facilitate the transport of compounds into the solvent at higher mass transfer rates (Eskilsson and Bjorklund 2000, Lopez-Avila 2000, Takeuchi et al. 2009, Veggi et al. 2013).

*Extraction time:* Compared to conventional extraction techniques, the exposure time of MAE processes is very short. For foods, the extraction times vary from 3 to 40 min, depending on the solid matrix and the compounds extracted. Higher extraction time usually tends to increase the extraction yield. However, this was influenced by the dielectric properties of the solvent. Solvents such as water, ethanol, and methanol may heat up tremendously on longer exposure, thus risking the future of thermolabile constituents. Sometimes, when longer extraction time is required, the samples are extracted in multiple steps using consecutive extraction cycles with fresh solvent. With this procedure, the extraction yield is enhanced, avoiding extensive heating of the same amount of solvent (Eskilsson and Bjorklund 2000, Lopez-Avila 2000, Takeuchi et al. 2009, Veggi et al. 2013).

*Water content:* The water content in the solid matrix is of great importance. The moisture in the matrix is heated, evaporated, and generates internal pressure in the cell, resulting in cell rupture, and hence improving the extraction yield. Water addition increases the polarity of the solvent and has a positive effect on the microwave-absorbing ability. Moreover, the additional water promotes hydrolyzation, and hence, the risk of oxidation of the compounds reduces (Leonelli et al. 2013, Takeuchi et al. 2009).

#### 6.6.1.4 Microwave-Assisted Extractors

There are two groups of MAEs, according to the way microwave energy is applied to the food sample: multimode and single mode, or focused systems. A multimode system allows dispersion of microwave radiation randomly in the microwave cavity, so that every sample and cavity region is irradiated. A single mode or focused system permits focusing of microwave radiation on a restricted region in which a stronger electromagnetic field is applied on the food (Chan et al. 2011, Destandau et al. 2013, Lopez-Avila 2000, Luque-Garcia 2005, Turner 2006).

Usually, multimode systems are of the closed-vessel type, where the MAE treatment is conducted under controlled temperature and pressure, whereas focused systems are of the open-vessel type, in which microwave radiation is applied at atmospheric pressure.

Both multimode and focused microwave apparatus consist of four major components (Figure 6.10) (Chan et al. 2011, Destandau et al. 2013, Lopez-Avila 2000, Luque-Garcia 2005):

1. The magnetron tube or microwave generator, where the microwave energy is generated at a fixed frequency.
2. The waveguide, used to transmit the microwaves from the source to the microwave cavity.
3. The applicator, where the sample is placed. It can be a multimode cavity where microwaves are randomly dispersed on the waveguide itself. In the latter case, the sample is introduced by wall slots and the waveguide is terminated by the matched load.
4. The circulator, which allows microwaves to pass only in the forward direction.

*Closed-vessel microwave systems*: The closed MAE system (Figure 6.10) is carried out in a sealed vessel with different modes of microwave radiation at high pressure and temperatures. The great advantage of this system

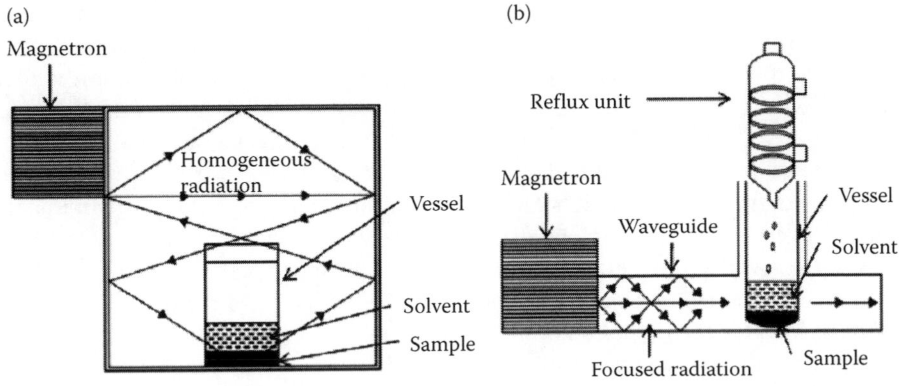


**FIGURE 6.10** (a) Closed vessel microwave system and (b) open vessel microwave system. (Adapted from Chan C.H. et al. 2011. *Journal of Chromatography A*. 1218: 6213–6225.)

is that the pressure inside the extraction vessel is controlled in order not to exceed the working pressure of the vessel, while the temperature can be regulated above the normal boiling point of the extraction solvent. Temperature and pressure increase accelerates MAE due to the ability of the solvent to absorb microwave energy. Although the closed-vessel system offers an efficient extraction with less solvent consumption, it is susceptible to losses of volatile compounds with limited sample throughput (Chan et al. 2011, Destandau et al. 2013, Lopez-Avila 2000, Luque-Garcia 2005, Turner 2006).

*Open-vessel microwave systems*: In open-vessel systems (Figure 6.10), the extraction process is performed at atmospheric pressure conditions and only a part of the vessel is directly exposed to the propagation of microwave radiation (mono-mode). This system is commonly named as focused microwave-assisted extraction (FMAE). The maximum temperature used in the system is approximately the normal boiling point of the solvent at atmospheric pressure. Open system is developed to counter the shortcomings of the closed system such as the safety issues and it is considered more suitable for extracting thermolabile compounds. This system has higher sample throughput and more solvent can be added to the system at any time during the process. Losses of the solvent are prevented by the presence of a cooling system on the top of the extraction vessel that condensates the solvent vapors (Chan et al. 2011, Destandau et al. 2013, Lopez-Avila 2000, Luque-Garcia 2005, Turner 2006).

### 6.6.2 Ultrasound-Assisted Extraction

In the food industry, UAE has been the subject of research and development and has been characterized as a green novel technology. Ultrasound comprises mechanical waves that need the material presence to spread. The difference between sound and ultrasound is the frequency of the wave; sound waves are at human-hearing frequencies (16 Hz to 16–20 kHz) while ultrasound has frequencies on the threshold of human hearing but below microwave frequencies (from 20 kHz to 10 MHz) (Azmir et al. 2013, Mason et al. 2005, Pingret et al. 2013, Takeuchi et al. 2009).

Ultrasounds are characterized by their frequency and wavelength that result in the wave speed through the medium. Amplitude or intensity of waves is also an important parameter and is used to classify the industrial application: low-intensity ultrasound (LIU) with <1 W/cm$^2$, and high-intensity ultrasound (HIU) with 10–1000 W/cm$^2$. HIU is applied at higher frequencies (up to 2.5 MHz) to modify processes or products by physical disruption of tissues and, among other applications, to speed up and improve the efficiency of sample preparation. LIU is used to control the quality of processes and products and provide information on the physicochemical properties of foods (e.g., firmness, ripeness, sugar content, and acidity) (Pingret et al. 2013, Takeuchi et al. 2009).

The principle of ultrasound has been attributed to the acoustic cavitation phenomenon. During the sonication process, when a sonic wave meets a liquid medium, longitudinal waves are created, thereby forming regions of alternating compression and rarefaction (expansion) waves induced on the molecules of the medium. The expansion process creates bubbles in a liquid and produces negative pressure that can reach a high local pressure of up to 50 MPa, intense heating with hot spots around 5000 K, and dauers a few microseconds. The extent of negative pressure depends on the nature and purity of liquid. At constant ultrasound intensity, dynamic equilibrium is established between the forming and the collapsing bubbles. When the cavitation bubbles collapse near cell walls, the high pressure and temperature generate microjets and shock waves directed toward the solid surface. As a result, there is an enhanced solvent penetration into the cells and an intensification of the mass transfer. These microjets can be explored in industrial processes enhancing the extraction of target compounds from plant materials; the cavitation bubble generated near the material surface (i) collapses during a compression cycle (ii) and a microjet directed toward the surface is created. Therefore, the cell walls of the plant matrix are destroyed and its content is released into the medium (Mason et al. 2005, Pingret et al. 2013, Takeuchi et al. 2009).

There are two forms of bubbles' cavitation: stable and transient. Stable cavitation bubbles have an existence of many acoustic cycles and oscillate nonlinearly around an equilibrium size, while the transient bubbles exist for one or at most a few cycles, during which time, they expand to at least double their initial size before collapsing violently into smaller bubbles. The dynamics of a transient cavitation bubble is expressed by the following equation of Rayleich–Plesset:

$$\rho\left[R\frac{d^2R}{dt^2}+\frac{3}{2}\left(\frac{dR}{dt}\right)^2\right]=\left(P_h-P_V+\frac{2\theta}{R_0}\right)\left(\frac{R_0}{R}\right)^{3k}-P_k-P_a+P_V-\frac{2\theta}{R} \qquad (6.15)$$

where

ρ is the solvent density

$R$ is the radius of the bubble
$P_h$ is the hydrostatic pressure
$P_a$ is the acoustic pressure
$P_V$ is the vapor or gas pressure
$P_k$ is the critical pressure of bubble nucleation
$k = C_D/C_V$ is the ratio of specific heats
$\theta$ is a parameter that depends on the viscosity and superficial tension of the liquid

The integration of the above equation allows the calculation of the size of a bubble and also the time of the implosion as a function of ultrasound frequency:

$$R_0 = \frac{4}{3W_a}(P_a - P_h)\left(\frac{2}{\rho P_a}\right)^{1/2}\left[1 + \frac{2(P_a - P_h)}{3P_h}\right]^{1/3} \tag{6.16}$$

where $W_a = 2\pi f_a$ and $f$ is the frequency.

$$t_i \cong 0.915 R_0 \sqrt{\frac{\rho}{(P_h + P_a - P_V)}} \tag{6.17}$$

The ultrasound frequency significantly affects the rate of extraction and kinetics. However, these influences are dependent on the structure of the material and on the compound to be extracted. The acceleration of the kinetics and of the extraction are obtained, probably as a result of the increase of the intraparticular diffusion of the solute that results from the disruption of the cell walls. In some cases, lower frequencies are required during the process to avoid degradation of bioactive compounds (Azmir et al. 2013, Takeuchi et al. 2009).

#### 6.6.2.1 Heat Transfer in UAE

The intensity applied is related to the effects produced by ultrasound in a mass transfer process. HIU enhances the mass transfer process by affecting internal and external resistance of the wall. Ultrasonic intensity (UI) can be determined by calorimetric methods and can be described by the following expression:

$$UI = \frac{P_o}{A_b} = \frac{\left(\frac{dT}{dt}\right)C_p m}{A_b} \tag{6.18}$$

where

$P_o$ is the average power, expressed in the function of $dT/dt$ that is the variation of temperature $T$ with the time $t$,
$C_P$ is the heat capacity of the liquid,
$m$ is the liquid mass added into the vessel, and
$A_b$ is the area of the reaction vessel's bottom.

#### 6.6.2.2 Important Parameters of UAE

The advantages of UAE compared to a conventional technique include reduction in extraction time, energy, and use of solvent. Ultrasound energy for extraction also allows easy–effective mixing, faster energy transfer, reduced thermal gradients and extraction temperature, selective extraction, reduced equipment size, faster response to process control, quick start-up, increased production, and even elimination process steps (Azmir et al. 2013, Mason et al. 2005, Pingret et al. 2013, Takeuchi et al. 2009).

The parameters that can be optimized in UAE are related to the ultrasonic equipment such as the frequency, wavelength and amplitude of the wave, the ultrasonic power, and intensity. Also, the temperature, pressure, time of sonication, solvent type and raw material's moisture content, milling degree, and particle size are governing factors for obtaining efficient and effective extraction and should be taken into consideration. The careful study of these parameters is of great importance to result in the highest extraction yield.

#### 6.6.2.3 Physical Parameters

The frequency, the wavelength, and the amplitude influence the cavitation bubbles and therefore the extraction. Frequency (f) is measured in hertz and expresses the number of cycles per second. Period (P) is the reciprocal of the frequency (1/f) and thus is the time of one cycle. Wavelength ($\lambda$) represents the distance of one cycle and it is determined by both the source of ultrasound and the medium. Amplitude (A) is the height of the wave and is measured in decibels (dB) or pascals (Pa). It has an effect on the formation of cavitation bubbles.

Ultrasonic power can be estimated by the direct or indirect measurement of the applied energy. The most common physical methods are the measurement of acoustic pressure using hydrophones or optical microscopes and the calorimetric method. Among the chemical methods, the indirect measurement of OH radicals formed by sonoluminescence is used. Generally, the highest efficiency of UAE can be achieved by increasing the ultrasound power, reducing the moisture of food matrices to enhance solvent–solid contact, and optimizing the temperature to allow a shorter extraction time.

Intensity can be expressed as energy transmitted per second and per square meter of the medium and is correlated with the amplitude of the sound wave; with an increase in the amplitude, bubble collapse will be more violent. The applied UI is expressed by Equation 6.18.

#### 6.6.2.4 Medium Parameters

The solvent choice is dictated by the solubility of the target compounds in the solvent and also by physical parameters including viscosity, surface tension, and vapor pressure of the medium. For cavitation bubbles to be effective, the negative pressure during the expansion cycle has to overcome the natural cohesive forces in the medium. The rise of viscosity increases the molecular interactions and, hence, the cavitation threshold rises significantly. Also, a high surface tension decreases cavitation phenomena. Therefore, a high intensity is advised to obtain the necessary mechanical vibrations that will cause cavitation.

The temperature increase generates the rise of the vapor pressure and the decrease of the viscosity and the surface tension, as more solvent vapors enter the bubble cavity, reducing the pressure difference between the inside and outside of the bubble. In extraction, a higher temperature results in a higher efficiency due to an increase in the number of cavitation bubbles and a larger solid–solvent area contact. However, it is possible to notice a decrease in the extraction yield as the temperature rises in the case of volatile compounds. Thus, the optimization of the temperature parameter can be performed to obtain the highest yield of the target compounds without degradation.

#### 6.6.2.5 Matrix Parameters

The matrix used can be fresh or dry depending on the target molecules. In the case of dry matrices, the absorption of the liquid could occur depending on the porosity of the material. Also, the solubility of the target compounds in the chosen solvent and the temperature can influence the extraction yield. Other parameters related to the solid–liquid extraction such as the solid/liquid ratio and the particle size of the matrix are relevant to extraction efficacy (Azmir et al. 2013, Mason et al. 2005, Pingret et al. 2013, Takeuchi et al. 2009).

### 6.6.3 Enzymic and Other Methods Enhancing Extraction

Enzymatic treatment of oilseed or flakes prior to the conventional extraction process has been investigated. Solvent extraction is commonly used in the production of edible oil from oilseeds. Mechanical extraction (by pressing or centrifugation) can also be used, that is, in olive oil production, but it is unlikely to obtain complete oil removal, since the solids retain significant quantities of oil inside the matrix. Then, a by-product is derived that can be utilized to obtain the rest of the oil by solvent extraction, that is, olive residue or olive–pomace containing about 10% oil is extracted to recover the olive–pomace oil (sansa oil). In any case, the solvent extraction of oil is significantly influenced by the structure of oilseed flakes as it includes oil dissolution and diffusion as well as viscosity flux of the solvent into the solid's capillary pipes. The incorporation of the enzymatic treatment prior to the conventional extraction process has been studied for canola seeds, soya beans, and sunflower. Enzymatic action, as it is known, damages cell walls (mainly composed of cellulose, associated with hemicelluloses, pectic substances, and lignin) favoring the permeability for oil. To enhance the extractability of oil from oilseeds, a number of enzymes are used: amylase, glucanase, protease, pectinase, as well as cellulolytic and hemicellulolytic enzymes. Nowadays, the aim is to incorporate enzymatic treatment in the industry without significant alteration of the conventional process. The interest of aqueous enzymatic extraction processes has been revived in view of seeking of environmentally friendly alternative technologies for oil industry (Rosenthal et al. 1996, 2001).

Aqueous enzymatic extraction is an alternative approach for protein recovery from proteinaceous solid materials. Enzymes can enhance the proteins extraction yield in several ways, such as carbohydrases, which liberate more protein from the matrix as they affect the cell wall components (Tang et al. 2002). Aqueous enzymatic protein extraction is an environmentally friendly, safe, and cheap alternative process to extract protein. Furthermore, this process avoids serious proteins degradation, improving their nutritional and functional properties. However, the enzymatic process shows some drawbacks, such as the extensive time required and the high cost of enzymes (Martínez-Maqueda et al. 2013).

Other methods suggested for enhancing the extraction of proteins are related to cell disruption. Owing to the fact that tissues are rich in proteases, plants are generally more problematic for protein extraction (Wang et al. 2008). Proteins are commonly found in protein bodies inside cell walls; thus, cell disruption is required before their extraction and many chemical and physical techniques can be used for this purpose. Mechanical homogenization is one of the best methods for plant tissues, while recently, most researchers use the ultrasonic method for protein homogenization achieving increased protein yield (Bodzon-Kulakowska et al. 2007, Martínez-Maqueda et al. 2013, Van Het Hof et al. 2000). High-pressure homogenization (HPH) has also been investigated generating at 40–80 MPa, approximately double the protein extraction compared to atmospheric pressure (Barbin et al. 2011, Martínez-Maqueda et al. 2013). Temperature treatments contain the use of freeze–thaw and heat treatments, where the lysis of cells or tissues is commonly achieved by flash freezing the cells in liquid nitrogen. Also, heat treatment of protein solutions usually enhances their solubility, emulsifying, and foaming properties, but the extraction of proteins is more difficult (Martínez-Maqueda et al. 2013). Finally, cell

**TABLE 6.2**
**Comparison of MAE and UAE**

| | Extraction Technique | |
|---|---|---|
| | MAE | UAE |
| Description | Sample is immersed in the solvent and submitted to microwaves | Sample is immersed in the solvent and submitted to ultrasound |
| Extraction time | 3–30 min | 10–60 min |
| Advantages | • Fast and multiple extractions<br>• Low solvent volumes<br>• Selectivity<br>• Elevated temperatures<br>• Controllable heating process | • Easy and multiple extractions<br>• Selective extraction<br>• Faster energy transfer<br>• Moderate solvent volumes<br>• Faster response to process control |
| Disadvantages | • Extraction solvent must be able to absorb microwaves<br>• Cleanup step is needed<br>• Waiting time for vessels to cool down | • Cleanup step is required<br>• Repeated extraction may be needed |

*Source:* Adapted from Eskilsson C.S., Bjorklund E. 2000. *Journal of Chromatography A*. 902: 227–250; Veggi P.C., Martinez J., Meireles M.A.A. 2013. In: Chemat F, Gravotto G (eds.). *Microwave-Assisted Extraction for Bioactive Compounds: Theory and Practice*. Springer, pp. 15–52.

lysis can be realized by osmotic shock by gentle shaking of hypertonic solution or by chemical treatment of antibiotics, chelating agents, detergents, and capable solvents (Martínez-Maqueda et al. 2013).

Concluding, in the past decade there have been many advances in the development of new extraction methods with a number of applications. Enzymic extraction in terms of solid material treatment by suitable enzymes is easy applicable and achieves increased product yields. In particular, for MAE and UAE methods, the mass and heat-transfer phenomena involved as well as the parameters that influence these processes have been extensively described and their advantages and disadvantages are presented in Table 6.2. Optimized operating parameters can improve the performance of these methods, thus concluding that MAE and UAE are considered as promising extraction techniques compared to conventional extraction.

## REFERENCES


Aguilera J.M. 2003. Solid–liquid extraction. In: Tzia C, Liadakis G (eds.). *Extraction Optimization in Food Engineering*. Marcel Dekker, New York, pp. 35–55.

Azadmard-Damirchi S., Alirezalu K., Fathi Achachlouei B. 2011. Microwave pretreatment of seeds to extract high quality vegetable oil. *World Academy of Science, Engineering and Technology*. 57: 72–75.

Azmir J., Zaidul I.S.M., Rahman M.M., Sharif K.M., Mohamed A., Sahena F., Jahurul M.H.A., Ghafoor K., Norulaini N.A.N., Omar A.K.M. 2013. Techniques for extraction of bioactive compounds from plant materials: A review. *Journal of Food Engineering*. 117: 426–436.

Badger W., Banchero J. 1955. Extraction. In: *Introduction to Chemical Engineering*. McGraw-Hill, New York, pp. 320–348.

Barbin D.F., Natsch A., Müller R.K. 2011. Improvement of functional properties of rapeseed protein concentrates produced via alcoholic processes by thermal and mechanical treatments. *Journal of Food Processing and Preservation* 35: 369–375.

Berk Z. 1992. Technology of production of edible flours and protein products from soybeans. *FAO Agricultural Services Bulletin 97*, FAO, Rome.

Berk Z. 2009. Extraction. In: *Food Process Engineering and Technology*. Academic Press, London, UK, pp. 259–272.

Bernardini E. 1976. Batch and continuous solvent extraction. *Journal of the American Oil Chemists' Society*. 53: 273–278.

Bernardini E. 1983. *Oilseeeds, Oils and Fats*. B.E. Oil Publishing House, Rome, pp. 361–363, 483–493.

Birch E.J., Ian D.W. 2000. Fats|extraction by solvent based methods. In: *Encyclopedia of Separation Science*. Academic Press, London, UK, pp. 2794–2801.

Bodzon-Kulakowska A., Bierczynska-Krzysik A., Dylag T., Drabik A., Suder P., Noga M., Jarzebinska J., Silberring J. 2007. Methods for samples preparation in proteomic research. *Journal of Chromatography B*. 849: 1–31.

Chakrabarty M.M. 2003. Extraction of fats and oils. In: *Chemistry and Technology of Oils and Fats*. Allied Publishers Pvt. Ltd., New Delhi, pp. 247.

Chan C.H., Yusoffa R., Ngoh G.C., Kung F.W.L. 2011. Microwave-assisted extractions of active ingredients from plants. *Journal of Chromatography A*. 1218: 6213–6225.

Christodoulou P. 2003. Sugars and carbohydrates. Chapter 8. In: Tzia C, Liadakis G (eds.). *Extraction Optimization in Food Engineering*. Marcel Dekker, New York, pp. 235–303.

Coulson J.M., Richardson J.F., Backhurst J.R., Harker J.H. 1993. Leaching. In: *Chemical Engineering*. Pergamon Press, New York, pp. 391–407.

Couper J., Penney W., Fair J., Walas S. 2012. Extraction and leaching. In: *Chemical Process Equipment*. BH Elsevier, Oxford, UK, pp. 487–528.

Destandau E., Michel T., Elfakir C. 2013. Microwave-assisted extraction. In: Rostagno MA, Prado JM, Kraus GA (eds.). *Natural Product Extraction: Principles and Applications*. RSC, Cambridge, UK, pp. 113–130.

Eggers R., Jaeger P. 2003. Extractions systems. In: Tzia C, Liadakis G (eds.). *Extraction Optimization in Food Engineering*. Marcel Dekker, New York, pp. 1–42.

Eskilsson C.S., Bjorklund E. 2000. Analytical-scale microwave-assisted extraction (review). *Journal of Chromatography A*. 902: 227–250.

Hamm W. 2003. Oil production and processing. In: *Vegetable Oils*. Elsevier Science Ltd., Boca Raton, pp. 5904–5915.

Harahsheh M.Al., Kingman S.W. 2004. Microwave-assisted leaching—A review. *Hydrometallurgy*. 73: 189–203.

Ibarz A., Barbosa-Cánovas G.V. 2003. *Operations in Food Engineering*. CRC Press, LLC, New York, pp. 773–810.

Jassie L., Revesz R., Kierstead T., Hasty E., Metz S. 1997. Microwave-assisted solvent extraction. In: Kingston HM, Haswell SJ (eds.). *Microwave-Enhanced Chemistry*. American Chemical Society, Washington, pp. 569.

Johnson L., Lusas E. 1983. Comparison of alternative solvents for oils extraction. *Journal of the American Oil Chemists' Society*. 60: 181–194.

Kemper T. 2005. Oil extraction. In: Shahidi F (ed.). *Bailey's Industrial Oil and Fat Products*. Vol. 5, John Wiley & Sons, Inc., New Jersey, pp. 57–97.

Leonelli C., Veronesi P., Gravotto G. 2013. Microwave-assisted extraction: An introduction to dielectric heating. In: Chemat F, Gravotto G (eds.). *Microwave-Assisted Extraction for Bioactive Compounds: Theory and Practice*. Springer, New York, pp. 1–10.

Liadakis G., Tzia C., Oreopoulou V., Thomopoulos C.D. 1995. Protein isolation from tomato seed meal, extraction optimization. *Journal of Food Science*. 60: 477.

Lopez-Avila V. 2000. *Microwave-Assisted Extraction*. Academic Press, New York, pp. 1389–1398.

Luque-Garcıa J.L. 2005. *Microwave-Assisted Solvent Extraction*. Elsevier Academic Press, New York, pp. 584–591.

Martínez-Maqueda D., Hernández-Ledesma B., Amigo L., Miralles B., Gómez-Ruiz J.A. 2013. Extraction/fractionation techniques for proteins and peptides and protein digestion. In: Toldrá F, Nollet LML (eds.). *Proteomics in Foods: Principles and Applications*. Springer, New York, pp. 21–50.

Mason T., Riera E., Vercet A., Lopez-Buesa P. 2005. Application of ultrasound. In: Sun DW (ed.). *Emerging Technologies for Food Process*. Elsevier Academic Press, Oxford, UK, pp. 323–342.

McCabe W., Smith J., Harriott P. 1993. Leaching and extraction. In: *Unit Operations of Chemical Engineering*. McGraw-Hill, New York, pp. 614–623.

Minkiewicz P., Dziuba J., Darewicz M., Iwaniak A., Dziuba M., Nalecz D. 2008. Food peptidomics. *Food Technology Biotechnology*. 46:1–10.

Moure A., Sineiro J., Domínguez H., Parajó J.C. 2006. Functionality of oilseed protein products: A review. *Food Research International*. 39: 945–963.

Oreopoulou V., Tzia C. 2007. Utilization of food-byproducts and treatment of waste in the food industry (Chapter 11). In: Oreopoulou V, Russ W (eds.). *Utilization of Plant by-Products for the Recovery of Proteins, Dietary Fibers, Antioxidants, and Colorants. Series: Integrating Safety and Environmental Knowledge into Food Studies towards European Sustainable Development, Vol. 3*. Springer Science + Business Media, LLC, New York, pp. 209–232.

Pingret D., Fabiano-Tixier A.S., Chemat F. 2013. Ultrasound-assisted extraction. In: Rostagno MA, Prado JM, Kraus GA (eds.). *Natural Product Extraction: Principles and Applications*. RSC, Cambridge, UK, pp. 89–105.

Prabhudesai R. 1997. Leaching. In: Schweitzer P (ed.). *Handbook of Separation Techniques for Chemical Engineers*. McGraw-Hill, New York, pp. 3–19.

Pramparo M., Gregory S., Mattea M. 2002. Immersion vs. percolation in the extraction of oil from oleaginous seeds. *Journal of the American Oil Chemists' Society*. 79: 955–960.

Ramaswamy H., Marcotte M. 2006. Separation and concentration. In: *Food Processing: Principles and Application*. CRC Press, Boca Raton, pp. 355–360.

Rao D.G. 2010. Extraction. In: *Fundamentals of Food Engineering*. PHI Learning Private Limited, New Delhi, pp. 365–371.

Rosenthal A., Pyle D.L., Niranjan K. 1996. Aqueous and enzymatic processes for edible oil extraction. Review. *Enzyme and Microbial Technology*. 19: 402–420.

Rosenthal A., Pyle D.L., Niranjan K., Gilmour S., Trinca L. 2001. Combined effect of operational variables an enzyme activity on aqueous and enzymatic extraction of oil and protein from soybean. *Enzyme and Microbial Technology*. 28: 499–509.

Sancey B. 2002. Extraction. In: *Encyclopedia of Science and Technology*. McGraw-Hill, New York, pp. 767–769.

Sattler K., Feidt H.J. 1995. Extraction. In: *Thermal Separation Processes*. VCH, New York, pp. 393–465.

Sharma G.M., Mundoma C., Seavy M., Roux K.H., Sathe S.K. 2010. Purification and biochemical characterization of Brazil nut (*Bertholletia excelsa* L.) seed storage proteins. *Journal of Agricultural and Food Chemistry*. 58: 5714–5723.

Spaninks J., Brun S. 1979. Mathematical simulation of the performance of solid–liquid extractors—Belt type extractors. *Chemical Engineering Science*. 34: 207–215.

Sumnu S.G., Sahin S. 2012. Microwave heating. In: Sun DW (ed.). *Thermal Food Processing: New Technologies and Quality Issues, Second Edition*. CRC Press, New York, pp. 555–560.

Suryanarayana A. 2002. Leaching. In: *Mass Transfer Operations*. New Age International (P) Limited Publishers, New Delhi, pp. 511–519.

Takeuchi T., Pereira C., Braga M., Marostica M., Leal P., Meireles A. 2009. Low-pressure solvent extraction (solid–liquid extraction, microwave assisted, and ultrasound assisted) from condimentary plants. In: Meireles A (ed.). *Extracting Bioactive Compounds for Food Products: Theory and Applications*. CRC Press, Boca Raton, pp. 140–171.

Tandon S., Rane S. 2008. Decoction and hot continuous extraction techniques. In: Handa SS, Khanuja SPS, Longo G, Rakesh DD (eds.). *An Overview of Extraction Techniques for Medical and Aromatic Plants*. ICS UNIDO, Trieste, Italy, pp. 93–105.

Tang S., Hettiarachchy N.S., Shellhammer T.H. 2002. Protein extraction from heat-stabilized defatted rice bran. Physical processing and enzyme treatments. *Journal of Agricultural and Food Chemistry*. 50: 7444–7448.

Thomas G., Veloso G., Krioukov V. 2007. Mass transfer modelling in counter-current crossed flows in an industrial extractor. *Food and Bioproducts Processing*. 85: 77–84.

Tiwari K.K. 1995. Extraction technologies related to food processing. In: Gaonkar AG (ed.). *Food Processing: Recent Developments*. Elsevier, New York, pp. 269–302.

Turner C. 2006. Overview of modern extraction techniques for food and agricultural samples. In: Turner C (ed.). *Modern Extraction Techniques*. ACS Symposium Series, Washington, pp. 3–19.

Tzia C. 2003. Optimization. Chapter 5. In: Tzia C, Liadakis G (eds.). *Extraction Optimization in Food Engineering*. Marcel Dekker, New York, pp. 137–180.

Van Het Hof K.H., De Boer B.C.J., Tijburg L.B.M., Lucius B.R.H.M., Zijp I., West C.E., Hautvast J.G.A.J., Weststrate J.A. 2000. Carotenoid bioavailability in humans from tomatoes processed in different ways determined from the carotenoid response in the triglyceride-rich lipoprotein fraction of plasma after a single consumption and in plasma after four days of consumption. *Journal of Nutrition* 130: 1189–1196.

Veggi P.C., Martinez J., Meireles M.A.A. 2013. Fundamentals of microwave extraction. In: Chemat F, Gravotto G (eds.). *Microwave-Assisted Extraction for Bioactive Compounds: Theory and Practice*. Springer, New York, pp. 15–52.

Vetter J. 1998. *Sugar Technology*. Verlag Bartens, Berlin, pp. 334–338.

Wakeman R. 1994. Extraction (liquid–solid). In: Kroschwitz J, Howe-Grant M (eds.). *Encyclopedia of Chemical Technology*. John Wiley & Sons, New Jersey, pp. 181–195.

Wang W., Tai F., Chen S. 2008. Optimizing protein extraction from plant tissues for enhanced proteomics analysis. *Journal of Separation Science*. 31:2032–2039.

Xu L., Diosady L. 2003. Fats and oils from plant materials. In: Tzia C, Liadakis G (eds.). *Extraction Optimization in Food Engineering*. Marcel Dekker, New York, pp. 1–35.

Zayas F. 1997. Solubility of proteins. In: Zayas F (ed.). *Functionality of Proteins in Food*. Springer, Germany, pp. 6–26.

# 7 Supercritical Fluid Extraction

*Epaminondas Voutsas*

## CONTENTS



## 7.1 INTRODUCTION

Supercritical fluid extraction (SFE) is a separation method that is based on the same basic principle with the more conventional separation processes of distillation and liquid solvent extraction, that is, they exploit phase equilibrium behavior between different states of matter at different operating conditions. SFE is essentially a typical solvent extraction with the basic difference to be that the supercritical (SC) solvent undergoes a change of state prior to the extraction process and changes the state again after the end of the process.

SFE takes advantage of the high dissolving ability of supercritical fluids (SCFs) under specific temperature and pressure conditions. The first observations and experiments on the solubility of substances in SCFs go back to the nineteenth century. The Irish chemist Thomas Andrews (Andrews, 1875) gave a lecture to the Royal Society, reporting for the first time that SCFs dissolve some nonvolatile solvents. He described the "critical point" as the end of the vapor pressure curve in a phase diagram, and measured with remarkable accuracy the critical properties of $CO_2$: critical temperature ($T_c = 30.92°C$) and critical pressure ($P_c = 73$ atm), which are very close to those widely accepted today, 30.95°C and 72.8 atm. Hannay and Hogarth (1879) and Hannay (1880) described the increased solubility of several inorganic salts in ethanol at elevated pressures and temperatures above the critical temperature of ethanol (234°C). By reducing the pressure, they observed the precipitation of the salts in the form of snow. Their experiments caused considerable debate in the scientific community and they strongly disputed, until finally scientists to be convinced that this is a new phenomenon and not simply an increase of the solubility due to the increase of temperature.

Research continued during the rest of the nineteenth century and up to the middle of the twentieth century, and there had been a significant number of experimental measurements in various systems. From 1950 to 1970, there has been a growing interest for the solubility of substances in SCFs. Francis (1954) presented an extensive list with the solubility of 261 substances in liquid $CO_2$ at 25°C (near critical), whose values could be used as an indication of the possibility of further increasing the solubility in SCF $CO_2$.

Despite the large volume of research work on SCFs, their industrial applications were extremely limited until the 1970s, which is mainly due to the high operating pressures of the SFE process that results in high installation costs. In particular, until the mid-1970s where the energy was cheap and the environmental and health regulations were more flexible, there was no incentive to exploit SFE. The first and the most classical applications of SFE using $SCCO_2$, were for the decaffeination of coffee in 1978 in Germany (Lack and Seidlitz, 1993) and 2 years later in Australia for the processing of hops (Clark and Mailer, 1983).

Although for many years, SFE was sidelined from the conventional separation methods—distillation, and liquid extraction—in the last two decades, an awareness of the potential of SCF processing as a viable component for "green" processing has arisen (King and Bott, 1993). SFE processes using $CO_2$ are especially advantageous compared to distillation processes, for example, in food, pharmaceutical, and cosmetic industries, when thermolabile compounds with low vapor pressures have to be separated.

The main reasons that gave a significant boost to the use of benign solvents, such as $SCCO_2$, are

- The high increase of the energy costs made traditional energy-intensive separation methods such as distillation, very expensive, and turned the interest toward others, which require less energy, such as SFE.
- The new health regulations and prohibitions imposed on the use of classical industrial solvents, and the new stricter environmental regulations for the

treatment and disposal of industrial wastes, have led to growing interest in the potential uses of $CO_2$ as a nontoxic solvent.
- The increasing consumer awareness of the identity and use of chemical solvents in food and natural products.

In this chapter, some of the basic fundamentals of SFE and their use for the extraction and fractionation of food-related materials are reviewed. The relevant physicochemical properties of SCFs, particularly those of $CO_2$, are first discussed, followed by a short discussion of solubility measurements in SC fluids. Then, two important parameters in SFE, namely, selectivity and cosolvent effect, are discussed. The following brief introduction to phase equilibrium modeling in SCFs that is a key factor for the successful design of SFE is presented. Three different SCF-processing schemes are discussed next, followed by mathematical modeling of the SFE process. Then, some safety and scale-up issues of the SCF technology are presented and, finally, some other applications of SCFs are briefly introduced.

## 7.2 SOME BASIC PROPERTIES OF SCFs

Table 7.1 presents the most common SCFs along with their critical properties—temperature and critical pressure. The following are observed:

- Most hydrocarbons have $P_c$ close to 50 atm.
- Critical temperatures of light hydrocarbons, such as ethylene and ethane, are close to that of the ambient temperature, while those of the cyclic and aromatic hydrocarbons are much larger.
- Carbon dioxide has a low $T_c$, while $P_c$ is somewhat higher.
- Ammonia and water have a high $T_c$ and $P_c$, because of their polarity and hydrogen bonding.

The most attractive and commonly used SCF at laboratory, pilot, and industrial scale is $CO_2$. It is a low-cost fluid, widely abundant in pure form, nonflammable, not toxic (although specific safety precautions are required), and environmentally friendly (Clavier et al., 1996). The critical point of $CO_2$ allows operations at moderate temperatures (typically 40–70°C) and accessible pressures; the maximum operating pressure is usually set at 300 bar for the most current applications (McHugh and Krukonis, 1994; Brunner, 1994). Therefore, with the use of $CO_2$ various problems of classical separation techniques such as distillation and liquid extraction, are avoided; problems such as (a) recycling and disposal of organic solvents, and (b) chemical and thermal deterioration of the product obtained when high temperatures and/or organic solvents are employed.

Another widely used SCF is water, despite its high values of $T_c$ and $P_c$ and the corrosive properties in some cases. Generally, water is a common solvent with many applications because of its nontoxicity. In SC conditions, water acquires some interesting properties. So, although as a liquid it exhibits a strong polar character, as an SCF it behaves almost like a nonpolar substance, while its acidity increases (Bellisent-Funel, 2001). This is because of the phenomenon of splitting into ions in

**TABLE 7.1**
**Critical Properties for Some SCFs**

| Fluid | Critical Temperature ($T_c$) (°C) | Critical Pressure ($P_c$) (atm) |
|---|---|---|
| Carbon dioxide | 31.1 | 72.8 |
| Ethane | 32.3 | 48.2 |
| Ethylene | 9.3 | 49.7 |
| Propane | 96.7 | 41.9 |
| Propylene | 91.9 | 45.6 |
| Cyclohexane | 280.3 | 40.2 |
| Nitrous oxide | 36.4 | 71.5 |
| Benzene | 289.0 | 48.3 |
| Toluene | 318.6 | 40.6 |
| p-Xylene | 343.1 | 34.7 |
| Chlorotrifluoromethane | 28.9 | 38.7 |
| Trichlorofluoromethane | 198.1 | 43.5 |
| Xenon | 16.9 | 58.0 |
| Ammonia | 132.5 | 111.3 |
| Water | 374.2 | 217.6 |

*Source:* Adapted from McHugh, M.A., Krukonis, V.J. 1986. *Supercritical Fluid Extraction*. Butterworths, USA.

the region near and above its critical point. Moreover, in this region, water presents better solvating capacity. These properties coupled with the fact that they can be significantly modified by small changes in pressure and temperature, allow the use of SC water as (a) solvent in separation processes (usually it is used at SC pressures, but at temperature less than its critical one; so, it is characterized as superheated water), (b) solvent in chemical reactions, which is the most common application, and (c) as acid catalyst in processes such as oxidation of organic toxic substances (Fukushima, 1999; Lang and Wai, 2001).

Other SC solvents that attract scientific interest are light hydrocarbons such as ethane, ethylene, propane, and so on (McHugh and Krukonis, 1994). The problem with them is that they are flammable, so their use requires special attention. The same applies for $N_2O$ (Vandana et al., 1996), which has been used as a polar solvent for extraction purposes, although it is not preferred because of the high risk of explosion inherent with its use. Regarding the halogenated hydrocarbons, they are expensive and harmful to the environment. Finally, xenon, an expensive fluid finding application in small-scale spectroscopy units because of its particular properties, and especially because it does not absorb in the range of infrared and ultraviolet (UV) radiation (Haake et al., 1998; Howdle et al., 1992; Smith, 1999).

In the phase diagram of a pure substance such as $CO_2$ (Figure 7.1), four distinct areas are shown: solid, liquid, gas, and the SCF, where the last area is extended over and on the right of the dashed lines, that is, at pressures and temperatures greater than the critical ones. In the region above the critical point, a substance exhibits

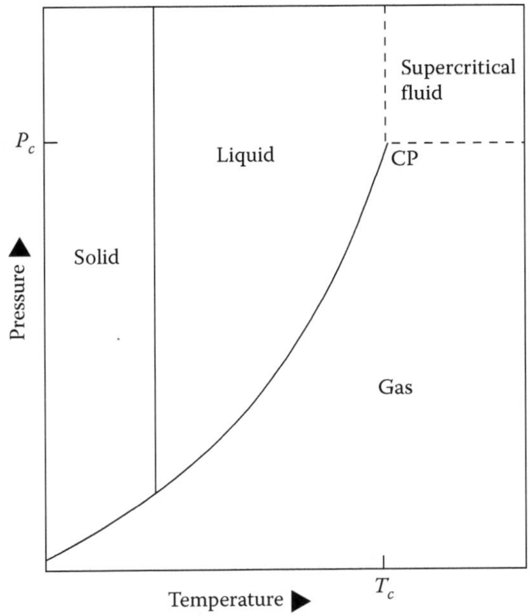


**FIGURE 7.1** Phase diagram of a pure substance.

properties similar to those of both liquid and gas, with the final result of the behavior to be a combination of these. This can be realized through the comparison of the values of some physicochemical properties, such as density, diffusivity, and viscosity, in the gas, liquid, and SC states shown in Table 7.2. It is evident that SCFs exhibit densities that correspond to those encountered in liquids, and diffusivities and viscosities similar to gases.

SCFs have increased dissolving power as compared to gases because they have much higher densities, which are comparable to those of liquids (about one-third to one-half of the liquids). At the near-SC region, that is, $T_r$ (= $T/T_c$) from 1 to 1.5 and

**TABLE 7.2**
**Comparison of Physicochemical Properties between Gas, Liquid, and SC States**

| State | Density (g/cm³) | Diffusion Coefficient (cm²/s) | Viscosity (cps) |
|---|---|---|---|
| Gas | $10^{-3}$ | $10^{-1}$ | $10^{-2}$ |
| Liquid | 1 | $<10^{-5}$ | 1 |
| SC | 0.3–0.8 | $10^{-4}$–$10^{-3}$ | $10^{-2}$–$10^{-1}$ |

*Source:* Adapted from Chordia, L., Martinez, J.L. 2002. *Lab. Focus,* 6(1).

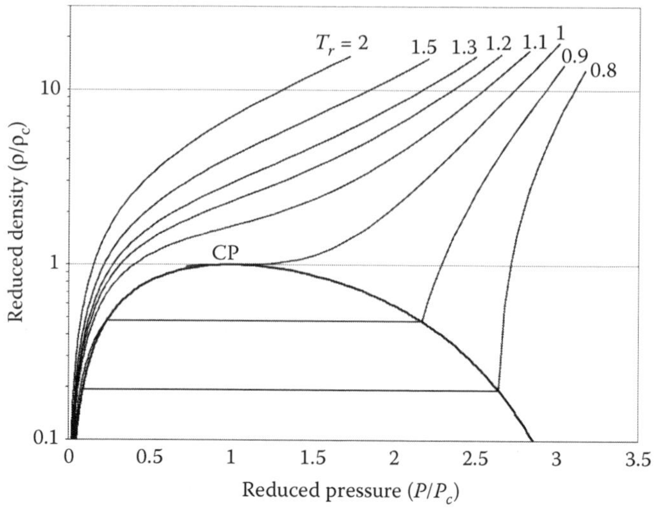


**FIGURE 7.2** Reduced pressure versus reduced density phase diagram of a pure substance.

$P_r$ (= $P/P_c$) from 1 to 3, a slight decrease in temperature at constant pressure causes a large increase in density, and a small increase in pressure at constant temperature causes a large increase in density (Figure 7.2).

Mainly, two factors influence the temperature dependence of the solubility of high-boiling compounds in SCFs—solids or liquids. On the one hand, the solubility rises with temperature because of the higher sublimation (or vapor) pressure. On the other, the density of the SCF is reduced at higher temperatures, which has an opposite effect on the solubility. So, at low pressures, the solubility is only influenced by the vapor pressure, which leads to a solubility decrease with decreasing temperature. At higher pressures, larger densities are obtained at lower temperatures, which lead to an improvement of the solvent power for the high-boiling compounds. This means that the isotherms will cross, as shown in Figure 7.3, for the experimental and calculated solubilities of benzoic acid in $CO_2$. Depending on the interactions, this can lead to higher solubilities at low temperatures at the pressures applied. However, at high pressures, after a second crossing of the isotherms, larger solubilities are obtained at high temperatures. Most likely, this is a result of the higher vapor pressure, while the temperature dependence of the interactions of the compounds should be considered as well. Also note that this complex behavior of the SCFs cannot be described by considering it as an ideal gas.

The influence of the other factors that affect solubility has a relatively more obvious interpretation. The increased diffusivities of an SCF as compared to those of liquid, results in high mass transfer rates. Self-diffusion coefficient of SC $CO_2$ is 1–2 orders of magnitude greater than those of dissolved substances in usual solvents (Kander and Paulaitis, 1983). In the same direction of increased solvating power drive, the low viscosity and almost zero surface tension of SCFs enable easy penetration of the fluid in porous solids. Furthermore, the viscosity of $CO_2$, even for the

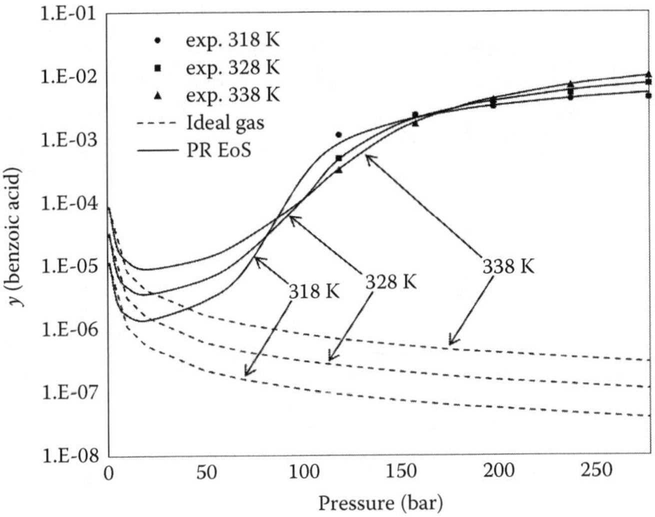


**FIGURE 7.3** Solubilities of benzoic acid in $CO_2$. Experimental data from Kurnik et al., 1981. Solid lines: PR EoS calculations with optimized binary interaction parameters ($k_{ij} = 0.025$ at 318 K, $k_{ij} = 0.016$ at 328 K and $k_{ij} = 0.0087$ at 338 K). Dashed lines: Solubilities predicted where the vapor phase is assumed to behave as an ideal gas.

high pressures is about 0.1 cps, one order of magnitude smaller than those of typical organic solvents (Schneider, 1978).

## 7.3 SOLUBILITY MEASUREMENTS IN SCFs

Solubility of the solute in the SC solvent is determined by measuring the amount of solute in the solution, giving the result in terms of mass fraction or mole fraction. Extensive reviews of experimental solubilities of solid and liquids in SCFs can be found in the literature (Bartle et al., 1991; Christov and Dohrn, 2002; Dohrn and Brunner, 1995; Dohrn et al., 2010; Fonseca et al., 2011; Paulaitis et al., 1982).

Two main methods for measuring the solubility of substances in SCFs exist, the static method and the dynamic method.

In the static method, equilibrium state between the solute and the solvent is achieved before any sample is taken out to analyze the solubility. Static method carries out the equilibrium process in many ways that include recirculation of the solvent, agitation by the magnetic stirrer, or simply trapping the solvent in the equilibrium cell for some time. A schematic diagram of an apparatus that operates with the static method is shown in Figure 7.4, in which known quantities of the substance to be dissolved and the SCF are fed to the desired pressure in a thermostated cell. The pressure is adjusted with an appropriate device and the whole system is agitated using a magnetic stirrer. The cell has an appropriate sapphire window through which it is possible to observe the phenomena within the cell with the help of a camera.

On the other hand, in the dynamic method, the solute that is the condensed phase remains in an equilibrium cell, while the SCF flows through the cell continuously.

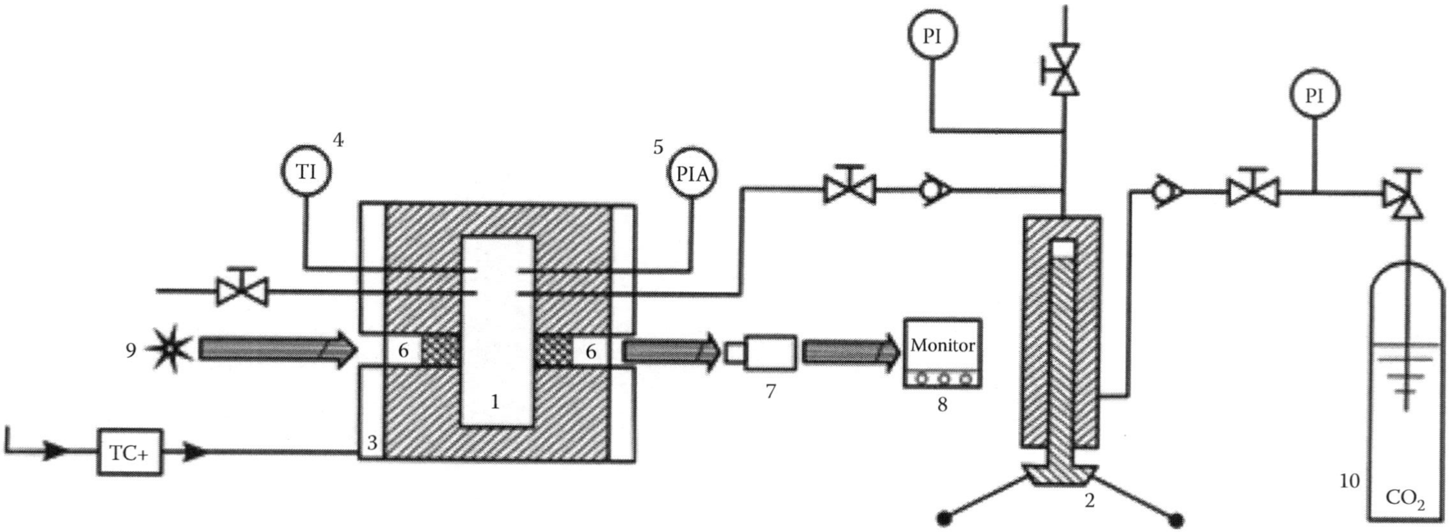


**FIGURE 7.4** Schematic diagram of an apparatus used for solubility measurements with the static method. (1) high-pressure optical cell; (2) manually operated high pressure pump; (3) thermostated heating jacket; (4) thermocouple; (5) pressure transducer; (6) sapphire windows; (7) camera; (8) monitor; (9) cold light source; (10) $CO_2$ cylinder. (Reprinted from *J. Supercrit. Fluids*, 41 (2), E. Bertakis et al. Measurement and thermodynamic modeling of solid-liquid-gas equilibrium of some organic compounds in the presence of $CO_2$, 240, Copyright (2007), with permission from Elsevier.)

SCF is contacted with the solute and equilibrium is achieved during the residence time of the SCF in the equilibrium cell. Then, the equilibrated solution is extracted and its equilibrium concentration is determined. As shown in the schematic diagram of an apparatus that operates with the dynamic method (Figure 7.5), a sample of the solute is placed into an extraction cup with a porous bottom and top lid. The cup is placed into the extractor, which is located inside the oven of the apparatus. When the system is being charged by means of a fluid pump, the carbon dioxide flows from the loop control valve through the recirculation pump and a six-port valve into the extractor, which is loaded with the solute. Once the desired pressure level has been reached, the extraction loop is isolated from the carbon dioxide supply tank, the recirculation pump is activated, and the fluid is allowed to flow through the closed loop. The progress of the extraction process is monitored by a detector, for example, a UV detector. When the absorbance reading becomes constant with time, the equilibrium between the solute and the gaseous phase has been reached. The injection of the sample extract, which is trapped inside a fixed volume sample loop, into the analysis system, for example, a high-performance liquid chromatography (HPLC) column, is accomplished using a six-port injection valve and an eight-port valve connected in tandem and controlled by a microprocessor.

Dynamic methods have the advantage over static methods in that they allow sequential solubility measurements without the need to interrupt the experiment to change the composition of the mixture.

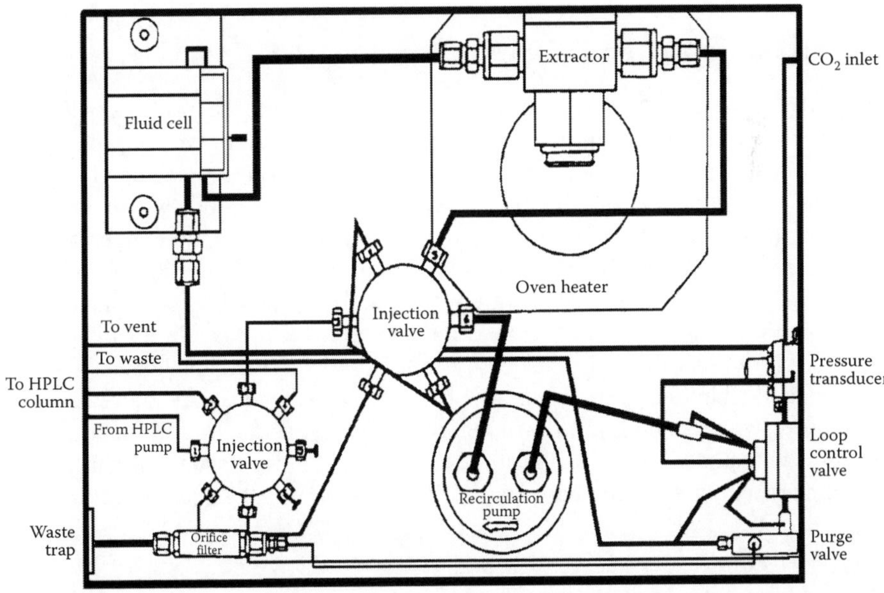


**FIGURE 7.5** Schematic diagram of an apparatus used for solubility measurements with the dynamic method. (Reprinted from *J. Chem. Eng. Data*, 40/4, Ph. Coutsikos, K. Magoulas and D. Tassios, Solubilities of phenols in supercritical carbon dioxide, 954, Copyright (1995), with permission from American Chemical Society.)

## 7.4 SOLVENT SELECTIVITY AND CO-SOLVENT EFFECT

One of the major advantages of SCFs is their selectivity, which can be defined as the ability of a solvent to dissolve the desired compound to a greater extent than the other constituents of the mixtures.

For example, as demonstrated by Schneider et al. (1980), from a mixture of benzoic acid and p-hydroxybenzoic acid, $SCCO_2$ prefers to dissolve benzoic acid at concentrations of about 1000 times greater than those of p-hydroxybenzoic acid. Consequently, the selection of a suitable solvent is very important for the effective separation of a mixture with SFE.

A significant increase of a compound solubility in an SCF may be achieved with the use of cosolvents, which are usually low-molecular-weight volatile compounds that are added to the SCF in small amounts, ca. up to 5% mole. The cosolvent effect is due to the strong interactions with the molecules of the solute, which results in a smaller quantity of the SCF needed for the extraction and, therefore, less energy for the SFE process.

McHugh and Krukonis (1986) showed that the addition of 5% (mole) of methanol in SC $CO_2$ increases the solubility of acridine almost 5 times at 50°C and 200 bar. Another example is the case of hydroquinone, whose solubility in pure $CO_2$ at 308 K is of the order of $10^{-6}$ to $10^{-5}$ in mole fraction basis (Coutsikos et al., 1995). The addition of only 2% in mole of tributyl phosphate ester increases the solubility of hydroquinone to the value of $10^{-3}$ (Lemert and Johnston, 1991). Also note that the use of a cosolvent may also lead to the increase of the selectivity of the SCFs.

## 7.5 PHASE EQUILIBRIUM IN SCFs

The design and optimization of SCF processes require the accurate knowledge of the solubility of the high-boiling compound in the SCF as a function of pressure and temperature. Of course, a comprehensive database with experimental solubility data would be desirable to fix optimal process conditions, for example, which gas or gas mixture provides the best solvent properties, which are the optimal temperature and pressure conditions, which cosolvent is most suitable for improving the solubility and selectivity, and so on. Unfortunately, the number of reliable experimental data is very limited. Moreover, the reliable measurement of the solubility of solids or high-boiling liquids in an SCF at high pressures requires sophisticated expensive experimental techniques. Since it is not possible to measure all the required solubilities for the various solutes in the different SCFs for a wide range of temperatures and pressures with and without cosolvents, thermodynamic models are applied.

The starting point for all phase equilibria calculations is the equality of fugacities of each species of the mixture $\left(\hat{f}_i\right)$ in every phase ($I, II, \ldots, P$):

$$\hat{f}_i^I = \hat{f}_i^{II} = \ldots = \hat{f}_i^P \quad i = 1,2,\ldots,c \tag{7.1}$$

The calculation of the fugacities depends on the phases involved. Three different cases of binary phase equilibrium in SCFs are considered here: (a) between an SCF and a solid phase, that is, solid–gas (SG) equilibrium, (b) between an SCF and a liquid phase, that is, gas–liquid (GL) equilibrium, and (c) the three-phase equilibrium between an SCF, a solid, and a liquid phase, that is, solid–liquid–gas (SLG) equilibrium.

### 7.5.1 Binary SG Equilibrium

If the high-boiling substance (component 2) is in the solid state, only the isofugacity condition for the solid component has to be taken into account because the solubility of the SC gas (component 1) in the solid phase can be neglected:

$$\hat{f}_2^s = \hat{f}_2^{scf} \tag{7.2}$$

where superscript *s* stands for the solid and *scf* stands for the SCF phase, respectively.

Owing to the assumption that the mole fraction of component 2 in the solid phase is equal to one, the fugacity of the high-boiling component in the solid phase is equal to that of the pure compound at the specific temperature and pressure, which is given by the following expression (McHugh et al., 1988):

$$\hat{f}_2^s(T,P) = P_2^{sat}\,\varphi_2^{sat}\exp\int_{P_2^{sat}}^{P}\left(\frac{v_2^s}{RT}\right)dP \tag{7.3}$$

where $P_2^{sat}$ is the vapor pressure of the solid (sublimation pressure) at the specific temperature, $\varphi_2^{sat}$ is the fugacity coefficient at the specific temperature and a pressure equal to $P_2^{sat}$, and $v_2^s$ is the molar volume of the solid at the specific temperature. The exponential term in Equation 7.3 is referred to as Poynting correction factor (*Poy*). The fugacity of the solute in the SC phase is calculated through its standard definition:

$$\hat{f}_2^{scf} = y_2\,\hat{\varphi}_2^{scf}\,P \tag{7.4}$$

where $\hat{\varphi}_2^{scf}$ and $y_2$ are the fugacity coefficient and the mole fraction (solubility) of the solid solute in the SCF phase, respectively.

By substituting Equations 7.3 and 7.4 into Equation 7.2, the solubility of the solid solute in the SCF is calculated:

$$y_2 = \frac{P_2^{sat}}{P}\,\frac{\varphi_2^{sat}(Poy)_2}{\hat{\varphi}_2^{scf}} \tag{7.5}$$

At low pressures, ideal gas behavior of the SCF phase can be assumed, and Equation 7.5 is reduced to

$$y_2^{IG} = \frac{P_2^{sat}}{P} \tag{7.6}$$

Equation 7.6 indicates that at low pressures, the solubility is only influenced by the sublimation pressure.

Equation 7.5 can be rewritten as

$$y_2 = y_2^{IG} E \tag{7.7}$$

where $E$ is referred as the enhancement factor:

$$E = \frac{\varphi_2^{sat}(Poy)_2}{\hat{\varphi}_2^{scf}} \tag{7.8}$$

Since solid vapor pressures are very low, $\varphi_2^{sat}$ is assumed to be equal to unity. Poynting correction $(Poy)_2$ represents the effect of pressure on the solid fugacity and it does not exceed the value of 2–3 even for very high pressures.

Fugacity coefficient, $\hat{\varphi}_2^{scf}$, represents the differences in the intermolecular forces between similar and dissimilar molecules. For SC conditions, it may have values of the order of $10^{-4}$ or lower, especially for the very high pressures. So, the value of $E$ becomes $10^4$ or higher, which leads to much higher solubility values than those that correspond to the ideal gas solubility. This is clearly shown in Figure 7.3 for the solubility of benzoic acid in $CO_2$. Ideal gas solubility of benzoic acid in $CO_2$ at 45°C and 80 atm is $y_2^{IG} = 10^{-7}$, while that in the real $SCCO_2$ is $y_2 = 10^{-5}$, that is, 100 times higher, that is, $E = y_2/y_2^{IG} = 100$. For pressures higher than 280 atm, $y_2^{IG} = 2 \times 10^{-8}$ and $y_2 = 2 \times 10^{-3}$, that is, $E = 10^5$.

For the calculation of fugacity coefficients in the SCF phase, $\hat{\varphi}_2^{scf}$, cubic equations of state (EoS) such as Soave–Redlich–Kwong (Soave, 1972), and Peng–Robinson (Peng and Robinson, 1976) are usually used. However, for the accurate description of the fugacity coefficients, binary interaction parameters in the combining rules of the attractive and covolume parameters of the EoS fitted to the experimental solubility data are needed (Coutsikos et al., 1995).

Although cubic EoS can be used only for correlation of experimental data, another possibility is the prediction of the solubility of solids or high-boiling compounds using the so-called predictive EoS/$G^E$ models, such as PSRK (Holderbaum and Gmehling, 1991), MHV2 (Dahl et al., 1991), LCVM (Spiliotis et al., 1994), or the more recent UMR-PRU (Louli et al., 2007; Voutsas et al., 2004a, 2006). These models combine a cubic EoS with a group contribution method of the UNIFAC type with the help of a suitable mixing rule. An example of the application of the UMR-PRU model in the prediction of phenanthrene solubility in pure $CO_2$ and $CO_2$ with cosolvent is shown in Figure 7.6.

### 7.5.2 Binary GL Equilibrium

If the high-boiling compound (2) is in the liquid state, the phase equilibrium can be calculated by the φ–φ approach starting from the isofugacity condition:

$$x_i \hat{\varphi}_i^L P = y_i \hat{\varphi}_i^{scf} P \quad i = 1,2 \tag{7.9}$$

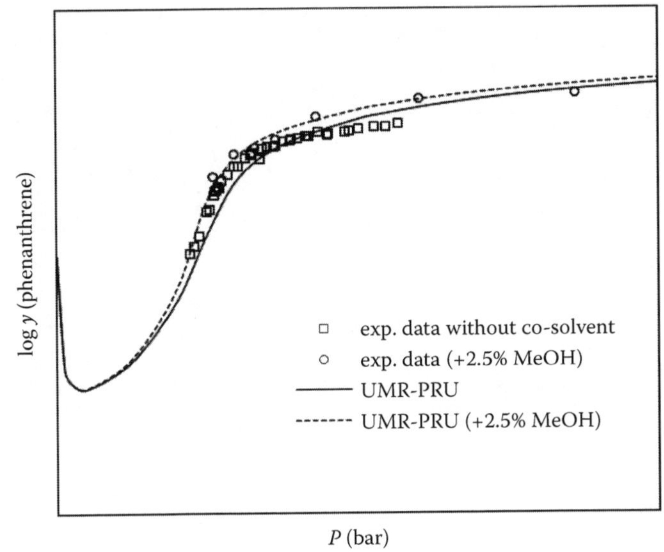


**FIGURE 7.6** Solubility of phenanthrene at 50°C in pure $CO_2$ and in $CO_2$ with methanol as co-solvent with the UMR-PRU model. (Reprinted from *Fluid Phase Equil.*, 241, E. Voutsas et al., Thermodynamic property calculations withe universal mixing rule for EoS/$G^E$ models: Results with the Peng-Robinson EoS and a UNIFAC model, 224, Copyright (2006), with permission from Elsevier.)

where superscript $L$ stands for the liquid. Note that $x_1$ is actually the solubility in mole fraction basis of the SCF in the liquid phase, and $y_2$ is the solubility of the high-boiling compound in the SCF phase.

For the calculation of fugacity coefficients, cubic EoSs are used, but only for correlation purposes, since binary interaction parameter in the combining rule of the attractive term parameter fitted to the experimental solubility data is needed, while when the mixture components are asymmetric with respect to their size, an extra adjustable parameter is required in the combining rule of the covolume parameter of the EoS (Voutsas et al., 2004b). The prediction of solubilities of high-boiling liquids can be achieved with EoS/$G^E$ models. UMR-PRU has been proven as a powerful predictive tool for this kind of phase equilibrium, regardless of the mixture asymmetry with respect to the size of the molecules involved (Voutsas et al., 2004a, 2006). Typical prediction results with the UMR-PRU model for the case of binary mixtures of $CO_2$ with toluene and comparison with experimental data are shown in Figure 7.7.

### 7.5.3 Binary SLG Equilibrium

Knowledge of the three-phase SLG equilibrium is key for the modeling and design processes for fine particle production, such as the technique of particle from gas-saturated solutions (Jung and Perrut, 2001).

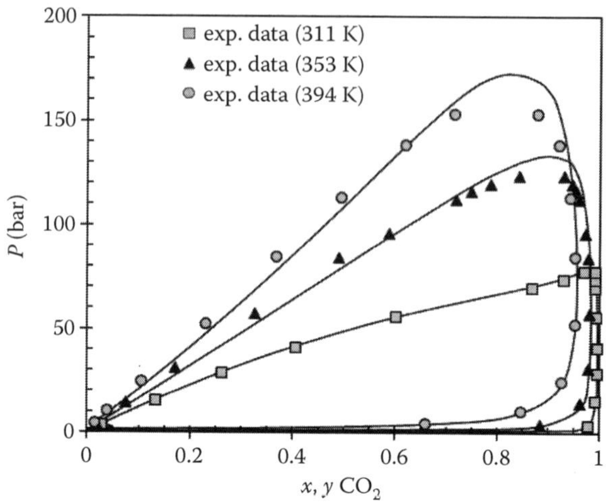


**FIGURE 7.7** P-x-y diagram with the UMR-PR model for the gas/liquid phase equilibrium of the $CO_2$/toluene binary mixture. (Reprinted from *Fluid Phase Equil.*, 241, E. Voutsas et al., Theromodynamic property calculations with the universal mixing rule for EoS/$G^E$ models: Results with Peng-Robinson EoS and UNIFAC model, 219, Copyright (2006), with permission from Elsevier.)

For the solution of the binary SLG equilibrium, the following procedure can be used (Bertakis et al., 2007). At a specified pressure, three unknowns exist: the temperature, the liquid, and vapor-phase compositions. These are calculated by solving the following three equations (subscript 1 refers to the SCF and 2 refers to the high-boiling-point compound):

$$\hat{f}_2^s = \hat{f}_2^G \tag{7.10}$$

$$\hat{f}_2^L = \hat{f}_2^G \Leftrightarrow x_2 \cdot \hat{\varphi}_2^L = y_2 \cdot \hat{\varphi}_2^G \tag{7.11}$$

$$\hat{f}_1^L = \hat{f}_1^G \Leftrightarrow x_1 \cdot \hat{\varphi}_1^L = y_1 \cdot \hat{\varphi}_1^G \tag{7.12}$$

The solution of Equation 7.10 for the specified pressure and an assumed temperature yields the composition of the gaseous phase. Equations 7.11 and 7.12 are then solved by guessing a liquid-phase composition and repeating until the equations are satisfied simultaneously. Next, the sum of the calculated liquid-phase molar fraction is checked. Depending on its value, the temperature is adjusted accordingly, and the calculations are repeated for the same pressure. If the sum of the mole fractions is greater than the sum of which the temperature is decreased, the opposite happens if it is less than one, while if the sum of the mole fractions is equal to unity, the result is recorded as a point on the SLG line.

Fugacity coefficients in the gas and the liquid phases are calculated by a cubic EoS or an EoS/G$^E$ model, while for the calculation of the fugacity of the solid phase, $\hat{f}_2^s$, the fugacity of a hypothetical subcooled liquid-phase approach, $f_1^{SCL}(T,P)$, is used as the reference of the solid-phase fugacity:

$$\hat{f}_1^s(T,P) = f_1^{SCL}(T,P) \cdot \exp\left\{\frac{\left(V_1^s - V_1^{SCL}\right)[P - P_{sub,1}(T)]}{RT} + \left[\frac{\Delta_{fus}h_1}{RT_{fus,1}}\left(1 - \frac{T_{fus,1}}{T}\right)\right]\right\} \quad (7.13)$$

where $\Delta_{fus}h_1, T_{fus,1}, V_1^s$, and $P_{sub,1}$ are the heat of fusion, the normal melting point temperature, the molar volume, and the sublimation pressure of the pure solid, respectively, while $V_1^{SCL}$ is the molar liquid volume of the hypothetical subcooled liquid. The fugacity of the solid in a hypothetical subcooled liquid phase, $f_1^{SCL}(T,P)$, is calculated by a cubic EoS or an EoS/G$^E$ model.

Bertakis et al. (2007) demonstrated that the UMR-PRU model coupled with the subcooled liquid reference stat allows reliable SLG equilibrium predictions.

## 7.6 SCF PROCESSING SCHEMES

### 7.6.1 Supercritical Fluid Extraction

SFE is one of the most widespread applications of SCFs, which is mainly applied in the food industry and also in the pharmaceutical and cosmetic industries. On the basis of the specific properties of SCFs, SFE process presents some important characteristic features, which are summarized below:

- Operating temperatures are close to the solvent critical temperature. Therefore, high-boiling, heat-sensitive components may be extracted at relatively low temperatures. So, SFE is suitable for processing thermolabile substances such as essential oils, antioxidants, and generally bioactive compounds, whose recovery by conventional methods is difficult or impossible because they cause decomposition, denaturation, or loss of the active properties of the compounds. Moreover, the low temperature involved leads to low-energy consumption as compared, for example, with the high energy demanding distillation.
- The selectivity and capacity of the process may be changed by varying the operating conditions, namely temperature and pressure, as well as the choice of solvent and/or the use of cosolvent.
- The recovery of the solute is accomplished in a simple depressurization/precipitation step, which is a distinct advantage over liquid extraction.
- Solute fractionation is possible.
- Solvents such as $CO_2$ are cheap, abundant, nontoxic, noncorrosive, nonflammable, and avoid environmental problems.

The characteristic advantages and disadvantages of SFE as compared to the classical separation methods of hydrodistillation and liquid extraction, usually used in the food industry, are summarized in Table 7.3. The table explains both the areas of application of SFE and the reasons for its limited commercial distribution.

A schematic diagram of the basic SFE process is shown in Figure 7.8. The substance chosen as the solvent is condensed from its gaseous state into a liquid, and then pressurized with a pump above its critical pressure threshold so that it is actually a high-pressure liquid. Finally, it is heated above the critical temperature to become an SC. The SCF then flows into an extractor containing the solute, where the desirable extraction occurs. The output stream from the extractor is expanded in a separator via a valve, where the pressure drop causes a solute solubility decrease, and hence, the solute is precipitated from the extract. Finally, the solvent is recirculated and the whole process is repeated until completion of the extraction.

The SFE process mainly using $CO_2$ has found various applications in the food industry starting from the first commercial application of the process in 1978 for the decaffeination of green coffee beans and 2 years later for the extraction of hop flavors. After that, SFE has been used in many fields of the food industry: extraction of cholesterol and other lipids from egg yolk, milk fat fractionation, extraction of lipids and cholesterol from meat and meat products, extraction of lipids and cholesterol from fish, extraction of natural colorings from foodstuffs, extraction–refining and fractionation of oils and vegetable fats, extraction and fractionation of natural

**TABLE 7.3**
**Pros and Cons Comparison of Hydrodistillation, Liquid Extraction, and SFE**

| Hydrodistillation | | Liquid Extraction | | SFE | |
|---|---|---|---|---|---|
| + | − | + | − | + | − |
| Low capital cost | High operating temperatures | Middle capital cost | Additional treatment for solvent recovery | Low operating temperatures | High operating pressures |
| Product purity | High energy consumption | Middle energy consumption | Presence of solvent in the final product | Low-energy consumption | High capital cost |
| | Thermal/chemical product deterioration | | Thermal/chemical extract deterioration | No thermal/chemical extract deterioration | |
| | | | High-energy consumption for final product purification | Extra pure product (solvent free) | |
| | | | | Extract fractionation | |

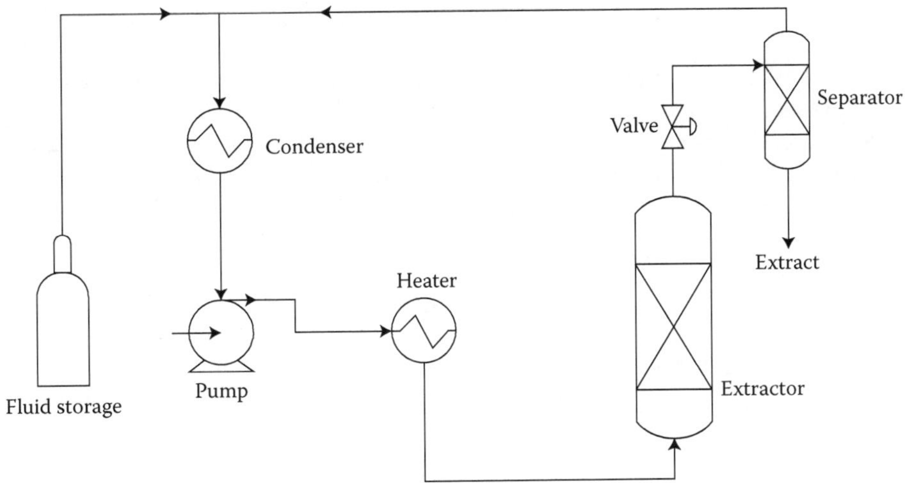


**FIGURE 7.8** Schematic diagram of a basic SFE process.

flavorings, and extraction of antioxidants. A review of SFE applications in food-related industries is given by Raventós et al. (2002).

SFE of solid raw materials is generally operated in batch or a semicontinuous mode of milled materials, pellets, or granulates. The equipment is composed of one or several extractors consisting of high-pressure autoclave(s) with fast-opening systems in which the material to be treated is not usually loaded directly in the extractor but inside a "basket" easy to put in and to remove from the extractor, and a separation section where extract and fluid are separated prior to fluid recycle. Some extractors are designed to directly load the raw material while the spent material is removed by a vacuum system. The operation of batch mode has strong limitations of the process, as it induces both investment costs for autoclave closing/opening systems and operating expenses for manpower and replacement of fluid losses.

In practice, more than one extractor and separator should be used. The use of more than one extractor aims to (a) increase the capacity as the plant can work during the time needed for unloading/loading one extractor, and (b) decrease of the fluid losses as it is possible to depressurize one extractor at the end of the extraction into another one (just loaded with raw material). The use of more than one separator allows the fractionation of the extract and, thus, the collection of products of different composition, with a simple change in the operational pressure and/or temperature.

The use of a continuous process rather than a batch or semibatch process may significantly improve the economics of an industrial-scale SFE process. Continuous filling and removal of large volumes of raw material from a high-pressure autoclave is very difficult and costly. However, advances in high-pressure technology may allow continuous feed of solid materials. An example is a plant of the Maxwell House Division of General Foods, in Houston, Texas, where coffee is decaffeinated by SC carbon dioxide using an almost continuous process. Since coffee is not a very valuable commodity, the only way the process can be made viable is by economies of scale and, therefore, the plant is very large with the extractor to be more than 60 $m^3$

in volume. The extractor has two smaller vessels above and below it, connected to it by large bore valves.

It should be pointed out that—as discussed in Section 7.4—apart from the SCF, it is possible to use a cosolvent as well, to increase the solubility of a substance and/or to control the selectivity of the process. This is a key feature of the method, which makes it flexible and broadens the range of its application.

### 7.6.2 Supercritical Fluid Fractionation

Fractionation of liquid mixtures with SC carbon dioxide in countercurrent columns is a process of the same family of extraction. This process generally operates in continuous mode with carbon dioxide as the solvent, possibly with a cosolvent added to increase its polarity. This represents a big advantage over extraction from solids, which is usually carried out in batch or in a semicontinuous mode. Fractionation with high-pressure carbon dioxide may be ideal for difficult separations in liquid mixtures. It has been applied for various separations of valuable substances from natural products, such as vegetable food oils, fish oils, wine and beer, and so on (Brunner, 2004; Reverchon and De Marco, 2006).

A typical process flow sheet that shows the general operation of SFF is presented in Figure 7.9. The feed is continuously fed to the fractionation column filled with a packing for ensuring a better contact between the two phases; the SCF is introduced through a sintered disk or perforated plate at the bottom of the column. Extract reflux must be applied when a high selectivity of the process is required. The extract exits at the top of the column and is partly depressurized and reheated to cause extract precipitation and collection through gravity or cyclonic separators. Extracts can be continuously recovered through a series of pressure locks working alternatively, avoiding fluid losses and pressure variation upward in the column. The liquid raffinate exits at the bottom of the column through a valve controlled by a level gauge located at a place chosen by the operator to settle the liquid–fluid interface (generally in the bottom head of the column). To avoid significant fluctuation of pressure inside the column, raffinate can be depressurized using a series of two or three settlers maintained at decreasing pressures, where part of the entrained fluid dissolved in the raffinate is being recovered and recycled. The fluid is recycled according to the classical way. Finally, in certain cases where part of the extract is a very volatile compound or mixture, it is necessary to trap these light ends by adsorption on an adsorbent bed, for example, activated carbon, swept by the compressed gas exiting the separators. The adsorbate can be later recovered by SFE. When operating two parallel adsorbent beds, this can be conducted periodically without stopping the main SFF operation.

### 7.6.3 Particle Design with SCFs

Except for the most common applications of SCFs in extraction and/or fractionation, particle formation processes using SCFs are now subjected to an increasing interest especially in the pharmaceutical and food industries. These processes mainly aim to increase bioavailability of poorly soluble molecules and to design sustained-release

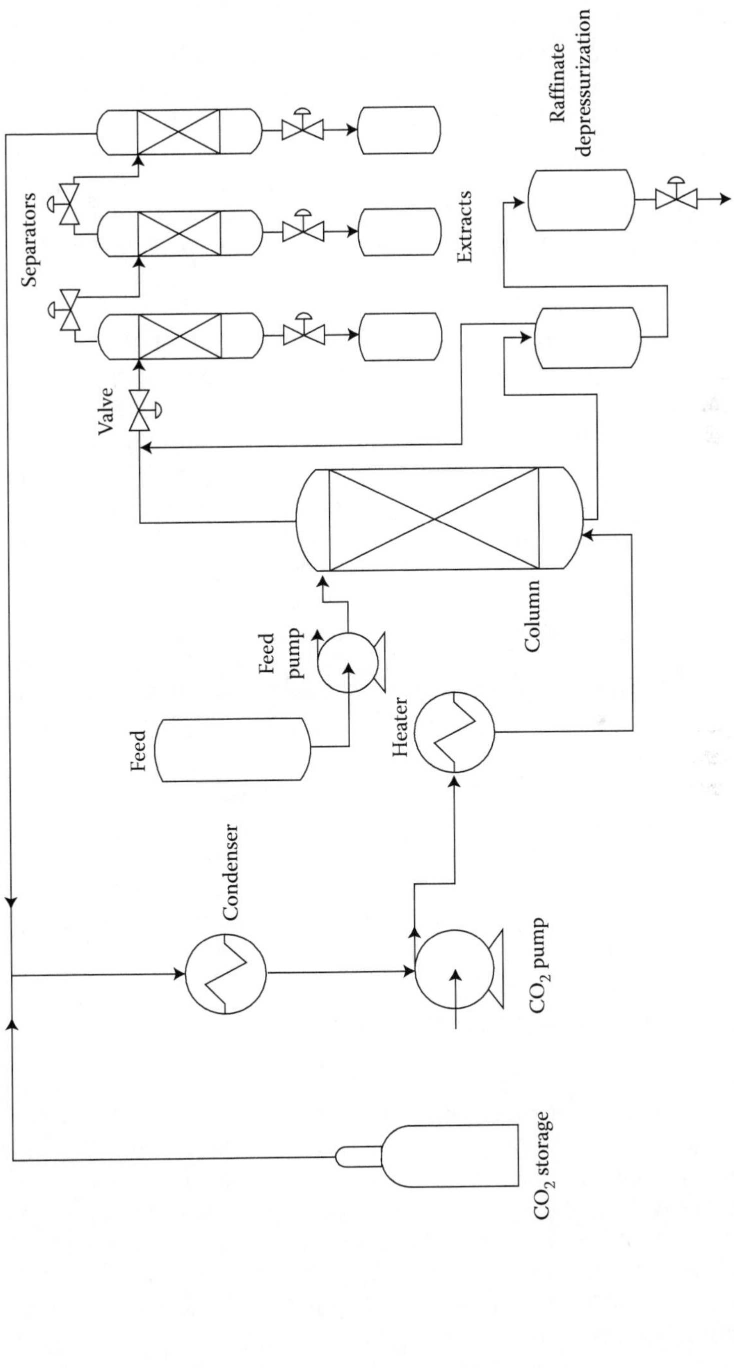


**FIGURE 7.9** Schematic diagram of a basic SFF process.

formulations. Several techniques have been proposed and investigated to produce particles from SCFs (Jung et al., 2001). Generally, the three following basic techniques are distinguished.

#### 7.6.3.1 Rapid Expansion of Supercritical Solutions

In the RESS technique, the substance is first dissolved in a SC solvent, after which the pressure is rapidly decreased in a specifically designed nozzle, resulting in supersaturation of the substance in the SC solvent and leading to the formation of small particles with a narrow particle size distribution. The RESS process could find valuable applications at a commercial scale when only product solubility in SC $CO_2$ is not too small at reasonable pressures, limiting the process application to nonpolar or low-polarity compounds (Tom and Debenedetti, 1991; Tom et al., 1993).

#### 7.6.3.2 Supercritical Antisolvent

In the SAS technique, the solid is dissolved in a conventional solvent and the solution is sprayed continuously through a nozzle into the SCF. The dispersion of solution in the SCF leads to an expansion of the droplets and at the same time, an extraction of the liquid into the fluid occurs. The solvent power of the conventional solvent decreases dramatically and supersaturation leads to the precipitation of particles. SAS has been successfully used to produce fine particles from polymers, proteins, superconductors, pigments, and pharmaceutical compounds. Moreover, a great variety of particle morphology can be obtained depending on operating parameters, nozzle shape, and material properties, leading to attractive "engineered" structures (Dixon et al., 1993; Reverchon, 1997; Reverchon and Perrut, 2000).

#### 7.6.3.3 Particle Generation from Supercritical Solutions or Suspensions

This process allows the formation of particles from substances that are insoluble in an SCF, but absorb a large amount of gas that either swells the substance or decreases its melting point (for polymers the glass transition temperature). In the PGSS technique, a saturated liquid solution or suspension of high-boiling substances with compressed gases is rapidly expanded at ambient pressure. Owing to the Joule–Thomson effect and/or the evaporation and volume expansion of the gas, the temperature decreases below the melting point temperature of the solute and fine particles are formed (Weidner et al., 1994).

## 7.7 MODELING OF THE SFE PROCESS

For the accurate design of the SFE process, special attention should be paid to determine optimal process conditions, such as pressure and temperature, use of cosolvent, solvent flow rate, mean particle diameter of the feedstock, and so on. For this purpose, at least some experimental data along with reliable mathematical models are needed.

The availability of flexible and accurate tools for the mathematical modeling of the SFE, are of great importance for the design and scale-up of the whole process. Such models have been proposed in the literature; here, we focus on mass balance models since they involve mass transfer coefficients, which are necessary for scale-up calculations.

Mass balance models are based on the following hypotheses: (i) extract is "a single compound"; (ii) the axial dispersion is negligible; (iii) the temperature, pressure, solvent density, and flow rate are constant along the SFE extractor; (iv) the solvent is solute free at the entrance to the extractor; and (v) the solid bed is homogeneous. According to the above factors, the following mass balance equations are derived:

$$\rho\varepsilon\left(\frac{\partial y}{\partial t} + u\frac{\partial y}{\partial z}\right) = J(x,y) \tag{7.14}$$

$$\rho_S(1-\varepsilon)\frac{\partial x}{\partial t} = -J(x,y) \tag{7.15}$$

where $\rho$ is the solvent density (kg/m$^3$), $\varepsilon$ is the void fraction on the bed, $y$ is the solute concentration in the fluid phase (kg solute/kg $CO_2$), $x$ is the solute concentration in the solid phase (kg solute/kg solute-free feed), $u$ is the interstitial velocity (solvent velocity/volume fraction of the fluid), $\rho_s$ is the solid density (kg/m$^3$), and $J$ is the solute exchange rate between the phases.

On the basis of these basic equations, different models with specific assumptions and different numbers of adjustable parameters, which must be determined by fitting laboratory SFE experimental data, can be derived. Some of them are presented below:

*Model I*

This is a simplified expression proposed by Reverchon and Sesti Ossèo (1994), which neglects the accumulation ($\partial y/\partial t$) of the solute in the fluid phase and assumes a uniform extraction along the bed ($\partial y/\partial z$ = constant), coupled with a linear equilibrium relationship. Its mathematical expression is as follows:

$$e = x_o[1-\exp(Ct)] \tag{7.16}$$

with $C = A\,K\,(1-B)$, $A = \alpha_0\,K_f\,\rho\,[\rho_s\,(1-\varepsilon)]^{-1}$, and $B = A/(\dot{q} + A)$

where $x_0$ is the initial total concentration of the solute in the solid (kg solute/kg solute-free feed), $\dot{q}$ is the specific mass flow rate of the solvent (kg solvent/s/kg solute-free feed), $\alpha_0$ is the specific interfacial area (m$^2$/m$^3$), $K_f$ is the overall mass transfer coefficient in the fluid ($f$) phase (m/s), and $K$ is the equilibrium constant, which was assumed to be equal to ($y_0/x_0$), where $y_0 = y_{t=0}$.

The independent adjustable parameters of the model are $x_0$, $y_0$, and $A$.

*Model II*

This model (Papamichail et al., 2000) differs from *Model I* in the equilibrium expression. Thus, it employs the equilibrium expression proposed by Perrut et al. (1997):

$$e = \begin{cases} y_0 \cdot A \cdot (1-B) \cdot t & x \geq \overline{x} \quad \text{or} \quad t \leq \overline{t} \\ x_0 - \overline{x}\exp[-A \cdot K \cdot (1-B) \cdot (t-\overline{t})] & x < \overline{x} \quad \text{or} \quad t > \overline{t} \end{cases} \tag{7.17}$$

where $\bar{x}$ is the solute concentration in the solid controlling the transition from the solubility-controlled regime to the diffusion-controlled one of the extraction, and $\bar{t} = (x_0 - \bar{x}) / [y_0 \cdot A \cdot (1 - B)]$ is the time that corresponds to the transition from one regime to another. In this case, the equilibrium constant $K$ was fitted to the experimental data and was not calculated using the assumption made in *Model I*.

As compared to *Model I*, two more parameters are required in this model: $\bar{x}$ and $K$.

*Model III*

This is an extended Lack's plug flow model proposed by Sovová (1994) and Sovová et al. (1994):

$$e = \begin{cases} qy_r[1 - \exp(-Z)] & , q < q_m \quad \text{(I)} \\ y_r[q - q_m \exp(z_w - Z)] & , q_m \le q \le q_n \quad \text{(II)} \\ x_0 - \dfrac{y_r}{W} \ln\{1 + [\exp(Wx_0 / y_r) - 1]\exp[W(q_m - q)]x_k/x_0\} & , q \ge q_n \quad \text{(III)} \end{cases} \tag{7.18}$$

with

$$q_m = (x_0 - x_k)/y_r Z \quad q_n = q_m + \frac{1}{W} \ln \frac{x_k + (x_0 - x_0)\exp(Wx_0/y_r)}{x_0}$$

$$\frac{z_w}{Z} = \frac{y_r}{Wx_0} \ln \frac{x_0 \exp[W(q - q_m)] - x_k}{x_0 - x_k}$$

$$Z = k_f a_0 \rho / [\dot{q}(1 - \varepsilon)\rho_s] \quad W = k_s a_0 / [\dot{q}(1 - \varepsilon)]$$

where $q$ is the specific amount of solvent passed through the extractor (kg solvent/kg solute-free feed), $y_r$ is the solubility of the solute in the solvent (kg solute/kg solvent), $Z$ is the dimensionless mass transfer parameter in the fluid phase, $q_m$ is the $q$ value when extraction begins inside the particles, $z_w$ is the dimensionless axial coordinate between fast and slow extraction, $q_n$ is the $q$ value when the easily accessible part of solute is all extracted, $W$ is the dimensionless mass transfer parameter in the solid phase, $x_k$ is the initial concentration of the difficult accessible solute in the solid (kg solute/kg solute-free feed), $k_f$ is the solvent-phase mass transfer coefficient (m/s), and $k_s$ is the solid-phase mass transfer coefficient (m/s).

Five adjustable independent parameters are involved in this model: $x_0$, $x_k$, $y_r$, $Z$, and $W$.

*Model IV*

This model employs the same assumptions with *Model III*, but it takes into account the accumulation term in the differential mass balance equations. The system of the differential equations (7.14) and (7.15), coupled with the expressions for $J(x, y)$ proposed by Sovová et al. (1994):

$$J(x > x_k, y) = k_f a_0 \rho (y_r - y) \text{ and } J(x < x_k, y) = k_s a_0 \rho_s x(1 - y/y_r) \tag{7.19}$$

is solved numerically by employing a fourth-order Runge–Kutta method and assuming that the solid bed is divided into n stages, where in each of them, the concentration is considered to be uniform. The fitted parameters of this model are the same with those of *Model III*.

*Model V*

The main assumptions involved in this model (Perakis et al., 2010) are the following: (i) the solid bed is homogeneous with respect to the particle size and the initial distribution of the solute, (ii) the temperature, pressure, solvent density, and flow rate are constant along the bed, (iii) the axial dispersion is negligible, and (iv) the mass transfer rate is controlled by the phase equilibrium and the solute diffusion in the solid. Also, the model takes into account the accumulation of the solute in the fluid phase ($\partial y/\partial t \neq 0$) and that the extraction is not uniform along the bed ($\partial y/\partial h \neq$ constant).

The basic model equations are the following:

$$\rho_f \varepsilon \frac{\partial y}{\partial t} = -U\rho_f \frac{\partial y}{\partial h} + \rho_s k_s \alpha_o \left( x - \frac{y}{K_{eq}} \right) \tag{7.20}$$

$$(1-\varepsilon)\frac{\partial x}{\partial t} = -k_s \alpha_o \left( x - \frac{y}{K_{eq}} \right) \tag{7.21}$$

where $\rho_f$ is the solvent density (kg/m$^3$), $\varepsilon$ is the void fraction on the bed, $y$ is the solute concentration in the fluid phase (kg solute/kg $CO_2$), $x$ is the solute concentration in the solid phase (kg solute/kg solute-free feed), $U$ is the superficial fluid velocity (m/s), $t$ is the extraction time (s), $\rho_s$ is the apparent solid density (kg/m$^3$), $h$ is the axial coordinate (m), $k_s$ is the mass transfer coefficient in the solid phase (m/s), and $\alpha_o$ is the specific interfacial area (m$^2$/m$^3$). Also, $K_{eq}$ is the partition coefficient of the solute between the fluid and the solid phase, and is defined as

$$K_{eq} = y_i / x_i \tag{7.22}$$

where $y_i$ and $x_i$ are the solute concentrations in the fluid and solid phase, respectively at their interface.

The initial and boundary conditions are

$$\begin{cases} y(t=0,h) = \dfrac{x_o K_{eq}}{1+BK_{eq}} \\ x(t=0,h) = \dfrac{x_o}{1+BK_{eq}} \\ B = \dfrac{\rho_f \varepsilon}{\rho_s (1-\varepsilon)} \end{cases} \tag{7.23}$$

$$y(t, h=0) = 0 \tag{7.24}$$

where $x_o$ is the initial total concentration of the extractable solute in the solid (kg solute/kg solute-free feed). As shown by Equation 7.23, it is assumed that a part of the solute has been already dissolved in the SC $CO_2$ at the beginning of the extraction. Actually, it is considered that the loading time was long enough so as to enable the fluid to reach equilibrium concentration before the extractions start. This assumption is justified by the favorable transport properties of SCFs that lead to high mass transfer rates, and has also been employed by other researchers (Reverchon and Marrone, 2001; Reverchon et al., 1999; Sovová, 2005).

Three adjustable independent parameters are involved in this model: $K_{eq}$, $k_s$, and $x_o$.

It should be noted that in all these models, the initial total concentration of the soluble components in the feedstock, $x_0$, can be determined experimentally by exhaustive extraction of the feedstock, thus, reducing the number of adjustable parameters (Louli et al., 2004).

Examples of experimentally determined pressure and temperature effect on the extraction of oil from dittany are shown in Figures 7.10 and 7.11, along with Model V results.

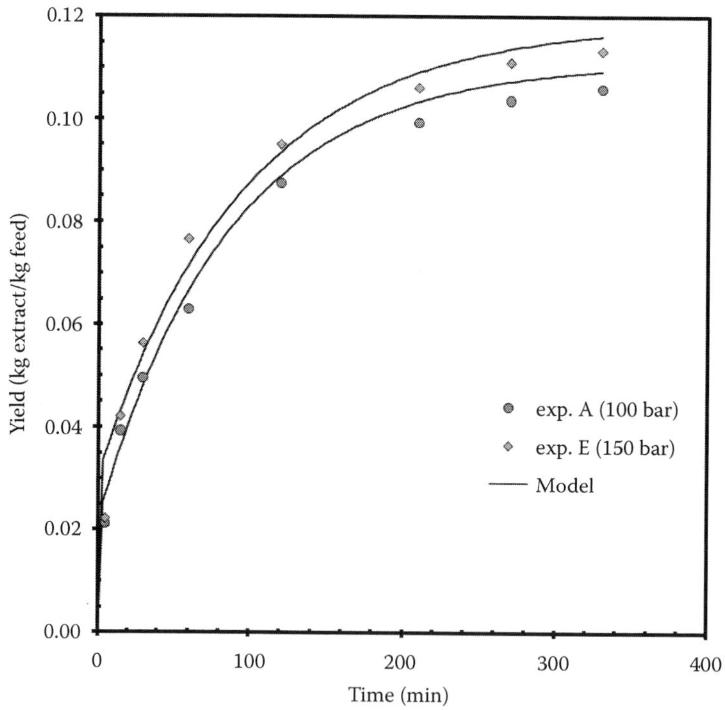


**FIGURE 7.10** Effect of pressure (all other parameters are constant) on the extraction yield of oil from dittanty (Origanum Dictamnus) using supercritical carbon dioxide versus time. Modeling results are also presented. (Reprinted from *J. Supercrit. Fluids*, 55, Ch. Perakis et al., Supercritical $CO_2$ extraction of dittany oil: Experiments and modelling, 576, Copyright (2010), with permission from Elsevier.)

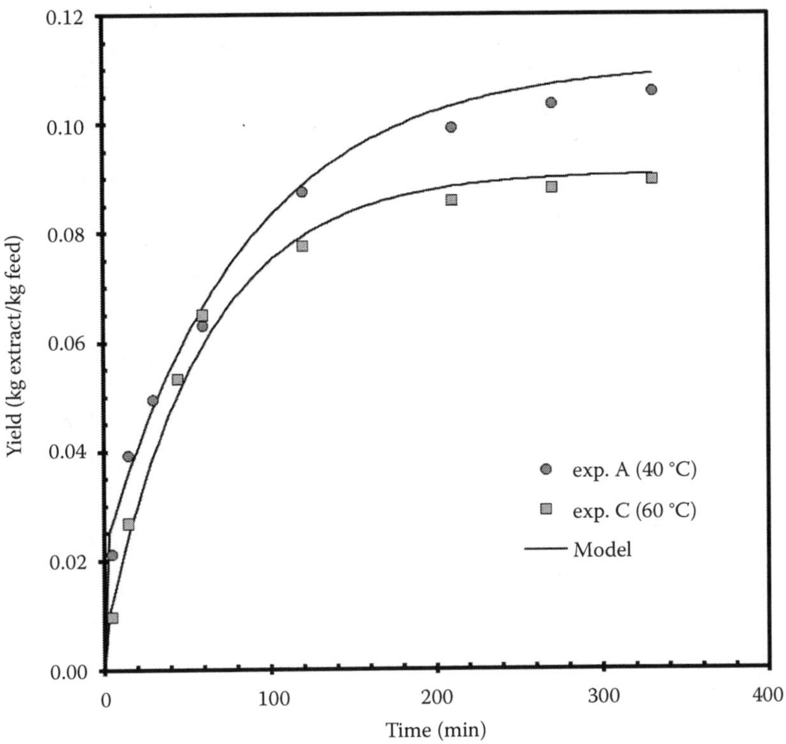


**FIGURE 7.11** Effect of temperature (all other parameters are constant) on the extraction yield of oil from dittany (Origanum Dictamnus) using supercritical carbon dioxide versus time. Modeling results are also presented. (Reprinted from *J. Supercrit. Fluids*, 55, Ch. Perakis et al., Supercritical $CO_2$ extraction of dittany oil: Experiments and modelling, 576, Copyright (2010), with permission from Elsevier.)

## 7.8 SAFETY AND SCALE-UP ISSUES

SCF technology maybe potentially hazardous and should not be used without proper safety precautions that must be taken into account at every stage of process design and operation, that is, in equipment design, building and installation, operation, inspection, and maintenance. Clavier and Perrut (1996) and Leboeuf and Deschamps (2010) have reported the hazards of high-pressure SCF extraction plants.

The main hazards associated with SCF applications are mechanical and chemical hazards. Industrial production on high-pressure equipment requires high reliability with drastic safety requirements as mechanical hazards, such as metal fatigue or brittle fracture, which must be eliminated. This requires a preventative maintenance, as many parts must be inspected and changed periodically. Moreover, a rigorous operation plan must be enforced to eliminate any risk of deterioration of the basic parts, and safety sensors must be continuously logged. Maintenance mainly concerns the high-pressure pumps, autoclaves, and baskets to avoid solvent bypass or sintered disks to detect deformation prior to rupture due to plugging.

Pressure vessels must be inspected and submitted to pressure tests according to official standards. Moreover, the main process valves must be often checked, as they are the key of safe operation during autoclave opening for raw material change. Sensors must be recalibrated periodically, in comparison with traceable reference sensors, and data logging validated.

Chemical hazards associated with SCF applications are particularly important when using flammable SCFs or cosolvents, which are used to increase SCF polarity and can be flammable or environmentally hazardous.

A very important remark concerning safety issues is that operators must be trained by specialists prior to work on SCF equipment. Furthermore, a detailed information exchange between the equipment supplier and user should be the key for a reliable and safe operation, both for laboratory- or pilot-scale equipment and for large-scale units as well.

As already mentioned, SFE and SFF processes have proven to be technically and economically feasible, and they have been implemented at a large industrial scale for more than 30 years (Brunner, 1994). When studying scale-up criteria for SFE and SFF, it is important to establish a methodology that allows predicting the behavior of the process at an industrial scale from laboratory or pilot unit data, considering the differences encountered in processes conducted in the equipment of significantly different sizes.

For SFE, scale-up criteria described in literature (Batista and Meireles, 2012; Leboeuf and Deschamps, 2010; Mezzomo et al., 2009; Moura et al., 2005) include kinetic parameters such as solvent residence time and superficial velocity, empirical equations based on the bed's geometry, and the use of mathematical models such as those presented in Section 7.7. When using scale-up criteria, it is necessary to assess their applicability to different types of raw materials, since the mass transfer mechanisms may differ among species and parts of the raw material used for extraction.

SFF industrial units are usually operated continuously in countercurrent columns. The methods used to design these types of columns are not essentially different from those for a standard liquid–liquid extraction unit, where the concept of using the theoretical or equilibrium stage approach is well known (Treybal, 1981; Perry, 1997). The number of theoretical stages necessary to perform a certain separation can be calculated from phase equilibrium data alone. Although mass transfer and capacity of the column are also to be considered to obtain the actual size of a separation device, knowledge of relevant phase equilibrium ratios is a prerequisite in a design calculation.

## 7.9 OTHER SCF APPLICATIONS

From the application of SCFs in extraction, fractionation, or particle formation processes mentioned above, the use of SCFs has also been examined in various fields. Among others, briefly, the following applications are reported: as reaction media for the hydrogenation of fats and oils (Brunner, 2004; King et al., 2001); for polymerization (Kazarian, 2000; Kemmere, 2005) and in enzymatic reactions (Halling, 1994; Voutsas et al., 2001); in preparative-scale SCF chromatography (Taylor and King, 2002; Rajendran, 2012); for paints and coatings (Brunner, 2010); for food preservation (Rawson et al., 2012); for ceramics (Bordet et al., 2002); for metal extraction and soil decontamination (Parka et al., 2013; Wai and Wang, 1997; Wang and Chiu, 2008); in nanotechnology (Brunner, 2010; Meziani et al., 2008); and for enhanced oil recovery (Baklid et al., 1996).

## 7.10 CONCLUDING REMARKS

SCF technology is an innovative and promising way for extraction and fractionation as well as for particle design. The main motivation for the application of SCFs, and especially $CO_2$, as an alternative to conventional extraction processes, is the possibility of exploiting the peculiar properties of SCFs, which are often described as intermediate between those of a liquid and a gas, and easily changeable with small modifications in operating conditions, namely pressure and temperature. Also, a significant increase in the solubilizing capacity of $CO_2$ can be achieved with the use of small amounts of appropriate cosolvents. Owing to these properties, SCFs are increasingly replacing organic solvents used in industrial purification and recrystallization operations because of increasingly stringent regulations regarding residual organic solvents in processed food products.

On the other hand, although extensive investigations have been carried out worldwide for more than three decades, SCF applications have still been rather limited. Although the efficacy and versatility of SFE and SFF are well recognized, there is a common apprehension in terms of capital investment costs, safety, and energy requirements. Probably, too optimistic forecasts had promised very attractive solutions, which, however, were far from economic and technical feasibility. SCF technology is comparatively more capital intensive than traditional separation technologies, due to the requirements for high-pressure operation and very accurate process control. The challenge here lies in proper process analysis and design, making it possible to optimize the capital investment cost and manufacture economically viable SCF plants. Concerning safety issues, accurate control over the whole process achieved with state-of-the-art instrumentation and control systems ensures high consistency, safety, and reliability in the performance of SCF plants. Finally, as for the energy requirements, the energy needed to attain the SC state is more than compensated by the negligible energy required for solvent recovery from the extract, which is a simple depressurization/precipitation step. Moreover, the relatively low temperatures involved in SCF applications lead to low-energy consumption as compared with high energy demanding separation methods such as distillation.

As a closing remark, I presume that the statement made in 1988 at the *First International Symposium on Supercritical Fluid* by the pioneer in the area of SCFs Dr. Val Krukonis "There is no point in doing something in a supercritical fluid just because it's neat. Using the fluids must have some real advantage," is still up to date.

## ACKNOWLEDGMENT

Dr. Vassiliki Louli and Professor Kostis Magoulas are gratefully acknowledged for fruitful discussions and/or advice and feedback.

## REFERENCES

Andrews, T. 1875. The Bakerian lecture: On the gaseous state of matter. *Proc. Roy. Soc. London*, 24:455.

Baklid, A., Korbøl, R., Owren, G.A. 1996. Sleipner Vest $CO_2$ disposal, $CO_2$ injection into a shallow underground aquifer. SPE no. 36 600, *SPE Annual Technical Conference*, Denver, USA.

Bartle K.D., Clifford A.A., Jafar S.A., Shilstone, G.F. 1991. Solubilities of solids and liquids of low volatility in supercritical carbon dioxide. *J. Phys. Chem. Ref. Data.* 20:713.

Batista, E.A.C., Meireles, M.A.A. 2012. Supercritical fluid extraction of grape seed: Process scale-up, extract chemical composition and economic evaluation. *J. Food Eng.* 109:249.

Bellisent-Funel, M. 2001. Structure of supercritical water. *J. Mol. Liq.* 90:313.

Bertakis, E., Lemonis, I., Katsoufis, S., Voutsas, E., Dohrn, R., Magoulas, K., Tassios, D. 2007. Measurement and thermodynamic modeling of solid–liquid–gas equilibrium of some organic compounds in the presence of $CO_2$. *J. Supercrit. Fluids* 41(2):238.

Bordet, F., Chartier, T., Baumard, J.-F. 2002. The use of co-solvents in supercritical debinding of ceramics. *J. Eur. Ceram. Soc.* 22:1067.

Brunner, G. 1994. *Gas Extraction*. Steinkopff, Darmstadt.

Brunner, G. 2004. *Supercritical Fluids as Solvents and Reaction Media*. Elsevier, Amsterdam, the Netherlands.

Brunner, G. 2010. Applications of supercritical fluids. *Annu. Rev. Chem. Biomol. Eng.* 1:321.

Chordia, L., Martinez, J.L. 2002. What's so hot about supercritical fluids? *Lab. Focus,* 6(1).

Christov, M., Dohrn, R. 2002. High pressure fluid phase equilibria: Experimental methods and systems investigated (1994–1999). *Fluid Phase Equil.* 202:153.

Clark, B.J., Mailer, R. 1983. How Carlton and United made hop extract history. *Brewer's Guardian*, 10:13–17.

Clavier, J., Perrut, M. 1996. Safety in supercritical fluid operations. In: *High Pressure Chemical Engineering*, eds., P.R. Rohr and C. Trepp, pp. 627–631. Amsterdam, The Netherlands, Elsevier.

Clavier, J.Y., Nicoud, R.M., Perrut, M. 1996. High pressure chemical engineering. In: *A New Efficient Fractionation Process: The Simulated Moving Bed with Supercritical Eluent*, eds., Ph. R. von Rohr and Ch. Trepp, pp. 429–434. Elsevier Science, London.

Coutsikos Ph., Magoulas K., Tassios, D. 1995. Solubilities of phenols in supercritical carbon dioxide. *J. Chem. Eng. Data*, 40:953.

Dahl, S., Fredenslund, Aa., Rasmussen, P. 1991. The MHV2 model: A UNIFAC-based equation of state model for prediction of gas solubility and vapor–liquid equilibria at low and high pressures. *Ind. Eng. Chem. Res.* 30:1936.

Dixon, D.J., Johnston, K.P., Bodmeier, R.A. 1993. Polymeric materials formed by precipitation with a compressed fluid antisolvent. *AIChE J.* 39:127.

Dohrn R., Brunner G. 1995. High pressure fluid phase equilibria: Experimental methods and systems investigated (1988–1993). *Fluid Phase Equil.* 106:213.

Dohrn, R., Peper, S., Fonseca, J.M.S. 2010. High pressure fluid phase equilibria: Experimental methods and systems investigated (2002–2004). *Fluid Phase Equil.* 288:1.

Fonseca, J.M.S., Dohrn, R., Peper, S. 2011. High pressure fluid phase equilibria: Experimental methods and systems investigated (2005–2008). *Fluid Phase Equil.* 300:1.

Fukushima, Y. 1999. Application of supercritical fluids. *R&D Rev. Toyota CRDL.* 35:1.

Haake, M., Goodson, B.M., Laws, D.D., Brunner, E., Cyrier, M., Havlin, R., Pines, A. 1998. NMR of supercritical laser-polarized xenon. *Chem. Phys. Lett.* 292:686.

Halling, P. 1994. Thermodynamic predictions for biocatalysis in nonconventional media: Theory, tests, and recommendations for experimental design and analysis. *Enzyme Microb. Technol.* 16:178.

Hannay, J.B. 1880. On the solubility of solids in gases II. *Proc. Roy. Soc. London.* 30:484.
Hannay, J.B., Hogarth, J. 1879. On the solubility of solids in gases. *Proc. Roy. Soc. London.* 29:324.
Francis, A.W. 1954. Ternary systems of liquid carbon dioxide. *J. Phys. Chem.* 58:1099.
Holderbaum, T., Gmehling, J. 1991. PSRK: A group-contribution equation of state based on UNIFAC. *Fluid Phase Equil.* 141:251.
Howdle, S.M., Jobling, M., Poliakoff, M. 1992. Spectroscopic investigations of organometallic photochemistry in supercritical fluids. *ACS Sympo. Ser.* 488:121.
Jung, J., Perrut, M. 2001. Particle design using supercritical fluids. Literature and patent survey. *J. Supercrit. Fluids,* 20:179.
Kander, R.G., Paulaitis, M.E., 1983. The adsorption of phenol from dense carbon dioxide onto activated carbon. In: *Chemical Engineering at Supercritical Fluid Conditions*, eds., M.E. Paulaitis, J.M.L. Penninger, R.D. Gray, and P. Davinson, Ann Arbor Science, Ann Arbor, Michigan.
Kazarian, S.G. 2000. Polymer processing with supercritical fluids. *Polym. Sci. Ser. C.* 42(1):78.
Kemmere, M.F. 2005. Recent developments in polymer processes. In: *Handbook for Polymer Reaction Engineering*, eds., Th. Meyer and J.T.F. Keurentjes, Wiley-VCH, Weinheim.
King, M., Bott, T. 1993. *Extraction of Natural Products Using Near Critical Solvents*. Blackie Academic, Glasgow.
King, J.W., Holliday, R.L., List, G.R., Snyder, J.M. 2001. Hydrogenation of vegetable oils using mixtures of supercritical carbon dioxide and hydrogen. *J. Am. Oil Chem. Soc.* 78(2):107.
Kurnik, R.T., Holla, S.J., Reid, R.C. 1981. Solubility of solids in supercritical carbon dioxide and ethylene. *J. Chem. Eng. Data,* 26:47.
Lack, E., Seidlitz, H. 1993. Commercial scale decaffeination of coffee and tea using supercritical carbon dioxide. In: *Extraction of Natural Products Using Near-Critical Carbon Dioxide*, eds., M.B. King and T.R. Bott, pp. 101–139. Blackie Academic, Glasgow.
Lang, Q., Wai, C.M. 2001. Supercritical fluid extraction in herbal and natural product studies—A practical review. *Talanta,* 53:771.
Leboeuf, F., Deschamps, F. 2010. Supercritical fluid processing in food and pharmaceutical industries: Scale-up issues. In: *Current Trends of Supercritical Fluid Technology in Pharmaceutical, Nutraceutical and Food Processing Industries*, eds., A.R.C. Duarte and C.M.M. Duarte, pp. 97–115. Bentham e-Books, the Netherlands.
Lemert, R.M., Johnston, K.P. 1991. Chemical complexing agents for enhanced solubilities in supercritical fluid carbon dioxide. *Ind. Eng. Chem. Res.* 30:1222.
Louli, V., Folas, G., Voutsas, E., Magoulas, K. 2004. Extraction of parsley seed oil by supercritical $CO_2$. *J. Supercrit. Fluids,* 30:163.
Louli, V., Boukouvalas, C., Voutsas, E., Magoulas, K., Tassios, D. 2007. Application of the UMR-PRU model to multicomponent systems: Prediction of the phase behavior of synthetic natural gas and oil systems. *Fluid Phase Equil.* 261:351.
McHugh, M.A., Krukonis, V.J. 1986. *Supercritical Fluid Extraction.* Butterworths, USA.
McHugh, M.A., Watkins, J.J., Doyle, P.T., Krukonis, V.J. 1988. High-pressure naphthalene–xenon phase behavior. *Ind. Eng. Chem. Res.* 27:1025.
McHugh, M.A., Krukonis, V.J. 1994. *Supercritical Fluid Extraction Principles and Practice,* 2nd edition. Butterworth-Heinemann, Boston, MA.
Meziani, M.J.m., Pathak, P., Sun, Y.-P. 2008. Supercritical fluid technology for nanotechnology in drug delivery. In: *Nanotechnology in Drug Delivery*, eds., M.M. de Villiers, P.A. Glen, and S. Kwon. Springer, New York, USA.
Mezzomo, N., Martínez, J., Ferreira, S.R.S. 2009. Supercritical fluid extraction of peach (*Prunus persica*) almond oil: Kinetics, mathematical modeling and scale-up. *J. Supercrit. Fluids,* 51:10.

Moura, L.S., Carvalho, Jr. R.N., Stefanini, M.B., Ming, L.C., Meireles, M.A.A. 2005. Supercritical fluid extraction from fennel (*Foeniculum vulgare*): Global yield, composition and kinetic data. *J. Supercrit. Fluids,* 35:212.

Papamichail, I., Louli, V., Magoulas, K. 2000. Supercritical fluid extraction of celery seed oil. *J. Supercrit. Fluids,* 18:213.

Parka, K., Lee, J., Sung, J. 2013. Metal extraction from the artificially contaminated soil using supercritical $CO_2$ with mixed ligands. *Chemosphere*, 91:616.

Paulaitis, M.E., Krukonis, V.J., Kurnik, R.T., Reid, R.C. 1982. Supercritical fluid extraction. *Rev. Chem. Engr,* 1(2):179.

Peng, D.Y., Robinson, D.B. 1976. A new two-constant equation of state. *Ind. Eng. Chem. Fundam.* 15:59.

Perakis, C., Louli, V., Voutsas, E., Magoulas, K. 2010. Supercritical $CO_2$ extraction of dittany oil: Experiments and modelling. *J. Supercrit. Fluids,* 55:573.

Perrut, M., Clavier, J.Y., Poletto, M., Reverchon, E. 1997. Mathematical modeling of sunflower seed extraction by supercritical $CO_2$. *Ind. Eng. Chem. Res.* 36:430.

Perry, R. 1997. *Perry's Chemical Engineers' Handbook*, 6th edition. McGraw-Hill, New York, USA.

Rajendran, A. 2012. Design of preparative–supercritical fluid chromatography. *J Chromatogr. A,* 1250:227.

Raventós, M., Duarte, S., Alarcón, R. 2002. Application and possibilities of supercritical $CO_2$ extraction in food processing industry: An overview. *Food Sci. Technol. Int.* 8:269.

Rawson, A., Tiwari, B.K., Brunton, N., Brennan, C., Cullen, P.J., O' Donnell, C.P. 2012. Application of supercritical carbon dioxide to fruit and vegetables: Extraction, processing, and preservation. *Food Rev. Int.* 28:253.

Reverchon, E., Sesti Ossèo, L. 1994. Modelling of supercritical extraction of basil oil. In: *Proceedings of the 3rd International Symposium on Supercritical Fluids*, eds., M. Perrut and G. Brunner, 189p. Strasbourg, France.

Reverchon, E. 1997. Supercritical fluid extraction and fractionation of essential oils and related products. *J. Supercrit. Fluids*, 10:1.

Reverchon, E., Daghero, J., Marrone, C., Mattea, M., Poletto, M. 1999. Supercritical fractional extraction of fennel seed oil and essential oil: Experiments and mathematical modeling. *Ind. Eng. Chem. Res.* 38:3069.

Reverchon, E., Perrut, M. 2000. *Proceedings of the 7th Meeting on Supercritical Fluids*, Antibes, December 6–8, 2000, eds., M. Perrut and E. Reverchon, pp. 3–20. Antibes, France. ISBN 2-905-267-33-10.

Reverchon, E., Marrone, C. 2001. Modeling and simulation of the supercritical $CO_2$ extraction of vegetable oils. *J. Supercrit. Fluids,* 19:161.

Reverchon, E., De Marco, I. 2006. Supercritical fluid extraction and fractionation of natural matter. *J. Supercrit. Fluids,* 38:146.

Schneider, G.M. 1978. Physicochemical principles of extraction at high pressures. *Angew. Chem. Int. Ed. Engl.* 17:716.

Schneider, G.M., Stahl, E., Wilke, G. 1980. *Extraction with Supercritical Gases*. Verlag Chemie, Deerfield Beach, FL.

Smith, R.M. 1999. Supercritical fluids in separation science—The dreams, the reality and the future. *J. Chromatogr. A,* 856:83.

Soave, G. 1972. Equilibrium constants from a modified Redlich–Kwong equation of state. *Chem. Eng. Sci.* 27:1197.

Sovová, H. 1994. Rate of the vegetable oil extraction with supercritical $CO_2$: I. Modelling of extraction curves. *Chem. Eng. Sci.* 49:409.

Sovová, H., Kučera, J., Jež, J. 1994. Rate of the vegetable oil extraction with supercritical $CO_2$—II. Extraction of grape oil. *Chem. Eng. Sci.* 49:415.

Sovová, H. 2005. Mathematical model for supercritical fluid extraction of natural products and extraction curve evaluation. *J. Supercrit. Fluids,* 33:35.

Spiliotis, N., Boukouvalas, C., Tzouvaras, N., Tassios, D. 1994. Prediction of vapor–liquid equilibrium with the LCVM model: A linear combination of the Vidal and Michelsen mixing rules coupled with the original UNIFAC and the t-mPR equation of state. *Fluid Phase Equil.* 92:75.

Taylor, S.L., King, J.W. 2002. Preparative-scale supercritical fluid extraction/Supercritical fluid chromatography of corn bran. *J. Am. Oil Chem. Soc.* 79(11):1133.

Tom, J.W., Debenedetti, P.G. 1991. *Proceedings of the Second International Symposium on Supercritical Fluids*, ed., M. Mc Hugh, pp. 229–232. Boston.

Tom, J.W., Lim, G.B., Denbenedetti, P.G., Prudhomme, R.K. 1993. Applications of supercritical fluids in the controlled released of drugs. In: *Supercritical Fluid Engineering Science*, eds., J.F. Brennecke and E. Kiran, pp. 238–257. ACS Symposium Service 514, American Chemical Society, Washington, DC.

Treybal, R.E. 1981. *Mass Transfer Operations*, 3rd edition. McGraw-Hill, Singapore.

Vandana, V., Teja, A.S., Zalkow, L.H. 1996. Supercritical extraction and HPLC analysis of taxol from *Taxus brevifolia* using nitrous oxide and nitrous oxide + ethanol mixtures. *Fluid Phase Equil.* 116:162.

Voutsas, E.C., Spiliotis, N., Tassios, D.P. 2001. Enzymatic reactions in non-conventional media: Prediction of solvent water content for optimum water activity. *Biocatal. Biotransf.* 19:99.

Voutsas, E., Magoulas, K., Tassios, D. 2004a. A universal mixing rule for cubic equations of state applicable to symmetric and asymmetric systems: Results with the Peng–Robinson equation of state. *Ind. Eng. Chem. Res.* 43:6238.

Voutsas, E., Coutsikos, P.h., Kontogeorgis, G. 2004b. Equations of state with emphasis on excess Gibbs energy mixing rules. In: *Computer Aided Property Estimation*, eds., R. Gani and G. Kontogeoris, pp. 75–111. Elsevier Science, the Netherlands.

Voutsas, E., Louli, V., Boukouvalas, C., Magoulas, K., Tassios, D. 2006. Thermodynamic property calculations with the universal mixing rule for EoS/$G^E$ models: Results with the Peng–Robinson EoS and a UNIFAC model. *Fluid Phase Equil.* 241:216.

Wai, C.M., Wang, S. 1997. Supercritical fluid extraction: Metals as complexes. *J. Chromatogr. A,* 785:369.

Wang, J.S., Chiu, K. 2008. Extraction of chromated copper arsenate from wood wastes using green solvent supercritical carbon dioxide. *J. Hazard. Mater.* 158:384.

Weidner, E., Knez, Z., Novak, Z. 1994. PGSS: A new process for powder generation. *In Proceedings of the 3rd International Symposium on SCF*, ISASF, 229p. Strasbourg.

# 8 Chilling and Freezing

*M. Giannakourou and Virginia Giannou*

## CONTENTS

## 8.1 INTRODUCTION

Food preservation at low temperatures has been applied since ancient times. In the early sixteenth century, selling of ice began in the large cities, while the first refrigeration machine was built in 1834 by Jacob Perkins. Since then, chilling or refrigeration has been widely used in household and industrial applications (transportation, distribution, and retailing) to prolong the shelf life of perishable foods, especially meat, fish, dairy products, fruit, vegetables, and ready-made meals.

From the technological point of view, chilling is considered to be a minimal processing method and concerns storage at temperatures above the freezing temperature and below 15°C. In general, chilled foods will lose value or, in several cases, will be destroyed if frozen (Maroulis and Saravacos, 2003).

Its purpose is to slow down

- The activity of microorganisms and enzymes
- The postharvest and postslaughter metabolic activities of plant and animal tissues, respectively
- The deteriorative chemical (e.g., oxidation, Maillard browning, and formation of off-flavors) and biochemical reactions (e.g., glycolysis, proteolysis, enzymatic browning, and lipolysis)
- Moisture loss or other physical changes resulting from the interaction of food components with the environment

This ensures that food retains its freshness for a longer time and extends retainability. However, chilling cannot improve the quality of a poor product; neither can it stop the processes of spoilage. Since it cannot be considered a reliable bactericidal step, it is often used in combination with other preservation methods, such as fermentation, irradiation, pasteurization, mild heat treatment, waxing, use of chemicals (acids or antioxidants), and controlled atmosphere, to ensure food safety and enhance quality (Drummond and Sun, 2010; Hui et al., 2004).

Chilling, is also used for nonpreservative purposes, such as crystallization, aging of meat, wine, and cheese, or to facilitate operations such as pitting of cherries and peaches, cutting of meat, and slicing of bread (Karel and Lund, 2003; Ramaswamy and Marcotte, 2006). However these are beyond the objectives of this chapter.

## 8.2 PRECOOLING

Good temperature management throughout the "field to fork" chain is the key for the preservation of quality. Precooling of most produce to a "safe" temperature is imperative in ensuring quality and increasing shelf life. It is a treatment administered prior to storage or long-distance shipping to promptly reduce the products' temperature from the levels that exist at the time of slaughter or harvest to an optimum chilling temperature. The purpose is to retard the products' degradation, reduce the rates of metabolic activities such as respiration, transpiration, and ethylene production, minimize the growth of decay microorganisms and enzymes' activity, and reduce the refrigeration capacity needed in the transport vehicle or storage facility. Improved flexibility in marketing is an additional benefit (Karel and Lund, 2003).

Harvested produce, in particular, contains a substantial amount of heat, known as field heat, associated with the products' temperature at harvest, which is considered a significant part of the cooling load. The amount of field heat necessary to be removed depends on the produce and the required storage temperatures. Since at the time of harvest, the products' temperature is same as that of the environment, it is considered important and advisable, wherever possible, to harvest goods when the ambient temperature is low during night, morning, or evening, to avoid high cooling loads (Mishra and Gamage, 2007).

Precooling usually precedes common cooling, and its purpose is to immediately extract the products' heat before transport, storage, and processing. It is especially needed for highly perishable products with short postharvest storage lives and for products where growth of microorganisms must be promptly retarded to avoid toxicological problems and quality defects (animal tissues). Vegetables included in this category are, for example: asparagus, broccoli, cauliflower, sweet corn, leafy vegetables (e.g., spinach), globe artichokes, Brussels sprouts, cabbage, celery, carrots, and radishes. Commercially, important fruits that need to be precooled immediately after harvest include apricots, avocados, most berries, cantaloupes, tart cherries, peaches, nectarines, plums, prunes, guavas, mangoes, papayas, and pineapples. Precooling, however, is not necessary for the more resistant fruits and vegetables or those designated for commercial curing or ripening (Becker and Fricke, 2002).

When precooling is recommended, it generally should be accomplished as rapidly as possible. It can be accomplished through cold/ambient air stream, water (hydrocooling), ice, room cooling, cryogenic cooling, or vacuum cooling by decreasing the pressure around the produce to a point at which the boiling point of water is reduced. However, refrigeration, under specific conditions, using appropriate equipment, is commonly required for ensuring short cooling time, which is critical in preventing the loss of quality. Different types of produce have different cooling requirements. The choice of a precooling method is also greatly influenced by the type of packaging used (if any). Precooling rates depend on the type of product, its size, weight, and the surface-to-volume ratio, and may affect the quality characteristics of fresh produce, especially their color (Brosnan and Sun, 2001; Mishra and Gamage, 2007; Vigneault et al., 2008).

## 8.3 CHILLING

Following precooling, it is important that the cold chain of the fresh product is maintained throughout transportation, storage, and preferably during retailing and domestic use. Typically, road and sea containers are refrigerated, as are the storage units at exporters, importers, and retail distribution centers. Air freight is rarely cooled and relies on adequate precooling, good pack insulation, and the speed of transport to maintain adequate quality. For most fruits and vegetables, this cold chain tends to be broken in the retail stores as the majority are rarely displayed in chilled cabinets (Aked, 2002).

Cooling can be divided into two distinct phases, (i) the chilling operation itself, in which the foodstuff is cooled from either an ambient temperature or a cooking temperature of over 70°C, and (ii) the chilled storage, at a closely controlled temperature of between 15°C and –2°C, depending on the foodstuff.

Chilling equipment and chilled storage equipment are quite different in their requirements and their design, and although some chilling equipment may be used for chilled storage, storage equipment is not designed to cool products, only to maintain temperature.

The main objectives of chilled storage are

- To extend the availability of fresh produce in the market,
- To ensure continuous supply of quality raw material to the processors,
- To extend the length of the processing season,
- To hold raw material obtained during favorable price situations,
- To condition certain commodities such as potatoes, onions, and garlic, and
- To ripen certain fruits such as mangoes and bananas (Mishra and Gamage, 2007).

Chilled foods can be grouped into three major categories according to their storage temperature range as follows:

1. – 1°C to 1°C (fresh fish, meats, sausages and ground meats, smoked meats, and breaded fish).
2. 0–5°C (pasteurized canned meat, milk, cream, yogurt, prepared salads, sandwiches, baked goods, fresh pasta, fresh soups and sauces, pizzas, pastries, and unbaked dough).
3. 0–8°C (fully cooked meats and fish pies, cooked or uncooked cured meats, butter, margarine, hard cheese, cooked rice, fruit juices, and soft fruits) (Fellows, 2000).

As mentioned before, the quality of chilled foods is primarily dependent on the overall quality of the raw materials employed, as cold storage cannot compensate for or improve the use of inadequate or substandard ingredients. The process of chilling foods under the indicated storage conditions causes little or no reduction in the eating quality or nutritional properties of food. However, it is important that the chilling rate enforced is according to the products' characteristics and storage is maintained at the chosen temperature with minimal fluctuation to prevent undesirable effects on the products' quality and sensory properties, as well as to avoid present microorganisms to grow or produce toxins (Drummond and Sun, 2010).

The most significant effect of chilling on the sensory characteristics of processed foods is hardening due to solidification of fats and oils. Chemical, biochemical, and physical changes during refrigerated storage may lead to loss of quality, and in many instances, it is these changes rather than microbiological growth that limit the shelf life of chilled foods. These changes include enzymic browning, lipolysis, color and flavor deterioration in some products, or moisture migration and retrogradation of starch leading to staling of baked products (which occurs more rapidly at refrigeration temperatures than at room temperature). Physicochemical changes also involve oil's migration, gel shrinking, and syneresis in sauces and gravies due to changes in starch thickeners or evaporation of moisture from unpackaged chilled meats and cheeses. Lipid oxidation is one of the main causes of quality loss in cook-chilled products, and cooked meats in particular rapidly develop an oxidized flavor termed "warmed-over flavor." Vitamin losses during chill storage of selected fresh and processed foods may also occur, while in cook–chill systems, nutritional losses are reported as well (Brown and Hall, 2000; Fellows, 2000). The effect of chilling, however, on different food categories is described in more detail below.

*Chilling rate—time*

The rate at which heat can be extracted during chilling is dependent on many factors. The size and shape of the pack or container will affect the rate of heat transfer to the cooling air (or, in some cases, water). The temperature and speed of the air will also affect this. Within the pack, the weight, density, water content, specific heat capacity, thermal conductivity, latent heat content, and initial food temperature will each play a part.

In the case of unpackaged foods, the factors leading to rapid cooling also lead to rapid loss of moisture; so, it may seem that slow cooling is better. Generally, this is not the case, as the extended cooling time is also an extended drying-out time. More rapid chilling is possible with thinner packs, with higher airspeeds, and with lower air temperatures. All these lead to higher operating costs; so, the equipment design has to be a compromise to give the best overall operating system (Heap, 2000).

To ensure that refrigeration is effective, it is often necessary to calculate processing times, product temperatures, heat loads, and water diffusion into and out of the product. These are influenced by both operating conditions (environmental temperature, humidity, air velocities, etc.) and product parameters (type, size, shape, and composition) (Pham, 2001). The simplest approach toward the calculation of chilling time is the application of Newton's law (Cortella, 2012). According to this, the change of temperature with respect to time is proportional to the difference between the temperature of the object and the temperature of its surroundings (Equation 8.1).

$$\frac{dT}{dt} = k(T_s - T) \tag{8.1}$$

where $T$ is the temperature of the object, $T_s$ is the temperature of the surroundings (e.g., cooling chamber), $k$ is a positive constant associated with the rate at which an object cools, and $t$ represents chilling time.

By integrating the above equation and setting that at $t = 0$ € $T = T_0$ (initial temperature of the product), the following relation (Equation 8.2) occurs:

$$T = T_s + (T_0 - T_s)e^{-kt} \tag{8.2}$$

The $k$ constant can be easily calculated (Equation 8.3) by measuring the time ($t_1$) required to reach a temperature ($T_1$) other than $T_0$.

$$k = \frac{1}{t_1} \ln\left(\frac{T_0 - T_s}{T_1 - T_s}\right) \tag{8.3}$$

More sophisticated approaches have been proposed in the literature for the calculation of processing time and heat load, most of which are based on the geometrical characteristics of the food product under refrigeration (Cleland, 1990; Heldman, 2011; Lin, 1994; Lin et al., 1996a,b; Lovatt et al., 1993; Pflug et al., 1965; Pham, 2002).

### 8.3.1 Chilling Effect on Microorganisms

Microbial growth in foods results in food spoilage with the development of undesirable sensory characteristics and off-flavors, and in certain cases, the food may become unsafe for consumption. The pathogenicity of certain microorganisms is a major safety concern in the processing and handling of foods in that they produce chemicals in foods that are toxic to humans (Rahman, 2007).

The activities of microorganisms, including growth are greatly affected by the chemical and physical state of their environment. Although several factors can potentially affect the growth of microorganisms (e.g., pressure, radiation, or presence of inhibitors/competing microorganisms), temperature, pH levels, water availability, and oxygen are considered the key elements that control the growth of all microorganisms. Out of these, temperature is probably the most important environmental factor (Madigan et al., 2012).

Each microorganism has (i) an optimum temperature at which it grows best, (ii) a minimum temperature below which growth no longer takes place, and (iii) a maximum temperature above which all development is suppressed. On the basis of temperature range for growth, they can be categorized into four broad groups (or five for other microorganisms, including "thermoduric," which can survive exposure to relatively high temperatures but will not grow at these temperatures) (Table 8.1).

Most spoilage and food-poisoning organisms grow rapidly at temperatures above 10°C. As the temperature is lowered, their growth rate decreases and generally reaches a limit below 3°C. Especially in cases of rapid chilling, some microbial injury and death is also likely to occur due to cold shock. However, a small increase in the refrigeration temperature may induce a significant increase in the growth rate of certain microorganisms, leading to a considerable decrease in the shelf life of foods. For example, the growth rate of some microorganisms doubles for each 3°C rise in temperature. Microbiological growth and transmission is also affected by the relative humidity of the environment. High humidity in cold rooms should be avoided since condensation that forms on the walls and ceiling creates the proper environment for mold growth and buildups.

The main microbiological concern in chilled foods is a number of pathogens that can grow during extended refrigerated storage below 5°C, or as a result of any

**TABLE 8.1**
**Cardinal (Minimum, Optimum, and Maximum) Growth Temperatures of the Basic Microorganisms' Categories**

| Category | Minimum Temperature | Optimum Temperature | Maximum Temperature |
|---|---|---|---|
| Thermophilic | 30–40°C | 55–65°C | ~100°C |
| Mesophilic | 5–10°C | 30–40°C | <45°C |
| Psychotropic | <0–5°C | 20–30°C | >25°C |
| Psychrophilic | <0–5°C | 12–18°C | <20°C |

*Source:* Hamad S.H. Factors affecting the growth of microorganisms in food (Chapter 20). In: *Progress in Food Preservation*. Eds., R. Bhat, A.K. Alias, G. Paliyath. 417–421, 2012. West Sussex: John Wiley & Sons Ltd, Copyright Wiley-VCH Verlag GmbH & Co. KGaA. Reproduced with permission; Adapted from Walker S.J., Betts G. 2000. In: *Chilled Foods* (2nd ed.). Eds. C. Dennis, M. Stringer. Chichester: Ellis Horwood Ltd.

increase in temperature (temperature abuse) and thus cause food poisoning. Examples of these pathogens are *Aeromonas hydrophilia*, *Listeria monocytogenes*, *Yersinia enterocolitica*, *Plesiomonas shigelloides*, *Clostridium botulinum*, some strains of *Bacillus cereus*, *Vibrio parahaemolyticus*, and enteropathogenic *Escherichia coli* (Brackett, 1992; Fellows, 2000).

Chilling prevents the growth of thermophilic and many mesophilic microorganisms; the latter may become a problem in situations of temperature abuse. The psychotropic or psychrophilic microorganisms are able to grow at commercial refrigeration temperatures and can cause off-flavors and physical defects in foods. The psychotropic organisms generally have a rapid growth above 0°C. They grow slowly up to about −10°C, below which there exists little or no growth. In fact, there can even be a slow death at temperatures below −10°C. However, in general, the activity of organisms of public health concern, such as those producing toxins, is largely limited below 4°C. Psychotrops include molds, yeasts (e.g., *Penicillium*, *Aspergillus*, *Geotrichum*, and *Botrytis*), many Gram-negative bacteria from genera *Pseudomonas*, *Achromobacter*, *Acinetobacter*, *Alcaligenes*, *Flavobacterium*, *Yersinia*, *Serratia*, and *Aeromonas*, and Gram-positive bacteria from genera *Leuconostoc*, *Lactobacillus*, *Bacillus*, *Clostridium*, *Streptococcus*, and *Listeria* (Ramaswamy and Marcotte, 2006; Ray and Bhunia, 2008; Spencer and Ragout de Spencer, 2001).

The characteristic of psychrophiles is that their transport processes function optimally at low temperatures. This is an indication that the cytoplasmic membranes of psychrophiles are structurally modified in such a way that low temperatures do not inhibit membrane functions. Cytoplasmic membranes from psychrophiles tend to have a higher content of unsaturated and shorter-chain fatty acids, which helps the membrane remain in a semifluid state at low temperatures. In addition, the lipids of some psychrophilic bacteria contain polyunsaturated fatty acids (increasing flexibility), something very uncommon in prokaryotes (Madigan et al., 2012). Most psychrophiles belong to the Gram-negative bacterial genera of *Aeromonas*, *Alcaligenes*, *Flavobacterium*,

*Pseudomonas*, and *Vibrio*. Psychrophilic yeasts such as *Candida gelida*, *Candida nivalis*, *Leucosporidium scottii*, and *Cryptococcus vishniacii*, are also considered important in food applications (Hamad, 2012; Herbert and Sutherland, 2000).

### 8.3.2 Chilling Effect on Plant Tissues

Fruits are commonly derived from an ovary and the surrounding parts, while vegetables are derived from different plant parts. Vegetables can be grouped into three categories, namely: (1) seeds and pods; (2) bulbs, roots, and tubers; and (3) flowers, buds, stems, and leaves (Tabil and Sokhansanj, 2001).

The most striking distinction between the phyto- and myosystems of foods is the presence of aerobic respiratory activity in the former. Respiration is a complex process of oxidation of organic matter (starch, sugars, acids, fats, proteins, etc.) to simpler, more useful to the cell, molecules such as $CO_2$ and $H_2O$, with the concurrent production of energy and other intermediate products. It is a basic reaction of plants that causes weight loss, excessive $CO_2$ buildup in storage, and triggers a host of other metabolic reactions such as ethylene synthesis, color, texture, and flavor changes (Singh, 2007).

Plant foods, during harvest, are cut off from their normal supply of water, minerals, and organic matter, which is normally translocated to them from other parts of the plant. The tissues, however, remain alive and are capable of continuing a wide range of metabolic activities, breaking down the organic matter to meet their energy requirements. Since the stored food is not replaced, senescence (progressive loss of membrane integrity) follows and the produce will eventually decay. Immature products such as peas and beans tend to have much higher respiration rates and short shelf lives caused by natural senescence whereas the opposite is true for mature storage organs such as potatoes and onions. Since oxygen is necessary for the respiration of fruits and vegetables, their storage life can be extended considerably by modifying the atmosphere in the cold-storage rooms. This can be accomplished for certain plant tissues by reducing the oxygen level from 1% to 5% while increasing the $CO_2$ level (Aked, 2002).

Another characteristic metabolic process in agricultural products is transpiration, which is associated with the rapid water loss of harvested produce from their surfaces. Transpiration is a major component of weight loss in fruits and vegetables and its rate varies with the environmental conditions, such as the temperature, relative humidity, and air motion. A 5–10% weight loss will cause significant wilting, shriveling, degraded texture, and poor taste. Moisture loss occurs rapidly in a warm, dry environment especially among injured products affected by commodity characteristics, such as surface-area-to-volume ratio, presence of waxy substances on the skin, tenderness of the skin, and presence of protrusions on the skin surface. The texture of the produce is adversely affected by excessive water loss, rendering the product unmarketable (Tabil, and Sokhansanj, 2001).

Part of the physiological activities is highly desirable, especially in some fruits, for the attainment of the desirable appearance and texture and optimum eating qualities ( e.g., the ripening of fruits such as banana, mango, papaya, pineapple, etc.). In others, they are less desirable and mostly deleterious. But, however undesirable they are, they will still have to be continued at some minimal level to maintain the quality of fruits and vegetables. It is this physiological activity that maintains the vigor of the tissue and provides a kind of defense against the attack of spoilage organisms (Francis et al., 1999; Ramaswamy and Marcotte, 2006).

On the basis of their respiratory activity, plant produce can be divided into climacteric and nonclimacteric. The first includes apple, apricot, avocado, banana, mango, peach, pear, plum, and tomato. They can be harvested unripe and ripened artificially at a later stage. During ripening, the respiration of climacteric fruits increases dramatically (usually until optimum ripeness) over a short period of time and without careful temperature control, the fruit will rapidly overripen and senesce, leading to internal tissue breakdown and the production of off-flavor volatiles will follow. Climacteric fruits also produce high levels of ethylene—a plant hormone that plays a key role in the ripening and senescence of fruits and vegetables—during the initiation of ripening. Nonclimacteric fruits include cherry, cucumber, fig, grape, grapefruit, lemon, pineapple, and strawberry. These fruits, in general, exhibit a decline in their respiratory activity during storage. Vegetables respire in a similar way to nonclimacteric fruits. However, many of them undergo an abrupt decline in the rate of respiration immediately after harvest followed by a more gradual decline during storage (Aked, 2002; Fellows, 2000; Karel and Lund, 2003).

Temperature reduction in fruits and vegetables is commonly used to slow the rate of aerobic respiration, thus delaying senescence and decay of plant tissues, and enables some fruits to be ripened at controlled rates. To achieve a maximum storage life of plant tissues at chilling temperatures, it is desirable that (1) aerobic respiration must be allowed at a slow rate so that the maintenance processes associated with life continue to function and the natural protective coating, which hinders invasion of microorganisms, remains intact and (2) the temperature must be suitably low so that major deteriorative reactions are slowed as much as possible (Karel and Lund, 2003).

The optimum storage temperatures for fruits and vegetables are determined by genetic parameters (species, cultivar, clone, etc.), their stage of development (maturation, stage of ripening, etc.), date of harvest, origin, and the pre- and postharvest conditions it has experienced. For example, optimum storage temperature is influenced by the ripeness of bananas, lemons, pineapples, and tomatoes, the variety of apples and pears, the harvest date of potatoes, and the origin of tomatoes, grapefruit, oranges, cucumbers, and green beans (Aked, 2000). Information on the optimum storage requirements, based on parameters such as the rate of respiration and ethylene generation or perishability for each fruit and vegetable can be found in several sources (Aked, 2002; Gross et al., 2004; Kader, 2002; McGregor, 1987; Nascimento Nunes, 2008; Salunkhe et al., 1991; Tabil, and Sokhansanj, 2001; Thompson, 1996, 2003).

However, during transportation or temporary storage in warehouses, it is usually not feasible to maintain the optimum environmental conditions for each product. In this case, it is suggested to provide two storage environments, one at 0°C with 90% relative humidity and the other at 10°C with 85–90% relative humidity. The first storage environment can facilitate the temporary storage of those fruits and vegetables that are not subject to chilling injury (see the paragraphs below) as well as the storage of eggs, milk, and animal tissues, if necessary. In this case, precautions should be taken to avoid storage of incompatible products (e.g., highly odoriferous products, such as fish and onions, with odor-absorbing products, such as butter) under the same room. The other storage environment (at 10°C) can accommodate those fruits and vegetables that are subject to chilling injury (Karel and Lund, 2003).

Fruits and vegetables are highly susceptible to postharvest damage caused by mechanical injury or injury from temperature effects. Losses by mechanical injury begin from the time of harvest until the produce reaches the consumers' table. Bruises, cuts, splits, and cracks are some of the manifestations of mechanical injury. These trigger an increase of physiological processes, which can be disastrous to the produce. Damage can also be caused by insects, which usually infest the crop while in the field, depositing their larva or egg on the produce, while fungi and bacteria infect the produce causing diseases. Pests and diseases can be controlled by quarantine treatments, employing physical and chemical methods or a combination of the two (Tabil and Sokhansanj, 2001).

Certain fruits and vegetables are particularly sensitive to temperature variations during handling and storage. Injury from temperature effects is caused by exposing the produce to extremes of temperature during postharvest handling. Heat injury is sustained by produce due to exposure to the sun after harvest, or with any warm surface such as the soil or a wall heated by the sun. Weight loss, softening, discoloration, and eventual desiccation of the affected tissue are its most common effects. Undesirable changes may also occur when the temperature is reduced below a specific optimum for the individual fruit. This is termed chilling injury and results in various physiological changes including water-soaked appearance, surface and internal discoloration, pitting, failure to ripen, uneven ripening, development of off-flavors, and heightened susceptibility to pathogen attack (Fellows, 2000; Kader, 2013). The reasons for this are not fully understood, but may include an imbalance in metabolic activity that results in the overproduction of metabolites that then become toxic to the tissues.

More specifically, there are 10 visual symptoms associated with chilling injury, namely:

1. Surface lesions—Pitting, large sunken areas, and discoloration
2. Water soaking of tissues—Disruption of cell structure and the accompanying release of substrates favoring growth of pathogens
3. Internal discoloration or browning of pulp, vascular strands, and seeds
4. Breakdown of tissues
5. Failure of fruits to ripen following removal from storage
6. Accelerated rate of senescence
7. Increased susceptibility to decay
8. Shortened shelf life due to one or more of the above responses
9. Compositional changes related to flavor and taste
10. Loss of growth capacity for stored propagules

Chilling injury can be found, for example, in apples (stored at <2–3°C), avocados (<13°C), bananas (<12–13°C), lemons (<14°C), mangoes (<10–13°C) and melons, pineapples, and tomatoes (each <7–10°C) (Fellows, 2000).

The early stages of chilling injury are believed to be reversible and some produce can tolerate chilling temperatures for short periods of time without development of symptoms. A range of methods is available to limit chilling injury. These include stepwise reduction in storage temperature, or intermittent warming during storage (e.g., nectarines and peaches). Some fruits may become less susceptible to chilling when held under appropriate modified atmospheres (MAPs), for example, mango, avocado (Aked, 2002).

The factors influencing the shelf life of plant tissues during chill storage are

- Produce type, variety, or cultivar
- The part of the crop selected (the fastest growing parts have the highest metabolic rates and the shortest storage lives)
- Food's status during harvesting (e.g., integrity/mechanical damage, microbial contamination, infestation of pests and diseases, and degree of maturity)
- The temperature of harvest, storage, distribution, and retail display
- The relative humidity of the storage atmosphere, which influences dehydration losses (Fellows, 2000; Tabil and Sokhansanj, 2001)

### 8.3.3 Chilling Effect on Animal Tissues

#### 8.3.3.1 Red Meat and Poultry

Regardless of the species, after an animal is slaughtered, blood circulation stops within the body and muscles lose their oxygen supply necessary for normal aerobic respiration (glucose + $O_2 \rightarrow CO_2 + H_2O$ + energy). Without oxygen, the muscle turns to anaerobic glycolysis (glycogen $\rightarrow$ glucose $\rightarrow$ lactic acid + energy) to generate energy. Anaerobic glycolysis produces energy to contract the muscle and it also produces lactic acid, which causes the muscle pH to fall from a physiological value (pH $\approx$ 7) to an ultimate value ranging from 5.1 to 6.5. As glycogen supplies are depleted, the energy for muscle contraction is lost and the actin and myosin filaments lock together in a permanent contraction called "rigor mortis" that may take up to 24 h to appear (Karel and Lund, 2003; North and Lovatt, 2012).

However, many of the biochemical reactions present in the living state retain some degree of activity in the nonliving state. These reactions are responsible for profound quality changes during storage. Postharvest quality is also affected by slaughter conditions or stress before death. The rate and extent of muscle postmortem metabolism is dependent on the availability of glycogen at slaughter, the temperature of the medium in which the reactions occur, and whether or not procedures intended to accelerate metabolic reactions have been applied. As mentioned, during rigor mortis, muscles become firm and inextensible but gain some softness after hanging and conditioning (aging) (Fellows, 2000; Kadim and Mahgoub, 2007; Rahman, 2007).

Aging is necessary as meat is often unacceptably tough immediately following rigor onset. It involves complex changes in muscle metabolism in the postslaughter period and is dependent on the type of meat, animal breed, metabolic status, and environmental factors such as rearing system and prior slaughter stress. During aging, the structure of the myofibrillar and other associated proteins undergoes some modifications, and collagen is weakened to a lesser extent. The proteolytic enzymes in meat (mainly cathepsins and calpains) play a significant role in improving meat tenderness during aging. However, they require specific conditions, such as temperature and pH, for optimal activity. Cathepsins are able to degrade myofibrillar proteins at the pH of meat. Calpains, on the contrary, require a higher pH (>6.6) for optimal activity and their maximum activity most likely occurs during the early postmortem stages. High-temperature conditioning may accelerate the aging process by keeping carcasses at temperatures of 15°C or greater. This type of conditioning may be applied in the pre- or postrigor state and is very effective in improving meat tenderness. Aging is also believed to increase flavor intensity, since postmortem processes, such as proteolysis and lipolysis, result in the development of flavor precursors (Kadim and Mahgoub, 2007).

The main factors that affect the shelf life of meat are microbial growth and deleterious chemical reactions. Although the internal musculature of a healthy animal is essentially sterile after slaughter, possibly due to the continued post-slaughter functioning of the immune system, all meat animals carry large numbers of different microorganisms on the outer surfaces of the body and in the alimentary tract (Gill, 1983). Meat animal carcasses and meat cuts can also be easily contaminated during slaughtering, dressing, chilling, and cutting processes when the muscles of animals are exposed to the environment. Contamination may vary depending on the characteristics of each animal, its geographic origin and the season of the year, as well as the sanitation and hygienic practices during the products' handling and processing.

Meat may support growth and serve as sources of various spoilage and pathogenic microorganisms. The spoilage microflora of fresh carcasses usually consists almost exclusively of Gram-negative rods (mainly Pseudomonads) and micrococci (mainly *Micrococcus* spp. and *Staphylococcus* spp.). In addition, Gram-negative bacteria such as *Acinetobacter*, *Alcaligenes*, *Moraxella*, and enterobacteriaceae, and Gram-positive species including spore-forming bacteria, lactic acid-producing bacteria (LAB), and *Brochothrix thermosphacta*, as well as yeasts and molds, may be present in small numbers (Borch et al., 1996; Jensen et al., 2004; Varnam and Sutherland, 1995).

The most important pathogens associated with meat include salmonellae, *Staphylococcus aureus*, verotoxigenic *E. coli*, *Clostridium perfringens*, *Campylobacter jejuni/coli*, *L. monocytogenes*, *Y. enterocolitica*, and *A. hydrophilia*. *Salmonella* spp., pathogenic *E. coli*, and *Campylobacter* are of enteric origin and are considered as the common food-borne pathogens in meat. In general, the presence of small numbers of pathogens is not a problem because meat is normally cooked before consumption. Adequate cooking will substantially reduce the numbers, if not completely eliminate all the pathogenic organisms present on the meat. Most meat-based food poisoning is associated with inadequate cooking or subsequent contamination after cooking (James and James, 2002; Jensen et al., 2004).

Lipid oxidation, as aforementioned, is the main chemical reaction responsible for quality deterioration and shelf-life reduction in meat and meat products. The susceptibility of different meat species depends on the level of unsaturated fatty acids in the tissue, their availability, and the presence of activators or inhibitors. Fish tissue is most affected, followed by poultry, pork, beef, and lamb (Brown and Hall, 2000; Drummond and Sun, 2010).

Historically, animal tissues have been preserved in a variety of ways, with the most common methods being drying, curing, smoking, heat processing, fermentation, irradiation, canning, packaging, and refrigeration. Of all these alternatives, refrigeration has the key benefit (along with irradiation, to some extent) in that it leaves the form of the meat product almost unchanged and (when carried out appropriately) almost indistinguishable from the original fresh product. By cooling the meat from the initial body temperature of the animal (about 38°C) to 10°C, the microbial growth rate drops by about 95%. The lowest cold-storage temperature for meat is –1.5°C while the minimum growth temperature of psychotropic bacteria is –3°C. Decreasing refrigeration temperatures decrease growth of spoilage microorganisms, and affect the composition of the bacterial flora. Moisture lost during chilling also tends to decrease microbial growth as a result of reduced surface water activity ($a_w$). Refrigeration is also effective in reducing deterioration due to chemical reactions (Borch et al., 1996; James and James, 2002; Jensen et al., 2004; North and Lovatt, 2012).

When meat is chilled at temperatures below 10°C (especially between 0°C and 5°C), before rigor mortis has occurred (at pH > 6.2), the excised muscles will contract, resulting in poor water-holding capacity and undesirable changes to its texture. This phenomenon, known as "cold shortening," entails hardening of meat and has been attributed to the negative impact of low chilling temperatures on the ability of the sarcoplasmic reticulum to regulate the distribution of calcium ions in muscles. Cold shortening has been extensively observed at prerigor excised muscle from poultry, beef, and lamb, while pork does not seem to be significantly affected. Therefore, it is essential to subsequently chill meat quick enough to ensure that bacterial growth remains to the minimum, and slow enough to encourage mild respiration for aging to take place and prevent cold shortening (Fellows, 2000; Karel and Lund, 2003; North and Lovatt, 2012).

More specifically, it is suggested to keep the carcasses after slaughter at ambient temperatures to accelerate rigor development and then chill rapidly to condition the meat. Postslaughter techniques are also used to facilitate the tenderizing process, such as electrical stimulation and pelvic suspension. Electrical stimulation by low voltage, either right before evisceration or directly after, causes a very rapid fall in meat pH and accelerates rigor mortis, allowing cooling to proceed without the development of cold shortening. Pelvic suspension, on the other hand, takes advantage of the fact that stretched muscles do not suffer from cold shortening during rigor mortis. It is based on hanging the carcass from the pelvic bone. The weight and the tension of the hind limb in this position leads to straightening of the vertebral column and prevention of muscle shortening due to skeletal restraint in both the back and the leg (Drummond and Sun, 2010; Lundesjo Ahnstrom, 2008). The tenderness of certain cuts of meat also depends on the location of the cut, the age, and the activity level of the animal. Cuts from the relatively inactive mid-backbone section of the animal such as short loins, sirloin, and prime ribs are, in general, more tender than those from active parts such as the legs and the neck. Meat cuts can be either chill stored/displayed unpacked or packed under air, vacuum, and MAP containing different levels of oxygen and carbon dioxide, balanced with inert nitrogen. Vacuum packaging is applied to constrain microbial growth as many of the organisms that influence meat spoilage require the presence of oxygen to grow. The vacuum-packed meat is also preserved from weight loss and discoloration. Packaging under various gaseous atmospheres has been used as an alternative to vacuum packing. The intention has been to preserve the fresh meat color and to prevent anaerobic spoilage by using high concentrations of oxygen (50–100%) along with 15–50% carbon dioxide to restrict the growth of Pseudomonads and related species. Pork is generally stored aerobically or in MAP, and beef is stored in a vacuum or MAP due to the need for tenderization during an extended storage (Borch et al., 1996; Kadim and Mahgoub, 2007).

The rate of spoilage depends on the numbers and types of organisms initially present, the conditions of storage (temperature and gaseous atmosphere), and characteristics (pH, $a_w$) of the meat. However, in general, there is little difference in the microbial spoilage of beef, lamb, pork, and other meat derived from mammals (Varnam and Sutherland, 1995). Enzymes also catalyze chemical reactions within the meat, and are able to produce chemical and physical changes that alter the quality characteristics of the meat. The first signs of putrefaction in meat appear with the development of off-odors when bacterial levels reach approximately $10^7$ cm$^{-2}$ of surface area. At a further 10-fold increase, slime begins to form on the surface. Discoloration and gas production may also occur (Borch et al., 1996; James, 2006).

Changes in appearance are normally the criteria that limit the display of unwrapped products, rather than microbiological considerations. The predominant type of spoilage in vacuum-packed chilled meat is souring. This is not normally detectable until bacterial numbers reach 8 $\log_{10}$ cfu cm$^{-2}$ or greater. The exact cause of such spoilage is unknown, but is assumed to result from lactic acid and other end products of fermentation by dominant LAB (James and James, 2002). The microflora of meat stored in commercially used MAP is in general similar to that of vacuum packs (Varnam and Sutherland, 1995). It is found that packages containing up to 80% oxygen and 20% carbon dioxide (high oxygen-MAP) will reduce the color deterioration of retail cuts of meat, but will only slightly increase the shelf life, compared to aerobic storage. The shelf life of meat increases in the order: air, high oxygen-MAP, vacuum, no oxygen-MAP, and 100% $CO_2$ (Borch et al., 1996). The spoilage of meat in MAP may involve souring similar to that in vacuum-packed meat. Other characteristics include "rancid" and "cheesy" odors. Chemical rancidity does not appear to be primarily involved and souring is probably caused by the metabolites of LAB or *Br. thermosphacta* (Varnam and Sutherland, 1995).

As far as the overall quality and acceptability of refrigerated meat products is concerned, it can be affected by several parameters. However, the most important of them are appearance and texture. The appearance of meat at its point of sale is the most prominent attribute governing its purchase. It is reflected by the ratio of fat to lean and the amount of marbled fat, but most importantly by the color of the meat. Myoglobin and hemoglobin are the muscle and blood pigments, respectively that determine the attractiveness of fresh red meat (James and James, 2002; Jensen et al., 2004).

Myoglobin is the primary meat pigment and exists as bright-red oxymyoglobin ($MbO_2$), purple–red deoxymyoglobin (Mb), or brown metmyoglobin (MetMb). Denatured globin hemichrome is the gray or tan oxidized pigment of cooked meats. The purple color of freshly cut meat is due to the deoxymyoglobin. On exposure to air, it is converted into the bright-red pigment oxymyoglobin, which gives fresh meat its normal desirable appearance. Hemoglobin, which is responsible for the color of blood, plays only a small role in the color of red meat, although it may be more significant in paler meat.

Oxymyoglobin is more stable at lower temperatures because the rate of oxidation of the pigment decreases. At low temperatures, the solubility of oxygen is greater and oxygen-consuming reactions are slowed down. There is also a greater penetration of oxygen into the meat and the meat is redder than at high temperatures. In general, conditioned meat is a brighter and more attractive red than unconditioned meat, but its color stability becomes progressively poorer the longer it is conditioned. During prolonged storage, oxymyoglobin oxidizes to brownish-green metmyoglobin, and discoloration occurs. Pigment changes in fresh meat are also amplified by bacterial activity that provokes the reduction of oxygen tension in surface tissues, eventually leading to the development of a variety of green pigments, including metsulfmyoglobin and cholemyoglobin. The discoloration rate, however, is different for different muscles (James and James, 2002; Jensen et al., 2004).

Meat texture, on the other hand, is the major characteristic of eating quality. Some of the factors that influence the toughness of meat are inherent in the live animal. It is now well established that it is the properties of the connective tissue proteins, and not the total amount of collagen in meat that largely determine whether meat is tough or tender. As the animal grows older, the number of immature reducible cross-links decreases. The mature cross-links result in a toughening of the collagen and this in turn can produce tough meat.

The tenderness of meat is affected by both chilling and storage. Under the proper conditions, tenderness is well maintained throughout the chilled storage life, but improper chilling can produce severe toughening and meat of poor eating quality. The rate of cooling, the length of time, and the temperature during conditioning are the most important refrigeration factors controlling the texture of meat (James and James, 2002).

Another important quality factor is weight loss by evaporation. Under typical commercial distribution conditions, it has been estimated that lamb and beef lose from 5.5% to 7% of their weight between slaughter and retail sale. Weight losses from pork are probably of the same magnitude. In addition to the direct loss in saleable meat, there are also secondary losses, since excessive drying during initial chilling and chilled storage can have a detrimental effect on meat juiciness and color, producing a dark unattractive surface on the meat (James and James, 2002). Color darkening can also be caused by rapid chilling of meat due to subtle changes in the rate and extent of pH decline during chilling (North and Lovatt, 2012). Chilling evaporation has prompted many plants to resort to water spraying of carcasses during chilling. However, the advantages of water spraying must be weighed against the risks of a higher surface water activity, its potential to increase bacterial loads, and its possible effects on carcass appearance (Drummond and Sun, 2010).

#### 8.3.3.2 Fish and Seafood

Fish is recognized as one of the most perishable food products and its rate of spoilage increases with contamination or damage to fish tissue during catching, transport, and processing. As with red meat and poultry, several changes take place shortly after the catch of fish and other seafood, affecting their quality (Medina et al., 2009). The quality of fishery products is primarily influenced by both intrinsic (species, size, sex, composition, spawning, viruses and parasites potential, and cultivation conditions) and extrinsic factors (location, season and methods of catch, on-board handling, and storage conditions). Environmental conditions, such as temperature, relative humidity, as well as direct exposure to sunlight and airflow induce the rate of quality deterioration and spoilage (Opara et al., 2007). By far, temperature is the most influential factor affecting the physiological and biochemical processes associated with tissue degradation. Fish and seafood, being high-water-content foods, are maintained better under conditions of high relative humidity. Excessive airflow around fresh fish should also be avoided as it contributes to the loss of surface moisture and undesirable changes in the skin and flesh quality. Mechanical handling, in addition, can result in physical injury, such as bruises, cuts, and abrasion, affecting both appearance and creating a favorable substrate for contamination. Immediate gutting of fish removes a major reservoir of microbial contamination; yet, it exposes internal surfaces to rapid spoilage. Filleting also exposes fish to microorganisms, enzymes, and oxygen, altering and accelerating spoilage (Drummond and Sun, 2010; Hall, 2011; Venugopal, 2006).

Deteriorative changes in fish and seafood are mainly attributed, at first, to autolytic enzymatic reactions and then followed by the action of microbial enzymes and the growth of microorganisms, eventually leading to fish spoilage. Fresh fish skin has a translucent appearance due to even scattering of incident light. Color and overall appearance of fish can be degraded due to the gradual disintegration of myofibrils, resulting in their wider and more random intracellular distribution. As a result, the incident light is unevenly scattered and the fish starts to look opaque. Flesh, on the other hand, becomes yellow due to oxidation of carotenoid pigments and lipids in tissues, and also reactions with carbonyl amines (Venugopal, 2006). In the crustacean (e.g., shrimps or lobsters), the most common problem is the development of black spots or melanosis due to a biochemical mechanism by which phenols are oxidized to quinones by the enzyme polyphenol oxidase (Hyldig and Nielsen, 2007). Fish meat texture can be affected by the rate and extent of rigor mortis progress and pH decline, as well as myofibrillar disintegration and weakening of connective tissue caused by proteolysis. Lipid oxidation is also a significant cause of deterioration (both in flavor and texture), often a determinant of products' shelf life, especially in fatty fish. The kind of lipid present is also important, since unsaturated fatty acids are less stable (Drummond and Sun, 2010).

Microorganisms such as *Shewanella putrefaciens*, *Pseudomonas* ssp., and *Photobacterium phosphoreum* are frequently responsible for quality changes and development of off-odors in fish, reducing their shelf life. The temperature characteristics of the associated flora are commonly affected by the temperature of the waters in which the fish live. More specifically, the bacterial flora of cold-water fish is predominantly Gram-negative and psychrotrophic, while in warm-water fish it is mainly represented by mesophilic, Gram-positive microorganisms (Drummond and Sun, 2010). The dominant microflora of cold-water fish species are *Pseudomonas*, *Alteromonas*, *Moraxella*, *Acinetobacter*, *Vibrio*, *Flavobacterium* and *Cytophaga*, while in warm-water fish, species of the Gram-positive *Micrococcus* and *Bacillus* are predominant.

In concern for food safety, pathogenic bacteria are considered the main hazards of fish and seafood. All aquatic environments (both marine and freshwater) can harbor spores of *C. botulinum*, while *V. parahaemolyticus* is the leading cause of food poisoning in raw fish products (Opara et al., 2007). *L. monocytogenes*, on the other hand, is of major concern to the fish-processing industry (Ben Embarek, 1994). Attention has also been drawn on the presence of algal toxins, especially in shellfish (Botana, 2008).

All these factors make refrigeration of fresh fish and fish products critical for maintaining their quality. Adequate chilling can slow down biochemical reactions and microbial growth, reduce mass losses, and maintain the associated quality aspects of fish. Rapid cooling and storing of fish, close to 0°C, are the most recommended conditions to reduce the deterioration rates and extend shelf life. Although from a microbial standpoint, storage at −2°C to −4°C is better than at 0°C; this is usually not preferred due to toughening and autolytic changes, such as lipid hydrolysis, which occur at the lower temperatures. The most common preservation systems available to the fish industry include the use of ice, chilled seawater, and slurry ice. During storage in ice, and depending on the species, fish can either gain or lose significant weight. Weight lost as drip is accompanied by leaching of proteins and flavor components, while weight gained usually incurs in salt absorption, discoloration, and off-odor production (Bellas and Tassou, 2005; Barros-Velazquez et al., 2008; Drummond and Sun, 2010; Medina et al., 2009).

A wide range of physical/chemical treatments have also been proposed by researchers for extending the shelf life of fish. These include electrical stimulation, decontamination with phosphate, hydrogen peroxide, chlorine, chlorine dioxide and ozone, surface treatment by organic acids, and the use of sorbate or sulfite. Coatings or films are finding increasing applications as well, as protective barriers against contact with air and water, delaying product deterioration during cold storage. Some are reported to additionally have antimicrobial, binding, and texturizing properties (Gelman et al., 2005; Rahman, 2007).

### 8.3.4 Chilling Effect on Processed Foods

Minimally processed, chilled foods were introduced in commercial catering and the retail food sector to satisfy the increasing consumers' demand for fresh-like, convenience products with superior sensorial and nutritional quality, less/no preservatives, and environmental friendly packaging. However, concerns arise regarding the microbiological safety and quality of cook-chilled products, due to their

1. Formulation, containing little or no preservatives and having a low acid and high moisture (high $a_w$) content
2. Minimal thermal processing
3. Packaging, which provides a favorable environment for anaerobic pathogens to grow and produce toxins (Juneja and Snyder, 2007)

Their production relies on cook–chill or cook–pasteurize–chill processes under the combination of minimal processing (65–100°C) and storage under controlled chill conditions to prevent the growth of pathogenic organisms. It includes pasteurizing (in cook–pasteurize–chill products) to a minimum temperature of 80°C for 10 min at the thermal center, rapidly cooling, and holding food under chilled conditions (0–3°C), before reheating (in most cases) and consumption. Heat treatment in cook-chilled products is designed to give a reduction by a factor of $10^6$ in the numbers of *L. monocytogenes* or *C. botulinum*, while chilling aims at preventing the survival of spore-forming bacteria, such as *C. botulinum*, *C. perfringens*, and *B. cereus* (Armstrong, 2004; Rodgers, 2003).

In an attempt to overcome the limitations of conventional cook–chill, enhancements to the process have been developed such as the creation of a barrier (usually a plastic pouch or tray), between the food and the surrounding environment, followed by extraction of the air surrounding the food, either prior to or immediately after thermal processing. The introduction of "sous vide" technology was another effort to prolong the shelf life of cook-chilled products. The term "sous vide" is used to describe the process of vacuum-packaging food, in a pouch or other suitable container, prior to the application of low-temperature thermal processing, to increase juiciness and maximize weight yield, and storage under chill conditions. The great advantage of the sous vide process is that it allows a delicate food to be cooked more gently at a precise, lower temperature than in conventional cooking (Juneja and Snyder, 2007; Roldan et al., 2013).

Various sous vide products require a preheat treatment to enable a browning or thickening effect, the maintenance of vegetable color (e.g., blanching), and/or the release of powerful flavors and aromas. This means that the ingredients can be marinated, trimmed, flavored, seared, grill marked, and/or partially cooked. Thereafter, they are vacuum packaged (preferably 99.9% removal of the air) and sealed before thermal processing, in a water bath or a steam combination oven. Heat treatment is performed by holding for a sufficient interval under appropriate time/temperature conditions, depending on the constituents. Rapid chilling is recommended within 30 min from the completion of thermal processing and aims at a core temperature of 0–3°C within a further 90 min to preserve quality and prevent the germination or growth of surviving *C. botulinum* spores (Armstrong, 2000; Baldwin, 2012).

Cook–chill products can be classified according to the level of microbial risk that they pose to consumers in:

**Class 1**: Foods containing raw or uncooked ingredients, such as salad or cheese as ready-to-eat (RTE) foods
**Class 2**: Products made from a mixture of cooked and low-risk raw ingredients
**Class 3**: Products first cooked and then packaged
**Class 4**: Products that are cooked after packaging

Each class of food products may require different handling and safety/hygiene precautions (Hill, 1987).

The shelf life of chilled processed foods is determined by

- The type of food
- The degree of microbial destruction or enzyme inactivation achieved during processing

- Packaging characteristics/properties
- Temperature conditions during processing, distribution, and storage (Fellows, 2000)

## 8.4 FREEZING

Freezing is one of the ancient methods of preservation, but its commercialization took place later than canning due to the lack of commercial refrigeration equipment (Maroulis and Saravacos, 2003). A major issue of the actual distribution chain is that a lot of products (mainly fresh fruits and vegetables) are seasonal and extremely perishable, with a short shelf life, limiting their availability to the market. Therefore, there is a need to use a preservation pattern, which extends food availability throughout the year, without compromising its quality and leading to a final product, similar to the "fresh" one, with a significantly longer shelf life. Among all preservation methods proposed to meet these requirements, freezing is one of the most popular. Nowadays, frozen foods have become part of the diet of developed countries. Today, in supermarkets, refrigerated and frozen foods occupy more than 50% of the floor space reserved for processed foods.

Freezing is a process of removing heat from a product, to bring down its temperature below its freezing point, accompanied by a phase change from water to ice. The common frozen storage temperature is –18°C or 0°F. There are two main factors that make freezing an appropriate preservation method (Reid, 1997): The first factor is related to the role of temperature in food stability. It is well established that, in general terms, reducing temperature leads to reduced rates of degradation phenomena and quality changes, thus stabilizing product quality. The second important factor is the change of water to ice, which implies the decrease of the concentration of the unfrozen phase. The liquid water notably promotes the microbiological, enzymatic, and chemical activity in the stored foods, reducing its storage life. The freezing of water to ice basically puts a stop to most of these activities.

Foods with reduced moisture levels have been proven to be less affected by microbiological deterioration; it is more stable concerning chemical and other degradation changes. When compared to refrigerated foods, it is the formation of ice crystals that cause frozen foods to keep much longer than refrigerated foods. Thus, a frozen system with ice has significantly reduced liquid moisture content (and much lower water activity, $a_w$), leading to a food matrix of enhanced stability.

The freezing of food is a complicated procedure that involves a variety of phenomena (thermodynamic, kinetic, etc.) and leads to more stable products with different properties. In this chapter, we will discuss the main scientific issues that have to be clarified to best comprehend and elucidate the freezing process.

### 8.4.1 The Freezing Process

During the freezing process, food undergoes three successive stages: the prefreezing period, the freezing stage, and the temperature decrease to the final storage temperature. Typically, in the first prefreezing process, there will be an initial drop in the temperature of the product until it reaches its initial freezing point, where the sensible heat of the product is removed. Then, the temperature of the product remains relatively steady as the latent heat is removed. At the final stage of deep freezing, the sensible heat of ice and that of the nonfrozen phase are removed until the final storage temperature is reached. At the same time, an amount of the latent heat is removed due to the freezing of additional water as a result of the decreasing temperature (Ramaswamy and Marcotte, 2006).

The freezing process for food products is much more complicated than that of pure water, as it is clearly shown in the following diagrams, where the freezing curve of pure water (Figure 8.1) is compared to that of a pure solution (Figure 8.2) and that of a common food product (Figure 8.3).

In the case of pure water, temperature is reduced from value $T_1$ to value $T_2$, since in a short time period, the latent heat of water (4.18 J/g°C) is removed until the initial freezing point of 0°C is reached ($T_2$). Before crystallization initiates, a temperature signaled as $S$ is reached, which is inferior to the real freezing point. When crystallization process begins, heat is abruptly released, leading to a sudden temperature increase to 0°C. This temperature remains constant until all water becomes ice. Phase change from water to ice is subsequently observed, where latent heat (333.2 J/g) is released. When solidification is done, further heat removal leads to temperature decrease with a rate of 1°C for every 2.1 J/g, until temperature $T_3$ is reached.

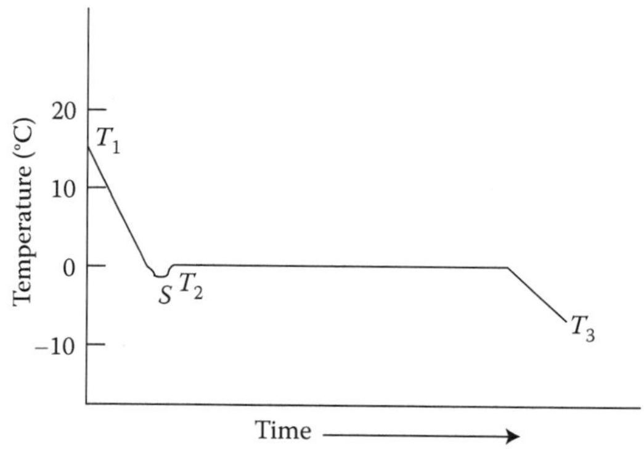


FIGURE 8.1 Freezing curve of pure water.

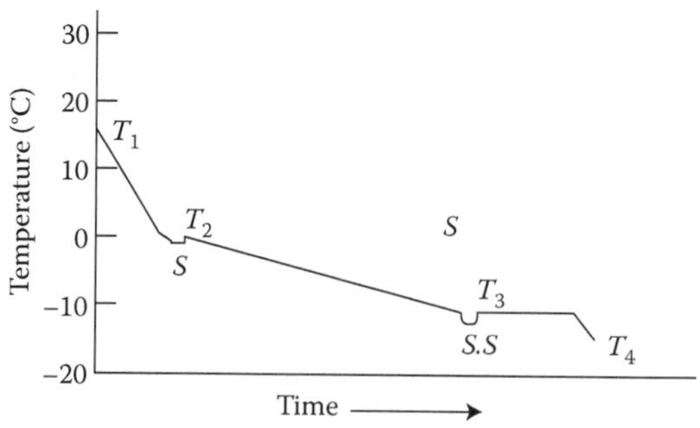


FIGURE 8.2 Freezing curve of a binary water–sucrose solution.

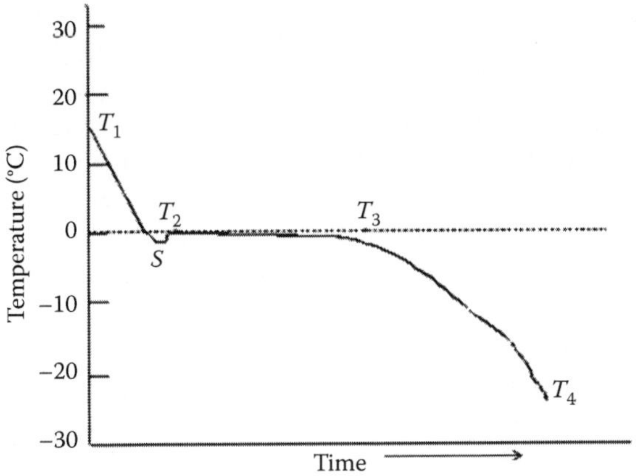


**FIGURE 8.3** Typical freezing curve of a high moisture food matrix.

Correspondingly, in Figure 8.2, the freezing curve for water–sucrose system is shown. When a pure solute is added to water, the freezing point of the pure solvent is decreased, allowing for the liquid and the solid phase to coexist in equilibrium at different temperatures, depending on the initial concentration of the solute and the amount of solvent that has been frozen. For an ideal, dilute solution, the freezing point can be determined using the following formula, based on Raoult's law:

$$\Delta T_f = \frac{R_g T_{Ao}^2 MW_A m}{1000L} \tag{8.4}$$

where $\Delta T_f$ is the freezing point depression, $R_g$ is the universal gas constant $(= 8.314\ J \cdot \text{mol}^{-1} \cdot \text{K}^{-1})$, $T_{Ao}$ is the freezing point of the pure solvent (absolute temperature), $MW_A$ is the molecular weight of the solvent, $m$ (molalilty) is the concentration of the solute (expressed in moles) per kilogram of the solvent, and $L$ is the latent heat of melting.

In Figure 8.2, after the initial freezing point of the solution $T_2$, an increased amount of the available water is gradually frozen, leading to a condensation of the solute in the unfrozen phase of the solution. As the solute concentration increases and more ice crystals are produced, there is a further freezing point depression. For any temperature in the range $T_2$–$T_3$, it is possible to estimate the concentration of the solute as well as the amount of ice related to the amount of the unfrozen phase using Equation 8.4.

The freezing process from temperature $T_2$ to temperature $T_3$ is realized through the release of the latent heat of melting and occurs in a slow rate, until the unfrozen state becomes saturated (or sometimes even oversaturated) and reaches its eutectic concentration (*S.S*), a value which is unique for every pure solute. Further, heat removal from this point leads to the production of mixed crystals with the participation of both ice and solute in a constant proportion. Therefore, at the saturation point, the amount of the unfrozen point is gradually reduced, but its composition remains constant, which is the reason why the temperature also remains constant during this phase.

Furthermore, freezing below $T_3$ occurs only if the solution has been totally solidified; in this case, freezing to the temperature $T_4$ would be extremely rapid, since only sensible heat is removed.

Finally, in the more complicated case of a food matrix, the real equilibrium taking place is much more complex and the freezing curves manage to give a rough approximation of the phenomenon. In Figure 8.3, where such a curve is depicted, it can be clearly observed that the straight line corresponding to the eutectic composition of the solute is not distinct, probably due to the presence of multiple cosolutes, with low initial concentrations and different eutectic points. As soon as each solute reaches its saturation point (eutectic point), its concentration within the solution will remain constant, whereas the concentration of the other cosolutes will gradually increase and, consequently, the freezing point will further reduce. The initial freezing point usually is observed for most foods in the range between –0.5°C and –3°C. In the temperature range $T_2$–$T_3$, the majority of water (at a percentage near 75%) is believed to have been transformed to ice. At the initial stages of the process, water is removed in the form of clear ice crystals, whereas at later stages, other solutes cocrystallize and mixed crystals are produced, giving birth to complicated solid structures. After temperature $T_3$ is reached, a small amount of available water remains unfrozen and in the case of any, even slight heat removal leads to a significant increase of the solute concentration in the unfrozen phase, causing a significant temperature decrease. Even at temperatures below –30°C, water continues to exist in food matrices in a liquid form (Rahman, 1995).

As an example of the significant effects of the special characteristics and composition of each food, in the following figure (Figure 8.4), the freezing curves of two different food matrices (sausage and Brussel sprout, respectively) are shown, obtained with the use of thermocouples and an automat data collection system (Data logger CR-10X, Campbell Scientific, Leicestershire, UK).

The mechanism through which ice crystals are formed during freezing (both size and number) is of great importance for the quality of the final product. There are two main aspects when considering the transformation of water to ice (Reid, 1997; Zaritzky, 2012): Thermodynamic factors that define the properties of the system under equilibrium conditions and kinetic factors that describe the rates at which this equilibrium might be approached.

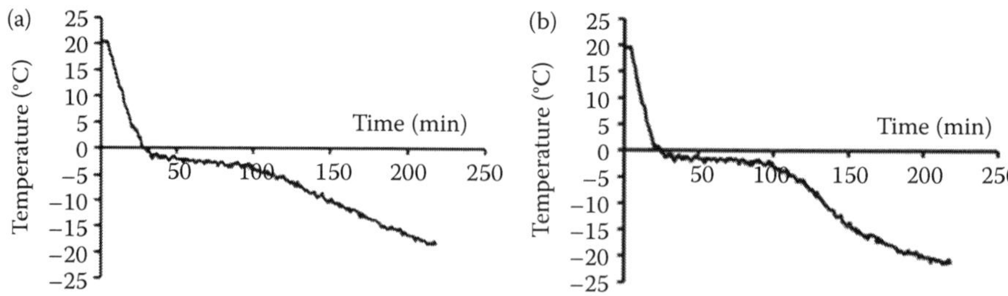


**FIGURE 8.4** Experimental freezing curves for the freezing of (a) Frankfurter and (b) Brussel sprout. The freezing process was realized in an horizontal closed freezer (Sanyo, MIR 553, Sanyo Electric Co, Ora-Gun, Gunma, Japan), set at –25°C, with air circulation (h ≃ 9.5 W/m$^2$ K, based on experimental data).

Freezing includes two successive processes when water is transformed in ice: the nucleation (formation of ice crystals) and the subsequent increase of the size of these crystals (crystal growth).

#### 8.4.1.1 Nucleation

Thermodynamics describes the equilibrium between a liquid and a solid phase. When ice and water coexist at atmospheric pressure, the temperature of the system reaches the freezing point of pure water ($T_f = 0°C$) and the amount of ice remains constant while no energy is either added or removed. Given that below 0°C, the equilibrium state is ice, one would expect that decreasing temperature below this point would automatically produce ice; this is not actually happening. It is necessary to get temperatures ($T$) substantially below the freezing point before ice begins to form due to supercooling (or undercooling phenomenon). Supercooling of pure water defined as $\Delta T_s = T_f - T$, is necessary for a nucleus, or seed to be produced before nucleation initiates. Nucleation can then be described as the process by which a minimum crystal is formed with a critical radius that can then expand and grow. Nucleation can either occur spontaneously (homogeneous nucleation) in water free from all impurities or it can take place on a foreign catalytic center (heterogeneous nucleation) when water molecules aggregate in a crystalline arrangement on nucleating agents such as active surfaces; this type of nucleation predominates in food systems. Homogeneous nucleation requires a higher supercooling than heterogeneous nucleation, and it practically does not take place until a temperature approaching −40°C is reached.

#### 8.4.1.2 Crystal Growth (Propagation)

Once nuclei have been formed, the crystals can then grow. Unlike nucleation, crystal growth can begin at temperatures just below the freezing point. At temperatures near the freezing point, crystal growth is favored to the creation of more nuclei. So, the small number of nuclei formed around the heterogeneous materials will begin to grow in size under these conditions. As the temperature is lowered, the rate of crystal growth increases. But at lower temperatures, nucleation overtakes crystal growth, with the result of many more tiny nuclei being formed before they begin to agglomerate into larger sizes. The rate of nucleation and crystal growth as a function of temperature are schematically shown in Figure 8.5 (Sahagian and Goff, 1996).

The size of ice crystals in the final frozen product significantly affects its quality and, therefore, is of great importance for consumer acceptability (Kiani and Sun, 2011). The formation of few nuclei leads to few crystals of big size, whereas the production of a high number of nuclei results in a correspondingly high number of small ice crystals when the freezing process is concluded. That implies that the size of ice crystals strongly depends on the nuclei formed, and consequently on the supercooling rate achieved as well as on the freezing rate. As it is clearly demonstrated in Figure 8.6, the rate of nuclei generation is suddenly increased when a critical value of supercooling is reached (point A), whereas the crystal growth increases smoothly as temperature is raised (Heldman and Singh, 1981). In this manner, if the rate of heat removal is slow and product temperature remains constant between 0°C and point A for a long period, each nucleus formed would have the possibility to grow

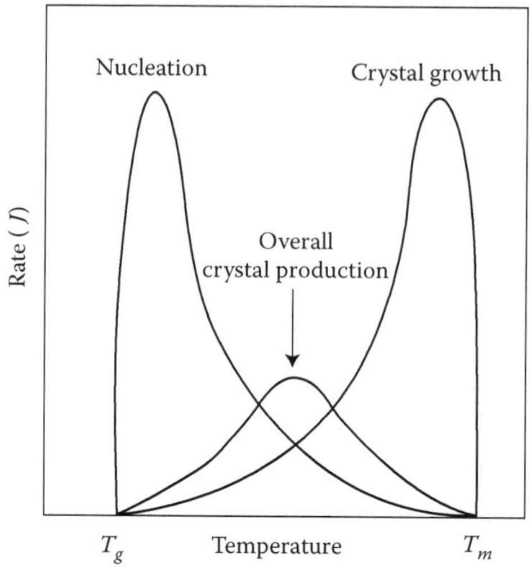


**FIGURE 8.5** Temperature effect on the rate of crystal production, where $T_g$ is the glass transition temperature and $T_m$ is the melting temperature.

significantly. On the other hand, when heat is rapidly removed, product temperature would be decreased below point A and an important number of nuclei would generate, without being capable of growing much. The conclusion is that the average size of ice crystals is adversely related to the number of nuclei formed; additionally, the number of nuclei generated can be controlled and manipulated by the rate of heat removal (and the freezing rate).

Another phenomenon strongly related to ice crystal formation is recrystallization, which is one of the most important degradation factors of frozen products and will be thoroughly studied in the following section.

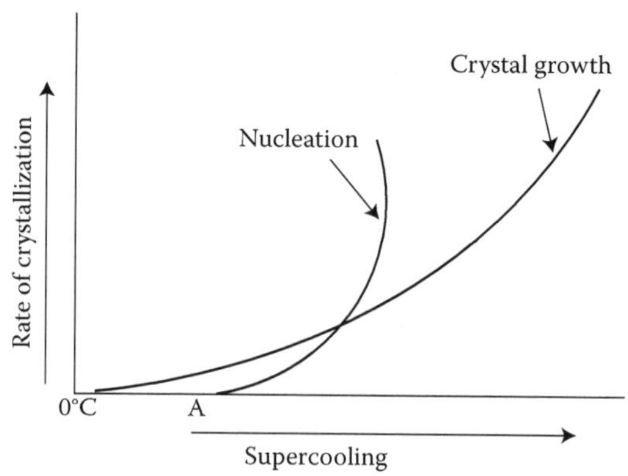


**FIGURE 8.6** Effect of supercooling on the rate of nucleation and crystal growth.

#### 8.4.1.3 Freezing Rate

The location, number, and size of the ice crystals formed determine the resulting texture of the frozen–thawed product. When the rate of heat removal is low, water can transfer from the interior of the cell fast, and the cell dehydrates, with water being incorporated into the external ice crystals. Therefore, slow freezing results in large ice crystal formation exclusively in the extracellular locations, which actually tend to squeeze the cell structures as they grow. This means that, upon thawing, they leave a product with severe textural breakdown. On the other hand, when there is a fast rate of heat removal, there is little water transfer from the interior of the cell, and numerous tiny ice crystals are formed, both intra- and extracellular, which do not grow appreciably in size. Hence, they do not significantly crush the cell structure, which should therefore retain a better texture upon thawing.

Besides the freezing rate, there are other factors that influence the quality of the final frozen–thawed product. Even though we prefer fast freezing procedures to attain the formation of numerous tiny nuclei in preference to larger nuclei, if the subsequent storage conditions are not the proper ones (large temperature fluctuations), these tiny ice crystals undergo recrystallization, and tend to merge, resulting in larger crystals with the result being that the advantage of fast freezing is lost. Therefore, appropriate temperature conditions during storage are as important as the freezing process itself.

## 8.5 MODELING OF FOOD FREEZING

### 8.5.1 Thermodynamics of Phase Change

The thermodynamics of food freezing are related to the changes in water within a food product as the freezing process proceeds (Heldman and Taylor, 1997; Singh and Heldman, 1981). As mentioned earlier, freezing of a food tissue is different from that of pure water in two important points: first, the temperature at which the initial ice crystals are formed is depressed below the temperature at which ice crystals begin to form in pure water (supercooling). The second differentiation is the gradual removal of latent heat of fusion as the product temperature decreases, giving a different shape to the freezing curve of food materials in comparison to that of pure water (Figures 8.1 through 8.3).

Another important characteristic of frozen foods is the relation between unfrozen water and temperature, which has a unique impact on the design of freezing equipment. For estimating this fraction, an assumption is usually adapted that pure ice crystals are formed during freezing and that all solutes are concentrated in the unfrozen water fraction.

### 8.5.2 Thermophysical Properties of Frozen Foods

When considering the freezing process, the most important properties include density, thermal conductivity, specific heat, enthalpy, latent heat, thermal diffusivity, and freezing point.

#### 8.5.2.1 Freezing Point

The equilibrium or initial freezing point is one of the most important physical properties of a food material because thermophysical properties change dramatically at the initial freezing point. Therefore, the initial freezing point is required for the prediction of thermophysical properties of frozen foods.

A freezing curve can be used to determine an initial freezing point for a food material. A typical freezing curve for most foods is shown in Figures 8.3 through 8.5. Alternatively, the onset, peak, and end of freezing can be determined from a heat flow exotherm, which can be developed using differential scanning calorimetry (DSC).

In theory, one cannot freeze all the water present in a food system, because there is always a certain fraction of it that is attached to the food constituents so tightly that it will not freeze. This constitutes "bound" water or unfreezable water. It is generally recognized that the majority of crystallization of ice takes place in the temperature range of –1 to –5°C. This temperature zone is therefore termed the zone of maximum ice crystal formation.

The initial freezing point of foods varies with water content, other nonwater components, component molecular weight, component interaction, and water-binding characteristics. It is obvious that the magnitude of freezing point depression is a function of product composition. The relationship between product composition and freezing temperature is most often explained in terms of the freezing temperature depression for a solution. The initial freezing point of an ideal, dilute solution can be calculated with the use of Raoult's equation, using Equation 8.4, as mentioned in an earlier section.

#### 8.5.2.2 Density

The overall influence of freezing on food density is relatively small, but an important change occurs just below the freezing point temperature (Heldman and Taylor, 1997). Above this temperature, the product density can be considered as constant. At the freezing point, as the freezing process proceeds, the fraction of frozen water in the product increases and the product density decreases rapidly.

The volume changes associated with foods undergoing freezing are, generally speaking, smaller than that associated with water. This can be due to several reasons (Ramaswamy and Marcotte, 2006). First, frozen food contains other components that reduce the weight percent of available water in the food. Since it is only water that is mostly responsible for expansion, the expansion will be much less when compared to that of pure water. Second, a certain fraction of water present in foods always remains as liquid water, which does not expand. And finally, there are several factors in the system (ice crystals, fat, and other components) that contract as temperature is lowered. Hence, the volume contracts as the temperature of the frozen product is lowered, which causes the density to increase.

#### 8.5.2.3 Heat Capacity

Heat capacity is the heat required to raise the temperature of a unit mass of water by a unit of temperature [$C_p = Q/(m\Delta T)$]. The heat capacity water is roughly constant between 0°C and 100°C (1 cal/(g°C) or 4.187 kJ/(kg°C)). Heat capacity of ice is 0.5 cal/(g°C), half that of water, and it goes down further as the temperature is lowered. The heat capacity of foods can generally be estimated based on the concentration of their components. Several empirical models exist in the literature (Heldman and Singh, 1981; Larkin et al., 1984; Martins and Silva, 2004; Ramaswamy and Marcotte, 2006).

#### 8.5.2.4 Enthalpy

Enthalpy is the heat content per unit mass of a food material with typical units of J/kg. Since it is difficult to define the absolute value of enthalpy, a zero value is usually arbitrarily defined at −40°C, 0°C, or any other convenient temperature. It is very convenient to use enthalpy for quantifying energy in frozen foods because it is difficult to separate latent and sensible heats in frozen foods as some unfrozen water exists in the foods even at very low temperature.

#### 8.5.2.5 Thermal Conductivity

The thermal conductivity of a food product depends on water content and its structure. The thermal conductivity of ice is about 4 times that of water. The thermal conductivity of ice increases at lower temperatures. Like heat capacity, thermal conductivity varies with temperature as with heat capacity, the thermal conductivity of foods is usually estimated based on its composition, and do not account for structural issues. To estimate the values of the thermal conductivity of frozen foods, some assumptions and approximations must be made about the structure of the food and the disposition of the various components dispersed in the food, including any air spaces in porous foods, and the direction (parallel or perpendicular) of heat flow relative to the layers of the components (Nesvadbha, 2008). To describe these properties, several simplified empirical models exist in the literature (Heldman and Singh, 1981; Ramaswamy and Marcotte, 2006).

#### 8.5.2.6 Thermal Diffusivity

The thermal property most often introduced into heat-transfer equations is thermal diffusivity and is numerically computed from thermal conductivity ($k$), heat capacity ($C_p$), and density ($\rho$) as shown below:

$$\alpha = \frac{k}{\rho C_p} \tag{8.5}$$

The ratio of the two (thermal conductivity divided by the volumetric heat capacity) determines the net ability of the material to respond to temperature changes. It also determines the ease at which it can undergo temperature changes (Ramaswamy and Marcotte, 2006).

## 8.6 PREDICTION OF FREEZING TIME

As already discussed, the rate of freezing is of great importance for food quality, with fast freezing having the least detrimental effects. The objective, therefore, especially where texture is the main quality attribute, is to choose a freezing method that assures a rapid freezing. To assess the rate of freezing, a proper definition must be provided for freezing time. After Hossain et al. (1992a,b), freezing time is the time required for the thermal center of the food (defined as the point that is the most difficult to freeze) to reach a predefined temperature. Some other important definitions proposed are (Delgado and Sun, 2001)

1. Nominal freezing time for a specific product, with known dimensions and a uniform initial temperature of 0°C is the time needed for its thermal center to reach a temperature 10°C below its initial freezing point (International Institute of Refrigeration, 1972).
2. The standard or effective freezing time is the total time needed for product temperature to reach from its initial value to a predetermined final value at the thermal center of the food. One end is fixed, in this case, but freezing time depends on initial temperature.
3. Another definition is the time taken to cross the zone of maximum ice crystal formation, that is, −1 to −5°C. This definition is somewhat similar to definition 1.

Freezing time can be either experimentally estimated with the freezing curve or theoretically calculated based on semiempirical or analytical models, or numerical methods. At this point, it should be stressed out that the exact prediction of freezing time is inevitable due to the complex effect of food thermophysical properties. Actually, the effectiveness of each prediction model strongly depends on the assumptions made and on how realistic these assumptions are. The main factors involved in those mathematical equations are the thermophysical properties of food and the parameters of the freezing process.

### 8.6.1 Analytical Solutions

These equations have been derived for a few ideal, simple phase change problems (Pham, 2012) and, therefore, are rarely of practical application, taking into account the complex geometry and behavior of food matrices. However, they can serve as initial estimations into numerical calculation methods for assessing a more accurate and realistic result.

The most popular analytical solution for freezing time is Plank's equation (Plank, 1913) or quasi-steady-state solution, which can be applied for different normal shapes (slab, sphere, and infinite cylinder). It is obtained by ignoring the specific heat of the product both above and below freezing and assuming the latent heat is released at a constant temperature $T_f$. The three main differential equations used refer to the successive stages of heat transfer during freezing, as depicted in Figure 8.7.

The freezing front $x$ gradually moves toward the center as latent heat is released and conducted away toward the surface. Heat transferred from the food interior to the surface by conduction is equal to the amount of heat removed toward the food environment by convection and, thus, the following equation can be obtained:

$$q = \frac{A(T_s - T_f)k}{x} = hA(T_\infty - T_s) \tag{8.6}$$

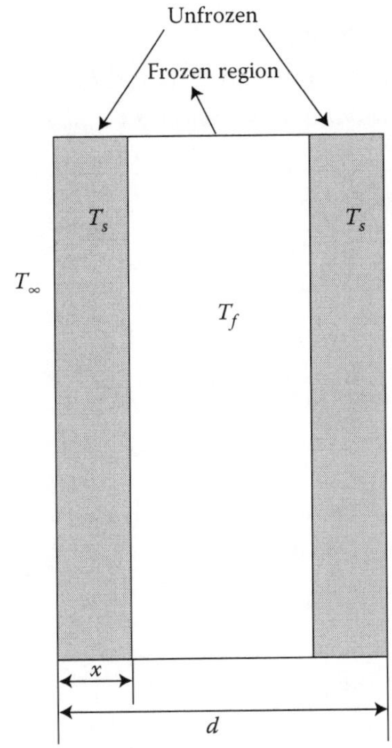


**FIGURE 8.7** Food slab during freezing. (Adapted from Singh R.P., Heldman D.R. 1981. *Food Process Engineering*. 158–215. Connecticut: AVI Publishing Company, Inc.)

where A is the product surface area, $T_s$ is the initial temperature of the product, $T_f$ is the product initial freezing point, $T_\infty$ is the temperature of the environment, $k$ is the food thermal conductivity, $x$ is the food slab thickness, and $h$ is the heat-transfer coefficient.

The rate of heat transfer from the freezing front can be described by the following equation ($L$ is the latent heat and $\rho$ is the product density):

$$q = AL\rho\frac{dx}{dt} \tag{8.7}$$

By equating Equations 8.6 and 8.7 and after integrating for normal geometrical shapes, Plank's equation is obtained, which calculates the freezing time:

$$t_F = \frac{\rho L}{T_f - T_\infty}\left[\frac{P\alpha}{h} - \frac{R\alpha^2}{k}\right] \tag{8.8}$$

where $P$ and $R$ are constants that vary depending on product geometry (Table 8.2) and $\alpha$ is product thickness in case of an infinite slab, diameter for an infinite cylinder, and diameter for a sphere.

**TABLE 8.2**
**Values for Plank's P and R Constants Depending on Shape**

| Shape | P | R |
|---|---|---|
| Infinite slab | 1/2 | 1/8 |
| Sphere | 1/6 | 1/24 |
| Infinite cylinder | 1/4 | 1/16 |

The most obvious difficulties are selection of a latent heat magnitude ($L$) and a value for thermal conductivity ($k$), which actually is not a constant value, since both temperature and food structure change during freezing. Except for the limitations and assumptions discussed earlier, the basic equation does not account for the time required to cool the product from its initial temperature to the freezing point and the time to cool to the final temperature after the completion of freezing (Heldman and Taylor, 1997).

Later in years, a lot of studies have been published proposing modifications of Plank's equation, mostly oriented in introducing the sensible heat removed from food until it reaches the initial freezing point (Cleland and Earle, 1976, 1977). Working with fish, Nagaoka et al. (1955), introduced in their equation both the precooling and freezing times, taking into account the initial product temperature above freezing point and its final temperature below this point. Their equation is the following one:

$$t_F = (C_{PU}(T_i - T_f) + X_W L + C_{PF}(T_f - T))(1 + 0.008(T_S - T_f)) \left(\frac{\rho}{(T_f - T_m)}\right)\left(\frac{P\alpha}{h} + \frac{R\alpha^2}{k}\right) \tag{8.9}$$

where $T_f$ = freezing point of the product, $X_W$ = water fraction, $T_s$ = product initial temperature, $T_m$ = temperature of the freezing medium, $T$ = final product temperature, $C_{PU}$ = heat capacity (unfrozen state), $C_{PF}$ = capacity (frozen state), $L$ = latent heat of fusion, and $\rho$ = density of the frozen product.

Finally, numerous work has been done on realistic geometrical shapes, which are based on modifications of the initial Plank's equation (Cleland and Earle, 1982; Hossain et al., 1992a,b; Ilicali et al., 1996; Lopez-Leiva and Hallstrom, 2003; Pham, 1991).

### 8.6.2 Other Methods for Calculating Freezing Time

Taking into account the laborious work required for analytical solutions and the rapid evolution of high-speed computers, numerical solutions have been developed with greater accuracy and flexibility. These techniques have the advantage of being applicable for nonstandard geometries, and for the heterogeneous food structure, taking into account the variable thermophysical properties of foods, and avoiding simplified assumptions. Usually, a finite element (Abdalla and Singh, 1985; Bonacina et al., 1973; Cleland et al., 1984), finite differences (Cleland and Earle, 1984; Sheen and Hayakawa, 1990), and a finite volume method are applied.

A more recent approach involves the use of artificial neural networks (ANNs) (Mittal and Zhang, 2000) or genetic algorithms (Goni et al., 2008) to assess the freezing time. Their principle is based on a batch of experimental data, used for "training" of the system to select the most appropriate mathematical model. The main advantage of this method is the speed, the rapid calculations achieved, and the easiness, which allow for immediate intervention in case some parameters need to be corrected.

## 8.7 GLASS TRANSITION IN FROZEN FOODS

Glass transition is not related to the release of latent heat; yet, it can be observed by dramatic changes in dielectric, mechanical, and physicochemical properties of frozen foods and is a property of the unfrozen, concentrated phase of frozen foods.

Glass transition is of high importance for frozen foods because it has been demonstrated (Goff, 1997) that several quality degradation mechanisms (enzymatic reactions, recrystallization, etc.) are highly reduced or even eliminated when the unfrozen phase is in the glassy state. Additionally, the kinetics of reactions affecting quality at temperatures close to the glass transition temperature (as will be dealt with later in this chapter) can be described by the Williams–Landel–Ferry (WLF) equation that the difference between storage temperature ($T$) and $T_g{}'$ is the crucial factor. Finally, when understanding the qualitative and quantitative effect of endogenous and environmental factors (e.g., temperature, freezing rate, composition, etc.) on glass transition phenomenon, it is possible to accordingly design its production or even selectively modify its composition to prolong its shelf life.

Glass transition is defined as the phenomenon observed when a glass is heated until it starts behaving as a supercooled melt (Levine and Slade, 1992; Roos, 1995; Sablani, 2012). The glass transition relates to the phenomena observed when a supercooled, malleable liquid, or rubbery material is changed into a disordered solid glass upon cooling, or conversely when a brittle glass is changed upon heating into a supercooled liquid or a rubbery material (Roudaut et al., 2004). This change is highly dependent on the temperature, time, food composition, and, where thermodynamics is concerned, it is a second-order transition. That means (Roos, 1995) that the second derivative of Gibbs energy is discontinued at the glass transition temperature, whereas enthalpy, entropy, and the volume of the two phases remain unchanged. Thus, the equation describing the main thermodynamic parameters is

$$\left(\frac{\partial^2 G}{\partial T^2}\right) = \frac{-C_p}{T}, \left(\frac{\partial^2 G}{\partial p \partial T}\right) = V\alpha, \left(\frac{\partial^2 G}{\partial p^2}\right) = -V\beta \tag{8.10}$$

where $\alpha$ is the constant of thermal change and $\beta$ is a measure of compressibility.

From this equation, it is obvious that $C_p$, $\alpha$, and $\beta$ are discontinuous, and their experimental determination can be used to trace and study this transition.

### 8.7.1 Definitions, Physical Parameters, and Phase Diagrams

During the freezing of foods, a two-step crystallization procedure is followed (nucleation and crystal growth) before ice crystals are formed. As pure water is frozen, the viscosity of the liquid phase rises. If the liquid is cooled very quickly, the viscosity may reach very high values that molecular rearrangements in the liquid become extremely slow avoiding ice crystallization. When the viscosity reaches $10^{11} - 10^{12}$ Pa · s, a solidification (vitrification) occurs, and the concentrated phase surrounding the ice crystals becomes a glass (Blond and Le Meste, 2004). The temperature at which this transition appears is called $T_g'$, the glass transition temperature of the maximally freeze-concentrated system. The liquid is in a metastable state until it gets below the glass transition temperature ($T_g'$) where the system is an amorphous solid or glass. A glass is defined as a non-equilibrium, metastable, amorphous, and disordered solid of very high viscosity (Sablani, 2012).

Glass transition is a change occurring in a temperature zone, rather than at a specific temperature point. When a material is gradually subjected to a temperature change in that temperature zone, the successive forms it assumes are

1. Glassy state: When referring to a glass, we assume an amorphous, non-crystallized solid, which is actually a supercooled liquid of high viscosity ($\eta > 10^{10} - 10^{14}$ Pa · s), which is met in a nonstable state that can support its own weight against flow, under gravity (Levine and Slade, 1992). When the material is in that state, changes occur in a very slow rate and are described with the term "physical aging." Food matrices that are in the glassy state are considered stable.
2. Glass transition range: The glass is transformed into a viscous, supercooled liquid, followed by an important change in the mechanical properties of the material. In that temperature zone, with a range of 10–30°C, even a slight temperature change can lead to significant changes of structure and stability of the food matrix. The main change observed refers to molecular mobility that practically ceases below $T_g$, due to "freezing" of molecules in the glassy state.
3. Rubbery zone: This zone usually initiates—with the exception of low molecular compounds such as sugars or water—with a subzone where the mechanical properties of the material remain constant. The range of this subzone depends on the molecular weight and is strongly influenced by the linearity of molecules. After this subzone of constant mechanical properties, there is another zone where plastic materials behave either as viscoelastic or as plastic liquids, depending on the experimental time. Within the zone of viscoelastic behavior, there is no possibility for whole chains to exhibit mobility, whereas segments of chain can still move.

4. Zone of liquid flow: In this zone, normal flow is observed and complex polymers can behave as viscous liquids (melts). Mechanical properties of semicrystallized polymers depend on the level of crystallinity. For $T_g < T < T_m$, increased hardness is observed before melting initiates at the melting point, $T_m$.

For frozen foods, as ice is gradually formed, soluble solids are concentrated in the nonfrozen phase. For each ratio of ice to the nonfrozen phase, there is a freezing temperature of equilibrium, which depends on the concentration of the solute. This equilibrium is depicted in a so-called phase diagram (Figure 8.7), and involves a liquidus curve, extending from the melting point of pure water $T_f$ (0°C), until the eutectic temperature ($T_e$) of the solute (which is the point where the solute has been maximally concentrated due to freezing (saturation point)) and the kinetically controlled glass transition curve ($T_g$ curve), are drawn. The curve $T_m$ represents the temperature at which ice begins to separate as a function of initial concentration, or the concentration of the freeze-concentrated phase as a function of temperature. For all points on this curve, the freeze-concentrated phase is in equilibrium with ice. For soluble solutes, the temperature of the glass transition depends on the water content; it decreases when the water content increases. The intersection of the two curves could provide a better estimation of $T_g'$ (Hatley et al., 1991). The freezing of water is stopped at this temperature; the concentration of the maximally freeze-concentrated phase is called $C_g'$ and its water content is considered the "unfreezable" water.

Below and to the right of the glass transition line, the solution is in the amorphous glassy state with or without the presence of ice (depending on the temperature and the freezing path followed), whereas above and to the left of the glass transition line, the solution is in the liquid state, with or without ice, depending on temperature.

For the carbohydrate solution depicted in Figure 8.8, point A shows the initial concentration of 40% at room temperature and the initial $T_g$ of this solution, at room temperature before phase separation is marked as point $A'$ (which would occur only if there was an immediate undercooling that would not allow for ice formation). However, in the case of slow cooling below its equilibrium freezing point, the crystallization process (nucleation and crystal growth) begins at point B, and water starts being removed as ice. As this freezing process continues, the solute concentration of the unfrozen phase gradually increases and, subsequently, the equilibrium freezing point of the unfrozen phase decreases following the path marked starting from point B. At the same time, the glass transition temperature $T_g$ of the unfrozen phase follows the path $A'$ of the glass transition line, leading to a rapid increase in viscosity, particularly in the late stages of the freezing procedure (Roos and Karel, 1991a,b). Since cocrystallization of the solute at the eutectic point $T_e$ is not likely to happen, freeze concentration continues into a nonequilibrium state, until a critical, solute-dependent concentration is reached. At this crucial point, the unfrozen phase has a very limited mobility and its physical state changes from a viscoelastic liquid to an amorphous solid glass (Goff, 1992, 1994). Point C (the intersection of the nonequilibrium section of the liquidus curve and the glass transition curve)

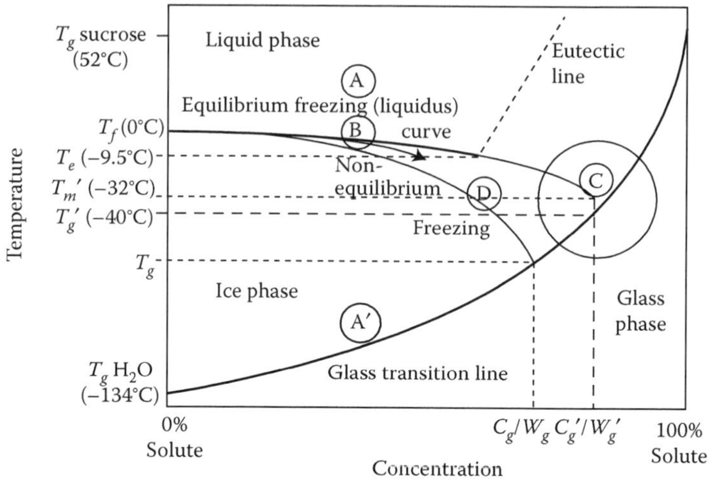


**FIGURE 8.8** Indicative state diagram for an aqueous sucrose solution showing the glass transition line, liquidus curve, theoretical eutectic line and the various physical states defined within the boundaries. Details of specific areas, points and paths are provided within the text. (Adapted from Goff H.D. 1997. *Quality in Frozen Food.* New York: Chapman & Hall.)

represents the glass transition temperature of the maximally concentrated phase ($T_g'$), where ice formation does not occur any more. The corresponding maximum concentrations of water and carbohydrate within the glass at $T_g'$ are denoted as $W_g'$ and $C_g'$, respectively.

Besides thermodynamic issues, freezing process is also controlled by important kinetic factors. At freezing temperatures, water removal in the form of ice is limited by the gradual increase of concentration and viscosity increase acts as the limiting factor for growth. Therefore, under special occasions, there is a possibility for nonequilibrium freezing to occur, resulting in a partial dilute glass (Roos and Karel, 1991a,b), depicted as the path D in the phase diagram, leading to a lower $T_g$ than $T_g'$ and higher water content in the glass ($W_g$) caused by excess undercooled water plasticized within the glass (Goff, 1997).

Levine and Slade (1988) have promoted the idea that the transition of a liquid to a glassy state for the maximally freeze-concentrated fraction, and consequently the temperature at which this glass transition takes place ($T_g'$), was the threshold of instability, and that the kinetics above this temperature were controlled by the difference between the storage temperature and $T_g'$. A good stability being related to the possibility of maintaining the product below $T_g'$, the formulation change could be a possible improvement. Therefore, knowing the effect of environmental or indigenous factors on the glass transition of the system, production, formulation, or storage of foods can be decided (e.g., increase of the $T_g'$ of a system) to enhance the stability of frozen foods (Champion et al., 2000). These phase diagrams constructed for several food components can be a useful tool for deciding the ideal conditions of processing and storage of frozen foods.

### 8.7.2 Reactions in Glassy/Rubbery Phase: Stability of Frozen Foods

Owing to significant modifications in the mechanical properties of the materials in the glassy or rubbery phase, it is expected to observe an influence of the physical state of the material on the rate of chemical reactions that depend on molecular mobility and diffusion. Bearing in mind the limited free volume, the increased viscosity, and the decreased molecular mobility, low rates of chemical reactions are expected in the glassy state, leading to a lesser extent of quality loss.

On the other hand, in the rubbery phase, due to an increase of the free volume and a viscosity decrease, an abrupt increase of chemical reaction rates is expected. As a result of this approach, several phenomena correlated earlier only to water activity issues, can now be addressed, also taking into account the important influence of the food physical state. More specifically, frozen food stability can be studied more profoundly using this approach.

The physical state of frozen foods is more sensitive to temperature changes when compared to other foods of low humidity, due to phase changes of water and solutes. At low temperatures, food solutes are concentrated in the concentrated amorphous phase, which is plasticized by the nonfrozen water. At typical temperatures of food storage, the physical state of the food matrix depends on properties of food components and their effect on ice melting (Roos, 1995).

When considering the main forms of quality deterioration of frozen foods, they are caused by the activity of the nonfrozen phase, through chemical or enzymatic reactions, leading in many cases to a significant propagation of ice crystals (Goff, 1992). Most raw vegetables can be preserved for a rather short time even at temperatures close to –20°C, due to degradation of texture, color, flavor, and nutritional value as a result of enzyme activity (for instance, pectic enzymes lead to texture degradation, polyphenoloxidase (PPO), and chlorophyllase, peroxidase may cause color changes, and ascorbic acid oxidase and thiaminase lead to nutritional loss, etc.) (Reid, 1990). Additionally, due to the concentration of the nonfrozen phase, its properties, such as the pH, titratable acidity, ionic strength, viscosity, initial freezing point, surface tension, and so on change significantly. Consequently, food may suffer from detrimental changes or even from an important structure collapse, due to crystals size. These crystals have been proven to be unstable and undergo major changes during frozen storage concerning their number, size, and shape (recrystallization phenomena), which strongly affect the overall quality of frozen food.

Mobility and activity of molecules that are responsible for food degradation reactions strongly depend on glass transition zone; therefore, this transition is of major importance for food stability, since below glass transition temperature, water can be considered as "frozen" and unavailable for any deterioration reaction. As mentioned earlier, storing frozen food below its glass transition point can stabilize its quality and extend its shelf life. Alternatively, since glass transition temperature of most raw materials is very low, there is the possibility to slightly modify food tissue to increase its glass transition temperature above the usual storage temperature.

In recent literature, a lot of material has been published on the effect of glass transition phenomena on the rate of deterioration reactions of frozen foods.

Extensive literature has been published on the effect of temperature on enzyme activity in amorphous systems (Burin et al., 2002; Champion et al., 2000; Chaudhary et al., 2013; Mazzobre et al., 1997a,b; Neri et al., 2010; Schebor et al., 1996, 1999; Terefe and Hendrickx, 2002; Terefe et al., 2004). In some of these publications, model systems are used to facilitate the control and measurement of several parameters. Although the focus of older literature was on the loss of enzymatic activity due to the reduction of the translational and rotational motions of molecules at the glassy state, recent research is oriented to a more complex explanation, based on the synergistic effect of glass transition phenomena, water activity reduction, viscosity changes, and even the special characteristics of the frozen matrix under study.

### 8.7.3 Quantifying the Effect of Temperature of Frozen Foods

Temperature is the most important factor that influences the rate of reactions occurring in frozen foods, and the Arrhenius law (Equation 8.11), derived from thermodynamic laws and statistical mechanics principles, is the most widely used (Arrhenius, 1889).

$$k = k_A \exp\left(-\frac{E_A}{RT}\right) \Rightarrow \ln k = \ln k_A - \frac{E_A}{R}\left(\frac{1}{T}\right) \tag{8.11}$$

with $k_A$ representing the Arrhenius equation constant and $E_A$, in joules or calories per mole, is defined as the activation energy, that is, the excess energy barrier that quality parameter A needs to overcome to proceed to degrade products. $R$ is the universal gas constant (1.9872 cal/mole.K or 8.3144 J/mole.K).

To estimate the temperature effect on the reaction rate of a specific quality deterioration mode, values of $k$ are estimated at different temperatures, in the range of interest, and lnk is plotted versus the term of $1/T$ in a semilog graph. A straight line is obtained with a slope of $-E_A/R$ from which the activation energy is calculated.

It should be noted that the Arrhenius equation implies that $k_A$ is the value of the reaction rate at 0 K that is of no practical interest. Alternatively, the use of a reference temperature, $T_{ref}$, is recommended, corresponding to a representative value in the temperature range of the process/storage of study. Equation 8.11 is then mathematically transformed as follows (Equation 8.12):

$$k = k_{ref} \exp\left[\frac{-E_A}{R}\left(\frac{1}{T} - \frac{1}{T_{ref}}\right)\right] \tag{8.12}$$

where $k_{ref}$ is the rate constant at the reference temperature $T_{ref}$. The value of $E_A$ is, in that case, calculated from the linear regression of lnk versus $(1/T - 1/T_{ref})$.

Values of $T_{ref}$ usually employed are 255 K for frozen, 273 K for chilled, and 295 K for ambient temperature-stored food products. Besides giving the constant a practical physical meaning, the above transformation of the Arrhenius equation provides enhanced stability during numerical parameter estimation and integration.

However, there are numerous cases published where there is an obvious deviation from the Arrhenius law. Examples of the formation of metastable glasses that deteriorate following a kinetic pattern that deviates from the Arrhenius law, include frozen carbohydrate-containing solutions or food products (Biliaderis et al., 1999; Blond and Simatos, 1991; Carrington et al., 1996; Champion et al., 2000; Furuki, 2002), spray-dried milk (Bushill et al., 1965), whey powder and dehydrated vegetables (Buera and Karel, 1993, 1995), osmotically dehydrofrozen fruits and vegetables (Chiralt et al., 2001; Torreggiani et al., 1999), and so on. In the systems that are subject to glass transition, due to drastic acceleration of the diffusion-controlled reactions above $T_g$, the dependence of the rate of a food reaction on temperature cannot be described by a single Arrhenius equation. In the rubbery state above $T_g$, the activation energy is not constant, but is rather a function of temperature. This behavior has been often described by an alternative equation, the WLF expression that empirically models the temperature dependence of mechanical and dielectric relaxations in the range $T_g < T < T_g + 100$:

$$\log\frac{k_{ref}}{k} = \frac{C_1(T - T_{ref})}{C_2 + (T - T_{ref})} \tag{8.13}$$

where $k_{ref}$ is the rate constant at the reference temperature $T_{ref}$ ($T_{ref} > T_g$) and $C_1$, $C_2$ are system-dependent coefficients.

Williams et al. (1955), assuming $T_{ref} = T_g$ and applying WLF equation for data available for various polymers, estimated mean values of the coefficients $C_1 = -17.44$ and $C_2 = 51.6$. However, the uniform application of these constants is often problematic (Buera and Karel, 1995; Peleg, 1992; Taoukis and Giannakourou, 2004; Terefe and Hendrickx, 2002) and the calculation of system-specific values, whenever possible, should be preferred, using Equation 8.13 with $T_{ref} = T_g$ and rearranging the mathematical expression to the following form (Equation 8.14):

$$\left[\log\frac{k_g}{k}\right]^{-1} = \frac{-C_2}{C_1(T - T_g)} - \frac{1}{C_1} \tag{8.14}$$

It has been stated that the Arrhenius model is more adequate for describing the temperature dependence of reactions within the glassy state of food matrices, and also at 100°C above the glass transition temperature, but is not applicable within the rubbery state (Slade et al., 1989). Terefe and Hendrickx (2002) studied the kinetics of the pectin methylesterase, catalyzed de-esterification of pectin in frozen food model in a wide temperature range (–24 to 20°C), and observed that a single Arrhenius plot could not describe the temperature dependence in the whole range. Owing to an important change of mechanical and mobility properties between glassy and rubbery systems, a break in the Arrhenius plot at or near the glass transition temperature of the system may be expected. At this point, it should be stressed that in some cases measuring the glass transition temperature may not be adequate for explaining the break in an

Arrhenius plot (Giannakourou and Taoukis, 2003a; Manzocco et al., 1999; Terefe and Hendrickx, 2002) and other changes should be taken into consideration.

### 8.7.4 Measuring Glass Transition of Frozen Foods

Since glass transition involves changes in thermal and mechanical properties of foods, thermal analysis techniques are commonly used to evaluate this process. DSC is the technique most often used to measure the glass transition temperature, in which the difference in energy inputs into the product under measurement and a reference material is measured as a function of temperature while both materials are subjected to a controlled temperature program. For frozen solutions of small solutes (sugars, polyols), the thermograms show a two-step endotherm (Figure 8.9).

As shown in the thermogram, second-order glass transitions are detected by endothermic step changes in heat flow, describing discontinuous changes to $C_p$. To trace the onset and endpoint of the glass transition, the baseline before and after the transition is used, as well as its straight line extrapolation.

Modulated DSC is usually applied where the temperature during the measurement is modulated by a controllable period and amplitude, leading to improved resolution (slow heating rates) and good sensitivity (instantaneous fast heating rates).

Other alternative techniques used are mechanical methods (thermomechanical analysis, dynamic mechanical thermal analysis (DMA, DMTA, etc.) (Goff, 1997), molecular mobility measurements (nuclear magnetic resonance (NMR), electron spin resonance, dielectric analysis, microscopy, etc.). Recently, moisture sorption isotherms are also proposed as a fast, quite accurate alternative to traditional thermal methods for glass transition determination, since new automatic isotherm generators can be used to produce high-resolution, dynamic isotherms (Carter and Schmidt, 2012).

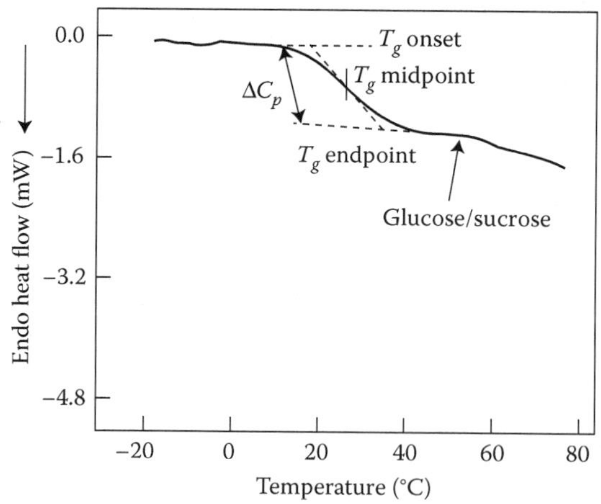


**FIGURE 8.9** Typical thermogram (temperature-heat flow profile) from DSC analysis of a frozen carbohydrate solution.

## 8.8 QUALITY OF FROZEN FOODS: PHYSICAL AND CHEMICAL CHANGES DURING FROZEN STORAGE

For each particular food, there is a finite time period and after its production, it will retain a required level of sensory and safety quality characteristic under stated conditions of storage (Taoukis and Labuza, 1997). This length of time can be generally defined as the shelf life of the food product or, alternatively, durability. Actually, there is not only one established and uniformly accepted definition for the term of "shelf-life," but a variety of approaches. The International Institute of Refrigeration (IIR) suggests two alternative definitions of shelf life in the case of frozen foods, HQL (high-quality life), which is the time from freezing of the product to the development of a just noticeable sensory difference (70–80% correct answers in a triangular sensory test) and PSL (practical storage life) that corresponds to the period of proper frozen storage after freezing of an initial high-quality product during which, the sensory quality remains suitable for consumption. A practical definition proposed by Fu and Labuza (1992) defines shelf life as "the time period within which the food is safe to consume and/or has an acceptable quality to consumers."

As for all food products, frozen foods are perishable, with a limited shelf life; during their life cycle from production to final consumption, they gradually deteriorate. Each type of frozen food loses its quality due to different mechanisms. Freezing extends food shelf life, maintaining at a satisfactory level its quality attributes. This is accomplished by a significant slowdown of most deterioration reactions that occur during storage of frozen foods. Nevertheless, quality does not cease to deteriorate during its commercial shelf life. There are numerous factors that affect frozen food quality that can be divided into two categories: processing and compositional factors. The quality factors associated with processing parameters are mostly related to the ice phase and the characteristics of the ice crystals. On the other hand, stability and quality issues associated with compositional factors mostly concern chemical reactions and affect sensory attributes, such as color, flavor, texture, as well as nutritional properties. For example, for frozen vegetables these factors are namely raw material, storage, and procedures before freezing, blanching, fluctuating temperatures, and packaging. Actually, time–temperature–tolerance (TTT) and product–processing–packaging (PPP) concepts are used to monitor and control the effects of temperature fluctuations on frozen food quality during production, distribution, and storage (Bogh-Sorensen, 1984).

In this section, we will summarize changes that occur due to freezing in different types of frozen food products.

### 8.8.1 Changes of Plant Tissue (Fruits and Vegetables) Due to Freezing

Vegetables are among the most important frozen commodities. The industrial freezing of untreated vegetables can cause significant tissue damage, especially affecting microstructural integrity of plant cells. Upon thawing, many vegetables exhibit significant off-flavors, off-odors, discoloration, and loss of the original texture. These detrimental effects have been shown to be a consequence of the disruption of membranes and the activation of enzyme systems within the cells. Consequently, many vegetables must be blanched prior to freezing (Reid, 1998). The influence of the freezing rate on plant tissues is of great importance (Fellows, 2000). When slow freezing is applied, ice crystals grow in intercellular spaces and may cause mechanical damage at adjacent cell walls. Owing to a water vapor gradient, water moves from the cells to the growing crystals. Cells become dehydrated and permanently damaged, ice crystals tend to grow to an undesirable size, and the overall cell structure tends to collapse. On thawing, cells do not regain their original shape and turgidity. As a consequence, food loses its original texture and cellular material leaks out from ruptured cells (termed "drip loss"). On the other hand, in fast freezing, smaller ice crystals form simultaneously within both cells and intercellular spaces, without forming water vapor gradients. The texture of the food is thus retained and there is little physical damage to cells.

As far as frozen storage is concerned, this step is very crucial for food quality and commercial viability. In general, the lower the temperature of frozen storage, the lower is the rate of microbiological and biochemical changes. However, freezing and frozen storage do not inactivate enzymes and do not extinguish microorganism risks. Therefore, physical and chemical changes take place within the product during its storage; in the case of frozen vegetables, at the temperatures concerned microbial alterations are of minor concern since very few bacteria can grow below −5°C and no fungi or bacteria have been reported to grow below −12.5°C (Parreno and Alvarez Torres, 2012).

Physical changes influencing frozen vegetable quality during storage are related to recrystallization and ice sublimation, both affecting ice crystals stability. Recrystallization consists of modifications in the number, size, shape, and orientation of the ice crystals, which are present within and on the surface of the product after freezing. In literature (Fellows, 2000) three main mechanisms of recrystallization are reported, namely isomass recrystallization (a change in surface shape or internal structure), accretive recrystallization (two adjacent ice crystals join together to form a larger crystal, leading to an overall reduction in the number of crystals in the food), and finally and most importantly, migratory recrystallization, which is actually an increase in the average size and a reduction in the average number of crystals, caused by the growth of larger crystals at the expense of smaller crystals. The latter is the major cause of tissue damage for frozen foods and it is strongly accelerated by fluctuations in temperature during frozen storage.

Sublimation of ice at the surface of the product can also take place when packaging is not efficient, leading to desiccation and water accumulation inside the packaging in the form of frost. The difference between the water vapor pressure on food surfaces and that in the surrounding atmosphere is the driving force for dehydration. Surface ice sublimes forming a porous dehydrated layer, whose thickness increases as time elapses (Campañone et al., 2005a,b). In addition to undesirable weight loss, a variety of oxidative and other deteriorating reactions may occur at the food surface, leading to extensive quality loss. Recrystallization and ice sublimation may cause a detrimental phenomenon, known as "freezer burn" that involves not only surface dehydration, but also degradation reactions in color, texture, and flavor at the surface during frozen storage.

Chemical changes, mainly observed as enzymatic and nonenzymatic reactions occur in plant tissue of frozen foods during freezing, and their effect on quality is very important, as they significantly alter the sensory attributes of the original plant tissue.

#### 8.8.1.1 Color Changes

Color is the most important attribute judged by the consumer, readily related to maturity, processing conditions, and the overall food quality. Frozen vegetables and fruits undergo several color changes during storage, due to alterations in natural pigments, chlorophylls, anthocyanins, carotenoids, or even by enzymatic browning. The characteristic green color of popular frozen vegetables, such as frozen beans, peas, spinach, broccoli, watercress, and so on gradually fades away, giving way to a brownish color during storage at temperatures near −18°C, due to the transformation of chlorophyll α and b into their corresponding pheophytins. This change is related to the substitution of the central ion of $Mg^{+2}$ of chlorophyll from two hydrogen ions and is catalyzed by endogenous acids in the plant tissue. Besides pheophytin formation, there are other paths, leading to loss of green color due to the activity of enzymes such as chlorophyllase or lipoxygenase-induced oxidation of polyunsaturated fatty acids in the presence of oxygen.

As it will be thoroughly explained in another chapter, blanching conditions influence to a great extent the rate of chlorophyll degradation (Cano, 1996, Kmiecik and Lisiewska, 1999; Martinez et al., 2013). When blanching involves high-temperatures-short time, it is found that chlorophyll pigment is retained better, whereas adding salts, such as NaCl, KCl, and $K_2SO_4$ into blanching water can reduce pheophytin formation in Brussel sprouts and spinach (Cano, 1996).

Another cause of color loss is the destruction of anthocyanins, which are water-soluble pigments, responsible for the red colors. Under certain conditions, they may be destroyed by the enzyme-induced oxidation of polyphenols, leading to significant color loss, during subsequent frozen storage. Cano (1996) states that the food pH is the most crucial factor affecting anthocyanin loss; at high pH values, with the presence of oxygen, the rate of color loss is rapid. Carotenoid oxidation, on the other hand, is another cause of color change, mainly due to acids, catalyzed by certain enzymes and is sensitive to light. These pigments, as well as anthocyanins (both categories being lipo-soluble pigments) are not affected during blanching; on the contrary, blanching protects these pigments due to the inactivation of enzymes, such as lipoxygenase, peroxidase, and so on that catalyze the oxidation of phenolic compounds.

In recent literature, there are numerous publications studying color changes of different frozen vegetables. A summarizing table follows (Table 8.3), showing some representative studies and their results.

Finally, another important cause of color degradation is enzymatic browning, when endogenous phenolic compounds are transformed into o-diphenols, and then into o-quinones, in the presence of oxygen and PPO. These quinones tend to condense and react with other phenolic compounds, amino acids, and so on, without the presence of enzymes, leading to the formation of complex brownish polymers. These phenomena occur especially in vegetables such as cauliflowers, potatoes, and

**TABLE 8.3**
**Characteristic Literature Studies about Color Changes of Frozen Vegetables**

| Food Matrix | Quality Index | Temperature Conditions | Kinetic Results | Reference |
|---|---|---|---|---|
| Chive (unblanched) | Chlorophyll | −20°C (measurement every 3 months)<br>−30°C (measurement every 3 months) | 25% loss<br>15% loss<br>(after 12 months) | Kmiecik and Lisiewska (1999) |
| Tomato | Carotenoids | −20°C (after 12 months)<br>−30°C (after 12 months) | 36% loss<br>16% loss | Lisiewska and Kmiecik (2000) |
| Tomato | Lycopene | −20°C (after 12 months)<br>−30°C (after 12 months) | 43% loss<br>19% loss | Lisiewska and Kmiecik (2000) |
| Parsley | Chlorophyll | −20°C (measurement every 3 months)<br>−30°C (measurement every 3 months) | 3.5% loss<br>2.5% loss<br>(after 9 months) | Lisiewska and Kmiecik (1997) |
| Mushrooms | Whiteness (*L, L, a, and b scale*) | −20°C (every 15 days) | 10.9% loss (after 90 days) | Czapski and Szudyga (2000) |
| Green beans | Chlorophyll α | −7°C, −15°C, and −30°C | $E_a = 48.73 \pm 1.27$ kJ/mol<br>$k(-15°C)*10^{-2} = 4.793 \pm 0.001(1/d)$ | Martins and Silva (2002) |
| Green beans | Chlorophyll b | −7°C, −15°C, and −30°C | $E_a = 45.60 \pm 1.94$ kJ/mol<br>$k(-15°C)*10^{-2} = 4.795 \pm 0.002(1/d)$ | Martins and Silva (2002) |
| Green beans | Color (*α, lab scale*) | −7°C, −15°C, and−30°C | $E_a = 103.05 \pm 0.07$ kJ/mol<br>$k(-15°C)*10^{-3} = 6.600 \pm 0.002(1/d)$ | Martins and Silva (2002) |
| Green beans | Color (*b, lab scale*) | −7°C, −15°C, and −30°C | $E_a = 55.33 \pm 1.79$ kJ/mol<br>$k(-15°C)*10^{-3} = 22.189 \pm 0.349(1/d)$ | Martins and Silva (2002) |
| Green beans | Total color $= \sqrt{(L-L_o)^2+(a-a_o)^2+(b-b_o)^2}$ | −7°C, −15°C, and −30°C | $E_a = 106.27 \pm 1.05$ kJ/mol<br>$k(-15°C)*10^{-3} = 5.999 \pm 0.000(1/d)$ | Martins and Silva (2002) |
| Green beans | Total color $= \sqrt{(L-L_o)^2+(a-a_o)^2+(b-b_o)^2}$ | Accelerated life testing (temperature steps at −30°C, −20°C, −10°C, and −5°C) | $E_a = 140.34 \pm 18.67$ kJ/mol<br>$k(-15°C)*10^{-3} = 135.976 \pm 15.981(1/d)$ | Martins et al. (2005) |
| Watercress | Hue angle $= \tan^{-1}\left(\frac{b_H}{a_H}\right)$ | −7°C, −15°C, and −30°C | $E_a = 168.65 \pm 28.18$ kJ/mol<br>$k(-15°C)*10^{-3} = 5.48 \pm 2.22(1/d)$ | Conçalves et al. (2009) |

mushrooms when there is a cutting or peeling step in the procedure. Blanching can protect plant tissues from enzymatic browning, since PPO enzyme is rather sensitive to temperature. The degree of this sensitivity depends on the pH of the blanching solution as well as the pH of the food matrix (Czapski and Szudyga, 2000).

#### 8.8.1.2 Ascorbic Acid Degradation

When blanching is applied before freezing, important losses of water-soluble ingredients are recorded, among which vitamin C is of great concern. On the other hand, in unblanched or insufficiently blanched plant tissues, ascorbate oxidase catalyzes the oxidation of ascorbic acid during frozen storage, especially when oxygen-permeable packaging is used. The retention of ascorbic acid is often used as an estimate of the overall nutrient loss of food products, as vitamin C is highly sensitive during processing and subsequent frozen storage (Patras et al., 2011). Vitamin C loss even at very low storage temperatures has been extensively studied in recent literature; a lot of studies focused on the role of blanching at the vitamin C retention, whereas others studied the ascorbic acid loss under fluctuating temperature conditions, which are more realistic than isothermal analysis (Concalves et al., 2011a,b; Cruz et al., 2009; Giannakourou and Taoukis, 2003b). Indicatively, in Table 8.4, some of these publications and the main findings regarding vitamin C loss are summarized.

As it will be thoroughly explained in another chapter, blanching (immersion in hot water or steaming at 100°C) is an important thermal treatment commonly applied in vegetables (and less in fruit tissue) mainly to inactivate enzymes responsible for the alterations mentioned above (at sensory or nutritional characteristics). Therefore, blanching is actually applied as a preliminary step before freezing, to decrease quality losses during storage. As far as blanching conditions are concerned, many processors select heat treatment sufficient to inactivate peroxidase, one of the more stable enzymes present, and one of the enzymes whose activity is relatively easy to measure. Since blanching is a heat treatment, changes associated with mild thermal processing can be expected, such as partial degradation of cell wall polymers, pigment degradation, thermally induced degradation of nutrients such as ascorbic acid, or even leaching of nutrients. Hence, a determination must be made as to whether blanching is necessary for a frozen product to retail a satisfactory quality level during its life, and, then a decision must be reached as to the extent and conditions of blanching needed to ensure optimum product quality.

### 8.8.2 Changes of Animal Tissue Due to Freezing

Freezing is one of the most important preservation methods for meat and meat products since, compared with other methods, it leads to a minimal loss of quality during long-term storage. The conditions for freezing meat appear to have little effect on the quality of the frozen product immediately after freezing (Reid, 1998). Freezing and subsequent thawing influence the water fraction of meat; during the freezing process, as water freezes, the concentration of the remaining solutes increases, thereby disrupting the homeostasis of the complex meat systems (Leygonie et al., 2012).

Frozen storage is used to retard undesirable biochemical reactions in meat, but there is some cell disruption and destruction of muscle fiber due to the formation

**TABLE 8.4**
**Characteristic Literature Studies about Vitamin C Loss in Frozen Vegetables**

| Food Matrix (after Blanching) | Temperature Conditions | Kinetic Result | Reference |
|---|---|---|---|
| Green peas | −20°C (measurement every 3 months) | 10% loss (after 12 months) | Favell (1998) |
| Green beans | −20°C (measurement every 3 months) | 20% loss (after 12 months) | Favell (1998) |
| Broccoli | −20°C (measurement every 3 months) | 2.7% loss (after 12 months) | Favell (1998) |
| Spinach | −20°C (measurement every 3 months) | 33.8% loss (after 12 months) | Favell (1998) |
| Green beans (packaged) | −22°C | 13% loss (after 12 months) | Oruna-Concha et al. (1998) |
| Green beans | −20°C | 33% loss (after 12 months) | Howard et al. (1999) |
| Chive | −20°C (every 3 months)<br>−30°C (every 3 months) | 34% loss<br>2.2% loss<br>(after 12 months) | Kmiecik and Lisiewska (1999) |
| Tomato | −20°C (at 12 months)<br>−30°C (at 12 months) | 70.3% loss<br>43.2% loss<br>(after 12 months) | Lisiewska and Kmiecik (2000) |
| Thermally induced degradation of nutrients such as ascorbic acid in broccoli | −20°C (every 3 months)<br>−30°C (every 3 months) | 17.6% loss<br>14.7% loss<br>(after 12 months) | Lisiewska and Kmiecik (2000) |
| Cauliflower | −20°C (every 3 months)<br>−30°C (every 3 months) | 8.8% loss<br>6.6% loss<br>(after 12 months) | Lisiewska and Kmiecik (1996) |
| Parsley | −20°C (every 3 months)<br>−30°C (every 3 months) | 35.8% loss<br>21.8% loss<br>(after 9 months) | Lisiewska and Kmiecik (1997) |
| Strawberry | −12°C (every 15 days)<br>−18°C (every 15 days)<br>−24°C (every 15 days) | 64.5% loss<br>10.7% loss<br>8.9% loss | Sahari et al. (2004) |
| Green peas | Storage at −3°C, −8°C, −12°C, and −20°C | $E_a = 97.9 \pm 9.6$ kJ/mol<br>$k(-20°C)*10^{-3} = 2.13$ (1/d) | Giannakourou and Taoukis (2003b) |
| Spinach | Storage at −3°C, −8°C, −12°C, and −20°C | $E_a = 112.0 \pm 23.2$ kJ/mol<br>$k(-20°C)*10^{-3} = 4.54$ (1/d) | Giannakourou and Taoukis (2003b) |
| Green beans | Storage at −3°C, −8°C, −12°C, and −20°C | $E_a = 101.5$ kJ/mol<br>$k(-20°C)*10^{-3} = 2.23$ (1/d) | Giannakourou and Taoukis (2003b) |
| Okra | Storage at −3, −8, −12 and −20°C | Ea = 105.9 kJ/mol<br>$k(-20°C)*10^{-3} = 1.05$ (1/d) | Giannakourou and Taoukis (2003b) |

of ice crystals (Soyer et al., 2010). During frozen storage, many reactions can occur between different meat components. These changes include oxidation of pigments and of lipids. There is also evidence for an insolubilization of proteins, which may contribute to the textural change. For example, fish and poultry meats are susceptible to oxidative reactions due to their high concentrations of oxidation catalysts (such as myoglobin and iron) and lipids. Lipid oxidation is the major form of deterioration in stored muscle foods. Oxidative reactions (lipid or phospholipid oxidation) in meat are the most important factor in quality losses, including flavor, texture, nutritive value, and color. Muscle cells contain high amounts of proteins, which can also be affected by oxidative reactions. Oxidation has been shown to induce a number of biological alterations such as protein solubility, protein fragmentation, or aggregation. Another factor that will shorten the storage life is the degree of comminution, which influences the access of oxygen. It accounts for the lower stability of ground beef.

Briefly, the main quality attributes affected by freezing are mentioned:

*Moisture loss:* When meat is frozen, its water-holding capacity is significantly reduced and the distribution of moisture is modified in meat tissues. This gives rise to "drip loss," or other similar phenomena, such as "thaw loss," "press loss," and "cooking loss." It has been suggested that the water-holding capacity is decreased due to the disruption of the muscle fiber structure and the modification/denaturation of endogenous proteins. Drip loss occurs in the postmortem phase due to a decrease of pH, and the loss of adenosine triphosphate (ATP), leading to a water release that was previously bound to proteins. Drip loss is apparent throughout the cold chain and is the main economic loss for the red meat industry, depending to a great extent on the species. In general, beef tends to lose more drip compared to pork and lamb (James and James, 2012) and important differences are reported even between different pig breeds. Moreover, small pieces of meat are found to drip more than large intact carcasses. Finally, the freezing rate is reported to significantly affect drip loss, with slow rates giving way to more drip during freezing, thawing, and cooking.

Excessive drip can affect the quality of meat after cooking, at the consumption phase, deteriorating meat juiciness, which is one of the most important attributes. When meat feels and/or tastes dry, this is related to lack of flavor and increased toughness.

*Tenderness and texture:* It is generally agreed in literature that meat tenderness increases with freezing and thawing. Other factors that play an important role at the toughness of the final product are the length of frozen storage and the degree to which the meat was aged prior to freezing (Leygonie et al., 2012). Different mechanisms are reported to contribute to texture changes: breakdown of muscle fibers by enzymes during proteolysis and loss of structural integrity caused by ice crystals formation.

*Oxidation of lipids and proteins:* The amount of unfrozen water into the frozen meat tissue, related to the final temperature of frozen storage, is of great importance for chemical reactions. Therefore, storing frozen meat at low temperatures, around −40°C decreases available water, decreasing the rate of chemical reactions. Additionally, the faction of unfrozen water influences oxidation processes, since primary lipid oxidation may initiate during frozen storage. This may lead to secondary lipid oxidation upon thawing, causing major undesirable changes in odor, flavor, color, and nutrition value. Actually, the formation of "off" or "rancid" flavors remains the main problem to extended storage of frozen meat. Oxidative rancidity is influenced by a variety of factors, such as species, type of muscle, and enzymes (intrinsic factors), and also by extrinsic factors, such as light, heat, mincing, addition of NaCl, and so on.

As far as protein oxidation is concerned, it is closely related to lipid oxidation, as it can be linked to any of the pro-oxidative factors, such as oxidized lipids, free radicals, and so on. This kind of oxidation severely alters juiciness, tenderness, flavor, and color of meat, also causing amino acid destruction, protein unfolding, increased surface hydrophobicity, fragmentation, and protein cross-linking, all leading to the formation of protein carbonyls (Leygonie et al., 2012). In general, it is stated in literature that oxidative changes to proteins decrease their water-holding capacity and, consequently, water leaches out of meat as exudates.

***Color and overall appearance***

During freezing, denaturation of the myoglobin may occur, increasing its susceptibility to autoxidation and subsequent loss of bright-red color. Species greatly affect myoglobin concentration, with beef and lamb containing more than pork and poultry, accounting for the different ("red" and "white" meat correspondingly) colors.

Another appearance problem related to freezing of meat tissues is "freezer burn," which occurs when a dry, spongy layer is formed at the surface, due to desiccation. It is a common problem in unpackaged or inadequately wrapped meat, and can be further aggravated if stored in areas of low-humidity and poor temperature control. If storage conditions are carefully monitored and controlled, this problem can be minimized, and the oxidation of oxymyoglobin to metmyoglobin sets the greater appearance issue. Storage temperature, light intensity on the display area, and type of packaging are of major concern for meat color and, therefore, for product acceptability by the consumers (Cornforth, 1994).

### 8.8.3 Changes of Fish–Seafood Tissue during Freezing

Fish is particularly unstable during frozen storage. There are several reasons for this instability, depending on fish species. High-fat fish tend to have a short shelf life in frozen storage due to lipid oxidation, which produces off-flavors. Freezing is a widely used preservation method for fish and other seafood, as it minimizes microbiological growth and some undesirable biochemical processes. Nevertheless, frozen muscle inevitably loses some of the special quality attributes of fresh fish, usually observed as a loss in juiciness and increase in toughness (Sigurgisladottir et al., 2000). Additionally, as freezing procedure cannot inactivate microorganisms, it is necessary to select the appropriate raw material with initial microbial load, as low as possible.

One of the main aspects related to the frozen fish deteriorative changes is the myofibrillar protein denaturation, which can lead to textural and functional changes in the frozen fish (Tironi et al., 2010). The production of free fatty acids (FFAs) and formaldehyde (FA), as well as storage temperature and time, have significant effect on protein changes during freezing and frozen storage. Myofibrillar protein denaturation and aggregation lead to loss of protein functionality and gel-forming ability in frozen fish. Factors influencing protein denaturation during freezing and frozen storage include salt concentration, pH, ionic strength, lipid oxidation, enzymatic reaction, surface tension, and the physical effects of freezing and dehydration (Shenouda, 1980). After being physically or chemically unfolded, proteins bind with each other forming aggregates, via hydrogen bonds, hydrophobic interactions, and disulfide bridges. This phenomenon leads to conformational changes and decreased protein solubility and extractability.

Other important aspects of deterioration include changes of enzyme activity and metabolite concentrations (aldehydes, amines, and nucleotide degradation) (Jaczynski et al., 2012). Ice crystal size, greatly depending on the freezing rate, is an important factor related to muscle deterioration, because the formation of large ice crystals leads to an extensive mechanical damage of cells; as a consequence, cellular components, such as lipids and proteins are free to interact with enzymes leading to protein denaturation and lipid degradation.

To counterbalance the damaging effects of freezing on proteins (denaturation and aggregation), it is suggested that cryoprotectants are used, such as sugars and other carbohydrates. Other compounds such as phosphates have also been used for fish tissue, in combination with sugar and sorbitol. Carbohydrates, polyols, some amino acids, and related compounds have shown the highest cryoprotective effectiveness. Nevertheless, many of them cannot be used for various reasons such as high cost, not permitted by food regulations, or adverse sensory properties (Herrera and Mackie 2004). The interest is now focused on novel, less-sweet cryoprotectants, which do not alter the sensory characteristics of fish; trehalose (Campo-Deano et al., 2010, 2011) combined with sugar or sorbitol is shown to offer a satisfactory alternative.

## REFERENCES


Abdalla H., Singh R.P. 1985. Simulation of thawing of foods using finite element method. *Journal of Food Process Engineering*, 7: 273–286.

Aked J. 2000. Fruits and vegetables (Chapter 11). In: *The Stability and Shelf-Life of Food*. Eds., D. Kilcast, P. Subramaniam. Boca Raton: Woodhead Publishing Limited.

Aked J. 2002. Maintaining the post-harvest quality of fruits and vegetables (Chapter 7). In: *Fruit and Vegetable Processing Improving Quality*. Ed. W. Jongen. 119–149. Boca Raton: Woodhead Publishing Ltd.

Arrhenius S. 1889. About the reaction rate of the inversion of non-refined sugar at souring. *Zeitschrift für Physikalische Chemie*, 4: 226–248.

Armstrong G.A. 2000. Sous vide products (Chapter 8). In: *The Stability and Shelf-Life of Food*. Eds., D. Kilcast, P. Subramaniam. Boca Raton: Woodhead Publishing Limited.

Armstrong G.A. 2004. Minimal thermal processing: Cook–chill and sous vide technology (Chapter 24). In: *Handbook of Preservation and Processing*. Eds., Y.H. Hui, S. Chazala, D.M. Graham, K.D. Murrell, W.-K. Nip. New York: Marcel Dekker, Inc.

Baldwin D.E. 2012. Sous vide cooking: A review. *International Journal of Gastronomy and Food Science*, 1: 15–30.

Barros-Velazquez J., Gallardo J.M., Calo P., Aubourg S.P. 2008. Enhanced quality and safety during on-board chilled storage of fish species captured in the Grand Sole North Atlantic fishing bank. *Food Chemistry*, 106(2): 493–500.

Becker B.R., Fricke B.A. 2002. Hydrocooling time estimation methods. *International Communications in Heat and Mass Transfer*, 29(2): 165–174.

Bellas I., Tassou S.A. 2005. Present and future applications of ice slurries. *International Journal of Refrigeration*, 28(1): 115–121.

Ben Embarek P.K. 1994. Presence, detection and growth of *Listeria monocytogenes* in seafoods: A review. *International Journal of Food Microbiology*, 23(1): 17–34.

Biliaderis C.G., Lazaridou A., Arvanitoyannis I. 1999. Glass transition and physical properties of polyol-plasticized pullulan–starch blends at low moisture. *Carbohydrate Polymers*, 40: 29–47.

Blond G., Le Meste M. 2004. Principles of frozen storage (Chapter 3). In: *Processing of Frozen Foods*. Eds., Y.H. Hui, P. Cornillon, I. Guerrero Legarretta, M.H. Lim, K.D. Murrell, W.-K. Nip. New York: Marcel Dekker, Inc.

Blond G., Simatos D. 1991. Glass transition of the amorphous phase in frozen aqueous systems. *Thermochimica Acta*, 175: 239–247.

Bogh-Sorensen L. 1984. The TTT–PPP concept. In: *Thermal Processing and Quality of Food*. Eds., P. Zeuthen, J.C. Cheftel, C. Eriksson, M. Lul, H. Leniger, P. Linko, G. Varela, G. Vos. 511–521. Essex: Elsevier Applied Science.

Bonacina C., Comini G., Fasano A., Primicero M. 1973. Numerical solution of phase change problems. *International Journal of Heat and Mass Transfer*, 16: 1825–1832.

Borch E., Kant-Muemans M.-L., Blixt Y. 1996. Bacterial spoilage of meat products and cured meat. *International Journal of Food Microbiology*, 33: 103–120.

Botana L.M. 2008. *Seafood and Freshwater Toxins: Pharmacology, Physiology, and Detection*. Boca Raton: Taylor & Francis Group, LLC.

Brackett R.E. 1992. Microbiological safety of chilled foods: Current issues. *Trends in Food Science and Technology*, 3: 81–85.

Brosnan T., Sun D.-W. 2001. Precooling techniques and applications for horticultural products—A review. *International Journal of Refrigeration*, 24: 154–170.

Brown H.M., Hall M.N. 2000. Non-microbiological factors affecting quality and safety (Chapter 9). In: *Chilled Foods—A Comprehensive Guide* (2nd ed.). Eds., M. Stringer, C. Dennis. 225–255. Cambridge: Woodhead Publishing Limited.

Buera M.P., Karel M. 1995. Effect of physical changes on the rates of nonenzymic browning and related reactions. *Food Chemistry*, 52: 167–173.

Buera P., Karel M. 1993. Application of the WLF equation to describe the combined effects of moisture, temperature and physical changes on non-enzymatic browning rates in food systems. *Journal of Food Processing and Preservation*, 17: 31–47.

Burin L., Buera M.P., Hough G., Chirife J. 2002. Thermal resistance of β-galactosidase in dehydrated dairy model systems as affected by physical and chemical changes. *Food Chemistry*, 76: 423–430.

Bushill J.H., Wright W.B., Fuller C.H.F., Bell A.V. 1965. The crystallization of lactose with particular reference to its occurrence in milk powders. *Journal of the Science of Food and Agriculture*, 16: 622.

Campañone L.A., Salvadori V.O., Mascheroni R.H. 2005a. Food freezing with simultaneous surface dehydration: Approximate prediction of weight loss during freezing and storage. *International Journal of Heat and Mass Transfer*, 48: 1195–1204.

Campañone L.A., Salvadori V.O., Mascheroni R.H. 2005b. Food freezing with simultaneous surface dehydration: Approximate prediction of freezing time. *International Journal of Heat and Mass Transfer*, 48: 1205–1213.

Campo-Deano L., Tovar C.A., Borderias J. 2010. Effect of several cryoprotectants on the physicochemical and rheological properties of suwari gels from frozen squid surimi made by two methods. *Journal of Food Engineering*, 97: 457–464.

Campo-Deano L., Tovar C.A., Borderias J., Fernandez-Martin F. 2011. Gelation process in two different squid (*Dosidicus gigas*) surimis throughout frozen storage as affected by several cryoprotectants: Thermal, mechanical and dynamic rheological properties. *Journal of Food Engineering*, 107: 107–116.

Cano M.P. 1996. Vegetables. In: *Freezing Effects on Food Quality*. Ed., L.E. Jeremiah. 247–297. New York: Marcel Dekker.

Carrington A.K., Goff H.D., Stanley D.W. 1996. Structure and stability of the glassy state in rapidly and slowly cooled carbohydrate solutions. *Food Research International*, 29(2): 207–213.

Carter B.P., Schmidt S.J. 2012. Developments in glass transitions determination in foods using moisture sorption isotherms. *Food Chemistry*, 132: 1693–1698.

Champion D., Blond G., Le Meste M., Simatos D. 2000. Reaction rate modeling in cryo-concentrated solutions: Alkaline phosphatase catalyzed DNPP hydrolysis. *Journal of Agriculture and Food Chemistry*, 48: 4942–4947.

Chaudhary V., Small D.M., Kasapis S. 2013. Effect of a glassy gellan/polydextrose matrix on the activity of α-D-glucosidase. *Carbohydrate Polymers*, 95: 389–396.

Chiralt A., Martinez-Navarette N., Martinez-Monzo J., Talens P., Moraga G., Ayala A., Fito P. 2001. Changes in mechanical properties throughout osmotic processes. Cryoprotectant effect. *Journal of Food Engineering*, 49: 129–135.

Cleland A.C. 1990. *Food Refrigeration Processes—Analysis, Design and Simulation*. London: Elsevier Applied Science.

Cleland A.C., Earle R.L. 1977. A comparison of analytical and numerical methods of predicting the freezing times of foods. *Journal of Food Science*, 42: 1390.

Cleland A.C., Earle R.L. 1982. Freezing time prediction for foods—A simplified procedure. *International Journal of Refrigeration*, 5: 134–140, 196.

Cleland A.C., Earle R.L. 1984. Assessment of freezing time prediction methods. *Journal of Food Science*, 49: 1034–1042.

Cleland A.C., Earle, R.L. 1976. A new method for prediction of surface heat transfer coefficients in freezing. Bull. I.I.R. Annexe: 567.

Cleland D.J., Cleland A.C., Earle R.L., Byrne S.J. 1984. Prediction of rates of freezing, thawing, and cooling in solids of arbitrary shape using the finite element method. *International Journal of Refrigeration*, 7: 6–13.

Concalves E.M., Abreu M., Brandao T.R.S., Silva C.L.M. 2011a. Degradation kinetics of colour, vitamin C and drip loss in frozen broccoli (*Brassica oleracea* L. ssp. Italica) during storage at isothermal and non-isothermal frozen conditions. *International Journal of Refrigeration*, 34: 2136–2144.

Concalves E.M., Cruz R.M.S., Abreu M., Brandao T.R.S., Silva C.L.M. 2009. Biochemical and colour changes of watercress (*Nasturtium officinale* R.Br.) during freezing and frozen storage. *Journal of Food Engineering*, 93: 32–39.

Concalves E.M., Pinheiro J., Abreu M., Brandao T.R.S., Silva C.L.M. 2011b. Kinetics of quality changes of pumpkin (*Curcubita maxima* L.) stored under isothermal and non-isothermal frozen conditions. *Journal of Food Engineering*, 106: 40–47.

Cornforth D. 1994. Color: Its basis and importance. In: *Advances in Meat Research Series, Vol. 9: Quality Attributes and Their Measurement in Meat, Poultry and Products*. Eds., A.M. Pearson, T.R. Dutson. 37–78. London: Blackie Academic and Professional.

Cortella G. 2012. Cooling (Chapter 1). In: *Operations in Food Refrigeration*. Ed., R.H. Mascheroni. Boca Raton: Taylor and Francis Group, LLC.

Cruz R.M.S., Vieira M.C., Silva C. 2009. Effect of cold chain temperature abuses on the quality of frozen watercress (*Nasturtium officinale* R.Br.). *Journal of Food Engineering*, 94: 90–97.

Czapski J., Szudyga K. 2000. Frozen mushrooms quality as affected by strain, flush, treatment before freezing, and time of storage. *Journal of Food Science*, 65(4): 722–725.

Delgado A.E., Sun D.-W. 2001. Heat and mass transfer models for predicting freezing processes—A review. *Journal of Food Engineering*, 47(3): 157–174.

Drummond L., Sun D.-W. 2010. Effects of chilling and freezing on safety and quality of food products (Chapter 11). In: *Processing Effects on Safety and Quality of Foods*. Ed., E. Ortega-Rivas. Boca Raton: Taylor and Francis Group, LLC.

Favell D.J. 1998. A comparison of the vitamin C content of fresh and frozen vegetables. *Food Chemistry*, 62(1): 59–64.

Fellows P. 2000. *Food Processing Technology, Principles and Practice* (2nd ed.). 387–405, 418–440. Cambridge: Woodhead Publishing Limited and CRC Press, LLC.

Francis G.A., Thomas C., O'Beirne D. 1999. The microbiological safety of minimally processed vegetables. *International Journal of Food Science and Technology*, 34: 1–22.

Fu B., Labuza T.P. 1992. Shelf-life testing: Procedures and prediction methods. In: *Quality in Frozen Food*. Eds., M.C. Erickson, Y.C. Hung. 377–415. New York: Chapman & Hall.

Furuki T. 2002. Effect of molecular structure on thermodynamic properties of carbohydrates. A calorimetric study of aqueous di- and oligosaccharides at subzero temperatures. *Carbohydrate Research*, 337: 441–450.

Gelman A., Sachs O., Khanin Y., Drabkin V., Glatman L. 2005. Effect of ozone pretreatment on fish storage life at low temperatures. *Journal of Food Protection*, 68(4): 778–784.

Giannakourou M.C., Taoukis P.S. 2003a. Stability of dehydrofrozen green peas pretreated with non-conventional osmotic agents. *Journal of Food Science*, 68(6): 2002–2010.

Giannakourou M.C., Taoukis P.S. 2003b. Kinetic modelling of vitamin C loss in frozen green vegetables at variable storage conditions. *Food Chemistry*, 83: 33–41.

Gill C.O. 1983. Meat spoilage and evaluation of the potential storage life of fresh meat. *Journal of Food Protection*, 46: 444–452.

Goff H.D. 1992. Low temperature stability and the glassy state in frozen foods. *Food Research International*, 25: 317–325.

Goff H.D. 1994. Measuring and interpreting the glass transition in frozen foods and model systems. *Food Research International*, 27: 187–189.

Goff H.D. 1997. Measurement and interpretation of the glass transition in frozen foods. In: *Quality in Frozen Food*. Eds., M.C. Erickson, Y.H. Hung. 29–51. New York: Chapman & Hall.

Goni S.M., Oddone S., Segura J.A., Mascheroni R.H., Salvadori V.O. 2008. Prediction of foods freezing and thawing times: Artificial neural networks and genetic algorithms approach. *Journal of Food Engineering*, 84: 164–178.

Gross K.C., Wang C.Y., Saltveit M. 2004. The commercial storage of fruits, vegetables, and florist and nursery stocks. *Agricultural Handbook Number 66*. Beltsville: USDA. (www.ba.ars.usda.gov/hb66/index.html).

Hall G.M. 2011. *Fish Processing—Sustainability and New Opportunities*. 77–81. Iowa: Blackwell Publishing Ltd.

Hamad S.H. 2012. Factors affecting the growth of microorganisms in food (Chapter 20). In: *Progress in Food Preservation*. Eds., R. Bhat, A.K. Alias, G. Paliyath. 417–421. West Sussex: John Wiley & Sons Ltd.

Hatley R.H.M., Van den Berg C., Franks F. 1991. The unfrozen water content of maximally freeze concentrated carbohydrate solutions: Validity of the methods used for its determination. *Cryo-Letters*, 12: 113–124.

Heap R.D. 2000. The refrigeration of chilled foods (Chapter 4). In: *Chilled Foods—A Comprehensive Guide* (2nd ed.). Eds., M. Stringer, C. Dennis. Cambridge: Woodhead Publishing Limited.

Heldman D.R., Taylor T.A. 1997. *Modeling of Food Freezing*. In: *Quality in Frozen Food*. Eds., M.C. Erickson, Y.C. Hung. 51–62. New York: Chapman & Hall.

Heldman D.R. 2011. *Food Preservation Process Design*. 111–138. San Diego: Elsevier Inc.

Heldman D.R., Singh R.P. 1981. *Food Process Engineering* (2nd ed.). Connecticut: AVI Publishing.

Herbert R.A., Sutherland J.P. 2000. Chill storage. In: *The Microbiological Safety and Quality of Food* (Vol. 1). Eds., B.M. Lund, T.C. Baird-Parker, G.W. Gould. 101–121. Gaithersburg: Aspen Publishers.

Herrera J.R., Mackie I.M. 2004. Cryoprotection of frozen–stored actomyosin of farmed rainbow trout (*Oncorhynchus mykiss*) by some sugars and polyols. *Food Chemistry*, 84: 91–97.

Hill M.A. 1987. The effect of refrigeration on the quality of some prepared foods. In: *Developments in Food Preservation* (Vol. 4). Ed., S. Thorne. 123–152. Essex: Elsevier Applied Science.

Hossain Md.M., Cleland D.J., Cleland A.C. 1992a. Prediction of freezing and thawing times of foods of regular multi-dimensional shape by using an analytically derived geometric factor. *International Journal of Refrigeration*, 15(4): 227–234.

Hossain Md.M., Cleland D.J., Cleland A.C. 1992b. Prediction of freezing and thawing times of foods of two-dimensional irregular shape by using a semi-analytical geometric factor. *International Journal of Refrigeration*, 15(4): 235–240.

Howard L.A., Wong A.D., Perry A.K., Klein B.P. 1999. β-carotene and ascorbic acid retention in fresh and processed vegetables. *Journal of Food Science*, 64(5): 929–936.

Hui Y.H., Lim M.-H., Nip W.-K., Smith J.S., Yu P.H.F. 2004. Principles of food processing (Chapter 1). In: *Food Processing—Principles and Applications*. Eds., J. Scott Smith, Y.H. Hui. Oxford: Blackwell Publishing.

Hyldig G., Nielsen D. 2007. Texture of fish, fish products, and shellfish (Chapter 42). In: *Handbook of Meat, Poultry and Seafood Quality*. Ed., L.M.L. Nollet. 549–561. Iowa: Blackwell Publishing.

Ilicali C., Cetin M., Cetin S. 1996. Methods for the freezing time of ellipses. *Journal of Food Engineering*, 28: 361–372.

International Institute of Refrigeration. 1972. *Recommendations for the Processing and Handling of Frozen Foods* (2nd ed.). Paris.

Jaczynski J., Tahergorabi R., Hunt A.L., Park J.W. 2012. Safety and quality of frozen aquatic products. In: *Handbook of Frozen Food Processing and Packaging*. Ed., D.-W. Sun. 343–385. Boca Raton: Taylor and Francis Group, LLC.

James S.J. 2006. Principles of food refrigeration and freezing (Chapter 112). In: *Handbook of Food Science, Technology, and Engineering* (Vol. 3). Ed., Y.H. Hui. Boca Raton: Taylor and Francis Group, LLC.

James S.J., James C. 2002. *Meat Refrigeration*. Boca Raton: Woodhead Publishing Ltd and CRC Press, LLC.

James S.S., James C. 2012. Quality and safety of frozen meat and meat products. In: *Handbook of Frozen Food Processing and Packaging*. Ed., D.-W. Sun. 303–323. Boca Raton: Taylor and Francis Group, LLC.

Jensen W.K., Devine C., Dikeman M. 2004. *Encyclopedia of Meat Sciences*. 249–255, 727–737, 1131–1161. Oxford: Elsevier Ltd.

Juneja V.K., Snyder O.P. 2007. Sous vide and cook–chill processing of foods: Concept development and microbiological safety. In: *Advances in Thermal and Non-Thermal Food Preservation*. Eds., G. Tewari, V.K. Juneja. 145–163. Iowa: Blackwell Publishing.

Kader A.A. 2002. *Postharvest Technology of Horticultural Crops* (3rd ed.). 39–47. Oakland: University of California.

Kader A.A. 2013. Postharvest technology of horticultural crops—An overview from farm to fork. *Ethiopian Journal of Applied Sciences and Technology*, 1: 1–8.

Kadim I.T., Mahgoub O. 2007. Postharvest handling of red meat (Chapter 7). In: *Handbook of Food Preservation* (2nd ed.). Ed., M.S. Rahman. 181–192. Boca Raton: Taylor & Francis Group, LLC.

Karel M., Lund D.B. 2003. Storage at chilling temperatures (Chapter 7). In: *Physical Principles Preservation of Food* (2nd ed.). 237–273. New York: Marcel Dekker, Inc.

Kiani H., Sun D.-W. 2011. Water crystallization and its importance to freezing of foods: A review. *Trends in Food Science and Technology*, 22: 407–426.

Kmiecik W., Lisiewska Z. 1999. Effect of pretreatment and conditions and period of storage on some quality indices of frozen chive (*Allium schoenoprasum* L.). *Food Chemistry*, 67: 61–66.

Larkin J.L., Heldman D.R., Steffe J.F. 1984. An analysis of factors influencing precision of thermal property values during freezing. *International Journal of Food Refrigeration*, 7(2): 86–92.

Levine H., Slade L. 1988. Principle of cryostabilization technology from structure/property relationships of carbohydrate/water systems. A review. *Cryo-Letters*, 9: 21–63.

Levine H., Slade L. 1992. Glass transitions in foods. In: *Physical Chemistry of Foods*. Eds., H.G. Schwartzberg, R.W. Hartel. 83–221. New York: Marcel Dekker, Inc.

Leygonie C., Britz T.J., Hoffman L.C. 2012. Impact of freezing and thawing on the quality of meat: A review. *Meat Science*, 91: 93–98.

Lin Z. 1994. *Prediction of Chilling Times of Foods* (PhD thesis). New Zealand: Massey University.

Lin Z., Cleland A.C., Sellarach G.F., Cleland D.J. 1996a. A simple method for prediction of chilling times for objects of two-dimensional irregular shape. *International Journal of Refrigeration*, 19: 95–106.

Lin Z., Cleland A.C., Sellarach G.F., Cleland D.J. 1996b. A simple method for prediction of chilling times: Extension to three-dimensional irregular shapes. *International Journal of Refrigeration*, 19: 107–114.

Lisiewska Z., Kmiecik W. 1996. Effect of level of nitrogen fertilizer, processing conditions and period of storage of frozen broccoli and cauliflower on vitamin C retention. *Food Chemistry*, 57(2): 267–270.

Lisiewska Z., Kmiecik W. 1997. Effect of freezing and storage on quality factors in hamburg and leafy parsley. *Food Chemistry*, 60(4): 633–637.

Lisiewska Z., Kmiecik W. 2000. Effect of storage and temperature on the chemical composition and organoleptic quality of frozen tomato cubes. *Food Chemistry*, 70: 167–173.

Lopez-Leiva M., Hallstrom B. 2003. The original Plank equation and its use in the development of food freezing rate predictions. *Journal of Food Engineering*, 58: 267–275.

Lovatt S.J., Pham Q.T., Loeffen M.P.F., Cleland A.C. 1993. A new method of predicting the time-variability of product heat load during food cooling. Part 1: Theoretical considerations. *Journal of Food Engineering*, 18: 13–36.

Lundesjo Ahnstrom M. 2008. *Influence of Pelvic Suspension on Beef Meat Quality (doctoral thesis)*. Uppsala: Swedish University of Agricultural Sciences.

Madigan M.T., Martinko J.M., Stahl D.A., Clark D.P. 2012. *Biology of Microorganisms* (13th ed.). 134–137. San Francisco: Pearson Education, Inc.

Manzocco L., Nicoli M.C., Anese M., Pitotti A., Maltini E. 1999. Polyphenoloxidase and peroxidase activity in partially frozen systems with different physical properties. *Food Research International*, 31(5): 363–370.

Maroulis Z.B., Saravacos G.D. 2003. Refrigeration and freezing (Chapter 5). In: *Food Process Design*. New York: Marcel Dekker, Inc.

Martinez S., Perez N., Carballo J., Franco I. 2013. Effect of blanching methods and frozen storage on some quality parameters of turnip greens (grelos). *LWT—Food Science and Technology*, 51: 383–392.

Martins R.C., Lopez I.C., Silva C.L.M. 2005. Accelerated life testing of frozen green beans (*Phaseolus vulgaris*, L.) quality loss kinetics: Colour and starch. *Journal of Food Engineering*, 67: 339–346.

Martins R.C., Silva C.L.M. 2002. Modelling colour and chlorophyll losses of frozen green beans (*Phaseolus vulgaris* L.). *International Journal of Refrigeration*, 25: 966–974.

Martins R.C., Silva C.L.M. 2004. Inverse problem methodology for thermal–physical properties estimation of frozen green beans. *Journal of Food Engineering*, 63: 383–392.

Mazzobre M.F., Buera M.P., Chirife J. 1997a. Protective role of trehalose on stability of lactase in relation to its glass and crystal forming properties and effect of delaying crystallization. *LWT—Food Science and Technology*, 30: 324–329.

Mazzobre M.F., Buera M.P., Chirife J. 1997b. Glass transition and thermal stability of lactase in low-moisture amorphous polymeric matrices. *Biotechnology Progress*, 13: 195–199.

McGregor B.M. 1987. *Tropical Products Transport Handbook, Agriculture Handbook No. 668*. 74–114. Washington: U.S. Department of Agriculture.

Medina I., Gallardo J.M., Aubourg S.P. 2009. Quality preservation in chilled and frozen fish products by employment of slurry ice and natural antioxidants. *International Journal of Food Science and Technology*, 44: 1467–1479.

Mishra V.K., Gamage T.V. 2007. Postharvest handling and treatments of fruits and vegetables (Chapter 3). In: *Handbook of Food Preservation* (2nd ed.). Ed., M.S. Rahman. Boca Raton: Taylor & Francis Group, LLC.

Mittal G.S., Zhang J. 2000. Prediction of freezing time for food products using a neural network. *Food Research International*, 33: 557–562.

Nagaoka J., Takaji S., Hohani S. 1955. Experiments on fish freezing in air blast freezers. *Proceedings of the 9th International Congress on Refrigeration*, Paris, paper 4.321.

Nascimento Nunes M.C. 2008. *Color Atlas of Postharvest Quality of Fruits and Vegetables*. Iowa: Blackwell Publishing.

Neri L., Pittia P., Bertolo G., Torreggiani D., Saccheti G. 2010. Influence of water activity and molecular mobility on peroxidase activity in salt and sorbitol–maltodextrin system. *Journal of Food Engineering*, 101: 289–295.

Nesvadbha P. 2008. Thermal properties and ice crystals development in frozen foods. In: *Frozen Food Science and Technology*. Ed., J.A. Evans. 1–26. Iowa: Blackwell Publishing.

North M.F., Lovatt S.J. 2012. Chilling and freezing meat (Chapter 19). In: *Handbook of Meat and Meat Processing* (2nd ed.). Ed., Y.H. Hui. 357–370. Boca Raton: Taylor & Francis Group, LLC.

Opara L.U., Al-Jufaili S.M., Rahman M.S. 2007. Postharvest handling and preservation of fresh fish and seafood (Chapter 6). In: *Handbook of Food Preservation* (2nd ed.). Ed., M.S. Rahman. 152–172. Boca Raton: Taylor & Francis Group, LLC.

Oruna-Concha M.J., Gonzalez-Castro M.J., Lopez-Hernandez J., Simal-Lozano J. 1998. Monitoring of the vitamin C content of frozen green beans and padron peppers by HPLC. *Journal of the Science of Food and Agriculture*, 76: 477–480.

Parreno W.C., Alvarez Torres M.D. 2012. Quality and safety of frozen vegetables. In: *Handbook of Frozen Food Processing and Packaging*. Ed., D.-W. Sun. 387–433. Boca Raton: Taylor & Francis Group, LLC.

Patras A., Tiwari B.K., Brunton N.P. 2011. Influence of blanching and low temperature preservation strategies on antioxidant activity and phytochemical content of carrots, green beans and broccoli. *LWT—Food Science and Technology*, 44: 299–306.

Peleg M. 1992. On the use of the WLF model in polymers and foods. *Critical Reviews in Food Science and Nutrition*, 32(1): 59–66.

Pflug I.J., Blaisdell J.L., Kopelman I.J. 1965. Developing temperature–time curves for objects that can be approximated by a sphere, infinite plate or infinite cylinder. *ASHRAE Transactions*, 71: 238–248.

Pham Q.T. 1991. Shape factors for the freezing time of ellipses and ellipsoids. *Journal of Food Engineering*, 13: 159–170.

Pham Q.T. 2001. Modelling thermal processes: Cooling and freezing (Chapter 15). In: *Food Process Modelling*. Eds., P. Tijskens, M. Hertog, B. Nicolai. Cambridge: Woodhead Publishing Limited.

Pham Q.T. 2002. Calculation of processing time and heat load during food refrigeration. AIRAH Conference *Food for Thought—Cool*, Sydney, 24 May 2002.

Pham Q.T. 2012. Mathematical modeling of freezing processes. In: *Handbook of Frozen Food Processing and Packaging*. Ed., D.-W. Sun. 147–184. Boca Raton: Taylor & Francis Group, LLC.

Plank R. 1913. Beitrage zur Berechnung und Bewertung der Gefriergeschwindigkeit von lebensmitteln. *Zeitschrift fur die gesamte Kalte Industrie Reihe* 3 Heft, 10: 1–16.

Rahman M.S. 2007. Food preservation: Overview (Chapter 1). In: *Handbook of Food Preservation* (2nd ed.). Ed., M.S. Rahman. Boca Raton: Taylor & Francis Group, LLC.

Rahman S. 1995. *Food Properties Handbook*. 1–137. Boca Raton: CRC Press, Inc.

Ramaswamy H., Marcotte M. 2006. Low-temperature preservation (Chapter 4). In: *Food Processing, Principles and Applications*. 169–232. Boca Raton: Taylor & Francis Group, LLC.

Ray B., Bhunia A. 2008. *Fundamental Food Microbiology* (4th ed.). 35–39, 60–61. Boca Raton: Taylor & Francis Group, LLC.

Reid D. 1998. Freezer preservation of fresh foods: Quality aspects (Chapter 14). In: *Food Storage Stability*. Eds., I.A. Taub, P.S. Reid. Boca Raton: CRC Press, LLC.

Reid D.S. 1990. Optimizing the quality of frozen foods. *Food Technology*, 44(7): 78–82.

Reid D.S. 1997. Overview of physical/chemical aspects of freezing. In: *Quality in Frozen Food*. Eds., M.C. Erickson, Y.-C. Hung. 10–28. New York: Chapman & Hall.

Rodgers S. 2003. Potential applications of protective cultures in cook–chill catering. *Food Control*, 14(1): 35–42.

Roldan M., Antequera T., Martin A., Mayoral A.I., Ruiz J. 2013. Effect of different temperature–time combinations on physicochemical, microbiological, textural and structural features of sous-vide cooked lamb loins. *Meat Science*, 93(3): 572–578.

Roos Y., Karel M. 1991a. Non equilibrium ice formation in carbohydrate solutions. *Cryo-Letters*, 12: 367–376.

Roos Y., Karel M. 1991b. Amorphous state and delayed ice formation in sucrose solutions. *International Journal of Food Science and Technology*, 26: 553–566.

Roos Y.H. 1995. *Phase Transitions in Foods*. New York: Academic Press, Inc.

Roudaut G., Simatos D., Champion D., Contreras-Lopez E., Le Meste M. 2004. Molecular mobility around the glass transition temperature: A mini review. *Innovative Food Science and Emerging Technologies*, 5: 127–134.

Sablani S.S. 2012. Glass transitions in frozen food systems. In: *Handbook of Frozen Food Processing and Packaging*. Ed., D.-W. Sun. 39–54. New York: Taylor & Francis Group, LLC.

Sahagian M.E., Goff H.D. 1996. Fundamental aspects of the freezing process. In: *Freezing Effects on Food Quality*. Ed., L.E. Jeremiah. 1–50. New York: Marcel Dekker, Inc.

Sahari M.A., Boostani M.F., Hamidi Z.E. 2004. Effect of low temperature on the ascorbic acid content and quality characteristics of frozen strawberry. *Food Chemistry*, 86: 357–363.

Salunkhe D.K., Bolin H.R., Reddy N.R. 1991. *Storage, Processing, and Nutritional Quality of Fruits and Vegetables —Fresh Fruits and Vegetables (Vol. 1)*. Boca Raton: CRC Press, LLC.

Schebor C., Del Pilar Buera M., Chirife J. 1996. Glassy state in relation to thermal inactivation of the enzyme invertase in amorphous dried matrices of trehalose, maltodextrin and PVP. *Journal of Food Engineering*, 30: 269–282.

Schebor C., Del Pilar Buera M., Karel M., Chirife J. 1999. Color formation due to nonenzymatic browning in amorphous, glassy, anhydrous, model systems. *Food Chemistry*, 65: 427–432.

Sheen S., Hayakawa K.I. 1990. Finite difference analysis for the freezing or thawing of an irregular food with volumetric change. In: *Engineering and Food—Preservation*

*Processes and Related Techniques* (Vol. 2). Eds., W.E.L. Spiess, H. Schubert. 426–441. Amsterdam: Elsevier Applied Science.

Shenouda S.Y. 1980. Theories of protein denaturation during frozen storage of frozen fish. *Advances in Food Research*, 256: 275–311.

Sigurgisladottir S., Ingrarsdottir H., Torrissen O.S., Cardinal M., Hafsteinsson H. 2000. Effect of freezing/thawing on the microstructure and the texture of smoked Atlantic salmon (Salmosalar). *Food Research International*, 33: 857–865.

Singh N.P. 2007. *Fruit and Vegetable Preservation*. Jaipur: Oxford Book Company.

Singh R.P., Heldman D.R. 1981. Thermodynamics of food freezing. In: *Food Process Engineering*. 158–215. Connecticut: AVI Publishing Company, Inc.

Slade L., Levine H., Finey J. 1989. Protein–water interactions: Water as a plasticizer of gluten and other protein polymers. In: *Protein Quality and the Effects of Processing*. Eds., R.D. Phillips, J.W. Finlay. 9–123. New York: Marcel Dekker.

Soyer A., Ozalp B., Dalmis U., Bilgin V. 2010. Effects of freezing temperature and duration of frozen storage on lipid and protein oxidation in chicken meat. *Food Chemistry*, 120: 1025–1030.

Spencer J.F.T., Ragout de Spencer A.L. 2001. *Food Microbiology Protocols*. 3–11. New Jersey: Humana Press Inc.

Tabil, Jr. L.G., Sokhansanj S. 2001. Mechanical and temperature effects on shelf life stability of fruits and vegetables (Chapter 2). In: *Food Shelf Life Stability Chemical, Biochemical, and Microbiological Changes*. Eds., N.A.M. Eskin, D.S. Robinson. Boca Raton: CRC Press, LLC.

Taoukis P.S., Giannakourou M.C. 2004. Temperature and food stability: Analysis and control. In: *Understanding and Measuring the Shelf-Life of Food*. Ed., R. Steele. 42–68. Cambridge: Woodhead Publishing Ltd.

Taoukis P.T., Labuza T.P. 1997. Kinetics of food deterioration and shelf life prediction. In: *Handbook of Food Engineering Practice*. Eds., K.J. Valentas, E. Rotsten, R.P. Sing. 361–403. Boca Raton: CRC Press.

Terefe N.S., Hendrickx M. 2002. Kinetics of the pectinmethylesterase (PME) catalysed de-esterification of pectin in frozen model systems. *Biotechnology Progress*, 18(2): 221–228.

Terefe N.S., Van Loey A., Hendrickx M. 2004. Modelling the kinetics of enzyme-catalysed reactions in frozen systems: The alkaline phosphatase catalysed hydrolysis of di-sodium-p-nitrophenyl phosphatase. *Innovative Food Science and Emerging Technologies*, 5: 335–344.

Thompson A.K. 1996. *Postharvest Technology of Fruits and Vegetables*. Oxford: Blackwell Science.

Thompson K. 2003. *Fruit and Vegetables: Harvesting, Handling and Storage* (2nd ed.). Oxford: Blackwell Publishing Ltd.

Tironi V., de Lamballerie M., Le-Bail A. 2010. Quality changes during the frozen storage of sea bass (*Dicentratchus labrax*) muscle after pressure shift freezing and pressure-assisted thawing. *Innovative Food Science and Emerging Technologies*, 11: 565–573.

Torreggiani D., Forni E., Guercilena I., Maestrelli A., Bertolo G., Archer G.P., Kennedy C.J., Bone S., Blond G., Contreras-Lopez E., Champion D. 1999. Modification of glass transition temperature through carbohydrates additions: Effect upon colour and anthocyanin pigment stability in frozen strawberry juices. *Food Research International*, 32: 441–446.

Varnam A.H., Sutherland J.P. 1995. *Meat and Meat Products*. London: Chapman & Hall.

Venugopal V. 2006. Postharvest quality changes and safety hazards (Chapter 2). In: *Seafood Processing: Adding Value through Quick Freezing, Retortable Packaging, Cook-Chilling, and Other Methods*. Boca Raton: Taylor & Francis Group, LLC.

Vigneault C., Rennie T.J., Toussaint V. 2008. Cooling of freshly cut and freshly harvested fruits and vegetables. *Stewart Postharvest Review*, 4(3): 1–10.

Walker S.J., Betts G. 2000. Chilled foods microbiology (Chapter 6). In: *Chilled Foods* (2nd ed.). Eds. C. Dennis, M. Stringer. Chichester: Ellis Horwood Ltd.

Williams M.L., Landel R.F., Ferry J.D. 1955. The temperature dependence of relaxation mechanisms in amorphous polymers and other glass forming liquids. *Journal of the American Chemical Society*, 77: 3701–3707.

Zaritzky N. 2012. Physical–chemical principles in freezing. In: *Handbook of Frozen Food Processing and Packaging*. Ed., D.-W. Sun. 3–37. New York: Taylor & Francis Group, LLC.

# 9 Drying of Foods

*Panagiotis A. Michailidis and Magdalini K. Krokida*

## CONTENTS



## 9.1 INTRODUCTION

Drying is a heat and mass transfer process to remove water or another solvent by evaporation, in most of the cases, from a solid, semisolid, or liquid. Solvent removal by sublimation takes place in freeze drying, while in osmotic dehydration, water is removed from a solid material as liquid due to osmotic pressure and hence concentration difference. It is used in the food, agricultural, ceramic, chemical, pharmaceutical, pulp and paper, mineral, polymer, and textile industries. Drying duration depends on the nature of the product, the drying method or technique applied, and the drying conditions. Drying is an operation with a heavy demand for energy, contributing ~12% of the total energy used in industrial sectors worldwide (Misha et al., 2012). According to Bórquez et al. (1999), drying of food materials is not just a heat and mass transfer process but also a technological process that has a significant effect on product quality as it contributes to preservation and often to the improvement of the technological properties of products.

Foods are dried, starting from their natural form (fruits, vegetables, grains, spices, milk) or after handling (e.g., instant coffee, soup mixes, and whey). The process is also used to obtain a desired physical form (e.g., powder, flakes, and granules) or texture and to create new products that would not otherwise be feasible. (Mujumdar, 1995). There are foods and agricultural materials, such as some fruits, that present an extremely short shelf life of 2–3 days. The application of some kind of postharvest processing is crucial for their preservation and expansion of their shelf life, and drying is the one used most widespread. Drying contributes to the devolution of the product in a stable and safe condition as the removal of water lowers its availability (water activity) and prevents microbiological deterioration or poisoning of the product. The moisture content for most foods must be lowered to below 50% wet basis (wb) to ensure protection against microorganisms (Labuza and Tannenbaum, 1972). A reduction in microorganism growth is best achieved when the water content of food materials is in the range of 15% or less, while there are chemical, enzymatic, and nonenzymatic reactions that are inhibited only when the moisture content decreases below 5% (Wiktor et al., 2013). In the case of meat and fish preservation and processing, moisture removal and enzyme inactivation by drying also prevents the growth and reproduction of microorganisms that cause decay and minimizes many of the moisture-mediated deteriorative reactions (Duan et al., 2004).

Although sun drying has been used from ancient times, the application of the method as a unit operation in food processing started in the eighteenth century and, initially, drying was used for the supply of troops during wars, such as the Crimea, the Boer, and the First World War (Van Arsdel and Copley, 1963). Among the first food materials processed by drying were potatoes, carrots, corn, beans, spinach, cabbage, turnips, celery, and soup mixtures. Although drying was initially adopted in the food industry for the extension of products shelf life, at the end of the twentieth century the retaining of quality attributes, which express the physical, chemical, microbial, and nutritional quality of the food product, such as color, visual appeal, shape, flavor, microbial load, retention of nutrients, porosity-bulk density, texture, rehydration properties, water activity, freedom from pests, insects, and other contaminants, preservatives, and freedom from taints and off-odors, has gained great importance for the development of high-quality dried foods. This makes food dehydration as one of the most challenging unit operations in food processing (Sablani, 2006, Perera, 2005).

The evolution of drying technology, aimed at the optimization of the process in terms of final food quality and energy consumption, can be divided into four generations:

- The first generation includes dryers that use hot air flowing over an extensive area of the product to remove water from the surface, such as cabinet- and bed-type dryers (tray, truck tray, rotary flow conveyor, tunnel, kiln, etc.). These dryers are mostly suitable for solid materials such as grains, sliced fruits and vegetables, or chunked products.
- The second-generation dryers are used mainly for the dehydration of slurries and purees to produce powders and flakes and include spray and drum dryers.
- The third generation includes drying methods such as freeze and osmotic drying, which facilitate water removal by methods based not on water

vaporization but on other principles, namely, sublimation and movement due to osmotic pressure differential.

- The fourth generation refers to methods, some of which have been recently developed and considered to be state-of-the-art dewatering methods in food processing. They aim at the acceleration of the drying rate, the extended preservation of bioactive components, the reduction of energy consumption, and the interaction of the drying medium with the inner molecules of the material. High-vacuum, fluidized bed, microwave, radio-frequency, and refractance window drying (RWD), among others are included in this generation (Vega-Mercado et al., 2001).

## 9.2 PSYCHROMETRY

Drying, humidification/dehumidification, and cooling (e.g., cooling towers) are processes that, along with vapor content measurements, include interfacial mass and energy transfer between a gas and a pure liquid when they are brought into contact with each other (Vega-Mercado et al., 2001). Especially drying in the presence of air as a medium of heating and carrying away the evaporated moisture from the product is the most common way to dewater food materials (direct or convective dryers). The knowledge of air properties such as its moisture content and temperature changes during its passage through the dryer is essential for the calculation of mass and energy balances for the streams involved in the process (Mujumdar and Devahastin, 2006). Psychrometry is the thermodynamic branch that deals with the determination of moist air properties (Maroulis and Saravacos, 2003). The ideal gas law is used to predict the behavior of air–water vapor mixtures, since air temperature is high enough and water vapor pressure low enough in relation to their respective saturation points.

The most important terms and properties associated with humid air are the following:

- Dry bulb temperature $T$ (°C): It is the temperature of the mixture measured by a common, nonmodified thermometer that is immersed in the mixture.
- Absolute humidity or humidity ratio $Y_v$ (kg water/kg dry air): It is the mass of moisture (water vapor) contained in the moist mixture per unit mass of dry air.
- Relative humidity or relative saturation $RH$ (%): It is the ratio of the partial pressure of water vapor in the system to the partial pressure of water vapor in a saturated condition (equilibrium vapor pressure) at the same temperature.
- Specific volume or humid volume $v$ (m$^3$/kg): It is the volume of unit mass of dry gas and its accompanying vapor at the mixture temperature and pressure.
- Humid heat $C_p$ (kJ/kg/°C): It is the amount of heat required to raise the temperature of unit mass dry air and its associated water vapor by 1°C.
- Enthalpy $\Delta H$ (kJ/kg): It is the internal energy and the flow work per unit mass and is expressed as the sum of enthalpies of the gas and its vapor content.

- Dew point temperature $T_d$ (°C): It is the dry bulb temperature at which an air–vapor mixture becomes saturated when cooled at constant total pressure out of contact with a liquid.
- Adiabatic saturation temperature $T_{as}$ (°C): It is the equilibrium gas temperature reached by unsaturated gas and vaporizing liquid under adiabatic conditions.
- Wet bulb temperature $T_w$ (°C): It is the steady-state nonequilibrium temperature that a small amount of water reaches when exposed to a continuous stream of gas under adiabatic conditions. Wet bulb temperature is defined thermodynamically as the temperature at which water, by evaporating into moist air at a given dry bulb temperature and moisture content, can bring the air to saturation adiabatically, while constant pressure is maintained and the latent heat required for evaporation is supplied at the expense of the liquid sensible heat resulting in its temperature decrease. Wet bulb temperature is determined with a thermometer on which the bulb has been covered with a wet cloth and is immersed in a rapidly moving air stream. Its value is lower than the dry bulb temperature if the air stream is unsaturated.
- Adiabatic saturation humidity $Y_{as}$ (kg water/kg dry air): It is the absolute humidity that corresponds to adiabatic saturation temperature ($T_{as}$) and lies on the saturation curve.
- Saturation: It is the state when a gas holds the maximum amount of vapor under the existing conditions of pressure and absolute humidity.
- Wet bulb depression (°C): It is the difference between dry and wet bulb temperature ($T - T_w$) and expresses a measurement of unsaturation of the air. It inclines to zero and then the air becomes saturated.
- Water activity $\alpha_w$ (–): It is the ratio of vapor pressure exerted by water in solid to that of pure water at the same temperature.

Maroulis and Saravacos (2003) presented a psychrometric model, which calculates the fore-mentioned properties and has the ability to handle even streams that expel water and become concentrated after changing one or more of their thermodynamic properties. A revised model is presented in Table 9.1. Table 9.2 presents the variables of the model that are not included in the properties presented above.

Equation P1 calculates the vapor pressure of water at saturation at a given temperature $T$. Equation P2 represents the absolute humidity of the air–water vapor at saturation for the given temperature and pressure $P$. It expresses the maximum amount of water that can be held by the unit mass of dry air at these conditions. Given an absolute humidity $Y$, Equation P3 presents its amount (per unit mass dry air) that remains in the gas phase, while Equation P4 calculates the amount that condenses and is removed from the air–vapor mixture. It is obvious that liquid water removal takes place only when the initial absolute humidity is higher than the corresponding value at saturation; otherwise, moisture content remains in the gas. Equation P5 presents the water activity of the mixture and Equation P6 gives its corresponding relative humidity. Equation P7 calculates the specific enthalpy of the mixture, taking into account the amount of water that may exist as liquid. If only the enthalpy of the gas phase is needed, the last term of the equation must not be taken into

**TABLE 9.1**
**Psychrometric Model**

$$P_s = 10^3 \exp\left(27.0214 - \frac{6887}{273.16 + T} - 5.32 \ln \frac{273.16 + T}{273.16}\right) \quad \text{(P1)}$$

$$Y_s = m\frac{P_s}{P - P_s} \quad \text{(P2)}$$

$$Y_V = \min(Y, Y_s) \quad \text{(P3)}$$

$$Y_L = Y - Y_V \quad \text{(P4)}$$

$$Y_V = m\frac{\alpha_w P_s}{P - \alpha_w P_s} \quad \text{(P5)}$$

$$RH = 100 \times \alpha_w = 100 \times \frac{P_v}{P_s} = 100 \times \frac{X_w}{X_s} \quad \text{(P6)}$$

$$H = C_{p_a} T + Y_V\left(\Delta H_o + C_{p_v} T\right) + Y_L C_{p_w} T \quad \text{(P7)}$$

$$v = \left(2.83 \times 10^{-3} + 4.56 \times 10^{-3} Y_V\right)\left(T + 273\right) \quad \text{(P8)}$$

$$\rho = \frac{P M_w}{R\left(T + 273.16\right)} \quad \text{(P9)}$$

$$P = 10^3 \exp\left(27.0214 - \frac{6887}{273.16 + T_b} - 5.32 \ln \frac{273.16 + T_b}{273.16}\right) \quad \text{(P10)}$$

$$\frac{Y_V P}{m + Y_V} = 10^3 \exp\left(27.0214 - \frac{6887}{273.16 + T_d} - 5.32 \ln \frac{273.16 + T_d}{273.16}\right) \quad \text{(P11)}$$

$$P_d = 10^3 \exp\left(27.0214 - \frac{6887}{273.16 + T_d} - 5.32 \ln \frac{273.16 + T_d}{273.16}\right) \quad \text{(P12)}$$

$$C_{p_a} = 1.00926 \times 10^3 - 4.0403 \times 10^{-2} T + 6.1759 \times 10^{-4} T^2 - 4.097 \times 10^{-7} T^3 \quad \text{(P13)}$$

$$C_{p_v} = 1.883 - 1.6737 \times 10^{-4} T + 8.4386 \times 10^{-7} T^2 - 2.6966 \times 10^{-10} T^3 \quad \text{(P14)}$$

$$C_{p_w} = 2.8223 + 1.1828 \times 10^{-2} T - 3.5043 \times 10^{-5} T^2 + 3.601 \times 10^{-8} T^3 \quad \text{(P15)}$$

$$C_p = C_{p_a} + Y_V C_{p_v} \quad \text{(P16)}$$

$$\Delta H_T = 2491.5 - 2.048 T - 0.0032 T^2 \quad \text{(P17)}$$

$$P_{as} = 10^3 \exp\left(27.0214 - \frac{6887}{273.16 + T_{as}} - 5.32 \ln \frac{273.16 + T_{as}}{273.16}\right) \quad \text{(P18)}$$

$$Y_{as} = m\frac{P_{as}}{P - P_{as}} \quad \text{(P19)}$$

*continued*

**TABLE 9.1 (continued)**
**Psychrometric Model**

$$C_{Pas} = 1.00926 \times 10^{3} - 4.0403 \times 10^{-2} T_{as} + 6.1759 \times 10^{-4} T_{as}^{2} - 4.097 \times 10^{-7} T_{as}^{3} \quad \text{(P20)}$$

$$C_{Pvs} = 1.883 - 1.6737 \times 10^{-4} T_{as} + 8.4386 \times 10^{-7} T_{as}^{2} - 2.6966 \times 10^{-10} T_{as}^{3} \quad \text{(P21)}$$

$$C_{ps} = C_{Pas} + Y_{as} C_{Pvs} \quad \text{(P22)}$$

$$\Delta H_{as} = 2491.5 - 2.048 T_{as} - 0.0032 T_{as}^{2} \quad \text{(P23)}$$

$$T_{as} = T + \frac{\Delta H_{as}}{C_{ps}} (Y_{as} - Y) \quad \text{(P24)}$$

$$P_{w} = 10^{3} \exp\left(27.0214 - \frac{6887}{273.16 + T_{w}} - 5.32 \ln \frac{273.16 + T_{w}}{273.16}\right) \quad \text{(P25)}$$

$$Y_{w} = m \frac{P_{w}}{P - P_{w}} \quad \text{(P26)}$$

$$C_{Paw} = 1.00926 \times 10^{3} - 4.0403 \times 10^{-2} T_{w} + 6.1759 \times 10^{-4} T_{w}^{2} - 4.097 \times 10^{-7} T_{w}^{3} \quad \text{(P27)}$$

$$C_{Pvw} = 1.883 - 1.6737 \times 10^{-4} T_{w} + 8.4386 \times 10^{-7} T_{w}^{2} - 2.6966 \times 10^{-10} T_{w}^{3} \quad \text{(P28)}$$

$$C_{pw} = C_{Paw} + Y_{w} C_{Pvw} \quad \text{(P29)}$$

$$\Delta H_{w} = 2491.5 - 2.048 T_{w} - 0.0032 T_{w}^{2} \quad \text{(P30)}$$

$$T_{w} = T + \frac{\Delta H_{w}}{h/M_{a}/k_{y}} (Y_{w} - Y) \quad \text{(P31)}$$

account. Equation P8 presents the specific volume of the air–vapor mixture (Treybal, 1980) and Equation P9 gives its density. Equation P10 calculates the boiling point of water, which is the temperature where its vapor pressure becomes equal to the total pressure of the system. Equation P11 is a combination of Equations P1 and P2 and represents the dew point temperature for the given pressure and absolute humidity. This is the lowest temperature in which the mixture could be cooled without changes in the amount of water vapor it holds. Equation P12 derives from Equation P1 and calculates the dew point pressure when the dew point temperature is known. Equations P13 through P15 calculate the specific heat of dry air, water vapor, and liquid water, respectively, Equation P16 gives the specific heat of moist air, while Equation P17 calculates the latent heat of water vaporization, as a function of temperature. Alternatively, the values of 1.01, 1.88, and 4.18 kJ/kg/°C can be used as mean values for heat capacities, respectively, and the value of 2501 kJ/kg can be used as the vaporization latent heat value at reference temperature 0°C. Equations P18 through P24 are used for the calculation of adiabatic saturation temperature and moisture of the mixture. The first six of them correspond to Equations P1, P2, P13, P14, P16, and P17, calculating the variable values at the adiabatic saturation point. This point has lower temperature and higher moisture compared to the initial

**TABLE 9.2**
**Variables Involved in the Psychrometric Model**

| | |
|---|---|
| $C_{pa}$ | Heat capacity of dry air (kJ/kg da/°C) |
| $C_{pw}$ | Heat capacity of water vapor (kJ/kg/°C) |
| $k_y$ | Convective mass transfer coefficient (kg mol/s/m$^2$ mol/frac) |
| $m$ | Air-to-water molecular weight ratio. It is equal to 0.622 |
| $M_a$ | Air molecular weight (gr/mol). It can be taken approximately as 29 gr/mol |
| $P$ | Absolute pressure of the mixture (bar) |
| $P_v$ | Partial pressure of water vapor (bar) |
| $P_s$ | Partial pressure of water vapor at saturation (bar) |
| $R$ | Universal gas constant, equal to 8.314 J/mol/°C |
| $X_w$ | mol fraction of water in the mixture (mol/mol) |
| $X_s$ | mol fraction of water in a saturated mixture at the same temperature (mol/mol) |
| $\Delta H_o$ | Latent heat of vaporization for water at temperature $T_o$ |
| $Y_s$ | Absolute humidity at saturation (kg moisture/kg dry air) |
| $Y_V$ | Absolute humidity (kg moisture/kg dry air) |
| $v$ | Specific volume (m$^3$/kg dry air) |

state (unless the initial stage corresponds to saturation). Equation P24 presents the adiabatic saturation temperature of the mixture. Equations P18 through P24 form a cyclic algebraic system, which is solved by a trial-and-error procedure using $T_{as}$ as trial variable. Equations P25 through P31 also form a cyclic algebraic system similar to that of Equations P18 through P24 for the calculation of wet bulb temperature. In this case, $T_w$ is the trial variable.

The term $h/M_a/k_y$, known as psychrometric ratio, ranges between 0.96 and 1.005 for air–water vapor mixtures. Since the heat capacity of the mixture has almost the same value, the adiabatic saturation and wet bulb temperatures are almost equal for the air–water vapor system. It worth mentioning that adiabatic saturation and wet bulb temperatures are conceptually quite different. The first one is a gas temperature and a thermodynamic entity while the latter is a heat and mass transfer rate-based entity and refers to the temperature of the liquid phase.

The partial pressure of the water vapor at saturated conditions can also be expressed by the following detailed equations (Treybal, 1980):

$$\ln P_s = \frac{-5674.53}{273+T} + 6.39 - 0.968\times10^{-2}\,T + 0.622\times10^{-6}\,T^2 + 0.208\times10^{-8}\,T^3 + 0.948\times10^{-12}\,T^4 + 4.16\ln T$$

for a temperature range of −100–0°C,

$$\ln P_s = \frac{-5800.22}{273+T} + 1.39 - 0.049T + 0.418\times10^{-4}\,T^2 - 0.145\times10^{-7}\,T^3 + 6.55\ln T$$

for a temperature range of 0–200°C.

In both these equations, the temperature is expressed in K and the pressure in Pa.

Two other useful properties of air–water vapor mixtures are dynamic viscosity $\eta_g$ (kg/m/s) and thermal conductivity $\lambda_g$ (W/m/°C), which are given by the following equations (Mujumdar, 1995; Pakowski et al., 1991):

$$\eta_g = 1.691 \times 10^{-5} + 4.984 \times 10^{-8} T - 3.187 \times 10^{-11} T^2 + 1.319 \times 10^{-14} T^3$$

$$\lambda_g = 2.425 \times 10^{-2} - 7.889 \times 10^{-5} T - 1.790 \times 10^{-8} T^2 - 8.570 \times 10^{-12} T^3$$

Water activity $\alpha_w$ is expressed by the equation

$$\alpha_w = \frac{P_v}{P_s} = \frac{Y_V P}{(m + Y_V) P_s}$$

which results in the useful expression

$$P_v = \frac{Y_V P}{m + Y_V}$$

The graphical representation of the psychrometric model constitutes the psychrometric chart for the air–water vapor mixture. It relates dry and bulb temperature, absolute and relative humidity, enthalpy and specific volume of the moist air, and is widely used due to its simplicity and the ease to rapidly perform the calculation of the values of these properties. A common way to specify an air–vapor mixture is by its dry bulb and wet bulb temperatures, which can be measured relatively easily. Other properties such as absolute humidity, relative humidity, dew point temperature, and enthalpy can be evaluated conveniently from the psychrometric chart. A psychrometric chart can be presented in a variety of forms among which the most common is the Grosvenor chart that shows the relationship between dry temperature (abscissa) and absolute humidity (ordinate) of humid air at a given absolute pressure (most commonly 1 atm). The most characteristic line in the chart is the saturation line, which is identified as 100% relative humidity. During air cooling, when this line is reached, the air becomes saturated, it is not able to retain or absorb more water, and the process moves along the saturation line. Other curves of importance are the adiabatic saturation curves. They represent the path followed by the air stream as it comes in contact with liquid and heat and mass transfer takes place among the two phases. Heat from the air stream is offered to the liquid, which vaporizes, increasing the moisture content of the air, while at the same time, a decrease in the temperature of the air is observed. It should be noted that when a drying process is analyzed, and heating of solids from the hot air stream or heat of wetting of solids, associated with additional energy required to remove bound moisture, is taking into account, the path of moist air on the chart is no longer parallel to an adiabatic saturation line or a line of constant enthalpy (Kemp, 2012). The dew point temperature, which is the lowest temperature of the mixture to retain its moisture content at a given pressure, can be determined from the psychrometric chart by drawing a straight line from a given point until the saturation line is reached and the corresponding dry bulb temperature presents the dew point. For many practical purposes, the adiabatic saturation curves of the psychrometric chart can be used for the determination of wet bulb temperature. In purely convective drying, drying surface reaches the wet bulb temperature during the constant rate period. Figure 9.1 presents an absolute humidity–dry bulb temperature psychrometric chart at a pressure of 540 torr, based on the mathematical model of Table 9.1. Gray lines refer to 760 torr (1 atm). It is evident that when pressure increases, the saturation line (along with lines corresponding to lower relative humidities) moves to increased temperatures (to the right of the chart). This indicates that moist air becomes saturated at lower temperature, for a given value of absolute humidity, or it

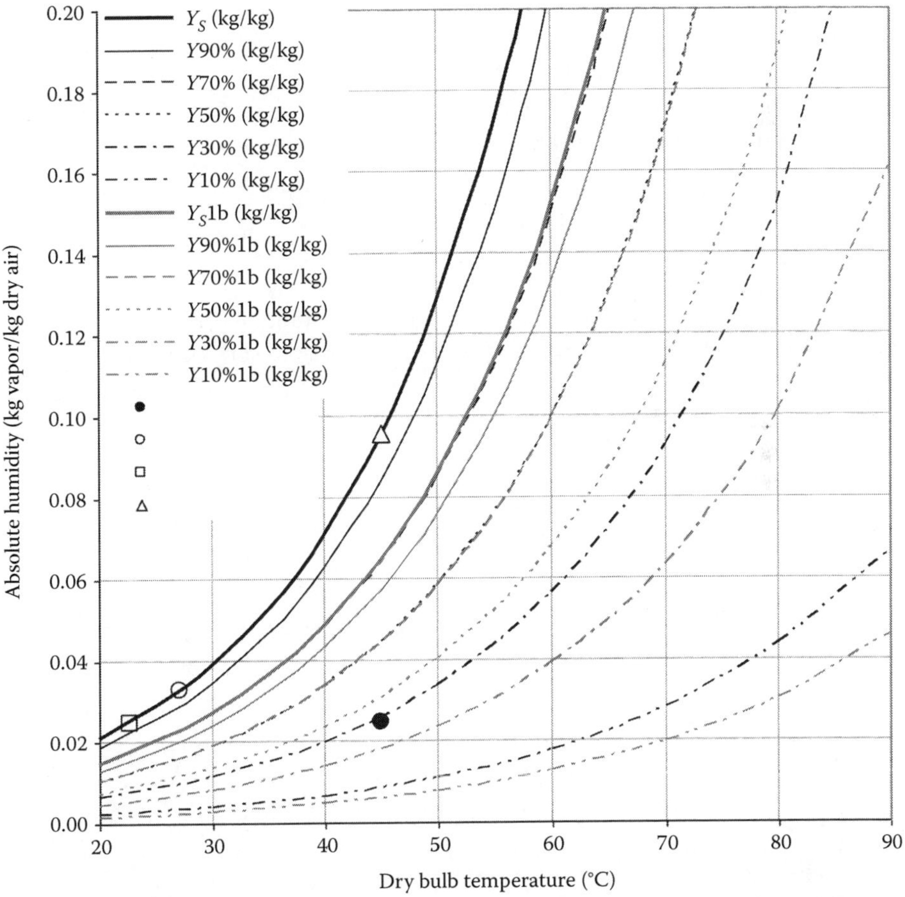


**FIGURE 9.1** Humidity–temperature psychrometric chart. Black lines refer to 540 torr and gray lines refer to 760 torr. Black solid marker represents a given air–vapor mixture, square marker its dew point, cyclic marker its adiabatic saturation point, and triangle marker its saturation point.

can hold a larger amount of vapor, for a given temperature, as the pressure decreases. In this figure, the black solid marker represents a given air–water vapor mixture, the square marker represents its dew point (approach to saturation at constant absolute humidity), the cycle marker represents its adiabatic saturation point (approach to saturation at constant enthalpy), and the triangle marker represents its saturation point (approach to saturation at constant dry bulb temperature). For air–water vapor mixtures, adiabatic saturation temperature is practically equal to wet bulb temperature (this is not valid for air mixtures with other compounds). It is obvious that not only temperature but pressure change as well can drive an unsaturated mixture to saturation or even to a state where liquid water will condense out. The Mollier chart, which presents enthalpy versus absolute humidity, is another common type. Figure 9.2 presents an enthalpy/dry bulb temperature–absolute humidity psychrometric chart (two ordinate chart) at a pressure of 760 torr based on the described psychrometric model. A detailed psychrometric chart is presented at the end of the chapter, while the reader can find a tool to calculate psychrometric properties of air–water vapor mixtures at the web address http://lpad.chemeng.ntua.gr/psychrometry.

## 9.3 DRYING MECHANISM

In the case of drying food materials under the influence of a fluid (usually air or inert gas), the main mechanisms of drying are

- Surface diffusion or liquid diffusion on the pore surfaces
- Liquid or vapor diffusion due to moisture concentration differences
- Capillary action in granular and porous foods due to surface forces

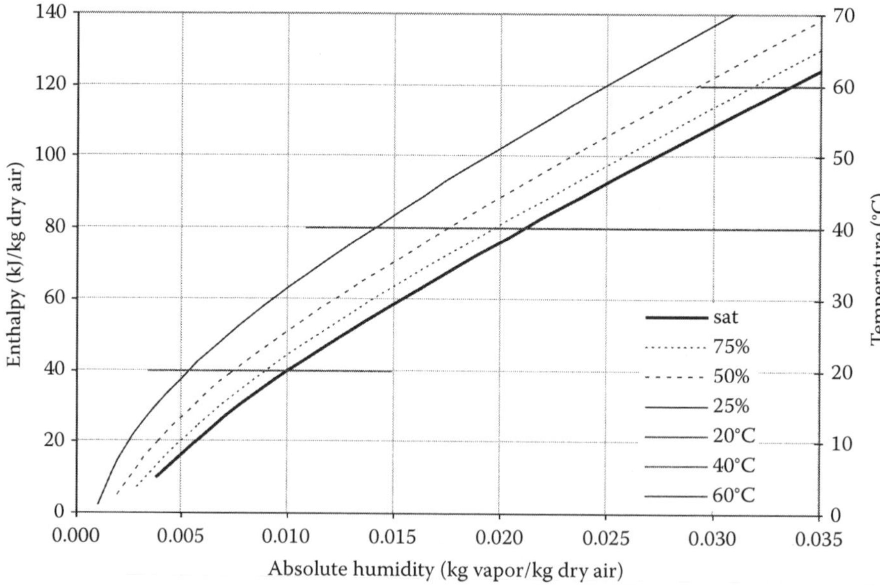


**FIGURE 9.2** Enthalpy/temperature–humidity psychrometric chart (pressure 760 torr).

Furthermore, thermal diffusion that is defined as water flow due to the vaporization–condensation sequence, and hydrodynamic flow that is defined as water flow due to shrinkage and pressure gradient may also take place during drying (Strumillo and Kudra, 1986). The dominant mechanism can change during the process. Diffusion is a function of the moisture content and the structure of the material and it determines the drying rate.

For hygroscopic products, generally, the product dries at a constant rate and subsequently, periods of falling rate; the process terminates when equilibrium is established. In the constant rate period, external conditions such as temperature and relative humidity of the drying medium, drying air velocity and flow direction, physical form of the product, the method of its supporting, and the desirability of agitation are the main parameters affecting drying, and the dominant diffusion mechanism is surface diffusion. As the constant rate period comes to an end, the moisture has to be transported from the inner layers of the solid to the surface by capillary forces. The drying rate may still be constant until the moisture content reaches the critical moisture content. At that stage, the surface moisture film has been reduced, dry spots appear on the surface, and the first falling rate period (unsaturated surface drying) begins. It is worth mentioning that the rate per unit wet solid surface area still remains constant, but as the rate is calculated with respect to the overall solid surface area, the drying rate seems to reduce (Mujumdar and Menon, 1995). Liquid diffusion due to moisture concentration difference is the dominant drying mechanism at that stage. The second falling rate period begins when the surface liquid film has entirely evaporated and the rate of the process is controlled by the vapor diffusion due to moisture concentration difference. Internal conditions such as temperature, moisture content, and structure of the product are the main factors affecting the water removal rate in both falling rate periods.

It has been observed that biological materials, although they have high moisture content, generally do not present a constant rate period during drying, and in some cases, when the initial moisture content is not very high, only the second falling rate period appears, as in the drying of grains or nuts (Bakshi and Singh, 1980; Parry, 1985).

## 9.4 DRYING KINETICS

Drying is a process governed by simultaneous and often coupled and multiphase heat, mass, and momentum transfer phenomena. That along with the different structure of food materials as well as the numerous physical, chemical, and biochemical transformations (sometimes desirable) that occur during drying add a significant complexity, which makes the application of many transport models unsuitable for a precise prediction of drying time. It is well known that physical changes such as glass transitions or crystallization during drying may result in changes in the mechanisms of mass transfer and rates of heat transfer within the material, often in an unpredictable manner (Mujumdar, 1995). Theoretical models are not practical and cannot unify drying calculations; this is a process not well understood at the microscopic level. To overcome this problem and to develop effective tools for the process design, optimization, control, and energy integration, thin-layer drying equations based on experimental measurements have been developed, which are practical and give sufficiently good results (Erbay and Icier, 2010; Kudra and Mujumdar, 2002). Thin-layer drying equations express the drying rate during the process.

Drying processes are mathematically modeled with two main models:

(i) Distributed models that consider simultaneous heat and mass transfer
(ii) Lumped parameter models that do not take into account the temperature gradient in the product and assume a uniform temperature distribution that equals the drying air temperature (Erbay and Icier, 2010)

Thin-layer drying generally refers to the drying as one thin layer of particles or slices where the temperature distribution can be assumed as uniform and the lumped parameter models can be applied (Akpinar, 2006a). Thin-layer models are theoretical and semi-theoretical or empirical. The first are derived from Fick's second law of diffusion and consider only the internal resistance to moisture transfer. The latter are derived from Fick's second law and modifications of its simplified forms, or by analogues with Newton's law of cooling, and take into account only the external resistance to moisture transfer between the product and the drying fluid. Semitheoretical and empirical models are valid only within the process conditions applied (Fortes and Okos, 1981). In the thin-layer drying concept, drying constant is the combination of drying transport properties such as moisture diffusivity, thermal conductivity, and interface heat and mass coefficients (Marinos-Kouris and Maroulis, 1995). Table 9.3 summarizes some of the common thin-layer drying models.

Other more complicated drying models have been introduced in the literature to simulate experimental drying data for different food materials. Adhami et al. (2013) proposed the following differential models for the description of the drying rate in the primary and secondary drying stages of quince undergoing freeze drying. These models embody the effect of freeze drier operating parameters, which are expressed through the imposed heat load $q$ (W/kg) and the sample temperature $T$. The model describing the kinetics of the primary drying stage is expressed as

$$\frac{dX}{dt} = \left(a + bq^{n}\right)\left(c + dT^{m}\right)\left(\frac{X}{X_o}\right)^{k}$$

For the secondary drying stage where the dominant mechanism is the removal of bounded water, the following equation is applied:

$$\frac{dX}{dt} = a'\left(\frac{X}{X_o}\right)^{n'} - b'\exp\left(\frac{c'}{T} - d'\right)\left(e' + q^{k'}\right)$$

The authors estimated the values of the parameters for quince freeze drying as $a = 0.0797$, $b = 0.0007$, $n = 0.9077$, $c = 1.1011$, $d = 4.4060$, $m = -0.3246$, $k = 0.7954$, $a' = 1.8224$, $n' = 1.3716$, $b' = 29.083$, $c' = 44.6125$, $d' = 8.7070$, $e' = 0.5844$, and $k' = 0.6116$.

## TABLE 9.3
## Thin-Layer Drying Models

**Models Derived from Newton's Law of Cooling**

Lewis (Newton) model, 1921 $X_t = X_e + (X_o - X_e)\exp(-kt)$

Page model, 1949 $X_t = X_e + (X_o - X_e)\exp(-kt^n)$

*Modified Page models*

Modified Page-I model (Overhults et al.), 1973 $X_t = X_e + (X_o - X_e)\exp(-kt)^n$

Modified Page-II model (White et al.), 1978 $X_t = X_e + (X_o - X_e)\exp - (kt)^n$

Modified Page-III model (Diamente and Munro), 1993 $X_t = X_e + (X_o - X_e)\exp - k(t/l^2)^n$

**Models Derived from Fick's Second Law of Diffusion**

Simplified Fick's second law of diffusion $X_t = X_e + (X_o - X_e) \cdot \frac{8}{\pi^2} \cdot \exp\left(-\frac{\pi^2 D_{eff}\, t}{4L^2}\right)$

Henderson and Pabis (single term) model, 1961 $X_t = X_e + a(X_o - X_e)\exp(-kt)$

Logarithmic (asymptotic) model (Chandra and Singh), 1995 $X_t = X_e + (X_o - X_e) \cdot [a\exp(-kt) + p]$

Midilli model, 2002 $X_t = X_e + (X_o - X_e) \cdot [a\exp(-kt) + b^*t]$

Modified Midilli model (Ghazanfari et al.), 2006 $X_t = X_e + (X_o - X_e) \cdot [\exp(-kt) + b^*t]$

Demir et al. model, 2007 $X_t = X_e + (X_o - X_e) \cdot [a\exp[(-kt)]^n) + b']$

Two-term model (Henderson), 1974 $X_t = X_e + (X_o - X_e) \cdot [a\exp(-k_1 t) + b\exp(-k_2 t)]$

Two-term exponential model (Sharaf-Eldeen et al.), 1980 $X_t = X_e + (X_o - X_e) \cdot [a\exp(-kt) + (1-a)\exp(-k\,a\,t)]$

Modified two-term exponential models (Verma model), 1985 $X_t = X_e + (X_o - X_e) \cdot [a\exp(-kt) + (1-a)\exp(-gt)]$

Diffusion approach model (Kaseem), 1998 $X_t = X_e + (X_o - X_e) \cdot [a\exp(-kt) + (1-a)\exp(-k\,b\,t)]$

Modified Henderson and Pabis (three-term exponential) model, 1999 $X_t = X_e + (X_o - X_e) \cdot [a\exp(-k_1 t) + b\exp(-k_2 t) + c\exp(-k_3 t)]$

**Empirical Models**

Thompson model, 1968 $t = g\ln\left(\frac{X_t - X_e}{X_o - X_e}\right) + h\left[\ln\left(\frac{X_t - X_e}{X_o - X_e}\right)\right]^2$

Wang and Singh model, 1978 $X_t = X_e + (X_o - X_e) \cdot [1 + a^*t + c^*t^2]$

Kaleemullah model, 2002 $X_t = X_e + (X_o - X_e) \cdot \left[\exp\left(-(d^*T + e^*)t^{(p^*T+m)}\right)\right]$

Weibull $X_t = X_e + (X_o - X_e)\exp\left[-\left(\frac{t}{w}\right)^v\right]$

Logistic $X_t = X_e + (X_o - X_e) \cdot \frac{g}{1 + h\exp(kt)}$

*Note:* $a$, $b$, and $c$ are defined as the indication of shape and generally named as model constant (dimensionless), $n, m, l, g, h, p, b', a^*, b^*\ c^*$ are empirical constants, $d^*$ is in °C$^{-1}$s$^{-(p^*T+m)}$, $e^*$ is in s$^{-(p^*T+m)}$, $p^*$ is in °C$^{-1}$, $L$ is the half-thickness of the material layer, $w$ is a scale parameter, $v$ is a shape parameter, and $T$ is the temperature (°C).

Drying constant $k$ can be used for the determination of effective moisture diffusivity $D_{eff}$, expressed in m²/s, which is a term that summarizes the effect of molecular diffusion, surface diffusion, capillary flow, Knudsen flow, and hydrodynamic flow, as a solid undergoes drying and moisture diffuses through its macromolecule structure. In general, the quantitative calculation of each of those effects separately is practically difficult.

$$D_{eff} = -\frac{A}{\pi^2}k$$

relates drying constant and diffusivity, where $A = 4L^2$ for an infinite slab, $A = 4r^2$ for a sphere, and $A = \left(L_1^2 + L_2^2 + L_3^2\right)^{-1}$ for a three-dimensional finite slab; $L$ is the half thickness of the slab if drying takes place from both sides, or the slab thickness if only one side is exposed to drying conditions; $r$ is the sphere radius; and $L_i$, $i = 1, 2, 3$ are the dimensions of the finite slab.

Moisture diffusivity is affected significantly by drying temperature and this effect is described by the following Arrhenius equation:

$$D_{eff} = D_o \exp\left(-\frac{E_a}{RT}\right)$$

where $D_o$ (m²/s) is the Arrhenius factor defined as the reference diffusion coefficient at infinitely high temperature, $E_a$ (kJ/mol) is the activation energy for diffusion, and $R$ (kJ/kmol/°C) is the universal gas constant.

Finally, a useful equation that connects moisture content in dry basis $X$, expressed in kg moisture/kg dry solids, to moisture content in wet basis $W$, expressed in kg moisture/kg of material (total), is the following:

$$X = \frac{W}{1 - W}$$

## 9.5 DRYING METHODS

Michailidis and Krokida (2013) presented an extensive description of various drying methods and techniques used in food processing. These include both extensively used methods and a few experimental methods. The main advantages, disadvantages and comments for each method are presented in Table 9.4.

In the following sections, further drying methods will be presented.

One of the most important features in drying technology is that prior to drying, a predrying process usually applies. Liquids are

- Vacuum concentrated, to reduce the food volume or strip aroma compounds in order to add them back to the dry powder
- Treated with enzymes, to reduce the viscosity, avoiding gelling and haze formation, or subjected to removal of some compounds to assure a natural color of the product

**TABLE 9.4**
**Description of Various Drying Processes**

| Dring Method | Advantages | Disadvantages | Comments |
|---|---|---|---|
| Hot-air drying | It characterized by simplicity and there is great engineering experience. | It presents high energy consumption, low energy efficiency (40–60%), and extensive reduction of product quality compared to original foodstuff (discoloration, browning, changes in physical structure, and functionality changes of biopolymers) due to extended exposure to oxygen at high temperatures. | A fluid (air in most cases) is used to supply heat for the vaporization of moisture, which removes from the dryer with the same fluid. It is the most common drying method and applies to a great variety of food materials, including fruits and vegetables, meat, seafood, grains, etc. |
| Vacuum drying | It can de used for heat-sensitive materials due to low temperatures applied and oxygen-deficient environment.<br>It presents high drying rates and induces limited degradation of nutritional compounds.<br>Allows solvent recovery. | The equipment is more complex compared to hot-air drying as the process takes place in reduced pressure. | Heat is supplied by conduction from the supporting elements and by radiation from the dryer walls. Moisture removes due to the pump used for the vacuum creation. |
| Microwave drying | It reduces drying time as heat is generated internally instead of requiring conduction from a hot surface or convection from the surrounding fluid.<br>It presents high rates of moisture removal, no overheating of the atmosphere or the surface, reduced heat losses, and compact equipment. | Uneven heating, underdrying or charring are common as microwave energy is not usually absorbed uniformly throughout the sample.<br>There is no common method to monitor or control the electromagnetic field distribution. | The effect of microwaves causes the heating and vaporization of water inside the material. The migration of solutes in the liquid phase is limited, preventing case hardening of the materials. The final product is less brown, odor is retained in higher degree, and the rehydration characteristics are better compared to hot-air drying. |

*continued*

**TABLE 9.4 (continued)**
**Description of Various Drying Processes**

| Dring Method | Advantages | Disadvantages | Comments |
|---|---|---|---|
| Freeze drying | It causes the minimum change of structure, texture, and appearance compared to other methods.<br>It preserves shape, taste, color, flavor, aroma, biological activity, and nutrient compounds and limits oxidation or modification reactions in the food material as it is processed near or under ambient temperature and in the absence of oxygen.<br>The final product presents high porosity leading to fast rehydration rate and high rehydration capacity. | Cost is the highest among the different drying methods due to the need for freezing the raw materials and operating under high vacuum.<br>The duration of the process is extensive (1–3 days). | Moisture of foods transforms in the state of ice during freezing and ice sublimation takes place under vacuum during drying. It is the most advanced drying method for the preservation of the initial characteristics of raw food material.<br>It can be used in the processing of products that are technically difficult to dehydrate at high temperature, for example, because of their high sugar content. |
| Spray drying | It is used for processing liquid feeds.<br>It is continuous and easy to be controlled.<br>Satisfies aseptic/hygienic drying conditions. | It presents relatively high cost (only freeze drying is more expensive drying process), low thermal efficiency, and large air volumes at low product hold-up, which increase the gas cleaning cost.<br>There is possibility of heat degradation of the product if high drying temperature is used. | It is a special process applied to transform a feed from a liquid state into a dried particulate form (powder, particles, granules, and agglomerates at different sizes).<br>It can be used in food encapsulation and microencapsulation to increase food shelf life during storage. |
| Osmotic dehydration | It is a simple technique that preserves texture and flavor, prevents discoloration, and increases microbiological stability. | It is used mainly as a pretreatment as it is inefficient and difficult to reduce the weight by more than 50% and it does not ensure product stability, as water activity is generally higher than 0.9. | It takes place by immersion of the food material in a sugar or salt solution. There is the ability for the introduction of a preservative agent, or any solute of |

| | | | |
|---|---|---|---|
| | | | nutritional interest, which improves sensory characteristics and reduces water activity of the product.<br>It is possible to change, to a certain extent, the nutritional and functional properties of a food without modifying its integrity. |
| Atmospheric freeze drying | It may apply without the need for an external condensed.<br>Energy cost can be lowered by ~30% compared to freeze drying. | It is much slower compared to freeze drying. | It is an alternative to freeze drying presenting similar quality properties.<br>An adsorbent can be used to enchain moisture. |
| Sonic drying | It is a simple, relatively cheap, and energy-saving technique. | | It relays on the pressure increase and decrease of the gas medium surrounding the dying material, which is caused by the energy generated from sound waves.<br>The sonic-assisted drying can reduce drying time to about half, depending on sound energy and frequency. |
| Heat pump drying | It increases energy saving and can be used in the cases of drying with inert gases for their recirculation after moisture removal. | It has increased capital cost and requirement for maintenance compared to hot-air drying.<br>Leakage of refrigerant to the environment may occur. | It is a variation of fluid drying in which, through the heat pump, heat from the exhaust drying fluid is recovered and offered again to the moist material or even the moisture from the exhaust drying fluid is removed and the fluid recirculates in the dryer. |
| Infrared drying | It presents relatively shortened drying time, high energy transfer rate, reduced energy consumption, efficient transmission through the air or evacuated space, low air flow through | | It is based on the interaction of infrared wavelength radiation with the internal structure of a food material. |

*continued*

**TABLE 9.4 (continued)**
**Description of Various Drying Processes**

| Dring Method | Advantages | Disadvantages | Comments |
|---|---|---|---|
| | the product, uniform temperature in the product, superior product quality, space saving, ease of automation, and clean working environment compared to other drying methods. | | |
| Superheated steam drying | The net energy consumption can be low enough. Drying medium does not contain oxygen and there is no risk for food substances oxidation.<br>There is preservation of color and nutrients.<br>Casehardened skin is unlikely to form. The treatment strips more of the acids that contribute to an undesirable taste or aroma of the products. | The equipment used is more complex than a hot air dryer and special care should be taken for the feed of raw materials and discharge of the products.<br>Heat sensitive materials may prone to damage. | Superheated steam is used as drying medium.<br>Low, near-atmospheric or high pressure (at ~5 bar, referred as high pressure superheated steam drying) operation is possible.<br>Low pressure superheated steam drying takes place in lower temperatures to avoid melting, glass transition or damage of the processing food. |
| Intermittent drying | It presents reduced energy consumption. Product quality is increased with no cracking of the material. | | It is a drying technique in which the heat is applied in a noncontinuous way. In the tempering period, moisture is redistributed inside the material resulting in the increase of drying rate when heat application is rebooted. |

| | | | |
|---|---|---|---|
| Instant controlled pressure drop drying | It presents reduction of energy consumption and overall production cost, controllability, increased safety and hygienics, and quality improvement in terms of sensorial, functional, and nutritional attributes. | | It is a treatment of high-temperature–short-time followed by an instant pressure drop toward a vacuum, which allows the water to autovaporize.<br>It can be applied for texturing of fruits, vegetables, and seaweeds.<br>In conjunction to hot-air drying, it can lead to lightly or highly expanded fruits and vegetables. |
| Sun drying | It is the oldest, simplest, and costless drying method. | Drying time is very protracted (up to 10 days) and there is need for extensive land.<br>Product quality is poor due to undesirable reactions, case hardening, and spoilage caused by unexpected rains, animals, and toxic fungi. | |
| Solar drying | It presents decreased drying time and harvest losses due to spoilage compared to sun drying and results in significant cost savings by reducing conventional fuel demand. | | It incorporates the sun energy by controlling the radiative heat. |
| Supercritical drying | | | It is an experimental method that leads to the retention of the original (micro) structure and the corresponding porous network functionality of the material. |

*continued*

**TABLE 9.4 (continued)**
**Description of Various Drying Processes**

| Dring Method | Advantages | Disadvantages | Comments |
|---|---|---|---|
| Flash Drying | It is an efficient technique that results in shortening the drying time. It has lower capital cost, simpler equipment, and less energy requirement compared to freeze drying. | | It is a drying technique based on the application of successive cycles of heating and vacuum pulses.<br>It can be used in the production of high-quality products even when they are sensitive to heat treatment. |
| Pulse drying | | | High-temperature short-time pulse drying is used to alter the structural properties of the dried material by changing the values of the drying parameters used in conventional hot-air drying. It has been used for the production of foods that present a highly porous structure and a formation of a crust on their external surface. |
| Pulse combustion drying | | | It is an experimental method based on intermittent combustion of a fuel, which generates pressure, velocity, and, to a certain extent, temperature waves that intensify heat and mass transfer. Product quality and unit cost is comparable to spray drying. |

Solids are

- Sulfated to retard nonenzymatic browning, reduce shrinkage and oxidation of liable food components, and improve rehydration
- Soaked in solutions of different compounds to retard nonenzymatic browning, and affect texture and shrinkage (calcium salts) or reduce bacterial load (acid solutions) or reduce drying time ($K_2CO_3$)
- Blanched to inactivate enzymes, change the tissue structure, reduce drying time, and increase firmness and microbiological quality
- Freezed to disorder tissue structure, increase diffusion, reduce drying time, and enhance rehydration
- Treated by high pressure to preserve color (Lewicki, 2006)

### 9.5.1 Drum Drying

Drum drying (DD) is one of the most energy-efficient methods in which drying takes place on the outer cylindrical surface of one or two slowly rotating drums, heated internally by steam, and is used for the treatment of sludge/slurries, suspensions, liquid solutions, pureed foods, and pastes over a wide viscosity range (Maroulis and Saravacos, 2003). Especially in the case of viscous materials (in their natural state or after concentration), drying in the form of very thin film is particularly effective (Falagas, 1985). The material is spread in the form of a thin layer and after about three quarters of a revolution from the point of feeding the product is considered dried and its removal, many times in the form of a thin sheet, is achieved by a scraper, knife, or blade. The applied film is not in motion relative to the drum because of rapid drying and solidification. Product temperature is in the range of 120–170°C for most of the applications. Exposure to high temperature is limited to a few seconds and the method can be applied even in the cases of heat-sensitive products. Feed temperature and concentration are frequently regulated using preheating and preconcentration to optimize the amount of heat to be transferred per unit weight of dry material. Preconcentration is desirable only up to the point that does not prevent uniform adherence of the product to the dryer surface (Moore, 1995). DD is regarded as conduction drying method where the required heat for the vaporization of moisture is being obtained by the heat transfer from the steam condensation inside the drum through its metallic wall to the material film spread over its external surface (Kostoglou and Karapantsios, 2003). Energy efficiencies in DD may range between 70% and 90%, compared to 40–60% for hot air drying, while the corresponding steam consumption is 1.2–1.5 kg per kg of water evaporated. Hot water under high pressure could be used as the heating medium when low energy consumptions are required. The method is considered as one of the most reliable among the various drying methods. On the other hand, DD at atmospheric pressure may result in extensive quality loss of heat-sensitive biocompounds due to the high temperature of the drum while the high cost of the exchange surface, which must be precisely manufactured to allow scraping, is another limitation of the method. There are many drum dryer types and configurations, which include a single drum dryer with or without applicator rolls, double-drum dryers where the drums rotate toward each other, twin-drum dryers where the drums rotate away from each other, and vacuum dryers that operate under reduced pressure. The latter presents the advantage of drying at lower temperatures but its high capital cost limits its applications. The drum(s) are mounted on a horizontal axis and mechanically rotated with variable speed control. Drum dryers may be dip or splash fed or equipped with applicator rolls, especially in the treating of highly viscous materials. The installation of a radiation heater in a zone near the sludge feeding point seems to improve drying performance (Islam et al., 2007). Common application of DD includes the production of flaky dry powder from thick suspensions of different types of starchy food materials. Drum dryers are commonly used for the industrial production of a variety of foodstuffs such as yeast creams, gelatin, breakfast cereal, fruit purees, fruit and vegetable pulp, applesauce, milk products, baby foods, mashed potatoes, dry soup mixtures, pregelatinized or cooked starches, fruit–starch mixtures, and so on (Bonazzi et al., 1996).

Gavrielidou et al. (2002) reported variables affecting drying rate in a double-drum dryer include steam pressure, drum clearance, pool level between the drums, as well as material physical characteristics, concentration, and the temperature the material reaches at the drum surface. Drying air velocity, drum rotational speed, and especially film thickness are also critical operating parameters affecting the performance of the dryer and controlling the final moisture content and productivity of the dryer (drying time) (Moore, 1995).

Kasiri et al. (2004) described the mechanism of DD analyzing the process into three segments. As the drum temperature is high due to the heating medium used inside the dryer drum, the feed temperature gradually increases. During the first segment, feed undergoes that increase in temperature up to the maximum slurry temperature when the water starts boiling. In the second segment, evaporation of liquid water takes place from the outer film surface, at the boiling point, due to the difference in vapor pressure between the drying air and the surface. The process is enhanced by the heat transfer from the high-temperature drum surface. Owing to the low thickness of the film and the abundance of moisture in the surface layer, superficial moisture loss is quickly replaced from the internal layers of the material to maintain a saturated surface. Moisture diffusion at this stage is not critical to this method as the drying material layer is thin. The evaporation rate remains constant due to the constant temperature. As the superficial moisture moves to the surrounding space, the material to be dried experiences a temperature increase from the water's boiling point to the final product temperature during the third segment. At the beginning of this segment, there is not enough liquid moisture in the main body to maintain the saturation conditions on the exposed drying surface (critical moisture content), and a dried solid film is formed and diffusion becomes the rate-limiting step, since moisture diffusion through the film is slower than surface evaporation.

The authors also examined the effect of a variety of factors on the drying time and the final moisture content of the product. The increase of the drying medium temperature increases the heat transfer rate. The increase of film thickness increases the time for the material to reach critical moisture content, while diffusion is slower at the final stage of the process. The increase of surrounding air velocity increases the mass transfer coefficient, although at some point, velocity reaches a saturation point above which an increase does not considerably affect the drying time. The increase of the drum rotation speed causes considerable increase in the final product moisture content due to a shorter drying time. The decrease of the surrounding pressure results in an increase of the mass transfer coefficient and the drying time.

### 9.5.2 Impingement Drying

Impingement drying or impingement jet drying (IJD) is a method in which the food material is stationary on a surface and hot air is fed to the drying chamber from a series of round nozzles at a short distance from the surface. The high-velocity turbulent jets of impinging air remove the boundary layer from the solid surface and the surrounding cold air creating a bed of hot air that surrounds the particles, they rapidly increase the temperature at the center of the product to the drying air temperature, and enhance the heat and mass transfer rate (Anderson and Singh, 2006). The pseudofluidization of the bulk material aims at the good contact between a gas and a solid. When the particulate material is pneumatically agitated, a random distribution of dry, partly wet, and wet material is expected, which forms a random distribution of transport resistances partly in series and partly in parallel (Xiao et al., 2010). The method ensures rapid dewatering of materials in the form of thin sheets or beds of coarse granules. Indeed, the drying procedure is very fast and the moisture content distributes more uniformly due to the pseudofluidized bed created by the high-velocity air flow that suspends and trembles the particles (Moreira, 2001). IJD is recommended only if the major fraction of the moisture to be removed is unbound. As the drying rate is very high, evaporative cooling in the constant rate period is also high and holds the product surface temperature below its degradation or ignition temperature. If the remaining moisture is in the form of bound water and/or the controlling mechanism of the process is the internal diffusion (drying shifts to falling rate period), product temperature may rise rapidly due to the high heat transfer rate, which will result in degradation of heat-sensitive products. IJD requires simple equipment is general, as no moving mechanical devices or need for maintenance are required (Bórquez et al., 1999).

In food processing, typical temperature and jet exit velocity values in IJD range from 100°C to 350°C and 10 to 100 m/s, respectively (Mujumdar, 1986). The maximum (center) velocity decreases with increasing distance from the nozzle exit. An impingement dryer consists of a single gas jet or an array of such jets, while there is a large variety of nozzles available (multizones). IJD has been applied to a variety of granular food materials such as coffee and cocoa beans, rice, bakery, snacks, and nuts (Moreira, 2001).

Superheated steam instead of air can also be used in this method to improve the texture of food products as steam usually causes changes in the texture that lead to a crispier final product after drying. Comparison of drying with superheated steam and air reveals that at elevated temperatures, superheated steam provides higher drying rates and there is more starch gelatinization compared to air drying. IJD can be combined with infrared radiation to reduce drying time and improve product quality as the infrared radiation energy can be absorbed directly by the material without significant loss to the environment.

Xiao et al. (2010a) investigated the IJD of carrot cubes ($1 \times 1 \times 1$ cm) with an average initial moisture content of 10.5 kg/kg dry basis (db) to a final value of

0.06 kg/kg db. As drying parameters, they explore air temperature (40°C, 50°C, and 60°C), air velocity (3.0, 8.0, and 13.0 m/s), and air relative humidity (10%, 20%, and 40%). The nozzle distance from the particles was 8 cm. Results indicated that the drying time of carrot cubes was more affected by drying air temperature, followed by air relative humidity and air velocity. The effective diffusivity of the moisture was calculated in the range $0.265 \times 10^{-9}$–$1.052 \times 10^{-9}$ $m^2/s$ and the activation energy as 20.17 kJ/mol. The color values L (white/dark), a (red/green), and b (yellow/blue) of the samples decreased from 50.02 to 32.71, 32.49 to 14.34, and 35.52 to 16.38, respectively, when the drying temperature increased from 40°C to 60°C, indicating that the nutrients of carrots were degraded more extensively at higher temperature, whereas the hue angle values had no obvious relationship with the drying temperature. The rehydration ratio of dried carrot was significantly affected by the drying temperature and relative humidity owing to the disruption of cellular integrity at high temperatures. Increased relative humidity improved the rehydration rate of the final product. Concerning its surface microstructure, it was noted that drying at 60°C and 10%RH led to the outer layers being more consolidated, rigid, and denser compared to drying at 40°C and 40%RH. Air velocity had little effect on the drying rate and color change and no obvious effect on the rehydration ratio.

Xiao et al. (2010b) investigated the IJD of Monukka seedless grapes at drying temperatures of 50°C, 55°C, 60°C, and 65°C and air velocity of 3, 5, 7, and 9 m/s. They concluded that the product hardness, which is an expression of its texture, increased from 9.53 to 17.16 N, showing a significantly increasing trend with an increase in temperature, while the air velocity had no significant effect. High temperatures cause higher water removal rate from the surface than the water migration rate from the interior and a hard layer containing previously dissolved solutes is formed on the surface. The retention ratio of vitamin C of the samples varied from 10.26% (0.57 mg vitamin C/100 g wb) to 39.73% (2.25 mg vitamin C/100 g wb) compared to the fresh one, corresponding to 50°C and 65°C, respectively. Air velocity did not significantly influence the retention of vitamin C. The loss of vitamin C is attributed to oxidation and thermal degradation as it is a heat-sensitive compound.

Bórquez et al. (1999) applied the IJD on pressed fish cake from mackerel previously cooked at 95°C for ~15 min. The cake had an initial moisture content of ~0.5 kg/kg. The operating conditions of the dryer were air temperature 32–51°C, wall temperature 40–62°C, air velocity 66 m/s, particle diameter 1.28 and 1.96 mm, specific surface area 5039 and 4233 $m^2/m^3$, and sphericity 0.62 and 0.52. They observed that the heat transfer coefficient increases by a factor of 2 when the particle diameter is reduced from 1.96 to 1.28 mm due to the increase in the specific surface area exposed to heat and mass transfer. The rate of oxidation did not increase with temperature at high values of moisture as increased water content presented a protective effect against oxidation at drying temperatures in the range of 32–62°C and also retarded lipid oxidation, probably due to a dilution of reactants and the promotion of nonenzymatic browning, enhanced by water and temperature, which resulted in the formation of antioxidant compounds. The progression of drying led to a substantial decrease of n-3 unsaturated fatty acids (lipids) concentration (C18:4, C20:5, and C22:6) and as the water content decreased from 13% to 5%, the losses of n-3 fatty acids increased from 2% to 8%.

Li et al. (1999) studied the IJD of tortilla chips at different superheated steam temperatures. They concluded that the microstructure of the product dried at higher steam temperature had more pores, showed a coarser appearance and a lower degree of shrinkage, and presented a higher modulus of deformation. Also, the products gelatinized less during drying at higher temperatures and had higher water absorption ability.

### 9.5.3 Impinging Drying

In impinging drying or impinging stream drying (ISD), the collision of two (or more) high-velocity gas streams (10–150 m/s) leads to the generation of a high shear rate and intense turbulence (in the case of turbulent impinging streams), which greatly enhances heat, mass, and momentum transfer in the impingement zone. The wet particles (0.2–3.0 mm diameter) are brought into an oscillatory motion by the impinging streams as they penetrate into the opposed stream(s) due to their inertia and decelerate due to the opposite flow of this opposed stream(s). At the end of the deceleration path, the particles accelerate and re-enter their original stream, and after performing several such oscillatory motions, their velocity eventually vanishes and the particles are withdrawn from the dryer. This unique flow pattern increases the residence time of particulate materials in the drying zone under intense heat and mass transfer conditions. Drying time in this method is limited to a few seconds (0.5–15 s). Hence, the method is suitable only for the removal of surface moisture or weakly bound water in the unhindered drying rate period, making the method an alternative to flash drying (Sathapornprasath et al., 2007). The equipment of ISD is relatively compact and simple in construction and operation, while the product quality is high. Pressure drop is an essential parameter of dryer operation and is in the range of 1–10 kPa, which is about 20 times lower than that of fluidized bed drying (Kudra et al., 1995). ISD, compared to other drying methods, presents a smaller footprint and high robustness due to the lack of moving parts (Choicharoen et al., 2010). The disadvantages of the method are the high energy consumption compared to traditional parallel-flow drying and the fact that it is not suitable for highly sensitive materials due to overheating (Kurnia et al., 2013).

There are many types (configurations) of impinging dryers, such as coaxial gas solid, two tangentially fed impinging stream, four impinging streams, multistage two impinging streams dryers, and others. Important operating parameters of the dryer performance include inlet air temperature and velocity, feed rate of particles, and spacing between the opposed inlet streams. Wet particles are generally carried to the dryer by one (or both) of the inlet streams.

Sathapornprasath et al. (2007) investigated the effect of inlet air temperature (110°C, 130°C, and 150°C), exit air velocity (16 and 45 m/s), particle flow rate (10 and 20 kg/h) and impinging distance between outlets of pipes of 0.025 m diameter (faces of the inlet pipes) (0.05 and 0.15 m), on the performance of a coaxial two-impinging stream dryer in terms of the volumetric heat transfer coefficient and volumetric water evaporation rate using a model material (resin) of initial moisture content 81–85% db, mean diameter of $0.5 \times 10^{-3}$ m, and bulk density of 1250 kg/m$^3$. The maximum pressure was 1800 mmH$_2$O and the flow rate was 5.6 m$^3$/min. They reported that the maximum volumetric water evaporation rate was in the range of

110 kg water/m$^3$h and the volumetric heat transfer coefficient was ~880 W/m$^3$/°C. The average particle residence time was ~2 s.

An increase in the inlet air temperature led to an increase of water removal as there was a higher temperature difference between the drying medium and the particle surface (at wet bulb temperature), which increased heat and mass transfer, especially in the unhindered rate drying period. An increase in air velocity led to an increase in the volumetric water evaporation rate as the higher velocity resulted in stronger collision between the opposed streams. This increased the shear rate and turbulence intensity within the impingement zone and decreased the boundary layer around the particles and consequently the resistance to heat and mass transfer in the particle–heating medium interface. An increase in the particle flow rate resulted in an increase in the volumetric water evaporation rate. The authors came to the conclusion that this trend must be related to the capacity of the dryer, which exceeded the load required to evaporate water from the particles entering the system, although the humidity of air in the dryer increased. The effect of the distance between the inlet pipes on the volumetric water evaporation rate is more complicated; as for high values of air velocity and particle flow rate, there was no significant effect, while for lower particle flow rates, the decrease of impinging distance led to higher volumetric water evaporation rate but only at the high inlet air temperatures. This attributed to the lower loading ratio that caused a higher shear rate and turbulence intensity while the higher temperatures offered higher energy to the particles. The combination of low inlet air velocity and high particle flow rate led to a lower volumetric water evaporation rate as impinging distance increases because the increased loading ratio in combination with the lower shear rate of the streams reduced the potential of inlet air streams to carry the particles to the impingement zone.

Choicharoen et al. (2010) investigated the drying of okara, which is a soy residue remaining as by-product in soymilk manufacturing. They recorded a maximum volumetric water evaporation rate of ~520 $kg_{water}$/m$^3$/h and a maximum volumetric heat transfer coefficient of ~4500 W/m$^3$/°C for a mean particle residence time of ~0.97–1.74 s. The lowest specific energy consumption was ~5.6 MJ/$kg_{water}$.

### 9.5.4 Refractance Window Drying

RWD is an emerging technology introduced by MCD Technologies, Inc. (Tacoma, WA) in 2000. The method was patented by Magoon in 1986 for the dehydration of heat-sensitive foods. The design innovation of this method focuses on the usage of hot water as the heat transfer medium and at temperatures just below boiling at standard conditions, making it the unique drying method that utilizes hot water for heating. The application of RWD requires the distribution of the product in a very thin and uniform layer on a conveyor belt (film), which floats on the hot water. The belt is constructed of a special plastic; it has special characteristics regarding refraction and acts not only as a support for the material but also as an interface relatively transparent to infrared radiation. As this plastic conveyor is very thin, it reaches the hot water temperature almost immediately. The system works at atmospheric pressure and the hot water, usually at 95–97°C, circulates beneath the belt on shallow troughs transmitting thermal energy for moisture evaporation from the wet solids by conduction and radiation with infrared transmission in the wavelength range that matches the absorption spectrum for water. During drying, infrared heat passes directly through the plastic belt to the product. RWD is a very efficient method as all heat transfer methods (conduction, convection, and radiation) take place simultaneously. Infrared radiation is the major heat transfer mechanism in the early drying stages as the moisture is quite high, while conduction becomes important when moisture no longer contacts with the belt. Heat losses are limited as the plastic belt is a poor heat conductor. Food materials dried with RWD retain their color and flavor to a great extent and are protected by oxidization as product temperature remains below the temperature of the circulating water beneath the belt. RWD is considered to be a state-of-the-art method and can be used for the treatment of high-value food products with superior quality, while drying time is very short. The rapid mass transfer from the wet product and the resulting high vapor saturation above it limits the product–oxygen interaction and contributes in maintaining product quality, while the intense evaporation cools the product to an actual temperature usually below 70°C. The equipment used is simple and relatively inexpensive and the operational cost is low (Kudra and Mujumdar, 2009; Nindo and Tan, 2007; Smith, 1994; Vega-Mercado et al., 2001). Products, which include juice, pulps, purees, suspensions, and sliced foods, are spread on the moving belt. Puree thickness is in the range of 1.0 mm and drying time may be less than 5 min. Thickness and moisture content are important parameters determining the absorbency of the puree. The recirculation and reuse of the hot water improves the thermal efficiency of the dryer. The cooling section at the discharge end of the dryer is intended to reduce the product temperature, preferably below its glass transition temperature, to facilitate product removal.

The method has the ability to handle a wide range of liquid food materials, and has been applied for the transformation of fruits (strawberries, mangoes, blueberries, avocadoes, and lingonberries), vegetables (carrots, squash, and asparagus), and herbs and spices into value-added concentrates and powders. Meat, fish, poultry, eggs, flavorings, starches and grains, dairy, cereals, and beverages, as well as spirulina, β-carotene, lycopene, barley grass, tea blends, and homogeneous dried materials from mixtures of foods or herb or food extracts, have also been successfully dried by RWD (Jones, 2006). Some of the most important applications of the method include scrambled egg mix, avocado fruits for dips, high carotenoid-containing algae for treating macular degeneration and cancer, herbal extracts and nutritional supplements for human use, food ingredients (e.g., spices), and nutritional supplements for shrimp farming. The method has also been used successfully for the dehydration of foods that are difficult to be spray dried without the addition of nonsugar carriers.

Ochoa-Martínez et al. (2012) investigated the drying of mango slices (1 and 2 mm thickness) by using the refractance window technique with the water bath temperature set at 92°C. They also used air drying, in the form of a tray dryer, at 62°C and air velocity of 0.52 m/s to compare the results. RWD demonstrated a significantly higher drying rate, resulting in a moisture content of 0.013 and 0.048 kg water/kg dry solids for 1 and 2 mm sample, respectively, after 1 h of drying. The corresponding values in the case of tray drying were 0.966 and 3.614 kg water/kg dry solid. Water activity was less than 0.5 for RWD and close to 0.97 for air drying, while air drying took 4 h to reach results similar to those of RWD at 1 h. The samples dried by RWD exhibited diffusivities of $4.40 \times 10^{-10}$ (for 1 mm) and $1.56 \times 10^{-9}$ (for 2 mm) $m^2/s$, compared to $2.08 \times 10^{-11}$ and $6.83 \times 10^{-11}$ $m^2/s$, respectively, for air-dried samples.

Abonyi et al. (2001) studied the drying of strawberry and carrot puree with an initial moisture content of 93.6% and 89.4% wb, respectively, using RWD (effective length 1.83 m, water temperature 95°C, belt speed 0.45–0.58 m/min, thickness of the puree 1 mm, air temperature 20°C, relative humidity 52% and velocity 0.7 m/s, final moisture content (fmc) of 6.1% wb for carrot, 9.9% wb for strawberry, and 5.7% wb for strawberry with maltodextrin carrier), SD (inlet air temperature of ~190°C, outlet air temperature of ~95°C, 70% maltodextrin (DE = 10) carrier was added to strawberry puree, fmc 2.3% wb), DD (counterrotating double-drum dryer of diameter 19 cm, rotational speed of 0.3 rpm, residence time of 3 min, surface temperature 138°C and fmc of 5.0% wb for carrot) and freeze drying (freezing temperature −35°C, absolute pressure 3.3 kPa, heating plate temperature 20°C, condenser temperature −64°C, drying time 24 h, fmc of 8.2% wb for carrot, 3.9% wb for strawberry with 70% maltodextrin as carrier and 12.1% wb for strawberry purees without carrier). They reported that RWD of carrot resulted in carotene losses slightly higher but not significantly different compared to freeze drying and in particular 8.7% (total carotene), 7.4% (α-carotene), and 9.9% (β-carotene) losses, while corresponding values for freeze drying were 4.0%, 2.4%, and 5.4%. DD caused extensive carotene reduction of 56.1% (total carotene), 55.0% (α-carotene), and 57.1% (β-carotene). The slight reduction in carotene content during freeze drying might have been caused by the formation of the cis isomer when exposed to a temperature close to that of the heating plate (20°C) for a long time (about 8 h), when drying in the secondary drying stage. The ascorbic acid (AA) concentration of strawberry puree reduced by 6.4% and 6.0% during freeze drying and RWD, respectively resulted in an insignificant difference between the two methods. Although RWD took place at around 70°C food temperature, which is significantly higher compared to that of freeze drying, the moisture reduction of the puree during the first minute in RWD was quite high, contributing to limited degradation of ascorbic acid due to the low partial pressure of oxygen close to the food surface. Concerning color, freeze drying produced carrots with a brighter color (highest L*) compared to DD and RWD. On the other hand, no significant difference was presented in b* values among samples produced by freeze drying and RWD and fresh samples, while DD resulted in a darkened product as indicated by high b*/a* and hue angle values. In the case of strawberry, RWD and freeze drying products were more red but slightly brighter than fresh puree as evidenced by the higher hue and L* values for samples both with and without additive. Among the strawberry purees with additive, SD gave samples with the lowest chroma, indicating a less saturation and hence a pale appearance. It should be noted that as maltodextrin has a white color, purees with the addition of this agent have higher L* values. Perhaps the most important finding of their research was the alteration of the overall flavor impression of dried strawberry by enriching heat-induced ketone- and aldehyde-flavor notes, when elevated temperatures such as in RWD and SD were employed during dehydration. RW-dried strawberry purees had less esters and alcohols. This was recorded by a flavor volatile analysis using the solid-phase microextraction (SPME) analysis with GC/MSD. Both RWD and SD presented a significant reduction in fruity and green ester notes compared to the control strawberry puree, while freeze drying retained the fruity and green compounds to a great extent. The nerolidol content was high in the control but there was a concentration decrease following the sequence of freeze drying, RWD, and SD products while the ethyl acetate content was decreased in RWD and SD compared to control and freeze-dried samples. Carvone, which is a heat-enriched compound, was detected in high concentration in RW-dried samples, whereas its concentration was low in control and spray-dried samples and was undetectable in freeze-dried ones.

### 9.5.5 Fluidized Bed Drying

Fluidized bed drying (FBD) is one of the most versatile and successful drying techniques and is used extensively for cost-effective industrial drying of a wide range of granular materials that can be fluidized. During the fluidization process, a layer of solid body particles acquires properties that make it similar to a liquid and the particles mix in a manner that is dependent on the velocity of the gas flow that causes the state of fluidization (Jaros and Pabis, 2006). The method presents advantages, such as intense material mixing resulting in uniformly dried material, substantial intensification of heat and mass transfer between drying gas (usually air) and the particles being dried, resulting in shortening of drying time, rapid equalization of the gas temperature in the entire volume of the dried product, to the extent that the process practically takes place in a gradient-free field of gas temperature, and easy material transport inside the dryer (Jaros and Pabis, 2006; Mujumdar, 2006).

Many different fluidized bed dryer designs are available, which range from small batch units to large continuous dryers capable of processing tens of tonnes of feedstock per hour, while advanced versions in which the particle motion is enhanced by stirring or vibration are also employed to extend the range of materials that can be fluidized by conventional means (Baker et al., 2006). The continuous fluidized bed dryers are classified as well-mixed and plug-flow based on residence time distribution (RTD) of the material. The well-mixed fluidized bed dryers present a wide distribution of residence time, moisture content, and temperature of the particles at its outlet and typically they consist of a cylindrical vessel with solids inlet and outlet ports. The plug-flow fluidized bed dryers usually feature a long, narrow channel through which the fluidized solids flow. The bed length may be up to 20 m and the length-to-width ratio is in the range of 4:1 to 30:1. In a fluidized bed dryer, the pressure drop across the distributor must be high enough to ensure good and uniform fluidization and for upwardly directed flow, the pressure drop across the distributor must exceed 30% of the pressure drop across the bed (Khanali et al., 2013; Yang, 2003).

Baker et al. (2006) examined the application of plug-flow fluidized bed dryers and concluded that drying time ranges from 1 min to 2 h, moisture content of the product depends only on the difference between the temperature and humidity of the exhaust air and the drying characteristics of the solids; the specific energy consumption increases with decreasing outlet moisture content.

A particular variant of FBD is the jet-tube drying. In the dryer used for this technique, hot air is directed from above through a series of long tubes onto an oscillating solid conveyor surface, creating a bed of air surrounding each particle. The air rises and exits from a cyclone, which separates out fine particles. In this way, a more uniform and controllable drying with less jet clogging is achieved while there is capability of creating multiple drying zones. Jet velocities up to 70 m/s can be attained and the production rate ranges between 90 and 41,000 kg/h (Pallas et al., 2012).

### 9.5.6 Electrohydrodynamic Drying (High Electric Field Drying)

Electrohydrodynamic drying (EHDD) is a novel nonthermal and low-energy-consuming drying method. It is a variant of high electric field drying methods that uses a point-to-plate electrode system (in general, the electric field used for drying consists of a point-to-plate or two flat electrodes) and it gained attention at the mid-1990s as having the ability to produce food materials of high quality. It can be applied successfully in the processing of thermally sensitive biological materials and can be used both for postharvest treatment of foods and during food processing. EHDD complies with high product quality and can be used for the production of high-value foods for the outdoors, instant meals, nutraceuticals, baby foods, and seasonal and perishable foods and herbs (Bajgai et al., 2006). It has also been applied in the cases of tomato, apple, potato, okara, spinach, scallop, and others.

Barthakur and Arnold (1995) studied the production of air ions by a single point-to-plane corona electrode system and their effect in the increasing of water evaporation rate. When a high-voltage field is created in the space between a pointed and an earthed electrode with a small radius of curvature, evaporation of water from a food material is observed. They measured the evaporation rate of 0.019 and 0.017 g/min at a 1 cm electrode gap for negative and positive air ions, respectively, and concluded that the principal driving force for the evaporation enhancement was an ion drag phenomenon (electric, ionic, or corona wind) that is governed by the principles of fluid movement under the effect of electrical forces. The corona wind is generated by gaseous ions of air components ($N_2^+$, $O_2^+$, $N^+$, $O^+$, $O_2^-$) and water vapor in the surrounding atmosphere that ionize and subsequently accelerate under the influence of the applied high-voltage electric field. These charged ions collide with noncharged molecules and transfer their momentum, which results in the increase of localized energy and in the aforementioned ion-drag phenomenon. The corona wind disturbs the saturated air layer, creates vortex motions (turbulence) in the water, produces impinges on the moist biological material, and leads to movement that forces water molecules and other components out of the biological matrix. In this way, the corona wind enhances mass transfer between the biological surface and the ambient air and accelerates evaporation. Furthermore, the water molecules within the material orient themselves in the direction of the electric field, which leads to lowering the entropy and results in subsequent lowering of the temperature where the material is being dried (Bajgai et al., 2006, Hashinaga et al., 1999). Foods are primarily composed of water while carbohydrates, proteins, vitamins, triglycerides, and minerals are the food's other main substances. When subjected to an electric field, polarization of dipole molecules and bulk movement of charge carriers, such as ions, induce a capacitive current and a resistive current. The rotational effect on the dielectric molecules caused by the electric field may also contribute to further enhancement of mass transfer from the product (Zang et al., 1994). The method can be carried out using either AC or DC high voltages.

Singh et al. (2012) presented an extensive review for this method. Some of their conclusions are the following:

- EHDD takes place under ambient temperature and pressure conditions.
- The corona current increases linearly with an increase in applied voltage for both positive and negative polarities and a negative corona discharge produces a larger corona current compared to a positive corona discharge.
- The specific energy consumption increases with an increase in applied voltage for both polarities but it is lower compared to conventional drying and it is also lower than that of the latent heat of vaporization, indicating water removal from the food surface by other means in addition to evaporation.
- Electrode configuration plays an important role in determining the efficiency of the process and multiple-needle electrode configurations present higher efficiency than wire and single-electrode configurations.
- The equipment is relatively simple does not include movable parts, and there are no wear and tear problems.
- The evaporation rate depends on current strength, electrode gap, as well as the properties of the fluid medium. During drying, no primary heat is involved although a slight temperature increase of the samples has been observed.

EHDD offers lower production costs along with superior quality in terms of properties such as color, shrinkage, flavor, and nutrient content compared to hot air drying. It also presents a simpler design and less energy consumption compared to freeze drying, thus showing great potential for bulk and industrial applications. Less shrinkage represents the retention of the natural structure of the material (Bajgai et al., 2006).

Hashinaga et al. (1999) investigated the appliance of EHDD in the case of apples (slices of 2–3 mm thickness and ~8.6 cm diameter) by using multiple point-to-plate electrodes with AC high voltages at ambient temperature (18°C). They concluded that the size of the point electrode, the electric field strength, the gap between electrodes, and the interelectrode distance were the parameters that affected drying. They determined the optimum electrode gap as 1.3 cm and the optimum electrode field strength as $4.4 \times 10^5$ V/m for copper electrode and $4.7 \times 10^5$ V/m for sewing needle. The use of multiple, independent point electrodes distributed randomly over the drying material increased the initial evaporation rate by almost 4.5 times compared to the control sample dried at ambient air. Regarding the electrode shape, it was reported that a thin and sharp pointed sewing needle was more efficient than thicker and blunter copper electrodes. EHDD resulted in product color closest to that of the original material compared to samples dried at ambient air or in oven at 55°C and the apple presented lighter colors with less browning, higher L and a values, and small change in b values. Another important result of the study was the authentication by using chromatographic analyses that EHDD does not contaminate the product with compounds produced by the electric wing such as ozone and nitrous oxide.

Li et al. (2005) studied the oven drying of okara cake at 105°C with and without the combined application of EHDD (–20 kV voltage, electrode gap 35, 50, and 65 mm, and number of needles in point electrodes 1 and 3). The final moisture content of the food material was 10% wb. They observed that the variation of the number of needles in the point electrodes and the electrode gap increased the initial drying rate by a factor of 1.7–3.2 times and reduced drying time by 15–40% when the samples were exposed to the electric field compared to the control sample. During the initial drying period of 180 min, the kinetics of single needle-to-plate drying was similar to that of a 3-needle-to-plate drying at 35 mm gap, while a divergence in the process became significant for the remaining drying period. When 3-needle-to-plate electrodes were used, the best results were reported for 50 mm gap. The dried cake retained its shape and no cracks were observed on its surface.

Bajgai and Hashinaga (2001) investigated the drying of the Japanese radish using an alternating current high electric field (ACHEF). The equipment included a multiple-point electrode and a grounded copper mesh and the electrical field strength was 430 kV/m. They found that after 7 h of drying, 87.5% of the total moisture was removed from the product slices while the corresponding reduction was 86.9% for drying in an oven at 60°C and 26.5% for air drying at ambient temperature (25°C). Compared to oven drying, the ACHEF-dried radish exhibited less shrinkage (50% in respect to 80%, which caused cellular collapse due to heat damage of cell walls and membranes), better rehydration (10% and 17% higher water absorbed at 25°C and 100°C, respectively), higher water absorption (365 g/100 g dry weight in respect to 340 g/100 g), more natural color as no discoloration was observed, and less solid loss (13.9 mg/g less loss in respect to oven drying).

### 9.5.7 Foam-Mat Drying

Drying of foamed materials dates back to 1917 when Campbell patented the method for the drying of foamed evaporated milk. Foam-mat drying (FMD) is a particularly helpful method for the processing of food materials that require long drying times or are heat sensitive, sticky, viscous, and generally difficult to dry. It has been applied in a variety of liquid and semiliquid materials, such as pastes (tomato), fruit pulps (mango, banana, guava, apple, pineapple, starfruit, cowpea, mandarin, bael, passion fruit, and papaya puree) and juices (tomato), coffee extract, soymilk, and egg melange drying (Hertzendorf et al., 1970). Foaming often converts a fluid material to a semisolid as it reduces its fluidity, altering the requirements for the support of the material during drying. Foam is a two-phase system having a dispersed phase (usually air), which is the larger one and is surrounded by a plateau border of a continuous phase. Foams are highly fragile and delicate in nature and the high level of surface energy at the air–water interface makes them thermodynamically unstable (Muthukumaran et al., 2008). During FMD, a liquid or semiliquid material is whipped to form stable foam and is dehydrated by thermal means. Foam density used in the method ranged between 300 and 600 kg/m$^3$. The layer of the foamed material was in the range of 0.003–0.01 m.

The chemical nature of the material to be dried, its soluble solids, the pulp fraction, and the type and concentration of the foaming agent and the foam stabilizer affect foam formation, density, and stability. The quality of the foam is described by properties such as foam expansion, stability, and density, which are determined by the type and concentration of each foaming agent. The main advantage of foam formation prior to dehydration is the increased drying rate mainly due to the expanded internal structure and increased exposure of the product surface area to the drying medium. Furthermore, the thin lamella walls of the foam bubbles promote liquid transfer to the surface of the material by capillary action, which is helpful, especially in the falling rate period. The very thin material layers that can be used and their porous structure accelerate the diffusion of the water to the surface and allow the implementation of relatively low temperatures, which along with short drying times result in products of high quality that present favorable (good) rehydration characteristics, retention of volatiles, impaired losses of bioactive compounds (e.g., vitamins), and controlled density (Morgan et al., 1961). The relatively simple and inexpensive equipment and control required can also spread the applicability of the method. A disadvantage of the method lies in the reduced density of foamed materials, which leads to limited throughput of the material to the dryer, for a given film thickness, and low material-to-dryer surface loading. A critical subject in the application of the method is the lack of stability of the foam during heating, which causes cellular breakdown, resulting in serious impairment of the operation. Furthermore, the very low thermal conductivity of foams and the expanded nature of the film tend to limit heat transfer through the food in the later stages of drying and cause the increase of its temperature close to that of the heating medium. In many cases, the dried product sticks to the drying surface and its removal becomes difficult. To prevent these undesired events, a nonstick food-grade Teflon sheet can be used for the coating of the drying surface. In regard to the dried material, the foamed structure of the final product is very conducive to reconstitution and the food material tends to disperse readily in water. During storage, the open structure of foam-dried products tends to present problems of stability as they are sensitive to oxidative deterioration. Although foamed foods is usually dried in a convectional manner (in a hot air dryer), FMD can be combined with infrared and microwave drying to induce convection with an irradiative supply of energy, while conduction heating can also be used. Research also includes the combination of FMD with SD and freeze drying (Ratti and Kudra, 2006; Rajkumar et al., 2007). Mean drying time of the method is ~10 min, while the introduction of vacuum can reduce drying time to 1.5 min (Sankat and Castaigne, 2004, Schoppet et al., 1970). Competitive methods that apply in the same kind of foods include spray drying and DD.

Rajkumar et al. (2007) examined the FMD of Alphonso mango pulp at drying temperatures of 60°C, 65°C, 70°C, and 75°C and layer thickness of 1, 2, and 3 mm. They also studied the effect of various food foaming agents on the process and the properties of the produced mango powder, obtained after 25 min whipping time. These included soy protein (0.25%, 0.5%, 1.0%, and 1.5%) with methyl cellulose (0.5%), glycerol mono stearate (0.5%, 1.0%, 2.0%, and 3.0%), and egg albumen (2.5%, 5.0%, 10%, and 15%) with methyl cellulose (0.5%). They recorded that the optimum concentrations of the foaming agent were 1% soy protein, 2% glycerol mono stearate, and 10% egg albumen. Drying time for foamed mango pulp was lower compared to nonfoamed pulp at all examined temperatures and thicknesses. It was observed that in the case of foamed mango pulp, drying time for the reduction of the moisture content from 393% to 5.8% db was 40 min for 1 mm foam thickness, 60 min for 2 mm foam thickness, and 80 min for 3 mm foam thickness, while in the case of fresh pulp, it took 100 min, 130 min, and 190 min, respectively, for the moisture to drop down to 6.6% wb from its initial value of 391%. Therefore, the decrease of foam thickness led to an increased reduction of the product moisture content. Drying at 60°C produced powder that retained a significantly higher content of biochemical compounds compared to the products dried at higher temperatures, which showed significant reduction of heat-sensitive compounds, such as ascorbic acid, β-carotene, and total soluble solids content. The concentration reduction of these bioactive molecules was greater for higher thicknesses due to the longer drying time. The optimum process conditions for the production of mango pulp with the highest nutritional quality is its initial foaming with 10% egg albumen and 0.5% methyl cellulose succeeded by drying of 1 mm thickness foam at 60°C. Other properties, including pH, acidity, and total sugar, were less affected by the increased temperature and foam thickness. It is worth noting that the retention of bioactive molecules in the foamed mango pulp was significantly higher than in the nonfoamed pulp, which is attributed to the longer drying time in the latter case; this finding has been verified in the FMD of other materials such as the apple.

Kudra and Ratti (2006) studied the convective drying of foamed and nonfoamed apple juice (19 mm thickness layer at air temperature 55°C and velocity 0.69 m/s) and observed that the drying time was reduced from 500 to 200 min in the case of the foamed juice due to its higher drying rate, while the drying rate was higher by a factor of ~1.2–1.6 at the beginning of the drying. The density of foamed apple juice was 210 kg/m$^3$. The structure of the foamed material presented significant porosity and moisture content at the level of the equilibrium, whereas the nonfoamed product was received in the form of viscous syrup under the same drying conditions. The drying kinetics for foamed juice was typical for capillary-porous solids. The energy consumption for drying of foamed apple juice was only 20% of that of nonfoamed juice. The authors also examined the drying of mango pulp (1 mm thickness layer at 60°C) and concluded that although foam material has a very low density, the increased drying time can lead to increased productivity of the process. In particular, the dryer yield was 0.83 for foamed and 0.68 kg/m$^2$/h for nonfoamed material, leading to a reduction of capital costs by ~11% and ~10% compared to a belt-conveyor dryer and a drum dryer, respectively.

Kandasamy et al. (2012) investigated the FMD of papaya (pulp of concentration 12, 11, 10, 9, 8, and 7°Brix) using a batch cabinet dryer at an air temperature of 60°C, 65°C, and 70°C, flow rate of 2.25 m$^3$/min, and thickness of 2, 4, 6, and 8 mm. Methyl cellulose (0.25%, 0.50%, 0.75%, and 1.00% w/w) was used as a foaming agent. They concluded that ascorbic acid (145 mg/100 g in the fresh pulp), β-carotene (4.056 mg/100 g) and total sugars (36.8 g/100 g) in the final product showed significant reduction at higher temperatures (65°C and 70°C) and thicknesses (6 and 8 mm) caused by the higher temperatures and increasing drying time, while pH (5.2) and acidity (1.4%) remained relatively constant. In the case of pulp with 13°Brix soluble solids, foam formation did not take place, possibly due to its high viscosity and consistency. Drying time of foamed papaya pulp from initial moisture content of ~844% to ~4.5% db at 60°C was 3, 4, 7, and 9 h for 2, 4, 6, and 8 mm foam thickness, respectively. In the case of nonfoamed pulp, the corresponding drying time was 6, 8, 10, and 12 h for a final moisture content ~18 ± 3% db. The product with the best color, flavor, taste, and overall acceptability was obtained at 60°C. The optimum conditions for the production of papaya powder was recorded as pulp of 9°Brix added with 0.75% methyl cellulose, whipped for 15 min, and dried with a foam thickness of 4 mm at 60°C.

### 9.5.8 Foam-Spray Drying

Foam-spray drying (FSD) was introduced in 1961 by Hanrahan and Webb for the production of acid cottage cheese whey, skim milk, and whole milk with the use of foamed feed. The method is based on the drying of purposely foamed droplets, which are being formed by the introduction of a gas (air, $CO_2$, $N_2$, $N_2O$) or by using highly volatile liquids that are completely miscible with the feed material or by chemical reaction. Foamed droplet formation can take place before, during, or after atomization and before or during drying itself. The method has been applied in the production of tea, coffee, whey, fat, and skimmed milk products. Gas bubbles in FSD are much smaller than in FMD. The expansion of gas bubbles during FSD results in large particles with an increased surface area, reduced bulk density and increased porosity. Other characteristics of the method are the improved dissolution and the enhanced retention of highly volatile substances. FSD is suitable for the production of free-flowing powders from materials that have a tendency to agglomerate and adhere to dryer walls. It can also be used for the dehydration of hard-to-dry food materials and mixtures of biological substances such as protein, fat, vitamins, and mineral components, which are common in meat, detergent, and pharmaceutical industry (Crosby and Weyl, 1977).

Frey and King (1986) studied the retention of volatiles during FSD using aqueous solutions of either sucrose or coffee-extract solids with ~600 ppm of various acetates added as model volatile aroma components, dried in a pilot-plant spray dryer. Solids content ranged from 15% to 60% w/w for sucrose solutions and from 15% to 50% w/w for coffee solutions. The inlet air temperature to the dryer was 200°C, the outlet air temperature ranged from 100°C to 150°C, the liquid feed temperature was 49.5°C and the atomization pressure ranged from 3.55 to 7.00 MPa. When the gas-desorption method was used, the ratio of the dissolved gas volume under standard conditions to the liquid volume was 1.0, while for gas-admixing method, the ratio of the gas injected volume under the same conditions to the liquid feed volume was 4.0. They established that the gas-admixing foaming method leads to reduction of volatiles losses from the liquid sheet emanating from the nozzle if undissolved foaming gas causes the earlier breakup of the sheet, while diffusional losses of volatiles from droplets are increased when bubbles are introduced into droplets. When the gas-desorption foaming method was applied, volatiles loss during atomization or in regions close to the atomizer did not affected as bubble growth rate is generally too small to influence transport processes in these regions. Nevertheless both methods increased the volatiles loss from droplets of sucrose solution in the late stages of drying.

Zbicinski and Rabaeva (2009) investigated the gas-admixing FSD of 20% maltodextrin solution with an addition of glyceryl monostearate as a surfactant in concentrations up to 2% by applying the shadowgraphy technique based on the visualization of particles by high-resolution imaging with pulsed backlight illumination, which allows the observation of particle sizes down to 5 μm. The experiments were carried out in a 9-m-long cocurrent spray-drying tower and nitrogen was used for feed foaming. They concluded that in the case of the foamed feed there was a decrease of the evaporation rate as a result of the decreasing coefficients of heat and mass transfer to larger particles caused by the expansion of gas bubbles trapped in the droplet during foaming and heating up during drying. This phenomenon also led to a clear trend in the increase of the Sauter mean diameter and wettability and the decrease of bulk, tap, and apparent density by increasing foaming gas flow rates due to the expansion of gas bubbles, especially when the foaming and atomization processes are stable. On the other hand, no significant effect of feed foaming on the solubility and the angle of repose of the powder was observed. The powder produced by foaming consisted of larger particles with visible voids, while 70–80% of the particles were not spherical. Their shape factor ranged between 0.6 and 0.75 and there was a minor influence of feed foaming on the shape of particles.

### 9.5.9 Desiccant Drying

Desiccant drying (DCD) refers to a convective (hot air) drying technique where the drying medium passes through an adsorption column or similar device (e.g., desiccant wheel) to lower its moisture content before entering the drying section. Moisture adsorption is accompanied by heat release due to the heat of adsorption that raises air temperature and increases the efficiency of the drying system. The introduction of hot air with minimum relative humidity increases the driving force of moisture removal and the water uptake capacity, accelerating the drying process or making it possible to maintain drying duration even at lower drying temperatures. This is often useful in food drying to protect heat-sensitive compounds and the overall functionality of the product. Desiccant materials are able to remove practically all the moisture content from an air stream and produce dry air when they operate before the breakthrough point of the column appears. DCD can increase the energy efficiency of convective dryers, which is around 50%, while it can be even less for dryers working at low inlet temperature.

Desiccant materials present advantages, such as low energy consumption, continuous operation, increased drying rate, increased product quality especially for heat-sensitive products, ease to design the system and replace the material after cycles of operations, and no requirement for maintenance over long periods. The disadvantages of this technique are the pressure drop in solid desiccants, the carry over in liquid desiccants by air stream, the low moisture adsorption capacity (heat pump drying can also be used for moisture removal from a drying medium stream) and especially the regeneration process, which requires the use of hot air of ~300°C. This hot air is necessary for the reuse of the saturated material as the use of heat to regenerate desiccant materials has limitations in energy saving. The use of solar energy or waste heat from industrial processes (e.g., flue gases of steam generators) for regeneration as well as the ability for further use of the air after regeneration for heat recovery is the most effective way to increase the efficiency of the dryer through heat integration with other unit operations, rendering the method attractive. Vapor compression systems (heat pumps), electrical heaters, ultrasonic technology assisted by hot air, and electro-osmosis have been considered as novel methods for regenerating desiccant materials, as they present low energy consumption. The use of adsorbents can reduce energy consumption up to 50% (Tsotsas and Mujumdar, 2012).

Desiccants can be solids such as silica gel, alumina silicate, activated carbon, and zeolite, or liquids such as $CaCl_2$, LiCl, LiBr, and KCOOH solution. Solid desiccants are used in the form of stationary or rotary wheel beds for packing the desiccant material and are more widely used in drying applications. Liquid desiccants are more complicated in handling and usage, although they are flexible and can be regenerated far away from the dryer, allowing localized dehumidification. Liquid desiccants also present relatively high moisture removal capacity, the ability to absorb organic and inorganic contaminants from the air, and regeneration at relatively lower temperatures compared to solid desiccants is possible (Gandhidasan, 2004; Misha et al., 2012). The method has been applied in the drying of herbs, cocoa, seeds, cereals, nuts, manure, and sludge, and the final products present good color, taste, and microbial stability. The efficiency of the regeneration step can be increased by using superheated steam (Bussmann, 2002) at a temperature level of 350°C, which may result in a regenerated adsorbent with a moisture content of 2%.

Figure 9.3 depicts a one-section–two-chamber desiccant dryer. The solids to be dried are fed in a conveyor belt, which is common for the two chambers, and come in contact with hot air for the removal of a percentage of their moisture. There are several arrangements in a dryer of such a type, including parallel-flow and counterflow of the air stream, recirculation (partial in general) of the hot air after its exit

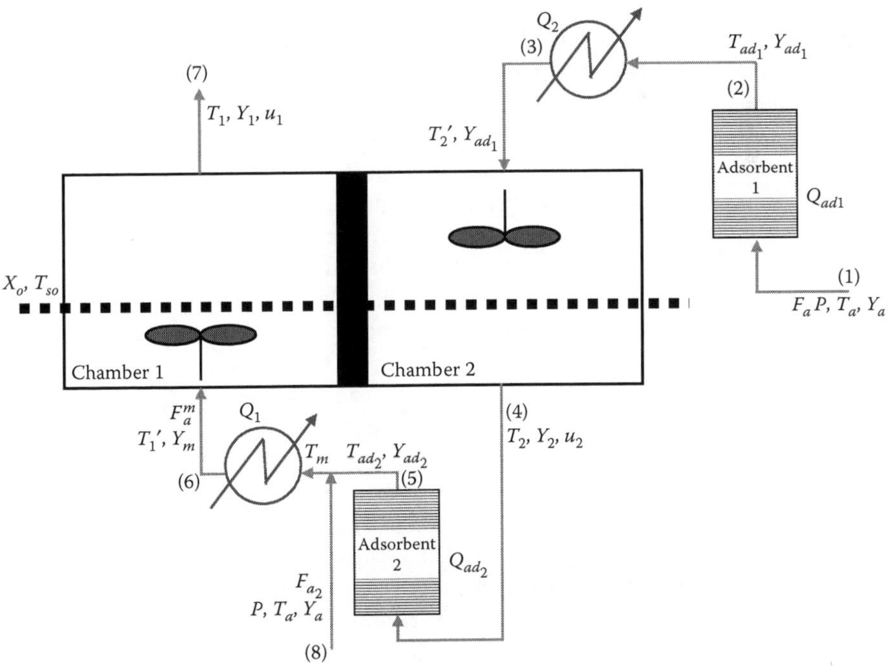


**FIGURE 9.3** A schematic representation of a one-section–two-chamber desiccant dryer.

from a chamber, and heat recovery by applying heat exchange and heat pumping. In the presenting dryer, the air is initially passed through an adsorption column (which usually has the form of a desiccant wheel) and loses humidity. At the same time, the releasing heat of adsorption results in the heating of air. Additional heating to the required inlet temperature is achieved by a heat exchanger (using steam as a heating medium in most cases). The hot dry air enters the second chamber, from which the final product is received, and its humidity increases as water is removed from the solids. At the same time, air temperature drops due to the heat of vaporization that is being offered to the solids for moisture removal, while additional amounts of heat are consumed for solids heating, removal of bound water and heat losses into the environment. The moist air is passed through a second adsorption column where again humidity is removed and its temperature increases. The pass through a heat exchanger and the first drying chamber are similar to that described before. An additional ambient air stream can be fed to the system after the second adsorption column as illustrated in Figure 9.3. This increases the controllability of the system as a certain value of humidity or velocity of air in the first chamber can be set. Different flowsheets are also available. A portion of air exiting chamber 2 may leave the drying system or it may be fed to chamber 1 bypassing adsorbent column 2.

The mathematical model of the described conveyor belt dryer is presented in Table 9.5. Equations Dl and D6 express the partial pressure of vapors at saturation for the air in chambers 1 and 2, respectively. They are the well-known Antoine equations, which give good results for engineering calculations at pressures close to ambient and temperatures encountered in food processing. Equations D2 and D7 calculate the water activity of the solid material as a function of pressure, temperature, and humidity of the drying air. Equations D3 and D8 give the equilibrium moisture content of the solid product, which is the minimum water content that remains in the product under the existing drying conditions, due to the equilibrium that takes place between the moisture of the solids and the humidity of the air. Equations D4 and D9 express the influence of drying parameters and solids characteristics in the drying process. High values of drying constant indicate easiness in the removal of moisture from the treated solids. Equations D5 and D10 are the mathematical expression of the drying kinetics and calculate the drying (or residence) time in each chamber for the desirable final moisture content of the product to be achieved. Since it is assumed that the two chambers are identical and the conveyor belt is common for the whole length of the section, drying time is equal for the two chambers. Equations D11, D12, and D13 calculate the specific enthalpy of the solids in the feed and the exit of the first and second chamber, respectively, as a function of the heat capacity of the dry solids and the water they contain. Equations D14, D16, D17, D20, and D21 give the specific enthalpy of the fresh (ambient) air, drying air fed in and exiting from the second chamber, and drying air fed in and exiting from the first chamber, respectively. Equations D15 and D18 calculate the specific enthalpy of drying air at the exit of adsorbent columns 1 and 2, respectively, while Equation D19 gives the specific enthalpy of air after the mixing point. Equations D22 and D27 present the amount of humidity remaining in the air exiting column 1 and 2, respectively, and Equations D23 and D28 express the flow rate of adsorbent in columns 1 and 2 compared to the air flow rate. In a way, this flow rate is an imaginary value. It relates to the size

**TABLE 9.5**
**Mathematical Model of a One-Section–Two-Chamber Desiccant Dryer**

$$P_{s_1} = \exp\left[a_1 - \frac{a_2}{a_3 + T_1}\right] \quad \text{(D1)}$$

$$a_{w_1} = \frac{Y_1 P}{(m + Y_1) P_{s_1}} \quad \text{(D2)}$$

$$X_{e_1} = b_1 \exp\left[\frac{b_2}{273 + T_1}\right]\left[\frac{a_{w_1}}{1 - a_{w_1}}\right]^{b_3} \quad \text{(D3)}$$

$$k_1 = k_o T_1^a Y_1^b u_1^c d_p^d \quad \text{(D4)}$$

$$t = -\frac{1}{k_1} \ln \frac{X_1 - X_{e_1}}{X_o - X_{e_1}} \quad \text{(D5)}$$

$$P_{s_2} = \exp\left[a_1 - \frac{a_2}{a_3 + T_2}\right] \quad \text{(D6)}$$

$$a_{w_2} = \frac{Y_2 P}{(m + Y_2) P_{s_2}} \quad \text{(D7)}$$

$$X_{e_2} = b_1 \exp\left[\frac{b_2}{273 + T_2}\right]\left[\frac{a_{w_2}}{1 - a_{w_2}}\right]^{b_3} \quad \text{(D8)}$$

$$k_2 = k_o T_2^a Y_2^b u_2^c d_p^d \quad \text{(D9)}$$

$$t = -\frac{1}{k_2} \ln \frac{X_2 - X_{e_2}}{X_1 - X_{e_2}} \quad \text{(D10)}$$

$$h_s = (C_{p_s} + X_o C_{p_w}) T_s \quad \text{(D11)}$$

$$h_{s_1} = (C_{p_s} + X_1 C_{p_w}) T_{s_1} \quad \text{(D12)}$$

$$h_{s_2} = (C_{p_s} + X_2 C_{p_w}) T_{s_2} \quad \text{(D13)}$$

$$h_a = (C_{p_a} + Y_a C_{p_v}) T_a + Y_a \Delta H_o \quad \text{(D14)}$$

$$h_{ad_1} = (C_{p_a} + Y_{ad_1} C_{p_v}) T_{ad_1} + Y_{ad_1} \Delta H_o \quad \text{(D15)}$$

$$h'_{a_2} = (C_{p_a} + Y_{ad_1} C_{p_v}) T'_2 + Y_{ad_1} \Delta H_o \quad \text{(D16)}$$

$$h_{a_2} = (C_{p_a} + Y_2 C_{p_v}) T_2 + Y_2 \Delta H_o \quad \text{(D17)}$$

$$h_{ad_2} = (C_{p_a} + Y_{ad_2} C_{p_v}) T_{ad_2} + Y_{ad_2} \Delta H_o \quad \text{(D18)}$$

$$h_a^m = (C_{p_a} + Y_m C_{p_v}) T_m + Y_m \Delta H_o \quad \text{(D19)}$$

*continued*

**TABLE 9.5 (continued)**
**Mathematical Model of a One-Section–Two-Chamber Desiccant Dryer**

$$h'_{a_1} = \left(C_{p_a} + Y_m C_{p_v}\right)T'_1 + Y_m \Delta H_o \quad \text{(D20)}$$

$$h_{a_1} = \left(C_{p_a} + Y_1 C_{p_v}\right)T_1 + Y_1 \Delta H_o \quad \text{(D21)}$$

$$Y_{ad_1} = f_{ad_1} Y_a \quad \text{(D22)}$$

$$F_{ad_1} = r_{ad_1} F_a \quad \text{(D23)}$$

$$F^w_{ad_1} = F_a\left(Y_a - Y_{ad_1}\right) \quad \text{(D24)}$$

$$Q_{ad_1} = F^w_{ad_1} \Delta H_{ad} \quad \text{(D25)}$$

$$T_{ad_1} = \frac{F_a (C_{p_a} + Y_a C_{p_v}) T_a + F_a (Y_a - Y_{ad_1}) \Delta H_o + F_{ad_1} C_{p_{ad}} T_{ad_{o1}} + Q_{ad_1}}{F_a (C_{p_a} + Y_{ad_1} C_{p_v}) + F_{ad_1} C_{p_{ad}}} \quad \text{(D26)}$$

$$Y_{ad_2} = f_{ad_2} Y_2 \quad \text{(D27)}$$

$$F_{ad_2} = r_{ad_2} F_a \quad \text{(D28)}$$

$$F^w_{ad_2} = F_a (Y_1 - Y_{ad_2}) \quad \text{(D29)}$$

$$Q_{ad_2} = F^w_{ad_2} \Delta H_{ad} \quad \text{(D30)}$$

$$T_{ad_2} = \frac{F_a (C_{p_a} + Y_2 C_{p_v}) T_2 + F_a (Y_2 - Y_{ad_2}) \Delta H_o + F_{ad_2} C_{p_{ad}} T_{ad_{o2}} + Q_{ad_2}}{F_a (C_{p_a} + Y_{ad_2} C_{p_v}) + F_{ad_2} C_{p_{ad}}} \quad \text{(D31)}$$

$$F^m_a = F_a + F_{a_2} \quad \text{(D32)}$$

$$F^m_a Y_m = F_a Y_{ad_2} + F_{a_2} Y_a \quad \text{(D33)}$$

$$F^m_a h^m_a = F_a h_{ad_2} + F_{a_2} h_a \quad \text{(D34)}$$

$$\dot{Q}_2 = F_a\left(h'_{a_2} - h_{ad_1}\right) \quad \text{(D35)}$$

$$\dot{Q}_1 = F^m_a\left(h'_{a_1} - h^m_a\right) \quad \text{(D36)}$$

$$F_{w_1} = F_s (X_o - X_1) \quad \text{(D37)}$$

$$F_{w_1} = F^m_a (Y_1 - Y_m) \quad \text{(D38)}$$

$$Q_{c_1} = F_s (h_{s_1} - h_{s_o}) + F_{w_1} (\Delta H_o + C_{p_v} T_{s_1} - C_{p_w} T_{s_o}) \quad \text{(D39)}$$

$$Q_{c_1} = F^m_a\left(h'_{a_1} - h_{a_1}\right) \quad \text{(D40)}$$

$$F_{w_2} = F_s (X_1 - X_2) \quad \text{(D41)}$$

$$F_{w_2} = F_a (Y_2 - Y_{ad_1}) \quad \text{(D42)}$$

**TABLE 9.5 (continued)**
**Mathematical Model of a One-Section–Two-Chamber Desiccant Dryer**

$$Q_{c_2} = F_s(h_{s_2} - h_{s_1}) + F_{w_2}(\Delta H_o + C_{p_v}T_{s_2} - C_{p_w}T_{s_1}) \tag{D43}$$

$$Q_{c_2} = F_a(h'_{a_2} - h_{a_2}) \tag{D44}$$

$$T_{s_1} = T_1 \tag{D45}$$

$$T_{s_2} = T_2 \tag{D46}$$

$$u_1 = \frac{F_a^m(1+Y_m)}{2\rho_a A_b} \tag{D47}$$

$$V_1 = \frac{F_s(1+X_o)t}{\rho_s(1-\varepsilon)} \tag{D48}$$

$$Z_1 = \frac{2V_1}{A_b} \tag{D49}$$

$$\Delta P_1 = f Z_1 u_1^2 \tag{D50}$$

$$E_{f_1} = 3.6\frac{\Delta P_1 F_a^m(1+Y_1)}{\rho_a} \tag{D51}$$

$$u_2 = \frac{F_a(1+Y_{ad_1})}{2\rho_a A_b} \tag{D52}$$

$$V_2 = \frac{F_s(1+X_1)t}{\rho_s(1-\varepsilon)} \tag{D53}$$

$$Z_2 = \frac{2V_2}{A_b} \tag{D54}$$

$$\Delta P_2 = f Z_2 u_2^2 \tag{D55}$$

$$E_{f_2} = 3.6\frac{\Delta P_2 F_a(1+Y_2)}{\rho_a} \tag{D56}$$

$$A_b = LD \tag{D57}$$

$$u_b = \frac{L}{2t} \tag{D58}$$

$$E_b = 3.6\, eLF_s(1+X_o) \tag{D59}$$

$$M_{s_o} = \frac{2F_s(1+X_o)t}{A_b} \tag{D60}$$

$$f_{\Delta T_1} = \frac{T_1' - T_1}{T_1' - T_{dp_1}} \tag{D61}$$

*continued*

**TABLE 9.5 (continued)**
**Mathematical Model of a One-Section–Two-Chamber Desiccant Dryer**

$$f_{\Delta T_2} = \frac{T_2' - T_2}{T_2' - T_{dp_2}} \quad \text{(D62)}$$

$$T_{dp_1} = \frac{a_2}{a_1 - \ln\dfrac{Y_1 P}{m + Y_1}} - a_3 \quad \text{(D63)}$$

$$T_{dp_2} = \frac{a_2}{a_1 - \ln\dfrac{Y_2 P}{m + Y_2}} - a_3 \quad \text{(D64)}$$

$$P_{s_a} = \exp\left[a_1 - \frac{a_2}{a_3 + T_a}\right] \quad \text{(D65)}$$

$$a_{w_a} = \frac{Y_a P}{(m + Y_a)P_{s_a}} \quad \text{(D66)}$$

$$b_{a_1} = b_o \exp\left(-\frac{E_a}{R(T_{ad_1} + 273)}\right) \quad \text{(D67)}$$

$$b_{a_2} = b_o \exp\left(-\frac{E_a}{R(T_{ad_2} + 273)}\right) \quad \text{(D68)}$$

$$X_{a1e}^{m} = X_a^{\max}\frac{b_{a_1}\alpha_{w_a}P_{s_a}}{1 + b_{a_1}\alpha_{w_a}P_{s_a}} \quad \text{(D69)}$$

$$X_{a2e}^{m} = X_a^{\max}\frac{b_{a_2}\alpha_{w_2}P_{s_2}}{1 + b_{a_2}\alpha_{w_2}P_{s_2}} \quad \text{(D70)}$$

$$X_{a1e} = \frac{F_{ad_1}^{w}}{F_{ad_1}} \quad \text{(D71)}$$

$$X_{a2e} = \frac{F_{ad_2}^{w}}{F_{ad_2}} \quad \text{(D72)}$$

and rotational speed of a desiccant wheel or the size of the adsorption column and the mass of adsorbent used before the column put for regeneration. Equations D24 and D29 calculate the humidity flow rate that is being removed from the air as it passes through the adsorbent columns 1 and 2. Equations D25 and D30 calculate the heat released due to the adsorption of air humidity on the adsorbent material, which is related to adsorbent nature. Typical values of heat of adsorption are 3200 kJ/kg adsorbed water for zeolite, 2900 kJ/kg adsorbed water for alumina, and 2300 kJ/kg adsorbed water for silica gel (Atuonwu et al., 2012). Equations D26 and D31 express the temperature of air at the exit of columns 1 and 2, respectively. These expressions are results of the equations

$$Q_{ad_1} = F_a(h_{ad_1} - h_a) + F_{ad_1}C_{Pad}(T_{ad_1} - T_{ad_{o1}}),$$

and

$$Q_{ad_2} = F_a(h_{ad_2} - h_{a_2}) + F_{ad_2}C_{Pad}(T_{ad_2} - T_{ad_{o2}})$$

by substituting specific enthalpy from the corresponding expressions of Table 9.5. Equations D32, D33, and D34 refer to the mixing of the air exiting from column 2 with fresh air to increase air flow rate in the first chamber, if this is needed. The first of these equations expresses the mass balance of dry air, the second gives the mass balance of air humidity, and the third is the enthalpy balance of the mixing. Equations D35 and D36 calculate the heat duty of the heat exchangers in the second and first chamber, respectively. Equations D37 and D41 give the moisture flow rate that removes from the solid material in the first and second chamber, respectively, while Equations D38 and D42 express the humidity flow rate that is received from the drying air at the corresponding chamber. In the case where heat losses are assumed to be negligible, each chamber works adiabatically. Equations D39 and D43 describe heat exchange in chamber 1 and 2, respectively, while Equations D40 and D44 calculate the effect of this exchange in the enthalpy of drying air in the corresponding chamber. Equations D45 and D46 express the assumption of the thermal equilibrium between the solids and the air as both exit each chamber. This assumption is quite realistic and very close to the experimental reality especially when the moisture content of the product has significantly reduced and its temperature has been shifted from the one of the wet bulb to the one of the dry bulb for the existing drying conditions. Equations D47 and D52 are expressions of the continuity equation of fluids in chambers 1 and 2. The factor 2 in the denominator of the equations is due to the fact that $A_b$ refers to the total belt surface of the dryer, including both chambers. Equations D48 and D53 are expressions for the specification of the bulk volume of the material transferred by the belt, assuming constant porosity of the bulk (the true volume generally decreases as the material losses water), while Equations D49 and D54 calculate the height (thickness) of the moist material in the entrance of each chamber. Again factor 2 is used in order to take into consideration that $V_i$ refers to each chamber, while $A_b$ to the whole section. Equations D50 and D55 express the pressure drop of drying air during its pass through the layer of the moist material and Equations D51 and D56 calculate the energy consumption of the funs to achieve the air circulation in the first and second chamber, respectively. Equation D57 expresses the total surface area of the conveyor belt; Equation D58 gives the velocity of the belt and Equation D59 calculates the consumption of electric energy in the motor for the belt movement. Typical values of constant $f$ in Equations D50 and D55 and for constant $e$ in Equation D59 are in the range of 2 (Maroulis and Saravacos, 2003). Equation D60 calculates the load per unit surface area in the feeding position of the belt (this specific load decreases along the belt due to water vaporization). The value of this variable relates to the durability of the conveyor belt. Equations D61 and D62 give the temperature drop of the drying air as it passes through the solids layer, which is transferred on the belt. The selection of this temperature drop is quite

**TABLE 9.6**
**Process Design Results of a One-Section–Two-Chamber Desiccant Dryer**

| Process Specification | | |
|---|---|---|
| Ambient pressure | $P$ (atm) | 1.0 |
| Ambient temperature | $T_a$ (°C) | 20 |
| Absolute humidity of ambient air | $Y_a$ (kg/kg da) | 0.005 |
| Solids flow rate | $F_s$ (kg db/s) | 0.08 |
| Initial solids water content | $X_o$ (kg/kg db) | 3.5 |
| Solids water content exiting chamber 2 | $X_2$ (kg/kg) | 0.5 |
| Initial solids temperature | $T_{so}$ (°C) | 20 |
| Exit air temperature in chamber 1 | $T_1$ (°C) | 52 |
| Air humidity content in chamber 1 (exit) | $Y_1$ (kg/kg) | 0.0009 |
| Drying air velocity in chamber 1 | $u_1$ (m/s) | 2.0 |
| Exit air temperature in chamber 1 | $T_2$ (°C) | 58 |
| Air humidity content in chamber 2 (exit) | $Y_2$ (kg/kg) | 0.0025 |
| Drying air velocity in chamber 2 | $u_2$ (m/s) | 2.0 |
| Product particle size | $d_p$ (m) | 0.02 |
| Dryer belt width | $D$ (m) | 1.4 |
| Adsorbent 1 initial temperature | $T_{oad_1}$ (°C) | 35 |
| Adsorbent 2 initial temperature | $T_{oad_2}$ (°C) | 20 |
| Fraction of moisture remains in air exiting adsorbent 1 | $f_{ad_1}$ (%) | 0.05 |
| Fraction of moisture remains in air exiting adsorbent 2 | $f_{ad_2}$ (%) | 0.05 |
| Air temperature drop in chamber 1 | $f_{\Delta T_1}$ (–) | 0.22 |
| Air temperature drop in chamber 2 | $f_{\Delta T_2}$ (–) | 0.15 |
| Adsorbent 1 "flow rate" to air flow rate ratio | $r_{ad_1}$ (kg/kg) | 0.06 |
| Adsorbent 2 "flow rate" to air flow rate ratio | $r_{ad_2}$ (kg/kg) | 0.07 |
| Technical data | | |
| Heat capacity of solid | $C_{ps}$ (kJ/kg/°C) | 2.1 |
| Heat capacity of dry air | $C_{pa}$ (kJ/kg/°C) | 1 |
| Heat capacity of water (liquid removed from the solid) | $C_{pw}$ (kJ/kg/°C) | 4.2 |
| Heat capacity of water vapor | $C_{pv}$ (kJ/kg/°C) | 1.9 |
| Heat capacity of adsorbent | $C_{pad}$ (kJ/kg/°C) | 0.836 |
| Air/water molecular weight ratio | $M$ (–) | 0.62198 |
| Air density | $\rho_a$ (kg/m$^3$) | 1.0 |
| Dry material density | $\rho_s$ (kg/m$^3$) | 1800 |
| Porosity of loading layer | $\varepsilon$ (–) | 0.45 |
| Specific drying constant | $k_o$ (h$^{-1}$) | 3.702E–04 |
| Temperature exponent in drying kinetic equation | $a$ (–) | 1.26 |
| Humidity exponent in drying kinetic equation | $b$ (–) | –0.26 |
| Velocity exponent in drying kinetic equation | $c$ (–) | 0.22 |
| Particle characteristic size exponent in drying kinetic equation | $d$ (–) | –0.63 |
| Antoine constant | $a_1$ (–) | 11.9 |
| Antoine constant | $a_2$ (–) | 3990 |
| Antoine constant | $a_3$ (–) | 234 |
| Constant of material isotherm equation | $b_1$ | 0.0005 |
| Constant of material isotherm equation | $b_2$ | 1850 |

**TABLE 9.6 (continued)**
**Process Design Results of a One-Section–Two-Chamber Desiccant Dryer**

| Process Specification | | |
|---|---|---|
| Constant of material isotherm equation | $b_3$ | 0.385 |
| Heat of vaporization of water | $\Delta H_w$ (kJ/kg) | 2500 |
| Air density | $\rho_z$ (kg/m$^3$) | 1.2 |
| Empirical constant of pressure drop equation | $e$ | 1.98 |
| Empirical constant of belt driver power equation | $f$ | 2.05 |
| Universal gas constant | $R$ (J/mol/°C) | 8.314 |
| Langmuir sorption constant | $b_o$ (–) | 5.62E–08 |
| Adsorbent activation energy | $E_1$ (J°C/mol) | –5.12E + 04 |
| Adsorbent capacity | $X_{zmax}$ (kg/kg db) | 0.1896 |
| Adsorbent latent heat of sorption | $\Delta H_{ad}$ (kJ/kg) | 3200 |
| Process design results | | |
| Specific enthalpy of ambient air | $h_a$ (kJ/kg) | 32.7 |
| Specific enthalpy of solids exiting chamber 1 | $h_{s1}$ (kJ/kg) | 382.2 |
| Vapor pressure at saturation of air in chamber 2 | $P_{s2}$ (atm) | 0.171 |
| Solids temperature exiting chamber 2 | $T_{s2}$ (°C) | 58 |
| Specific enthalpy of solids exiting chamber 2 | $h_{s2}$ (kJ/kg) | 243.6 |
| Absolute humidity of drying air exiting adsorbent 1 | $Y_{ad_1}$ (kg/kg da) | 0.00025 |
| Dew point temperature of air exiting chamber 2 | $T_{dp_2}$ (°C) | –4.7 |
| Air temperature entering chamber 2 | $T_2'$ (°C) | 69.1 |
| Specific enthalpy of air entering chamber 2 | $h_{a2}'$ (kJ/kg) | 69.72 |
| Rate of water removed from the solids in chamber 2 | $F_{w2}$ (kg/s) | 0.060 |
| Drying air flow rate in chamber 2 | $F_a$ (kg db/s) | 26.1 |
| Heat exchanged in chamber 2 | $Q_{c2}$ (kW) | 132.4 |
| Specific enthalpy of air exiting chamber 2 | $h_{a2}$ (kJ/kg) | 64.65 |
| Moisture removal rate in absorbent 1 | $F_{ad1}^{w}$ (kg/s) | 0.124 |
| Heat released in adsorbent 1 | $Q_{ad_1}$ (kW) | 396.8 |
| Adsorbent 1 flow rate | $F_{ad_1}$ (kg/s) | 1.566 |
| Drying air temperature exiting adsorbent 1 | $T_{ad_1}$ (°C) | 46.7 |
| Specific enthalpy of air exiting adsorbent 1 | $h_{ad_1}$ (kJ/kg) | 47.31 |
| Relative humidity of air exiting chamber 2 | $a_{w_2}$ (–) | 0.024 |
| Equilibrium moisture content in chamber 2 | $X_{e2}$ (kg/kg) | 0.032 |
| Drying kinetic constant in chamber 2 | $k_2$ (h$^{-1}$) | 3.99 |
| Heat duty of heat exchanger in chamber 2 | $Q_2$ (kW) | 585.3 |
| Solids water content exiting chamber 1 | $X_1$ (kg/kg) | 1.25 |
| Absolute humidity of drying air exiting adsorbent 2 | $Y_{ad2}$ (kg/kg da) | 0.00013 |
| Moisture removal rate in absorbent 2 | $F_{ad_2}^{w}$ (kg/s) | 0.063 |
| Heat released in adsorbent 2 | $Q_{ad_2}$ (kW) | 202.2 |
| Drying air temperature exiting adsorbent 2 | $T_{ad_2}$ (°C) | 69.2 |
| Specific enthalpy of air exiting adsorbent 2 | $h_{ad_2}$ (kJ/kg) | 69.52 |
| Adsorbent 2 flow rate | $F_{ad_2}$ (kg/s) | 1.827 |
| Ambient air flow rate to the mixing point | $F_{a_2}$ (kg db/s) | 0.00 |
| Air temperature after the mixing | $T_m$ (°C) | 69.2 |

*continued*

**TABLE 9.6 (continued)**
**Process Design Results of a One-Section–Two-Chamber Desiccant Dryer**

| Process Specification | | |
|---|---|---|
| Drying air flow rate in chamber 1 | $F_a^m$ (kg db/s) | 26.1 |
| Specific enthalpy of air after the mixing point | $h_a^m$ (kJ/kg) | 69.52 |
| Vapor pressure at saturation of air in chamber 1 | $P_{s_1}$ (atm) | 0.129 |
| Solids temperature exiting chamber 1 | $T_{s_1}$ (°C) | 52 |
| Specific enthalpy of solids at initial stage | $h_{so}$ (kJ/kg) | 336.0 |
| Dew point temperature of air exiting chamber 1 | $T_{dp_1}$ (°C) | −17.3 |
| Air temperature entering chamber 1 | $T_1'$ (°C) | 71.5 |
| Specific enthalpy of air entering chamber 1 | $h_{a_1}'$ (kJ/kg) | 71.88 |
| Rate of water removed from the solids in chamber 1 | $F_{w_1}$ (kg/s) | 0.180 |
| Heat exchanged in chamber 1 | $Q_{c_1}$ (kW) | 456.4 |
| Specific enthalpy of air exiting chamber 1 | $h_{a_1}$ (kJ/kg) | 54.40 |
| Relative humidity of air exiting chamber 1 | $a_{w_1}$ (−) | 0.012 |
| Equilibrium moisture content in chamber 1 | $X_{e_1}$ (kg/kg) | 0.03 |
| Drying kinetic constant in chamber 1 | $k_1$ ($h^{-1}$) | 4.53 |
| Heat duty of heat exchanger in chamber 1 | $Q_1$ (kW) | 61.7 |
| Material bulk volume in chamber 1 | $V_{b_1}$ ($m^3$) | 0.31 |
| Material maximum height in chamber 1 | $Z_1$ (m) | 0.024 |
| Pressure drop of air flowing through the material in chamber 1 | $\Delta P_1$ (bar) | 0.20 |
| Fan electrical load in chamber 1 | $E_{f_1}$ (kW) | 18.52 |
| Material bulk volume in chamber 2 | $V_{b_2}$ ($m^3$) | 0.06 |
| Material maximum height in chamber 2 | $Z_2$ (m) | 0.004 |
| Pressure drop of air flowing through the material in chamber 1 | $\Delta P_2$ (bar) | 0.04 |
| Fan electrical load in chamber 2 | $E_{f_2}$ (kW) | 3.39 |
| Dryer length | $L$ (m) | 18.7 |
| Belt velocity | $u_b$ (m/h) | 39.7 |
| Belt driver electrical load | $E_b$ (kW) | 48.3 |
| Maximum specific load of the belt | $M_{so}$ (kg/$m^2$) | 23.3 |
| Vapor pressure at saturation of ambient air | $P_{sa}$ (atm) | 0.022 |
| Relative humidity of ambient air | $a_{wa}$ (−) | 0.360 |
| Langmuir sorption constant of adsorbent 1 | $b_{z_1}$ (−) | 1344.8 |
| Langmuir sorption constant of adsorbent 2 | $b_{z_2}$ (−) | 377.9 |
| Maximum moisture loading of adsorbent 1 | $X_{a1e}^m$ (kg/kg db) | 0.173 |
| Maximum moisture loading of adsorbent 2 | $X_{a_2e}^m$ (kg/kg db) | 0.115 |
| Adsorbent loading of adsorbent 1 | $X_{a_1e}$ (kg/kg db) | 0.079 |
| Adsorbent loading of adsorbent 2 | $X_{a_2e}$ (kg/kg db) | 0.035 |

important for the efficient operation of the dryer. When drying rate is constant, it is desirable to have a large temperature drop during the pass of air through the chamber in order to achieve high efficiency of the dryer. In the case of drying in the falling rate period, a significant temperature drop inside the chamber leads to a decrease of the drying rate and an increase of the drying time according to the general drying kinetics expression. Equations D63 and D64 calculate the dew point temperature of air at the exit of chambers 1 and 2, respectively. They are not used directly in the model solution, but they ensure that as drying air cools during its pass through the material, its final temperature will be above the dew point under the conditions of working pressure and exit humidity. Equations D65 and D66 calculate the vapor pressure at saturation and water activity of ambient air. Equations D67 and D68 express the Langmuir sorption constant of the adsorbent material in columns 1 and 2, respectively, and Equations D69 and D70 give the maximum moisture loading of the corresponding adsorbents, which is a physical property of the adsorbent material and is affected by the temperature and relative humidity of the column or desiccant wheel. Equations D71 and D72 calculate the specific air humidity removed by the adsorbent in columns 1 and 2, respectively. This is the amount of water vapor that is adsorbed by the unit mass of dry adsorbent. It is obvious that the amount of vapor adsorbed must be lower than the maximum amount the adsorbent can hold under its working conditions.

Table 9.6 presents the application of the described model through its results. In this specific solution, the flow rate of ambient air in the mixing point is equal to 0, which is the simplest solution, but other solutions may be investigated by the reader.

Figure 9.4 illustrates the path of drying air through the drying system (drying chambers and adsorbents) on a Grosvenor psychrometric chart. It is evident that the process occurs in humidity well below its ambient value. Section 2–3 of the path refers to air heating in the second chamber, while path 5–6 is the corresponding heating in the first chamber heat exchanger. It is obvious that the latter heat amount is very low as the air is heated significantly in the adsorbent 2. The exit stream (point 7) maintains a temperature much higher than ambient and can be used for heat

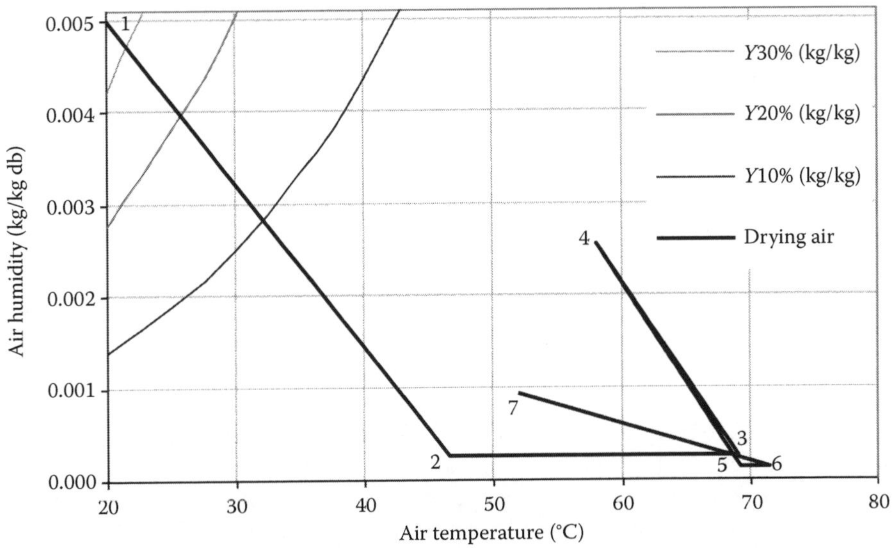


**FIGURE 9.4** The path of drying air on the psychrometric chart (numbers related to positions in Figure 9.3).

integration. DCD is an interesting option for minimizing heat requirements of the dryer in the cases where there is sufficient amount of heat rejected to environment (e.g., from steam generation units using fossil fuel or natural gas), which is necessary for adsorbent regeneration.

## 9.6 HYBRID DRYING

The presentation of a variety of drying methods and techniques in the previous sections reveals that there is no ideal method that prevails in engineering practice. Drying time, product quality, and operating cost are considered the most important factors and the recent trend in drying research is the application of combined or hybrid drying in order to exploit the best characteristics of each individual method. The selection of the most appropriate methods for each material, the sequence of their application, the operating parameters, and the duration of each method have to be established through experimental procedures and theoretical analyses. The utilization of hybrid methods may include their sequential or simultaneous application. Some of the most recent examples of hybrid drying are presented in this section.

Bai et al. (2012) examined the combination of EHDD (45 kV at temperature 18°C and relative humidity 45%) with vacuum freeze drying (EHD-freeze drying) for the dehydration of sea cucumbers. They came to the conclusion that the combined EHD-freeze drying presented less drying time and had lower energy consumption than freeze drying (113.2 MJ electric energy/kg water, in comparison with 168 MJ/kg water for freeze drying, while EHDD was the most efficient method). In the EHDD process, the needle electrode produced corona wind (resembling a round jet) that impinged and removed moisture from the surface of the product. No heat energy was lost and the energy use was very efficient. During the application of the process, the samples were dried initially with EHDD to reduce the moisture content to ~40% and the remaining water removed using freeze drying to a final water content of ~12%. The drying rate of freeze drying (20.5 h) was the slowest, while EHDD (12 h) had the fastest rate. EHD-freeze drying took 17.5 h for the production of the dried material. The product processed by the hybrid method presented improved quality compared to EHD-dried products, which included lower shrinkage (4.10% for freeze drying, 10.75% for EHD-freeze drying, and 16.8% for EHDD), higher rehydration rate and capability compared to EHDD, higher protein content (~44.7 g/100 g db compared to ~41.2 g/100 g db for EHDD), and better overall sensory qualities. freeze drying led to the highest quality, attributed to the lower working temperature of the method, followed by EHD-freeze drying and EHDD.

Zheng et al. (2011) investigated the dehydration of the blackcurrant berry, a fruit with very high vitamin C content, using microwave-assisted foam mat drying (MW-FMD). They used a household MW oven and heated air as the drying medium. The concept to combine the two methods was the volumetric heat generation and water vaporization inside the materials caused by microwave radiation, which can confine the poor heat transfer of air wrapped in the foamed materials. They described the beneficial effect of vapors, which strip up bubbles of the material, increasing the evaporation surface area and the instantaneous removal of water from the surface layer of the bubbles before their collapse. This hybrid method resulted in an increased drying rate combined with superior quality in terms of color and appearance. In general, the increase of microwave power and the decrease of pulp load accelerated the dehydration of the pulp. The authors determined that pulp thickness is the most important factor for the retention of vitamin C followed by pulp load and drying time, while microwave power is the least important parameter. On the other hand, pulp thickness was the least important in regard to moisture content, which was affected mainly by microwave power, followed by drying time and pulp load. In the case of anthocyanin content, pulp thickness is again the most important factor, followed by pulp load and microwave power, while drying time is the least important. Furthermore, microwave power and pulp load had a positive effect on both vitamin C and anthocyanin up to a certain level (420–490 W and 350–420 W and 57.5–65.0 gr and 35.0–42.5 gr, respectively), while the further increase of power resulted in degradation of these biomolecules. The optimum operating conditions for drying were microwave power of 560 W, pulp load of 65 g, drying time of 8 min, and pulp thickness of 4.46 mm. Under these conditions, vitamin C content was ~1.37 mg/g (compared to the maximum of ~1.40 mg/g when only vitamin C was the compound of interest), anthocyanin content was ~27.7 (color value) (compared to the maximum of ~33.57 when only anthocyanin was the compound of interest) and moisture content was ~0.197 kg moisture/kg db.

Muthukumaran et al. (2008) combined freeze drying and foaming methods to increase freeze-drying rate, taking advantage of the increased surface area made available by foaming. Xanthan gum (XG) at 0.125% concentration was used as a stabilizer. They concluded that the moisture reduction was rapid during the first 6 h of drying followed by a lower and almost constant drying period up to 24 h. Moisture content of egg white foamed with XG and without XG reduced from the initial value of ~733% db to the final value of 8.34% and 9.25% db, respectively. In the case of egg white without foaming (with the same volume as foamed egg white), moisture reduction was lower due to higher initial mass, and it took nearly 15 h to remove 50% of ice compared to just 6 h when a foamed sample was used. On the other hand, egg white without foaming presented the same drying trend as that of foamed egg white when the two samples had the same initial mass. The drying rate of egg white foam treated with 0.125% XG and without XG was 1.59 and 1.19 g/h, respectively, during the initial stage of drying, and reduced to 0.35 and 0.32 g/h, respectively, during the final stage of the process. That difference was attributed to the stabilized foamed surface during drying.

Supmoon and Noomhorm (2013) examined the combination of air impinging jet drying with infrared drying (IJ-IRD) for the production of potato chips, which were traditionally produced with deep-fat frying, leading to considerable oil uptake and they compared this hybrid method to stand-alone IJD and hot air drying (air velocity 1 m/s). IJ-IRD took place in an air temperature of 85°C, air velocity of 5, 10, and 15 m/s, and infrared intensity of 0.16, 0.27, and 0.33 W/cm$^2$. The infrared emission spectrum was in the range of 3–10 μm and the maximum power was 500 W. The nozzle was an 18-mm-diameter circular jet tube. Slices of initial moisture content of 5.5–7.4 kg water/kg db and 1 mm thickness, blanched in hot water at 85°C for 3 min, were used. The final moisture content was 0.035 kg water/kg db. The authors concluded that the drying rate increased with an increase in infrared intensity, as more energy was available for the increase of temperature and vapor pressure inside the product. IJ-IRD decreased the drying time by ~35%, 50%, and 50% at an intensity of 0.16, 0.27, and 0.33 W/cm$^2$, respectively, compared to IJD both applied at 5 m/s air jet velocity. At high intensities, the effect was not significant presumably due to the influence of the air velocity, which seemed to control moisture removal at each velocity level. The total specific energy consumption (SEC) decreased as the air velocity was increased and infrared intensity was decreased, although the drying time decreased with an increase of air velocity and infrared intensity. The lowest value of SEC of 452 MJ/kg water was observed at an air velocity of 15 m/s and infrared intensity of 0.16 W/cm$^2$, while the highest value of 988 MJ/kg water was observed at an air velocity of 5 m/s and an infrared intensity of 0.33 W/cm$^2$. Hot air drying presented the lower SEC of only 24.6 MJ/kg water, while IJD (5 m/s) had a consumption of 59.7 MJ/kg water. Drying time was 26.0 and 11.4 min, respectively. The lowest drying time of 3.3 min was achieved under the highest air velocity and infrared intensity, with a total SEC of 619 MJ/kg water. The quality of potato chips was affected mostly by the infrared intensity and slightly by the air jet velocity. Conclusively, shrinkage in IJ-IRD was slightly lower than that in IJD and significantly lower than that in hot air drying. Furthermore, IJ-IRD and IJD caused almost uniform shrinkage in all directions, in contrast with hot air drying, which resulted in bending in the direction perpendicular to the air flow. High air velocities tend to lead in relatively lower volume shrinkage due to case hardening. The increase of infrared intensity caused a decrease of volume shrinkage as infrared radiation increased the product temperature. IJ-IRD led to increased lightness and lighter color compared to HA-dried samples. The latter stands for IJD as well. It should be noted that the lightness of the potato decreased significantly after blanching and drying possibly due to starch gelatinization (Pimpaporn et al., 2007). Air velocity had no statistically significant difference on the lightness at the same infrared intensity level. IJ-IRD resulted in a product with lower hardness compared to that produced by IJD and hot air drying. Furthermore, hardness was decreased with increase of infrared intensity especially at low air velocities. This was attributed to the puffing effect caused by the rapid evaporation of water in the inner layers of the tissue due to infrared energy absorption.

Cui et al. (2008) investigated the hybrid drying of carrot (8-mm wafers with a diameter of ~40 mm of imc 88.2%) and apple (20 × 20 × 8 mm of imc 87.3%) chips with the application of microwave-vacuum drying (MW-VD) followed by freeze drying (MW-V-freeze drying). The combination of these particular methods aimed at the minimization of two main problems: at the latter stages of MWD, the food material temperature tends to rise rapidly as its moisture becomes quite low, while microwave–freeze drying suffers from uniform heating and possibility of corona discharge in an absolute pressure of 1–5 mbar. Both these incidents have undesirable effects on the product structure and degrade its quality. To overcome these problems, the authors applied MW-VD (power output 400 W, absolute pressure 2.5 kPa, rotation speed of the turntable 5 rpm, and initial incident microwave power density 2 W/g) as an initial stage, followed by freeze drying (freezing temperature –25°C for 20 min, heating plate temperature 30°C, and pressure 0.2 kPa). The moisture content after MW-VD was ~48% for carrot slices and ~37% for apple slices, while the final moisture content was ~7%. They concluded that the total carotene losses were 5.1% for MW-V-freeze drying, 5.3% for MW-VD (drying time 0.5 h for carrot slices and 0.6 h for apple slices), 4.6% for freeze drying, and 29.4% for hot air drying (temperature 60–65°C, velocity 0.5 m/s, drying time ~5.5 h). The results confirmed that shorter drying times and low oxygen concentration minimize carotene degradation. The retention of vitamin C was 91.8%, 89.0%, 97.0%, and 63.7%, respectively. Freeze-dried carrot presented the highest lightness, redness, and yellowness, MW-V-F-dried had slightly lower values, while HA-dried showed the least lightness and red and yellow hues. It is worth noting that color attributes relate not only to the presence of carotenes but also to the density of the product (Lin et al., 1998). The freeze-dried apple caused the lowest changes of L, a, and b values, MW-V-F-dried had a low alternation of their values, while HA-dried presented the greatest decrease of L and increase of a and b values, due to browning of the product caused by higher temperature and oxygen presence. The shrinkage ratio of the carrot was 59.59%, 81.6%, 26.81%, and 9.65% for MW-V-freeze drying, freeze drying, MW-VD, and hot air drying, respectively. The corresponding values for the apple were 60.65%, 83.50%, 40.87%, and 10.50%. Freeze drying and MW-V-freeze drying resulted in products with almost the same rehydration capacity presumably due to similar surface microholes and capillaries created during drying. MW-VD produced foodstuff with the highest rehydration capacity possibly due to its puffed texture and hot air drying produced products with the lowest water absorption due to sealing of surface capillaries and long drying duration (related to irreversible cell destruction), which is an inherited effect of the method.

Duan et al. (2007) studied the microwave freeze drying (MFD) in the case of cabbage and they noted that this hybrid method reduced drying time by more than half compared to freeze drying. The quality of the product of the two methods was quite the same. Cavity pressure had no significant effect in the preservation of vitamin C while microwave power significantly affected its preservation. Vitamin C content was highest when microwave power of 700 W was applied, while its content decreased at lower power levels due to the extensive duration of the process as well as at higher power levels due to the high temperature achieved. MFD also has a sterilization effect. The combination of advanced product quality and low power consumption led to the selection of 700 W microwave power and over 100 Pa pressure to be adopted. The quality of the product may be enhanced by applying relatively high microwave power and cavity pressure during the sublimation phase and relatively lower values of the forementioned parameters during desorption phase.

Wang et al. (2011) studied a hybrid unit for rapid drying applications consisting of a recirculating heat pump, which acts as refrigerant circuit, integrated with a rotary dehumidifier (desiccant wheel). Refrigerant R134a was used in the heat pump. The regeneration of the desiccant wheel is achieved with the heat dissipated by the heat pump condenser. Dehumidification of processed air was achieved by cooling the air below its dew point and removing humidity in the form of condensation from the evaporator and by adsorption on the desiccant material. The dryer presented efficient capability and supplied low dew point temperatures (–10°C to –20°C) and dry bulb temperatures of 20–30°C. The energy savings of the hybrid system was up to 60% compared to dryers consisting of a single refrigerant circuit or desiccant wheel. The addition of the desiccant wheel improved the specific moisture extraction rate of the dryer by up to 20%, and provided additional sensible heating to the air without the incorporation of an auxiliary heater. The regulation of the airflow through the control damper was an effective method to adjust drying air humidity while enhancing the amount of energy recovered from the air to minimize or even avoid the need for cooling and reheating air. The increase of air flow rate over the cooling coil and desiccant wheel was found to improve the performance of the dryer in terms of moisture removal capacity, which was increased from 7.3 to 11 kg/h when the air flow rate increased from 1000 to 1500 $m^3/h$. This type of dryer can find applications in food and beverage packaging industry (bottles, cans, and food packets and pouches).

## 9.7 EFFECT OF DRYING PROCESS IN BIOACTIVE COMPOUNDS OF FOODS

The selection of a drying method has direct impact on the deterioration of bioactive molecules present in the foodstuff. Different drying methods and conditions may have a great effect on the degree of preservation of individual valuable bioactive compounds. There are key components that are associated with the quality and the preservation of bioactive compounds that include ascorbic acid, β-carotene, chlorophylls, lysine, and others. In the following, some recent research efforts concerning the preservation of bioactive compounds are presented. They clearly denote that the selection of the drying method and its operating conditions is essential for presenting the best possible quality and nutritional value in the production of foods.

Igual et al. (2012) studied the change in organic acids, phenolic compounds, and antioxidant activity of dried apricots when hot air drying combined or not when microwave (MW) energy was used. They concluded that the industrial processing of dried apricots may be improved by using microwave energy, as the fruit presented a higher phenolic content, particularly of chlorogenic acid, catequin, and epicatequin while drying time was considerably reduced and the antioxidant capacity was maintained. Drying led to a significant decrease in malic, caffeic, and tartaric acid and although hot air drying did not affect the citric acid content, the application of microwaves caused a significant decrease in the citric acid. Also, hot air drying, with or without MW assistance, significantly decreased the amount of gallic acid to a greater extent than when only MWD was employed. MWD significantly increased the epicatequin and catequin content. The combination of hot air drying–MWD increased catequin to a greater extent than MWD alone, while epicatequin decreased to the same degree as when hot air drying was used. All the drying methods led to a total loss of kaempferol. Higher temperatures contributed to a greater total phenolic content, while the application of MW heated the material even more extensively. The authors note the fact that this might be due to the more complete extraction of phenols after extensive heating or due to the sulfite pretreatment. DPPH scavenging activity (antioxidant capacity was assessed using the free radical-scavenging activity of the samples evaluated with the stable radical DPPH*) was increased to 3.5–3.8% for dried product compared to 2.4% for a frozen one.

Ortiz et al. (2013) investigated the air drying of Atlantic salmon (*Salmo salar* L.) fillets at 40°C, 50°C, and 60°C and the influence of temperature on biochemical characteristics. They observed that the content of palmitic acid, eicosapentaenoic acid (EPA—an omega-3 fatty acid), and docosahexaenoic acid (DHA—an omega-3 fatty acid) were significantly lower in the dried fish fillets compared to the fresh fish. Palmitic acid decreased by 20% compared to fresh fish while a temperature increase caused a decrease of its content. Concerning the content of the typical fish fatty acids, EPA remained practically unchanged and DHA decreased slightly with increased temperature. The tocopherol content decreased during drying by 40%. α-Tocopherol decreased significantly after drying at all three temperatures and especially at 40°C, although at 50°C, the highest content was observed, possibly due to the restricted flow out of melted fat from the fish muscle to the surface. This led to less loss of the liposoluble α-tocopherol due to the viscosity decrease of the lipid phase with increasing temperature. γ-Tocopherol decreased significantly only during drying at 40°C, maybe due to the longer drying time at that temperature. Astaxanthin content, which is a dietary carotenoid and the main pigment responsible for the pink color of the fillets, was practically the same as that of fresh fish after drying at 40°C, while it was significantly increased in the dried products processed at 50°C and 60°C. This indicates the concentrating effect of the process in regard to this compound as well as the stability of astaxanthin in the applied conditions. They also detected secondary lipid oxidation compounds formed during drying, which were measured by the thiobarbituric acid index (TBA-i). This index can be used for the evaluation of the effect of drying. An increase of TBA-i was recorded after drying at all temperatures, while the index increase was lower at higher drying temperatures, resulting in shorter drying time. Anisidine value was also used for the measurement of secondary oxidation products. Drying significantly increased the anisidine values of the products, confirming the rapid decomposition of hydroperoxides into secondary oxidation products (alcohols, aldehydes, ketones, acids, dimers, trimers, polymers, cyclic compounds, and free radicals) due to hydrolytic and oxidative degradation at high temperatures. The anisidine value of the dried products increased less with increasing drying temperature, leading to the conclusion that fish degradation is more relevant to longer drying times than higher drying temperatures.

Siriamornpun et al. (2012) studied the effect of hot air drying (at air temperature of 60°C and velocity of 1.0 m/s for 4 h), freeze drying (at pressure of 0.14 mbar for 48 h after frozen at −40°C), and combined far-infrared radiation with hot air convection (FIR-hot air drying) (at intensities of 5 kW/m$^2$, wavelength of 15–100 μm, air temperature of 40°C, and velocity of 1 m/s for 60 min) on carotenoids and phenolic compounds of marigold flowers dried to a final moisture content of ~7% db. The Marigold flower is used as food (tea, food colorant, and in cooking), as an ingredient in animal feed, and as a medicinal plant. The product treated with FIR-hot air drying presented the highest amount of lycopene (58.7 mg/100 g DW), while the concentrations for hot air drying, freeze drying, and fresh material were 51.2, 48.7, and 38.1 mg/100 g DW, respectively. This increase in bioaccessible lycopene is attributed to the release of phytochemicals from the matrix, which were accessible for extraction. Hot air drying and FIR-hot air drying gave powder with the highest level of β-carotene (15.5 and 12.8 mg/100 g DW) compared to 9.8 mg/100 g DW content of the fresh sample. On the other hand, freeze drying led to a decrease of that compound (9.0 mg/100 g DW). The nonthermal treatment during freeze drying is considered responsible for that result. Freeze drying and FIR-hot air drying led to products with the highest level of lutein (283.2 and 278.9 mg/100 g DW), which are by far greater than those of the fresh product (53.8 mg/100 g DW) and the HA-dried one (110.4mg/100 g DW). This makes the Marigold flower a very important ingredient as lutein is not synthesized by humans and must be obtained by the ingestion of foods such as fruits and leafy green vegetables. FIR-hot air drying was responsible for the higher values of DPPH radical scavenging (85%) compared to freeze drying (67%), fresh petals (65%), and hot air drying (52.4%). This was attributed to the infrared radiation that causes molecular vibrations within the food material, which heats causing complex molecular structures to break and release antioxidant compounds. Furthermore it liberates and activates low-molecular-weight natural antioxidant compounds. The FRAP (ferric reducing ability of plasma) values, which indicate the total level of redox-active compounds in a solution, showed that FIR-hot air drying presented the greatest reducing power (972.7 Lmol $FeSO_4$/g DW), followed by fresh material (821.0 Lmol $FeSO_4$/g DW), freeze drying (811.0 Lmol $FeSO_4$/g), and hot air drying (730.7 Lmol $FeSO_4$/g DW) due to the higher levels of total phenolic content (TPC), total flavonoid content (TFC), lycopene, β-carotene and lutein. FIR-hot air drying products showed the highest TFC (133.1 mg RE/g DE), followed by fresh (107.8 mg RE/g DW), freeze drying (105.5 mg RE/g DW), and hot air drying-treated samples (69.0 mg RE/g DW). The same trend was also presented in the TPC with values 60.0, 55.8, 53.6, and 43.6 mg GAE/g DW, respectively. The phenolic acid with the highest content was *p*-coumaric acid (ranging from 591 lg/g DW in freeze-dried product to 1803 lg/g DW in FIR-HA-dried product), while other acids, including gallic, protocatechuic, caffeic, syringic, and ferulic acids, showed the highest content in the flower dried with FIR-hot air drying. *p*-Hydroxybenzoic and sinapic acid were the only two found in higher content in the HA-dried products. Rutin content, which was detected as the most important flavonoid in the flower had the following contents: 282, 1388, 1546, and 1950 lg/g DW for freeze drying, hot air drying, fresh, and FIR-hot air drying, respectively.

Desobry et al. (1997) investigated the degradation of β-carotene, which implicates as an anticancer compound, antioxidant and free radical quencher, encapsulated in 25 dextrose equivalent maltodextrin by three drying methods (SD, freeze drying, and DD) during drying and storage. The drying (at 11% and 32% RH and 25°C, 35°C, and 45°C) and encapsulation process led to an 8% degradation of β-carotene with freeze drying, 11% with SD, and 14% with DD. No significant influence of relative humidity was observed on the retention of β-carotene. Although DD caused more initial loss of the compound, the lower amount of surface carotenoids and larger particle size (average 105 μm compared to 80 μm for freeze drying and 30 μm for SD) resulted in greater stability compared to the other methods as the high ratio of surface/volume for a sphere and the large amount of small spheres favor the oxidation of β-carotene. Moreover, in a 15-week storage test, DD gave the best β-carotene preservation for all temperatures with at least ~47% retention during storage, retaining 40% of the original compound. On the other hand, spray-dried powder showed the fastest degradation kinetics (27% total retention after 12 weeks storage) attributed to the smallest particle size of the material and the highest carotenoid surface content among the three methods. The freeze-dried powder showed retention of ~35% after 15 weeks. The smaller diameter of the spray-dried capsules increases the exposure of surface β-carotene to oxygen, while the bulk density (817 kg/m$^3$ for SD, 699 kg/m$^3$ for DD, and 677 kg/m$^3$

for freeze drying) favors faster oxidation of freeze-dried samples since it has a lower value, indicating the least collapse.

Dev et al. (2011) studied the drying of *Moringa oleifera* pods, which are rich in bioactive molecules and present medicinal properties, using microwave-assisted hot air drying (MWHAD) and conventional hot air drying. The samples were dried to a final moisture content of 13% wb at three different temperatures (50°C, 60°C, and 70°C), while microwave power density was 1 W/g. It was found that MWHAD reduced the loss of volatiles (analyzed using an electronic nose for the quantitative classification of the aromatic quality of the products) and preserved most of the bioactive molecules (analyzed using gas chromatography–mass spectrometry). β-Sitosterol (a phytosterol) was considered the key volatile component for the classification of the treatments. The samples dried at 50°C using MWHAD presented the best quality attributes as they retained ~67% of the original concentration of β-sitosterol, which was nearly 13% more than in 60°C. On the other hand, MWHA-dried samples processed at 70°C lost a great amount of that compound and only 23% of the original concentration remained. The rapid moisture migration occurred at 70°C with the effect of microwaves resulting in a greater diffusion of the volatiles into the drying air. Among the bioactive compounds found in the product, 2,3-dihydro-3,5-dihydroxy-6-methyl-4H-pyran-4-one, which presents hypotensive characteristics, *n*-hexadecanoic acid, and 9,12-octadecadienoic acid, which have antioxidant behavior and reduce the free radicals in the body, were the most important. MWHA-dried samples retained much greater quantities of these molecules and presented higher overall quality.

Pallas et al. (2012) investigated the jet tube fluidized bed drying for the production of high-quality sweetened and nonsweetened blueberries ($a_w < 0.55$) at 99°C, 107°C, and 116°C. Osmotic dehydration as a pretreatment was also examined. Drying time was 50–100 min depending on the conditions. Total monomeric anthocyanins (TMA) level, decreased as the drying temperature increased and ranged from 4.07 down to 1.51 mg cyanidin-3-*O*-glucoside equivalents (C3G eq)/g extract for dried blueberries while the initial value was 7.65 mg C3G eq for frozen blueberries/g extract. At 99°C, 107°C, and 116°C, TMA was 4.07, 3.32, and 2.52 mg C3G eq/g extract, respectively. OD followed by hot air drying at 107°C and 116°C resulted in the greatest loss in TMA as high temperatures and long drying periods relate to the decomposition of anthocyanins. The total phenolic content increased during drying for noninfused blueberries, with highest levels of 31.6 mg gallic acid equivalents (GAE)/g extract for samples dried at 107°C, while the initial value was 24.9 mg GAE/g extract. Blueberries without osmotic pretreatment exhibited 20–30% higher phenolic content compared to treated ones at the same temperatures possibly due to migration of some compounds into the sugar solution although sugar may have protection against thermal degradation. With the exception of sugar-infused berries dried at 107°C and 116°C, the dried blueberries maintained or demonstrated slightly increased hydrophilic oxygen radical absorbance capacity values (H-ORACFL), indicating that their antioxidant capacity was retained upon drying. Blueberries dried at 107°C possessed the greatest preference scores and best retention of blueberry flavor and required a relatively short drying time.

Lycopene is a pigment presenting very important antioxidant properties and gives the red color to watermelons, yet is very sensitive to light, oxygen, and heat.

Watermelon pomace prior to treatment presents high lycopene content (0.201 mg/g), about 4.5 times higher than fresh watermelon. Arocho et al. (2012) investigated the usage of watermelon pomace as a food ingredient after drying, using a cabinet dryer and a drum dryer. Cabinet drying did not expose samples to light, in contrast with DD, which exposes them to high heat and light, causing higher loss of the examined compound. Lycopene content of fresh material was in the same order as cabinet-dried samples. Cabinet drying preserved not only the lycopene content but also color, although drying time was much longer than during DD. Nevertheless, cabinet drying for extended periods affected the sugar composition of the product, causing browning and stickiness.

Ghandi et al. (2012) studied the influence of SD conditions, such as inlet air temperature (130–200°C), outlet air temperature (38–65°C), drying medium (air and nitrogen), and milk-derived protectants (10%, 15%, and 25% lactose; 5% and 10% sodium caseinate; 10%, 25%, and 35% lactose/sodium caseinate (Lac:NaCas, 3:1)) on the survival of *Lactococcus lactis* ssp. *cremoris*. An inlet and outlet air temperature of 130°C and 65°C, respectively, maintained high survival of the bacteria without sacrificing low moisture content. A mixture of Lac:NaCas (3:1) showed a better protective effect on bacteria survival than lactose and sodium caseinate individually, and this effect increased with an increasing amount of protectant. The results were generalized by substituting whey protein isolate for sodium caseinate. The positive effect of oxygen elimination was demonstrated both by replacing air with nitrogen and adding ascorbic acid as an oxygen scavenger to improve the survival of the bacteria. In fact, the addition of an oxygen scavenger seems a better candidate for industrial application considering the potential high cost of manufacturing if nitrogen was used as the atomization and/or drying medium.

Apricots (especially some variations) are suitable for drying to give a high calorific-value product enriched with vitamin and minerals. Karatas and Kamışlı (2007) studied the drying of "near-ripe" and "ripe" apricot by applying IRD and MWD and the variations of vitamins A, C, and E and malondialdehyde (MDA) with moisture removal. These compounds are sensitive to the intensity and duration of the applied heat and radiation. They concluded that for both methods and both ripeness states, the values of vitamins A, C, and E and MDA increased with moisture removal. The values of vitamins A and E and MDA of the ripe apricot were higher than those of the near-ripe one in the examined range of moisture removal, while the near-ripe apricot presented higher values of vitamin C. Furthermore, the values of the examined vitamins and MDA in the products dried by microwave were higher than those dried by infrared radiation and, consequently, MWD is more efficient than IRD in terms of fewer losses of bioactive compounds as well as in drying duration and preservation of the original color. The difference in vitamins and MDA losses in the two drying methods can be attributed to the drying time, which is a lot shorter in MWD (at a range of 3–8 min) compared to the 20–90 min required for IRD according to the final moisture content.

Wilford et al. (1997) studied the kinetics of changes in the four major carbohydrates (glucose, fructose, sucrose, and sorbitol) during the dehydration of d'Agen prunes at 70–90°C using high-performance liquid chromatography and gas chromatography–mass spectrometry. They recorded a series of reactions that can be described as follows: Initially, acid hydrolysis of sucrose takes place over a period of

1.5–6 h, depending on the drying temperature, and corresponds to a moisture loss of 40–50% resulting in a rapid decrease in the amount of sucrose, while the amounts of the hydrolysis products (fructose and glucose) increase proportionally. In a succeeding stage, degradation of these products occurs via Maillard-type reactions with nitrogen-containing compounds, such as proteins or free amino acids, which proceed effectively at temperatures greater than 50°C. Sorbitol on the other hand suffers little change in its amount as its structure protects it from Maillard reactions due to the lack of carbonyl group and its content remains relatively constant up to ~10 h of drying at 80°C. In the final stage of drying at 80°C, taking place after 8–10 h, quite a rapid loss of fructose, glucose, and sorbitol is observed due to the onset of caramelization reactions, which only occur at very low moisture. At higher temperatures, sorbitol starts to decrease only after 3–4 h of drying, while sucrose disappears completely by 6–7 h at 70°C (40% moisture reduction) and by 2 h at 90°C (30% moisture reduction). Furthermore, it was established that low pH (~3.65 initially and 3.77 at 50% weight loss) and high moisture content, for example, in the initial stage of the process, favor sucrose hydrolysis.

## REFERENCES

Abonyi, B.I., Feng, H., Tang, J., Edwards, C.G., Chew, B.P., Mattinson, D.S. and Fellman, J.K. 2001. Quality retention in strawberry and carrot purees dried with refractance window system. *Journal of Food Science*, 67(2): 1051–1056.

Adhami, S., Rahimi, A. and Hatamipour, M.S. 2013. Freeze drying of quince (*Cydonia oblonga*): Modelling of drying kinetics and characteristics. *Korean Journal of Chemical Engineering*, 30(6): 1201–1206.

Akpinar, E.K. 2006a. Determination of suitable thin layer drying curve model for some vegetables and fruits. *Journal of Food Engineering*, 73: 75–84.

Anderson, B.A. and Singh, R.P. 2006. Modeling the thawing of frozen foods using air impingement technology. *International Journal of Refrigeration*, 29: 294–304.

Arocho, Y.D., Bellmer, D., Maness, N., McGlynn, W. and Rayas-Duarte, P. 2012. Watermelon pomace composition and the effect of drying and storage on lycopene content and color. *Journal of Food Quality*, 35: 331–340.

Atuonwu, J.C., van Straten, G., van Deventer, H.C. and van Boxtel, A.J.B. 2012. On the controllability and energy sensitivity of heat-integrated desiccant adsorption dryers. *Chemical Engineering Science*, 80: 134–147.

Bai, Y., Yang, Y. and Huang, Q. 2012. Combined electrohydrodynamic (EHD) and vacuum freeze drying of sea cucumber. *Drying Technology*, 30: 1051–1055.

Bajgai, T.R. and Hashinaga, F. 2001. High electric field drying of Japanese radish. *Drying Technology*, 19(9): 2291–2302.

Bajgai, T.R., Raghavan, G.S.V., Hashinaga, F. and Ngadi, M.O. 2006. Electrohydrodynamic drying—A concise overview. *Drying Technology*, 24(7): 905–910.

Baker, C.G.J., Khan, A.R., Ali, Y.I. and Damyar, K. 2006. Simulation of plug flow fluidized bed dryers. *Chemical Engineering and Processing*, 45: 641–651.

Bakshi, A.S. and Singh, R.P. 1980. Drying characteristics of parboiled rice, In: *Drying'80*, Mujumdar, A.S. ed., Hemisphere Publishing Company, Washington DC.

Barthakur, N.N. and Arnold, N.P. 1995. Evaporation rate enhancement of water with air ions from a corona discharge. *International Journal of Biomaterials*, 39(1): 29–33.

Bonazzi, C., Dumoulin, E., Raoult-Wack, A., Berk, Z., Bimbenet, J.J., Courtois, F., Trystram, G., and Vasseur, J. 1996. Food drying and dewatering. *Drying Technology*, 14(9): 2135–2170.

Bórquez, R., Wolf, W., Koller, W.D. and Spieß., W.E.L. 1999. Impinging jet drying of pressed fish cake. *Journal of Food Engineering*, 40: 113–120.

Bussmann, P.J.T. 2002. Method for drying a product using a regenerative adsorbent. Patent EP 1503854 A1 (WO03/097231 A1).

Choicharoen, K., Devahastin, S. and Soponronnarit, S. 2010. Performance and energy consumption of an impinging stream dryer for high-moisture particulate materials. *Drying Technology*, 28: 20–29.

Crosby, E.J. and Weyl, R.W. 1977. Foam spray drying: General principles. *American Institute of Chemical Engineers (AIChE), Symposium Series*, 73(163): 82–94.

Cui, Z.-W., Li, C.-Y., Song, C.-F. and Song, Y. 2008. Combined microwave-vacuum and freeze drying of carrot and apple chips. *Drying Technology*, 26: 1517–1523.

Desobry, S.A., Netto F.M. and Labuza T.P. 1997. Comparison of spray-drying, drum-drying and freeze-drying for β-carotene encapsulation and preservation. *Journal of Food Science*, 62(6): 1158–1162.

Dev, S.R.S., Geetha, P., Orsat, V., Gariépy, Y. and Raghavan G.S.V. 2011. Effects of microwave-assisted hot air drying and conventional hot air drying on the drying kinetics, color, rehydration, and volatiles of *Moringa oleifera*. *Drying Technology*, 29: 1452–1458.

Duan, X., Zhang, M. and Mujumdar, A.S. 2007. Studies on the microwave freeze drying technique and sterilization characteristics of cabbage. *Drying Technology*, 10(25): 1725–1731.

Duan, Z., Zhang, M. and Tang, J. 2004. Thin layer hot-air drying of bighead carp. *Fisheries Sciences*, 23: 29–32.

Erbay, A. and Icier, F. 2010. Thin-layer drying behaviors of olive leaves (Olea Europaea L.). *Journal of Food Process Engineering*, 33(2): 287–308.

Falagas, S. 1985. *Drying Agricultural Products*, pp. 80–83, Athens: ELKEPA (in Greek).

Fortes, M. and Okos, M.R. 1981. Non-equilibrium thermodynamics approach to heat and mass transfer in corn kernels. *Transactions of American Society of Agricultural Engineers (ASAE)*, 22: 761–769.

Frey, D.D. and King, C.J. 1986. Experimental and theoretical investigation of foam-spray drying. 2. Experimental investigation of volatiles loss during foam-spray drying. *Industrial & Engineering Chemistry Fundamentals*, 25(4): 730–735.

Gandhidasan, P. 2004. A simplified model for air dehumidification with liquid desiccant. *Solar Energy*, 76: 409–416.

Gavrielidou, M.A., Vallous, N.A., Karapantsios, T.D. and Raphaelides, S.N. 2002. Heat transport to a starch slurry gelatinising between the drums of a double drum dryer. *Journal of Food Engineering*, 54: 45–58.

Ghandi, A., Powell, I.B., Chen X.D. and Adhikari, B. 2012. The effect of dryer inlet and outlet air temperatures and protectant solids on the survival of *Lactococcus lactis* during spray drying. *Drying Technology*, 30: 1649–1657.

Hashinaga, F., Bajgai, T.R., Isobe, S. and Barthakur, N.N. 1999. Electrohydrodynamic (EHD) drying of apple slices. *Drying Technology*, 17(3): 479–495.

Hertzendorf, M.S., Moshy R.J. and Seltzer, E. 1970. Foam drying in the food industry. *CRC Critical Reviews in Food Technology*, 1(1): 25–70.

Islam, Md. R., Thaker, K.S. and Mujumdar, A.S. 2007. A diffusion model for a drum dryer subjected to conduction, convection, and radiant heat input. *Drying Technology*, 25: 1043–1053.

Igual, M., García-Martínez, E., Martín-Esparza, M.E. and Martínez-Navarrete, N. 2012. Effect of processing on the drying kinetics and functional value of dried apricot. *Food Research International*, 47: 284–290.

Jaros, M. and Pabis, S. 2006. Theoretical models for fluid bed drying of cut vegetables. *Biosystems Engineering*, 93: 45–55.

Jones, K. 2006. Dehydration of food combinations. US Patent US 2006/0112584 A1.

Kandasamy, P., Varadharaju, N., Kalemullah, S. and Moitra R. 2012. Production of papaya powder under foam-mat drying using methyl cellulose as foaming agent. *Asian Journal of Food and Agro-Industry*, 5(5): 374–387.

Kasiri, N., Hasanzadeh, M.A. and Moghadam, M. 2004. Mathematical modeling and computer simulation of a drum dryer. *Iranian Journal of Science & Technology, Transaction B*, 28(B6): 679–687.

Karatas F. and Kamışlı, F. 2007. Variations of vitamins (A, C and E) and MDA in apricots dried in IR and microwave. *Journal of Food Engineering*, 78: 662–668.

Kemp, I.C. 2012. Fundamentals of energy analysis of dryers. In: *Modern Drying Technology, Vol. 4: Energy Savings*, 1st ed., Tsotsas, E. and Mujumdar, A.S. eds., Wiley-VCH Verlag GmbH and Co. KGaA, Weinheim, Germany.

Khanali, M., Rafiee, S, Jafari, A. and Hashemabadi, S.H. 2013. Experimental investigation and modeling of plug-flow fluidized bed drying under steady-state conditions. *Drying Technology*, 31: 414–432.

Kostoglou, M. and Karapantsios, T.D. 2003. On the thermal inertia of the wall of a drum dryer under a cyclic steady state operation. *Journal of Food Engineering*, 60: 453–462.

Kudra, T. and Mujumdar, A.S. 2002. Part I. General discussion: Conventional and novel drying concepts. In: *Advanced Drying Technologies*, pp. 1–26, Marcel Dekker, New York.

Kudra, T. and Mujumdar, A.S. 2009. *Advanced Drying Technologies*, 2nd ed., CRC Press, Boca Raton, FL.

Kudra, T., Mujumdar, A.S. and Meltser, V. 1995. Impinging stream dryers. In: *Handbook of Industrial Drying*, Mujumdar, A.S. ed., pp. 539–566, Marcel Dekker, New York.

Kudra, T. and Ratti, C. 2006. Foam-mat drying: Energy and cost analyses. *Canadian Biosystems Engineering/Le génie des biosystèmes au Canada*, 48: 3.27–3.32.

Kurnia, J.C., Sasmito, A.P., Tong, W. and Mujumdar, A.S. 2013. Energy-efficient thermal drying using impinging-jets with time-varying heat input—A computational study. *Journal of Food Engineering*, 114: 269–277.

Labuza, T.P. and Tannenbaum, S.R. 1972. Nutrient losses during drying and storage of dehydrated foods. *CRC Critical Reviews in Food Technology*, 3(2): 217–240.

Lewicki, P.P. 2006. Design of hot air drying for better foods. *Trends in Food Science and Technology*, 17: 153–163.

Li, F.D., Li, L.T., Sun, J.F. and Tatsumi, E. 2005. Electrohydrodynamic (EHD) drying characteristic of okara cake. *Drying Technology*, 23(3): 565–580.

Lin, T.M., Durance, T.D. and Scaman, C.H. 1998. Characterization of vacuum microwave, air and freeze dried carrot slices. *Food Research International*, 31(2): 111–117.

Marinos-Kouris, D. and Maroulis, Z.B. 1995. Transport properties in the drying of solids. In: *Handbook of Industrial Drying*, 2nd ed., pp. 113–160. Mujumdar, A.S. ed., Marcel Dekker Inc., New York.

Maroulis, Z.B. and Saravacos, G.D. 2003. *Food Process Design*, Marcel Dekker, Inc., New York.

Michailidis, P.A. and Krokida, M.K. 2013. Drying and dehydration processes in food preservation and processing. In: *Conventional and Advanced Food Processing Technologies*. Bhattacharya, S. ed., Wiley-Blackwell (John Wiley and Sons, Ltd.), UK.

Misha, S., Mat, S., Ruslan, M.H. and Sopian, K. 2012. Review of solid/liquid desiccant in the drying applications and its regeneration methods. *Renewable and Sustainable Energy Reviews*, 16: 4686–4707.

Moore, J.G. 1995. Drum dryers. In: *Handbook of Industrial Drying*, 2nd ed., Mujundar, A.S. ed., Vol. 1, pp. 249–262, Marcel Dekker, Inc., New York.

Moreira, R.G. 2001. Impingement drying of foods using hot air and superheated steam. *Journal of Food Engineering*, 49(4): 291–295.

Morgan, A.I., Graham R.P., Ginnette, L.F. and Williams, G. 1961. Recent developments in foam-mat drying. *Food Technology*, 15: 37–39.

Mujumdar, A.S. 1986. Impingement drying. In: *Handbook of Industrial Drying*, Mujumdar, A.S. ed., pp. 498–502, Marcel Dekker, New York.

Mujumdar, A.S. 1995. *Handbook of Industrial Drying*, 2nd ed., Marcel Dekker, New York.

Mujumdar, A.S. 2006. *Handbook of Industrial Drying*, Marcel Dekker Inc., New York.

Mujumdar, A.S. and Devahastin, S. 2006. Fundamental principles of drying. In: *Handbook of Industrial Drying*, 3rd ed., Taylor & Francis Group, Boca Raton, FL.

Mujumdar, A.S. and Menon, A.S. 1995. Drying of solids: Principles, classification, and selection of dryers. In: *Handbook of Industrial Drying*, Mujumdar, A.S. ed., 2nd ed., pp. 1–40, Marcel Dekker Inc., New York.

Muthukumaran, A., Ratti, C. and Raghavan, V.G.S. 2008. Foam-mat freeze drying of egg white and mathematical modeling. Part I: Optimization of egg white foam stability. *Drying Technology*, 26(4): 508–512.

Muthukumaran, A., Ratti, C. and Raghavan, V.G.S. 2008. Foam-mat freeze drying of egg white and mathematical modeling. Part II: Freeze drying and modeling. *Drying Technology*, 26(4): 513–518.

Pallas, L.A., Pegg, R.B., Shewfelt, R.L. and Kerr, W.L. 2012. The role of processing conditions on the color and antioxidant retention of jet tube fluidized bed-dried blueberries. *Drying Technology*, 30: 1600–1609.

Li, Y.B., Seyed-Yagoobi, J., Moreira, R.G. and Yamsaengsung, R. 1999. Superheated steam impingement drying of tortilla chips. *Drying Technology*, 17(1&2): 191–213.

Nindo, C.I. and Tan, J. 2007. Refractance window dehydration technology: A novel contact drying method. *Drying Technology*, 25: 37–48.

Ochoa-Martínez, C.I., Quintero, P.T., Ayala, A.A. and Ortiz, M.J. 2012. Drying characteristics of mango slices using the Refractance Window™ technique. *Journal of Food Engineering*, 109: 69–75.

Ortiz, J., Lemus-Mondaca, R., Vega-Gαlvez, A., Ah-Hen, K., Puente-Diaz, L., Zura-Bravo, L. and Aubourg S. 2013. Influence of air-drying temperature on drying kinetics, colour, firmness and biochemical characteristics of Atlantic salmon (*Salmo salar* L.) fillets. *Food Chemistry*, 139: 162–169.

Pakowski, Z., Bartczak, Z., Strumillo, C. and Stenstrom, S. 1991. Evaluation of equations approximating thermodynamic and transport properties of water, steam and air for use in CAD of drying processes. *Drying Technology*, 9: 753–773.

Parry, J.L. 1985. Mathematical modeling and computer simulation of heat and mass transfer in agricultural grain drying. *Journal of Agricultural Engineering Research*, 54: 339–352.

Perera, C.O. 2005. Selected quality attributes of dried foods. *Drying Technology*, 23(4): 717–730.

Pimpaporn, P., Devahastin, S. and Chiewchan, N. 2007. Effects of combined pretreatments on drying kinetics and quality of potato chips undergoing low-pressure superheated steam drying. *Journal of Food Engineering*, 81(2): 318–329.

Rajkumar, P., Kailappan, R., Viswanathan, R., Raghavan, G.S.V. and Ratti, C. 2007. Foam-mat drying of Alphonso mango pulp. *Drying Technology*, 25(2): 357–365.

Ratti, C. and Kudra, T. 2006. Drying of foamed biological materials: Opportunities and challenges. *Drying Technology*, 24(9): 1101–1108.

Sablani, S.S. 2006. Drying of fruits and vegetables: Retention of nutritional/functional quality. *Drying Technology*, 24: 123–135.

Sankat, C.K. and Castaigne, F. 2004. Foaming and drying behaviour of ripe bananas. *Lebensmittel-Wissenschaft and Technologie*, 37: 517–525.

Sathapornprasath, K., Devahastin, S. and Soponronnarit, S. 2007. Performance evaluation of an impinging stream dryer for particulate materials. *Drying Technology*, 25: 1121–1128.

Schoppet, E.F., Aceto, N.C., Craig, J.C. and Sinnamon, H.I. 1970. Continuous vacuum drying of whole milk foam, IV. Pilot plant. *Journal of Dairy Science*, 58(1): 56–62.

Singh, A., Orsat, V. and Raghavan V. 2012. A comprehensive review on electrohydrodynamic drying and high-voltage electric field in the context of food and bioprocessing. *Drying Technology*, 30: 1812–1820.

Siriamornpun, S., Kaisoon, O. and Meeso, N. 2012. Changes in colour, antioxidant activities and carotenoids (lycopene, β-carotene, lutein) of marigold flower (*Tagetes erecta* L.) resulting from different drying processes. *Journal of Functional Foods*, 4: 757–766.

Smith, T.M. 1994. Heat transfer dynamics. *TAPPI Journal*, 77: 239–245.

Strumillo, C. and Kudra, T. 1986. Heat and mass transfer in drying processes. In: *Drying: Principles, Applications and Design*. Gordon and Breach Science Publishers, Montreux.

Supmoon, N. and Noomhorm, A. 2013. Influence of combined hot air impingement and infrared drying on drying kinetics and physical properties of potato chips. *Drying Technology*, 31(1): 24–31.

Treybal, R.E. 1980. *Mass-Transfer Operations*, 3rd ed., McGraw-Hill Chemical Engineering Series. McGraw-Hill, New York.

Tsotsas, E. and Mujumdar, A.S. 2012. *Modern Drying Technology, Energy Savings. Vol. 4*. Wiley-VCH Verlag and Co, Germany.

Van Arsdel, W.B. and Copley, M.J. 1963. *Food Dehydration. Vol. 1—Principles*. The AVI Publishing Company, Inc., Westport, Connecticut.

Vega-Mercado, H., Góngora-Nieto, M.M. and Barbosa-Cánovas, G.V. 2001. Advances in dehydration of foods. *Journal of Food Engineering*, 49(4): 271–289.

Wang, W.C, Calay, R.K. and Chen, Y.K. 2011. Experimental study of an energy efficient hybrid system for surface drying. *Applied Thermal Engineering*, 31(4): 425–431.

Wiktor, A., Iwaniuk, M., Śledź, M., Nowacka, M., Chudoba, T. and Witrowa-Rajchert, D. 2013. Drying kinetics of apple tissue treated by pulsed electric field. *Drying Technology*, 31: 112–119.

Wilford, L.G., Sabarez, H. and Price, W.E. 1997. Kinetics of carbohydrate change during dehydration of d'Agen prunes. *Food Chemistry*, 59(1): 149–155.

Xiao, H.-W., Gao, Z.-J., Lin, H. and Yang W.-X. 2010a. Air impingement drying characteristics and quality of carrot cubes. *Journal of Food Process Engineering*, 33(5): 899–918.

Xiao, H.-W., Pang, C.-L., Wang, L.-H., Bai, J.-W., Yang, W.-X., Gao, Z.- J. 2010b. Drying kinetics and quality of Monukka seedless grapes dried in an air-impingement jet dryer. *Biosystems Engineering*, 105(2): 233–240.

Xiao, H-W., Gao, Z.-J., Lin, H. and Yang, W.-X. 2010. Air impingement drying characteristics and quality of carrot cubes. *Journal of Food Process Engineering*, 33(5): 899–918.

Yang, W.C. 2003. *Handbook of Fluidization and Fluid-Particles Systems*, Marcel Dekker: New York.

Zang, Q., Barbosa-Cánovas, G.V. and Swanson, B.G. 1994. Engineering aspects of electric field pasteurization. *Journal of Food Engineering*, 25: 261–281.

Zbicinski, I. and Rabaeva, J. 2009. Analysis of gas admixing foam spray drying process. *Drying Technology*, 28(1): 103–110.

Zheng, X.-Z., Liu C.-H. and Zhou H. 2011. Optimization of parameters for microwave-assisted foam mat drying of blackcurrant pulp. *Drying Technology*, 29(2): 230–238.

# 10 Fluidized Bed, Spouted Bed, and In-Store Drying of Grain

*Robert H. Driscoll and George Srzednicki*

**CONTENTS**

## 10.1 PREFACE

Two-stage drying is the concept that optimal drying of a product may sometimes be best achieved by breaking down the drying process into two stages, a high-moisture stage and a low-moisture stage. The reasons for doing this vary, and include thermal efficiency, greater throughput, and improved quality. Different forms of two-stage systems have been developed over the last few decades for the main grain crops. This chapter describes the equipment components for two-stage drying systems.

The two-stage drying system has been developed especially for conditions characterized by high temperature and relative humidity at the time of harvest, common in subtropical and tropical regions of the world. However, with some variations, such systems can also apply to conditions of short summers with temperate climate followed by winter with several months of subzero temperatures. The concept of two-stage drying breaks down the moisture reduction operation into two separate stages. As grain after harvest is subject to rapid deterioration within the first 12 h, the first stage involves initial high-speed removal of water, reducing the product water activity to a level sufficient for medium-term storage. The second stage involves slow continued moisture removal to a content suited for long-term storage using near-ambient air in-bin mechanical aeration under conditions suited to a safe rate of moisture reduction at the lowest price. Two-stage drying interfaces best with bulk handling, and requires a sufficient volume of grain, careful management, and trained operators.

The first stage should be designed to reduce the water activity to below 0.80 or about 18% moisture content wet basis (w.b.). Equipment suitable for the first stage includes any form of a fast dryer, and especially the fluidized bed or spouted bed dryers developed through a series of international research projects and now commercialized. The second stage allows further reduction in water activity to about 0.70 and has been primarily implemented through in-store dryers (ISD). These dryers consist of large bins in which the bulk product at several meters depth (typically 3 m for tropical conditions) is aerated with ambient air that can be slightly heated, if required for reduction of its relative humidity. The main incentive for two-stage drying is recognition of the drying mechanics of grains (especially rice), which are soft above about 18% moisture content and can be dried rapidly without damage. In fact, if rice grain is dried in a suitable temperature range, its surface hardness may increase, reducing damage when subsequently milled. In contrast, below 18%, rapid drying may induce fissuring, causing a predisposition to fracture during milling. Gentle, slow drying allows moisture and thermal re-equilibration within the grain to relax drying stresses and improve milling yield. Moreover, near-ambient, in-store drying benefits from the natural drying capacity of the air, reducing the energy consumption and hence environmental impact.

The chapter first introduces some important concepts of drying. Next, two forms of first-stage drying are considered, the fluidized bed and the spouted bed, which are actually variants of the same concept. These dryers will quickly bring the product down to a safe moisture range. The last section considers in-store dryers, an option for gentle slow-drying, which reduces energy use and operates at low temperatures, helping to protect both the product and the environment.

## 10.2 THIN-LAYER DRYING CURVES

Before studying specific dryers, we need to consider how a particle dries. The rate of moisture loss from a particle depends on many factors:

- Availability of water for evaporation from the surface.
- Surface area.

- Air temperature, which provides the driving force for convective heat transfer into the surface of the product.
- Air humidity, which affects the capacity and rate of moisture uptake from the product surface.
- Air speed, which has an effect of the thickness of the air boundary layer between the external air stream and the product surface. The thickness of this relatively stationary layer affects the rate of heat and moisture diffusion to the surface, and higher air speed reduces its thickness, increasing transfer rates.

When we look at a thin layer of particles drying in an air stream of constant speed, temperature, and humidity, we observe four distinct periods (see Figure 10.1). These are discussed in the following section.

In considering drying, it is important to measure state variables such as moisture in a consistent way. To avoid error, engineers find it better to express many quantities on a dry basis. For example, we can define moisture content as mass of water over mass of product, which gives us the correct compositional proportion of water. However, during drying, the total product weight changes, and so our basis for measurement changes. If I dry a product from 50% to 20%, I do not lose 30% of the product weight but 37%. So, for drying, the moisture content of the air and the product are best expressed on a dry basis. I will use the symbol M to indicate dry basis product moisture measurement, and H for dry basis air moisture.

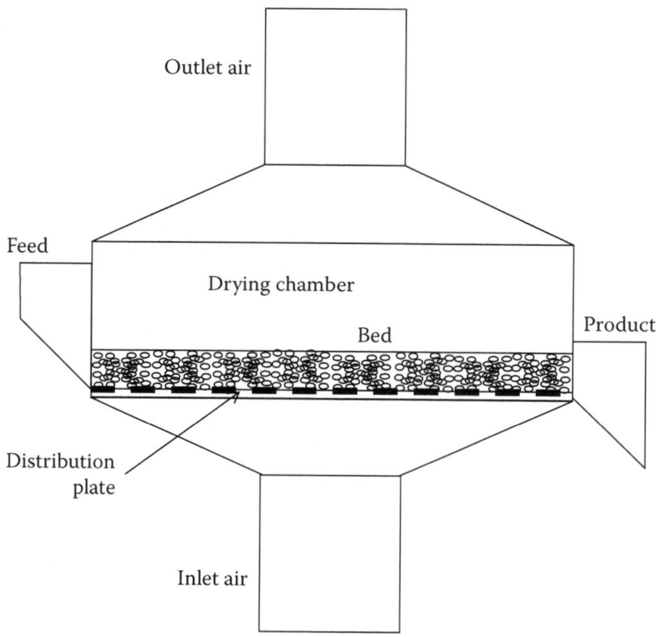


**FIGURE 10.1** Fluidized bed dryer.

### 10.2.1 Initial Transient Region

Initially, the heat content of the product, being more concentrated, has a dominant effect on the air stream. For example, a very cold product may initially cause the drying air to cool to the point where moisture condenses onto the product surface, for example, as observed when putting very cold tomatoes into a hot oven.

This region is transient, meaning that it is only a very short-term effect and the change in product moisture is normally insignificant. The initial transient ends when the product has achieved the same specific enthalpy as the air, so that any further exchanges of energy between the product and the air are balances: rate of heat loss through evaporation becomes equal to heat gain by convection.

A cold product entering the dryer must gain heat to achieve thermal equilibrium with the drying air, and so will gain a small amount of surface moisture. A hot product (at a temperature higher than the wet bulb temperature of the air) will initially lose heat, and thus will lose a small amount of moisture, drying quickly for a short time. In this way, a cold product entering the dryer wastes some the air heat energy, but a preheated product results in a higher thermal efficiency for the dryer.

### 10.2.2 Constant Rate Period

After thermal equilibrium between the product and air has been achieved, the rate of drying depends primarily on the availability of surface moisture. Provided that the air conditions and the size of the particles do not change, the rate of evaporation will depend solely on the rate of heat transfer to the surface, which will be constant. The product will be at a water activity of exactly one during this period, and the situation is identical to drying from a free water surface. The product will be at the wet bulb temperature of the air.

In practice, few food products exhibit a constant rate period (CRP). Very few products have a water activity of exactly one, and so do not have free water available on the surface, and so there will be a diffusional component to the rate of evaporation. The CRP may be approximated for high-moisture products such as slurries and pastes, provided that the proportion of soluble solids is small. Generally though, this region is not of great importance in studying food drying.

For the CRP to be sustained, water evaporated from the surface needs to be replenished (from pores or capillaries, or by diffusion), so that the surface of the product continues to act like a free water surface. After a sufficient period of time, these pores and capillaries will also become depleted, and the product surface will start to dry. The point at which this happens will depend on both the geometry and structure of the product, and also on the rate of drying to that point. This dependence is because the rate of surface moisture replacement may be sufficient at slow drying rates, but not at higher drying rates.

Assuming that shrinkage during drying can be neglected, the time required to dry a product in the CRP can be calculated as follows:

$$\frac{dM}{dt} = -k_o \tag{10.1}$$

where $k_o$ is the drying rate constant, in inverse time units. Integrating,

$$t_o = k_o/(M_o - M_c) \tag{10.2}$$

The only difficulty with this approximation is that the initial transient has not been separated from the CRP, and this will lead to a small error. For example, for a cold product, this equation will under estimate the time required to dry to $M_c$.

### 10.2.3 Transition Region

As the surface starts to dry, two mechanisms become important for moisture removal, surface evaporation, and internal moisture diffusion. During the transition period, some of the surface continues to act as a free water surface for evaporation, while the remainder of the surface becomes increasingly limited by internal diffusion, and the overall drying rate starts to drop. In commercial practice, this region is not significant, and is normally not modeled.

### 10.2.4 Falling Rate Period

During the falling rate period (FRP), the rate of evaporation is controlled by the rate of internal moisture diffusion. As the local moisture gradients within the product decrease, the rate of diffusion also decreases. Historically, this period is often modeled as an exponential decay equation:

$$\frac{dM}{dt} = -k_1(M - M_e) \tag{10.3}$$

So

$$MR = \frac{M - M_e}{M_c - M_e} = \exp(-k_1 t) \tag{10.4}$$

Equation 10.4 allows prediction of the FRP drying time in terms of the required final moisture content. Here, the main drying constant is $k_1$, in inverse time units. $MR$ is called the moisture ratio, and varies from 1 at the start of the FRP to 0 at complete equilibration with the drying air.

### 10.2.5 Summary of Drying Rates

From a food product point of view, the most important drying period is the FRP. Understanding the transient response period is also helpful, as this period may affect the thermal efficiency of the dryer. The other two regions (constant rate and transitional periods) are rarely observed in foods.

The fluidized bed dryer (FBD) is more effective at high water activities and less effective as the water activity drops, as the air can no longer absorb as much moisture. Thus, for food products, there is typically a window of moisture contents where the fluidized bed is effective. Once the surface moisture has been removed, the FBD loses its advantage of high contact surface area. Since fluidized bed dryers have advantages in terms of saving space, they may still be the preferred choice of dryer for a particular product, but the use of tempering bins to allow moisture equilibration may then be advisable.

## 10.3 THEORETICAL PREDICTIONS OF DRYING RATE

Any real model of the drying process must take into account the following effects:

- Diffusion within the product particles, based on Fick's equations
- Modeling of the surface resistance to moisture and heat transfer
- Isotherm relationship between product moisture and product water activity
- Particle and air properties
- Dryer geometry
- Particle geometry

Most methods for predicting drying rate of particles in any form of dryer consider only one or two factors from this list, and most predictions of particle drying rates are accordingly poor approximations to reality. For example, many drying models make the following assumptions:

- Fickian diffusion
- Ideal particle shape (sphere, cylinder, thin slab)
- Isotropic and homogeneous
- Constant air conditions
- Negligible surface resistance (to both heat and mass transfer)

After fitting to experimental data, these models may be used to estimate an effective mass diffusivity of the product. A different set of assumptions might yield a different diffusivity, and even predict a different shape to the drying curve.

### 10.3.1 Modeling Drying at Surface

Heat and mass transfer at the surface can be modeled as

$$\dot{Q} = \lambda_T \frac{dm_w}{dt} = \lambda_T k_y A(H_s - H_a) = h_c A(T_a - T_s) \tag{10.5}$$

The convective heat transfer coefficient $h_c$ in a fluidized bed is typically around 5–20 W/m$^2$.K (Chen 2003). Ciesielczyk (1996) analyzed heat and mass transfer for the CRP specifically.

The surface water activity is a function of both the surface moisture concentration and its temperature, a unique function required for each product. This relationship is called an isotherm:

$$a_w = f(M_s, T_s) \tag{10.6}$$

This equation can be determined by imposing equilibrium between the air and the product. A sample of the product in a jar will equilibrate over time with the conditions inside the jar, so that the temperature of the sample will match the temperature of the headspace in the jar (thermal equilibrium) (note that headspace is the space above the product and below the lid), and the water activity of the product will match the water activity of the air (concentration equilibrium). Air water activity is called relative humidity and is easy to measure, so that by determining the temperature and relative humidity of the air above the sample, the product temperature and water activity are also determined. The sample can then be oven-dried to determine its moisture content. Many standard procedures for doing this are published in various compendiums of methods. The resulting data can be fitted to equations such as the Guggenheim–Anderson–de Boer equation (GAB model).

Thus modeling of surface evaporation requires

- Knowledge of air properties and equations for their calculation
- Calculation of the surface heat transfer coefficient, for example, by Reynolds–Prandtl correlations with the particle Nusselt number as published in many chemical engineering texts
- Determination of the product isotherms
- Application of the wet bulb Equation 10.5 for rates of evaporation, modified for the effect of surface water activity
- Solution of the heat and mass balance equations for an element

### 10.3.2 Modeling Drying within Product

Modeling of diffusion within the product requires

- Fick's transport equation, in an integrated form for the particular geometry. Solutions to spheres, cylinders, and thin plates exist (Crank), but these models are not realistic for food products due to their irregular shapes and nonuniform properties. Finite difference, volume, or element solutions are rarely used for dryer modeling, but would allow this geometry limitation to be reduced.
- Modeling of heat flow within the particle, using Fourier's heat transfer equation and conservation of energy, or by simply assuming temperature is uniform within the product.

The solution for a sphere of radius $r$ is shown below.

$$MR = 1 - \frac{6}{\pi^2}\sum_{n=1}^{\infty}\frac{1}{n^2}exp\left(-\frac{D_m n^2\pi^2 t}{r^2}\right) \quad (10.7)$$

where $M$ is the average particle moisture content and $D_m$ is the effective diffusivity of the material, defined as the equivalent diffusivity of a smooth-surfaced sphere of equal mass.

The theoretical model is generally a poor approximation to the drying rate because it assumes ideal conditions that are rarely valid. Most food products are not regular shapes, and so the constants in Equation 10.7 will not be correct. Nor are food products isotropic or homogeneous, but may have distinct layers, or internal directional structure, which affects the rates of diffusivity.

Nevertheless, the form of equation is instructive, and most practical drying models adopt the concept of a number of terms involving exponential decay. Empirical models based on this concept have been successful with both short-term and long-term drying periods.

### 10.3.3 Empirical Models of Product Drying

A simple but reasonably effective prediction of the drying rate is derived by analogy to Newton's law of cooling:

$$\frac{dM}{dt} = -k_1(M - M_e) \tag{10.8}$$

Separating parts and integrating gives the familiar exponential form of Equation 10.4, which can now be recognized as the first exponential term of Crank's form of solution to Fick's equation.

Does this give a theoretical justification for Equation 10.4? Not at all, for Fick's equation is only valid for internal resistance (and ignores the surface resistance of the product, as described earlier). The derivation ignores geometry, structure, heterogeneity, and does not model the initial transient. In addition, Equation 10.7 could only be valid for constant air conditions, since its derivation involves an integration over constant boundary conditions.

But it works! Despite this solution having no reasonable theoretical basis, Equation 10.4 has been used for thin layers of products drying under constant aeration conditions with excellent results over short time periods. The reality is that drying does approximate an exponential curve, since the moisture is changing from some initial state to an equilibrium state, conditions well described by exponential decay.

In terms of agreement with data, the most successful model for thin-layer drying under constant conditions is based on Equation 10.7 but uses two terms:

$$\frac{M - M_e}{M_o - M_e} = Ae^{-k_1 t} + Be^{-k_2 t} \tag{10.9}$$

Two concerns should be addressed before applying Equation 10.8 or 10.9 to describe fluidized bed drying. First, was the model derived under thin-layer drying conditions? This would assume the product is stationary, and would give a slower drying rate than would occur in the fluidized bed dryer due to the increased surface area exposed to the air during fluidization. It is important that the constants for the equation are derived from fluidization drying runs.

Second, the above equation is only valid for constant conditions, and should not be used when conditions are changing. Specifically, the temperature and humidity of the inlet air must remain constant, or the equation will not be reliable as a predictor of moisture content.

Many other empirical equations for predicting drying rate exist, but have not been extensively tested or used with fluidized bed drying.

## 10.4 MODELING OF DRYERS

The previous section considered the drying rate of an individual particle under constant conditions. This gives us an equation for the drying rate, $dM/dt$, which can be used to estimate the rate of moisture loss at a particular point in a dryer. This equation is one of four fundamental equations describing the interaction between the product and the drying air.

Define dry basis product mass flow as

$$G_p = \frac{\rho_b v_p A}{1+M} \tag{10.10}$$

Similarly, define air mass flow as

$$G_a = \frac{\rho_a v_s A}{1+H} \tag{10.11}$$

where the densities $\rho_b$ and $\rho_a$ are the product bulk density and air density, respectively. The four equations for the complete air/product system for a static bed are

- Mass balance:

$$G_p\left(\frac{dM}{dt} + v_p\frac{dM}{dx}\right) = -G_a\left(\frac{dH}{dt} + v_s\frac{dH}{dx}\right) \tag{10.12}$$

- Heat balance:

$$G_p\left(\frac{dc_pT_p}{dt} + v_p\frac{dc_pT_p}{dx}\right) = -G_a\left(\frac{dh}{dt} + v_s\frac{dh}{dx}\right) \tag{10.13}$$

- Mass transfer:

$$\left(\frac{dM}{dt}\right)_{disp} = D_{ac}\frac{\partial^2 M}{\partial x^2} - \left(\frac{dM}{dt}\right)_{dry} \tag{10.14}$$

- Heat transfer:

$$\left(\frac{dc_pT_p}{dt}\right)_{disp} = D_{ac}\frac{\partial^2 h_p}{\partial x^2} + \frac{h_c A_p(1-\in)(T_a - T_s)}{\rho_b} - \left[\lambda_T + c_v(T_a - T_p)\right]\left(\frac{dM}{dt}\right)_{dry} \tag{10.15}$$

$(dM/dt)_{dry}$ is the empirical drying rate equation discussed in the previous section (see Equations 10.8 and 10.9), and $(dM/dt)_{disp}$ is the drying rate at a particular position. The terms involving $D_{ac}$ represent movement of moisture due to moisture diffusion, and are generally neglected. In fact, many researchers simplify these equations by ignoring some of the space and time derivatives given here.

There are some assumptions in the equations:

- Shrinkage is negligible.
- No dry solids are lost from the system.
- The dryer is adiabatic.
- The dryer can be considered as one dimensional.
- Axial (conduction) heat flow is negligible.

Note that in Equation 10.15, the total surface area $A_p$ in a layer is required, which can be calculated using the individual (averaged) particle surface area. However, in practice, Equation 10.15 is rarely used in simulation, since heat equilibration is typically 30–60 times faster than mass equilibration. Thus, the particle temperature will normally be close to the air wet bulb temperature. For in-store drying, Equation 10.15 is thus not required. However, for fluidized bed drying, the greater air speeds that are used invalidate this assumption, and the rate of heat transfer becomes important.

The above equations have been written in terms of a static product bed, and so are immediately applicable to in-store drying. If the product is moving through the dryer, as, for example, in a continuous fluidized bed dryer, the equations need to be modified by replacing $\rho_b \frac{d}{dt}$ by $\rho_b v_p \frac{d}{dx}$, where $v_p$ is the particle speed through the dryer.

A number of other equations are required to build up a complete simulation of the interaction of the product with the air. These are the property equations, consisting of

- Air psychrometric relations, such as equations linking temperature, absolute humidity, and relative humidity.
- Product property equations, such as the isotherm equation linking product moisture, temperature and water activity, the shape factor relating surface area to volume, the latent heat of wetting (required for greater precision at low moisture contents), equations for bulk and particle density in terms of moisture, and specific heat. The thin-layer drying Equation 10.12 could also be included in this group.

The air properties are available in textbooks on psychrometrics. The product property equations may be available in published literature or may need to be determined experimentally, a topic beyond the scope of this chapter.

The heat and mass balance approach described above is adequate for in-store dryer simulation, but requires some modification for fluidized bed drying, and more will be said on this topic in Sections 10.12 and 10.13.

## 10.5 FLUIDIZATION

A static bed of granular material may behave like a liquid under certain conditions, for example, when a fluid stream is passed through the bed. The fluid stream creates frictional drag forces that support the weight of the particles, reducing the contact forces that would normally prevent particles in a granular mass from moving relative to each other. Fluidization allows each particle to continuously change orientation, ideal for enhanced heat and mass transfer. In addition, fluidization allows transport of the product through a bed. Good general texts on fluidization include McCabe et al. (2001), Holdich (2012), Yang (2003), Yates (1983), and Chung and Mujumdar (2007).

The first industrial use of a fluidized bed was for coal gasification in 1922 (Tavoulareas 1991). Over time, the concept was refined to allow continuous recycling, combustion, spouting, and drying applications. Fluidized and spouted beds are now a well-established technology, having been used commercially for many decades. They offer a fast, uniform drying option suitable for powders, grains, pellets, and other particulate materials, a group called granular materials.

A granular material is composed of individual particles in contact with each other, allowing fluids to flow through the space between the particles, called interstitial space. Grains and powders are good examples of granular materials. This material is essentially in a different state from both solids and liquids, possessing some properties of both. At rest, a granular material appears solid, and the particles may partially lock together due to surface resistance, irregular shape, or cohesive forces. However, if sufficient kinetic energy is transferred to a granular material, it may flow like a liquid, for example, a conical pile of grain may be stable, yet the grain can be poured down a chute.

The solid/liquid duality of grains is a useful property for drying, as the product may be poured into a container, and then heated air passes through the connected interstitial space to remove moisture. Drying in bins or silos has in consequence been a successful and economical method of drying; a store of granular materials in a silo can be aerated, fumigated, and cooled with great uniformity due to this property of allowing air flow.

To facilitate aeration, the product is placed on a porous grid called a distribution plate, designed to carry the weight of the bed, but allow aeration from a plenum chamber below the bed. This may be simply a metal mesh, hemispherical ducting, mesh-covered channels, or some other form. The bed of materials is called a packed bed.

Air passing through the space between the particles must continuously change direction as it follows the natural tortuosity of the connected air space. The change in direction requires a momentum change, which is transferred to the particles over time as a force on the particle. The net force if aerated vertically from below will be upwards, supporting some of the weight of the product. As the aeration speed increases, the particle becomes lighter to the point where (assuming cohesive forces are minimal) the particle can start to move independently (Wen and Yu 1966). This is the point at which the full weight of the particle is being supported by the rate of momentum transfer from the air, although the exact air speed for fluidization will depend on other forces such as product weight above the particle and buoyancy.

During this time, the bed moves from stationary particulate (uniform) fluidization, then to bubbling, with air circulation bubbles picking up, cycling, then dropping particles back to the base of the bed. Particles suited for fluidization are typically in the range of 70 μm to 2 mm.

During the onset of fluidization, the bed expands and its porosity increases. With additional increases in air speed, there will be a greater transfer of energy to the particles, to the point where they are entrained in the air flow. Except when used for recirculating product within a fluidized bed, we want to prevent entrainment, so the Stokes settling speed of the particle should be greater than the air speed. This results in a net downward speed toward the supporting pressure distribution plate. Effectively, Stokes' equation sets a lower limit on particle sizes.

If we plot the air speed against the pressure drop for a bed of grain (see Figure 10.2), we would see a parabolic pressure change (Ergun 1952) up to the point of fluidization, at which point the pressure becomes constant as the particles start moving freely. Further increase in air speed has little effect on the pressure drop over the bed. Depending on the type of particle (see Geldart's classification later) and its compaction, this onset may be gradual, sudden, or with a small initial overshoot in pressure if particles are cohesive or the bed is tightly packed.

Fluidization starts when the bed mass per unit area is supported by the drag force imposed by the air, and this point defines a minimum fluidization speed, $u_{mf}$. As the normal force between particles reduces, a bed of particles will start to expand, typically doubling or quadrupling in size as the interstitial space increases.

Contact between the particles may affect the minimum fluidization speed. This may be due to

- Surface cohesion between particles, for example, particles high in sugar may have a sticky surface that increases the bond energy between particles. As particles dry, cohesiveness due to moisture decreases, allowing fluidization at lower speeds.

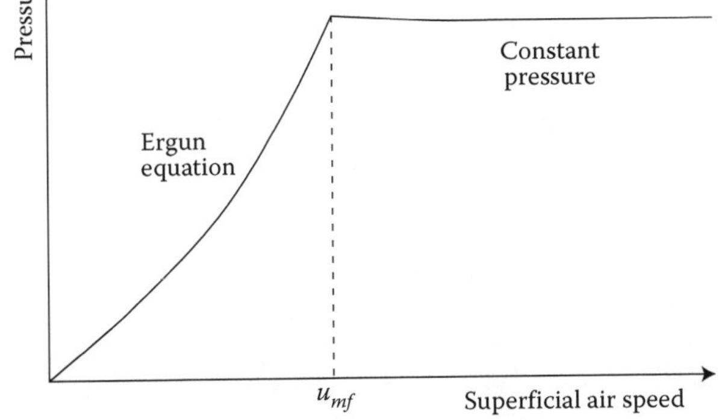


**FIGURE 10.2** Pressure across fluidized bed.

- Geometry of particles, as spherical particles have a small contact area, but milled particle may have a degree of interlocking, which prevents particles sliding easily over each other at the point of incipient fluidization.
- Uniformity of particles, for example, fine particles may reduce air flow in areas of accumulation, and large particles have a higher area-to-volume ratio and so are not as easily lifted by air pressure and have a higher settling speed.
- Density of material, which increases the gravitational body force on the particle, increasing the required fluidization speed.
- Depth of material, as pressure between particles tends to increase with depth, so that only the top portion of a bed may fluidize. Choice of bed depth is therefore an important parameter for ensuring uniform drying.
- Fluid properties such as viscosity and density (both affected by temperature).

With increasing air speed above $u_{mf}$, the pressure drop across the bed changes very little, but the bed flow characteristics change. Just above the minimum fluidization speed, the bed will fluidize smoothly and uniformly. At around twice the minimum fluidization speed, a specific form of fluidization called bubbling occurs.

Bubbling fluidization creates the best conditions for heat and mass transfer because of its high rate of mixing and its peculiar form of particle transport. There are four distinct zones in a bubble, which are a dense phase close to the pressure plate, a dilute phase of rising and descending particles, clear air bubbles with few particles, and a transition zone on top of the bed where particles reach the end of their trajectories and fall back toward the bed (Rowe et al. 1964). Thus, fluidized bed density is a maximum near the plate, dilute through the bubble zones, and decreasing at the top of the bed. This upper region is also called the freeboard region, and the height above which particles disengage from the air is called the transport disengagement height (TDH), determined by the distribution of particle speeds and the difference between the air speed and the particle settling speed (Stokes' equation).

Other forms of fluidization exist, and have been categorized by several authors. These include channeling, where air flows through relatively fixed channels in the product bulk, with consequently reduced air mixing, and expanding bubbles zones. Also with increased air speed, turbulent mixing occurs with rapid formation and dissolution of bubbles and eddies of product. This leads to further expansion of the bed, greater entrainment, and then finally as air speed increases, the onset of pneumatic conveying of the product mass.

Since fine particles have a lower settling speed, even during bubble fluidization they will tend to be entrained to greater heights, and may even be pneumatically conveyed out of the dryer, a potential hazard if the air is recirculated without some form of filtration such as cyclone separation. Also, since finer particles spend longer time in the freeboard region, their drying characteristics will be different, reducing the uniformity of the drying. However, this is a minor consideration compared with the effect of particle size on drying rates, which is considered later. Entrainment can be used to remove contaminants such as insects from grain (Ghaly and Sutherland 1984) or to separate fine from coarse materials for grading purposes.

When used for drying, a fluidized bed creates an ideal medium for particulate solids to receive rapid and uniform heating, and to release moisture uniformly and quickly from the particle surface. This is a major reason for considering a fluidized bed dryer. The dryer has an energy efficiency comparable with large multidirectional column dryers for grains, yet due to its high mass and heat transfer rates, requires only a small volume and footprint for operation. The smaller size allows fast changes in aeration conditions, and the product hold-up volume is small. Typically, the drying time is of the order of 10 min to an hour.

## 10.6 PRESSURE DROP DURING FLUIDIZATION

The pressure drop through a packed bed is modeled by the Ergun (1952) equation, valid over a wide range of face velocities $v_s$ (Looi et al. 2002):

$$\frac{\Delta P}{L} = \frac{150\mu v_s}{D^2}\frac{(1-\in)^2}{\in^3} + \frac{1.75\rho v_s^2}{D}\frac{1-\in}{\in^3} \tag{10.16}$$

where $\Delta P/L$ is the pressure drop, $D$ is the equivalent spherical diameter, $\in$ is the void fraction (porosity) of the bed, $\rho$ is the fluid density, and $\mu$ is the fluid viscosity. For Reynolds numbers (Re) below 20, only the first term (viscous energy) is required. For Reynolds numbers greater than 1000, only the second term (kinetic energy) need be considered. This is often true for the larger, grain-sized particles of concern for this chapter.

To find the minimum fluidization speed $v_{sm}$, let the particle density be $\rho_p$ and assume the bed expands to a porosity $\in_m$ at $v_{sm}$, then

$$\frac{\Delta P}{L} = (\rho_p - \rho)(1-\in_m)g \tag{10.17}$$

Equating the two expressions at $v_{sm}$, $\in_m$ allows the minimum speed for fluidization to be calculated. Although this works well for fine spherical particles, experimental verification is important for larger particles, especially for grains where experimental agreement is poor.

For Reynolds numbers less than 1, Stokes' equation provides a way of estimating the maximum fluidization speed for a given particle size:

$$v_s = \frac{D^2 g(\rho_f - \rho_p)}{18\mu} \tag{10.18}$$

Since Stokes' equation is derived for spherical particles in a laminar flow regime, the effective diameter $D$ must be calculated based on the particle size and shape factor, and the flow type checked using the Reynolds number for the flow. Finally, the viscosity at fluidization for Equation 10.18 must be corrected for bed porosity:

$$\mu = \mu_o \in^{4.6} \tag{10.19}$$

which is valid for Re < 0.2 (Richardson and Zaki 1954). Other correlations exist at higher values of the Reynolds number. From these equations, for example, a 200-μm-diameter particle would fluidize around 0.02 m/s air speed and have a settling speed of 37 m/s (assumed particle density 1400 kg/m$^3$, porosity 0.5).

## 10.7 COMPONENTS OF A FLUIDIZED BED DRYER

A fluidized bed dryer consists of the following components: an air source, a heater, a pressure plate, and a drying chamber.

### 10.7.1 Air Source

The air source consists of a fan of sufficient capacity to produce the air speed necessary to fluidize the product, and is normally a centrifugal fan. The fan inlet may be supplied with fresh air from a source away from the air outlet, or may be connected to the air outlet to allow recirculation of air. This allows more efficient use of the exit air, but the air would normally be cleaned by cyclone separation and/or filtration to avoid product contamination.

### 10.7.2 Heater (Burner)

The air is heated to a temperature, which allows rapid convective heat transfer to the product. The heater will typically be a gas burner, but could also be steam or a biomass combustor.

### 10.7.3 Pressure Plate

Air leaving the burner will not flow uniformly. This is due to the following:

- A centrifugal fan creates a pulsating flow.
- Air ejected from the fan is pushed to the outside of the fan casing, creating a higher pressure on the outside of a duct. This higher pressure may induce the air to "bounce" from one side to another, creating a degree of turbulence.
- Air in contact with the duct sides is stationary, so that there is typically a distribution of speeds across a duct from a maximum at the center to a minimum (zero) at the duct walls.
- Any corners, constrictions, expansions, or joins in the duct may be a source of nonuniformity in the air flow, caused by the momentum of the air.

The pressure plate is designed to create a uniform air flow through the drying chamber, and also to prevent product from falling out of the chamber. The plate holes are chosen with a diameter sufficient to prevent particles escaping, yet with a sufficient density to minimize the pressure drop across the plate, so its design is an important part of the dryer. The plate should create a sufficient pressure drop to make the air flow through the drying chamber as uniform and vertical as possible, yet not so high as to create an increased fan size requirement that would add to the drying cost. Generally, the plate should contribute about 30% of the pressure drop across the bed. Distribution baffles below the plate can also help to ensure a uniform and vertical air supply to all parts of the drying bed.

Particles that fall through the pressure distribution plate may reach the heating unit, lodging there and creating a fire hazard. However, even particles that do not completely fall through may create a problem by blocking holes and reducing the uniformity of fluidization, which can lead to some particles not being adequately dried, or may even be retained in the chamber and overdried to the point of combustion. Caps over the holes may prevent this, at the expense of a greater pressure drop and reduction of the vertical component of the air velocity. A plate may also have multiple layers, a coarse perforation thicker plate for providing strength, and a finer mesh to retain finer particles but with a reduced pressure drop due to its thinness.

### 10.7.4 Drying Chamber

The drying chamber supports and contains the product. A fluidized bed dryer may be operated in batch mode or continuous mode, and this affects the chamber design. For batch mode, the product may be loaded as a static bed (nonfluidized), and then become a fluidized stationary bed. For continuous operation, a weir must be designed to hold sufficient product in the chamber to allow adequate particle retention time in the drying air. The chamber may also be designed to allow circulation of product in some cases, particularly in the chemical industry where entrainment of reactants may be used to recirculate product within the dryer. Generally for the food industry, the dryer will be run as batch or continuous with no product recirculation.

The shape of the drying chamber should be conducive to good air flow and good product flow. The air must be streamlined and without flow obstacles such as sudden constrictions, which would increase the pressure drop through the chamber. The product must flow in such a way that the residence time is as invariant as possible, and to avoid dead spots where product might accumulate, overdry, and present a contamination or a fire risk.

## 10.8 PARTICLE COHERENCE

Particles in a granular mass may adhere to each other for several reasons, and there will be an energy associated with this. The geometry and material of the surface may also create a friction reaction force that opposes movement, allowing strain energy to be stored in the particles. Even particles that are relatively inert may not slide easily past each other due to locking of particles into stable packing configurations so as to minimize the gravitational potential energy. Whichever mechanism or combination of mechanisms is involved, the particles are restrained from moving relative to each other until the local energy density is sufficiently high to provide individual particles with energy in excess of their local stored energy, and at this point the particles are enabled to move past each other. If I allow a stream of granular particles to fall out of a hopper onto a flat surface, they will act as a fluid as they fall, exchanging potential energy of height for kinetic energy. However, as the particles hit the ground and lose energy by friction and impact, they will form a conical pile on the surface, packing together in configurations that resist movement. The angle of the conical pile is a critical measure of interparticle resistance, characteristic of a granular material, and is called the angle of repose.

Since interparticle friction is a major mechanism preventing fluidization, the application of sufficient pressure to a bed will prevent fluidization. Thus, when a fast air stream is applied to a deep bed of materials, only the top layer may fluidize, as the lower layers carry the weight of the upper layers and so are under higher pressure. Conversely, increasing the density of the interstitial fluid will increase particle buoyancy, decrease interparticle normal forces, so reducing friction and allowing fluidization at lower flow rates.

## 10.9 RESIDENCE TIME DISTRIBUTIONS

Residence time distributions are frequency plots of the time spent by each particle in a system. Often, they approximate normal (Gaussian) distributions, and provide information on

- The average time a particle spends in the equipment
- The minimum time, which is important to ensure all particles receive sufficient drying
- The maximum time, important to determine damage to product quality (e.g., by loss of volatiles, production of cooked notes, and overdrying)
- The variation in drying time, to determine the uniformity of the final product

The difference between the maximum and minimum residence time is relevant. Generally, the plant operator prefers a sharp residence time distribution, so that each particle receives a uniform treatment. If all particles have the same residence time, the flow is called plug flow, and each particle will receive the same treatment. However, the design of a fluidized bed favors backward and forward mixing, and so tends to promote the opposite extreme, called a well-mixed bed. Particles can move quickly through the bed, or can move backward and forward randomly for some time before being ejected. Thus, the variation in time in the dryer is large.

Note that the fluidized bed dryer is best suited to the removal of surface water only. Excess drying time past the time required to remove this water has little effect in the short times generally suited to fluidized bed drying. So the problem of large residence time variation is not as serious for a fluidized bed dryer as for other processing operations. However (as discussed later) there are designs of fluidized bed dryers which favor plug flow over well-mixed flow, and generally such designs are favored so as to reduce thermal damage for particles with a longer residence time.

A drying chamber with length about the same as the width tends to be well mixed, and is called a back-mixed FBD, whereas a chamber with a long length and short width gives a more narrow distribution, closer to plug flow. The choice of which design is better suited depends on the specific application, and sometimes a degree of mixing can be an advantage. If the wet particles are sticky and adhere, the presence of drier particles will tend to allow the bed to fluidize properly, by reducing the average interparticle cohesion, and provided only surface moisture is being removed, this should not affect the final product quality excessively. If particle fluidization is not difficult, a longer narrower bed is better suited as it will give a more uniform treatment for the same bed area. Another option if product variation must be minimized would be to use a batch fluidized bed dryer, accepting the reduced thermal efficiency.

## 10.10 EFFECTS OF MAIN PARAMETERS ON A FLUIDIZED BED DRYER

Many factors, such as chamber geometry, distribution plate, and air temperature, will affect the performance of a fluidized bed dryer. In this section, the main factors affecting dryer design are discussed.

### 10.10.1 Air Speed

Assuming the air speed is adequate for fluidization (about twice the minimum fluidization speed), yet well below the Stokes settling speed, it will give a basic operating configuration for the dryer to run correctly, as discussed earlier. Air speed will also affect the rate of drying. However, the effect is not linear. The rate of evaporation from the surface in a fluidized bed dryer is not controlled by the air speed, but is limited by the rate of heat flow by convection to the surface, with the convective heat transfer coefficient, $h$, being the critical term. The dependence of $h$ on air speed is disputed, varying from Chen (2003) who predicts a variation with $v^{1.3}$ (see Equation 10.4), to heat transfer texts, which suggest the variation is $v^{0.37}$. Note that convection heat transfer is the rate-limiting step regardless of whether internal diffusion or surface evaporation is the main resistance to mass flow.

### 10.10.2 Particle Size

If surface resistance is the main barrier to mass transfer, drying time will be in proportion to particle size, determined by the volume-to-area ratio for the particle. Conversely, if internal moisture diffusion is the rate-limiting step, doubling the particle size will quadruple drying time. Most materials will have a low initial resistance to surface evaporation, as moisture is freely available near the surface. However, once the surface moisture is removed, diffusion becomes the rate-limiting step. As a result, for most products the effect of doubling particle size will be from doubling or quadrupling the drying time.

Particle size also has an effect on the type of fluidization, as discussed earlier (see Geldart's classification). Increased air speed may also allow fluidization of a deeper bed of material.

### 10.10.3 Air Temperature

Increasing the air temperature increases the drying rate; again the effect is not proportional but depends on a complex combination of factors. If surface resistance is dominant, increased air temperature increases the surface mass transfer coefficient

$k_y$, and so increases the rate of heat transfer to the surface. According to the Clausius–Clapeyron equation, the capacity of the air to hold moisture also increases exponentially with temperature. Finally, increasing the air temperature will increase the product temperature, which in turn increases the mass diffusivity $D_m$ of the product, resulting in faster release of internal moisture to the surface. The effects will combine to give rapid increases in drying rates with temperature. Unfortunately, high temperatures will also significantly reduce the quality of the product, and so a balance between rate of drying and quality must be carefully maintained.

There is an increase in product surface exposed to the drying air due to fluidization. This initially allows a relatively higher air temperature to be used than static bed drying, since evaporation occurs from a greater product surface area and heat can be transferred more quickly back into the product. In terms of resistance to mass flow and heat transfer, we have reduced the surface resistance, but the internal diffusion resistance. The higher air temperature used in fluidized bed drying does not necessarily affect the product temperature, since the rate of heat loss through evaporation will come to equilibrium with the rate of convective heat transfer at a temperature close to the air wet bulb temperature.

This demonstrates a limitation of the fluidized bed dryer. Although addressing surface flux limitations, its ability to remove moisture trapped internally within the product is limited. In fact, excessive heating to the outer layers of a material may affect its quality, so that once surface moisture is depleted, the product temperature will rise and the product may start to deteriorate if left in the dryer. In this way, the advantages of an FBD in allowing fast, uniform drying may become severe quality disadvantages. To protect the product, the air temperature must now be reduced. Moreover, the thermal efficiency of the fluidized bed dryer starts to reduce with the depletion of surface moisture, and since for drying the exit air relative humidity cannot exceed the product surface water activity, the air leaves unsaturated. Once the product surface water activity has dropped to about 80% (0.8), there is no longer a reason for using the fluidized bed dryer.

In practice, this dilemma is often solved by means of tempering the product moisture in conditioning bins. After tempering (typically for about 8 h), the product is readmitted to the fluidized bed dryer and dried more quickly than without tempering, because the internal moisture has been given time to reequilibrate, specifically to reprovision the surface moisture.

Does this extend the drying time? The product must sit in storage for the tempering period, and so from the product point of view the time required for drying is longer. However, from the dryer's perspective, its thermal efficiency has been improved. If we then provide sufficient fresh or tempered product to keep the dryer running continuously, its utilization is 100% and we must be drying more product than without tempering. Successful operation of the dryer becomes a study in management of fresh and partially dried product through the dryer.

It is from this concept of tempering that the spouted bed dryer (SBD) makes its appeal. If the tempering could be in-built with the dryer, management might be easier. The concept is to fluidize just a small part of the total product mass, allowing an annular outer region to temper (under some aeration) before re-entering a spouting region.

### 10.10.4 The Shape of the Drying Chamber

Early designs of fluidized beds had rectangular drying chambers, a design best suited to easily manufacture. Streamlining the chamber to allow better air flow allows the same drying speeds at reduced fan size and expense. Long narrow drying chambers give better plug-flow properties. Increasing the bed cross-sectional area with height allows precise control of the height of the fluidized bed and reduces entrainment.

## 10.11 PRODUCT CLASSIFICATION FOR FLUIDIZED BEDS

Fluidization is affected by the particle size and particle buoyancy. Based on very careful tests for a range of particle sizes and densities, Geldart (1973) proposed four characteristic groups (in order of particle size):

**Group C**: Very difficult to fluidize. Particles are 20–30 μm and/or cohesive. These are difficult to fluidize as they stick together. Large air bubbles (slugs) move upward through the bed causing the bed to rise and collapse as each air bubble bursts. Fine flour powders behave in this way.

**Group A**: Easy to fluidize. Particles are small (20–100 μm), density <1400 kg/m$^3$. These fluidize easily with air, first by a period of expansion called particulate fluidization, followed by bubbling fluidization (with small uniform air cells).

**Group B**: Easy to fluidize. Particles are 40–500 μm, density 1400–4000 kg/m$^3$. Bubbling starts as soon as the air speed reaches the critical fluidization speed and is vigorous (like boiling), for example, sand particles. At higher speeds, slugging occurs.

**Group D**: Difficult to fluidize. Particles are larger than 600 μm, often with higher densities (density >1400 kg/m$^3$). High fluid speeds are required, which may cause particles to abrade each other as a result. Use of shallow beds is recommended, or an SBD. During fluidization, channeling may occur, and solids do not mix well. Grains and lentils fall in this category.

Note that density in the above categorization refers to relative density, that is, the difference between the true particle density and the fluid density.

Geldart's classification is used extensively in commercial applications of fluidization. Note that grains fall in the group D category: particles that are difficult to fluidize and so require high air speeds to achieve good bubbling fluidization. For this reason, grains are normally fluidized in relatively shallow beds of around 5–10 cm unfluidized (which gives a bed depth of around 20 cm when fluidized). Another way to circumvent the problem of fluidizing grains is to use a spouted bed. Since grains are fluidized at high speeds, each particle will have a large amount of energy, and so there may be impact damage or attrition between the particles (Soponronnarit et al. 2001).

## 10.12 FLUIDIZED BED DRYING

At low speeds, the packed bed is fixed, and air passes through paths of measurable tortuosity, impinging primarily on one surface of the product. However, in a fluidized bed, the particles tumble, presenting different surfaces to the drying effect. Since more surface area is exposed to drying, the rate of heat extraction from the air is also increased, so heat can be supplied to the drying air at a greater rate. The result is a faster and more uniform drying for each particle. In addition, since particles are cycling from top to bottom within the bed, there is more uniform drying over the whole bed. This greater contact between the heating fluid and the bed is the main advantage of fluidized bed drying.

Since a high air speed is required for fluidization, the contact time between the air and particles is comparatively short, so that for drying purposes, the exit air may not be used to its full efficiency. Also, the high air speeds require a larger fan, so that a fluidized bed dryer has a high energy cost. This problem can be reduced by recirculating the exit air. Typically, about 50–90% of the air is recirculated, and 10% is discarded. The exit air may need to be cleaned before recirculation, for example, using a cyclone precipitator to remove dust particles originating from the product.

When air interacts with a product, two fundamental forms of transfer occur simultaneously, heat and mass. For a dryer, mass transfer (specifically loss of moisture from the product) is the desired outcome. For evaporation to occur, the latent heat of evaporation must be provided, and this heat is carried away from the product with the escaping water vapor. No more evaporation can occur unless the heat being removed is continuously replenished to the drying surface. By removing the latent heat from the product surface, a temperature difference is created between the air and the surface. Thus, the heat and mass transfer processes are coupled.

Initially, these mechanisms are not in balance. Consider a cold product that has just entered a dryer, and is still at room temperature. As it comes in contact with hot drying air, heat transfer to the surface will dominate over evaporation of water. As the temperature of the product increases, mass transfer rates increase until the rate of heat transfer to the surface matches the rate of heat loss through evaporation. The product temperature cannot rise higher, since the two rates are now in equilibrium. Similarly, with a hot product, heat transfer would decrease until a balance was reached between heat loss by evaporation and heat gain by convection, based on the convective heat transfer equation:

$$\dot{Q} = hA(T_s - T_\infty) \tag{10.20}$$

A useful correlation for estimating $h$ is that of Kunii and Levenspiel (1991) valid at low $Re$:

$$Nu_g = 0.03Re_p^{1.3} \tag{10.21}$$

For the particular example of a completely saturated surface, the product temperature depends on the air properties only, and is called the wet bulb temperature. If the product is not completely saturated and has a lower water activity, the surface temperature will be higher than the air wet bulb temperature, and closer to the air temperature. Hence, heat transfer will be reduced and the rate of drying decreases as the product dries. This effect can be modeled as (Chen 2003)

$$k_y A(p_v - p_a) = hA(T_a - T_p) \tag{10.22}$$

Here, $p_v$ is the vapor pressure of water at the surface:

$$p_v = a_w p_s \tag{10.23}$$

where $a_w$ is the product water activity and $p_s$ is the saturation vapor pressure at the product surface temperature. If the product water activity is 1, and we make the approximation that air humidity varies linearly with water vapor pressure, this equation becomes the wet bulb equation. For this reason, the product temperature in this equation is sometimes called the pseudo-wet bulb temperature. If $a_w$ is less than 1, the product temperature will be greater than the air wet bulb temperature at the same air conditions.

The initial equilibration period is normally very rapid, since heat transfer is generally much quicker than mass transfer. For this reason, it is called an initial transient. Note that some moisture transfer will occur in the initial transient, until the heat exchange by evaporation equilibrates with heat exchange due to convection. For most practical drying situations, this is a small amount.

There are several consequences of the reduced product temperature due to thermal equilibration:

- The product temperature is less than the inlet air temperature during drying. A wet product comes close to the wet bulb temperature of the air, which at high temperatures is well below the dry bulb air temperature. Thus, high air drying temperatures can be used for food products. For example, for a product that is thermally damaged at 45°C, it may still be possible to use ambient air heated to over 100°C, if the product is at a high water activity (close to 1) and the air wet bulb temperature is below 45°C.
- As the product dries, its water activity decreases, so the rate of evaporation also decreases. From Equation 10.22, this means that the product temperature must increase, and the product may enter a temperature region where thermal damage occurs. Knowing the product and its drying characteristics can help select the maximum safe temperatures to use at different periods of drying. For batch drying, it is necessary to successively reduce temperatures during drying, the overall efficiency of the dryer reducing in proportion. For continuous dryers, the temperature of the product leaving the dryer should be monitored and the air temperature reduced to keep the product within safe thermal limits.
- For a cold product, the exit air will initially be cooled and wetted as it passes over the product. During the initial transient, this air will not be used effectively, but once the product has reached thermal equilibrium with the air, the cooling effect on the air will reduce and its capacity to carry moisture increase. Thus, the exit air will tend to become warmer and more humid over the period of the initial transient.

- Modeling a drying curve that includes data obtained during the initial thermal equilibration period will be incorrect, since for the same drying conditions the initial transient varies depending on the initial product thermal state—a cold product will initially pick up surface moisture, and a hot product will lose a small amount. Forcing an empirical thin-layer drying model to pass through the initial point $M = M_i$ is therefore incorrect unless the product has first been heated to the wet bulb temperature.

After the initial transient, the rate of drying is limited by

- The rate of heat transfer to the surface
- The capacity of the air (especially during the CRP)
- Diffusion of moisture to the surface

### 10.12.1 Surface versus Internal Resistance

The surface and the interior are distinct separate barriers to heat and mass transfer. Each constitutes a resistance to moisture movement, and it is the sum of the two resistances that determines the rate of drying. In most cases, one of the two barriers will be dominant, and we can ignore the other. If the rate of evaporation from the surface is the rate-limiting step, the rate of surface evaporation must be increased. If diffusion within the product is the problem, there are a different set of solutions available. Therefore, it is difficult to solve problems with drying rates without understanding these two resistances in more detail.

Food products typically do not exhibit a CRP (where the product water activity is 100%). However, the fluidized bed dryer operates best when the moisture is available at the surface, and becomes increasingly inefficient during the FRP, which is diffusion-controlled. In effect, the fluidized bed dryer is designed to reduce surface resistance, but cannot affect internal resistance to mass transfer.

Since food products are heat labile, the product cannot be allowed to exceed a certain temperature—time regime. While the product water activity is close to 100%, evaporative cooling brings the product to the air wet bulb temperature, allowing high air temperatures to be used. Once internal resistance starts to limit the availability of water to the surface, the product temperature rises above the air wet bulb temperature, and the air temperature must be successively reduced. The coupling of these two effects (internal resistance and reduced air temperature) results in slower drying and loss of thermal efficiency.

Note the difference in limiting factors for the two important rate periods. As a consequence, during the early stages of drying (but after the initial transient), external convective resistance is dominant, and efforts to improve the drying rate during this period should target known factors of surface evaporation, such as even aeration and increased air speed (see the next section). However, during the FRP, these factors will have less effect.

In practice, the problem of reduced drying rates can be reduced by means of tempering the product moisture in conditioning bins. After tempering (typically for about 4–8 h), the product is readmitted to the fluidized bed dryer and dries more quickly than without tempering, because the internal moisture has reprovisioned the surface moisture. The utilization factor of the dryer is improved at the expense of a greater hold-up volume of the product. The spouted bed version solves the internal resistance problem by incorporating a tempering bin into its design.

### 10.12.2 Decreasing Surface Resistance

Heat must arrive at the surface of the product so as to be available for providing the latent heat of water molecule evaporation. Surface water evaporates at a rate proportional to its availability and the surface area, and recondenses back to the surface. This suggests the following factors are of importance in determining surface resistance to mass transfer:

- The convective heat transfer between the air and the product surface, which is affected by the choice of drying fluid (assumed here to be air), the speed of the air (although this has a relatively small effect above a minimum critical speed, which is normally exceeded in a fluidized bed dryer due to the requirements of fluidization), the temperature of the air, and the size of the particles.
- The temperature difference between the air and the product surface. The rate of heat transfer is proportional to this difference, which is normally the difference between the air temperature and that predicted by Equation 10.22. Generally, dehumidifying the air or increasing the air temperature will affect the temperature driving force.
- The mass transfer coefficient for evaporation from the surface (see Equation 10.22).
- Surface area, which will not normally be a regular shape for food products, such as used by the ideal solutions to Fick's equation for diffusion. Generally the area-to-volume ratio predicted by these assumptions will be 10–50% lower than the actual ratio. One of the main reasons for selecting a fluidized bed dryer is to maximize air/surface contact.
- Water availability at the surface, which is affected by molecular binding, surface polarity, difference between rate of condensation and evaporation, presence of pores or capillaries near the surface and dissolved solutes.

Methods to increase the rate of surface evaporation focus on removing surface water as quickly as possible by increasing air/surface contact, increased temperature, heating, and drying the air. Increasing the air speed has some effect on the drying rate (the mass transfer coefficient increases less than linearly with air speed, but the required fan size increases with the cube of the speed, as can be seen from the kinetic energy term in Bernoulli's equation).

### 10.12.3 Decreasing Internal Resistance

Once the surface has dried, the steps described above become ineffective. Thus, increasing the internal rate of moisture movement is also important. The significant factors are

- The size of the product. The time for diffusion increases between linear and quadratic dependency on the particle size. Thus, doubling the particle thickness may almost quadruple the required drying time. Heat must have time to penetrate into the product, but this is normally fast. The rate-controlling step is normally the water molecules diffusing through pores and capillaries, or through vapor diffusion, or through diffusion–evaporation mechanisms. Some water is occluded within a product in a way that makes it difficult to move, and some is bound chemically to the food matrix.
- The properties of the food that determine the rate of diffusion are summarized in a single term, called the mass diffusivity of the material. To assume one dominant mass diffusivity to characterize a product is unwise, since most food materials have complex nonisotropic (directional) and nonuniform (compositional) structures. For example, a rice grain is normally dried in its hull as "rough grain." The hull itself provides little or no barrier to moisture movement, but below the hull is a bran layer of variable thickness, which is rich in hydrophobic oils, and provides a low mass diffusivity for this one thin layer of the whole product. The average diffusivity of a rice grain is far different, and the varying thickness of this layer causes drying difference within the product, which may lead to cracking of the rice grain during milling.
- Temperature has a strong effect on the diffusivity, which shows a clear Arrhenius temperature dependence in most cases. Increasing the temperature increases the rate of diffusion, so that as with the surface resistance, the operator of the dryer would love to increase the drying rate by increasing the air temperature. However, this must not be at the expense of product quality, which is also strongly affected by temperature.

This suggests the following steps for increasing drying rates during the diffusion-controlled period. The air temperature should be as high as is compatible with the requirements of product quality, taking into account the temperature of the product toward the end of drying when the evaporative cooling effect is at its lowest. The product should be as fine as possible, within the limits of the capacity of the dryer to safely fluidize the product. Resting the product at various intervals is also a sensible technique, allowing time for some of the internal moisture to move to the surface due to internal moisture gradients, a process called tempering. This increases both the utility and the economy of the dryer, since the resting product is held in separate bins and fresh product can be introduced.

## 10.13 ANALYSIS OF FLUIDIZED BED DRYERS

Various techniques exist for analyzing a fluidized bed dryer, ranging from simple to complex. The following approaches have been used:

(1) The continuum model, where the particles and interstitial air are modeled as a uniform continuum. The main equations used are heat and mass

balances, and some form of an empirical drying rate model for individual particles. Two forms of this model are common, one assuming a stationary bed (batch drying), the other assuming both product and air flows through the dryer (Tumambing 1993; Tumambing and Driscoll 1993).

(2) Two-phase analysis, where discrete bubbles are assumed to exist in an otherwise uniform bed of the granular product.

(3) Forces analysis (discrete particle modeling), where thousands of particles are modeled in order to study bubble and slug formation in a two-dimensional fluidized bed (Hoomans et al. 1996). This considers wall and interparticle interactions.

Further details on these models are available in Mujumdar (2006) and other references.

## 10.14 TYPES OF FLUIDIZED BED DRYERS

Designating the class form as a well-mixed fluidized bed dryer, variations from the classic form include batch/continuous, plug flow, pulsed air flow, superheated steam, vacuum, paste and liquid, multistage, rotating, inert gas, microwave, and combination dryers.

### 10.14.1 Classic Form (Well Mixed)

The product enters the drying chamber onto a static mesh screen, which allows the fluidized product to spread out evenly like a liquid on the bed (see Figure 10.3). A gate or weir at one end constrains the product, and the residence time distribution tends to be quite large, as particles can move freely around the bed. The whole bed is at a uniform temperature, and the particle residence time distribution has a large variance, so that products exit with a wide range of moisture contents.

Fresh feed replaces product escaping over the weir, so that the total load on the distribution plate stays constant. The dry particles tend to have lower interparticle

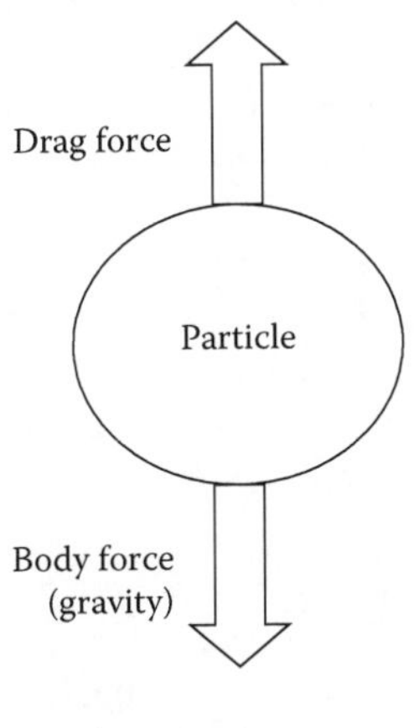


**FIGURE 10.3** Stokes particle separation speed.

cohesiveness than the fresh feed, allowing the bed to fluidize more easily at the expense of the high variability in final product moisture.

Wet feed is often sprayed onto the top of the bed to allow good initial air contact and dispersal among the drier product.

For many years, it was believed that fluidized beds were unsuited to drying paddy due to the high temperatures and propensity to cracking during milling. However, following detailed studies, modeling, and pilot trials, a region of safe operation was found. For paddy drying at air temperatures of around 140–150°C, milling yields were shown to be similar to the paddy dried with conventional low-temperature long-time techniques, such as solar or in-store dryers, and in fact may even increase due to partial gelatinization. This has led to commercial development in Southeast Asia (Soponronnarit and Prachayawarakorn 1994; Soponronnarit et al. 1996a,b; Wangji 1996).

### 10.14.2 Plug Flow

Channels and baffles are added to allow the product to flow along a defined path, giving a more uniform drying effect to the product. The residence time distribution has a lower variance. For this reason, the method of analysis is different than for the classic fluidized bed dryer. A well-mixed dryer can be analyzed by assuming the same conditions throughout the bed (hence the name), but plug flow means that conditions vary along the length of the bed in the product flow direction.

Typical arrangements of guide baffles are

1. Spiral for a circular fluidized bed
2. Rectangular channels for a rectangular or square bed

The plug flow dryer is a better design from the product's point of view, as it gives a more uniform treatment time to each particle. However, from an energy utilization perspective, the air efficiency will vary from high utilization near the product inlet, but lower energy efficiency near the product outlet where the air leaves less saturated. This effect might be reduced by recirculation of the air.

An example of this design was constructed for the Sri Lankan rice industry, and became adopted as the industry standard. The original design was for a fluidized bed dryer with a square fluidization floor, and this was modified to force the product to go through channels added to the square floor. The product uniformity improved, and milling yields, which are sensitive to individual particle moisture content, also improved (Taweerattanapanish et al. 1999; Tirawanichakul et al. 2004).

Sutherland and Ghaly (1992) also analyzed the application of fluidized bed drying for handling second season crops, typically harvested in rainy weather. Their work also studied removal of insects and short-term treatments for paddy arriving at large mills with limited drying capacity.

### 10.14.3 Air Recirculating Dryer

For most fluidized bed dryer designs, the capacity to recirculate the air improves the efficiency of the dryer. Soponronnarit et al. (1996b) showed that recirculating

70–80% of the air improved the dryer thermal efficiency by 50%. Since a fluidized bed dryer operates at high superficial air speeds, the air exiting the drying chamber may not have time to saturate. Mixing fresh ambient air with the exit air recovers some of the heat energy added by the burner, allowing a smaller burner to do the same job and hence saving energy. This should be done with care, since mixing warm moist exit air with cold humid ambient air may lead to supersaturation, allowing moisture to be precipitated and accumulated within the equipment. However, during normal fluidized bed operation, the high air speed and low levels or exit air saturation make this a safe and preferred option for operation.

Since the exit air must be fed back to the burner, the direction of the air needs to be changed from vertically upwards to a direction that matches the fan inlet. This creates a pressure drop, requiring a slightly larger pump to achieve the same air flow and offsetting some of the advantages of recirculation. This is especially true for fluidized bed dryers, where the high air speeds result in high pressure drops with corners, and careful streamlined flow design is necessary.

The recirculated air should also be cleaned, either through a bag filter or by means of a cyclone separator. Again, the design of the cyclone is critical to minimize air resistance, for example, by having large inlet and outlet ducting, and the hold-up volume to allow settling of the particles within the traverse time of the cyclone. This prevents the exit air from contaminating the product, or accumulating fine materials within the equipment.

### 10.14.4 Product Recirculating Dryer

The granular product can be recirculated from the product exit back to the inlet, reducing the average moisture content of the bed, increasing the ease of fluidization for sticky products, and decreasing the final product moisture content. Although not common for food products, recirculation is very important in the chemical industry.

A significant variation on product recirculation is to use sufficiently high air speeds so that the product is entrained by the air circulating through the dryer and fan continuously. Complete entrainment recirculation requires that the fan and heater are not affected by the product, and that no dead spots exist in the recirculation system, which would allow dried product accumulation. For food products, this is rarely the case, and so recirculation by entrainment is rare. However, for a feed with a variety of particle sizes and shapes, lighter particles may be entrained at lower speeds. This can be used to separate finer materials from a mixture, but unless anticipated and controlled, it can cause contamination or even hazardous situations such as accumulation of dried fine material near the heat source.

### 10.14.5 Vibratory Fluidized Bed Dryer

The bed is vibrated using an offset cam driven by a small motor, with the bed mounted on springs on two, three, or all corners to allow vibration of the bed without affecting the rest of the dryer. Vibration allows particles with lower bulk densities to be effectively fluidized at a lower air speed. This technique, for example, is used with drying of tea in India, drying of milk, whey, and cocoa. The effect of vibration is to add mechanical energy to the particles so that they can move independently of each other at lower air speeds, useful for cohesive particles. Vibration also facilitates transport of particles, and for fragile particles allows shallower beds to be used.

This arrangement is also better suited to a feed with a wide range of particle sizes. Normally, fine particles would be entrained in the air flow required to fluidize the larger particles, and vibrating the bed allows a suitable speed for the fine particles to be used, while still providing adequate mechanical energy to mobilize larger particles.

Vibrating the bed may also help to protect fragile products from being damaged by the agitation of the bed due to complete fluidization—the movement of the product will be more gentle, so less attrition. Conversely, vibration may be used to help break up clumps of particles, allowing more uniform treatment. The effect of vibration on minimum fluidization speed has been studied by Gupta and Mujumdar (1983, p. 218).

A variation of the vibratory fluidized bed dryer is the agitated FBD, for example, the Niro swirl fluidizer, which can dry pastes, sludges, and liquids. A rotor in the bed breaks up the feed as it dries, creating a dried powder, which is entrained in the exit air. Instead of the air being pumped vertically through the bed, the air inlet is tangential to create a swirling effect. The final power can be extracted from the drying air using a cyclone or bag filter. The dryer operates continuously.

### 10.14.6 Pulsed Fluidized Bed Dryers

The air flow through a fluidized bed dryer may be pulsed, for example, by having a rotating baffle in the duct work leading to the drying chamber or at the fan outlet. This gives a nonuniform air flow through the chamber, which will lead to rapid bed collapse and re-establishment. For example, rough rice (paddy) grains at high moistures tend to agglomerate, and since moisture transfer for a paddy is dominated by diffusion, this reduces drying effectiveness. Pulsation may improve fluidization quality by allowing lower air speeds (economy) and breaking up air bubbles and conglomerates (for improved mixing). The mechanical effect is stronger than just increasing air speed. Extreme pulsation causes part of the bed to act as a packed bed, and partly as a fluidized bed, at any one time (Gawrzyński and Glaser 1996; Gawrzyński et al. 1996; Moussa and Fowle 1985).

A comparison of the performance of a pulsed and a conventional fluidized bed dryer was made by Prachayawarakorn et al. (2005). A single fluidized bed dryer was constructed, and then a rotating baffle added, which allowed the dryer to be switched between normal and pulsed flow. Various parameters, especially drying rates and energy consumption, were then measured for both configurations. From previous pilot and mathematical modeling studies, the researchers had demonstrated that a fluidized bed dryer could be successfully used to dry rough rice (paddy), and had set limits on the temperature regime, which would deliver an acceptable product. They showed that for a paddy under 28% moisture content, air temperatures above 145°C resulted in significant and observable discoloration and loss of whiteness of the milled rice. Higher temperatures also affected the hardness of the rice after cooking. This determined the conditions used for comparison testing of pulsed versus nonpulsed air.

At a superficial air speed of 1.85 m/s, they found that pulsed air flow was more economical, with an energy saving of ~30%.

Prachayawarakorn et al. (2005) and Nie and Liu (1998) modeled pulsed fluidized beds.

### 10.14.7 Fluidized Bed Superheated Steam Drying

Provided the product is not too heat sensitive, steam may be a suitable way of fluidizing and heating the product. Steam has a large thermal capacity and condenses, releasing this heat very efficiently. Contamination of the steam by the product is a problem, requiring effective cleaning of entrained solids and possibly volatiles from the steam with each cycle. Potentially, the use of steam has a further advantage in that drying produces steam as water vapor from the product, and so a source of steam in the product itself is available, which could supplement direct steam injection if cleaned and repressurized (by centrifugal fan or steam injector).

A generic study of superheated steam fluidized bed drying was made by Tatermoto et al. (2002). Soponronnarit's team studied steam drying in fluidized beds (Taechapairoj et al. 2003a,b, 2004).

Fluidized bed drying was found to be thermally efficient only for a very high moisture paddy, and once the paddy moisture was reduced to about 20%, the FBD became less economic to use. Since thermal damage to the rice at high temperatures was now proven to do little damage, the possibility of using superheated steam for drying was considered, and in 2005, some initial results were published.

The study considered dryer performance and paddy quality aspects, including drying rates, energy consumption, head rice yield, whiteness, pasting properties of rice flour, and visual cell examination. The dryer used a rewetted paddy that was run as a batch dryer.

The initial results did not show a marked improvement in drying characteristics. There were lower drying rates due to initial steam condensation. The head rice recovery was found to be higher due to partial gelatinization, but this also resulted in reduced whiteness due to Maillard reaction browning. The critical air speed required for fluidization was higher, increasing from 1.65 m/s for hot air to 2.6 m/s for steam. This correlates to higher fan power requirements as well.

### 10.14.8 Fluidized Bed Drying Under Vacuum

For heat-sensitive products, the drying chamber, fan, and heat source can operate under a partial vacuum, increasing the fluidization speed to provide the same buoyancy effect on the particles. This allows drying at a lower temperature. Kozanoglu et al. (2006, 2010) showed that decreasing chamber pressure increased the drying rate, and that this effect was more pronounced with increasing bed porosity. Majid and Wahab (2009) studied the rate of drying of an ion exchange resin and pharmaceuticals in a vacuum fluidized bed dryer looking at air pressures from 13 to 65 kPa and temperatures of 30–50°C, demonstrating the strong effect of the product properties (especially size) on drying rates. They concluded that the fluidized bed dryer under vacuum had potential with pharmaceuticals and foods, particularly when the presence of an organic solvent could increase the explosion risk within the equipment, and when the lower temperatures would decrease the thermal damage to a product.

Although reduced pressure increased the minimum fluidization speed, the bed voidage during fluidization was not significantly affected (Llop et al. 1996).

### 10.14.9 Application to Drying Pastes

High-moisture pastes can be difficult to dry if the paste does not readily disperse into particulate solids. The paste could be dried slowly in trays in a belt dryer, or could be dissolved into a liquid feed for a spray dryer. However, it is possible to adapt fluidized bed dryer technology to meet this requirement. The paste is fed into the base of the product bed and broken up by rotating paddles before being picked up by the incoming heated air, carried up the dryer and exiting through the exhaust. The air speed is designed to carry the particles with a sufficient residence time to dry by the time they reach the drying chamber exit. This is considered a form of fluidized bed dryer, since a fluidized bed of particulate solids is formed from the paste by agitation and aeration, but it also combines elements of pneumatic flash drying (Ormós and Blickle 1980).

### 10.14.10 Application to Drying Liquids

SBDs (see Section 10.16) can be modified to handle a liquid feed, and have been used for drying a number of liquid products. Well-mixed fluidized beds can also be modified to dry liquids. The liquid is sprayed onto an inert bed of spherical particles. During fluidization, the liquid coats the particle surfaces and dries. Attrition between the particles then creates a fine powder much like a ball mill, which is then carried away by the fluidizing air and can be separated using a cyclone.

As with SBDs, fluidized beds can also be used to agglomerate powders, producing larger particles but retaining good wetting properties useful for redissolving the powder in water.

### 10.14.11 Multiple-Stage FBDs with Tempering

Fluidized bed dryers are most effective at high moisture contents. As a result, methods for resting (or *tempering*) the product have been studied, for example, the SBD. The tempering process allows moisture from within the product to diffuse top the surface, replacing water evaporated previously, and so increasing the utilization of the dryer. Multiple fluidized bed dryers allow each drying stage to be configured in the best possible arrangement (temperature and air speed) to improve thermal efficiency and reduce thermal damage. Between each stage, the product is stored in tempering bins for several hours, allowing moisture to redistribute. The tempering bins may be gently aerated to cool and keep the product surface skin dry. For details of studies and applications, see Poomsa-ad et al. (2002) and Prachayawarakorn (2004).

### 10.14.12 Rotating Fluidized Bed Dryers

The fluidization speed of a granular material is affected by gravity. Gravity can be artificially increased by arranging the bed around a rotational axis, the fluidizing air being fed to the outside radius of the bed, then inwards toward the central axis (Chen et al. 1999). Centripetal forces may then create a higher body force on the particles than can be achieved by gravity, allowing the fluidization conditions to be adjusted. Higher air flows can be used, without channeling or excessive bubbling.

Chilies were dried in a rotating fluidized bed dryer (Dongbang et al. 2010) in 4 cm layers using air speeds of 1.8 m/s and temperatures of 70–120°C. Although the capsaicin content decreased, the final product quality was comparable with sun-dried chilies. The principle has also been applied to soybeans, green beans, and rice (Chen et al. 1999). Daud (2008) noted that centrifugal fluidized bed dryers are not yet commercially available.

### 10.14.13 Combination Spray and Fluidized Bed Dryers

A liquid feed is distributed using a high-pressure nozzle or rotating atomizer as fine droplets into a concurrent hot air flow, then onto a fluidized bed, where particle agglomeration is allowed. The air flow leaving the bed flows countercurrently through the sprayed particles falling onto the bed, facilitating agglomeration. Complete units are available commercially. The fluidized bed may be nonaerated, using vibration to fluidize.

Granulation or agglomeration between relatively inert particles can be achieved using the combination spray and fluidized bed dryer. The inert particles are fluidized in a stationary bed, and a mist of sprayed liquid used to coat the particles, making the surfaces sticky. As the particle collides with other particles, they stick together, forming larger particles. The flow rate of sprayed liquid must be faster than the air capacity to evaporate moisture.

The system can also be used for particle coating, where high spray flow rates are used to create a uniform coating on the surface of the particles suspended in the bed, which is then dried by the fluidizing hot air flow. To prevent agglomeration, the bed walls are inclined, so that particles move up into the fluidizing zone, then fall back into the drying zone.

### 10.14.14 Fully Enclosed Units Using Inert Gases

Since granular materials have high surface-area-to-volume ratios, food products may be susceptible to deterioration from contact with air, for example, through lipid oxidation, which can produce rancid flavors. Fully enclosed air recirculation fluidized bed dryers can be used, which replace the internal air with inert nitrogen, which has similar moisture and heat transfer properties to air.

### 10.14.15 Microwave Fluidized Bed Dryers

A patent for a combination microwave fluidized bed dryer (CMFD) was submitted in 1968 and subsequently granted. The fluidized bed drying chamber is configured to act as a wave guide for microwave energy, allowing rapid drying times. Normal fluidized beds use convection and conduction (between particles) as the modes of heat transfer. Microwave-assisted FBDs allow quicker drying, so require the product to remain hot for a shorter period of time. In addition, the microwave energy heats the enter product, not just the surface, and so increases the rate of moisture diffusion within the product. Kaensup (2004) studied CMFD drying of peppercorns, finding an increased rate of drying over an FBD by a factor of 10. Better color was retained, and the internal structure was also better preserved, due to the reduced exposure time of the surface to high temperatures in much the same way that microwaved dough is not discolored by the cooking process in the way that a load of bread is browned in an oven.

Emel et al. (2005) studied the drying of macaroni, a form of pasta, which is difficult to dry because of the danger of cracking due to nonuniform drying. In a theoretical study, Chen et al. (2001) studied the effect of microwave distribution patterns on product heating, finding that the shape and distribution of wave patterns affected the drying performance. Creating a uniform microwave field would thus be of greatest importance in ensuring control and final product uniformity. Doelling and Nash (1992) studied the application of microwave FBDs to pharmaceuticals, also finding a highly reduced drying time (sixfold), where the microwave provided over two-thirds of the total energy input.

### 10.14.16 Drying Seed in Fluidized Beds

Seed drying has specific requirements. Grains are dried as seed to provide for the next season's crop, so must be preserved in a way that guarantees viability. Viability can be defined as the fraction of seeds that germinate under certain conditions, but the effect is confounded to some extent by seed dormancy, which means that some form of initial seed activation may be required. Thus, in practice, a viable seed may not always germinate, unless its dormancy has been broken. Generally, viability is strongly affected by heat, and many seeds start to lose viability above 43°C, with the germ itself being vulnerable to moderate to high temperatures. Two-stage drying has been practiced in the seed industry for decades, with Thai companies using slow first-stage drying and fast finishing drying (the reverse of modern practice) since the 1980s. A recent study by Jittanit et al. (2010a) looked at the application of two-stage drying to corn, rice, and wheat, and showed that germination levels were sustained by a combination of either fluidized or spouted bed dryer for the first stage, drying the grains to around 18%, followed by in-store drying at 18–30°C for the second stage. For the higher temperature first stage, temperatures of 40°C were acceptable for rice and corn, but up to 60°C could be used for wheat before significant reduction in germination was observed (Jittanit et al., 2010b). Forms of Giner's model (Jittanit et al. 2009) for germination were found to be a satisfactory description of temperature effects in the fluidized and spouted bed dryers.

### 10.14.17 Other Configurations

Many other modifications of fluidized bed dryer designs exist (see Mujumdar 2000, Soponronnarit 2005, wan Daud 2008), and are mentioned briefly here:

- Supplementary heaters within the fluidization region
- Multiple bed inputs: A fluidized dryer must have a uniform inlet air flow across the whole bed, so that all particles receive the same aeration. To reduce the problem of variation in air flow, distributor plates are often placed below the bed, which with the distributor plate that supports the product, helps to reduce air flow variation. An alternative is to have multiple bed air inlets, with a battery of fans supplying air to multiple stages.
- Floating beds: The air speed through the chamber for a specific air volume flow rate will depend on the bed cross-sectional area. If the bed width is kept constant, the air speed will stay constant, but by sloping the sides of the chamber so that cross-sectional area increases with distance from the distributor plate will cause the air speed to reduce as the air rises above the bed. This can be used to control the bed shape. Near the bed, where the air speed is lowest, particles may tend to move with the air, especially lighter particles. However, as the distance from the bed increases, the difference between the air speed and the Stokes settling speed will increase, allowing the particles to fall back toward the distributor plate again. This will reduce entrainment of lighter particles with the air leaving the drying chamber.

## 10.15 ENERGY EFFICIENCY

Fluidized bed dryers are more efficient when the inlet fluid temperature is high and the exhaust air is close to saturated. For food products, these goals are difficult to be achieved. High temperatures compromise the product by evaporating volatiles (such as aromas and flavors) and by causing chemical reactions (such as protein denaturation, caramelization, and nonenzymic browning), which may give cooked flavors to the product. Thus, efficiency must be balanced against quality, and pilot plant studies coupled with sensory analysis are required to find the best solution. Essence recovery systems such as packed beds may be required, and will reduce the impact of high-temperature drying, but with increased capital and processing cost. Fluidized bed dryers may not be suitable for complete drying of a high-value product, but may still be suitable for the initial phase of drying at high water activity.

In the same way, it is difficult to ensure that the exit air is saturated, although the SBD has a considerable advantage in this area if a high proportion of the inlet air travels through the annular region. Generally, the fluidized bed is too shallow and the inlet air moving too fast to achieve high relative humidities. However, by recirculating 50% of the air, a higher degree of saturation can occur, lowering the volume of air required and reducing the heater size. The air in the dryer cannot get to a higher relative humidity than the product water activity, but air that is close to saturation may cause condensation problems in cooler parts of the equipment, so achieving perfectly saturated air is not a safe goal.

Additional heat can be supplied to a fluidized bed dryer through steam rings, fin and tube heat exchangers, or radiation. Since the product in a fluidized bed is in a highly agitated state, transfer of heat from the heat exchanger surfaces to the product is relatively fast; many FBDs in the chemical industry have this type of facility.

## 10.16 SPOUTED BED DRYERS

### 10.16.1 Introduction

Since coarse particles such as grains are difficult to fluidize, a different approach was developed in the 1950s, called spouted-bed drying. The concept was to create a region of fluidization leading to entrainment in one specific area, called the spout. The rest of the product receives relatively low flow rates, but is still partially fluidized by gravity fall through an annular region, before being directed back to the spout. Thus, an overall circular motion is generated by the injected air.

This led to many potential advantages over other forms of drying:

1. The travel time for particles in the spout is very short, minimizing thermal damage to the product.
2. Time in the annular outer region allows some tempering of the product, so that the time in the spout is used to greatest energy efficiency. This means that larger particles (with a greater mass transfer resistance due to their size) can be efficiently fluidized.
3. Liquids can be dried in an SBD.
4. All of the advantages of the fluidized bed dryer also pertain to the SBD: excellent air product contact for maximum heat and mass transfer, product uniformity, and very small equipment footprint.
5. Since only a portion of the product bed is fluidized, there is a savings in fan power requirement.

For food products, which dry almost exclusively in the FRP, the advantages are clear. By allowing time for the main bed to temper, higher air temperatures can be used in the inlet air, and so higher thermal efficiencies should be achievable. The product is protected from the higher air temperatures by the faster evaporation rates resulting from the tempering process (Becker and Sallans 1961).

The concept was invented and developed by Mathur and Gishler (Mathur 1953, see also Mathur 1974). In a fluidization experiment, most of the distribution plate was blocked off, and a zone of stable fluidization was observed, with the formation of a stable spout that allowed the circulation of material.

SBDs have proved difficult to commercially apply due to one major problem, which is the difficulty of scaling up a laboratory or pilot plant dryer to industrial proportions (Nemeth et al. 1983). Spout instabilities have led to numerous failed dryer designs around the world. Part of the problem is that as the bed depth increases, the pressure of grain on the spout reduces its stability, and the spouted bed becomes effectively a fluidized bed.

A further problem revealed by recent high-speed camera studies was that the residence time distribution in a continuous spouted bed can be highly variable, with some particles being cycled many times through the spout and annulus, and others only once or twice. Modern designs try to alleviate this problem by either high internal circulation rates or specific geometries.

### 10.16.2 Design

The original design for a SBD consisted of four main elements, a heated air supply system, an annular product region, which may be cylindrical or conical-cylindrical (Figure 10.4), a central spout where hot air and product are mixed, and a fountain region above the annular region where product and air separate. Later modifications to this design included rectangular and triangular bins, designed with the intention of reducing the scale-up problems.

As a further addition, solutions may be injected into the annular region and dried as coatings or broken into dried powders. Most commonly, though, they are used for drying of particulate solids.

SBDs may be operated as batch or continuous units. For batch mode, the wet product is loaded, the dryer sealed, and the process begins. Heated air entering the chamber picks up product from the outer region (the dense phase). This creates a spout that reaches from the air inlet to the top of the annular region (the dilute phase). Particles in the spout are carried upward by the air, tumbling and so exposing all surfaces to the drying air. The air speed required to create stable spouting characteristics will be a function of the height of spout required, the inlet air pressure, densities of the dense and dilute phases, and geometric factors that have to do with the ease of a particle entrance to the spouting region. For this (and other) reasons, it is very difficult to scale up from a small working pilot plant dryer to a full commercial

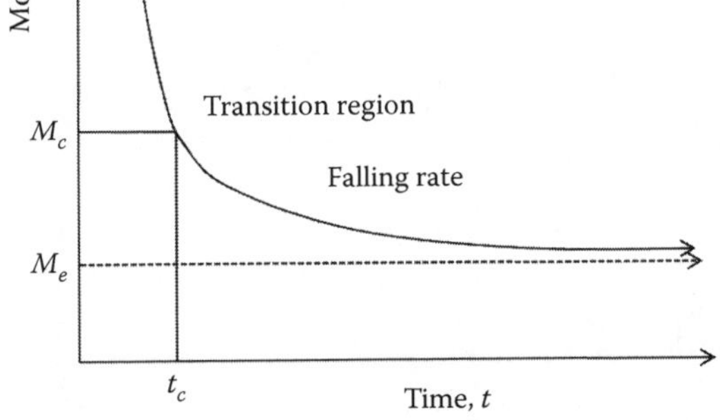


**FIGURE 10.4** Thin layer drying curve.

unit. The drying effect relates to area, the spout and annular regions to volume, and the entrance dimensions to distance, so that doubling one —the area, volume, or distance—does not increase all effects equally.

#### 10.16.2.1 Design Modifications

To improve the original cylindrical chambered design, several design modifications have been proposed over recent years (Mujumdar 2006, Passos et al. 1989).

##### *10.16.2.1.1 Spout Modifications*

The bed height can be doubled to over 2 m by using draft tubes above the drying air inlet, which separate the spout from the product in the annular region and provide a constant shape. Draft tubes also improve circulation and reduce pressure drop (Claflin and Fane 1984).

Slots to the draft tube can be used to modify the proportion of air that goes to the annular outer region, allowing a better balance to be achieved between tempering time in the annular region and drying time in the spout by increasing the drying effect in the annulus.

The air required for the spout can be introduced at an angle, which reduces the pressure drop on the spout, allowing higher units to be constructed (Evin et al. 2008). Similarly, air may be introduced tangentially, creating a vortex near the inlet, which reduces pressure drop and also increases spout stability.

##### *10.16.2.1.2 Drying Chamber Modifications*

Many modifications to the shape of the drying chamber have been suggested. The spouted bed can effectively be made two dimensional by changing the chamber shape to a thin rectangular chamber (side ratio of greater than 10), with the spout at one end, or sloping down to a spout in the middle. The floor slopes downward toward the spout, allowing the product to circulate. The resulting design behaves better in scale-up (Kalwar et al. 1991; Kalwar and Raghavan 1992, 1993). Scale-up is solved by designing one such unit, then combining several units together. Madhiyanon and Soponronnarit (2005) investigated the effect of geometry on the drying rates in different regions of the rectangular unit.

A second such solution was proposed by Mujumdar, which was to use triangular spouted beds, with the spout at one corner, and the annular region sloping down toward the spout as for the rectangular spouted bed. These units can be combined into a hexagonal configuration with the spouts at the center. The smooth walls provided by the drying chamber facilitate good behavior in the spout, and all of the heat is provided to the center of the equipment, improving thermal efficiency (Hung-Nguyen et al. 2001; Go et al. 2007).

##### *10.16.2.1.3 Other Modifications*

The recirculating gas may be inert, for example, nitrogen. This reduces problems of oxygen causing product deterioration.

The dryer may be run in a batch or continuous mode, and product recirculation may be used (at the expense of increasing the residence time distribution of particles in the dryer).

#### 10.16.2.2 Drying Zones

The pressure drop across a spouted bed is less than for a fluidized bed, as not all of the bed needs to receive buoyancy from the inlet air. However, during start-up, the initial pressure will still tend to be very large until the spout can form. Not all of the inlet air travels upward through the spout. A significant amount (which can be controlled by the chamber geometry) travels through the annular region, and may contribute substantially to the drying.

Early studies (Mathur and Gishler 1953) of temperature profiles in a spouted bed showed that the main region for heat uptake of the product was in the spout (especially the base of the spout), as would be expected since the product in the spout is as well mixed with the air similar to a fluidized bed dryer. However, only moisture available to the product surface would be removed in this zone, and the contact time is very short. In the annular region, the heated product comes in contact with lower air speeds, but diffusion at the bed temperature allows the moisture to travel to the surface of the product and then evaporate either in the annular region or at the bottom of the next zone. The proportion of time spent in each region is then a critical function of the rate of diffusion within the product.

Drying models for SBDs must take the two main zones into account—the high-temperature, high-area spouted region behaves differently to the slow air speed, limited contact area, diffusion-controlled annular region. Since the conditions are continually varying for the product, a thin-layer drying model based on constant conditions is not strictly valid, but may provide a useful indication of the drying rate.

#### 10.16.2.3 Applications

SBDs are well suited to larger particles such as grain and lentils. They were first successfully developed for wheat, but due to the problems of scale-up, SBDs have not proved very successful with grain. However, with the geometric solutions now proposed, new research is finding satisfactory solutions and field trials of new designs are again being done. For example, Hung-Nguyen et al. (2001) looked at the development of a field trial SBD using hexagonally arranged triangular chambers for paddy drying in Vietnam, encountering practical difficulties with weather-proofing the unit, but demonstrating the feasibility of the concepts.

Liquids can be dried using a spouted bed. The main chamber is filled with spherical particles that are inert, for example, nylon beads. The size of the particles is chosen as providing good spouting characteristics. The liquid is sprayed into the chamber as a liquid stream or as fine droplets, which coalesce on the surface of the beads forming a thin coating. As the liquid dries on the bead surface, attrition between particles produces a fine dried powder, which is entrained in the fluidizing air and can then be separated by cyclone or bag filter.

A modification of this concept can be used for granulation, which is the combination of fine product particles into larger particles. A spray of liquid coats the surfaces, making them sticky so that particles adhere.

In a similar way, the SBD may be used for coating. The product itself replaces the inert spheres and is coated, where by minimizing attrition and reducing the drying effect, the coated particles become stable after being sprayed.

#### 10.16.2.4 Comparison with Fluidized Bed Dryers

Sometimes, both the fluidized and the more specialized spouted bed dryer are suited to a specific product and choosing which to use may not be obvious. A study on the difference between the FBD and SBD was made by Jittanit et al. (2013) for drying high-moisture seeds. When harvested at around 25–30%, the grain moisture must be reduced quickly to prevent significant deterioration.

The purpose of the study was to find the best option for two-stage drying of seeds, the second stage being an energy-efficient slow method such as in-store drying. The final goal is to reduce the grain to below 14% for 6 month storage or around 12% for 12 months or longer storage. Fast dryers such as FBDs are energy efficient when moisture is available close to the product surface, but are inefficient and can damage lower-moisture grains; seed viability is strongly dependent on kernel temperature being kept low during drying. Studies on head rice recovery in northeast China demonstrated improved milling recovery with the gentler two-stage drying system.

The transition point between the two drying stages is normally chosen around 18% (w.b.), which works well for many grains (those low in oil content). Above 18%, grains deteriorate rapidly, but at around 18%, deterioration has slowed to the point where the grain can be kept for several weeks, and so the slower drying regime is appropriate.

For the moisture range from around 18% and upward, both FBD and SBD dryers proved to be reasonably efficient and comparable with continuous column dryers and cross-flow dryers used for commercial grain storage systems. The study by Jittanit et al. (2013) considered engineering (drying rate and specific energy consumption) and quality aspects (germination rates or viability) of using both. They found that the SBD had a lower specific energy consumption. However, there was little difference in terms of final product viability.

## 10.17 IN-STORE DRYERS

### 10.17.1 History of the Development of In-Store Dryers

In-store drying has been practiced in the United States since the 1950s among maize farmers in the north central region of the country and rice farmers in Texas (Chung et al. 1986). The technique was used mostly on an empirical basis and sometimes resulted in grain deterioration due to air flow rates being too low or drying air temperatures being too high.

In order to determine the appropriate conditions for the use of in-store drying, systematic studies were conducted, resulting in the development of a mathematical predictive model (Thompson 1972) called the "near equilibrium" model. The model could predict changes in grain moisture content, grain temperature, and dry matter deterioration. It took into account factors such as heat transfer through the walls of the bin, respiration of the grain mass, and changes occurring in the grain through continuous aeration. The approach consisted of calculating heat and mass balances across a single layer of drying material. A deep bed was considered a superposition of a number of single layers of material subjected to a stream of drying air perpendicular to each layer. This work was initially done on maize, a major commercial crop in the United States. Farmers were required to reduce the moisture content of maize to 15.5%.

Any deviation from that moisture content attracted a penalty. Maize with a higher moisture content was subjected to a risk of deterioration due to mold and insect activity. Computer simulations based on this model using weather data collected over a long period of time and including temperature and relative humidity from various locations in the maize-growing areas were used to predict the drying time for maize in these locations.

Since in-store drying operates at near-ambient conditions, minor changes in these conditions can have large effects on drying rates and, therefore, careful research was required to ensure that drying air of adequate quality was available each year to be able to move the drying front through the grain bed before a significant deterioration took place in the top layer. As a result of this research, areas in the United States were zoned according to the aeration conditions suitable for in-store drying (Thompson 1972). Schedules for different locations were specified so that grain could be quickly cooled after harvesting and was receiving sufficient aeration for drying. Various systems for improvement in the drying process have been developed over time, particularly for prevention of overdrying the bottom layer, for the reduction of a moisture gradient between the top and bottom layer, for efficient loading and unloading, and also for rewetting of the inlet layer.

In-store drying was later extended to other countries and crops grown in temperate climates. These included wheat in the United States (Barrett et al. 1981), barley in the United Kingdom (Smith and Bailey 1983), and canola in Canada (Muir et al. 1991). Considerable research was undertaken in those countries involving the analysis of weather data and appropriate air flow rates in order to design adequate in-store drying facilities. In Australia, in-store drying was adopted by the rice industry soon after its introduction in the United States (Bramall 1986). It was a logical addition to the conversion from bag to bulk handling in order to deal with the problem of high moisture content at harvest and sun cracking. The usual practice in Australia is to harvest the paddy at about 21% moisture content w.b. and dry it in storage to below 14% for storage and milling. The ambient conditions in the rice-growing region in Australia are favorable for aeration drying and require no supplementary heating. Initially, round bins with central ducting providing radial air distribution and mesh walls were used. They were later replaced by horizontal warehouses (see Figure 10.3) with in-floor or on-floor below were introduced subsequently, for example, layer filling, computer-controlled bins, supplementary heating, and use of computer models in the management of drying and storage operations.

### 10.17.2 Design

A factor favoring the adoption of mechanical drying was the increasing use of large-scale bulk handling equipment for grain. The various types of in-store dryers operating at high or near-ambient air temperature with low energy inputs, but longer drying time are closely integrated in the overall handling capacity of the processing plant. There are various types of in-store dryers (McLean 1989) such as

- Warehouses with in-floor or above-floor aeration ducts (see Figure 10.5)
- Ventilated silos with perforated floors and vertical air flow (round or square) (see Figures 10.6 and 10.7)
- Ventilated silos with separate inlet and exhaust ducts
- Ventilated silos with vertical ducts and radial air flow (see Figure 10.8)

The two-stage drying systems have been developed to take into account different drying rates of grain at various moisture contents. Normally, the first drying stage involves high-temperature, fast drying in order to reduce the moisture content from its harvest level so that the product's water activity falls below 0.80. The moisture targeted by fast drying is concentrated at and near the product surface and can be removed easily. Once the first-stage drying has been completed, the grain is transferred to a storage bin where it is cooled down to ambient temperature and dried further to a safe storage level using air at near-ambient temperature. The second stage of drying is targeting moisture at the center of the kernel and thus this process is mostly diffusion driven. The drying rates are significantly lower than during the first-stage drying, the second-stage drying taking place in the "falling rate" period.

In-store drying (Srzednicki and Driscoll 1994) is synonymous with in-bin drying using air at near-ambient temperature. This technique is used when grain remains in store until being milled or exported, or if drying is considered as the primary purpose of the equipment, with the dried grain being transferred to another bin for aerated storage. The advantage of in-store drying is the increase in throughput and reduction of capital cost per unit dried. There may be a situation when the trader needs a fast turnover of grain to purchase a fresh stock.

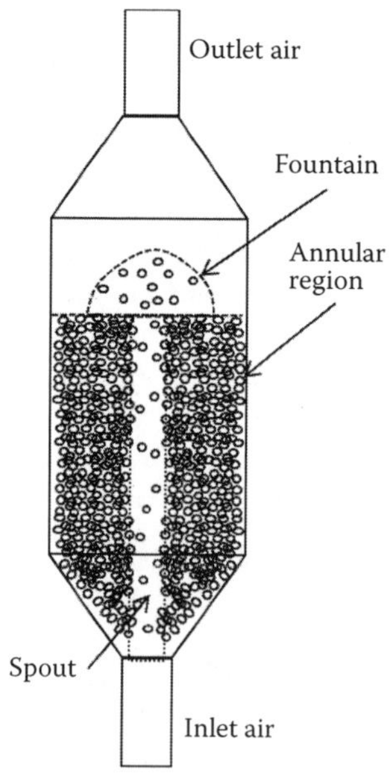


**FIGURE 10.5** Spouted bed dryer.

FIGURE 10.6 Warehouse with on-floor ductings for in-store drying.

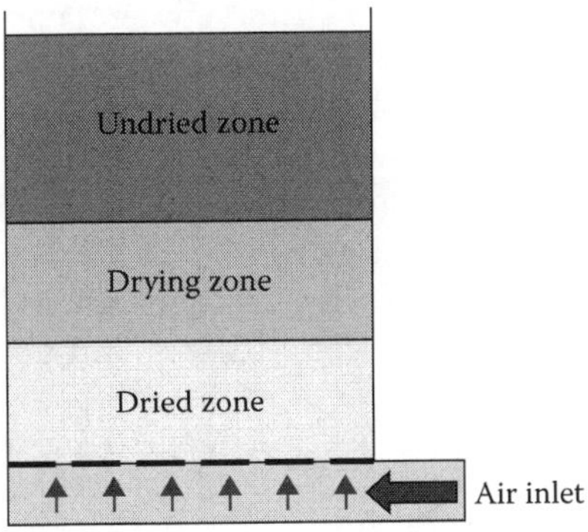


FIGURE 10.7 Silo with perforated floor for in-store drying.

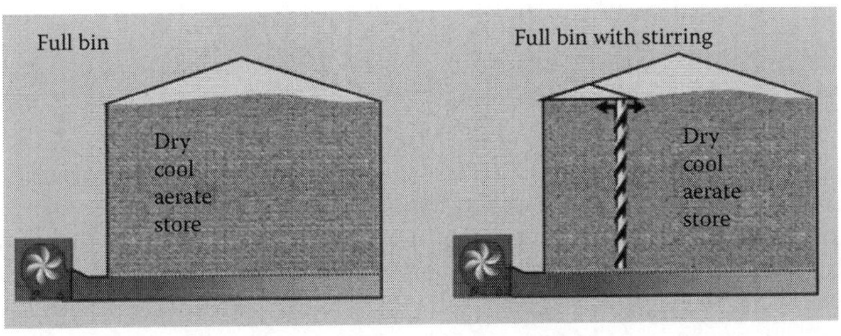


FIGURE 10.8 Silos with vertical airflow for in-store drying.

Two-stage drying has the following advantages (Morey et al. 1981):

- Reduced energy requirements
- Increased drying system capacity
- Improved grain quality

The reduced energy requirement over conventional drying methods is due to the increased air efficiency compared with continuous-flow dryers, so that less heat is vented to the atmosphere. The second point is related to the capacity of the first-stage dryer, since discharge of the grain at higher moisture content before cooling will free the dryer for the next batch of high-moisture grain, which is where the high-temperature dryers are more efficient. The third point relates to the relaxation time given the grain during the second stage of drying that allows moisture gradients within the grain to dissipate, preventing the outer layer of the grain from being overdried and hence made brittle and susceptible to damage.

*Accessories:* Various accessories and operating modes have been developed in order to improve the performance of in-store dryers, generally focusing on grain quality. These include

*Stir augers:* Stir augers are devices for vertical mixing of the grain (see Figure 10.7). They rotate slowly around the bin, taking 1–2 days to complete a circuit. Major advantages are the reduction of moisture gradient between bottom and top layers, and a decrease in compaction of the bottom layers by the top ones, resulting in an improved air flow pattern. The disadvantages are high cost, wear on augers if used with paddy, mechanical damage to grain, and reduced thermal efficiency.

*Dehumidification:* Dehumidification is used with recirculation. There are various methods ranging from cooling the exit air in order to remove moisture by condensation to using chemical desiccants. However, except for high-value crops such as seed, dehumidification has generally not been adopted for grain due to high equipment cost.

*Grain recirculation:* This technique consists of moving partially dried grain from one bin to another or from bottom to top of the bin. There are obvious advantages from this process in terms of tempering, mixing, repackaging, and reduction of compaction, leading to a better air flow through the grain bed.

### 10.17.3 Applications

The weather conditions in humid tropical climates are generally less suitable for in-store drying than in temperate climatic zones. Therefore, initially, the principles of in-store drying were used for tempering of paddy dried in high-temperature dryers by aerating bulk grain in a holding bin with ambient air. However, since the late 1970s, researchers in subtropical and tropical countries began investigating conditions for successful adoption of in-store drying for paddy and subsequently maize, peanuts, and soybeans. Techniques initially developed for temperate countries, particularly optimization achieved by the use of computer models, have been extended to tropical conditions. In order to compensate for high daily relative humidity values, higher drying air temperatures and air flow rates have been introduced.

Extensive research on in-store drying of paddy, based on fundamental studies on deep-bed drying of granular solids conducted in the rice-growing areas since the 1970s, resulted in improved strategies for paddy drying in Australia (Bramall 1986). The following conditions were found to be essential for successful drying:

- Segregation of procured paddy according to moisture contents
- Use of low-speed fans
- Use of aeration strategies taking into account daily fluctuations in weather conditions

These studies were validated on an industrial scale using bins with a capacity of 3000–5000 t and bed heights of 4–7 m.

In order to test in-store drying techniques developed in Australia, collaborative research has been conducted since the early 1980s with scientific partner organizations in Thailand, the Philippines, and Malaysia (Srzednicki and Driscoll 1995). Later, this research was extended to Vietnam, various regions of China and Northeast India. The main objectives of this research were

- To determine thermophysical data for the main grain crops with the aim of designing adequate drying systems for these crops
- To collect weather data in order to be able to simulate a range of drying scenarios occurring in the main grain-producing regions
- To study the effects of various drying strategies on the grain quality

A computer simulation model based on the thermophysical data obtained above was developed by the research team at the University of NSW in Sydney. The model is based on thermodynamic balances between air and grain. Different strategies can be simulated for in-store drying, among them constant aeration, relative humidity control, time control, and modulated burner control. The model makes provisions for options such as recirculation of air, stirring of grain, dehumidification, and heat losses through the walls.

At a later stage, significant research was conducted in northeastern China where wet frozen maize is stored in winter until it can be dried after thawing in spring. Weather data for Northeast China and also the major grain-growing areas of the country were obtained. Thermophysical properties for frozen maize and later for frozen paddy varieties grown in that part of China were determined and weather data obtained (Borompichaichartkul et al. 2003, 2004; Jittanit et al. 2013; Srzednicki et al. 1999, 2012).

Two-stage drying concept has also been adapted to seed drying due to the high germination rate of dried seeds. Nonetheless, the drying temperature must be carefully selected. A drying temperature of 40°C was clearly safe for all samples, whereas more than 90% of wheat seeds still germinated after drying at 60°C in FBD. Furthermore, drying seeds with an initial moisture content of 18% w.b. by ISD was safe under specified drying conditions (Jittanit et al. 2010).

## 10.18 CONCLUSIONS

Two-stage drying is a system that takes into account the different drying behavior of grains at different moisture contents. The first stage of drying involving high-temperature dryers evolved toward use of grain fluidization, with fluidized and spouted bed dryers. There has also been increased use of recirculation of the drying air. More recently, biomass furnaces have reduced the impact of fuel cost on the overall cost of drying. For the second stage of drying, in-store dryers are recommended, which combine drying and storage. Recent work in China has shown that new configurations of in-store dryer design are possible, allowing flexible adaptation to the needs of industry, with radial air distribution and flexible ducting making the system applicable for both silos and horizontal warehouses.

## NOMENCLATURE

| | |
|---|---|
| $a_w$ | water activity |
| c | specific heat, J/kg.K |
| D | particle diameter, m |
| $D_{ac}$ | mass diffusivity between product layers, $m^2/s$ |
| $D_m$ | mass diffusivity, $m^2/s$ |
| g | gravity, 9.81 $m/s^2$ |
| h | specific enthalpy, J/kg |
| $h_c$ | convective heat transfer coefficient, $W/m^2.K$ |
| k | constant |
| $k_y$ | mass transfer coefficient, $kg/m^2.s$ |
| H | air humidity (dry basis), kg/kg d.a. |
| L | distance, length, m |
| m | mass, kg |
| M | moisture (dry basis, i.e., mass water over solids mass), fraction |
| MR | dimensionless moisture ratio |
| P | pressure, Pa |
| Q | heat |
| r | radius, m |
| t | time, s |
| T | temperature, °C |
| $v_p$ | superficial face speed of product through drying chamber, m/s |
| $v_s$ | superficial face speed of air through drying chamber, m/s |
| x | position, m |
| $\epsilon$ | porosity |
| $\rho$ | density (dry basis, i.e., mass of dry solids per unit volume), $kg/m^3$ |
| $\mu$ | viscosity, Pa.s |
| $\lambda_T$ | latent heat at temperature T, kJ/kg |

### SUBSCRIPTS

| | |
|---|---|
| a | air |
| b | bulk |

c critical
e equilibrium
f fluid
o initial
p product or particle
s saturation
w water

### Accents

Dot over symbol rate (e.g., $\dot{Q}$ is rate of heat flow)

## REFERENCES


Barrett, J. R., Okos, M. R. and Stevens, J. B. 1981. Simulation of low temperature wheat drying. *Transactions of the American Society of Agricultural Engineers*, 24, 1042–1048.

Becker, H. A. and Sallans, H. R. 1961. Drying wheat in a spouted bed. *Chemical Engineering Science*, 13(3), 97–112.

Borompichaichartkul, C., Srzednicki, G. S. and Driscoll, R. H. 2003. Drying at sub-zero temperatures: Case study on in-store drying of grains. *Drying Technology*, 21, 4, 735–754.

Borompichaichartkul, C., Moran, G., Srzednicki, G. S. and Driscoll, R. H. 2004. Studies of physical state of water in maize from Northeast China (cv. Huangmo 417) during drying at subzero temperatures. *Drying Technology*, 22, 1&2, 295–305.

Bramall, L. D. 1986. Paddy drying in Australia. *Proceedings of an International Seminar 'Preserving Grain Quality by Aeration and in-store Drying'*, Kuala Lumpur, Malaysia, October 9–11, 1985, eds B. R. Champ and E. Highley, ACIAR Proceedings No 15, 219–223, ACIAR, Canberra.

Chen, G., Wang, W., Yan H. and Wang, Z. 1999. Experimental research on mass transfer in a centrifugal fluidized bed dryer. *Drying Technology*, 17, 1845–1857.

Chen, G., Wang W. and Mujumdar A. S. 2001. Theoretical study of microwave heating patterns on batch fluidized bed drying of porous material. *Chemical Engineering Science*, 56(24), 6823–6835.

Chen, J.C. 2003. Heat transfer. *Handbook of Fluidization and Fluid Systems*, Chapter 10, ed. W.C. Yang, Marcel Dekker, New York.

Chung D. S., Kanuyoso, B., Erickson, L. and Chong, H. L. 1986. Grain aeration and in-store drying in the USA. *Proceedings of an International Seminar 'Preserving Grain Quality by Aeration and Instoredrying'*, Kuala Lumpur, Malaysia, October 9–11, 1985, eds B. R. Champ and E. Highley, ACIAR Proceedings no 15, 67–80, ACIAR, Canberra.

Chung, L. L. and Mujumdar, A. S. 2007. Fluidized bed dryers. In *Handbook of Industrial Drying*, 3rd Ed., ed A. S. Mujumdar, Taylor & Francis, New York, 173–202.

Ciesielczyk, W. 1996. Analogy of heat and mass transfer during constant rate period in fluidized bed drying. *Drying Technology*, 14(2), 217–230.

Claflin, J. K. and Fane, A. G. 1984. Fluid mechanics, heat transfer, and drying in spouted beds with draft tube. In *Drying '84*, ed A. S. Mujumdar, Hemisphere Publ. Corp., New York, NY, USA, 137–141.

Doelling, M. K. and Nash, R. A. 1992. The Development of a microwave fluid-bed processor. II. Drying performance and physical characteristics of typical pharmaceutical granulations. *Pharmaceutical Research*, 9(11), 1493–1551.

Dongbang W., Pirompugd, W. and Triratanasirichai, K. 2010. The drying kinetics of chillies using a rotating fluidized bed technique. *American Journal of Applied Science*, 7(12), 1599–1606.

Emel, I. G., Sumnu, G. and Esin, A. 2005. Effect of microwave on fluidized bed drying of macaroni beads. *Journal of Food Engineering*, 66 (4), 463–468.

Ergun, S. and Orning, A. A. 1952. Fluid flow through packed columns. *Chemical Engineering Progress*, 48, 89–94.

Evin, D. I., Gu, H. and Tanyildizi, V. 2008. Grain drying in a paraboloid-based spouted bed with and without draft tube. *Drying Technology*, 26(12), 1577–1583.

Gawrzyński, Z. and Glaser, R. 1996. Drying in a pulsed fluid bed with relocated gas stream. *Drying Technology*, 14(5), 1121–1172

Gawrzyński, Z., Glaser, R., Zgorzalewicz, J., Pelech, Z., Stanislawski, J., Rogula, G. and Pieczaba, B. 1996. Operational tests of a pulsed fluid bed dryer/cooler for granulated sugar. In *Proceedings of the10th International Drying Symposium*, ed. C. Strumillo and Z. Pakowski, Krakow, Poland, 30 July–2 August, 771–777.

Geldart, D. 1973. Types of gas fluidization. *Powder Technology*, 7(5), 285–292. http://dx.doi.org/10.1080/07373939908917656.

Ghaly, T. F. and Sutherland, J. W. 1984. Heat damage to grain and seeds. *Journal of Agricultural Engineering Research*, 30, 337–345.

Go, A., Das, S. K., Srzednicki, G. and Driscoll, R. H. 2007. Modeling of moisture and temperature changes of wheat during drying in a triangular spouted bed dryer. *Drying Technology*, 25(4), 575–580.

Gupta, R. and Mujumdar, A. S. 1983. Recent developments in fluidized bed drying. In *Advances in Drying*, Vol. 2, ed A. S. Mujumdar. Hemisphere Publishing, New York, NY, USA, 155–192.

Holdich, R. 2012 *Fundamentals of Particle Technology*, Midland Information Technology and Publishing, West Bridgford, Nottingham, UK. http://www.particles.org.uk/particle_technology_book/

Hoomans, B. P. B., Kuipers, J. A. M. Briels, W. J., and van Swaaij, W. P. M. 1996. Discrete particle simulation of bubble and slug formation in a two-dimensional gas-fluidised bed: A hard-sphere approach. *Chemical Engineering Science*, 51(1), 99–118.

Hung-Nguyen, L., Driscoll, R. H. and Srzednicki, G. 2001. Drying of high moisture content paddy in a pilot scale triangular spouted bed dryer. *Drying Technology*, 19(2), 375–387.

Jittanit, W., Srzednicki, G. and Driscoll, R. H. 2009. Germination models for seeds dried in fluidized and spouted bed dryers. *Seed Science and Technology*, 37, 180–191.

Jittanit, W., Srzednicki, G. and Driscoll, R. H. 2010a. Corn, rice, and wheat seed drying by two-stage concept. *Drying Technology*, 28(6), 807–815, doi: 10.1080/07373937.2010.485081

Jittanit, W., Srzednicki, G. and Driscoll, R. H. 2010b. Seed drying in fluidized and spouted bed dryers. *Drying Technology*, 28(10), 1213–1219.

Jittanit, W., Srzednicki, G. and Driscoll, R. H. 2013. Comparison between fluidized bed and spouted bed drying for seeds. *Drying Technology*, 31(1), 52–56.

Kalwar, M. I., Kudra, T., Raghavan, G. S. V. and Mujumdar, A. S. 1991. Drying of grains in a drafted two dimensional spouted bed. *Journal of Food Process Engineering*, 13, 321–332.

Kalwar, M. I. and Raghavan, G. S. V. 1992. Spouting of two-dimensional beds with draft plates. *Canadian Journal of Chemical Engineering*, 70, 887–894.

Kalwar, M. I. and Raghavan, G. S. V. 1993. Batch drying of shelled corn in two-dimensional spouted beds with draft plates. *Drying Technology*, 11(2), 339–354.

Kaensup, W. and Wongwises, S. 2004. Combined microwave/fluidized bed drying of fresh peppercorns. *Drying Technology*, 22 (4), 779–794.

Kozanoglu, B. U., Davila Nava, I. R. and WeltiChanes, J. 2006. Influence of effective particle porosity on vacuum fluidized bed drying. *Chemical Engineering and Technology*, 29, 384–389, http://dx.doi.org/10.1002/ceat.200500021

Kozanoglu B. U., J. A. Vilchez, J. Casal and J. Arnaldos. 2010. Drying of solids in vacuum fluidized bed. *The Canadian Journal of Chemical Engineering*, 80(3), 376–385.

Kunii, D. and Levenspiel, O. 1991. *Fluidization Engineering*, 2nd Ed., Butterworth-Heinemann, Boston, MA, USA.

Llop, M.F., Madrid, F., Arnaldos, J. and Casal, J. 1996. Fluidization at vacuum conditions. A generalized equation for the prediction of minimum fluidization velocity. *Chemical Engineering Science*, 51(23), 5149–5157.

Looi, Y. A., Mao, Q. M. and Rhodes, M. 2002. Experimental study of pressurized gas-fluidized bed heated transfer. *International Journal of Heat and Mass Transfer*, 45, 255–265.

Madhiyanon, T. and Soponronnarit, S. 2005. High temperature spouted bed paddy drying with varied down comer air flows and moisture contents: Effects on drying kinetics, critical moisture content, and milling quality. *Drying Technology*, 23(3), 473–495.

Majid I., Wahab, A. and Rheima, K. H. 2009. Drying of solid materials by vacuum fluidized bed dryer. *Iraqi Journal of Chemical and Petroleum Engineering*, 10(4), 1–11.

Mathur, K. B. and Epstein, P. E. 1974. *Spouted Beds*, Academic Press, New York.

Mathur, K. B. and Gishler, P. E. 1953. Wheat drying using the spouted bed technique. Unnumbered National Research Council of Canada Report, Ottawa, Ontario.

McCabe, W. E, J. C. Smith and P. Harriott. 2001. *Unit Operations of Chemical Engineering*. McGraw-Hill, New York, NY, USA.

McLean, K. A. 1989. *Drying and Storing Combinable Crops*, 2nd Ed., Farming Press Books, Ipswich, UK.

Morey, V. R., Gustafson, R. J. and Cloud, H. A. 1981. Combination high-temperature ambient-air drying. *Transactions of the American Society of Agricultural Engineers*, 21, 509–512.

Moussa, N. A. and Fowle, A. A. 1985. Exploratory study of pulsed atmospheric fluidized-bed combustion. *Proceedings of Eighth International Conference on Fluidized-Bed Combustion*, 3, 1300–1310, Texas.

Muir, W. E., Sinha, R. N., Zhang, Q. and Tuma, D. 1991. Near-ambient air drying of canola. *Transactions of the American Society of Agricultural Engineers*, 34, 2079–2084.

Mujumdar (Ed), A. S. 2006. *Handbook of Industrial Drying*, 3rd ed., CRC Press, Boca Raton, Florida, 453.

Mujumdar, A. S. and Suvachittanont, S. 2000. *Developments in Drying*, Vol. 2, Kasetsart University Press, Bangkok, Thailand.

Nemeth J., Pallai, E. and Aradi, E. 1983. Scale-up examination of spouted bed dryers. *Canadian Journal of Chemical Engineering*, 61, 419–425.

Nie, Y. and Liu, D. 1998. Dynamics of collapsing fluidized beds and its application in the simulation of pulsed fluidized beds. *Powder Technology*, 99, 132–139.

Ormós, Z. and Blickle, T. 1980. *Drying of Pastes in Fluidized Bed (The Fluidized Bed Grinding-Drier), Drying 80, Vol.1., Developments in Drying*. Hemisphere Publ. Corp., Washington, New York, London, 200–204.

Passos, M. L., Mujumdar, A. S., Vijaya, G. and Raghavan, V. G. S. 1989. Spouted and spouted-fluidized beds for grain drying. *Drying Technology*, 7(4), 663–696.

Poomsa-ad, N., Soponronnarit, S., Prachayawarakorn, S. and Terdyothin, A. 2002. Effect of tempering on subsequent drying of paddy using fluidisation technique. *Drying Technology*, 20(1), 195–210.

Prachayawarakorn, S., Soponronnarit, S., Wetchacama, S. and Chinnabun, K. 2004. Methodology for enhancing drying rate and improving maize quality in a fluidized-bed dryer. *Journal of Stored Products Research*, 40(4), 379–393.

Prachayawarakorn, S., Tia, W., Poopaiboon, K. and Soponronnarit, S. 2005. Comparison of performances of pulsed and conventional fluidized-bed dryers, *Journal of Stored Products Research*, 41, 479–497.

Richardson, J. F. and Zaki, W. N. 1954. Sedimentation and fluidisation. Part 1. *Transactions of the Institution of Chemical Engineers*, 32, 35–53.

Rowe, P. N., Partridge, B. A. and Lyall, E. 1964. Cloud formation around bubbles in gas fluidized beds. *Chemical Engineering Science*, 19, 973–985.

Smith, E. A. and P. H. Bailey. 1983. Simulation of near-ambient grain drying. II: Control strategies for drying barley in Northern Britain. *Journal for Agricultural Engineering Research*, 28, 301–317.

Soponronnarit, S. and Prachayawarakorn, S. 1994. Optimum strategy for fluidized bed paddy drying. *Drying Technology*, 12(7), 1667–1686.

Soponronnarit, S., Prachayawarakorn, S., and Sripawatakul, O. 1996a. Development of cross-flow fluidized bed paddy dryer. *Drying Technology*, 14, 2397–2410.

Soponronnarit, S., Prachayawarakorn, S. and Wangji, M. 1996b. Commercial fluidized-bed paddy dryer. In: *Proceedings of the 10th International Drying Symposium*, eds C. Strumillo and Z. Pakowski, Vol. A, Krakow, Poland, 30 July–2 August 1996, 638–644.

Soponronnarit, S., Wetchacama, S., Trutassanawin, S. and Jariyatontivait, W. 2001. Designing, testing and optimization of vibro-fluidized paddy dryer. *Drying Technology*, 19, 1891–1908.

Srzednicki, G. S. and Driscoll, R. H. 1994. Advances in research on in-store drying, *Proceedings of the 6th International Working Conference on Stored-Product Protection*, Canberra, April 17–23, 1994, eds E. Highley, E. J. Wright, H. J. Banks and B. R. Champ, CAB International.

Srzednicki, G. S. and Driscoll, R. H. 1995. Adoption of in-store drying technology in Southeast Asia. *Proceedings of the 17th ASEAN Technical Seminar on Grain Postharvest Technology*, Lumut, Malaysia, 25–27 July 1995, 8pp.

Srzednicki, G., Driscoll, R. and Niu, X. 1999. In-store drying under different climatic conditions in China. *Proceedings of the 19th ASEAN (Association of South-East Asian Nations) 1st APEC Seminar on Postharvest Technology*, Ho Chi Minh City, Vietnam, November 9–12, 1999, eds G. I. Johnson, L. V. To, N. D. Duc, and M. C. Webb, 296–302.)

Srzednicki G., Driscoll, R. H., Niu, X. and Guo, D. 2012. Strategies for in-store drying based on the analysis of weather data—Case study: Key grain storing regions of China *Drying Technology*, 30(16), 1863–1869, Doi: 10.1080/07373937.2012.690802.

Sutherland, J. W. and Ghaly, T. F. 1992. Rapid fluid-bed drying of paddy rice in the humid tropics. *Proceedings of the 13th ASEAN Conference on Grain Post-harvest Technology*, Brunei Darussalam.

Tavoulareas, S. 1991. Fluidized-bed combustion technology. *Annual Reviews Inc.*, 16, 25–27.

Taechapairoj, C., Prachayawarakorn, S., and Soponronnarit, S. 2003a. Superheated steam fluidized bed paddy drying. *Journal of Food Engineering*, 58, 67–73.

Taechapairoj, C., Dhuchakallaya, I., Soponronnarit, S., Wetchacama, S., and Prachayawarakorn, S. 2003b. Superheated steam fluidized bed paddy drying. *Journal of Food Engineering*, 58, 67–73.

Taechapairoj, C., Prachayawarakorn, S., and Soponronnarit, S. 2004. Characteristics of rice dried in superheated-steam fluidized-bed. *Drying Technology*, 22(4), 719–743.

Tatermoto, Y., Mawatari, Y., Sakurai, K., Noda, K. and Komatsu, N. 2002. Drying characteristics of porous material in superheated steam fluidized bed. *Proceedings of the 13th International Drying Symposium*, Vol. A, 624–633, Beijing, China.

Tirawanichakul, S. Prachayawarakorn, S. Varanyanond, W. Tungtrakul, P. Soponronnarit, S. 2004. Effect of fluidized bed drying temperature on various quality attributes of paddy. *Drying Technology*, 22(7), 1731–1754.

Taweerattanapanish, A., Soponronnarit, S., Wetchacama, S., Kongseri, N. and Wongpiyachon, S. 1999. Effect of drying on rice yield using fluidization technique. *Drying Technology*, 17(1&2), 345–353.

Thompson, T. L. 1972. Temporary storage of high moisture shelled corn using continuous aeration. *Transactions of the American Society of Agricultural Engineers*, 15, 119–129.

Tumambing, J. A. 1993. *Mathematical modelling and design of a continuous fluidized bed dryer for pre-drying of paddy in the humid tropics*. Ph.D. Thesis, The University of New South Wales, Sydney, Australia.

Tumambing, J. A. and Driscoll, R. H. 1993. Modelling the performance of continuous fluidized bed paddy dryer for rapid pre-drying of paddy. *Proceedings of the 14th ASEAN Seminar on Grain Post-harvest Technology*, 193–213, Manila, Philippines.

wan Daud, W. R. 2008. Fluidized bed dryers—Recent advances. *Advanced Powder Technology*, 19(5), 403–418.

Wangji, M. 1996. *Development of industrial scale fluidized bed paddy dryer*. Master's Thesis, King Mongkut's University of Technology, Thonburi, Bangkok, Thailand, 197.

Wen, C. Y. and Yu, Y. H. 1966. *Mechanics of Fluidization*. Chem. Eng. Prog. Symp. Series, 62, 100–111.

Yang, W. C. 2003. *Handbook of Fluidization and Fluid–Particle Systems*, Marcel Dekker, New York, NY, USA.

Yates, J. G. 1983. *Fundamentals of Fluidized-Bed Chemical Processes*. Butterworths, London, UK.

# 11 Fermentation and Enzymes

*Constantinos Katsimpouras, Paul Christakopoulos, and Evangelos Topakas*

## CONTENTS



## 11.1 INTRODUCTION

Fermented foods are food substrates that have been subjected to the action of edible microorganisms whose enzymes hydrolyze the polysaccharides, proteins, and lipids to nontoxic products with flavors, aromas, and textures pleasant and attractive to the human consumer (Steinkraus 2002). The term "fermentation," from the Latin word *fervere*, in a more rigorous way describes a form of energy-yielding metabolic process from organic compounds, usually carbohydrates, without the involvement of an exogenous oxidizing agent (Bourdichon 2012). Fermentation is one of the oldest forms of food preservation as the origins of fermented foods in our diets date back many thousands of years (Campbell-Platt 1994). Chemical evidence enables the production of a fermented beverage from the Neolithic village of Jiahu in China to be traced back as far as 7000 BC (McGovern et al. 2004). The earliest evidence of wine production, dated to 5400–5000 BC, was found at the Hajji Firuz Tepe site in the northern Zagros Mountains in Mesopotamia (This et al. 2006). Similar evidence comes from the remains of juice grape extraction in the Neolithic site of Dikili Tash in Greece at the same period (Valamoti et al. 2007). Systematic analysis by microscopy of Egyptian bread loaves derived from several sites that span approximately 2000–1200 BC reveals their similarity to that of modern cereal foods (Samuel 1996). Fermentation processes, over the past half century, have improved to a great extent and have reached a point where desired compounds can be produced by carefully selecting starter bacteria and acquiring deep knowledge of the procedure and, hence, regulation of conditions (van Boekel et al. 2010).

Enzymes are commonly used in the production of food ingredients and in food processing. The industrial production of such enzymes dates back to 1874, where Christian Hansen, a Danish scientist, extracted rennin (chymosin) from calves' stomachs for use in the manufacture of cheese (Nielsen et al. 1994). Nowadays, chymosin is produced by recombinant bovine prochymosin expressed by *Escherichia coli* K-12 bacterium. This enzyme was the first recombinant approved for use in food processing by the US Food and Drug Administration (FDA) (Flamm 1991). Another group of enzymes called pectinases have been used for juice clarification since 1930s, while invertase was used during World War II for the production of sugar syrup, applying for the first time in history immobilization techniques on an industrial scale (Vasic-Racki 2006). The industrial use of enzymes for food processing was established in the 1960s, when the traditional starch acid hydrolysis was substituted by biochemical hydrolysis using amylases and amyloglucosidases together with glucose isomerases (see Fernandes 2010 for a review). The present review is aimed at a brief overview of both the fermentation processes and enzymes used in food processing and engineering.

## 11.2 FERMENTATION

Among the first processed foods, produced and consumed by humans, were fermented foods. Their popularity stems from the desirable features they present such as extended shelf life, improved safety, enhanced nutritional value, and special sensory characteristics. However, nowadays, fermentation is more about attributing unique flavors to foods in a totally natural way. In the literature, there are several classifications of fermented foods: according to the microorganism involved, the fermentation type, and so on, depicting this way authors' different points of view.

Lactic acid bacteria (LAB) are the main microorganisms employed in most fermentation processes such as dairy products, vegetables, and meat fermentations. Yeasts are also involved especially in alcoholic beverage fermentations. Moreover, microorganisms, defined as probiotics, have been included in many fermented foods, also known as fermented functional foods, due to their health-promoting effects.

Recently, as the application of new tools such as high-throughput sequencing technologies, phylobiomics, metagenomics, and metatranscriptomics contribute to acquiring a deeper knowledge of food fermentation and fermenting microbial ecosystems; the modern food industry is capable of up-scaling and designing more efficient fermentation processes.

### 11.2.1 Classification of Fermented Foods

Fermented foods can be classified in different ways, for example, by categories according to the kind of microorganism involved (Yokotsuka 1982), by classes, by commodity, and so on. Sudanese traditionally classify their foods on a functional basis such as porridges and breads (Kissar), sauces and relishes for the staples (Milhat), alcoholic drinks (marayiss), and food for special occasions (Akil-munasabat) (Dirrar 1993). Campbell-Platt (1987) classified fermented foods by classes such as beverages, dairy products, meat and fish products, vegetables and fruits, cereal products, and starch crop products. Steinkraus (2002) suggested the following classification based on fermentation: (i) protein-rich meat substitutes produced from vegetables, such as Indonesian tempeh an ontjom, (ii) high-salt, meat-flavored sauces and pastes, such as Chinese soy sauce, (iii) lactic acid-fermented foods, such as pickles, sauerkraut, kefir, table olives, and so on, (iv) alcoholic beverages, such as various types of wines and beers, Japanese sake, and so on, (v) alkaline-fermented foods, such as Japanese natto, (vi) vinegars, (vii) leavened breads, such as Western yeast and sourdough breads, and (viii) flat unleavened breads.

### 11.2.2 Microorganisms Used in Fermentation in Food Industry

#### 11.2.2.1 Dairy Products

The fermentation of dairy products represents one of the most important segments of the food industry. From milk, yogurt, cheese, and artisanal dairy products to the probiotic ones, fermentation prolongs shelf life, improves the taste, texture, flavor, and nutritional properties, and benefits human health.

LAB such as *Lactobacillus*, *Lactococcus*, *Leuconostoc*, *Pediococcus*, and *Streptococcus* are usually present in fermented dairy products and can be divided into two groups: mesophilic and thermophilic bacteria, based on their optimum growth conditions. LAB owe their designation to their ability to ferment sugars primarily into lactic acid. A scientific explanation for the beneficial effects of LAB was first given by the Russian bacteriologist Eli Metchnikoff who supported that LAB resulted in the displacement of toxin-producing bacteria normally present in intestinal flora (Lourens-Hattingh and Viljoen 2001).

One of the most popular dairy products is yogurt. Mixed cultures of *Streptococcus thermophilus* and *Lactobacillus bulgaricus* are responsible for yogurt production by the fermentation of pasteurized milk. Another fermented dairy product is kefir, a traditional acidic-alcoholic beverage that was originated in the Balkans, Eastern Europe, and the Caucasus (Fontán et al. 2006). Kefir is produced by inoculating milk with kefir grains (Magalhães et al. 2010). Kefir grains are irregular granules, with sizes that vary from 3 to 35 mm in diameter, where the polysaccharide kefiran plays the role of a matrix on which LAB and yeast are retained. The microflora present in kefir grains depends mainly on their source. It has been reported that kefir grains contain lactobacilli, lactococci, leuconostoc, yeasts (*Kluyveromyces lactis, Kluyveromyces marxianus, Torula kefir* and *Saccharomyces cerevisiae*) and sometimes acetic acid bacteria (Güzel-Seydim et al. 2005).

LAB are also employed in cheese making, taking part in both the fermentation process and cheese ripening. They are called starter and nonstarter LAB, respectively. Mesophilic species such as *Lactococcus lactis* and *Leuconostoc* spp. and thermophilic species such as *S. thermophilus*, *Lactobacillus delbrueckii*, and *Lactobacillus helveticus* are among starter LAB, while a more heterogeneous pattern appears for those belonging to nonstarter LAB (Settani and Moschetti 2010). Some varieties of cheese owe their unique characteristics to molds, yeasts, and bacteria other than LAB. A typical example is the saprophytic fungi *Penicillium roqueforti* that is used in blue-veined cheese production (Caplice and Fitzgerald 1999).

#### 11.2.2.2 Meat Products

Meat products have been consumed for centuries in many different countries all over the world and are a major part of the human diet. However, fresh meat is a highly nutritious medium that supports the growth of various microorganisms if not appropriately preserved, resulting in rapid deterioration. One of the first developed ways of preserving meat was drying, with smoking and the addition of salt following (Hutkins 2006). Moreover, another efficient way of meat preservation that has a very old history is fermentation.

The first fermented sausages originated in the Mediterranean region where the climate provides the favorable conditions for the process, such as the proper temperature and relative humidity (Lücke 1994). Sausages were initially invented as a means of exploiting meat leftovers. Fermented sausages nowadays are often called *salami* after the ancient Greek town of Salamis on the Cyprian coast (Zeuthen 2007). Chopped or ground lean meat, comminuted fat, salt, curing agents, sugar, and spices are the main ingredients of fermented sausages. The mixture consisting of the mentioned ingredients is stuffed into casings and is allowed to undergo a lactic fermentation during a drying process (Hugas and Monfort 1997). Quality, consistency, extended shelf life, enhanced sensory characteristics, and safety are crucial elements for the modern meat industry, necessitating the use of starter cultures in order to standardize the fermentation process. Starter cultures containing LAB and coagulase negative cocci (CNC) are commercially available and are the two main groups of bacteria that are considered technologically important in the fermentation of sausages. Additionally, LAB along with *Staphylococcus* are used when nitrite is utilized as the sole curing agent (Lindner et al. 2013). The main LAB species isolated from fermented sausages are *Lactobacillus*, *Pediococcus*, *Leuconostoc*, *Weisella*, and *Enterococcus* (Ammor and Mayo 2007) with *Lactobacillus plantarum*, *Lactobacillus sakei*, *Lactobacillus curvatus*, and *Staphylococcus xylosus* being the most well adapted to the sausage fermentation ecosystem (Rantsiou and Cocolin 2006; Leroy et al. 2006).

### 11.2.2.3 Vegetable and Fruits

Vegetables and fruits have a high nutritional value and are sources of vitamins, minerals, dietary fibers, phytosterols, and phytochemicals for the human diet. The majority of vegetables and fruits is consumed fresh or minimally processed as they are perishable and susceptible to spoilage microorganisms (Di Cagno et al. 2013). Fermentation is among the other methods of vegetable processing such as canning, freezing, drying, and chemical preservation and is considered a simple and valuable method to improve safety and shelf life, and in some cases contribute to the sensory and nutritional characteristic enhancements of vegetables. Most vegetable fermentations occur as a consequence of providing the appropriate environmental conditions necessary for LAB growth (Caplice and Fitzgerald 1999).

Fermented vegetables that are of great interest for the food industry and are produced on a large-scale basis are sauerkraut, pickles, olives, and Korean kimchi. Cabbage is widely used as a fermentation substrate with sauerkraut and kimchi being in the limelight. Sauerkraut, meaning "sour cabbage" in German, is manufactured from fine shredded, salted, white cabbage through lactic acid fermentation (Wacher et al. 2010). It is also well known for its antiscurvy properties as cabbage is rich in vitamin C (Hutkins 2006). Two distinguished stages are responsible for the final result in sauerkraut fermentation: an early heterofermentative stage, which is followed by a homofermentative one. The main species that are present in sauerkraut fermentation are *Leuconostoc mesenteroides*, *Pediococcus pentosaceus*, *Lactobacillus brevis*, and *L. plantarum* (Plengvidhya et al. 2004).

The Korean version of sauerkraut is called kimchi. Kimchi is a traditional Korean food, fermented from Chinese cabbage and radish, and is usually processed with various seasonings such as red pepper powder, garlic, ginger, green onion, fermented seafood, and salts (Jung et al. 2011). It is also remarkable that in Korea, the per capita consumption of kimchi is more than 43 kg per year (Hutkins 2006). Kimchi belongs to lactic acid-fermented vegetable products and is considered to be an important source of potentially beneficial and useful LAB (Lee et al. 2011). *L. mesenteroides* starts the fermentation, and as lactic acid increases, it gives its place to more acid-tolerant species such as *L. brevis*, *L. plantarum*, and *L. sakei* (Di Cagno et al. 2013). Other species that are responsible for kimchi fermentation are the *Leuconostoc* species *Leu. kimchii*, *Leu. citreum*, *Leu. gasicomitatum*, and *Leu. gelidum*, *L. curvatus*, *Lc. lactis*, *P. pentosaceus*, *Weissella confuse*, *Weissella kimchii*, and *Weissella koreensis* (Jung et al. 2011).

Another popular fermented food is pickles, which generally refer to any vegetable or fruit that is preserved by using salt or acid. However, most of the time, the term "pickles" refers to pickled cucumbers and can be separated into fermented or brined pickles and fresh-pack or quick-process pickles. Fermented pickles undergo lactic acid fermentation while fresh-pack pickles are manufactured by acidification, usually by the addition of vinegar (Wacher et al. 2010). Microorganisms mainly involved in pickles fermentation are *L. mesenteroides*, *S. cerevisiae*, *L. brevis*, and *L. plantarum*. Several bacteriocins and antimicrobial peptides that are produced by LAB during fermentation contribute to spoilage bacteria inhibition (Di cagno et al. 2013).

Table olives are also a fermentation product with high economic relevance and appear to be a main part of the Mediterranean diet along with olive oil. Leading producers of olives are Spain, Greece, Italy, and Turkey (Panagou et al. 2008). Among hundreds of methods of processing, only three are economically relevant: the Spanish-style green olives, the Greek-style naturally black olives, and the Californian-style black ripe olives (Brenes 2004). Processing consists of many steps, and several parameters such as salt content, pH, aerobic/anaerobic conditions, and temperature have to be controlled in order to obtain a consistent and high-quality product. The fermentation processes not only eliminate the bitterness due to oleuropein but also ensure the preservation and improvement of the final product through the action of LAB. *L. plantarum* and *Lactobacillus pentosus* are the main species found in most fermentations along with yeasts (Hurtado et al. 2012). The relative population of LAB and yeasts plays an important rolc in a product's characteristics with the LAB's predominance being desirable in olive fermentation. Thus, pure starter cultures are employed by the modern table olive industry in order to achieve the appropriate fermentation process (Panagou et al. 2008).

#### 11.2.2.4 Alcoholic Beverages

Fermented and distilled beverages are the two major categories of alcoholic beverages. Two of the most popular fermented alcoholic beverages are wine and beer. Wine from the Latin word *vinum* is a beverage resulting from fruit juice fermentation by yeasts after necessary processing and additions. The main raw material for wine production is grapes, but other fruits such as apples and berries could also be used for the same purpose. Stainless-steel tanks, open wooden vats, or wine barrels are playing the role of the fermenter in the wine fermentation process (Tamang 2010). The principal organisms employed in most alcoholic beverage fermentations are yeasts, which are able to produce ethanol through metabolism of low-molecular-weight sugars. Glucose and fructose are the two main soluble sugars present in grape must that are fermented to ethanol and carbon dioxide (Berthels et al. 2004). The principal yeast utilized in wine fermentation is *S. cerevisiae* but *S. bayanus* and *S. paradoxus* have also been reported in some cases (Tamang 2010). Furthermore, immobilized cell systems may be used for alcoholic fermentation at low temperatures and also result in a product of high quality with low concentration of higher alcohols such as propanol-1 and isobutyl alcohol (Kourkoutas et al. 2001).

Beers are a type of fermented alcoholic beverage and typically fall into two categories, lagers and ales. The fermentation process employs only one pure strain for beer production and is called monofermentation. The main raw material for beer production is barley (*Hordeum vulgare*) but other cereal grains such as wheat, corn, and rice could be used as well. Barley is germinated and kilned (malting) and then it has to be converted to fermentable carbohydrates and other nutrients for yeast metabolism through a process called mashing. Yeasts utilized in beer fermentation are mainly strains of *S. cerevisiae* (Tamang 2010). Brewing yeasts can be separated into top-fermenting and bottom-fermenting yeasts according to certain characteristics of the fermentation process they promote. *S. cerevisiae* is a top-fermenting yeast while *Saccharomyces pastorianus*, formerly known as *Saccharomyces carlsbergensis* (Kurtzman and Robnett 2003) belongs to bottom-fermenting yeasts, typically used to produce lager-type beers. Beer owes its bitterncss and aroma to hops. Hops are derived from the plant *Homulus lupulus* and are responsible not only for the flavor and aroma of beer but also for enhancing its preservation and prolonging its shelf life (Hutkins 2006).

### 11.2.3 Benefits from Fermentation

Fermentation plays an important role in food processing, improving not only the sensory characteristics but also the nutritional value, safety, and shelf life of foods and also contributes to fuel economy by reducing cooking times.

Preservation or shelf life is an important feature of fermented foods as before the usage of modern methods for food preservation, such as freezing and canning, even a short extension in shelf life would be crucial. It is achieved through the formation of inhibitory metabolites such as organic acids, ethanol, carbon dioxide, bacteriocins, and so on, usually in combination with a decrease in water activity by either drying or the use of salt (Gaggia et al. 2011). These end products control the growth of food spoilage microorganisms providing longer shelf life to perishable foods.

The fact that fermented foods are consumed by hundreds of millions of people every day reveals that fermentation provides a means for producing safe and well-preserved foods. The detoxification as well as low pH and elimination of antinutrients during food fermentation processing contribute to food safety. In addition, foods containing desirable microorganisms become resistant to invasion by spoilage, toxic or food poisoning, as less desirable organisms find it difficult to compete (Bourdichon et al. 2012).

Fermentations also show a significant potential in the improvement and design of the nutritional quality and health effects of foods and ingredients (Poutanen et al. 2009). Fermentation results in a lower proportion of dry matter in the food and the concentrations of vitamins, minerals, and proteins appear to increase when measured on a dry-weight basis (Adams 1990). For example, natural lactic fermentation has been shown to improve the nutritional value of grains such as wheat and rice principally by increasing the content of the essential amino acids such as lysine, methionine, and tryptophan (Adams 1990). In the Indonesian tapé ketan fermentation, rice starch is hydrolyzed to reducing sugars and fermented to ethanol, resulting in a loss of starch solids, which translates into higher protein content on a dry-weight basis (Cronk et al. 1977).

Decreased cooking times is also an advantage of fermented foods as most of them can be consumed in their current state, a fact that translates into fuel economy, which is crucial, especially, in the developing world. For example, soybeans can be transformed by the Indonesian tempeh fermentation to a product that requires about 97% less cooking time (Steinkraus 2002).

### 11.2.4 Types of Fermentation

Solid-state fermentation (SSF) is defined as the fermentation involving solids in the absence or near absence of free water (Pandey 2003). However, the substrate must possess enough moisture to support growth and metabolism of microorganisms, as below a moisture level of approximately 12% all biological activities cease (Binod et al. 2013). It is a system consisting of three phases: the continuous gas phase, the liquid film, and the solid phase. The low moisture content means that fermentation can only be carried out by a limited number of microorganisms, mainly yeasts and fungi, but genetically improved or genetically modified bacterial strains may be made to suit SSF processes (Couto and Sanromán 2006; Bhargav 2008). Almost all the fermentation processes used in ancient times were based on the principles of SSF as it stimulates the growth of microorganisms in nature on moist solids. This type of fermentation is considered environmentally friendly as it has lower energy requirements and lesser wastewater production.

Two major categories of SSF could be distinguished based on the type of microorganisms involved: natural or indigenous SSF and pure culture SSF, where individual strains or a mixed culture is used, respectively (Pandey et al. 2000). In industrial SSF processes, pure cultures are frequently used, as they lead to optimum substrate utilization. SSF systems could be also classified depending on the type of solid phase used. Cultivation on a natural material appears to be the most frequently used system, unlike the less frequent cultivation on an inert support impregnated with liquid medium (Ooijkaas et al. 2000).

The potential applications of SSF are the production of several metabolites of great interest to the food industry, such as enzymes (lipases, pectinases, fructosyl transferases, α-amylases), flavoring compounds, gums (xanthan gum from the bacterium *Xanthomonas campestris*), and organic acids (lactic acid, citric acid, etc.) (Couto and Sanromán 2006). Typical examples of SSF are traditional fermentations such as the Japanese koji, the Indonesian tempeh, and the French blue cheese. SSF has also been used in China to produce brewed foods such as soy sauce and vinegar since ancient times. Cheese is produced from mixed fermentation using *Lc. lactis* and *Streptococcus*. Many of these fermentations have been applied to more sophisticated, large-scale processes (Raimbault 1998).

On the other hand, most industrial fermentations are carried out by submerged fermentation (SmF), where the liquid medium remains in contact with the microorganism necessitating an oxygen supply, using a wide range of microorganisms. SmF has been employed extensively in industries for large-scale production of alcohol, organic acids, enzymes, vitamins, and amino acids (Kaur et al. 2013). The fermentation process is divided into two operations: upstream and downstream. The former describes all the operations before starting the fermenter, such as preparation and sterilization of the culture media, sterilization of the reactor, preparation and growth of inoculums of microbial strains, and so on, while the latter includes all the other operations after the fermentation such as filtration, sedimentation, evaporation, distillation, and so on (Binod et al. 2013).

In recent years, several studies for many substances of interest to the food industry have shown that SSF can give higher yields or better product characteristics than SmF (Couto and Sanromán 2006). Other advantages of SSF over SmF are better oxygen circulation, simpler technology with scarce operational problems, easier downstream processing, and resemblance to the natural habitat for several microorganisms. However, there are few designs available in the literature for bioreactors operating in solid-state conditions, as considerable drawbacks such as transfer resistance, steep gaseous concentration, and heat gradients that develop within the medium bed seem to adversely affect solid-state fermenter performances. In addition, a difference in sensory characteristics between SmF and SSF has been reported. For example, Chinese soy sauce, which can be classified into high-salt SmF and low-salt SSF soy sauce according to the fermentation process, demonstrates significant differences in flavor quality, having a richer aroma profile in the case of liquid-state fermentation (Feng et al. 2014).

### 11.2.5 Types of Fermenters

A fermenter is defined as a vessel for the growth of microorganisms, which, while not permitting contamination, enables the provision of conditions necessary for the maximal production of the desired products (Okafor 2007). Fermenters are also known as bioreactors. Batch, continuous, and fed-batch are the most common modes of operation of fermenters (Rani and Rao 1999). The relationship of substrate consumption to the biomass and products is the decisive factor in which the above fermentation modes should be used (Raj and Karanth 2006). Most of the fermenters consist of the following main components: (a) the body of construction, which is usually made up of stainless steel but it can also be made up of glass (lab fermenters), concrete, or wood, (b) the stirrers with impellers, which mix the gases and liquid media, (c) the baffles, which are used to prevent a vortex and to improve the aeration and stirrers' efficiency, (d) the sparger, a device for introducing air below the liquid level in the fermenter, (e) the jacket or coils, which are responsible for the cooling or heating of the media inside the fermenter, (f) the various valves and steam traps, and (g) the different kinds of probes that are connected to measure and control various process parameters such as pH, temperature, dissolved oxygen, and so on (Kaur et al. 2013).

The bioreactors used to perform SSF processes can be distinguished by the type of aeration or mixed system involved and they aim to create the appropriate conditions for SSF. The four major types of SSF bioreactors are the following: (a) the tray bioreactors, where the substrate is spread onto each tray forming a thin layer, (b) the packed-bed bioreactors, which are mainly composed of a column of glass or plastic and a perforated base where the solid substrate is retained, (c) the horizontal drum bioreactor, where mixing is performed by a rotating drum allowing better aeration and mixing of the substrate, and (d) the fluidized bed, where pneumatic agitation is applied in order to avoid the aggregation of substrate particles (Couto and Sanromán 2006).

On the other hand, SmFs are carried out in either shaker flasks (laboratory-scale fermentation) or stainless-steel tank fermenters (large-scale fermentations). Typical examples of bioreactors used in large-scale fermentations are the stirred tank fermenter (STF), the tower fermenters, which are modified stirred tank reactors, the airlift fermenter, the bubble column and fluidized bed reactors, which are similar to each other, and finally the membrane bioreactor (Kaur et al. 2013).

A very crucial element for the industry is product quality. Quality control during upstream processing of fermentation is required in order to ensure that each batch of the fermented products has a uniform quality and meets the design specifications. A step toward this direction is the careful selection of fermentation conditions and the continuous monitoring of all the upstream stages of fermentation.

### 11.2.6 Human Health Benefits

Fermented foods have been widely associated with human well-being, as they present various health benefits through manifold ways. Fermentation processes involve among others the production of secondary metabolites, such as B-complex vitamins and bioactive peptides that present great interest due to their properties (van Boekel et al. 2010). Moreover, several bacteria and fungi that are responsible for many fermentation processes are considered to be nutritious biomass associated with health benefits (Lindner et al. 2013). Therefore, fermented foods can be considered functional foods, as they promote human health by providing health benefits beyond basic nutrition. However, there is no strict definition of functional foods and although some prominent types of them exist, it is more a matter of marketing. Fortified foods, enriched foods, altered products, and enhanced commodities are, according to Spence (Spence 2006), the main types of functional foods.

The two main ways through which fermented functional foods express their health benefits include: the direct one through the interaction of ingested microorganisms with the consumer and the indirect one as a consequence of the consumption of metabolites produced by microorganisms involved in the fermentation process (Stanton et al. 2005). The direct one is also referred to as the probiotic effect. Probiotics is a term that is used to describe "live microorganisms which when administered in adequate amounts confer a health benefit on the host" (FAO/WHO report 2001). The most commonly used probiotics belong to the genera of *Lactobacillus* and *Bifidobacterium*, but also strains of *Bacillus*, *Pediococcus*, and some yeasts have been reported that could serve the same purpose (see Soccol et al. 2010 for a review). Health benefits derived from probiotics consumption are multiple such as gastrointestinal tract health improvement, enhancement of immune system and nutrients bioavailability, alleviation of lactose intolerance, diarrhea, gastroenteritis, irritable bowel syndrome, inflammatory bowel disease, *Helicobacter pylori* infections, hepatic diseases, hyperlipidemia, hypertension, and even reduction of risk of certain cancers (Parvez et al. 2006).

A microorganism, in order to be used as probiotics, should fulfill certain criteria such as having a beneficial effect on the host, maintaining proper viability throughout product's shelf life, withstanding transit through the gastrointestinal tract, producing inhibitors toward pathogens, and stabilizing intestinal microflora (Parvez et al. 2006). However, many fermented foods can be highly acidic and as a result probiotics may perform poorly or show limited stability. A method to overcome this drawback is encapsulation, as encapsulated bacteria can be used in many fermented products and protect them from bacteriophage and tough conditions such as freezing and gastric solutions (Krasaekoopt et al. 2003). Probiotics also need a delivery method to the host. Several foods are used as matrices for probiotics delivery with dairy products considered as an ideal vehicle. Typical examples are fermented milks, cheese, and yogurt. The probiotic bacteria are usually used either as a starter culture or as an addition to other starter cultures (Soccol et al. 2010).

### 11.2.7 Application of Molecular and Sequencing Technologies to Fermentation

The microbial consortia participating in food fermentations range from simple to very complex and vary in abundance and diversity during the processing. The application of molecular technologies complements culturing studies in order to facilitate the monitoring of fermentation ecosystems and characterization of the various microbial species. The fact that gives birth to this necessity is that culturing studies alone often offer a filtered view of the microbial diversity and potentially misleading information. Novel sequencing technologies along with the associated postsequencing analyses contribute to understanding the microbiota and their relation to fermentations (van Hijum et al. 2013). Furthermore, the large variations between industrial strains or environmental isolates and the model strains for which a complete genome is known suggests the approach of comparative genomics (De Vos 2001).

Amplification by PCR of 16S rRNA gene regions of total DNA in combination with denaturing gradient gel electrophoresis (DGGE) is one form of molecular community profiling. DGGE of PCR-amplified rDNA fragments is a genetic fingerprinting technique that provides a profile of the genetic diversity in a microbial community (Muyzer and Smalla 1998). Randazzo and coworkers described the change in diversity of the microbial communities throughout the stages of an artisanal cheese fermentation by using PCR and DGGE (Randazzo et al. 2002). Since then, many sequencing approaches were used to profile and characterize microorganisms participating in fermentations and emphasize the inadequacy of culturing studies alone (van Hijum et al. 2013).

Another method used for microbial community profiling in fermentations are amplified ribosomal DNA restriction analysis (ARDRA), which is a restriction analysis of 16S–23S rDNA fragments that can be analyzed on agarose gel (Markiewicz et al. 2010). According to Plengvidhya and coworkers, random amplified polymorphic DNA PCR (RAPD-PCR) can be used to investigate the growth, survival, and predominance of a starter culture in sauerkraut fermentations (Plengvidhya et al. 2004). Vogel et al. (2011) isolated and sequenced the genome of *Lactobacillus sanfranciscensis* by a combination of Sanger and 454 pyrosequencing revealing the features, which contribute to the microorganism's ability to outcompete other bacteria in the traditional sourdough fermentation. Genome-probing microarrays (GPMs) have been used to monitor the population dynamics of LAB during fermentation of kimchi (Bae et al. 2005). However, GPMs cannot reveal the real dynamics of microbial communities during fermentation as only the total number of microorganisms can be determined by DNA-based analyses but not their biological activities. Therefore, a metatranscriptomic analysis should be employed in order to address this issue. Nam et al. (2009) were the first to show that metatranscriptomic analysis can be used to monitor microbial dynamics during kimchi fermentation.

Furthermore, the development of next-generation DNA sequencing technologies, such as 454 pyrosequencing and metagenomic approaches based on the random sequencing of environmental DNA can be used in order to elucidate the gene content, metabolic potential, and the function of microbial communities. The first application of metagenomic approaches using pyrosequencing was conducted by Jung et al. (2011) to kimchi and provides information about the microflora composition, overall features, and the metabolic potential throughout the fermentation process.

## 11.3 ENZYMES

The industrial biotechnology sector became a very competitive market since the first industrial use of the enzymes in food processing, such as in the manufacture of meat, vegetables, fruit, baked goods, milk products, and both alcoholic and nonalcoholic beverages (Norus 2006). Industrial-scale production of enzymes used for food and feed applications include those for baking, beverages and brewing, dairy, dietary supplements, as well as fats and oils (Berka and Cherry 2006; Kirk et al. 2002). In 2006, a survey on world sales of enzymes attributed 31% for food enzymes, 6% for feed enzymes, and the remaining for technical enzymes used for detergent, personal care, leather, textile, and pulp and paper industries (Berka and Cherry 2006). This competitive market involves a relatively large number of companies located in the United States, Europe, and Japan. According to Novozymes A/S report (2011), Novozymes occupied 47% of the global market followed by DuPont (21%) and DSM (6%), while enzymes for food and beverage processing represented 29% of the global enzyme business (Novozymes A/S report 2012). The food and beverage processing enzymes represented $1220 million in 2010 that correspond to 36.5% of the total world industrial enzyme demand ($3345 million; Miguel et al. 2013). In addition, the demand for food and beverage enzymes will reach 40.1% of the global share in 2020, accounting for $2520 million of the $6280 million of the world enzyme market (Miguel et al. 2013). In the last decade, the rising demand for novel biocatalysts resulted in an increase in the number of published scientific articles describing the use of enzymes for process optimization in food and beverage manufacture. This section aims to provide an updated and concise overview on the enzymes that are being used in food processing for ingredient production and texture modifications applied in starch, dairy, bakery, brewery, beverage, and meat industries (Table 11.1).

### 11.3.1 Recombinant DNA Technology for Enzyme Production

Most of the enzymes that are currently used in food processing are produced by recombinant microorganisms. There are a number of reasons why enzyme companies take advantage of this technology with the most profound being the high amount of production and engineered enzymes, which is possible even for enzymes that originate from microorganisms unable to be cultivated on an industrial scale. Protein engineering of food processing enzymes makes them compatible with food processing conditions. For example, commonly used sweeteners, such as glucose or fructose syrups, are typically produced from corn starch using $\alpha$-amylase by heating at 105°C for 2–5 min followed by 1–2 h at 90–100°C (Olempska-Beer et al. 2006). This became possible by increasing the thermostability of the enzyme, through DNA modification of the corresponding gene and thus changing its amino acid sequence.

The commercially available enzyme preparations typically contain several added compounds such as preservatives and stabilizers that are needed for increasing the shelf life of the biocatalyst. Moreover, enzyme preparation may also carry other enzymes and metabolites that come from the production microorganism, wild type (WT) or recombinant, and residues from the growth medium and/or the following downstream process. Therefore, each enzyme preparation should follow the current

**TABLE 11.1**
**Examples of Food Processing Enzymes Used in Industry**

| Enzyme | EC Number | CAZy Family | Industry | Application |
|---|---|---|---|---|
| | | **Oxidoreductases** | | |
| Glucose oxidase | 1.1.3.4 | AA3 | Bakery | Enhance properties of wheat flour dough, increase shelf life |
| Laccases | 1.10.3.2 | AA1 | Bakery<br>Brewery<br>Winery | Improvement of gluten-free products<br>Prevention of chill haze formation and creation of off-flavor compounds<br>Wine stabilization |
| Tyrosinase | 1.14.18.1 | – | Dairy | Texture engineering of milk products, formation of mik gels |
| | | **Transferases** | | |
| Transglutaminases | 2.3.2.13 | – | Dairy<br>Meat | Modification of gel and stability of dairy products<br>Improvement of functional characteristics (e.g., texture, flavor), reconstruction |
| | | **Hydrolases** | | |
| α-Amylases and glucoamylases | 3.2.1.1 and 3.2.1.3 | GH13 and 14 | Bakery<br>Brewery<br>Starch | Bread color and flavor, dough properties (e.g., viscosity, gas retention)<br>Light beers<br>Starch liquefaction and saccharification |
| Asparaginase | 3.5.1.1 | – | Bakery | Prevention of acrylamide formation during baking |
| β-Galactosidases | 3.2.1.23 | GH42 | Dairy | Delactose milk products |
| Lipases | 3.1.1.3 | – | Bakery<br>Dairy | Emulsifiers, flavor development<br>Flavor development during cheese ripening |
| Pectinases[a] | 3.2.1.15 or 4.2.2.10 or 3.1.1.11 or other | GH28, PL1, CE8 | Fruit and juice | Juice clarification |
| Proteases | 3.4.X | – | Bakery<br>Brewery<br>Dairy<br>Meat | Regulate gluten strength and texture in bread, improve flavor<br>Prevention of chill haze formation<br>Flavor improvement in cheese products<br>Meat tenderizer |
| Pullulanases | 3.2.1.41 | GH13 | Starch | Starch liquefaction |

*continued*

**TABLE 11.1 (continued)**
**Examples of Food Processing Enzymes Used in Industry**

| Enzyme | EC Number | CAZy Family | Industry | Application |
|---|---|---|---|---|
| Xylanases | 3.2.1.8 | GH8 and 11 | Bakery | Change of rheological and organoleptic properties of bread |
| | | **Lyases** | | |
| Acetolactate decarboxylase | 4.1.1.5 | – | Brewery | Beer maturation (acetoin production) |
| | | **Isomerases** | | |
| Xylose isomerase | 5.3.1.5 | – | Starch | Production of high-fructose corn syrup (HFCS) |

[a] Enzyme mixture consisted of polygalacturonase, pectin lyase, pectin methylesterase, and other pectinolytic activities.

good manufacturing practice (cGMP), resulting in a high-purity product compatible for use in the food industry (Olempska-Beer et al. 2006). The safety of the enzymes that are produced by recombinant DNA technology has been extensively discussed in the literature (Pariza and Johnson 2001). The key issues that are important for the evaluation of enzyme safety are the assessment of the pathogenic and toxigenic potential of the production host strain. Even if the enzyme expression host is not considered as pathogenic or toxigenic, certain microorganisms under specific fermentation conditions might produce low levels of secondary metabolites that might be harmful to human health (Olempska-Beer et al. 2006). In the last two decades, the FDA has approved the use of a number of enzymes that are generally recognized as safe (GRAS) for food processing. According to Regulation (EC) No 1332/2008, all food processing enzymes currently on the EU market, as well as novel ones, are subjected to a safety evaluation by the European Food Safety Authority (EFSA) and approval via a Union list. The list of these commercial enzyme preparations produced from WT or recombinant microorganisms can be found on the websites of the Enzyme Technical Association (http://www.enzymetechnicalassoc.org) and the Association of Manufacturers and Formulators of Enzyme Products (http://www.amfep.org).

### 11.3.2 Enzymes in Starch Process

Acid hydrolysis of starch for the production of glucose was developed by a Russian chemist named Kirchoff in 1811, while de Saussure commercialized for the first time in 1815 this kind of production for syrups and crude sugars. Acid-catalyzed hydrolysis of starch was an efficient reaction; however, the lack of sweetness compared to sucrose and the formation of reversion products limited its use for the production of glucose syrups. The enzyme-catalyzed reaction revolutionizes starch hydrolysis, reaching yields up to 90% of glucose by the synergistic action of the enzymes α-amylase and amyloglucosidases or α-glucosidases. Starch consists of two fractions, amylose that is a long linear polysaccharide with

D-glucose units linked by α (1 → 4) glycosidic bonds and amylopectin that is highly branched and has both α (1 → 4) and (1 → 6) glycosidic bonds (Finn 1987). Therefore, for the efficient degradation of both starch fractions, the addition of a debranching enzyme such as pullulanase (EC 3.2.1.41) or isoamylase (EC 3.2.1.68) is necessary to attack the α (1 → 6) bonds (Guzman-Maldonado and Paredes-Lopez 1995). The glucose or maltose syrup production of starch is a two-step process. In the liquefaction step, the concentrated slurry of starch granules (30–40%, w/v) is gelatinized at elevated temperatures (90–110°C), adding a thermostable endoamylase (EC 3.2.1.1) to control the rapid increase of solution viscosity (Guzman-Maldonado and Paredes-Lopez 1995). By the action of endoamylase, a solution of dextrins having different polymerization degrees is made. The hydrolysis step is extended at a lower temperature (90°C), leading to the complete hydrolysis of maltooligosaccharides into glucose or maltose units. Commercial α-amylases used for starch liquefaction and saccharification usually show optimal activity between pH 5.5 and 6.0, requiring the addition of $Ca^{+2}$ for stability, while glucoamylases are active at low pH values around 4.5 (Guzman-Maldonado and Paredes-Lopez 1995). As starch slurry pH is approximately at 4.5, the whole process needs sequential pH adjustments using HCl and NaOH solutions, increasing the production cost and the creation of undesirable NaCl. The downstream process is further charged by anion-exchange chromatography for the removal of stabilizing $Ca^{+2}$ because the latter inhibits xylose isomerase (EC 5.3.1.5; glucose isomerase synonym) that catalyzes the conversion of glucose to fructose units. This step is commonly used for the production of high-fructose corn syrup (HFCS), where the isomerase converts a part of glucose into fructose to produce a desired sweetness. This isomerization step takes place in a reactor system employed with immobilized glucose isomerase prior to the posttreatment step, including evaporation to obtain the end product.

The complete saccharification of the produced maltodextrins is accomplished by glucoamylases (EC 3.2.1.3) that are inactivated at temperatures above 60°C (Zdzieblo and Synowiecki 2002), requiring the rapid cooling of the liquefacted solution. Therefore, the need of thermostable glucoamylases is also essential for further reducing the production costs, such as α-glucosidases from thermophilic archaeal sources such as *Pyrococcus furiosus*, *Sulfolobus solfataricus*, *Thermococcus hydrothermalis*, and others reaching optimal activity above 100°C (see Leveque et al. 2000 for a review).

### 11.3.3 Processing Enzymes Used in Dairy Industry

For many years, the only technical enzyme used in cheese making, apart from the sheep and goat gland esterases, was rennet obtained from the stomach of calves. Owing to the huge demand for cheese and dairy products worldwide, proteases such as chymosin (known also as rennin), have been used as substitutes for animal rennet that are produced by genetically engineered microorganisms. *E. coli* K-12, which set the basis for a safe chymosin production, has been used as a laboratory microorganism for over 30 years, being one of the most extensively studied bacteria without reported incidents of infection (Olempska-Beer et al. 2006).

Proteinases or peptidases (including general and specific aminopeptidases, carboxypeptidases, and oligopeptidases) have been used in the dairy industry for different applications (Stepaniak 2004). Such enzymes are commercially available; however, the preparations do not include pure enzymatic activities due to the high cost involved in downstream process. For the production of enzyme-modified cheeses (EMCs) that are rich in bioactive peptides with improved functional properties and reduced allergenicity, proteinases are the key enzymes described by an extensive literature. Commercial proteinases from microbial sources that also have the potential use for accelerated cheese ripening are derived from *Bacillus*, *Aspergillus* spp., or *Rhizomucor niveus* (Stepaniak 2004). Specific activity of such proteinases is crucial for reducing the allergenicity of milk proteins (Kilcawley et al. 2002). On the other hand, aminopeptidases (EC 3.4.11) and carboxypeptidases (EC 3.4.16–3.4.18) are used for the creation of flavor and debittering of protein hydrolysates, as well as for the production of EMCs. Such enzymes that catalyze the hydrolysis of proline-containing bonds are also commercially available from *Lc. lactis* and *Flavobacterium* spp. (Kilcawley et al. 1998, 2002; Raksakulthai and Haard 2003).

Lipolytic enzymes are also used in flavor development and enhancement of dietary milk products, such as in the manufacture of cheese products and EMCs, or in the general acceleration of cheese ripening. Lipolytic enzymes, except glycosidases, are hydrolases recognizing long-chain fatty acid esters separating sugars from glycolipids. More specifically, lipases or triacylglycerol hydrolases catalyze the hydrolysis of long-chain triglycerides into free fatty acids and glycerols. For the cheese manufacturing industry, several fungal lipases are used from *Mucor miehei*, *Aspergillus niger*, and *Aspergillus oryzae* among others (Hasan et al. 2006) that are robust enzymes compared to plant or animal homologs exhibiting extraordinary pH values and temperatures (Gandhi 1997). Lipases together with aminopeptidases are used for flavor development during cheese ripening, as free fatty acids and soluble peptides and amino acids are released that are effective flavors or flavor precursors. Different lipases with distinct substrate specificity releasing short-chain (C2–C6), medium-chain (C8–C14), and long-chain (>C16) fatty acids produce cheesy, buttery, and soapy flavors, respectively, since these fatty acids are converted by the microbial population of cheese into flavor components such as acetoacetate, esters, lactones, beta-keto acids, and methyl ketones (Beermann and Hartung 2012; Hasan et al. 2006). On the other side, uncontrolled conditions may form undesirable rancid flavors in dairy products such as milk (de Felice et al. 1991). In the same direction, aminopeptidases from bacteria or mold, such as *Lactococcus* and *Aspergillus* species, release single amino acids supporting flavor formation and acceleration in the manufacture of cheese and dairy products (Vishwanatha et al. 2010).

Enzymatic cross-linking of proteins in food processing is a method that received increasing attention in the last 20 years. The most known cross-linking enzyme that catalyzes the covalent bond formation between protein molecules and is commercially available is transglutaminase (EC 2.3.2.13) (Dickinson 1997). This enzyme was first identified by Heinrich Waelsch more than 40 years ago as a liver enzyme that incorporates amines into proteins (Fesus and Piacentini 2002). Transglutaminases originate from various animal tissues or organs and plants (Özrenk 2006). This enzyme catalyzes the acyl-transfer reaction in which the $\gamma$-carboxamide groups of peptide-bound glutaminyl residues are the acyl donors (Özrenk 2006). The enzyme exchanges a wide variety of primary amines and ammonia at the carboxamide groups of glutamine residues. The peptide-bound lysine residues may also function as acyl receptors, resulting in the formation of ε-(γ-glutaminyl) lysine isopeptide bonds creating polymers with high molecular weight. In addition to the cross-linking ability of transglutaminase, these enzymes could also catalyze the deamidation and amine incorporation, as in the absence of primary amines, water becomes the acyl acceptor and the γ-carboxamide groups of glutamines are deaminated (Zhu et al. 1995). The interest of these enzymes for use in dairy products dates back to the early 1980s. Since these enzymes show similarities with the catalytic triad and mechanism of papain (EC 3.4.22.2) and papain-like cysteine proteases, they are classified within the same superfamily in the Structural Classification of Proteins (SCOP) database (Fesus and Piacentini 2002). Transglutaminases act on milk caseins that are particularly good substrates, since reactive groups are exposed due to low degree of tertiary structure, a flexible, random-coil arrangement and absence of disulfide bonds in the $\alpha_{s1}$ and β-caseins (O'Connell and Kruif 2003). The application of transglutaminase in dairy products has the potential of increasing gel strength, surface viscosity, water-holding capacity, stability, rennetability, and mechanical stability while decreasing permeability (Özrenk 2006). Tyrosinase (EC 1.14.18.1) is an alternative to the transglutaminase cross-linking enzyme that catalyzes the *ortho*-hydroxylation of monophenols, such as tyrosine and the subsequent oxidation to quinones. This cross-linking activity could be applied from small molecules, such as tyrosine, to polymeric molecules, such as protein molecules, resulting in the formation of tyrosine–tyrosine, tyrosine–cysteine, and tyrosine–lysine protein cross-links (Beermann and Hartung 2012). However, the substrate specificity of different tyrosinases affects their cross-linking ability on different dairy products. For example, a tyrosinase from *Agaricus bisporus* was effective in cross-linking whey, while a different tyrosinase from *Pycnoporus sanguineus* preferred casein (Beermann and Hartung 2012). The cross-linking ability of tyrosinase could be exploited in texture engineering of milk products, for example, for the production of milk gels.

A very interesting sector of the dairy industry is the enzymatic reduction of lactose in milk and dairy products using several β-galactosidases (EC 3.2.1.23) or β-glucosidases (EC 3.2.1.21) showing β-galactosidase activity originating from microorganisms, plants, and animal tissues (Shukla 1975). This process is very important, as lactose-intolerant individuals have insufficient levels of lactase, an enzyme that catalyzes the hydrolysis of lactose found in milk and to a lesser extent in milk-derived dairy products, into glucose and galactose in their digestive system. β-Galactosidases exhibit product inhibition that is competitive with D-galactose and noncompetitive with D-glucose (Jurado et al. 2004; Ladero et al. 2002). In addition to product inhibition, the formation of galacto-oligosaccharides is achieved due to the trans-galactosylation activity of β-galactosidase, further decreasing the final yield of delactose process as this side activity competes with the hydrolysis of lactose (Mahoney 1998). On the other hand, galacto-oligosaccharides promote the growth of desirable intestinal microflora, such as bifidobacteria and lactobacilli, claiming specific prebiotic health benefits (Beermann and Hartung 2012). In order to enhance lactose hydrolysis, the use of a continuous-flow reactor with immobilized enzyme bring up conformational changes in the tertiary structure of β-galactosidase, reducing its transferase activity and enhancing the final hydrolysis yield (Greenberg and Mahoney 1981; Ladero et al. 2002). Different immobilization techniques were applied for β-glucosidase, including ionic adsorption or covalent attachment of the enzyme (Greenberg and Mahoney 1981). The use of thermostable immobilized β-galactosidases will allow longer operation times in elevated temperatures and the reduction of the reaction system contamination by the growth of mesophilic microorganisms (Synowiecki et al. 2006).

### 11.3.4 Enzymes in Bakery

Industrial enzymes played a key role in the baking evolution, representing a significant part of the industry, since bakery-processing enzymes market is expected to increase from $420 million in 2010 to $900 million in 2020 (Miguel et al. 2013). Enzymes are added for various reasons depending on the final product, for example, to modify dough rheology, crumb softness, and gas retention in bread making, dough rheology in pastry and biscuits making, change softness in the production of cakes, and reduce acrylamide formation (Cauvain and Young 2006). The most common enzymes added are hydrolases (Table 11.1), which are supplemented into flour during the mixing step of the bread-making process. For example, the industry uses α- and β-amylases belonging to GH13 and 14 of the Carbohydrate Active Enzymes (CAZy) database (www.cazy.org), respectively, as well as glucoamylases of GH14, pullulanases and isoamylases of GH13 (see Section 11.3.2). For the increase of fermentable and reducing sugars of the flour, microbial and malt α-amylases are widely used. α-Amylases catalyze the breakdown of starch particles into dextrins of low molecular weight during the dough stage, while endogenous β-amylases produce maltose from these oligosaccharides, which is exploited by yeast or sourdough microorganisms (Goesaert et al. 2005; Synowiecki et al. 2006). Reducing sugars are very important for the formation of Maillard products, bringing crust color and bread flavor (Miguel et al. 2013). Product volume and softness is affected by the action of amylases through the reduction of dough viscosity during starch gelatization, also improving gas retention of the fermented dough (Cauvain and Young 2006; Goesaert et al. 2009).

Another class of hydrolases that are being used in baking products are proteases, which in wheat and rye flours are endogenous corresponding to aspartic proteases; however, the proteolytic activity of ungerminated grain is normally low (Linko et al. 1997; Miguel et al. 2013). Bakery processing proteases are common enzymes used in the industry to reduce dough consistency, reduce mixing time, regulate gluten strength and texture in bread, assure dough uniformity, and improve flavor (Goesaert et al. 2005; Miguel et al. 2013). As the main goal of proteases is to weaken the gluten network; these enzymes are replacing bisulfite that was previously used in the control of consistency through the reduction of disulfide bonds of glutens (Linko et al. 1997). Proteases could be used in the blend or added to dough preparations in order to change dough rheology and bread quality, possibly due to their effect in gluten network or gliadin (Salleh et al. 2006).

Hemicellulases are also used in food processing, especially in the preparation of baked products. These enzymes contain a large amount of different activities, with the most prevalent being represented by endo-1,4-β-xylanases (EC 3.2.1.8), enzymes with a great potential in different industrial applications (see Topakas et al. 2013 for a review). For example, for the complete hydrolysis of arabino-glucuronoxylans, a battery of enzymes are needed, including endo-1,4-β-xylanases, β-D-xylosidases (EC 3.2.1.37), α-L-arabinofuranosidases (EC 3.2.1.55), feruloyl (EC 3.1.1.73) and acetyl (EC 3.1.1.72) esterases, and α-D-glucuronosidases (EC 3.2.1.131). These enzymes are preferably needed for the hydrolysis of water-unextractable arabinoxylan (WU-AX), in order to solubilize polysaccharides that interfere with the creation of the gluten network (Courtin and Delcour 2001; Rouau et al. 1994). According to CAZy database, xylanases may belong to glycoside hydrolase GH families 5, 7, 8, 10, 11, and 43 based on their physicochemical properties, structure, mode of action, and substrate specificities (Topakas et al. 2013). Xylanases of GH8 and GH11 families, such as a psychrophilic xylanase from *Pseudoalteromonas haloplanktis* and *Penicillium occitanis*, respectively, were used as processing enzymes for increasing loaf volume and moisture content, as well as enhancing sensory and textual properties (Collins et al. 2006; Driss et al. 2013).

Lipases (EC 3.1.1.3) that hydrolyze triacylglycerols to monoacylglycerols were recently introduced in the baking industry, starting their use back in the 1990s. Nowadays, engineered enzymes belonging to the third generation of lipases show lower affinity for short-chain fatty acids, reducing the risk of off-flavor formation of baked foods when butter or milk fat is used. In addition, these enzymes increase the expansion of gluten network and wall thickness, enhancing the volume and crumb structure of high-fiber white bread (van Oort 2010). Hydrolysis products mediated by lipases (mono-, diacylglycerols, monoacylgalactolipids, and lysophospholipids) and their role in bread making along with their interaction with gluten have been reviewed elsewhere (Miguel et al. 2013; Pareyt et al. 2011). Lipases in bakery products may act as emulsifiers, as found by the comparison of two lipases and diacetyl tartaric esters of monoglycerides (DATEM) emulsifier, where the enzymes improved dough handling properties to a similar or better extent compared to the emulsifier DATEM (Colakoglu and Ozkaya 2012; Miguel et al. 2013). These enzymes may contribute to the development of particular flavors in baking and in a similar way with dairy products (Miguel et al. 2013; Olesen et al. 2000).

The extensive research on bakery processing enzymes has included oxidoreductases, such as glucose oxidase (EC 1.1.3.4), produced from fungal sources that catalyze the oxidation of β-D-glucose to D-glucono-δ-lactone and hydrogen peroxide. As a glucose oxidoreductase, the enzyme removes residual oxygen and glucose, enhancing the shelf life of food and beverages, while the produced hydrogen peroxide may increase antimicrobial protection (Kirk et al. 2002; Sisak et al. 2006; van Oort 2010). The hydrogen peroxide may also promote the creation of disulfide bonds or dityrosine cross-links in gluten network, enhancing the properties of wheat flour dough (machinability, gas retention, colume, crumb structure) through gluten matrix modifications (Bonet et al. 2006; Rasiah et al. 2005). The remaining hydrogen peroxide could be removed by the action of catalase (EC 1.11.1.6) toward the production of water and oxygen.

Close to oxidoreductases, laccases (EC 1.10.3.2) are copper-containing enzymes that catalyze the oxidation of phenolic compounds through one-electron removal creating reactive phenolic radicals with cross-linking abilities (Bourbonnais et al. 1995). This cross-linking property is important for the formation of strong arabinoxylan networks by connecting ferulic acid moieties of decorated polysaccharides of dough during baking, resulting in better strength and stability properties, as well as stickiness reduction (Selinheimo et al. 2006). Owing to the growing awareness of celiac disease, there is considerable focus in the preparation of gluten-free baked products. Celiac disease is an autoimmune disorder of the small intestine that occurs in genetically predisposed people of all ages from middle infancy upward, which is triggered by the ingestion of gluten contained in many cereal flours such as wheat, rye, and barley. For this purpose, cereal flours, such as rice, potato, oats, starches, and corn are used for the production of gluten-free baked products; however, such flours and starches lack the protein matrix responsible for dough formation and other characteristics important in the baking process (Gallagher 2009). Gluten-free products made from oat flour is possible using enzymes, such as laccases, improving the texture quality together with increase of loaf volume, while decreasing crumb hardness and chewiness. Chemical analysis proved that β-glucan depolymerization and polymerization resulted in improved properties of oat flour bread (Renzetti et al. 2010).

Enzymes are not only used for processing food but also for protecting the consumer's health. An example of such an enzyme is asparaginase (EC 3.5.1.1) that is believed to have a high potential of reducing the formation of acrylamide during baking, a possible human carcinogen that is created via the Maillard reaction between asparagine and carbonyl groups (Claus et al. 2008). A fungal asparaginase from *A. niger* has the potential of broad commercialization as it is considered safe (Olempska-Beer 2008).

### 11.3.5 Enzymes in Brewing

The brewing process consists of several steps, including mashing, lautering, wort boiling, fermentation, maturation, filtration, and clarification. In the first step of the malting process, endogenous enzymes of barley (amylases, glucanases, proteases, and hemicellulases) are secreted to the medium; therefore, there is no need for adding exogenous enzymes. However, in case of unmodified malt or low-quality heterogeneous malt, the addition of enzymes is needed. After the fermentation stage where fermentable sugars in the wort are metabolized into ethanol, the maturation stage is intended for the adjustment of the flavor, carbon dioxide, haze stability, and other properties (James et al. 1996).

In the fermentation step, exogenous enzymes such as β-glucanases are used to degrade residual glucans, which are the main cause of filter blockage (Schmedding and van Gestel 2002). At the maturation stage, a number of conversions that aim in flavor changes are taking place. One of the main goals of this step is the oxidative transformation of α-acetolactic acid into diacetyl, which is considered an off-flavor in lager beer, and the subsequent reduction of diacetyl to acetoin by yeast reductases, a rate-limiting step for the overall maturation process (James et al.

1996). The formation of acetoin is important, as it is a major flavor component in beer. Microbial α-acetolactate decarboxylases (EC 4.1.1.5) are used for the transformation of α-acetolactic acid directly into acetoin. Commercially, a recombinant α-acetolactate decarboxylase originating from *L. brevis* that is expressed by *Bacillus subtilis* is proven to be a safe enzyme for use in the brewing industry (De Boer et al. 1993).

Another important aspect in beer manufacture is the removal of chill haze, since beer contains complexes of proteins with carbohydrates and tannin that become insoluble when cooled, while redissolved by warming beer at room temperature or above. However, sulfydryl groups replace phenolic rings after extended periods of time, leading to permanent haze that does not redissolve at room temperature (Minussi et al. 2002). Haze formation is stimulated by proanthocyanidin polyphenols that are naturally present in small quantities (Mathiasen 1995) and depends on the starting materials and processing variables used for the manufacture of beer (Finley et al. 1979). In order to avoid chill haze, proteolytic enzymes, such as papain immobilized on chitin, are appropriate for hydrolyzing proteins involved in haze creation (Finley et al. 1979). Moreover, proanthocyanidin polyphenols could be removed by the use of laccases, retaining haze stability (Mathiasen 1995; Rossi et al. 1988). These enzymes have also been used for removing oxygen at the end of beer manufacture, enhancing its storage life (Mathiasen 1995). A commercial laccase preparation from Novozymes A/S named "Flavourstar" is being used in brewing to prevent the creation of off-flavor compounds, such as *trans*-2-nonenal, by scavenging the oxygen that would otherwise form off-flavor precursors by reacting with fatty acids, proteins, and alcohol (Olempska-Beer 2004).

Enzymes are also used for the preparation of low-calorie beer production. As malt-derived dextrinase is heat sensitive, a large amount of nonfermentable by yeast dextrins are present in finished beer, contributing to its calorific value together with proteins. Therefore, low-calorie beer is produced by adding microbial amyloglucosidase during fermentation that will degrade dextrins to fermentable sugars, resulting in a normal alcohol beer; however, wort containing a reduced amount of extract is necessary. In addition, low-calorie, nonalcoholic beer could be created by adding one more step in the production process, involving the removal of alcohol by vacuum distillation (Schmedding and van Gestel 2002).

### 11.3.6 Processing Enzymes in Beverage Industry

Commercial sources of fungal enzymes have been used for fruit juice processing since the 1930s for the clarification of the juices, including pectinase, cellulase, and hemicellulase activities, among others. Enzymes are used in the fruit and juice production industry for enhancing the separation of juice from the fruit cells and to help the clarification of juice by the removal of pectin and naturally occurring polysaccharides, which contribute to the cloudy appearance, undesired viscosity, and poor filtration (see Kashyap et al. 2001 for a review). Pectinases are an integral part in this sector, being responsible for the degradation of the long and complex molecules of pectin that occur as structural polysaccharides and confer rigidity on cell walls. In unripe fruits, the insoluble pectin is bound to cellulose microfibrils, while during ripening the structure of pectin is broken down by endogenous enzymes, resulting in a more soluble form. All pectic substances, including protopectin, pectic acid, pectining acid, and pectin, account for about 0.5–4% of fresh material (Kashyap et al. 2001). After ripening, a part of pectin goes to the liquid phase, creating a cloudy preparation with an increase of viscosity, while the remaining amount is bound to cellulose fibrils, facilitating water retention (Pifferi et al. 1989). The cloud particles have a protein nucleus that is charged positive, which is coated by pectin polysaccharides that have negative charge (Pilnik and Voragen 1993). These negatively charged particles repel one another; therefore, the use of pectinases releases the positive part of the protein beneath, enhancing aggregation by the reduction of the electrostatic repulsion. The prepared juice drops in viscosity, the pressability of the pulp improves, the jelly structure disintegrates, and the fruit juice is obtained in higher yields (Kashyap et al. 2001). Pectinases are distinguished by one of two main groups, the acidic and alkaline pectinases. Acidic pectinases are widely used in the fruit juice industry and wine making, and are mainly derived from fungal sources such as *A. niger.* These enzymes are employed for the production of sparkling juices (apple, pear, and grape juices), cloudy juices (citrus, prune, tomato, and nectar juices), and unicellular products (Kashyap et al. 2001). Glucose oxidase coupled with catalase reported in cases of bakery products is also used for removing oxygen in bottled drinks, in order to reduce the nonenzymatic browning that might occur due to oxidation.

Laccases have been used in wine stabilization as an alternative to physical–chemical absorbents (Minussi et al. 2002). Wines are complex systems composed of ethanol, organic acids, inorganics, and phenolic compounds that give color and taste. Laccase-mediated polyphenol removal should be resistant to the acidic pH of wine and the presence of sulfite that may play the role of reversible inhibitor (Tanriöven and Ekşi 2005).

### 11.3.7 Meat Industry

Meat production, including pork, beef, poultry, goat, and mutton, has increased to cover the global demand that is estimated to reach 40 kg per capita per annum in 2020 (Delgado et al. 1998). Consumers rate tenderness as the most important attribute of meat quality. This characteristic is a result of the interaction of the actomyosin effect of myofibrillar proteins, the fat density effect, and the background effect of connective tissue (Castro Marques et al. 2010). Proteolytic enzymes are used for the tenderization of meat, in comparison with chemical or physical tenderization that mainly reduces the amount of connective tissue without degrading myofibrillar proteins (Qihe et al. 2006). Papain, one of the most popular proteases used in the meat industry, is a nonspecific thiol protease and a major protein constituent of latex in the tropical plant *Carica papaya*. However, papain has a tendency to overtenderize meat surface, resulting in a "mushy" product limiting its use as a commercial meat tenderizer (Han et al. 2009). Other plant-originating proteases called bromelin (EC 3.4.4.24) that are present in large amounts in fruit, leaves, and stems of the Bromeliaceae family such as pineapple (*Ananas comosus*) also degrade myofibrillar proteins and collagen, often resulting in overtenderization of meat (Melendo et al.

1996). The ideal proteolytic enzyme for the tenderization of meat would be a protease with specificity for collagen and elastin of the connective tissue that would have high activity at the low pH of meat and low temperature during storage (Gerelt et al. 2000). On the other hand, proteases can be exploited in waste valorization of meat industrial by-products. For instance, keratinases are capable of degrading hard and insoluble keratin proteins for the conversion of chicken feather waste into highly digestable animal feed (Khardenavis et al. 2009). Processes, such as the production of bioactive peptides against hypertension or the reduction of allergenics in meat foods, were also catalyzed by using proteases (Balti et al. 2010).

Transglutaminase (EC 2.3.2.13) that was mentioned in cases of dairy products is an enzyme with great potential in the meat industry, improving the functional characteristics of meat products, such as texture, flavor, and shelf life. The enzyme has the ability of adhering to the bonding surfaces of meat, fish, eggs, and even vegetables by cross-linking proteins. A variant of *Streptoverticillium mobaraense* is the main source of industrial transglutaminase, showing a pH optimum in the area between 5 and 8 and temperature at 50°C. However, the enzyme is also active in chilling temperatures, a property that it is exploited for the manufacture of restructured meat products (Castro-Briones et al. 2009). The enzyme is widely applied for the production of different meat products, including ham, beef and chicken sausages, fish, and others (Ahhmed et al. 2007; Romero de Ávila et al. 2010). Gelation in meat foods is induced by the presence of transglutaminases that catalyze the interconnections of myofibrils making a protein–protein network and thus improving gel elasticity of meat protein (Ahhmed et al. 2007).

## REFERENCES

Adams, M.R. 1990. Topical aspects of fermented foods. *Trends in Food Science and Technology* 1: 141–4.

Ahhmed, A.M., S. Kawahara, K. Ohta, K. Nakade, T. Soeda, and M. Muguruma. 2007. Differentiation in improvements of gel strength in chicken and beef sausages induced by transglutaminase. *Meat Science* 76: 455–62.

Ammor, M.S., and B. Mayo. 2007. Selection criteria for lactic acid bacteria to be used as functional starter cultures in dry sausage production: An update. *Meat Science* 76: 138–46.

Bae, J.W., S.K. Rhee, J.R. Park et al. 2005. Development and evaluation of genome-probing microarrays for monitoring lactic acid bacteria. *Applied and Environmental Microbiology* 71: 8825–35.

Balti, R., N. Nedjar-Arroume, A. Bougatef, D. Guillochon, and M. Nasri. 2010. Three novel angiotensin I-converting enzyme (ACE) inhibitory peptides from cuttlefish (*Sepia officinalis*) using digestive proteases. *Food Research International* 43: 1136–43.

Beermann, C., and J. Hartung. 2012. Current enzymatic milk fermentation procedures. *European Food Research and Technology* 235: 1–12.

Berka, R.M., and J.R. Cherry. 2006. Enzyme biotechnology. In *Basic Biotechnology*, ed. C. Ratledge and B. Kristiansen, 477–98. Cambridge: Cambridge University Press.

Berthels, N.J., R.R. Cordero Otero, F.F. Bauer et al. 2004. Discrepancy in glucose and fructose utilization during fermentation by *Saccharomyces cerevisiae* wine yeast strains. *FEMS Yeast Research* 4: 683–9.

Bhargav, S., B.P. Panda, M. Ali, and S. Javed. 2008. Solid-state fermentation: An overview. *Chemical and Biochemical Engineering Quarterly* 22: 49–70.

Binod, P., R. Sindhu, and A. Pandey. 2013. Upstream operations of fermentation processes. In *Fermentation Processes Engineering in the Food Industry*, ed. C.R. Soccol, A. Pandey, and C. Larroche, 75–87. USA: CRC Press.

Bonet, A., C.M. Rosell, P.A. Caballero, M. Gomez, I. Perez-Munuera, and M.A. Lluch. 2006. Glucose oxidase effect on dough rheology and bread quality: A study from macroscopic to molecular level. *Food Chemistry* 99: 408–15.

Bourbonnais, R., M.G. Paice, I.D. Reid, P. Lanthier, and M. Yaguchi. 1995. Lignin oxidation by laccase isozymes from *Trametes versicolor* and role of the mediator 2′,2′-azinobis (3-ethylbenzthiazoline-6-sulphonate) in kraft pulp depolymerisation. *Applied and Environmental Microbiology* 61:1876–80.

Bourdichon, F., S. Casaregola, C. Farrokh et al. 2012. Food fermentations: Microorganisms with technological beneficial use. *International Journal of Food Microbiology* 154: 87–97.

Brenes, M. 2004. Olive fermentation and processing: Scientific and technological challenges. *Journal of Food Science* 69: 33–4.

Campbell-Platt, G. 1987. *Fermented Foods of the World: A Dictionary and Guide*. London, 290. UK: Butterworths.

Campbell-Platt, G. 1994. Fermented foods—A world perspective. *Food Research International* 27: 253–7.

Caplice, E., and G.F. Fitzgerald. 1999. Food fermentations: Role of microorganisms in food production and preservation. *International Journal of Food Microbiology* 50: 131–49.

Castro Marques, A., M.R. Maróstica Jr, and G.M. Pastore. 2010. Some nutritional, technological and environmental advances in the use of enzymes in meat products. *Enzyme Research* Article ID 480923, http://dx.doi.org/10.4061/2010/480923

Castro-Briones, M., G.N. Calderón, G. Velazquez, M.S. Rubio, M. Vázquez, and J.A. Ramirez. 2009. Mechanical and functional properties of beef products obtained using microbial transglutaminase with treatments of pre-heating followed by cold binding. *Meat Science* 83: 229–38.

Cauvain, S., and L. Young. 2006. Ingredients and their influences. In *Baked Products*, ed. S. Cauvain, and L. Young, 72–98. Oxford: Blackwell Publishing.

Claus, A., R. Carle, and A. Schieber. 2008. Acrylamide in cereal products: A review. *Journal of Cereal Science* 47: 118–33.

Colakoglu, A.S., and H. Ozkaya. 2012. Potential use of exogenous lipases for DATEM replacement to modify the rheological and thermal properties of wheat flour dough. *Journal of Cereal Science* 55: 397–404.

Collins, T., A. Hoyoux, A. Dutron, J. Georis, B. Genot, T. Dauvrin, F. Arnaut, C. Gerday, and G. Feller. 2006. Use of glycoside hydrolase family 8 xylanases in baking. *Journal of Cereal Science* 43: 79–84.

Courtin, C.M., and J.A. Delcour. 2001. Relative activity of endoxylanases towards water-extractable and water-unextractable arabinoxylan. *Journal of Cereal Science* 33: 301–312.

Couto, S.R., and M.Á. Sanromán. 2006. Application of solid-state fermentation to food industry: A review. *Journal of Food Engineering* 76: 291–302.

Cronk, T.C., H. Steinkraus, L.R. Hackler, and L.R. Mattick. 1977. Indonesian tape ketan fermentation. *Applied and Environmental Microbiology* 33: 1067–73.

De Boer, A.S., R. Marshall, A. Broadmeadow, and K. Hazelden. 1993. Toxicological evaluation of acetolactate decarboxylase. *Journal of Food Protection* 56: 510–7.

De Felice, M., T. Gomes, and T. de Leonardis. 1991. Addition of animal and microbial lipases to curd. Effects on free fatty acid composition during ripening. *Lait* 71: 637–43.

De Vos, W.M. 2001. Advances in genomics for microbial food fermentations and safety. *Current Opinion in Biotechnology* 12: 493–8.

Delgado, C.L., C.B. Courbois, and M.W. Rosegrant. 1998. Global food demand and the contribution of livestock as we enter the new millennium. Markets and Structural Studies Division. International Food Policy Research Institute.

Di Cagno, R., R. Coda, M. De Angelis, and M. Gobetti. 2013. Exploitation of vegetables and fruits through lactic acid fermentation. *Food Microbiology* 33: 1–10.

Dickinson, E. 1997. Enzymic crosslinking as a tool for food colloid rheology control and interfacial stabilization. *Trends in Food Science and Technology* 10: 333–9.

Dirrar, M.A. 1993. *The Indigenous Fermented Foods of the Sudan: A Study in African Food and Nutrition*. Wallingford: CAB International.

Driss, D., F. Bhiri, M. Siela, S. Bessess, S. Chaabouni, and R. Ghorbel. 2013. Improvement of bread making quality by xylanase GH11 from *Penicillium occitanis* Pol6. *Journal of Texture Studies* 44: 75–84.

FAO/WHO 2001. Report on joint FAO/WHO expert consultation on evaluation of health and nutritional properties of probiotics in food including powder milk with live lactic acid bacteria, URL: http://www.who.int/foodsafety/publications/fs_management/en/probiotics.pdf

Feng, Y., Y. Cai, G. Su, H. Zhao, C. Wang, and M. Zhao. 2014. Evaluation of aroma differences between high-salt liquid-state fermentation and low-salt solid-state fermentation soy sauces from China. *Food Chemistry* 145: 126–34.

Fernandes, P. 2010. Enzymes in food processing: A condensed overview on strategies for better biocatalysts. *Enzyme Research* Article ID 862537, http://dx.doi.org/10.4061/2010/862537.

Fesus, L., and M. Piacentini. 2002. Transglutaminase 2: An enigmatic enzyme with diverse functions. *Trends in Biochemical Sciences* 27: 534–9.

Finley, J.W., W.L. Stanley, and G.G Walters. 1979. Chill proofing beer with papain immobilized on chitin. *Process Biochemistry* 7: 12–3.

Finn, R.K. 1987. Conversion of starch to liquid sugar and ethanol. In *Biotec 1. Microbial Genetic Engineering and Enzyme Technology*, ed. C.P. Hollenberg, and H. Sahm, 101–7. New York: Gustav Fisher.

Flamm, E.L. 1991. How FDA approved chymosin: A case history. *Bio/Technology* 9: 349–51.

Fontán, M.C.G., S. Martínez, I. Franco, and J. Carballo. 2006. Microbiological and chemical changes during the manufacture of Kefir made from cows' milk, using a commercial starter culture. *International Dairy Journal* 16: 762–7.

Gaggia, F., D. Di Gioia, L. Baffoni, and B. Biavati. 2011. The role of protective and probiotic cultures in food and feed and their impact in food safety. *Trends in Food Science and Technology* 22: S58–S66.

Gallagher, E. 2009. Improving gluten-free bread quality through the application of enzymes. *Agro Food Industry Hi-Tech* 20: 34–7.

Gandhi, N.N. 1997. Applications of lipase. *JAOCS* 74: 621–34.

Gerelt, B., Y. Ikeuchi, and A. Suzuki. 2000. Meat tenderization by proteolytic enzymes after osmotic dehydration. *Meat Science* 56: 311–8.

Goesaert, H., K. Brijs, W.S. Veraverbeke, C.M. Courtin, K. Gebruers, and J.A. Delcour. 2005. Wheat flour constituents: How they impact bread quality, and how to impact their functionality. *Trends in Food Science and Technology* 16: 12–30.

Goesaert, H., L. Slade, H. Levine, and J.A. Delcour. 2009. Amylases and bread firming—An integrated view. *Journal of Cereal Science* 50: 345–52.

Greenberg, N.A., and R.R. Mahoney. 1981. Immobilization of lactase (β-galactosidase) for use in dairy processing. A review process. *Process Biochemistry* 16: 2–8.

Güzel-Seydim, Z., J.T. Wyffels, A.C. Seydim, and A.K. Greene. 2005. Turkish kefir and kefir grains: Microbial enumeration and electron microscopic observation. *International Journal of Dairy Technology* 58: 25–9.

Guzman-Maldonado, H., and O. Paredes-Lopez. 1995. Amylolytic enzymes and products derived from starch: A review. *Critical Reviews in Food Science and Nutrition* 35: 373–403.

Han J., J.D. Morton, A.E.D. Bekhit, and J.R. Sedcole. 2009. Pre-rigor infusion with kiwifruit juice improves lamb tenderness. *Meat Science* 82: 324–30.

Hasan, F., A.A. Shah, and A. Hameed. 2006. Industrial applications of microbial lipases. *Enzyme and Microbial Technology* 39: 235–51.

Hugas, M., and J.M. Monfort. 1997. Bacterial starter cultures for meat fermentation. *Food Chemistry* 59: 547–54.

Hurtado, A., C. Reguant, A. Bordons, and N. Rozès. 2012. Lactic acid bacteria from fermented table olives. *Food Microbiology* 31: 1–8.

Hutkins, R.W. 2006. *Microbiology and Technology of Fermented Foods*, 233–59. Ames: Blackwell Publishing.

James, J., B.K. Simpson, and M.R. Marshall. 1996. Application of enzymes in food processing. *Critical Reviews in Food Science and Nutrition* 36: 437–63.

Jung, J.Y., S.H. Lee, J.M. Kim, M.S. Park, J. Bae, Y. Hahn, E.L. Madsen, and C.O. Jeon. 2011. Metagenomic analysis of kimchi, a traditional Korean fermented food. *Applied and Environmental Microbiology* 77: 2264–74.

Jurado, E., F. Camacho, G. Luzon, and J.M. Vicaria. 2004. Kinetic model of lactose hydrolysis in a hollow-fibre bioreactor. *Chemical Engineering Science* 59: 397–405.

Kashyap, D.R., P.K. Vohra, S. Chopra, and R. Tewari. 2001. Applications of pectinases in the commercial sector: A review. *Bioresource Technology* 77: 215–27.

Kaur, P., A. Vohra, and T. Satyanarayana. 2013. Laboratory and industrial bioreactors for submerged fermentations. In *Fermentation Processes Engineering in the Food Industry*, ed. C.R. Soccol, A. Pandey, and C. Larroche, 165–78. USA: CRC Press.

Khardenavis, A.A., A. Kapley, and H.J. Purohit. 2009. Processing of poultry feathers by alkaline keratin hydrolyzing enzyme from *Serratia* sp. HPC 1383. *Waste Management* 29: 1409–15.

Kilcawley, K.N., M.G. Wilkinson, and P.F. Fox. 1998. Enzyme-modified cheese. *International Dairy Journal* 8: 1–10.

Kilcawley, K.N., M.G. Wilkinson, and P.F. Fox. 2002. Properties of commercial microbial proteinase preparations. *Food Biotechnology* 16: 29–55.

Kirk, O., T.V. Borchert, and C.C. Fuglsang. 2002. Industrial enzyme applications. *Current Opinion in Biotechnology* 13: 345–51.

Kourkoutas, Y., M. Komaitis, A.A. Koutinas, and M. Kanellaki, 2001. Wine production using yeast immobilized on apple pieces at low and room temperatures. *Journal of Agricultural and Food Chemistry* 49: 1417–25.

Krasaekoopt, W., B. Bhandari, and H. Deeth. 2003. Evaluation of encapsulation techniques of probiotics for yoghurt. *International Dairy Journal* 13: 3–13.

Kurtzman, C.P. and C.J. Robnett. 2003. Phylogenetic relationships among yeasts of the '*Saccharomyces* complex' determined from multigene sequence analyses. *FEMS Yeast Research* 3: 417–432.

Ladero, M., M.T. Perez, A. Santos, and F. Garcia-Ochoa. 2002. Hydrolysis of lactose by free and immobilized β-galactosidase from *Thermus* sp. strain T2. *Biotechnology and Bioengineering* 81: 241–52.

Lee, H., H. Yoon, Y. Ji, H. Kim, H. Park, and J. Lee. 2011. Functional properties of *Lactobacillus* strains from kimchi. *International Journal of Food Microbiology* 145: 155–61.

Leroy, F., J. Verluyten, and L. De Vuyst. 2006. Functional meat starter cultures for improved sausage fermentation. *International Journal Food Microbiology* 106: 270–5.

Leveque, E., S. Janecek, B. Haye, and A. Belarbi. 2000. Thermophilic archaeal amylolytic enzymes. *Enzyme and Microbial Technology* 26: 3–14.

Lindner, J., A.L.B. Penna, I.M. Demiate, C.T. Yamaguishi, M.R.M. Prado, and J.L. Parada. 2013. Fermented foods and human health benefits of fermented functional foods. In *Fermentation Processes Engineering in the Food Industry*, ed. C.R. Soccol, A. Pandey, and C. Larroche, 264–91. USA: CRC Press.

Linko, Y.-Y., P. Javanainen, and S. Linko. 1997. Biotechnology of bread baking. *Trends in Food Science and Technology* 8: 339–44.

Lourens-Hattingh, A., and B.C. Viljoen. 2001. Yogurt as probiotic carrier food. *International Dairy Journal* 11: 1–17.

Lücke, FK. 1994. Fermented meat products. *Food Research International* 27: 299–307.

Magalhães, K.T., M.A. Pereira, A. Nicolau, G. Dragone, L. Domingues, J.A. Teixeira, J.B. de Almeida Silva, and R.F. Schwan. 2010. Production of fermented cheese whey-based beverage using kefir grains as starter culture: Evaluation of morphological and microbial variations. *Bioresource Technology* 101: 8843–50.

Mahoney, R.R. 1998. Galactosyl-oligosaccharide formation during lactose hydrolysis: A review. *Food Chemistry* 63: 147–54.

Markiewicz, L.H., E. Biedrzycka, E. Wasilewska, and M. Bielecka. 2010. Rapid molecular identification and characteristics of Lactobacillus strains. *Folia Microbiology* 55: 481–8.

Mathiasen, T.E. 1995. Laccase and beer storage, PCT Int. Appl. WO 9521240 A2.

McGovern, P.E., J. Zhang, J. Tang et al. 2004. Fermented beverages of pre- and proto-historic China. *PNAS USA* 101: 17593–8.

Melendo, J.A., J.A. Beltrán, I. Jaime, R. Sancho, and P. Roncalés. 1996. Limited proteolysis of myofibrillar proteins by bromelain decreases toughness of coarse dry sausage. *Food Chemistry* 57: 429–33.

Miguel, A.S.M., T.S. Martins-Meyer, E.V. da Costa Figueiredo, B.W.P. Lobo, and G.M. Dellamora-Ortiz. 2013. Enzymes in bakery: Current and future trends. In *Food Industry*, ed. I. Muzzalupo, InTech, http://www.intechopen.com/books/food-industry/enzymes-in-bakery-current-and-future-trends.

Minussi, R.C., G.M. Pastore, and N. Durán. 2002. Potential applications of laccase in the food industry. *Trends in Food Science and Technology* 13: 205–16.

Muyzer, G., and K. Smalla. 1998. Application of denaturing gradient gel electrophoresis (DGGE) and temperature gradient gel electrophoresis (TGGE) in microbial ecology. *Antonie van Leeuwenhoek* 73: 127–41.

Nam, Y.D., H.W. Chang, K.H. Kim, S.W. Roh, and J.W. Bae. 2009. Metatranscriptome analysis of lactic acid bacteria during kimchi fermentation with genome-probing microarrays. *International Journal of Food Microbiology* 130: 140–6.

Nielsen, P.H., H. Malmos, T. Damhus et al. 1994. Enzyme applications (industrial). In *Encyclopedia of Chemical Technology*, Vol. 9, ed. Kirk-Othmer, 567–620. New York: John Wiley and Sons, Inc.

Norus, J. 2006. Building sustainable competitive advantage from knowledge in the region: The industrial enzymes industry. *European Planning Studies* 14: 681–96.

Novozymes A/S: The Novozymes Report 2011. (http://report2011.novozymes.com/, accessed October 15, 2013).

Novozymes A/S: The Novozymes Report 2012. (http://report2012.novozymes.com/, accessed October 15, 2013).

O'Connell, J.E., and C.G. Kruif. 2003. β-casein micelles: Cross-linking with transglutaminase. *Colloids and Surfaces A: Physicochemistry Engineering Aspects* 216: 75–81.

Okafor, N. 2007. *Modern Industrial Microbiology and Biotechnology.* USA: Science Publishers.

Olempska-Beer, Z. 2004. Laccase from *Myceliophthora thermophila* expressed in *Aspergillus oryzae*. Chemical and Technical Assessment (Cta) for 61st JECFA.

Olempska-Beer, Z. 2008. Asparaginase from *Aspergillus niger* expressed in *A. niger*. Chemical and Technical Assessment (Cta) for 69st JECFA.

Olempska-Beer, Z.S., R.I. Merker, M.D. Ditto, and M.J. DiNovi. 2006. Food-processing enzymes from recombinant microorganisms—A review. *Regulatory Toxicology and Pharmacology* 45: 144–58.

Olesen, T., J. Qi Si, and V. Donelyan. 2000. Use of lipase in baking. US Patent Application; 2000. US 6110508.

Ooijkaas, L.P., F.J. Weber, R.M. Buitelaar, J. Tramper, and A. Rinzema. 2000. Defined media and inert supports: Their potential as solid-state fermentation production systems. *Trends in Biotechnology* 18: 356–60.

Özrenk, E. 2006. The use of transglutaminase in dairy products. *International Journal of Dairy Technology* 59: 1–7.

Panagou, E.Z., U. Schillinger, C. Franz, and G.H. Nychas. 2008. Microbiological and biochemical profile of cv. Conservolea naturally black olives during controlled fermentation with selected strains of lactic acid bacteria. *Food Microbiology* 25: 348–58.

Pandey, A. 2003. Solid-state fermentation. *Biochemical Engineering Journal* 13: 81–4.

Pandey, A., C.R. Soccol, P. Nigam et al. 2000. Biotechnological potential of coffee pulp and coffee husk for bioprocesses. *Biochemical Engineering Journal* 6: 153–62.

Pareyt, B., S.M. Finnie, J.A. Putseys, and J.A. Delcour. 2011. Lipids in bread making: Sources, interactions, and impact on bread quality. *Journal of Cereal Science* 54: 266–79.

Pariza, M.W., and E.A. Johnson. 2001. Evaluating the safety of microbial enzyme preparations used in food processing: Update for a new century. *Regulatory Toxicology and Pharmacology* 33: 173–86.

Parvez, S., K.A. Malik, S.Ah. Kang, and H.Y. Kim. 2006. Probiotics and their fermented food products are beneficial for health. *Journal of Applied Microbiology* 100: 1171–85.

Pifferi, P.G., G. Buska, I. Manenti, A. Lo Presti, and G. Spagna. 1989. Immobilization of pectinesterase on gamma alumina for the treatment of juices. *Belgian Journal of Food Chemistry and Biotechnology* 44: 173–82.

Pilnik, W., and A.G.J. Voragen. 1993. Pectic enzymes in fruit juice and vegetable juice manufacture. In *Food and Science Technology, Enzymes in Food Processing*, ed. G. Reeds, 363–99. New York: Academic Press.

Plengvidhya, V., Jr.F. Breidt, and H.P. Fleming. 2004. Use of RAPD-PCR as a method to follow the progress of starter cultures in sauerkraut fermentation. *International Journal of Food Microbiology* 93: 287–96.

Poutanen, K., L. Flander, and K. Katina. 2009. Sourdough and cereal fermentation in a nutritional perspective. *Food Microbiology* 26: 693–9.

Qihe, C., H. Guoqing, J. Yingchun, and N. Hui. 2006. Effects of elastase from a *Bacillus* strain on the tenderization of beef meat. *Food Chemistry* 98: 624–9.

Raimbault, M. 1998. General and microbiological aspects of solid substrate fermentation. *EJB Electronic Journal of Biotechnology* 1(3), ISSN: 0717-3458.

Raj, E.A., and G.N. Karanth. 2006. Fermentation technology and bioreactor design. In *Food Biotechnology*, ed. K. Shetty, G. Paliyath, A. Pometto, and R.E. Levin. USA: CRC Press.

Raksakulthai, R., and N.F. Haard. 2003. Exopeptidases and their application to reduce bitterness in food: A review. *Critical Reviews in Food Science and Nutrition* 43: 401–45.

Randazzo, C.L., S. Torriani, A.DL. Akkermans, W.M. de Vos, and E.E. Vaughan. 2002. Diversity, dynamics, and activity of bacterial communities during production of an artisanal Sicilian cheese as evaluated by 16S rRNA analysis. *Applied and Environmental Microbiology* 68: 1882–92.

Rani, K.Y., and V.S.R. Rao. 1999. Control of fermenters: A review. *Bioprocess Engineering* 21: 77–88.

Rantsiou, K. and L. Cocolin. 2006. New developments in the study of the microbiota of naturally fermented sausages as determined by molecular methods: A review. *International Journal of Food Microbiology* 108: 255–267.

Rasiah, I.A., K.H. Sutton, F.L. Low, H.M. Lin, and J.A. Gerrard. 2005. Crosslinking of wheat dough proteins by glucose oxidase and the resulting effects on bread and croissants. *Food Chemistry* 89: 325–32.

Renzetti, S., C.M. Courtin, J.A. Delcour, and E.K. Arendt. 2010. Oxidative and proteolytic enzyme preparations as promising improvers for oat bread formulations: Rheological, biochemical and microstructural background. *Food Chemistry* 119: 1465–73.

Romero de Ávila, M.D., J.A. Ordóněz, L. de la Hoz, A.M. Herrero, and M.I. Cambero. 2010. Microbial transglutaminase for cold-set binding of unsalted/salted pork models and restructured dry ham. *Meat Science* 84: 747–54.

Rossi, M., G. Giovanelli, C. Cantarelli, and O. Brenna. 1988. Effects of laccase and other enzymes on barley wort phenolics as a possible treatment to prevent haze in beer. *Bulletin de Liaison- Groupe Polyphenols* 14: 85–8.

Rouau, X., M.L. El-Hayek, and D. Moreau. 1994. Effect of an enzyme preparation containing pentosanases on the breadmaking quality of flours in relation to changes in pentosan properties. *Journal of Cereal Science* 19: 259–72.

Salleh, A.B., C.N.A Razak, R.N.Z.R.A Rahman, and M. Basri. 2006. Protease: Introduction. In *New Lipases and Proteases*, ed. A.B. Salleh, R.N.Z.R.A. Rahman, and M. Basri, 23–39. New York: Nova Science Publishers.

Samuel, D. 1996. Investigation of ancient Egyptian baking and brewing methods by correlative microscopy. *Science* 273: 488–90.

Schmedding, D.J.M., and M.J.M.C van Gestel. 2002. Enzymes in brewing. In *Enzymes in Food Technology*, ed. R.J. Whitehurst, and B.A. Law, 57–75. Sheffield: CRC Press.

Selinheimo, E., K. Kruus, J. Buchert, A. Hopia, and K. Autio. 2006. Effects of laccase, xylanase and their combination on the rheological properties of wheat doughs. *Journal of Cereal Science* 43: 152–9.

Settani, L., and G. Moschetti. 2010. Non-starter lactic acid bacteria used to improve cheese quality and provide health benefits. *Food Microbiology* 27: 691–7.

Shukla, T.P. 1975. Beta-galactosidase technology: A solution to the lactose problem. *CRC Critical Reviews in Food Technology* 5: 325–56.

Sisak, C., Z. Csanádi, E. Rónay, and B. Szajáni. 2006. Elimination of glucose in egg white using immobilized glucose oxidase. *Enzyme and Microbial Technology* 39: 1002–7.

Soccol, C. R., L.P.S. Vandenberghe, M.R. Spier et al. 2010. The potential of probiotics: A review. *Food Technology and Biotechnology* 48: 413–34.

Spence, J.T. 2006. Challenges related to the composition of functional foods. *Journal of Food Composition and Analysis* 19: S4–S6.

Stanton, C., R.P. Ross, G.F. Fitzgerald, and D. Van Sinderen. 2005. Fermented functional foods based on probiotics and their biogenic metabolites. *Current Opinion in Biotechnology* 16: 198–203.

Steinkraus, K.H. 2002. Fermentations in world food processing. *Comprehensive Reviews in Food Science and Food Safety* 1: 23–32.

Stepaniak, L. 2004. Dairy enzymology. *International Journal of Dairy Technology* 57: 153–71.

Synowiecki, J., B. Grzybowska, and A. Zdziebło. 2006. Sources, properties and suitability of new thermostable enzymes in food processing. *Critical Reviews in Food Science and Nutrition* 46: 197–205.

Tamang, J.P. 2010. Diversity of fermented beverages and alcoholic drinks. In *Fermented Foods and Beverages of the World*, ed. J.P. Tamang, and K. Kailasapathy, 85–125. USA: CRC Press.

Tanriöven, D., and A. Ekşi. 2005. Phenolic compounds in pear juice from different cultivars. *Food Chemistry* 93: 89–93.

This, P., T. Lacombe, and M.R. Thomas. 2006. Historical origins and genetic diversity of wine grapes. *Trends in Genetics* 22: 511–9.

Topakas, E., G. Panagiotou, and P. Christakopoulos. 2013. Xylanases: Characteristics, sources, production and applications. In *Bioprocessing Technologies in Biorefinery for Sustainable Production of Fuels, Chemicals, and Polymers*, ed. S.-T. Yang, H.A. El-Enshasy, and N. Thongchul, 147–66. New Jersey: John Wiley and Sons.

Valamoti, S.M., M. Mangafa, C. Koukouli-Chrysanthaki, and D. Malamidou. 2007. Grape-pressings from northern Greece: the earliest wine in the Aegean? *Antiquity* 81: 54–61.

Van Boekel, M., V. Fogliano, N. Pellegrini et al. 2010. A review on the beneficial aspects of food processing. *Molecular Nutrition and Food Research* 54: 1215–47.

van Hijum, S. AFT., E.E. Vaughan, and R.F Vogel. 2013. Application of state-of-art technologies to indigenous food fermentations. *Current Opinion in Biotechnology* 24: 178–86.

van Oort, M. 2010. Enzymes in bread making. In: *Enzymes in Food Technology*, ed. R.J. Whitehurst, and M. van Oort, 103–43. Chichester: Wiley-Blackwell.

Vasic-Racki, D. 2006. History of industrial biotransformations-dreams and realities. In *Industrial Biotransformations*, ed. A. Liese, K. Seelbach, and C. Wandrey, 1–35. Weinheim: Wiley-VCH.

Vishwanatha, K.S., A.G.A. Rao, and S.A. Singh. 2010. Production and characterization of a milk-clotting enzyme from *Aspergillus oryzae* MTCC 5341. *Applied Microbiology and Biotechnology* 85: 1849–59.

Vogel, R.F., M. Pavlovic, M.A. Ehrmann, A. Wiezer, H. Liesegang, S. Offschanka, S. Voget, A. Angelov, G. Böcker, and W. Liebl. 2011. Genomic analysis reveals *Lactobacillus sanfranciscensis* as stable element in traditional sourdoughs. *Microbial Cell Factories* 10: S6.

Wacher, C., G. Díaz-Ruiz, and J.P. Tamang. 2010. Fermented vegetable products. In *Fermented Foods and Beverages of the World*, ed. J.P. Tamang, and K. Kailasapathy, 149–90. London: CRC Press.

Yokotsuka, T. 1982. Traditional fermented soybean foods. In *Fermented Foods*, ed. A.H. Rose. UK: Academic Press.

Zdzieblo, A., and J. Synowiecki. 2002. New source of thermostable α-glucosidase suitable for single step starch processing. *Food Chemistry* 79: 485–91.

Zeuthen, P. 2007. A historical perspective of meat fermentation. In *Handbook of Fermented Meat and Poultry*, ed. F. Toldra, 3–8. UK: Blackwell Publishing.

Zhu, Y., A. Rinzema, J. Tramper, and J. Bol. 1995. Microbial transglutaminase-a review of its production and application in food processing. *Applied Microbiology and Biotechnology* 44: 277–82.

# 12 Fluid and Species Transfer in Food Biopolymers

*Pawan S. Takhar*

## CONTENTS



## 12.1 INTRODUCTION

Most foods exhibit a hierarchy of spatial scales (micro, meso, and macro), where physicochemical interactions between fluids, biopolymers, and dissolved species occur at various scales (Singh et al., 2003a). For example, during frying of foods, oil transport occurs at macroscale, water transport occurs at all three scales, and chemical reactions occur below microscale. The state of the food (glassy, rubbery, or glass transition) also affects the quality changes in foods and nature of fluid transport (Fickian and non-Fickian). Several experimental studies have shown that fluid transport follows Fick's law of diffusion only in glassy and rubbery regimes. Near glass transition, the conformational changes in polymers lead to an additional time-dependent stress term in the transport equation (Singh et al., 2003b), which makes fluid transport non-Fickian (or non-Darcian). This transport cannot be described using Fick's law even when the coefficient of diffusivity is defined as a function of fluid concentration (Thomas and Windle, 1980). Despite the complexity of these interactions, the overall system can be described using continuum mechanics to develop chemical potential, pressure, thermomechanical stress, generalized Darcy's law, and Fick's law-based relations, and coupling them to kinetic equations representing chemical reactions (Singh et al., 2003a,b,c). Modeling of fluid and species transport is needed for several food processing applications such as drying, frying, salt transport, extrusion, flavors release, encapsulation, and so on.

## 12.2 FLUID TRANSPORT IN FOODS

In a porous food matrix, Fick's law has been used extensively to model diffusion of fluids. This law states that velocity of fluid diffusing through the matrix is directly proportional to the concentration gradient. Fick's second law can be written as

$$\frac{\partial C}{\partial t} = D\nabla^2 C \tag{12.1}$$

where $C$ is the concentration of fluid, $D$ is the coefficient of diffusivity, and $t$ is time.

Polymer science studies have shown that fluid transport does not follow Fick's law near glass transition. Fick's law is followed only when the food is in rubbery or glassy states, away from the glass-transition regime (Singh et al., 2003b).

The transport becomes non-Fickian (or non-Darcian) near glass transition (Alfrey et al., 1966; Astarita and Sarti, 1978; Peterlin, 1979; Thomas and Windle, 1980, 1982). The fluid content versus radial cross-section profiles become sharp near glass transition (profiles are round for Fickian transport). These studies concluded that a time-dependent stress term resulting from viscoelastic relaxation of polymers is responsible for causing non-Fickian transport near glass transition.

In foods undergoing glass transition in drying temperature and fluid content range (e.g., soybeans, corn, pasta, and potatoes), as water evaporates from the outside, a glassy layer is formed near the surface that separates from the rubbery interior by a glass-transition region (Figure 12.1d). In the glassy region, a small amount of moisture diffuses by following Fick's law of diffusion. In the glassy region, moisture creates space by pushing apart and disentangling the polymer chains and the polymers relax at a very slow rate as compared to the diffusion rate (relaxation time of the order of infinity). The stress in glassy polymers is not a function of time and it causes only a baseline shift to the diffusivity values.

In the glass-transition region, moisture content is of intermediate magnitude, greater than the glassy region and less than the rubbery region. In this band, polymers undergo viscoelastic relaxation at a time scale on the order of diffusion time.

The energy transferred from the fluid to the polymers is dissipated with time, which is of the order of the diffusion time. This is exhibited as time-dependent stress relaxation in polymers. The polymer–fluid interaction results in an additional stress term that opposes or enhances the fluid movement depending upon the state of the food and direction of fluid motion (Singh et al., 2004a). It is well recognized that the extra stress term is responsible for anomalous (non-Fickian) transport. During drying, the core of the food may contain a higher amount of moisture and exist in a rubbery state. The flow behavior in the rubbery state is also Fickian, because the polymers relax at a significantly faster rate as compared to the diffusion rate. Therefore, in the rubbery state, too, the viscoelastic relaxation of polymers does not result in a time-dependent stress term. When the moisture distribution is plotted across the cross section of a food matrix undergoing glass transition, sharper profiles are obtained (Figure 12.1a). When fluid transport takes place in predominantly

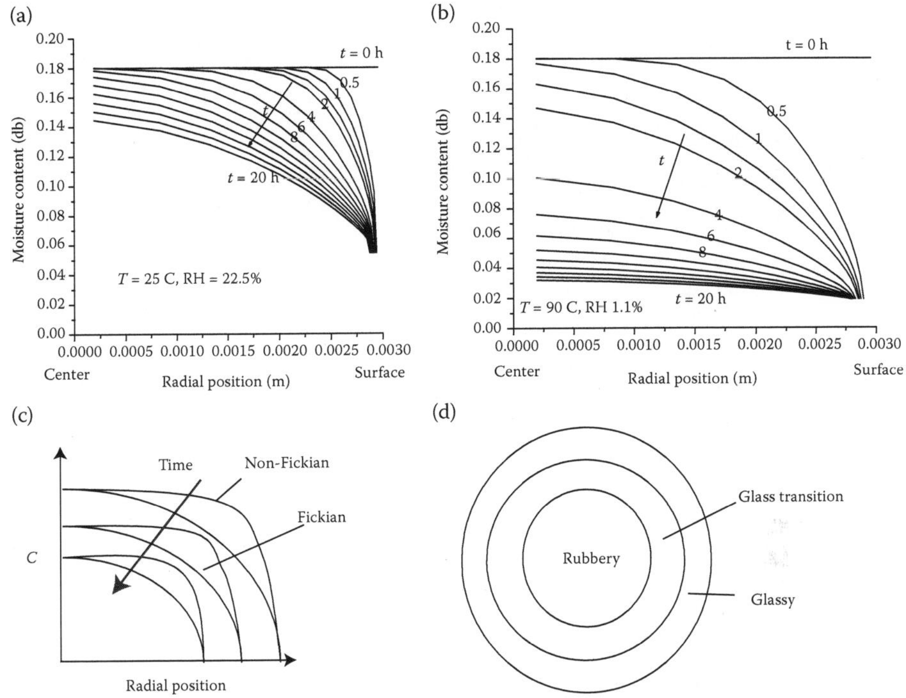


**FIGURE 12.1** (a) Sharper moisture profiles in a soybean seed during drying in non-Fickian regime as calculated by Singh et al. (2004a), (b) round moisture profiles in a soybean seed drying under Fickian transport conditions in a predominantly glassy region (Singh et al., 2004a), (c) round profiles become sharp when viscoelastic relaxation opposes fluid flow during drying, and (d) schematic of a spherical food particle showing glassy, transition, and rubbery regions during drying.

glassy (or rubbery) regimes, Fickian transport occurs that result in round moisture profiles (Figure 12.1b).

Figure 12.1c shows that round concentration versus radial cross-section profiles become sharp due to time-dependent opposing forces applied by the bending of polymers during drying.

Figure 12.2 shows a comparison of normalized stress profiles during sorption and drying of soybeans obtained by Singh et al. (2004a). Soybeans undergo glass transition near 25°C and 10% moisture content (Bruni and Leopold, 1991). By comparing the moisture and stress profiles, it was concluded that during sorption, the falling part of the stress curve (near the surface) aids in moisture transport and the rising part of the curve (near the center) opposes moisture transport (Figure 12.2a). During drying, the stress profiles decrease from a maximum at the surface toward the center (Figure 12.2a). A higher magnitude of stress relaxation function near the surface in dried region caused higher stress values near the surface during drying. This demonstrated that mechanical stress in biopolymers adds a significant component that opposes or aids moisture transport during glass transition.

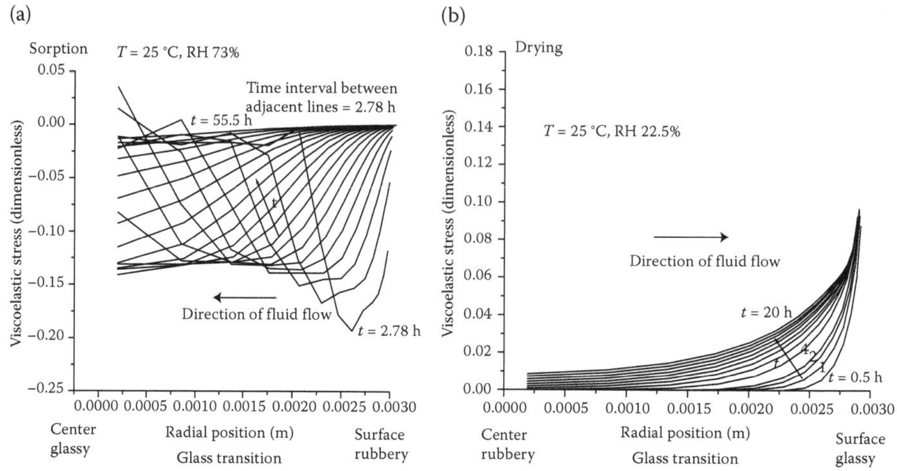


**FIGURE 12.2** Normalized stress profiles in soybeans during (a) sorption and (b) drying as calculated by Singh et al. (2004a).

### 12.2.1 Multiscale Transport Theory

Singh et al. (2003a,b,c) showed that the effect of viscoelastic relaxation on fluid transport can be included using the multiscale hybrid mixture theory (HMT). HMT is a systematic framework for developing governing equations, which describe the behavior of porous media such as soils, polymers, starch foams, cells, and living tissues. The media may consist of multiple phases, multiple dissolved constituents (e.g., biochemicals, salts, and proteins), multiple spatial scales (micro, meso, and macro), and interfaces between different phases (Achanta et al., 1994; Bennethum and Cushman, 1996; Gray and Hassanizadeh, 1998). HMT is a hybridization of classical volume averaging of field equations (laws of mass, momentum, and energy conservation) with a classical theory of mixtures (Bowen, 1971). The constitutive theory is developed by imposing the second law of thermodynamics via entropy inequality (Coleman and Noll, 1963) and following constitutive axioms (pertaining to constitutive theory) of continuum mechanics (Eringen, 1980). Applying this method to a wide class of materials and transport problems has shown that the systems that are thermodynamically viable are also physically viable (Eringen, 1980). The approach leads to the development of physically viable equations in a straightforward manner instead of adjusting their terms after years of experimentation.

For two-scale material, the results of HMT include governing equations applicable at the mesoscale (e.g., the scale of cell cytoplasm) as opposed to variables measured at the microscale (e.g., the scale of cell walls, protein bodies, and starch granules). For three-scale material, the resulting governing equations are applicable at the macroscale (e.g., the scale of several cells grouped together). The resulting equations at higher scales also include the effects of physical phenomena at the lower scales (micro and meso).

The three-scale fluid transport equation of Singh et al. (2003b) includes the effect of viscoelastic relaxation taking place at the scale of polymers on the macroscale movement of fluids within the food matrix. In vectorial notation, this equation can be written as (Singh et al., 2004b)

$$\frac{D^s\varepsilon^f}{Dt} + (\varepsilon^f - 1)\nabla \cdot (D\nabla\varepsilon^f) - (\varepsilon^f - 1)\nabla \cdot \left[\int_0^t B_c G(t-\tau)\nabla \frac{d\varepsilon^f(\tau)}{d\tau}d\tau\right] = 0, \quad (12.2)$$

where $\varepsilon^f$ is the volume fraction of the fluid and $\nabla$ is the gradient operator in spatial dimensions. The $D^s\varepsilon^f/Dt$ is the material time derivative calculated with respect to the solid-phase particles. The first two terms are similar to the classical Fickian equation. The integral term is novel, which showed up while following the procedure of HMT for viscoelastic solid matrix interacting with the viscous fluid phase at lower scales. The experimental properties needed to solve Equation 12.2 are the Fickian diffusion coefficient $D$, the stress relaxation function $G(t)$, and the relaxation parameter $B_c$. Since the equation was obtained after upscaling (or volume averaging) the microscale field equations, the variables of Equation 12.2 are defined at the macroscale.

Therefore, the classical techniques commonly used in engineering can be used to measure these properties. The values published in rheological and transport literature on foods can also be utilized. $D$ can be obtained using invasive or noninvasive experimental techniques. $G(t)$ can be measured by performing the classical stress relaxation tests using a universal testing machine such as Instron (Instron, Norwood, MA). $B_c$ $[\text{kg}/(\text{m}^3\text{s})]^{-1}$ is a parameter that links the effect of polymer relaxation with the fluid movement. In other words, the product $B_cG(t)$ accounts for the effect of stress relaxation in polymers on fluid transport. As shown in Singh et al. (2004a), $B_c$ can be determined by superimposing the $\varepsilon^f$ values (obtained using Equation 12.2 over the experimentally measured drying (or sorption) curves (Singh et al., 2004a) or using the asymptotic behavior of $D/G(t)$ in glassy and rubbery states (Takhar, 2011).

## 12.3 GENERALIZED DARCY'S LAW

While generalized transport equation can describe the diffusion of fluids in food systems, transport can also be pressure driven in applications such as rapid drying, frying, polymer expansion, and membrane transport. Transport in foods can involve both concentration and pressure-driven flow occurring in the same matrix. Takhar (2014) used HMT to develop the following generalized Darcy's law that combines both concentration and pressure-driven flow:

$$v^{f,s} = -\left[\frac{K^f}{\mu^f}\varepsilon^f\nabla p^f + \varepsilon^f D^f\nabla\varepsilon^f + \varepsilon^f\frac{K^f}{\mu^f}B^f\nabla\dot{\varepsilon}^f\right] \quad (12.3)$$

where $v^{f,s}$ is the velocity of the fluid phase relative to the solid phase, $K^f$ is the permeability of the fluid, $\mu^f$ is the dynamic viscosity of the fluid, $D^f$ is the coefficient of diffusivity, and $B^f$ is the mixture viscosity (Achanta et al., 1994; Singh et al., 2003a).

The generalized Darcy's law equation can be combined with the following two-scale mass balance equation (Achanta et al., 1994) to calculate the fluid volume fraction distribution in the food matrix:

$$\frac{D^s(\varepsilon^f\rho^f)}{Dt} + \nabla\cdot(\varepsilon^f\rho^f v^{f,s}) - \varepsilon^f\rho^f\frac{\dot{\varepsilon}^s}{\varepsilon^s} = \hat{e}^f, \tag{12.4}$$

where the material time derivative is taken with respect to the solid phase. The solid-phase volume fraction $\varepsilon^s$ can be obtained from its relation with the porosity, $\varepsilon^s = 1 - \phi$. $\dot{\varepsilon}^s$ represents the material time derivative of the solid-phase volume fraction. The right-hand side represents the source/sink term due to mass exchange between the phases during phenomena such as condensation or evaporation. Using $f = w, v$, the mass balance equations for water and vapor phases can be written. The water and vapor phase equations would be coupled using right-hand side of Equation 12.4, $\hat{e}^w = -\hat{e}^v$, as mass lost by one phase is gained by the other phase.

## 12.4 GENERALIZED FICK'S LAW FOR MASS TRANSPORT

Various phases in a food matrix can be composed of dissolved constituents (species) such as salts, biochemicals, flavors, and so on. The species can diffuse from a point of higher concentration to a point of lower concentration. The mechanism of transport for species is molecular motion of one species relative to the surrounding phases (Bird et al., 2006). The driving force for mass transport is species concentration represented using mole fraction or mass fraction. The reader is referred to Bird et al. (2006) for various forms of Fick's law. In this chapter, a more general thermodynamic form, which involves chemical potential gradient as the driving force, is presented.

When the framework of continuum thermodynamics is used for deriving equations, the chemical potential gradient shows up naturally as the driving force for mass transport in multicomponent systems (Singh et al., 2003a). Using continuum mechanics-based framework, the following relation for chemical potential ($\mu^{f_j}$) can be derived (Singh et al., 2003b):

$$\mu^{f_j} I = A^{f_j} I - \frac{t^{f_j}}{\rho^{f_j}} \tag{12.5}$$

where $A^{f_j}$ represents the Helmholtz free energy of the "jth" species dissolved in fluid $f$, $t^{f_j}$ is the mechanical stress, and $\rho^{f_j}$ is the density of the species. In multicomponent systems, $j$ can vary from 1 to $N$, where $N$ is the number of components comprising a phase.

To solve mass transfer problems, the two-scale species mass balance equation of Achanta et al. (1994) and Bennethum and Cushman (1996) can be combined with the generalized Fick's law. The species mass balance equation for the transport of the jth species is

$$\frac{D^{f_j}(\varepsilon^f\rho^{f_j})}{Dt} + \varepsilon^f\rho^{f_j}\nabla\cdot u^{f_j} = \hat{e}^{f_j} + \hat{r}^{f_j} \tag{12.6}$$

Here, $D^{f_j}/Dt$ represents material time derivative with respect to a jth particle dissolved in phase $f$ and $u^{f_j}$ represents the diffusive velocity of component $j$. The first term on the right-hand side of Equation 12.6 represents the mass gained by a species from other species and $\hat{r}^{f_j}$ represents the production of species $j$ due to chemical reactions. To calculate the diffusive flux of component $j$, the following generalized Fick's law can be used (Singh et al., 2003a):

$$R^j u^{f_j} = -\varepsilon^f \rho^{f_j} \nabla \mu^{f_j} + \varepsilon^f \rho^{f_j} g^{f_j} \tag{12.7}$$

where $R^j$ is the near-equilibrium material coefficient and $g^{f_j}$ is the gravitational or body force acting on a species.

## 12.5 CONCLUSIONS

Although polymer and food science literature has realized the importance of glass transition on the non-Fickian nature of fluid transport, its mathematical representation has not been complete. Continuum mechanics-based modeling approaches are a promising method to include such phenomenon. The continuum mechanics-based framework resulted in an integro-differential equation that can be solved for predicting moisture and stress profiles in foods. The equation predicts that Fick's law is valid only in the glassy and rubbery states. In the vicinity of glass transition, the transport becomes non-Fickian due to viscoelastic relaxation in polymers at a time scale of the order of diffusion or experimental time. Since many foods undergo glass transition in the processing temperature and moisture content range, its effect needs to be incorporated for enhancing the accuracy of predictions, for improved understanding of transport mechanisms, and for predicting the associated quality changes in the food. The continuum thermodynamics framework also yields generalized Fick's law and chemical potential relations, which can be used for studying species transport in food systems.

## REFERENCES

A. Cemal Eringen. *Mechanics of Continua* R. E. Krieger Pub. Co., Huntington, New York, 1980.

A. Peterlin. Diffusion with discontinuous swelling. V. Type II diffusion into sheets and spheres. *Journal of Polymer Science: Part B Polymer Physics*, 17(10):1741–1756, 1979.

B. D. Coleman and W. Noll. The thermodynamics of elastic materials with heat conduction and viscosity. *Archives of Rational Mechanical Analysis*, 13:167–178, 1963.

F. Bruni and A. C. Leopold. Glass transition in soybean seed: Relevance to anhydrous biology. *Plant Physiology*, 96(2):660–663, 1991.

G. Astarita and G. C. Sarti. A class of mathematical models for sorption of swelling solvents in glassy polymers. *Polymer Engineering Science*, 18:388–395, 1978.

L. S. Bennethum and J. H. Cushman. Multiscale, hybrid mixture theory for swelling systems. Balance laws. *International Journal of Engineering Science*, 34(2):125–145, 1996.

N. L. Thomas and A. H. Windle. A deformation model for case-II diffusion. *Polymer*, 21(6):613–619, 1980.

N. L. Thomas and A. H. Windle. A theory of case-II diffusion. *Polymer*, 23(4):529–542, 1982.

P. P. Singh, D. E. Maier, J. H. Cushman, and O. Campanella. Effect of viscoelastic relaxation on moisture transport in foods. Part II: Sorption and drying of soybeans. *Journal of Mathematical Biology*, 49(1):20–35, 2004a.

P. P. Singh, D. E. Maier, J. H. Cushman, K. Haghighi, and C. Corvalan. Effect of viscoelastic relaxation on moisture transport in foods. Part I: Solution of general transport equation. *Journal of Mathematical Biology*, 49(1):1–19, 2004b.

P. P. Singh, J. H. Cushman, and D. E. Maier. Multiscale fluid transport theory for swelling biopolymers. *Chemical Engineering Science*, 58(11):2409–2419, 2003b.

P. P. Singh, J. H. Cushman, and D. E. Maier. Three scale thermomechanical theory for swelling biopolymeric systems. *Chemical Engineering Science*, 58:4017–4035, 2003c.

P. P. Singh, J. H. Cushman, L. S. Bennethum, and D. E. Maier. Thermomechanics of swelling biopolymeric systems. *Transport in Porous Media*, 53(1):1–24, 2003a.

P. S. Takhar. Hybrid mixture theory based moisture transport and stress development in corn kernels during drying: Coupled fluid transport and stress equations. *Journal of Food Engineering*, 105(4):663–670, 2011.

P. S. Takhar, Unsaturated fluid transport in swelling poroviscoelastic biopolymers. *Chemical Engineering Science*, 109: 98–110, 2014.

R. B. Bird, W. E. Stewart, and N. L. Lightfoot. *Transport Phenomena*, volume 2. John Wiley and Sons, New York, 2nd edition, 2006.

R. M. Bowen. *Theory of Mixtures*, volume 3. Academic Press, New York, 1971.

S. Achanta, J. H. Cushman, and M. R. Okos. On multicomponent, multiphase thermomechanics with interfaces. *International Journal of Engineering Science*, 32(11):1717–1738, 1994.

T. Alfrey, E. F. Gurnee, and W. G. Lloyd. Diffusion in glassy polymers. *Journal of Polymer Science*, C12:249–261, 1966.

W. G. Gray and S. M. Hassanizadeh. Macroscale continuum mechanics for multiphase porous-media flow including phases, interfaces, common lines and common points. *Advances in Water Resources*, 21(4):261–281, 1998.

# 13 Encapsulation of Food Ingredients

## *Agents and Techniques*

*Charikleia Chranioti and Constantina Tzia*

**CONTENTS**

## 13.1 INTRODUCTION

### 13.1.1 Basic Principles of Encapsulation

Encapsulation is defined as a process in which the core material (liquid droplet, solid particle, or gas compound) is entrapped (coated or embedded) into a food-grade wall material to give an encapsulated product with many useful properties. The core material may be composed of just one or several ingredients/materials and is referred to by various names such as coated, active or entrapped material, payload, fill, or internal phase. On the other hand, the wall material may be single or in blends/mixtures and is also called an encapsulating agent, coating material, carrier, membrane, matrix, or shell (Risch, 1995; Gharsallaoui et al., 2007). The size of particles (capsules) formed through encapsulation may be classified as macro (>5000 mm); micro (1.0–5000 mm); and nano (<1.0 mm) (Couvreur et al., 2002; Jafari et al., 2008; Legrand et al., 1999; Muller et al., 2000; Shapiro, 2004).

A conventional encapsulation process includes two basic steps, the production of a dispersion/suspension or emulsion of the core and the wall material and its subsequent drying attained by a typical drying method or an alternative encapsulation technology. In such a process, the retention of the ingredient–core material into the encapsulating agent is governed by factors related to the chemical nature of the core (such as molecular weight, chemical functionality, solubility, polarity, and volatility), to the physicochemical properties of the agent (such as solubility, stability, molecular weight, conformation, and physical state), as well as to the nature and the parameters of the applied encapsulation technology (Madene et al., 2006).

The development of the encapsulation process historically started in the 1950s in the research into pressure-sensitive coatings for the manufacture of carbonless copying paper (Green and Scheicher, 1955). Nowadays, encapsulation technology is well developed and accepted within the printing, chemical, cosmetic, pharmaceutical, and food industries (Augustin et al., 2001; Heinzen, 2002).

Concerning food industries, well-defined encapsulation methods and techniques have been applied, the most commonly used of which are discussed in this chapter. Food components that may benefit from the encapsulation process belong to different categories including preservatives, flavoring agents (aroma compounds, oleoresins), colorants, sweeteners, enzymes, antioxidants, vitamins, minerals, and lipids/fats (Dziezak, 1988; Jackson and Lee, 1991; Shahidi and Han, 1993). The above food components could be added or incorporated into a food system in an encapsulated form, thus leading to the production of novel and functional foods.

### 13.1.2 Benefits of Encapsulation in Food Industry

Encapsulation is a beneficial process proposed to be used to solve various problems such as

- To protect nutritive ingredients against loss,
- To add nutrients-encapsulated ingredients to food products after processing,
- To mask the undesirable taste/flavor of some ingredients, thus enabling the production of functional foods with desirable sensory properties (LaBell, 1999),
- To protect sensitive food components from undesirable interactions with the other components during storage (Reineccius, 1991),
- To guard against light-induced reactions and/or oxidation,
- To allow a controlled release (Tari and Singhal, 2002), and
- To provide an additional attractiveness for the display (greater flexibility and control) and merchandizing of food products, thus better meeting the expectations of today's consumers (Ré, 1998).

Therefore, encapsulation can offer numerous benefits to the materials/ingredients being encapsulated, particularly in the case of functional food ingredients. More specifically, various properties of active materials may be modified or even changed by encapsulation. For example, handling and flow properties can be improved by converting a liquid into a solid-encapsulated form. Hygroscopic materials can be protected from moisture. The stability of functional ingredients and bioactive compounds that are volatile or sensitive to heat, light, or oxidation can be protected, thereby extending their shelf life. Materials that are otherwise incompatible can be mixed and utilized together safely (DeZarn, 1995; Versic, 1998). One of the major benefits from the use of food components (such as flavoring agents) in an encapsulated form is the extended shelf life that can be achieved.

### 13.1.3 Encapsulation Process: Design: Evaluation

#### 13.1.3.1 Purpose/Aim of an Encapsulation Process

The design of an encapsulation system requires a clear understanding of the purpose of encapsulation. A rational compromise between the large numbers of physical, chemical, and biological factors influencing the performance of both the active ingredient and the food matrix is necessary. Therefore, in designing an encapsulation process, the following issues should be taken into consideration (Shahidi and Han, 1993):

1. The functionality that the encapsulated ingredients should provide to the final product
2. The type of the food-grade coating material that should be selected (physical properties such as elasticity, permeability, and adhesive characteristics)
3. The processing conditions that the encapsulated ingredient may be subjected before releasing its content
4. The optimum concentration of the core material in the microcapsule
5. The mechanism by which the ingredient will be released from the microcapsule
6. The desired properties (e.g., particle size, wettability, solubility, density, and final physical form) and the stability requirements for the encapsulated ingredient
7. The cost constraints of the encapsulated ingredient

The encapsulation technology provides a variety of engineering techniques and scientific disciplines, thus offering the food technologists greater flexibility and control in developing foods that are more flavorful, nutritious, and may provide physiological benefits to human health (functional foods). However, the final choice of the technique and the coating material applied is based on a fundamental understanding of the physical and chemical phenomena determining stability, release, and perception (Ubbink and Kruger, 2006). Therefore, to design a successful encapsulated product, a proper selection of the food matrix and the process, is needed.

#### 13.1.3.2 Encapsulated Product's Yield

There are various ways to describe the retention of the ingredient–core material into the encapsulating agent, among which the most commonly used are identified below.

#### 13.1.3.3 Microencapsulation Yield

The microencapsulation yield (MEY) is defined as the ratio of core material in the final dried microcapsules to that in the emulsion (Zilberboim et al., 1986) and calculated as follows:

$$\text{MEY} = \frac{\text{Core material in microcapsules (g/100 g solids)}}{\text{Core material in emulsion (g/100 g solids)}}$$

#### 13.1.3.4 Microencapsulating Efficiency

The microencapsulating efficiency (MEE) is calculated by using the formula such as (Pauletti and Amestoy, 1999)

$$\text{MEE} = \frac{(\text{Total core material} - \text{Extractable core material})}{\text{Total core material}} \times 100$$

or

$$\text{MEE} = \frac{\text{Encapsulated core material}}{\text{Total core material}} \times 100$$

where the total core material is the initial core content added to the system before applying any encapsulation method and extractable core material is the actual core content, extracted from the final dried product, and their difference is the encapsulated core material content.

#### 13.1.3.5 Percentage Retention

The percentage retention is calculated by using the formula such as (Cai and Corke, 2000)

$$\text{Retention}\,\% = \frac{\text{Analyte at } X \text{ storage time}}{\text{Analyte at zero storage time}} \times 100$$

##### *13.1.3.5.1 Encapsulated Product's Structure: The Art of Architecture in Encapsulation*

Microencapsulation technology is sometimes considered more of an art than science. Depending on the physicochemical properties of the core, the wall composition, and the microencapsulation technique used, different types (Figure 13.1) of particles can be obtained:

- *Simple sphere surrounded by a coating of uniform thickness*; this classification is known as matrix encapsulation. This is the simplest structure in which a sphere is surrounded by a wall or membrane of uniform thickness, resembling that of a hen's egg. In this design, the core material is buried to varying depths inside the shell. This microcapsule has also been termed as a single-particle structure
- Particle containing an irregular-shaped core
- *Several core particles embedded in a continuous matrix of wall material*; this type of design is also termed as the aggregate or matrix structure. All the particles in the aggregate structure need not be the same material. In the case of aggregate capsule structures, a degree of particle size control can be achieved (Bakan, 1973)

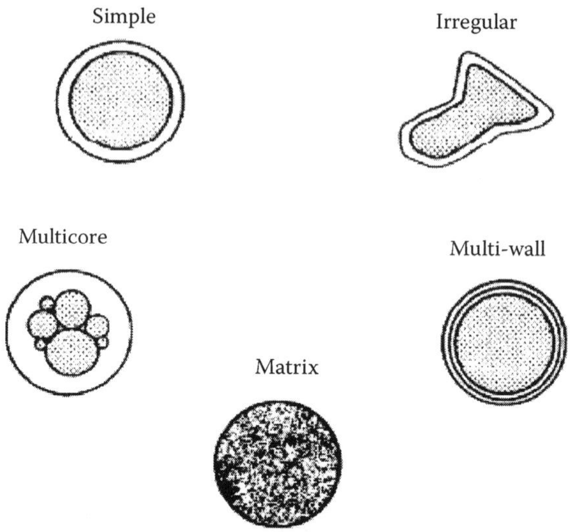


**FIGURE 13.1** The morphology of different types of microcapsules. (From Gibbs, B.F. et al. 1999. *International Journal of Food Sciences and Nutrition*, 50, 213–224.)

- *Several distinct cores within the same capsule*; this type of design is also termed as a multicore structure
- *Multiwalled microcapsules*; another well-known design for a microcapsule is to form a multiwalled structure, in which, the different concentric wall layers can have the same or quite different compositions. In this case, the multiple walls are placed around a core to achieve multiple purposes related to the manufacture of the capsules, their subsequent storage, and controlled release. This design is particularly important for controlled-release encapsulation systems using nanospheres containing an active ingredient (Shefer and Shefer, 2003)

## 13.2 ENCAPSULATING AGENTS

To encapsulate a core material/ingredient, the first requirement is the selection of an appropriate wall material or encapsulating agent. An ideal agent should have the following properties (Balssa and Fanger, 1971; Bangs, 1985; Madene et al., 2006; Reineccius, 1994):

1. Good rheological properties at high concentrations, meaning low viscosity at high-solids levels and low hygroscopicity
2. Be easy to work with it during the process of encapsulation
3. The capability to disperse or emulsify the active material/ingredient and stabilize the emulsion produced; therefore, it should have adequate emulsifying properties and be a good film former
4. Nonreactivity with the material to be encapsulated both during processing and upon prolonged storage

5. The capability to seal and hold the active material within its structure during processing and storage
6. The capability to offer maximum protection to the active material against environmental conditions (e.g., oxygen, heat, light, and humidity)
7. Solubility in solvents acceptable by the food industry (e.g., water and ethanol)
8. Specified or desired solubility properties of the capsules and release properties of the active material/ingredient from the capsule when incorporated in a product
9. Bland in taste
10. Stable in supply
11. Low in cost

Thus, the criteria for selecting an encapsulating agent are mainly based on its physicochemical properties (solubility, crystallinity, diffusibility, and redispersability), film forming, emulsifying, and drying properties (Gharsallaoui et al., 2007). Finally, a number of factors including the nature of the core material/ingredient and the process of encapsulation, as well as economics will be taken into account (Chranioti and Tzia, 2013). Since no single encapsulating agent meets all the criteria listed above, in practice, agents are employed either in blends/mixtures or with modifiers such as oxygen scavengers, antioxidants, chelating agents, and surfactants (Pegg and Shahidi, 2007).

The encapsulating agents can be selected from a wide variety of natural or synthetic polymers, depending on the core material/ingredient to be coated and the characteristics desired in the final encapsulated products. The materials mainly used in the food industry as encapsulating agents include carbohydrate polymers, proteins, lipids, and gums, with carbohydrate polymers being the most broadly used (Murugesan and Orsat, 2012). An overview of the various materials used as encapsulating agents is given in Table 13.1.

### 13.2.1 Carbohydrate Polymers

#### 13.2.1.1 Maltodextrins and Modified Starches

Maltodextrins and modified starches are widely used as encapsulation agents (Reineccius, 1988). Maltodextrins are formed by partially hydrolyzing corn flour

**TABLE 13.1**
**Overview of Various Materials Used as Encapsulating Agents**

| Category | Encapsulating Agent |
|---|---|
| Carbohydrates | Starch, modified starches, maltodextrins, cyclodextrins, cellulose, polysaccharides, and chitosan |
| Proteins | Gluten, isolates (pea, soy), caseins, whey proteins, and gelatin |
| Lipids | Fatty acids, alcohols, glycerides, waxes, and phospholipids |
| Gums | Seaweed (carrageenans, alginates), seeds (locust bean gum, guar gum), exudates (arabic gum, gum karaya, and mesquite gum), and microbial biosynthesis (xanthan gum, gellan gum) |

with acids or enzymes and they are supplied as dextrose equivalents (DEs); the DE value is a measure of the degree of starch polymer hydrolysis. Maltodextrins possess a good compromise between cost and effectiveness (Apintanapong and Noomhorm, 2003). They provide a good oxidative stability to encapsulated oily ingredients but exhibit poor emulsifying capacity, emulsion stability, and low oil retention (Buffo and Reineccius, 2000). The retention of volatile flavor compounds increases with an increase of the maltodextrin DE (Anandaraman and Reineccius, 1987), suggesting the importance of DE to the functionality of the encapsulating agent.

Chemically modified starches are considered good encapsulating agents because they exhibit low viscosities at high-solids concentrates and good water solubility (Gharsallaoui et al., 2007). Also, due to their surface-active properties, they can stabilize emulsions toward flocculation and coalescence (Dalgleish, 2006). However, an undesirable off-taste and a not so good protection to oxidizable flavorings are some of their disadvantages (Qi and Xu, 1999). To give some emulsifying capabilities to starch molecules, side chains of lipophilic succinic acid are inserted into starch to produce chemically modified starches. Various forms of modified starches are used for flavor and oil encapsulation such as Capsul, N-lok, Encapsul, and Hi-cap (Jafari et al., 2008).

#### 13.2.1.2 Chitosan

Chitosan is a polycationic polymer commercially obtained by alkaline deacetylation of chitin (an N-acetylglucosamine polymer). Chitin is the second most abundant natural polymer in nature after cellulose and it is found in the structure of a wide number of invertebrates (crustaceans, exoskeleton insects, and cuticles among others) (Kumar, 2000). Chitosan has attracted a great deal of attention due to its nontoxicity, biodegradability, biocompatibility, and mucus adhesiveness properties (Alishahi et al., 2011a, b; Kumar, 2000). Because of its useful characteristics, such as antibacterial and antioxidant properties, film-forming ability (Shahidi et al., 1999), gel enhancement, emulsifying properties (Rodríguez et al., 2002), and encapsulating capacity, it has also been used as an encapsulating agent (Alishahi and Aïder, 2012; Estevinho et al., 2013). In particular, it has been used as a support agent in microencapsulation processes for controlled release of bioactive compounds (Ko et al., 2003).

#### 13.2.1.3 Cyclodextrins

Cyclodextrins are chemically and physically stable molecules formed by the enzymatic modification of starch using cyclodextrin glycosyltransferase. The cyclodextrins contain six, seven, or eight glucose monomers; referring to as α-, β-, and γ-cyclodextrin, respectively. The glucose monomers are joined to one another in a double nut-shaped ring, giving the cyclodextrins a molecular structure that is relatively rigid and has a hollow cavity of specific diameter and volume. While the outer surfaces (top and bottom) are hydrophilic, the internal cavity has a relatively high electron density and is hydrophobic in nature due to the hydrogen and glycosidic oxygen atoms being oriented to the interior of the cavity (Marques, 2010). They have an ability to form complexes with a wide variety of organic compounds within their ringed structure, most notably flavor compounds (Reineccius et al., 2002). The ability of these unusual molecules to form inclusion complexes, which can change the physical and chemical properties of guest molecules, offers a variety of potential uses to the food industry.

Organic molecules of suitable size, shape, and hydrophobicity are able to interact noncovalently with cyclodextrins to form stable complexes. Several forces, such as van der Waals, hydrophobic interaction, and dipole–dipole interaction, are involved in the binding of guest molecules to the cyclodextrin cavity (Hedges et al., 1995). Much research has focused on the ability of cyclodextrins to prevent the volatilization of flavors and essences from spices, flavor extracts, and lipids (Nagatomo, 1985; Pagington, 1986), as well as to protect bioactive compounds from oxidation, light-induced reactions, thermal decomposition, and evaporation loss (Szente and Szejtli, 2004).

### 13.2.2 Gums

One class of material often exploited for its encapsulating capabilities is that of hydrocolloids, or more commonly known as, gums. These compounds are long-chain polymers that dissolve or disperse in water to give a thickening or viscosity-building effect (Glicksman, 1982). Gums are generally used as texturing agents, but their secondary effects include encapsulation capabilities (Carroll et al., 1984), stabilization of emulsions, control of crystallization, and inhibition of syneresis—the release of water. Additionally, a few gums are also capable of forming gels (Glicksman, 1982).

Food gums are obtained from a variety of sources. Although most gums are obtained from plant materials, such as seaweed (e.g., carrageenans and alginates), seeds (e.g., locust bean gum and guar gum), and tree exudates (e.g., gum acacia or arabic gum); others are products of microbial biosynthesis (e.g., xanthan gum and gellan gum); and still others are produced by chemical modification of natural polysaccharides. One of the most commonly used gum as a coating material for food encapsulation is discussed below.

#### 13.2.2.1 Gum Arabic

Gum arabic has been the agent of choice for many years since it is nontoxic, odorless, tasteless, with excellent emulsification capacity, low viscosity in aqueous solution, and very good volatile retention properties (Gabas et al., 2007). It is described as a highly branched arabinogalactan protein (Phillips and Williams, 1995), while both protein and polysaccharide moieties are fundamental to the functional properties of this polysaccharide (Dickinson, 2009). Gum arabic has some carboxylate groups as part of its structure, but it is not highly (negatively) charged, a globular-like, random coil structure, plus a considerable amount of water occluded between its polysaccharide chains (Burgess and Singh, 1993).

Therefore, its quite unique behavior is attributed not only to its molecular structure, but also to its conformation, so that its stabilizing mechanism can be comparable to very few other polysaccharide–protein systems (Jafari et al., 2012). Notwithstanding, due to unpredictable fluctuations in price, supply, and quality of gum arabic (Seisun, 2002), its total or partial replacement by alternative materials remains a continuing source of interest to the food industry (Chranioti and Tzia, 2013, 2014; Kshirsagar and Singhal, 2007; McNamee et al., 1998).

### 13.2.3 Proteins

Proteins possess many desirable functional properties, which allow them to be good candidates as coating material for the encapsulation of food ingredients, even though food hydrocolloids are mostly used. Gelatine is the most commonly used protein for this purpose, but other protein products such as sodium caseinate, whey protein, soy protein concentrates, and soy protein isolates have also been utilized (Hogan et al., 2001).

#### 13.2.3.1 Gelatine

Gelatine is a water-soluble protein derived from collagen and is a valuable encapsulating agent partially because it is nontoxic, inexpensive, and commercially available, with good film-forming properties. Gelatine is often used in combination with arabic gum to form coating films. When the pH is lower than its isoelectric point, gelatine becomes polycationic, and hence, there is an interaction between polycationic gelatine and polyanionic arabic gum, resulting in the formation of a complex (coacervate). The type of gelatine and gum selected and the formation and fixing procedures employed ultimately influence coating the final permeability (Hogan et al., 2001).

### 13.2.4 Use of Polymer Blends/Combinations as Encapsulating Agents

As no single encapsulating agent possesses all the properties required for an ideal one, approaches have been focused on mixtures of different agents, thus exploiting the individual properties of each material (Kaushik and Roos, 2007). For example, concerning flavor encapsulation, gum arabic can be used as an encapsulating agent combined with maltodextrin (McNamee et al., 2001; Soottitantawat et al., 2003; Xiang et al., 1997), modified starch (Jung and Sung, 2000), or with chitosan (Moschakis et al., 2010) for volatile substances. Combinations of gum arabic, modified starch, and maltodextrin have also been reported for the encapsulation of cardamom, cumin, turmeric, and fennel oleoresin (Chranioti and Tzia, 2013, 2014; Kanakdande et al., 2007; Krishnan et al., 2005; Kshirsagar et al., 2009). Moreover, proteins have also been used in an encapsulation process of oily or volatile substances, together with other coating materials such as carbohydrates (Hogan et al., 2001; Ono, 1980; Ono and Aoyama, 1979).

## 13.3 ENCAPSULATION METHODS

Many different methods have been proposed for the production of encapsulated products that can be divided into three main groups as (Shahidi and Han, 1993):

1. Physical methods: Spray drying, freeze drying, spray chilling/cooling, fluidized-bed coating, extrusion, and cocrystallization
2. Chemical methods: Molecular inclusion and interfacial polymerization
3. Physicochemical methods: Simple/complex coacervation and liposome entrapment

The choice of the encapsulation method for a specific application is based on parameters including physical–chemical properties of both core and coating material, applications of the final encapsulated material/ingredient, desired release mechanisms, industrial manufacturing, and processing cost. According to the encapsulation method selected, the final products will present various shapes (films, spheres, and irregular particles), various structures (porous or compact), and various physical structures (amorphous or crystalline dehydrated solid, rubbery, or glassy matrix). All the above properties will influence the diffusion of the core ingredients or the external substances (oxygen, solvent), as well as influencing the food product stability during storage (Madene et al., 2006).

### 13.3.1 Spray Drying

Spray drying accounts for the majority of commercially encapsulated materials/ingredients in food products (Shefer and Shefer, 2003). It is one of the oldest and well-known encapsulation techniques and is a standard technology used for flavor encapsulation (Mermelstein, 2001). In the 1930s, spray drying was employed to prepare the first encapsulated flavors using gum acacia as the coating agent (Blenford, 1986). Although it is most often considered as a dehydration process and is used in the preparation of dried materials, it can also be used as an encapsulation process for many food ingredients, some of which are listed in Table 13.2.

The process is conducted in a spray dryer and the basic steps involved have been discussed thoroughly in Chapter 6. In brief, the process involves the dispersion of the substance to be encapsulated in a carrier material, followed by atomization and spraying of the mixture into a hot chamber (Dziezak, 1988; Watanabe et al., 2002). The resulting microcapsules are then transported to a cyclone separator for recovery. In most cases, the encapsulated substances/materials can be released upon contact of the product with water, which dissolves the spray-dried products (Shefer and Shefer, 2003).

Spray drying is an economical and flexible encapsulation method as it offers substantial variation in the choice of the encapsulating agents. Moreover, it is adaptable to the commonly used processing equipment and can produce final encapsulated products (particles or capsules) of good quality (Meyers, 1995; Taylor, 1983). Also, the most important is that its processing cost is lower than those associated with most other methods of encapsulation (Dzondo-Gadet et al., 2005).

However, spray drying sets limitations concerning the concentration of both core and agent material, the requirement of high temperature for drying, and of a feed solution of low viscosity (Byun et al., 1998).

### 13.3.2 Freeze Drying

The freeze-drying technique or lyophilization is one of the most useful processes for drying thermo-sensitive substances that are unstable in aqueous solutions. In this process, frozen material is subjected to a pressure below the triple point (at 0°C, pressure: 610 Pa) and heated to cause ice sublimation to vapor. The basic principles of this technique, the stages, as well as the available freeze-drying systems have been discussed in detail in Chapter 6.

**TABLE 13.2**
**Encapsulation of Some Different Food Ingredients by Spray Drying**

| Food Category | Core Material | Wall Material | Air Inlet Temperature (°C) | Air Outlet Temperature (°C) | References |
|---|---|---|---|---|---|
| Flavors (Component —essential oils —oleoresins) | Orange peel oil | Maltodextrin and mesquite gum | 195–205 | 105–115 | Beristain et al. (1999) |
| | Caraway (*Carum carvi* L.) essential oil | Whey protein isolate, skimmed milk powder, waxy corn starch, and maltodextrin | 175–185 | 85–95 | Bylaitë et al. (2001) |
| | Rosa mosqueta (*R. rubiginosa*) oleoresin | Gelatine and starch | 95–105 (gelatin) 145–155 (starch) | 60–70 (gelatin) 65–75 (starch) | Robert et al. (2003) |
| | Sumac flavor (Sumac berries extract) | Sodium chloride, sucrose, glucose, and starch | 200 | 100 | Bayram et al. (2005) |
| | D-limonene | Gum arabic, maltodextrin, and modified starch | 200 | 105–115 | Soottitantawat et al. (2005a) |
| | L-menthol | Modified starch and gum arabic | 180 | 95–105 | Soottitantawat et al. (2005b) |
| | Cardamom oleoresin | Gum arabic, maltodextrin, and modified starch | 176–180 | 115–125 | Krishnan et al. (2005) |
| | La France pear aroma model (mixing of five kinds of esters and hexanal) | α-Cyclodextrin, gum arabic, and soybean-soluble polysaccharide | 150 | 96 | Tobitsuka et al. (2006) |
| | Black pepper oleoresin | Gum arabic and modified starch | 176–180 | 105–115 | Shaikh et al. (2006) |
| | Oregano (*Origanum vulgare* L.), citronella (*Cymbopogon nardus* G.), and marjoram (*Majorana hortensis* L.) | Whey protein concentrate and skimmed milk powder | 185–195 | 85–95 | Baranauskiene et al. (2006) |

| | | | | | |
|---|---|---|---|---|---|
| | Cumin oleoresin | Gum arabic, maltodextrin, and modified starch | 158–162 | 115–125 | Kanakdande et al. (2007) |
| | Red chili oleoresin | Mesquite gum, maltodextrin (DE 10), and whey protein concentrate | 1665–175 | 90–100 | Perez-Alonso et al. (2008) |
| | Turmeric oleoresin | Gum arabic and maltodextrin (DE 18) | 138–142 | 84–92 | Kshirsagar et al. (2009) |
| | Paprika oleoresin | Gum arabic and soy protein isolate | 155–165<br>175–185<br>195–205 | 105–115 | Rascón et al. (2011) |
| Colorants—pigments | *Amaranthus* | β-cyanin pigments Maltodextrins (DE 10, 20, and 25), maize starch, and modified starch | 150–210 | 87–115 | Cai and Corke (2000) |
| | Black carrot (*Daucus carota* L.) juice | Maltodextrins with different DE values | 160, 180, and 200 | 107, 118, and 131 | Ersus and Yurdagel (2007) |
| | Bayberries (*Myrica rubra* Sieb. et Zucc.) juice | Maltodextrins (DE 12 and 19) | 140–160 | 65–85 | Gong et al. (2008) |
| | Beetroot juice | Gum arabic | 175–185 | 80–90 | Pitalua et al. (2010) |
| | Muitle (*Justicia spicigera*)—Mexican native plant | Gum arabic and mesquite gum | 130–140 | 75–85 | Pavón-García et al. (2011) |
| | Roselle calyces (*Hibiscus sabdariffa* L.) | Maltodextrin (DE 11–15), gum arabic, and soluble starch | 150 | 100 | Idham et al. (2012) |
| Vitamins | β-carotene (vitamin A) | Maltodextrin (DE 25) | 150 | 100 | Desobry et al. (1997) |
| | Ascorbic acid | Gum arabic, rice starch, and gelatin | 150 | 80 | Trindade and Grosso (2000) |
| | Ascorbic acid | Starch and β-cyclodextrin | 200–300 | 70–95 | Uddin et al. (2001) |
| | Cashew apple juice (ascorbic acid) | Maltodextrin and cashew tree gum | 185 | 90 | de Oliveira et al. (2009) |
| Enzymes | α-Amylase invertase | Chitosan | 105–120 | 65–85 | Gonzalez Siso et al. (1997) |
| Oils/fatty acids (Ω-3/Ω-6) | Soya oil | Sodium caseinate and carbohydrate | 180 | 95 | Hogan et al. (2001) |
| | Fish oil | Sugar beet pectin and glucose syrup (DE 38) | 170 | 70 | Drusch (2007) |
| | Fish oil | Alginate and starch | 150 | 80 | Tan et al. (2009) |

This method is mainly used for the production of high-quality dried products, which contain heat-sensitive components such as aromas, vitamins, antibiotics, and microbial culture (Rahman and Perera, 2007). Apart from the dehydration of almost all heat-sensitive components, freeze drying has also been used to encapsulate water-soluble essences and natural aromatic substances (Kopelman et al., 1977a, b) as well as drugs compounds (Kirby and Gregoriadis, 1984).

Freeze drying is a simple technique that is particularly suitable for the encapsulation of flavor-aromatic substances. It is believed that freeze drying should provide a high retention of volatile compounds, since the entire dehydration process is carried out at low temperature and low pressure (Flink and Karel, 1970). Therefore, the virtual absence of air in combination with the low temperature prevents the deterioration due to oxidation or chemical modification of the product. Moreover, the application of freeze drying leads to the production of very porous products, which results in high rehydration rates. It has also been shown that the freeze-drying process can maintain the shape of the capsules because of fixation by freezing (Nagata, 1996).

However, freeze drying is a slow and expensive process. The dehydration period required for the process is long (commonly 20–24 h). A long processing time demands additional energy to operate the compressor and refrigeration units, which makes the process very expensive for commercial use. Its cost can be up to 50 times higher than spray drying (Desobry et al., 1997) and the storage and transport of particles produced is extremely expensive (Jacquot and Pernetti, 2003). Therefore, it is mainly used for high-value products (Cohen and Yang, 1995). Some examples of freeze-dried encapsulated food ingredients are listed in Table 13.3.

### 13.3.3 Basic Steps of Encapsulation Process for Spray and Freeze Drying

The use of spray- or freeze-drying techniques for encapsulation purposes involves three basic steps: the preparation of a dispersion or emulsion, the homogenization of the emulsion, and the subsequent drying of the emulsion that contains the core material/ingredient with the encapsulating agent (Lim et al., 2011; Madene et al., 2006).

#### 13.3.3.1 Preparation of the Dispersion or Emulsion

Once an encapsulating agent or a combination/mixture has been selected, it must be hydrated. It is desirable to use a particular in-feed solids level that is optimum for each encapsulating agent or the combination chosen. Research has shown that the solids level is the most important determinant during the encapsulation process (Reineccius, 1991). It has been found that there is an optimum in-feed solids level that is unique for each encapsulating agent (Bangs, 1985; Leahy et al., 1983; Reineccius and Bangs, 1985). Once the agent or mixture has been solubilized (with or without heating), the ingredient to be encapsulated is added to the mixture and then thoroughly dispersed throughout the system. At this point, an important parameter is the ratio of encapsulating agent to core material/ingredient, which depends on the application.

#### 13.3.3.2 Homogenization

The mixture is thereafter homogenized to create small droplets of the core ingredient within the encapsulating agent solution. The production of a finer emulsion

**TABLE 13.3**
**Encapsulation of Some Different Food Ingredients by Freeze Drying**

| Food Category | Core Material | Wall Material | Operating Conditions | References |
|---|---|---|---|---|
| Flavors (Component — essential oils — oleoresins) | Methyl linoleate | Lactose and gelatin | Freezing for 24 h at −30°C<br>Freeze drying (P < 0.1 mbar) | Shimada et al. (1991) |
| | La France pear aroma model (mixing of five kinds of esters and hexanal) | α-Cyclodextrin, gum arabic, and soybean-soluble polysaccharide | Freezing at −80°C<br>Freeze drying (25°C and 20 Pa) | Tobitsuka et al. (2006) |
| D-Limonene | Gum arabic, Sucrose, and gelatin | Freezing for 12 h at –20°C<br>Freezing for 24 h at –80°C<br>Freeze drying<br>(T < –45°C and P < 0.1 mbar) | | Kaushik and Roos (2007) |
| D-Limonene | Native potato starch, b-cyclodextrin, maltodextrin, acid-treated potato starch, and succinated potato starch | Freezing for 24 h at –70°C<br>Freeze drying for 48 h (–50°C and 1.33 Pa) | | Lee et al. (2009) |
| | Flavor oil (R-carvone) | Maltodextrin (DE 20)<br>Chitosan | Freezing at −80°C<br>Freeze drying for 24 h (−100°C and 1.33 Pa) | Kaasgaard and Keller (2010) |
| | Fennel oleoresin | Modified starch (octenyl-succinylated starch), maltodextrin (21DE), and chitosan | Freezing for 24 h at −30°C<br>Freeze drying (−100°C and 0.017 mbar) | Chranioti and Tzia (2013) |
| | Fennel oleoresin | Gum arabic, modified starch (octenyl-succinylated starch), maltodextrin (21DE), and chitosan | Freezing for 24 h at −30°C<br>Freeze drying (−100°C and 0.017 mbar) | Chranioti and Tzia (2014) |

*continued*

**TABLE 13.3 (continued)**
**Encapsulation of Some Different Food Ingredients by Freeze Drying**

| Food Category | Core Material | Wall Material | Operating Conditions | References |
|---|---|---|---|---|
| Colorants—pigments | *Amaranthus* | β-cyanin pigments Maltodextrins (DE 10, 20, and 25), maize starch, and modified starch | Freezing with liquid nitrogen<br>Freeze drying for 24 h | Cai and Corke (2000) |
| | Saffron (crocins) | Pullulan and two polyvinylpyrrolidone | Freezing<br>Freeze drying | Selim et al. (2000) |
| | Carotenoid (carrot pulp waste) | Sucrose and gelatin | Freeze drying for 24 h (0.4 mm Hg) | Tang and Chen (2000) |
| | Beetroot pigment | Maltodextrin (DE 5, 20) and pullulan | Freezing<br>Freeze drying | Serris and Biliaderis (2001) |
| | Potassium norbixinate and curcumin | Maltodextrin (DE 20) | Freezing with liquid nitrogen<br>Freeze drying (at –50°C for 24 h) | Sousdaleff et al. (2013) |
| Oils/fatty acids (Ω-3/Ω-6) | Rapeseed oil | Sucrose, maltodextrin (DE 10), and gelatine | Freezing for 15 h at –50°C<br>Freeze drying for 48 h (0.1 mbar) | Orlien et al. (2000) |

increases the retention of the core ingredient during the drying process (Andres, 1977). The addition of an emulsifier is often required while the type and amount of which should be determined. Another important aspect is the selection of the homogenization process; considerable process variation exists within the industry. Among the most common homogenization processes are the use of a typical homogenizer as well as the application of an ultrasonic treatment (Chemat et al., 2011). The homogenization process applied will subsequently influence the properties of the emulsion including emulsion droplet size, viscosity, and stability (Jafari et al., 2008).

#### 13.3.3.3 Drying of the Emulsion

During the final step of drying, when spray drying is applied, parameters that should be taken into account and optimized in each application involve the inlet and outlet temperatures, in-feed temperature, airflow and humidity, as well as pump rate. Research has shown that the above operating conditions greatly affect the quality properties of the final encapsulated product (Gharsallaoui et al., 2007; Jafari et al., 2008). When freeze drying is applied as an encapsulation method, freezing of the emulsion prior to drying is necessary. Therefore, the significant parameters to consider are the time and temperature of freezing that probably influence the properties of the final product (Kaasgaar and Keller, 2010) as well as the operating conditions of the applied freeze-drying system.

### 13.3.4 Spray Cooling/Chilling

Spray cooling/chilling is the least-expensive encapsulation technology and is routinely used for the encapsulation of a number of organic and inorganic salts as well as textural ingredients, enzymes, flavors, and other functional ingredients to improve heat stability, delay release in wet environments, and/or convert liquid hydrophilic ingredients into free-flowing powders (Gouin, 2004). Typically, the coating material used both in spray cooling and chilling technology is a fat.

Spray cooling and chilling are two commercially practiced encapsulation processes similar to spray drying in which both involve dispersing the core material into a liquefied coating material and spraying through heated nozzles into a controlled environment (Bakan and Anderson, 1970). Unlike spray drying, however, there is no water to be evaporated since the coating material is a fat. Other principal differences between these processes and spray drying lie first in the temperature of the air used in the drying chamber and second in the type of coating material applied. Spray drying employs hot air to volatilize the solvent from a coating dispersion; in contrast, spray cooling and spray chilling use air cooled to ambient or refrigerated temperatures. The core and lipid wall mixture are atomized in the chilled air that causes the fat to solidify around the core, thereby forming a crude encapsulated product (Pegg and Shahidi, 2007). Concerning the type of coating agent applied, in spray cooling, it is typically a vegetable oil or one of its derivatives, fat and stearin with melting points of 45–122°C (Risch, 1995) as well as hard mono- and diacylglycerols with melting points of 45–65°C (Taylor, 1983), whereas in spray chilling, the encapsulating agent is typically selected from a range of fractionated vegetable oils with comparatively lower melting points of 32–42°C.

A common myth concerning spray cooling/chilling is that the release of the active ingredient is thought to occur upon the melting of the encapsulating agent (fat matrix). This has been proven to be inaccurate since it is believed that a significant number of active ingredient particles may be located at the surface of the microcapsules or have direct access to the environment. Furthermore, other reasons such as osmotic forces, slow diffusion of water through the shell imperfection, or mechanical disruption of the particles, may influence the release mechanisms of the process. Overall, irrespectively of the payload or the composition of the fat matrix, it is very difficult to achieve a delayed release of a water-soluble ingredient over 30 min in a foodstuff of relatively high water activity by spray cooling/chilling (Gouin, 2004).

However, improvement of the release behavior can be achieved by modifying the crystalline structure of the encapsulating agent used. For instance, additives and/or mixtures of glycerides can be used to preferably obtain one crystalline form over another and affect the diffusion rate as well as slow down or speed up the release kinetic of the active ingredient. Even though the spray cooling/chilling process does not lead to a prefect product, its obtained properties are often sufficient to achieve the desired delayed release of the active ingredient. Therefore, this method of encapsulation can be used for controlled-release purposes (Gouin, 2004).

Besides the use for controlled-release purposes, spray cooled/chilled products have applications in bakery products, dry soup mixes, and foods containing high levels of fat (Blenford, 1986). Moreover, since spray cooled/chilled products are insoluble in water due to the hydrophobic coating (lipid matrix), they tend to be utilized for encapsulating water-soluble core materials (such as minerals, water-soluble vitamins, enzymes, acidulants, and spray-dried flavors) that may otherwise be volatilized or damaged during thermal processing (Augustin et al., 2001; Shahidi and Han, 1993). Spray chilling is also employed for the encapsulation of solid food additives, such as ferrous sulfate, acidulants, vitamins, minerals, and solid flavors, as well as for those materials that are not soluble in typical solvents (Taylor, 1983).

Finally, the major drawbacks of spray cooling/chilling process include interactions between the fat and the active ingredient, volatilization of lipid-soluble materials over time, loss of volatile materials during processing (Shefer and Shefer, 2003), as well as the requirement of special handling and storage conditions due to the nature of the encapsulating agent (Taylor, 1983).

### 13.3.5 Fluidized-Bed Coating

Fluidized-bed coating, also referred to as air suspension coating, spray coating, or the Wurster process, is a common technique used for commercial production of encapsulated products. It is already widely employed in the pharmaceutical and cosmetic industry, but its use in the food industry has also been studied (Madene et al., 2006). In particular, a number of food ingredients have been encapsulated by fluidized-bed coating, such as ascorbic acid (Knezevic et al., 1998), acidulants for processed meat (Weiss and Reynolds, 1989), leavening agents (Balassa and Brody, 1968), as well as flavor ingredients (Dezarn, 1998; Lee and Krochta, 2002).

#### 13.3.5.1 Basic Concept

Fluidized-bed coating was developed by D.E. Wurster in the 1950s. This technique relies upon a nozzle spraying of the coating material into a fluidized bed of core particles in a hot environment. The size of the resulted product varies from 0.3 to 10 mm (Panda et al., 2001).

#### 13.3.5.2 Fluidized-Bed Encapsulation Process

Solid particles to be sprayed are suspended in an upward-moving column of air in a fluidized-bed chamber at a controlled temperature and humidity. Depending on the specific application, the airflow may be heated or cooled (Bakan and Anderson, 1970). Once the moving fluid bed of particles has reached the prescribed temperature, the encapsulation-coating material is introduced into the system. Great variations exist as to the type of the wall material chosen. Cellulose derivatives, dextrins, emulsifiers, lipids, protein derivatives, and starch derivatives are examples of typical coating systems, and they may be used in a molten state or dissolved in an evaporable solvent. The coating is atomized through binary or pneumatic spray nozzles at the top of the chamber, whose droplets are of smaller size than the substrate being coated. The atomized particles travel down to the particle stream and deposit as a thin layer on the surface of the suspended core material. The turbulence of the air column is sufficient to keep the coated particles suspended, thereby allowing them to tumble and become uniformly coated. Upon reaching the top of the air stream, the particles move into the outer, downward-moving column of air, which returns them to the fluidized bed with their coating nearly dried (Figure 13.2). The particles pass through the coating cycle many times per minute (Sparks, 1981). With

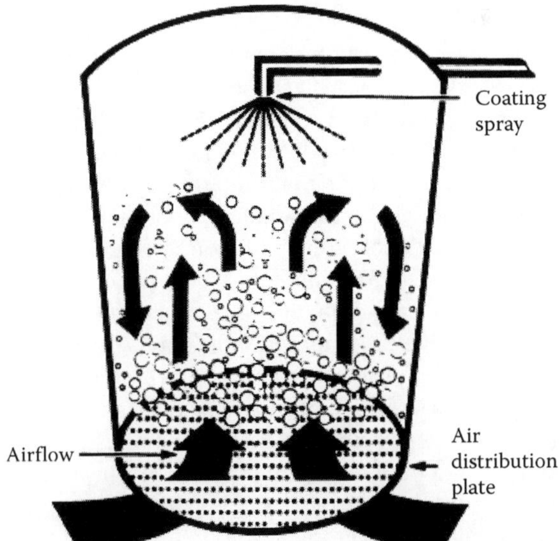


**FIGURE 13.2** A schematic representation of a conventional air suspension system. (From Dziezak, J.D. 1988. *Food Technology*, 42 (4), 136–151.)

each successive pass, the random orientation of the particles further ensures their uniform coating. In the case of hot melts, the coating is hardened by solidification in cool air. In the case of solvent-based coatings, the coating is hardened by evaporation of the solvent in hot air. The amount of coating applied can be regulated by controlling the length of time (residence time) that the particles are in the chamber. To achieve a good degree of coating, the process takes place from 2 to 12 h to complete. After this period, only 0.2–1.5% of the particles remain uncoated (Jacquot and Pernetti, 2003).

#### 13.3.5.3 Benefits of Fluidized-Bed Coating

- It is the most suitable method for encapsulating spray-dried flavors because the wall materials used in flavor systems are readily dissolved and form strong interparticle bridges on redrying (Buffo et al., 2002). Moreover, further coating by the fluidized bed of flavor ingredients with a fat layer, for instance, can also impart better protection and shelf life (Gouin, 2004).
- Interestingly, fluidized-bed technology is one of the few advanced technologies capable of coating particles with basically any kind of shell material (hydrocolloids such as gums and proteins, ethanolic solutions of synthetic polymers, and melted fats/waxes). Therefore, the controlled-release possibilities are considerably more versatile with the fluidized-bed technology than with any other technologies.
- Fluidized-bed spray granulation is a more recent technology that allows specific particle size distributions from 0.2 to 1.2 mm with low porosities to be designed for the product. As with spray drying, continuous spray granulation starts off with an aqueous emulsion. Repeated spraying, applying, and drying steps in a fluidized-bed encapsulation process form granules with an onion-like structure.
- A particular advantage offered by granulation technology is the possibility of producing large flavored particles of uniform particle size and shape without the need for any additional production steps (Uhlemann et al., 2002).
- The use of melted fats, waxes, or emulsifiers as shell materials is a relatively new, yet a very promising and interesting concept. From an industrial point of view, the inherent advantage of hot-melt fluidized-bed coating lies in the fact that the coating formulation is concentrated (no solvent as in aqueous-based coating formulation), which means dramatically shorter processing times. The energy input is also much lower than with aqueous-based formulation since no evaporation needs to be done (Jozwiakowsksi et al., 1990).
- This technology allows specific particle size distribution and low porosities to be designed into the product (Uhlemann and Morl, 2000).
- Other advantages of a high fluidized bed include
    - High drying rates because of good gas-particle contact, leading to optimal heat and mass transfer rates
    - Smaller flow area
    - High thermal efficiency

    - Lower capital and maintenance costs
    - Ease of control (Mujumdar and Devahastin, 2000)
- According to Chua and Chou (2003), the fluidized-bed dryer is a low-cost drying technology for developing countries compared with some state-of-the-art drying equipment, such as spray dryers.
- High production volumes and well-developed technologies have made a number of encapsulated food products standard items and are available at cost-effective prices (Dewettinck and Huyghebaert, 1999).

### 13.3.6 Extrusion

Extrusion was first patented in 1957 and further developed by the group that originally patented the technique (Swisher, 1957). This process is particularly useful for heat-labile compounds/bioactives and has been used to encapsulate flavors, vitamin C, and colorants (Shahidi and Han, 1993). Moreover, extrusion microencapsulation has been used almost exclusively for the encapsulation of volatile and unstable flavors in glassy carbohydrate matrices (Gouin, 2004).

#### 13.3.6.1 Basic Concept

Extrusion, mostly applied to flavor encapsulation, is a relatively low-temperature entrapping method that involves the forcing of a core material, which is dispersed in a molten carbohydrate mass, through a series of dies into a bath of dehydrating liquid. The pressures and temperatures employed are typically 100 psi and seldom exceed 115°C, respectively (Reineccius, 1994). Upon contact with the liquid, the coating material, which forms the encapsulating matrix, hardens and entraps the core material. Isopropyl alcohol is the most common liquid employed for the dehydration and hardening process. The extruded filaments or strands are then broken into small pieces, dried to mitigate hygroscopicity (an anticaking agent such as calcium triphosphate can facilitate this), and sized.

### 13.3.7 Extrusion Methods

Several methods of extrusion have been developed; the most commonly employed are described below.

#### 13.3.7.1 Simple Extrusion

A volatile compound is dispersed in a matrix polymer at 110°C. This mixture is then forced through a die and the filaments obtained are plunged into a desiccant liquid that, by hardening the extruded mass, traps the active substances (Crouzet, 1998; Rizvi et al., 1995). The most common liquid used for the dehydration and hardening process is isopropyl alcohol. The strands or filaments of hardened material are broken into small pieces, separated, and dried (Risch, 1995). Several factors have been reported to improve the quality of capsules including the DE of the corn syrup, emulsifier and flavor oil content, and emulsification pressure (Crocker and Pritchett, 1978).

#### 13.3.7.2 Double-Capillarity Extrusion Devices

##### *13.3.7.2.1 Coaxial Double-Capillary Device*

The core substance and the carrier material are fed through, respectively, the inner and outer opening of a coaxial double capillary. The core is usually a liquid and the polymer may be applied as a solution or as a melt (the core and the coat fluids must be immiscible). At the tip of the coaxial nozzle, the two fluids form a unified jet flow, which breaks up to form the corresponding microdroplets.

##### *13.3.7.2.2 Centrifugal Extrusion Device*

This process utilizes nozzles located on the outer circumference of a rotating cylinder. The liquid flavor is pumped through the inner orifice and the liquid shell material is pumped through the outer orifice, forming a coextruded rod of flavor components surrounded by the wall material. As the device rotates, the extruded rod breaks into droplets that form capsules (Schalmeus, 1995).

#### 13.3.7.3 Recycling Centrifugal Extrusion

Centrifugal extrusion is another encapsulation technique that has been investigated and used by some manufacturers. A number of food-approved coating systems have been formulated to encapsulate products such as flavorings, seasonings, and vitamins. These shell materials include gelatine, sodium alginate, carrageenan, starches, cellulose derivatives, gum acacia, fats/fatty acids, waxes, and polyethylene glycol (Pegg and Shahidi, 2007).

The technology of recycling centrifugal extrusion is combined with a facility for recycling of the excess coating fluid. The core material is dispersed in the carrier material. The suspension is extruded through the rotating disk in such a way that the excess coating fluid is atomized and separated from the coated particles. Excess coating fluid is then recycled, while the resulting microcapsules are hardened by cooling or solvent extraction.

#### 13.3.7.4 Advantages and Disadvantages of Extrusion Encapsulation

The extrusion encapsulation offers several advantages such as:

- The primary advantage of extrusion is unquestionably the outstanding stability offered to flavors against oxidation. For example, shelf life of up to 5 years has been reported for extruded flavor oils, compared to typically 1 year for spray-dried flavors and a few months for unencapsulated citrus oils (Gouin, 2004).
- This technique produces larger particles that can be used when visible flavor pieces are desirable (Risch, 1995).
- Extrusion remains the most suitable process for the use of glassy carbohydrates as shell materials. Carbohydrate matrices in the glassy states have very good barrier properties, enabling the encapsulation of flavors in them. The very long shelf life imparted to normally oxidation-prone flavor compounds (such as citrus oils) is due to the very slow diffusion rate of atmosphere gases through the hydrophilic glass of the matrix, thus providing an almost impermeable barrier against oxygen (Gouin, 2004).

- Optimization of the process parameters, including cooking temperature, emulsifier concentration, and pressurization of the cooking vessel, can result in improved encapsulation efficiency at high flavor loadings (Miller and Mutka, 1986).

However, based on the research conducted mainly on flavor encapsulation, extrusion encapsulation presents some disadvantages such as:

- Extrusion is considerably more expensive than spray drying; process costs are estimated to be nearly double (Shahidi and Han, 1993).
- The payload in the extrusion systems remains in very low levels (around 8%) (Shahidi and Han, 1993). Higher payloads lead to unstable systems; leaking out and fast oxidation of the sensitive flavor oil. Such low payloads in flavor microcapsules are very unattractive, from an industrial point of view, because
    - The cost in use becomes unacceptable
    - The substantial amount of carbohydrate added to the foodstuff along with the flavor often requires an undesirable adjustment of the recipe and might not be appropriate for sugar-free or savory products (Gouin, 2004)
- It is a high-temperature batch process; the flavors must be able to tolerate temperatures in the range of 110–120°C for a substantial period of time without deterioration (Shahidi and Han, 1993).
- Structural defects such as cracks, thin wall, or pores formed during or after processing can enhance the diffusion of flavors from the extruded carbohydrates (Villota and Hawkes, 1994; Wampler, 1992).
- One of the drawbacks of this technology is the rather large particles formed by extrusion (typically 500–1000 mm), which limit the use of extruded flavors in an application where mouthfeel is a crucial factor (Gouin, 2004).
- A very limited range of shell material is available for extrusion encapsulation. Most of the applications use carbohydrates of various DE, starch, and mixture of additives (Gouin, 2004).
- The major problem still facing this process is related to emulsion stability, which is difficult to obtain in extremely viscous carbohydrate melts (Risch, 1988).

However, some of the drawbacks of this technology can be confronted. For instance, the use of hydrophobically modified starches, which have very good amphiphilic properties, instead of simple carbohydrates would allow the payload to be increased up to 40% while preserving the attractive shelf life (Mutka and Nelson, 1988). A payload of 40% is more than double the payload of typical spray-dried flavors and while extrusion is a slightly more expensive process than spray drying, the much better shelf life and resistance to oxidation might compensate and give a comparable or even lower cost in use (Gouin, 2004).

Moreover, the replacement of simple carbohydrates such as sucrose with modified starches can result in a product that is "sugar free" and this might have some advantages in marketing of the final food product. The use of modified starches can also provide greater flexibility to manufacturers since simple carbohydrates such as sucrose will invert to glucose and fructose at low pH values and high temperatures, leading to a final product that would be more hygroscopic and readily participate in nonenzymatic browning reactions (Barnes and Steinke, 1987; Beck, 1972).

Finally, research has also been conducted in developing lower-temperature processes that would be a better alternative for encapsulation of sensitive flavor compounds while maintaining a high payload (Gouin, 2004).

### 13.3.8 Cocrystallization

Cocrystallization is an encapsulation process utilizing sucrose as a matrix for the incorporation of core materials. It involves spontaneous crystallization of sucrose (above 120°C) that produces aggregates of very small crystals ranging from 3 to 30 μm that incorporate the core materials either by inclusion within the crystals or by entrapment (Chen et al., 1988; Mullin, 1972). The use of the cocrystallization process allows many types of food ingredients, either single ingredients or combinations of ingredients, to be incorporated, thus providing interesting and useful characteristics (Shahidi and Han, 1993). However, few studies have been published on the encapsulation by cocrystallization process mostly referring to encapsulated products such as fruit juices, essential oils, flavors, and brown sugar (Beristain et al., 1994, 1996; Chen et al., 1988).

A typical procedure of cocrystallization could be described as follows: sucrose syrup is concentrated to the supersaturated state and maintained at a temperature high enough to prevent crystallization. A predetermined amount of the core material is then added to the concentrated syrup with vigorous mechanical agitation, thus providing nucleation for the sucrose–ingredient mixture to crystallize. As the syrup reaches the temperature at which transformation and crystallization begin, a substantial amount of heat is emitted. Agitation is continued to promote and extend crystallization until the agglomerates are discharged from the vessel. The encapsulated products are then dried to a desirable moisture (if necessary) and screened to a uniform size (Chen et al., 1982a, b). It is very important to properly control the rates of nucleation and crystallization as well as thermal balance during the various phases. The agglomerates form a loose network, bonded together by point contacts. The encapsulated materials are primarily located in the interstices between crystals. Owing to the porosity of the agglomerates, it is easy for an aqueous solution to rapidly penetrate the agglomerate and release core materials for dispersion and dissolution (Shahidi and Han, 1993).

Therefore, to achieve encapsulation, the structure of sucrose must be modified from a single-solid, dense-perfect crystal with a limited surface area to a micro-sized, irregular, agglomerated form, so as to increase void space and surface area and, therefore, be suitable as an encapsulating agent to incorporate the core materials (Chen, 1988; Awad and Chen, 1993).

Overall, the cocrystallization process offers several advantages such as:

- *It offers an economical and flexible alternative*: Compared with various encapsulation processes, cocrystallization is an economical and flexible technique as the procedure involved is relatively simple (Chen, 1994; Jackson and Lee, 1991).

- *It can be employed as a drying method*: In the highly saturated solution, nucleation and crystallization proceed at a rapid rate and the resulting heat of crystallization can be utilized to affect particle dehydration by evaporation. Therefore, by means of the cocrystallization process, core materials in a liquid form can be converted into a dry powdered form without additional drying (Shahidi and Han, 1993).
- *It enhances the stability of core material*: There is no tendency for the core material to separate from or settle out during handling, packaging, or storage since it is well entrenched in the modified sucrose matrix (Chen et al., 1988; Mullin, 1972).
- *It leads to products of good quality properties:* The resulted granular products have a low hygroscopicity, good flowability and dispersion properties (LaBell, 1991; Quellet et al., 2001), as well as free-flowing property (Beristain et al., 1996).
- *It provides significant advantages to the candy and pharmaceutical industries*: All cocrystallized sugar/flavor products offer direct tableting characteristics due to their agglomerated structure (Rizzuto, et al., 1984).
- It offers good retention of the core material (Beristain et al., 1996).

However, based on the research conducted mainly on flavor encapsulation, it presents some disadvantages as well such as:

- Not good protection against heat degradation
- Need for addition of a strong antioxidant to retard oxidation (Bhandari et al., 1998)

### 13.3.9 Coacervation: Phase Separation

Coacervation, also called phase separation, is a process in which a homogeneous polymer solution is converted into two phases. One is a polymer-rich phase, called a coacervate; and the other is a polymer-poor phase, that is, a solvent (Yan and Jin, 2005). This technique was the first encapsulation process studied and was initially employed by Green and Scheicher (1955) to produce pressure-sensitive dye microcapsules for the manufacturing of carbonless copying paper. Because of the very small particle sizes attainable with this process (ranging from a few submicrons to 6 mm), coacervation is regarded by many as the original and true microencapsulation technique (Dziezak, 1988).

According to Blenford (1986), the technology has also been used in fragrances that are applied in the form of "scratch and sniff" strips in promotional literature. The technology is typically used in flavor encapsulation as well as in the storage and delivery of additives (Korus, 2001; Korus et al., 2003; Passin, 1969, Soper, 1995). However, it can also be adapted for the encapsulation of other food ingredients including fish oils (Lamprecht et al., 2001), nutrients, vitamins (Junyaprasert et al., 2001), preservatives, and enzymes (Dubin et al., 1998).

#### 13.3.9.1 Basic Concept

The concept behind simple or complex coacervation is the phase separation of one or many hydrocolloids from the initial solution and the subsequent deposition of the newly formed coacervate phase around the active ingredient suspended or emulsified in the same reaction media. An appropriate chemical or enzymatic cross-linker can be added, if needed (Gouin, 2004). In general, batch-type simple coacervation processes consist of three steps, as summarized below, and are carried out under continuous agitation (Todd, 1970).

##### *13.3.9.1.1 Formation of a Three-Immiscible Chemical Phase*

In the first step, a three-phase system consisting of a liquid-manufacturing vehicle phase, a core material phase, and a coating material phase is formed by one of the following methods: (i) direct addition: In the direct addition approach, the coating-insoluble waxes, immiscible polymer solutions, and insoluble liquid polymers are added directly to the liquid-manufacturing vehicle, provided that it is immiscible with the other two phases and is capable of being liquefied and (ii) the *in situ* separation technique: In the *in situ* separation technique, a monomer is dissolved in the liquid vehicle and then subsequently polymerized at the interface.

##### *13.3.9.1.2 Deposition of the Coating*

Deposition of the liquid polymer coating around the core material is accomplished by controlled physical mixing of the coating material (while liquid) and the core material in the manufacturing vehicle. Deposition of the liquid polymer coating around the core material occurs if the polymer is sorbed at the interface formed between the core material and the liquid vehicle phase; this sorption phenomenon is a prerequisite to effective coating. Continued deposition of the coating is promoted by a reduction in the total free interfacial energy of the system brought about by a decrease of the coating material surface area during coalescence of the liquid polymer droplets.

##### *13.3.9.1.3 Solidification of the Coating*

Solidification of the coating is achieved by thermal, cross-linking, or desolvation techniques, and forms a self-sustaining microcapsule entity. The microcapsules are usually collected by filtration or centrifugation, washed with an appropriate solvent, and subsequently dried by standard techniques such as freeze- or fluidized-bed drying to yield free-flowing, discrete particles.

### 13.3.10 Coacervation Systems

Coacervation can be simple or complex. Simple coacervation deals with systems containing only one type of polymer (e.g., gelatine) with the addition of strongly hydrophilic agents to the solution. By strict definition, however, complex coacervation involves two or more polymers with opposite charges forming a complex coacervate as a result of ionic interaction at the interface. In either case, the coacervation mixture must be continually stirred. Addition of a suitable droplet stabilizer may also be necessary to avoid coagulation of the resulting microcapsules products (Arshady, 1999). A very large number of hydrocolloid systems have been evaluated for coacervation microencapsulation, but the most studied and well-understood coacervation system is probably the gelatine/gum acacia system (Arneodo, 1996; Jegat and Taverdet, 2000, 2001; Rabiskova and Valaskova, 1998). However, there are other coacervation systems such as gelatine/pectin, gelatine/carboxymethylcellulose, and whey protein/gum arabic (Yan and Jin, 2005) that also exhibit very good properties.

#### 13.3.10.1 Advantages and Disadvantages of Coacervation

The very high payloads achievable (up to 99%) and the controlled-release possibilities based on mechanical stress, temperature, or sustained release render coacervation a unique and promising microencapsulation technology. Although coacervation is a very efficient encapsulation technique, major problems face the food scientists when it comes to commercializing a coacervated food ingredient (Gouin, 2004). First of all, the cost of the process is very high, setting restrictions in its application (Versic, 1988). Moreover, the resulted complex coacervates are highly unstable and the chemical agents such as glutaraldehyde that are necessary to stabilize them are toxic (Sanchez and Renard, 2002) and therefore must be carefully used according to the country's legislation. Other reasons for its limited use include the optimization of wall material concentration (Nakagawa et al., 2004), the evaporation of volatile active compounds, the oxidation of the final products (Flores et al., 1992), the low availability of suitable encapsulating agents/materials that are food approved, as well as the difficulty in dealing with materials having both aqueous and lipid solubility properties (Shefer and Shefer, 2003).

However, some of the problems encountered during a typical complex coacervation system can be overcome. For instance, the processing cost can be dramatically decreased by applying at the end of the coacervation (third step) an economical isolation procedure such as spray drying (Pearl et al., 1993; Porzio and Madsen, 1996) instead of the typically used freeze- or fluidized-bed drying. Moreover, problems related to harmful chemical cross-linkers (glutaraldehyde) could eventually be solved by using enzymatic cross-linkers in contrast. Although the enzymatic cross-linking of coacervate is not, at the moment, a viable industrial option, it is believed that more suitable enzymes are being or will be developed shortly in the future (Gouin, 2004).

### 13.3.11 Liposome Encapsulation or Entrapment

Liposome encapsulation has been used mostly in pharmaceutical and cosmetic applications. However, the technology has evolved in recent years to the point that it is now conceivable for liposome encapsulation to become a routine process in the food industry (Gouin, 2004). For instance, liposomes have been prepared to encapsulate vitamins, food preservatives (liposome-entrapped nisin), meat tenderizers (bromelain-loaded liposome), as well as enzymes (Benech et al., 2002; Fresta and Puglisi, 1999; Kirby et al., 1991; Lee et al., 2000).

#### 13.3.11.1 Basic Principle

Liposomes are basically vesicles composed of one or more phospholipid bilayers encapsulating a volume of aqueous media. The mechanism of liposome formation is essentially based on the unfavorable interactions occurring between phospholipids and water molecules, where the polar headgroups of phospholipids are exposed to the aqueous phases (inner and outer), and the hydrophobic hydrocarbon tails are forced to face each other in a bilayer (Jesorka and Orwar, 2008). When phospholipids, such as lecithin, are dispersed in an aqueous phase, the liposomes form spontaneously. Owing to the possession of both lipid and aqueous phases, liposomes can be utilized in the entrapment, delivery, and release of water-soluble, lipid-soluble, and amphiphilic materials (Fang and Bhandari, 2010).

### 13.3.11.2 Formation of Liposomes: Physicochemical Aspect

From a physicochemical point of view, the formation of liposome structures may be illustrated by phase diagrams. A simplified phase diagram of the 1,2-dipalmitoyl phosphatidylcholine–water system is depicted in Figure 13.3. Addition of water decreases the transition temperature of the phospholipid to a limiting value ($T_c$), which is the minimum temperature required for water to penetrate between the layers of lipid molecules. When the system is cooled below $T_c$, the hydrocarbon chains adopt an ordered packing. The structure of this phase, known as the gel, is lamellar and the hydrocarbon chains are extended (Chapman et al., 1967). Each type of phospholipid molecule is characterized by a phase transition temperature. Below $T_c$, its fatty acyl chains are in a quasicrystalline array; while above $T_c$, the chains are in a fluid-like state. There are two principal requirements for liposome microencapsulation. First,

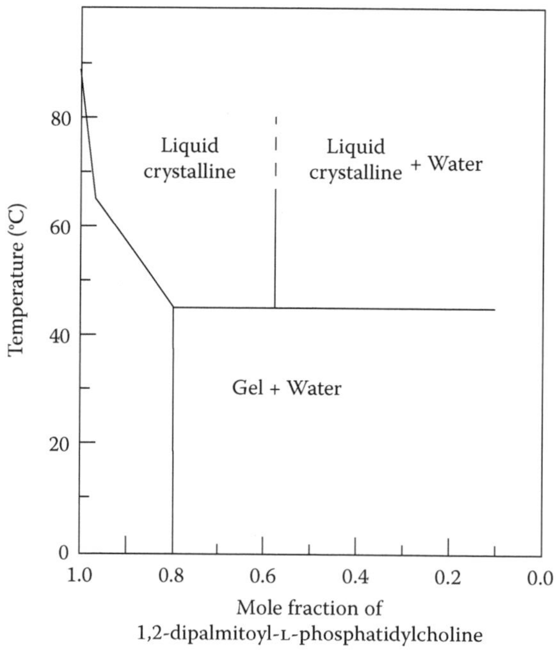


**FIGURE 13.3** Phase diagram of the 1,2-dipalmitoyl-L-phosphatidylcholine–water system. (From Shahidi, F. and Han, X.Q. 1993. *Critical Reviews 456 in Food Science and Nutrition*, 33, 501–547.)

the lipid of choice must have a negative Gibb's free energy value (ΔG) for bilayer structure formation, because a negative ΔG value between two states of the system indicates a favorable reaction. Second, sufficient energy must be put into the system to overcome the energy barrier. Close to room temperature, the value of ΔG for the formation of liposomes is always negative and, therefore, favorable. Even though the thermodynamics are favorable, this does not mean that the reaction will proceed automatically; it is usually necessary to overcome an energy barrier to initiate a reaction. Different lipids and types of energy input may be used to produce different varieties of liposomes for specific purposes. Numerous methods of liposome entrapment have been developed (Gregoriadis, 1987; Kim and Baianu, 1991; Reineccius, 1995), and some of the most commonly employed are described below.

#### *13.3.11.2.1 Microfluidization*

The microfluidization technique is based on the dynamics in specially designed microchannels. The resulting momentum and turbulence allow the lipid emulsion to overcome the energy barrier. An air-driven microfluidizer operates at pressures of up to 10,000 psi. A pump driven by compressed air is used to pump the aqueous emulsion of lipids, and the single-feed stream is split into two fluidized streams. The two flows interact with one another at ultrahigh velocities and in precisely defined microchannels. By means of microfluidization technology, small (0.1 μm in diameter) liposomes with high encapsulation efficiency can be easily formed (Mayhew et al., 1987). The advantages of the microfluidization technique include: (i) a large volume of liposomes can be formed in a continuous and reproducible manner; (ii) the average size of the liposomes can be adjusted; (iii) very high encapsulation efficiencies (larger than 75%) can be obtained; (iv) the solutes to be encapsulated are not exposed to sonication, detergents, or organic solvents; and (v) the resulting liposomes appear to be stable and do not aggregate or fuse. Therefore, the microfluidization technique has been shown to be an effective, cost-effective, and solvent-free continuous method for the production of liposomes with high encapsulation efficiency. This method can process a few hundred liters per hour of aqueous liposomes on a continuous basis (Maa and Hsu, 1999; Vuillemard, 1991; Zheng et al., 1999).

#### *13.3.11.2.2 Ultrasonication*

Ultrasonic dispersion is often employed for the preparation of small unilamellar vesicles (SUVs); the lipid emulsion overcomes the energy barrier through ultrasound absorption. In one approach, phospholipids are sonicated by immersing a metal probe directly into a suspension of large liposomes. In the second method, the lipid dispersion is sealed in a glass vial, which is then suspended in an ultrasonic cleaning bath. Bath sonication requires longer periods (up to 2 h) than probe sonication (only a few minutes), but it has the advantage that it can be carried out in a closed container under nitrogen or argon and does not contaminate the lipid with metal from the probe tip (Deamer and Uster, 1983).

#### *13.3.11.2.3 Reverse-Phase Evaporation*

This technique has been developed for the preparation of large unilamellar vesicles (LUVs), in which lipids in mixed aqueous–nonpolar solvents form inverted micelles (i.e., the lipid tails are inserted into the nonpolar phase and the head groups surround water droplets). When the nonpolar solvent is removed by rotary evaporation under vacuum, the gel-like intermediate phase changes to large unilamellar and oligolamellar vesicles. This technique produces liposomes of uniform size, ranging from 0.1 to 1.0 μm in diameter, with high encapsulation efficiency of up to 65% in low ionic strength media. Its disadvantage, however, is that components are exposed to both organic solvents and sonication. This may result in the denaturation of proteins and other molecules of similar stability (Szoka and Papahadjopoulos, 1980). However, it is important to note that LUVs are the most appropriate liposomes for the food industry because of their high encapsulation efficiency, their simple production methods, and good stability over time (Gouin, 2004).

#### 13.3.11.3 Advantages and Disadvantages of Liposome Encapsulation

The liposome encapsulation offers several advantages such as:

- *Stability in high $a_w$ environments*: The great advantage of liposomes over other encapsulation technologies is the stability offered to water-soluble material in high water activity application. For instance, spray dried, extruded, and fluidized-bed technologies impart great stability to food ingredients in the dry state, but release their content readily in high $a_w$ environments, giving up all protection properties (Gouin, 2004).
- *Special delivery properties*: Another unique property of liposomes is the targeted delivery of their content in specific parts of the foodstuff (Fresta and Puglisi, 1999) as well as at a specific and well-defined temperature. In particular, the liposome bilayer is instantly broken down at the transition temperature of the phospholipids, typically around 50°C, at which temperature of the content is immediately released (Gouin, 2004).

However, the main issues in liposome encapsulation for the food industry are:

- The *scaling up* of the process at *acceptable cost-in-use* levels: The recent advances in liposome encapsulation such as microfluidization technology have most probably solved this problem.
- The *delivery form* of the liposome-encapsulated ingredients:
  - *Aqueous* form: If the liposomes are kept and used as an aqueous suspension, this might be a very serious drawback for the large- scale production, storage, and shipping of encapsulated food ingredients. In particular, stabilization with antimicrobials is necessary but might disrupt the liposome structure while storage and shipping at low temperature adds an extra cost that might not be acceptable for most low-cost food ingredients.
  - *Dry* form: Dry liposome microencapsules can be obtained by freeze drying the liposome suspension, but the considerable cost of large-scale freeze-drying processes can only be carried out by high-price- encapsulated ingredients in a niche market. Moreover, not all liposome formulations can be freeze dried and the reconstitution of the wet formulation

is not always straightforward and usually requires complex steps and processes (Lasic, 1993).

Overall, the development of a cost-effective drying method for liposome microcapsules would enable them to readily reconstitute upon rehydration, which would ensure a promising future to liposome encapsulation of food ingredients.

#### 13.3.11.4 Inclusion Encapsulation: Molecular Inclusion

Molecular inclusion, unlike the other encapsulation processes, takes place at a molecular level using cyclodextrins (typically β-cyclodextrin) as the encapsulating agents (Reineccius and Risch, 1986). As already mentioned, cyclodextrins are six- (α-cyclodextrin), seven- (β-cyclodextrin), or eight-membered (γ-cyclodextrin) cyclic glucose molecules, which are produced from starch by enzymatic process using cyclodextrin glycosyltransferase (Hedges and McBride, 1999).

Cyclodextrins are hollow-truncated cone-shaped molecules having an inner diameter of approximately 5–8 A ($10^{-10}$ m), sufficient to hold typically 6–17 molecules of water. Their molecular dimensions allow total or partial inclusion of a wide range of compounds. The central cavity of the molecule creates a relatively hydrophobic environment, whereas its external surface has a hydrophilic character. This unique conformation is largely responsible for the characteristic physicochemical properties of cyclodextrins (Shieh and Hedges, 1996; Steinbock et al., 2001) that affect the formation of complexes with the different food compounds.

#### 13.3.11.5 Formation of β-Cyclodextrin Complexes

The β-cyclodextrin molecule forms inclusion complexes with core compounds that can dimensionally fit into its central cavity. These complexes are formed in a reaction that takes place only in the presence of water. Molecules that are less polar than water (i.e., most flavor substances) and have suitable molecular dimensions to fit inside the cyclodextrin interior can be incorporated into the molecule. In an aqueous solution, the slightly nonpolar cyclodextrin interior is occupied by water molecules. This situation is energetically unfavorable and, therefore, the sites occupied by water are readily substituted by the less-polar guest molecules. The resulting cyclodextrin complexes are relatively stable and due to their reduced solubility in aqueous solutions, they readily precipitate out of the solution and can be recovered simply by filtration. It must be noted that the formation of the cyclodextrin complexes can be influenced to a greater or lesser extent by several factors including the molecular weight and shape, steric hindrance, chemical functionality, polarity, and volatility of the core material (Goubet et al., 1998; Hedges et al., 1995).

The complexing of a cyclodextrin with a guest compound can be accomplished by three basic methods (Pagington, 1986) as described:

1. Stirring or shaking the cyclodextrin and guest molecules in an aqueous solution and filtering off the precipitated complex. However, in some cases, complexation of an insoluble guest can only be accomplished through dissolution of the guest in a water-soluble solvent.

2. Blending/kneading of solid cyclodextrin and guest molecules with water to form a paste. (Solvent should not be used. This method is particularly applicable for oleoresins.)
3. Forcing a gas through the solution for complexation to occur. This method is seldom employed.

However, it should be emphasized that there are several variations to these basic techniques, but still, in all methods, both the cyclodextrin and the guest molecules must be solubilized. If the guest material is insoluble in water, it is necessary to dissolve it in another solvent such as alcohol (Shahidi and Han, 1993). Irrespective of the applied technique, the encapsulated core compound is kept within the cyclodextrin's cavity by hydrogen bonding, van der Waals forces, or by the entropy-driven hydrophobic effect. Moreover, once the complex is formed, the presence of water or high temperature is required so as to liberate the guest molecules (Reineccius et al., 2002).

A typical application of the molecular inclusion process is the protection of unstable and high value-added speciality flavor chemicals (Uhlemann et al., 2002). However, there are substantial data in the literature documenting excellent protection for food ingredients that have been treated with cyclodextrins (Lindner et al., 1981; Loftsson and Kristmundsdottir, 1993; Reineccius and Rish, 1986; Reineccius et al., 2002; Szente and Szejtli, 1988; Versic, 1988; Westing et al., 1988), including mostly flavor components, hydrophobic vitamins (A, E, or K) and pigments. Overall, the cocrystallization process offers several advantages such as:

- Unique release characteristics
- *Thermal stability*: Cyclodextrin complexes are proven to be thermodynamically stable up to 200°C (Pagington, 1985)
- *Variable binding properties*: Cyclodextrins appear to have variable affinity for different guest compounds that can act, for instance, to selectively remove bitter compounds from orange and grapefruit juices (Shaw et al., 1984)
- *Stability against evaporation*: The cyclodextrin–guest complex formed is very stable to evaporation (Szente and Szejtli, 1988)
- *Oxidative stability*: Many reports have demonstrated that inclusion complexes are quite stable to oxidation (Westing et al., 1988; Szente and Szejtli, 1988)
- *Flowability*: Pagington (1986) and Bhandari et al. (1999) showed that the produced β-cyclodextrin–lemon oil complex was a very fine powder with an excellent flowability
- *Suitable for laboratory evaluation*: Qi and Hedges (1995) provided experimental details of a coprecipitation method deemed to be appropriate for laboratory use
- *Convenient for large-scale production*: The slurry or paste method seem to be appropriate for large-scale production since less water has to be removed during drying (Qi and Hedges, 1995)

- *Good retention during thermal processes*: β-cyclodextrin can retain some aroma compounds in food matrices during cooking and pasteurization (Jouquand et al., 2004)

However, based on the research conducted mainly on flavor encapsulation, it presents some disadvantages as well such as:

- *Low payload*: When applying molecular inclusion, a low payload is typically achieved. For instance, b-cyclodextrin, the most studied cyclodextrin for encapsulation, has a molecular weight of 1135 g/mol and the size of its cavity is such that it could hold a compound of about 150 g/mol. This means a theoretical maximum payload of 11% w/w. Payload between 6% and 15% has been reported for artificial flavor compounds (Heath and Reineccius, 1985).
- *Segregation problems*: It is important to note that any compounds (e.g., flavor) significantly larger than the cavity would not "fit in" and would therefore be left unencapsulated. This leads to segregation problems during the encapsulation process, where some components are preferably encapsulated over others.
- *Expensive method*: Another drawback is the rather high cost of the cyclodextrins. A recent report suggests that the cost of cyclodextrin could never reach a level below $6/kg (Gouin, 2004).
- *Variable binding properties*: They can also be a disadvantage when it comes to the encapsulation of flavor compounds. The variable inclusion properties of cyclodextrins can result in a dry flavor quite different from that of the original when the flavor comprises a broad range of molecules, for instance, an artificial flavor (Reineccius and Risch, 1986).
- As with all processes, there are limits to the application of cyclodextrin in the formation of flavor complexes (Szente et al., 1988) and these include the following:
    - There is a *limited amount of flavor*, which can be incorporated into a formulation (average 9–14% by weight).
    - The size and polarity of flavors to be complexed limit the usefulness of the process.
    - Cyclodextrin can act as an artificial enzyme, sometimes enhancing the rate of hydrolysis of some ester-type flavor components. This can result in undesirable adulteration of the flavor.
    - The water solubility of β-cyclodextrin flavor complexes is generally much lower than that of spray-dried and other microencapsulated samples.

### 13.3.12 Interfacial Polymerization

Interfacial polymerization happens when two different polymeric solutions are brought together. These two reactive polymeric species, each solubilized in a different liquid, react with one another when one liquid is dispersed in the other. The polymerization reaction takes place at the interface between the two polymeric liquids. The interfacial polymerization process can be employed to encapsulate solutions or dispersions of hydrophobic substances. It can also be used to encapsulate aqueous solutions or dispersions of hydrophilic substances. In the interfacial polymerization microencapsulation process, both the dispersed and continuous phases serve as a source of reactive polymeric species (Shahidi and Han, 1993).

In general, an interfacial polymerization reaction proceeds at a rapid rate that results in the formation of a very thin film having physical property characteristics of a semipermeable membrane. Properties of the film are markedly influenced by the reaction time (Kondo, 1979). The ultimate capsule size of interfacial polymerization is determined by the size of the first monomer. In general, the capsule size ranges from ~1 μm to several millimeters. This capsule size is a direct function of the agitation rate (Kondo, 1979). It is found that an increase in the concentration of the emulsifier yields a narrow size distribution range and a reduction of the average particle size. The patent application for the microencapsulation process utilizing the principle of interfacial polymerization was filed by IBM (*S/N*: 813,425) in 1959 (Kondo, 1979). However, the use of interfacial polymerization for food systems is limited, as most coatings are not food grade (Shahidi and Han, 1993).

## REFERENCES


Alishahi, A. and Aïder, M. 2012. Applications of chitosan in the seafood industry and aquaculture: A review. *Food and Bioprocess Technology*, 5, 817–830. DOI 10.1007/s11947011-0664-x.

Alishahi, A., Mirvaghefi, A., Tehrani, M.R., Farahmand, H., Koshio, S., Dorkoosh, F.A. and Elsabee M.Z. 2011a. Chitosan nanoparticle to carry vitamin C through the gastrointestinal tract and induce the nonspecific immunity system of rainbow trout (*Oncorhynchus mykiss*). *Carbohydrate Polymers*, 86, 142–146.

Alishahi, A., Mirvaghefi, A., Tehrani, M.R., Farahmand, H., Shojaosadati, S.A., Dorkoosh, F.A. and Elsabee M.Z. 2011b. Shelf life and delivery enhancement of vitamin C using chitosan nanoparticles. *Food Chemistry*, 126, 935–940.

Anandaraman, S. and Reineccius, G.A. 1987. Analysis of encapsulated orange peel oil. *Perfumer and Flavourist*, 12, 33–39.

Andres, C. 1977. Encapsulation ingredients: I. *Food Process*, 38(12), 44.

Apintanapong, M., and Noomhorm, A. 2003. The use of spray drying to microencapsulate 2-acetyl-1-pyrroline, a major flavour component of aromatic rice. *International Journal of Food Science and Technology*, 38, 95–102.

Arneodo, C.J.F. 1996. Microencapsulation by complex coacervation at ambient temperature. FR patent 2732240 A1.

Arshady, R. 1999. Microspheres, microcapsules and liposomes. In: *Preparation and Chemical Applications* (edited by R. Arshady), Vol. 1, pp. 279–322, London, UK: Citus Books.

Augustin, M.A., Sanguansri, L., Margetts, C. and Young, B. 2001. Microencapsulation of food ingredients. *Food Australia*, 53, 220–223.

Awad, A. and Chen, A.C. 1993. A new generation of sucrose products made by cocrystallization, *Food Technology*, 47(1), 146.

Bakan, J.A. 1973. Microencapsulation of foods and related products. *Food Technology*, 27(11), 34.

Bakan, J.A. and Anderson, J.L. 1970. *Microencapsulation, the Theory and Practice of Industrial Pharmacy* (edited by L. Lachman, H.A. Lieberman, and J.L. Kang), p. 384, Philadelphia, PA: Lea and Febiger.

Balassa, L.L. and Brody, J. 1968. The Balchem way. *Food Engineering*, 40, 89–91.

Balssa, L.L. and Fanger, G.O. 1971. Microencapsulation in the food industry. *Critical Review of Food Technology*, 2, 245.

Bangs, W.E. 1985. Development and characterization of wall materials for spray dried flavorings production, *Dissociation of Abstract International B,* 46(4), 1011.

Baranauskiene, R., Venskutonis, P.R., Dewettinck, K. and Verhé, R. 2006. Properties of oregano (*Origanum vulgare* L.), citronella (*Cymbopogon nardus* G.) and marjoram (*Majorana hortensis* L.) flavors encapsulated into milk protein-based matrices. *Food Research International*, 39, 413–425.

Barnes, J.M. and Steinke, J.A. 1987. Encapsulation matrix composition and encapsulate containing same, U.S. patent 4,689,235.

Bayram, Ö.A., Bayram, M. and Tekin, A.R. 2005. Spray drying of sumac flavour using sodium chloride, sucrose, glucose and starch as carriers. *Journal of Food Engineering*, 69, 253–260.

Beck, E.E. 1972. Essential oil composition and method of preparing the same, U.S. patent 3,704,137.

Benech, R.O., Kheadr, E.E., Laridi, R., Lacroix, C. and Fliss, I. 2002. Inhibition of *Listeria innocua* in cheddar cheese by addition of Nisin Z in liposomes or by *in situ* production in mixed culture. *Applied Environmental Microbiology*, 68, 3683–3690.

Beristain, C.I., Garcia, H.S. and Vernon-Carter, E.J. 1999. Mesquite gum *(Prosopis juliflora)* and maltodextrin blends as wall materials for spray-dried encapsulated orange peel oil. *Food Science and Technology International*, 5, 353–356.

Beristain, C.I., Mendoza, R.E., Garcia, H.S. and Vaequez, A. 1994. Co-crystallization of Jamaica (*Hibiscus sabdarifa* L.) granules. *Lebensmittel-Wissenschaft und-Technologie*, 27, 347–349.

Beristain, C.I., Vazquez, A., Garcia, H.S. and Vernon-Carter, E.J. 1996. Encapsulation of orange peel oil by cocrystallization. *Lebensmittel-Wissenschaft und-Technologie*, 29, 645–647.

Bhandari, B.R., Datta, N., D'Arcy, B.R., and Rintoul, G.B. 1998. Co-crystallization of honey with sucrose. *Lebensmittel-Wissenschaft und Technologie*, 31, 138–142.

Bhandari, B.R., D'Arcy, B.R. and Padukka, I. 1999. Encapsulation of lemon oil by paste method using b-cyclodextrin: Encapsulation efficiency and profile of oil volatiles. *Journal of Agricultural and Food Chemistry*, 47, 5194–5197.

Blenford, D. 1986. Fully protected. *Food Flavor Ingredient Packaging Process*, 8(7), 43.

Burgess, D. and Singh, O. 1993. Spontaneous formation of small sized albumin/acacia coacervate particles. *Journal of Pharmacy and Pharmacology*, 45(7), 586–591.

Buffo, R. and Reineccius, G.A. 2000. Optimization of gum acacia/modified starches/maltodextrin blends for the spray drying of flavours. *Perfumer and Flavourist*, 25, 37–49.

Buffo, R.A., Probst, K., Zehentbauer, G., Luo, Z. and Reineccius, G.A. 2002. Effects of agglomeration on the properties of spray-dried encapsulated flavours. *Flavour and Fragrance Journal*, 17, 292–299.

Bylaitė, E., Rimantas Venskutonis, P. and Mapdpierienė, R. 2001. Properties of caraway (*Carum carvi* L.) essential oil encapsulated into milk protein-based matrices. European *Food Research and Technology*, 212, 661–670.

Byun, M.H., Choi, M.J., Lee, S., and Min, S.G. 1998. Influence of freezing rate on the aroma retention in a freeze drying system. *Korean Journal for Food Science of Animal Resources*, 18, 176–184.

Cai, Y.Z. and Corke, H. 2000. Production and properties of spray dried Amaranthus betacyanins pigments. *Journal of Food Science*, 65, 1248–1252.

Carroll, T.J., Feinerman, D., Huzinec, R.J., and Piccolo, D.J. 1984. Gum composition with plural time releasing flavors and method of preparation, U.S. patent 4,485,118.

Chapman, D., Williams, R.M. and Ladbrook, B.D. 1967. Physical studies of phospholipids: VI. Thermotropic and lyotropic mesomorphism of some 1,2 diacylphosphatidylcholines (lecithin). *Chemistry Physics Lipids*, 1, 445.

Chemat, F., Zill-e-Huma, and Khan, M.K. 2011. Applications of ultrasound in food technology: Processing, preservation and extraction. *Ultrasonics Sonochemistry*, 18, 813–835.

Chen, A.C. 1994. Ingredient technology by the sugar cocrystallization process. *International Sugar Journal*, 96, 493–494.

Chen, A.C., Veiga, M.F. and Rizzuto, A.B. 1988. Cocrystallization: An encapsulation process. *Food Technology*, 42(11), 87.

Chen, A.C.C., Lang, C.E., Graham, Jr., C.P. and Rizzuto, A.B. 1982a. Crystallized, readily water-dispersible sugar product, U.S. patent 4,338,350.

Chen, A.C.C., Lang, C.E., Graham, Jr., C.P. and Rizzuto, A.B. 1982b. Crystallized, readily water-dispersible sugar product containing heat sensitive, acidic or high invert sugar substances, U.S. patent 4,362,757.

Chranioti, C. and Tzia, C. 2013. Binary mixtures of modified starch, maltodextrin and chitosan as efficient encapsulating agents of fennel oleoresin. *Food and Bioprocess Technology*, 6, 3238–3246. DOI 10.1007/s11947-012-0966-7

Chranioti, C. and Tzia, C. 2014. Arabic gum mixtures as encapsulating agents of freeze-dried fennel oleoresin products. *Food Bioprocess Technology*, 7, 1057–1065. DOI 10.1007/s11947-013-1074-z.

Chua, K.J. and Chou, S.K. 2003. Low-cost drying methods for developing countries. *Trends in Food Science and Technology*, 14, 519–528.

Cohen, J.S. and Yang, T.C.S. 1995. Progress in food dehydration. *Trends in Food Science Technology*, 6, 20–25.

Couvreur, P., Barratt, G., Fattal, E., Legrand, P. and Vauthier, C. 2002. Nanocapsule technology: A review. *Critical Reviews in Therapeutic Drug Carrier Systems*, 19(2), 99–134.

Crocker, G.C. and Pritchett, D.E. 1978. Improved encapsulated citrus oils. *Food Technology*, 32, 36–45.

Crouzet, J. 1998. Aromes alimentaires (Food flavorings). In: Techniques de l'ingenieur, *Agroalimentaire F*, 4100, pp. 1–16, Paris

Dalgleish, D.G. 2006. Food emulsions—Their structures and structure-forming properties. *Food Hydrocolloids*, 20, 415–422.

de Oliveira, M.A., Maia, G.A., Figueiredo, R.W., Souza, A.C.R., Brito, E. S. and Azeredo, H.M.C. 2009. Addition of cashew tree gum to maltodextrin-based carriers for spray drying of cashew apple juice. *International Journal of Food Science and Technology*, 44, 641–645.

Deamer, D.W. and Uster, P.S. 1983. *Liposome Preparation: Methods and Mechanisms, Liposomes* (edited by M.J. Ostro), p. 27, New York, NY: Marcel Dekker, Inc.

Desobry, S.A., Netto, F.M. and Labuza, T.P. 1997. Comparison of spray-drying, drum-drying and freeze-drying for β-carotene encapsulation and preservation. *Journal of Food Science*, 62, 1158–1162.

Dewettinck, K. and Huyghebaert, A. 1999. Fluidized bed coating in food technology, *Trends in Food Science Technology*, 10, 163.

DeZarn, T.J. 1995. Food ingredient encapsulation: An overview. *Encapsulation and Controlled Release of Food Ingredients* (edited by S.J. Risch and G.A. Reineccius), p. 74, ACS Symposium Series No. 590, Washington, DC: American Chemical Society.

DeZarn, T.J. 1998. Food ingredient encapsulation. In: *Encapsulation and Controlled Release of Food Ingredients* (edited by S.J. Risch and G.A. Reineccius), pp. 74–86, ASC Symposium Series 590, Washington, DC: American Chemical Society.

Dickinson, E. 2009. Hydrocolloids as emulsifiers and emulsion stabilizers. *Food Hydrocolloids*, 23, 1473–1482.

Drusch, S. 2007. Sugar beet pectin: A novel emulsifying wall component for microencapsulation of lipophilic food ingredients by spray-drying. *Food Hydrocolloids*, 21, 1223–1228.

Dubin, P. L., Muhoberac, B. B., and Xia, J. 1998. Preparation of enzyme-polyelectrolyte coacervate complexes and their properties. US patent 5834271 A.

Dziezak, J.D. 1988. Microencapsulation and encapsulated ingredients. *Food Technology*, 42 (4), 136–151.

Dzondo-Gadet, J.M., Nzikou, A., Etoumongo, M.L., and Desobry, S. 2005. Encapsulation and storage of safou pulp oil in 6DE maltodextrins. *Process Biochemistry*, 40, 265–271.

Ersus, S. and Yurdagel, U. 2007. Microencapsulation of anthocyanin pigments of black carrot (*Daucus carota* L.) by spray drier. *Journal of Food Engineering*, 80, 805–812.

Estevinho, B.N., Rocha, F., Santos, L. and Alves, A. 2013. Microencapsulation with chitosan by spray drying for industry applications. A review. *Trends in Food Science and Technology*, 31, 138–155.

Fang, Z. and Bhandari, B. 2010. Encapsulation of polyphenols: A review. *Trends in Food Science and Technology*, 21, 510–523.

Flink, J. and Karel, M. 1970. Retention of organic volatiles in freeze-dried solution of carbohydrates. *Journal of Agricultural Food Chemistry*, 18, 295.

Flores, R.J., Wall, M.D., Carnahan, D.W. and Orofino, T.A. 1992. An investigation of internal phase losses during the microencapsulation of fragrances. *Journal of Microencapsulation*, 3, 287–307.

Gabas, A.L., Telis, V.R.N., Sobral, P.J.A., and Telis-Romero, J. 2007. Effect of maltodextrin and arabic gum in water vapour sorption thermodynamic properties of vacuum dried pineapple pulp powder. *Journal of Food Engineering*, 82(2), 246–252.

Gharsallaoui, A., Roudaut, G., Chambin, O., Voilley, A., and Saurel, R. 2007. Applications of spray-drying in microencapsulation of food ingredients: An overview. *Food Research International*, 40(9), 1107–1121.

Gibbs, B.F., Kermasha, S., Alli, I. and Mulligan, C.N. 1999. Encapsulation in the food industry: A review. *International Journal of Food Sciences and Nutrition*, 50, 213–224.

Glicksman, M. 1982. Background and classification. In: *Food Hydrocolloids* (edited by M. Glicksman), Vol. 1, p. 3, Boca Raton, FL: CRC Press.

Gong, Z., Zhang, M., Mujumdar, A.S. and Sun, J. 2008. Spray drying and agglomeration of instant bayberry powder. *Drying Technology: An International Journal*, 26, 116–121.

Gonzalez Siso, M.I., Lang, E., Carreno-Gomez, B., Becerra, M., Otero Espinar, F. and Blanco Mendez, J. 1997. Enzyme encapsulation on chitosan microbeads. *Process Biochemistry*, 32(3), 211–216.

Goubet, I., Le Quere, J.L. and Voiley, A.J. 1998. Retention of aroma compounds by carbohydrates: Influence of their physicochemical characteristics and of their physical state. A review. *Journal of Agricultural and Food Chemistry*, 46, 1981–1990.

Gouin, S. 2004. Micro-encapsulation: Industrial appraisal of existing technologies and trends. *Trends in Food Science and Technology*, 15, 330–347.

Green, B.K. and Scheicher, L. 1955. Pressure sensitive record materials, U.S. patent 418, 2, 217–507, Ncr C.

Gregoriadis, G. 1987. Encapsulation of enzymes and other agents in liposomes. In: *Chemical Aspects in Food Enzymes* (edited by A.J. Andrews), p. 97, London: The Royal Society of Chemistry.

Heath, H.B. and Reineccius, G. 1985. *Flavor Production in Flavor Chemistry and Technology*. Westport Conn: AVI Publishing Company, Inc., Chapter 11.

Hedges, A. and McBride, C. 1999. Utilization of b-cyclodextrin in food. *Cereal Foods World*, 44, 700–704.

Hedges, A.R., Shieh, W.J. and Sikorski, C.T. 1995. Use of cyclodextrins for encapsulation in the use and treatment of food products. In: *Encapsulation and Controlled Release of Food Ingredients* (edited by S.J. Risch and G.A. Reineccius), p. 60, ACS Symposium Series No. 590, Washington, DC: American Chemical Society.

Heinzelmann, K. and Franke, K. 1999. Using freezing and drying techniques of emulsions for the microencapsulation of fish oil to improve oxidation stability. *Colloids and Surfaces B: Biointerfaces*, 12, 223–229.

Heinzen, C. 2002. Microencapsulation solve time dependent problems for foodmakers. *European Food and Drink Review*, 3, 27–30.

Hogan, S.A., McNamee, B.F., O'Riordan, E.D. and O'Sullivan, M. 2001. Emulsification and microencapsulation properties of sodium caseinate/carbohydrate blends. *International Dairy Journal*, 11, 137–144.

Idham, Z., Muhamad, I.I. and Sarmid, M.R. 2012. Degradation kinetics and color stability of spray-dried encapsulated anthocyanins from *Hibiscus sabdariffa* L. *Journal of Food Process Engineering* 35, 522–542.

Jackson, L.S. and Lee, K. 1991. Microencapsulation in the food industry. *LWT—Food Science and Technology*, 24, 289–297.

Jacquot, M. and Pernetti, M. 2003. Spray coating and drying processes. In: *Cell Immobilization Biotechnology* (edited by U. Nedovic and R. Willaert), pp. 343–356, Series: Focus on biotechnology, Dordrecht: Kluwer Academic Publishers.

Jafari, S.M., Assadpoor, E., He, Y. and Bhandari, B. 2008. Encapsulation efficiency of food flavours and oils during spray drying. *Drying Technology*, 26, 816–835.

Jegat, C. and Taverdet, J.L. 2000. Stirring speed influence study on the microencapsulation process and on the drug release from microcapsules. *Polymer Bulletin*, 44, 345–351.

Jegat, C. and Taverdet, J.L. 2001. Microencapsulation by complex coacervation method: Influence of some parameters on the particles morphology. *Annales des Falsifications de l'Expertise Chimique*, 94, 103–113.

Jesorka, A. and Orwar, O. 2008. Liposomes: Technologies and analytical applications. *Annual Review of Analytical Chemistry*, 1, 801–832.

Jouquand, C., Ducruet, V. and Giampaoli, P. 2004. Partition coefficients of aroma compounds in polysaccharide solutions by the phase ratio variation method. *Food Chemistry*, 85, 467–474.

Jozwiakowsksi, M.J., Jones, D. and Franz, R.M. 1990. Characterization of a hot melt fluid bed coating process from fine granules. *Pharmaceutical Research*, 7, 3–10.

Jung, J.M. and Sung, T.K. 2000. A new method for analysis of capsaicinoids content in microcapsules. *Korean Journal of Food Science and Technology*, 32, 42–49.

Junyaprasert, V.B., Mitrevej, A., Sinchaipanid, N., Boonme, P., and Wurster, D.E. 2001. Effect of process variables on the microencapsulation of vitamin A palmitate by gelatin-acacia coacervation. *Drug Development and Industrial Pharmacy*, 27, 561–566.

Kaasgaard, T. and Keller, D. 2010. Chitosan coating improves retention and redispersibility of freeze-dried flavour oil emulsions. *Journal of Agricultural and Food Chemistry*, 58, 2446–2454.

Kanakdande, D., Bhosale, R. and Singhal, R.S. 2007. Stability of cumin oleoresin microencapsulated in different combination of gum arabic, maltodextrin and modified starch. *Carbohydrate Polymers*, 67, 536–541.

Kaushik, V. and Roos, Y.H. 2007. Limonene encapsulation in freeze-drying of gum arabic–sucrose–gelatin systems. *LWT—Food Science and Technology*, 40, 1381–1391.

Kim, H.-H.Y. and Baianu, I.C. 1991. Novel liposome microencapsulation techniques for food applications. *Trends in Food Science Technology*, 2, 55.

Kirby, C.J. and Gregoriadis, G. 1984. A simple procedure for preparing liposomes capable of high encapsulation efficiency under mild conditions. In: *Liposome Technology* (edited by G. Gregoriadis), Vol. 1, p. 19, Boca Raton, FL: CRC Press.

Kirby, C.J., Whittle, C.J., Rigby, N., Coxon, D.T. and Law, B.A. 1991. Stabilization of ascorbic acid by microencapsulation. *International Journal of Food Science and Technology*, 26, 437–449.

Knezevic, Z., Gosaki, D., Hraste, M. and Jalsenjak, I. 1998. Fluid-bed microencapsulation of ascorbic acid. *Journal of Microencapsulation*, 15, 237–252.

Ko, J.A., Park, H.J., Park, Y.S., Hwang, S.J. and Park, J.B. 2003. Chitosan microparticle preparation for controlled drug release by response surface methodology. *Journal of Microencapsulation*, 20(6), 791–797.

Kondo, A. 1979. Microencapsulation by interfacial polymerization. In: *Microcapsule Processing and Technology* (edited by J. Wade van Valkenburg), p. 35, New York, NY: Marcel Dekker, Inc..

Kopelman, I.J., Meydav, S. and Wilmersdorf, P. 1977a. Freeze drying encapsulation of water-soluble citrus aroma. *Journal of Food Technology*, 12, 65.

Kopelman, I.J., Mangold, D.J. and Weinberg, S. 1977b. Storage studies of freeze-dried lemon crystals. *Journal of Food Technology* 12, 403.

Korus, J. 2001. Microencapsulation of flavours in starch matrix by coacervation method. *Polish Journal of Food and Nutrition Sciences*, 10(51), 17–23.

Korus, J., Tomasik, P., and Lii, C.Y. 2003. Microcapsules from starch granules. *Journal of Microencapsulation*, 20, 47–56.

Krishnan, S., Bhosale, R., and Singhal, R.S. 2005. Microencapsulation of cardamom oleoresin: Evaluation of blends of gum arabic, maltodextrin and a modified starch as wall materials. *Carbohydrate Polymers*, 61, 95–102.

Kshirsagar, A.C. and Singhal, R.S. 2007. Optimization of starch oleate derivatives from native corn and hydrolyzed corn starch by response surface methodology. *Carbohydrate Polymers*, 69, 455–461.

Kshirsagar, A.C., Yenge, V.B., Sarkar, A. and Singhal, R.S. 2009. Efficacy of pullulan in emulsification of turmeric oleoresin and its subsequent microencapsulation. *Food Chemistry*, 113, 1139–1145.

Kumar, M.N.V. 2000. A review of chitin and chitosan applications. *Reactive Functional Polymer*, 46, 1–27.

LaBell, F. 1991. Co-crystallization process aids dispersion and solubility. *Food Processing*, 52, 60–63.

Lamprecht, A., Schafer, U., and Lehr, C.M. 2001. Influences of process parameters on preparation of microparticle used as a carrier system for W-3 unsaturated fatty acid ethyl esters used in supplementary nutrition. *Journal of Microencapsulation*, 18, 347–357.

LaBell, F. 1999. Energy drinks with electrifying flavour—Research efforts of Forbitech Inc. in developing vitamin-fortified juice drinks. *Preparation of Foods*, 171(7), 28.

Lasic, D.D. 1993. *Liposomes: From Physics to Applications*. London: Elsevier.

Leahy, M.M., Anandaraman, S., Bangs, W.E. and Reineccius, G.A. 1983. Spray drying of food flavors: II. A comparison of encapsulating agents for the drying of artificial flavor. *Perfect Flavor*, 8(5), 49.

Lee, D.H., Jin, B.H., Hwang, Y.I. and Lee, S.C. 2000. Encapsulation of bromelain in liposome. *Journal of Food Science and Nutrition*, 5, 81–85.

Lee, S.W., Kang, S.Y., Han, S.H. and Rhee, C. 2009. Influence of modification method and starch concentration on the stability and physical properties of modified potato starch as wall materials. *European Food Research and Technology*, 228, 449–455.

Lee, S.Y. and Krochta, J.M. 2002. Accelerated shelf life testing of whey protein coated peanuts analysed by static gas chromatography. *Journal of Agricultural and Food Chemistry*, 50, 2022–2028.

Legrand, P., Barratt, G., Mosqueira, V., Fessi, H. and Devissaguet, J.P. 1999. Polymeric nanocapsules as drug delivery systems—A review. *Pharmaceutical Sciences*, 9(5), 411–418.

Lim, H.K., Tan, C.P., Bakar, J., and Ng, S.P. 2012. Effects of different wall materials on the physicochemical properties and oxidative stability of spray-dried microencapsulated red-fleshed pitaya (*Hylocereus polyrhizus*) seed oil. *Food and Bioprocess Technology*, 5, 1220–1227. DOI 10.1007/s11947-011-0555-1.

Lindner, K., Szente, L. and Szejtli, J. 1981. Food flavoring with b-cyclodextrin-complexed flavour substances. *Acta Aliment*, 10, 175.

Loftsson, T. and Kristmundsdottir, T. 1993. Microcapsules containing water-soluble cyclodextrin inclusion complexes of water insoluble drugs. In: *Polymeric Delivery Systems* (edited by M.A. El-Nokaly, , D.M. Piatt, and B.A. Charpentier), pp. 168 189, Washington, DC: AMS.

Maa, Y.F. and Hsu, C. 1999. C performance of sonication and microfluidization for liquid–liquid emulsification. *Pharmaceutical Development Technology*, 4, 233–240.

Madene, A., Jacquot, M., Scher, J. and Desobry, S. 2006. Review flavour encapsulation and controlled release—A review. *International Journal of Food Science and Technology*, 41, 1–21.

Marques, H.M.C. 2010. A review on cyclodextrin encapsulation of essential oils and volatiles. *Flavour Fragrance Journal*, 25, 313–326.

Mayhew, E., Conroy, S., King, J., Lazo, R., Nikolopoulus, G., Siciliano, A. and Vail, W.J. 1987. High pressure continuous-flow system for drug entrapment in liposomes. In: *Drug and Enzyme Targeting, Part B* (edited by R. Green and K.J. Widder), p. 64, New York, NY: Academic Press, Inc.

McNamee, B.F., O'Riordan, E. and O'Sullivan, M. 1998. Emulsification and microencapsulation properties of gum arabic. *Journal of Agricultural and Food Chemistry*, 46, 4551–4555.

McNamee, B.F., O'Riordan, E.D. and O'Sullivan, M. 2001. Effect of partial replacement of gum arabic with carbohydrates on its microencapsulation properties. *Journal of Agriculture and Food Chemistry*, 49, 3385–3388.

Mermelstein, N.H. 2001. Spray drying. *Journal of Food Technology*, 55(4), 92.

Miller, D.H. and Mutka, J.R. 1986. Preparation of solid essential oil flavor composition, U.S. patent 4,610,890.

Moschakis, T., Murray, B.S., and Biliaderis, C.G. 2010. Modifications in stability and structure of whey proteincoated o/w emulsions by interacting chitosan and gum arabic mixed dispersions. *Food Hydrocolloids*, 24, 8–17.

Mujumdar, A.S. and Devahastin, S. 2000. Fluidized bed drying. In: *Developments in Drying: Food Dehydratation* (edited by A.S. Mujumdar and S. Suvachittanont), Vol. 1, pp. 59–111, Bangkok: Kasetsart University Press.

Muller, C.R., Bassani, V.L., Pohlmann, A.R., Michalowski, C.B., Petrovick, P.R. and Guterres, S.S. 2000. Preparation and characterization of spray-dried polymeric nanocapsules. *Drug Development and Industrial Pharmacy*, 26(3), 343–347.

Mullin, J.W. 1972. *Crystallisation Kinetics*, p. 174, London: Crystallisation, Butterworth & Co. Ltd.

Murugesan, R. and Orsat, V. 2012. Spray drying for the production of nutraceutical ingredients: A review. *Food and Bioprocess Technology*, 5, 3–14. DOI 10.1007/s11947-011-0638-z.

Mutka, J.R. and Nelson, D.B. 1988. Preparation of encapsulated flavors with high flavor level. *Food Technology*, 42, 154–157.

Nagata, T. 1996. Techniques et application of electron microscopic radioautography. *Journal of Electron Microscopy*, 45, 258–274.

Nagatomo, S. 1985. Cyclodextrins: Expanding the development of their functions and applications. *Chemistry of Economics English Review*, 17, 28.

Nakagawa, K., Iwamoto, S., Nakajima, M., Shono, A., and Satoh, K. 2004. Microchannel emulsification using gelatine and surfactant-free coacervate microencapsulation. *Journal of Colloids and Interface Science*, 278, 198–205.

Ono, F. 1980. New encapsulation technique with protein–carbohydrate matrix. *Journal of Japanese Food Science Technology*, 27, 529.

Ono, F. and Aoyama, Y. 1979. Encapsulation and stabilization of oily substances by protein and carbohydrate. *Journal of Japanese Food Science Technology*, 26, 13.

Orlien, V., Andersen, A.B., Sinkko, T. and Skibsted, L.H. 2000. Hydroperoxide formation in rapeseed oil encapsulated in a glassy food model as influenced by hydrophilic and lipophilic radicals. *Food Chemistry*, 68, 191–199.

Pagington, J.S. 1986. b-Cyclodextrin and its uses in the flavor industry. In: *Developments in Food Flavors* (edited by G.G. Birch and M.G. Lindley), p. 131, London: Elsevier Applied Science Publishers Ltd.

Panda, R.C., Zank, J. and Martin, H. 2001. Modelling the droplet deposition behaviour on a single particle in fluidized bed spray granulation process. *Powder Technology*, 115, 51–57.

Passin, J.Z. 1969. Encapsulation process. U.S. patent 868619.

Pauletti, M.S. and Amestoy, P. 1999. Butter microencapsulation as affected by composition of wall material and fat. *Journal of Food Science*, 64, 279–282.

Pavón-García, L.M.A., Pérez-Alonso, C., Orozco-Villafuerte, J., Pimentel-González, D.J., Rodríguez-Huezo, M.E., and Vernon-Carter, E.J. 2011. Storage stability of the natural colourant from Justicia spicigera microencapsulated in protective colloids blends by spraydrying. *International Journal of Food Science and Technology*, 46, 1428–1437.

Pearl, T.T., Soper, J.C., and Wamper, D.J. 1993. Heatstable and fracturable spray-dried free flowing flavor oil capsules, method of making and using in food PCT. WO 93/19622 A2.

Pegg, R.B. and Shahidi, F. 2007. Encapsulation, stabilization, and controlled release of food ingredients and bioactives. In: *Handbook of Food Preservation*, 2nd edition (edited by M. Shafiur Rahman), Boca Raton, FL: CRC Press, Taylor & Francis Group.

Perez-Alonso, C., Cruz-Olivares, J., Barrera-Pichardo, J.F., Rodrıguez-Huezo, M.E. Baez-Gonzalez, J.G. and Vernon-Carter, E.J. 2008. DSC thermo-oxidative stability of red chilli oleoresin microencapsulated in blended biopolymers matrices. *Journal of Food Engineering*, 85, 613–624.

Phillips, G.O. and Williams, P.A. 1995. Interaction of hydrocolloids in food systems. In: *Ingredient interactions: Effect on food quality* (edited by A.G. Gaonkar), pp. 131–170, New York, NY: Marcel Dekker, Inc.

Pitalua, E., Jimenez, M., Vernon-Carter, E.J. and Beristain, C.I. 2010. Antioxidative activity of microcapsules with beetroot juice using gum arabic as wall material. *Food and Bioproducts Processing*, 8, 253–258.

Porzio, M.A. and Madsen, M.G. 1996. Double encapsulation process and flavorant compositions prepared thereby. PCT WO 96-US16124.

Qi, Z.H. and Hedges, A.R. 1995. Use of cyclodextrins for flavors. In: *Flavor Technology: Physical Chemistry, Modification and Process* (edited by C.T. Ho and C.H. Tong), pp. 231–243, ACS Symposium Series 610, Washington, DC: American Chemical Society.

Qi, Z.H. and Xu, A. 1999. Starch based ingredients for flow encapsulation. *Cereal Food World*, 44, 460–465.

Quellet, C., Schudel, M., and Ringgenberg, R. 2001. Flavors and fragrance delivery systems. *Chimia*, 55, 421–428.

Rahman, M.S. and Perera, C.O. 2007. Drying and food preservation. In: *Handbook of Food Preservation*, 2nd edition, Boca Raton, FL: CRC Press, Taylor & Francis Group.

Rascón, M.P., Beristain, C.I., García, H.S. and Salgado, M.A. 2011. Carotenoid retention and storage stability of spray-dried encapsulated paprika oleoresin using gum arabic and soy protein isolate as wall materials. *LWT—Food Science and Technology*, 44, 549–557.

Rabiskova, M. and Valaskova, J. 1998. The influence of HLB on the encapsulation of oils by complex coacervation. *Journal of Microencapsulation*, 15, 747–751.

Ré, M.I. 1998. Microencapsulation by spray drying. *Drying Technology*, 16(6), 1195–1236.

Reineccius, G.A. 1988. Spray drying of food flavours. In: *Flavour Encapsulation* (edited by G.A. Reineccius and S.J. Risch), pp. 55–66, Washington, DC: American Chemical Society.

Reineccius, G.A. 1991. Carbohydrates for flavor encapsulation. *Food Technology*, 454 (45), 144–147.

Reineccius, G.A. 1994. *Part II-Flavor encapsulation, Source Book of Flavors*, 2nd edtition (edited by G.A. Reineccius), p. 605. New York, NY: Chapman and Hall.

Reineccius, G.A. 1995. Liposomes for controlled release in the food industry. In: *Encapsulation and Controlled Release of Food Ingredients* (edited by S.J. Risch and G.A. Reineccius), p. 113. ACS Symposium Series No. 590, Washington, DC: American Chemical Society .

Reineccius, G.A. and Bangs, W.E. 1985. Spray drying of food flavors: III. Optimum infeed concentrations for the retention of artificial flavors. *Perfect Flavor*, 10(1), 27.

Reineccius, G.A. and Risch, S.J. 1986. Encapsulation of artificial flavors by b-cyclodextrin. *Perfect Flavor*, 11(4), 1.

Reineccius, T.A., Reineccius, G.A. and Peppard, T.L. 2002. Encapsulation of flavours using cyclodextrins: Comparison of flavour retention in alpha, beta and gamma types. *Journal of Food Science*, 67, 3271.

Risch, S.J. 1995. Encapsulation: Overview of uses and techniques. In: *Encapsulation and Controlled Release of Food Ingredients* (edited by S.J. Risch and G.A. Reineccius), p. 2, ACS Symposium Series No. 590, Washington, DC: American Chemical Society.

Risch, S.J. 1988. Encapsulation of flavours by extrusion. In: *Flavour Encapsulation* (edited by S.J. Risch and G.A. Reineccius), pp. 103–109. ACS Symposium Series 370. Washington, DC: American Chemical Society.

Rizvi, S.S.H., Mulvaney, S.J. and Sokhey, A.S. 1995. The combined application of supercritical fluid and extrusion technology. *Trends in Food Science and Technology*, 6, 232–240.

Rizzuto, A.B., Chen, A.C. and Veiga, M.F. 1984. Modification of the sucrose crystal structure to enhance pharmaceutical properties of excipient and drug substances. *Pharmaceutical Technology*, 8(9), 32.

Robert, P., Carlsson, R., Romero, N. and Masson, L. 2003. Stability of spray-dried encapsulated carotenoid pigments from rosa mosqueta (*Rosa rubiginosa*) oleoresin. *Journal of the American Oil Chemists' Society*, 80, 1115–1120.

Rodríguez, M.S., Albertengo, L. and Agulló, E. 2002. Emulsification capacity of chitosan. *Carbohydrate Polymers*, 48, 271–276.

Sanchez, C. and Renard, D. 2002. Stability and structure of protein–polysaccharide coacervates in the presence of protein aggregates. *International Journal of Pharmaceutics*, 242, 319–324.

Schalmeus, W. 1995. Centrifugal extrusion encapsulation. In: *Encapsulation and Controlled Release of Food Ingredient* (edited by S.J. Rish and G.A. Reineccius), pp. 96–103, Washington, DC: American Chemical Society.

Seisun, D. 2002. Market overview. In: *Gums and Stabilisers for the Food Industry* (edited by P.A. Williams and G.O. Phillips), Vol. 11, pp. 3–9, Cambridge: Royal Society of Chemistry.

Selim, K., Tsimidou, M. and Biliaderis, C.G. 2000. Kinetic studies of degradation of saffron carotenoids encapsulated in amorphous polymer matrices. *Food Chemistry*, 71, 199–206.

Serris, G.S. and Biliaderis, C.G. 2001. Degradation kinetics of beetroot pigment encapsulated in polymeric matrices. *Journal of Science of Food and Agriculture*, 81, 691–700.

Shahidi, F., Arachchi, J.K.V. and Jeon, Y.J. 1999. Food applications of chitin and chitosans. *Trends in Food Science and Technology*, 10, 37–51.

Shahidi, F. and Han, X.Q. 1993. Encapsulation of food ingredients. *Critical Reviews 456 in Food Science and Nutrition*, 33, 501–547.

Shaikh, J., Bhosale, R., and Singhal, R. 2006. Microencapsulation of black pepper oleoresin. *Food Chemistry*, 94, 105–110.

Shapiro, Y.E. 2004. Nanoencapsulation of bioactive substances. In: *Dekker Encyclopedia of Nanoscience and Nanotechnology* (edited by J.A. Schwarz, C.I. Contescu, and K. Putyera), pp. 2339–2352, New York, NY: Marcel Dekker, Inc.

Shaw, P.E., Tatum, J.H. and Wilson, C.W. 1984. Improved flavor of navel orange and grapefruit juices by removal of bitter components with b-cyclodextrin polymer. *Journal of Agricultural Food Chemistry*, 32, 832.

Shefer, A. and Shefer, S. 2003. Novel encapsulation system provides controlled release of ingredients. *Food Technology*, 57(11), 40.

Shieh, W.J. and Hedges, A.R. 1996. Properties and applications of cyclodextrins. *Journal of Macromolecular Science*, 33, 673–683.

Shimada, Y., Roos, Y. and Karel, M. 1991. Oxidation of methyl linoleate encapsulated in amorphous lactose-based food model. *Journal of Agricultural and Food Chemistry*, 39(4), 637–641.

Soottitantawat, A., Bigeard, F., Yoshii, H., Furuta, T., Ohkawara, M. and Linko, P. 2005a. Influence of emulsion and powder size on the stability of encapsulated D-limonene by spray drying. *Innovative Food Science and Emerging Technologies*, 6, 107–114.

Soottitantawat, A., Takayama, K., Okamura, K., Muranaka, D., Yoshii, H., Furuta, T., Ohkawara M., and Linko. P. 2005b. Microencapsulation of l-menthol by spray drying and its release characteristics. *Innovative Food Science and Emerging Technologies*, 6, 163–170.

Soottitantawat, A., Yoshii, H., Furuta, T., Ohgawara, M., and Linko, P. 2003. Microencapsulation by spray drying: Influence of emulsion size on the retention of volatile compounds. *Journal of Food Science*, 68, 2256–2262.

Soper, J.C. 1995. Utilization of coacervated flavours. In: *Encapsulation and Controlled Release of Food Ingredients* (edited by S.J. Risch and G.A. Reineccius), p. 104, ACS Symposium Series No. 590, Washington, DC: American Chemical Society.

Sousdaleff, M., Baesso, M.L., Neto, A.M., Nogueira, A.C., Marcolino, V.A. and Matioli, G. 2013. Microencapsulation by freeze-drying of potassium norbixinate and curcumin with maltodextrin: Stability, solubility, and food application. *Journal of Agricultural Food Chemistry*, 61, 955 – 965.

Sparks, R.E. 1981. Microencapsulation. *Kirk-Othmer Encyclopedia of Chemical Technology*, 3rd edition, (edited by M. Grayson and E. David), Vol. 15, p. 470, New York, NY: Wiley.

Steinbock, B., Vichailkul, P.P. and Steinbock, O. 2001. Nonlinear analysis of dynamic binding in affinity capillary electrophoresis demonstrated for inclusion complexes of b-cyclodextrin. *Journal of Chromatography A*, 943, 139–146.

Swisher, H.E. 1957. Solid flavoring composition and method of preparing the same, U.S. patent 2,809,895.

Szente, L., Gal-Fuzy, M. and Szejtli, J. 1988. Tea aromatization with b-cyclodextrin complexed flavors. *Acta Aliment*, 17, 193.

Szente, L. and Szejtli, J. 1988. Stabilization of flavors by cyclodextrins. In: *Flavor Encapsulation* (edited by S.J. Risch and G.A. Reineccius), p. 148, ACS Symposium Series No. 370, Washington, DC: American Chemical Society

Szente, L. and Szejtli, J. 2004. Cyclodextrins as food ingredients. *Trends in Food Science and Technology*, 15, 137–142.

Szoka, F. and Papahadjopoulos, D. 1980. Comparative properties and methods of preparation of lipid vesicles (liposomes). *Annual Review of Biophysics Bioengineering*, 9, 467.

Tan, L.H., Chan, L.W. and Heng, P.W.S. 2009. Alginate/starch composites as wall material to achieve microencapsulation with high oil loading. *Journal of Microencapsulation: Micro and Nano Carriers*, 26, 263–271.

Tang, Y.C. and Chen, B.H. 2000. Pigment change of freeze-dried carotenoid powder during storage. *Food Chemistry*, 69, 11–17.

Tari, T.A. and Singhal, R.S. 2002. Starch based spherical aggregates: Reconfirmation of the role of amylose on the stability of a model flavoring compound, vanillin. *Carbohydrate Polymers*, 50, 279–282.

Taylor, A.H. 1983. Encapsulation systems and their applications in the flavor industry. *Food Flavor Ingredient Packaging Process*, 5(9), 48.

Tobitsuka, K., Miura, M. and Kobayashi, S. 2006. Retention of a European pear aroma model mixture using different types of saccharides. *Journal of Agricultural and Food Chemistry*, 54, 5069–5076.

Todd, R.D. 1970. Microencapsulation and the flavor industry, *Flavor Industry*, 1, 768.

Trindade, M.A. and Grosso, C.R.F. 2000. The stability of ascorbic acid microencapsulated in granules of rice starch and in gum arabic. *Journal of Microencapsulation: Micro and Nano Carriers*, 17, 169–176.

Ubbink, J. and Krüger, J. 2006. Physical approaches for the delivery of active ingredients in foods. *Trends in Food Science and Technology*, 17, 244–254.

Uddin, M.S., Hawlader, M.N.A. and Zhu, H.J. 2001. Microencapsulation of ascorbic acid: Effect of process variables on product characteristics. *Journal of Microencapsulation: Micro and Nano Carriers*, 18, 199–209.

Uhlemann, H. and Morl, L. 2000. *Wirbelschicht-Spruhgranulation*. Berlin: Springer.

Uhlemann, J., Schleifenbaum, B. and Bertram, H.-J. 2002. Flavor encapsulation technologies: An overview including recent developments. *Perfumed Flavor*, 27(5), 52.

Versic, R.J. 1988. Flavor encapsulation: An overview. *Flavor Encapsulation* (edited by G.A. Reineccius and S.J. Risch ), p. 1, ACS Symposium Series No. 370, Washington, DC: American Chemical Society.

Villota, R. and Hawkes, J.G. 1994. Flavoring in extrusion. An overview. In: *Thermally Generated Flavors* (edited by T.H. Parliament, M.J. Morello, and R.J. McGorrin), p. 280, Washington, DC: American Chemical Society.

Vuillemard, J.C. 1991. Recent advances in the large scale production of lipid vesicles for use in food products: Microfluidization. *Journal of Microencapsulation*, 8, 547–562.

Wampler, D.J. 1992. Flavor encapsulation: a method for providing maximum stability for dry flavor systems. *Cereal Foods World*, 37, 817–820.

Watanabe, Y., Fang, X., Minemoto, Y., Adachi, S. and Matsuno, R. 2002. Suppressive effect of saturated L ascorbate on the oxidation of linoleic acid encapsulated with maltodextrin or gum arabic by spray-drying. *Journal of Agricultural and Food Chemistry*, 50, 3984–3987.

Weiss, H.D. and Reynolds, R.G. 1989. Method of acidulating a comminuted meat product, U.S. patent 4,803,092.

Westing, L.L., Reineccius, G.A. and Caporaso, F. 1988. Shelf life of orange oil: Effects of encapsulation by spray-drying, extrusion, and molecular inclusion. In: *Flavor Encapsulation* (edited by S.J. Risch and G.A. Reineccius), p. 110, ACS Symposium Series No. 370, Washington, DC: American Chemical Society.

Xiang, Y.F., Yang, J.Z., Li, P.F., Wang, L.Q., and Cheng, M. 1997. Microencapsulation of capsicum oleoresin. *Food Science China*, 18(11), 27–30 (Cited from FSTA 1998-09-T0626).

Yan, N. and Jin, Y. 2005. Microcapsules having multiple shells and method for the preparation thereof, U.S. patent application, U.S. 2005/0067726 A1.

Zheng, S., Alkan-Onyuksel, H., Beissinger, R.L. and Wasan, D.T. 1999. Liposome microencapsulations without using any organic solvent. *Journal of Dispersion Science Technology*, 20, 1189–1203.

Zilberboim, R., Kopelman, I.J. and Talmon, Y. 1986. Microencapsulation by a dehydrating liquid: Retention of paprika oleoresin and aromatic esters. *Journal of Food Science*, 51, 1301–1306.

# 14 Multiphysics Modeling of Innovative and Traditional Food Processing Technologies

*Kai Knoerzer and Henry Sabarez*

## CONTENTS



## 14.1 INTRODUCTION

The food industry is an increasingly competitive and dynamic arena with consumers being more aware of what they eat and, more importantly, what they want to eat. Important food quality attributes such as taste, texture, appearance, and nutritional content are strongly dependent on the way foods are processed (Knoerzer et al., 2011b).

In recent years, a number of innovative food processing technologies, also referred to as "emerging" or "novel" technologies have been proposed, investigated, developed, and implemented with the aim to improve or replace conventional processing technologies. These technologies take advantage of other physics phenomena such as static high hydrostatic pressure or dynamic pressure waves, or electric and electromagnetic fields, and provide the opportunity for the development of new foods and also for improving the quality of established food products through gentle processing. The physical phenomena utilized by these technologies can potentially reduce energy and water consumption and, therefore, can play an important role toward environmental sustainability of food processing and global food security by expanding the shelf stable product spectrum (Knoerzer et al., 2011b). Although developing new technologies is crucially important, incremental improvements by optimization of the design and operation of the traditional processing technologies are still readily embraced by the food industry due to reduced inherent risks compared to implementing new technologies.

Apart from the underlying thermo- and fluid-dynamic principles of conventional processing, innovative technologies incorporate additional Multiphysics dimensions, for example, pressure waves, and electric and electromagnetic fields, among others. To date, they still lack an adequate and complete understanding of the basic principles of intervening in temperature and flow evolution in product and equipment during processing (Barbosa-Canovas et al., 2011). The development and optimization of suitable equipment and process conditions that provide the adequate uniformity still remains a challenge. Computational fluid dynamics (CFD) is an established tool for characterizing, improving, and optimizing traditional food processing technologies; the partial differential equations (PDEs) solved are the ones describing the conservation of mass, momentum, and energy (i.e., continuity, Navier–Stokes, and Fourier equations). In addition, innovative technologies provide additional complexity and challenges for modelers because of the concurrent interacting multiphysics phenomena; further PDEs need to be solved simultaneously, such as the Maxwell's and the constitutive equations for problems involving electromagnetics (e.g., microwave and radiofrequency processing), the charge conservation (e.g., pulsed electric field (PEF) processing), and the wave equations (e.g., Helmholtz wave equations) for ultrasonic and megasonic processing (Knoerzer et al., 2011b). These equation systems can increase in complexity when not only the process variables are to be predicted but also process outcomes such as microbial/enzyme inactivation or food matrix modification. In such cases, the numerical problem is coupled to equations describing the dynamics of occurrence of such phenomena, for example, inactivation, and the development of (acoustic) forces leading to transport of concentrated species, among others.

## 14.2 SIMULATING INNOVATIVE AND TRADITIONAL FOOD PROCESSING TECHNOLOGIES

A common problem of innovative, but also traditional, food processing technologies is the nonuniformity of the treatment, which can be caused by gradients of process variables such as temperature, fluid flow, electric field strength, or sound pressure fields in the processing chambers. A nonuniform distribution of a certain process variable leads to nonuniformities in the resulting outcomes of the process (e.g., microbial inactivation and also product quality, such as color, texture, flavor, and nutritional content).

While trial-and-error optimization is always an option to improve equipment and process design, it is the least preferred way, as it is very cost-, labor-, and time-intensive, and a good performance may be missed, as not all possibilities can be tested following this approach. On the other hand, numerical modeling using CFD can be used exactly for this purpose at reduced costs and time of equipment use. This way, advantages and disadvantages of the respective technology can be identified and either utilized or minimized.

Numerical modeling studies have been reported across the range of food processing technologies, such as drying, mixing, extrusion, emulsification, microwave, radiofrequency, ultraviolet light, high-pressure (thermal), PEF, and ultrasonics/megasonics processing.

The following sections will highlight the latest advances in modeling high-pressure thermal, PEF, ultrasonics/megasonics processing, as well as industrial drying operations.

### 14.2.1 High-Pressure Thermal Processing

High-pressure thermal processing is a technology that is not only effective in inactivating vegetative organisms but also microbial spores, due to the elevated temperatures involved. Because the process time can be reduced compared to thermal-only processing through rapid compression heating and decompression cooling, quality attributes, such as color, nutrients, flavor, and texture can be better retained (Olivier et al., 2011).

A number of studies have been reported on the utilization of Multiphysics modeling for equipment and process characterization in terms of process temperature and flow field distributions (Knoerzer et al., 2007; Knoerzer and Chapman, 2011), prediction of microbial (spore) inactivation (Juliano et al., 2009), and equipment optimization (Knoerzer et al., 2010) (Figure 14.1).

Knoerzer et al. (2007) reported on the use of a numerical model to describe temperature and flow distribution in a 35 L pilot-scale high-pressure sterilization system (Avure Technologies Inc., Seattle, WA, USA). They evaluated the differences of the process variables for a number of different product carriers made of metal and insulating plastic

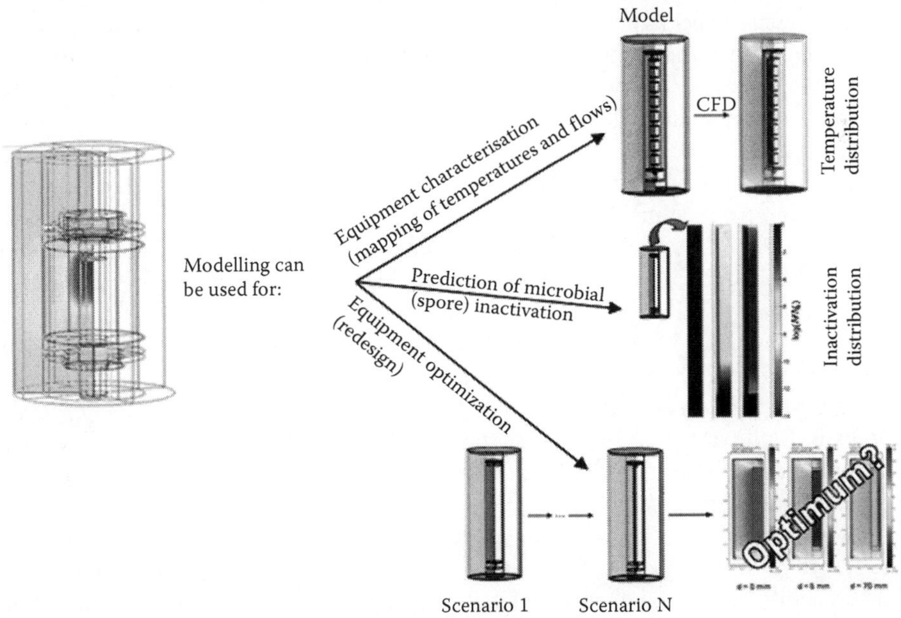


**FIGURE 14.1** Applications of numerical models describing high-pressure thermal processing.

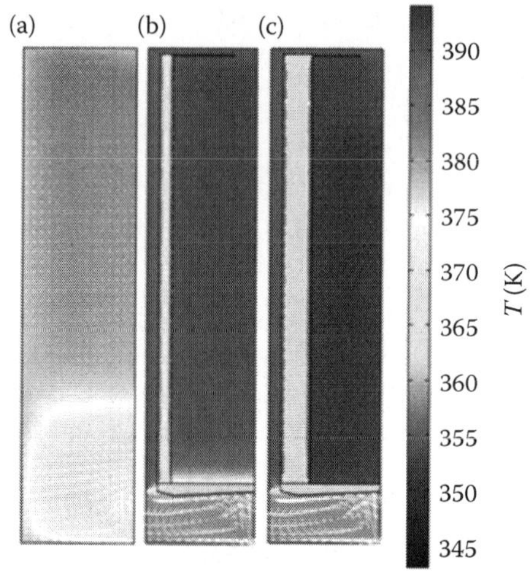


**FIGURE 14.2** The temperature distribution in an axis-symmetric section of the cylindrical high-pressure vessel: (a) no carrier in the vessel, (b) inclusion of cylindrical metal carrier, and (c) inclusion of cylindrical PTFE carrier; at the end of pressurization.

material. Figure 14.2 shows the temperature distributions in three investigated scenarios at the end of pressurization, namely, a cylindrical steel high-pressure vessel without carrier, one with a metal carrier and one scenario where a carrier made from insulating polytetrafluoroethylene (Teflon) (PTFE) was placed into the vessel.

As shown in the figure, the temperature distribution achieved in scenario (a) shows nonuniformities and relatively low temperatures compared to the scenarios where carriers are included, which avoid pronounced cooling down caused by the incoming pressurization fluid. Temperatures in scenario (b) are more uniform but still lower than in scenario (c). Furthermore, during pressure hold time, scenarios (a) and (b) exhibit pronounced heat losses, whereas the PTFE carrier in scenario (c) was able to retain the heat inside the carrier.

As expected, only the insulated carrier provided process conditions feasible for sufficient and uniform product sterilization through microbial spore inactivation (Figure 14.3).

Juliano et al. (2009) applied more detailed models (Figure 14.4a) describing the process variables (i.e., pressure, temperature, and flow) and evaluated the differences in the extent and distribution of predicted inactivation of *Clostridium botulinum* spores in food packages. In the first step, the CFD models were able to show heat retention inside the food packs and temperature magnitudes of ~121°C during pressure hold time (Figures 14.4b,c).

The predicted transient temperature distributions were then coupled to selected predictive microbial inactivation models, namely, the commonly known log-linear model, an nth order model, and a Weibull distribution model. The different inactivation models predicted very different levels of spore inactivation for the same pressure and temperature conditions. For example, the log-linear model predicted inactivation of *C. botulinum* spores in the order of 16 $\log_{10}$ after 3 min processing at 600 MPa

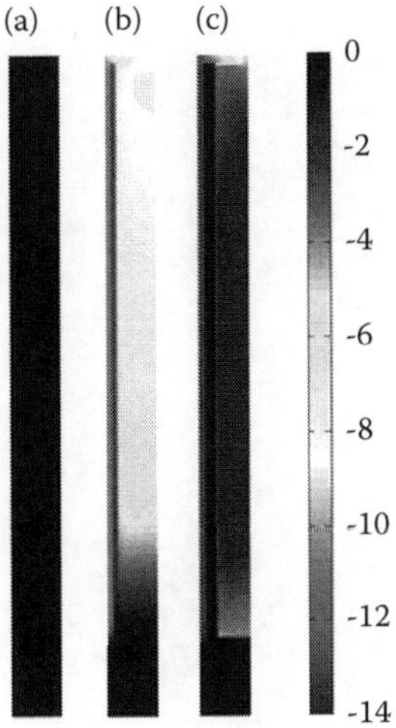


**FIGURE 14.3** The distribution of *Clostridium botulinum* spore inactivation as predicted by the log-linear model in (a) the vessel without product carrier, (b) the vessel including a steel carrier, and (c) the vessel including a PTFE carrier.

and 121°C (Figure 14.5a) whereas the Weibull model indicated spore inactivation of only 9 $\log_{10}$ for the same process (Figure 14.5b).

Knoerzer et al. (2010) then used a modified version of the model to optimize the wall thickness of the insulating carrier while increasing product load capacity. The carrier supplied by the manufacturer was designed with a wall thickness such that sufficient heat retention was ensured during processing. The authors derived a dimensionless parameter, referred to as integrated temperature distributor value

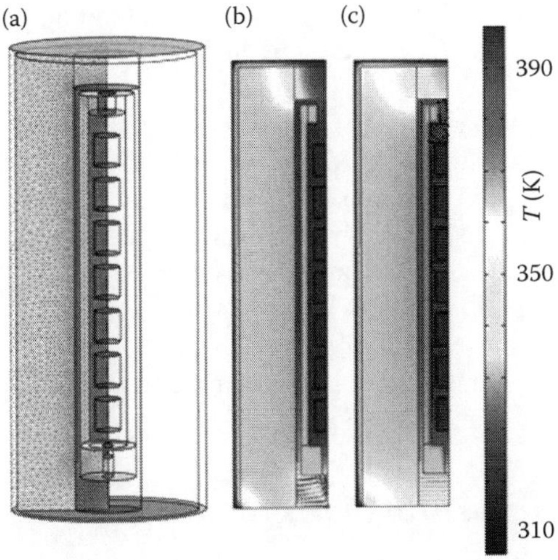


**FIGURE 14.4** Model geometry (a) and predicted temperature distributions at the end of pressurization (b) and pressure hold time (c).

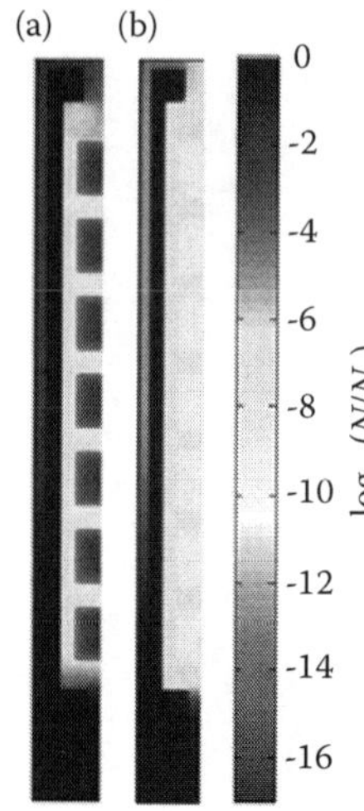


**FIGURE 14.5** An indication of inactivation of *Clostridium botulinum* spores as predicted by the log-linear model (a) and the Weibull distribution model (b).

(Equation 14.1) to evaluate temperature uniformity, and the temperature magnitude expressed relative to a target temperature and heat retention during processing

$$ITD = \frac{\int_{r_{min}}^{r_{max}} \int_{z_{min}}^{z_{max}} 10^{\left(\left(\left(\int_0^t T(t)dt/t_{process}\right)-T_{target}\right)/z_T\right)} dr\,dz}{(r_{max} - r_{min}) \cdot (z_{max} - z_{min})} \tag{14.1}$$

where $r_{min}$, $r_{max}$, $z_{min}$, $z_{max}$ cover the region of interest (the carrier volume), $t_{process}$ is the process time of interest (in this case, the pressure holding time where most of the heat loss is expected), and $T_{target}$ is the targeted holding temperature of the process under pressure.

An iterative strategy was applied, which consisted of a model that automatically changed the carrier wall thickness in a range of 0–70 mm, and evaluated the temperature performances and load capacities for the respective scenarios. Figure 14.6 shows the modified model geometry with variable carrier wall thickness and the predicted temperature distributions at the end of pressure hold time for a wall thickness of 0, 5, and 70 mm.

The study showed that the wall thickness can be reduced from 28 mm to ~4 mm without compromising temperature performance, leading to an increase of carrier load capacity by more than 100% (Figure 14.7).

### 14.2.2 Pulsed Electric Field Processing

PEF processing is a technology that can be applied for cold or low-temperature pasteurization of liquid products. It is able to inactivate vegetative microorganisms through the application of electric fields in the order of several ten thousand volts per

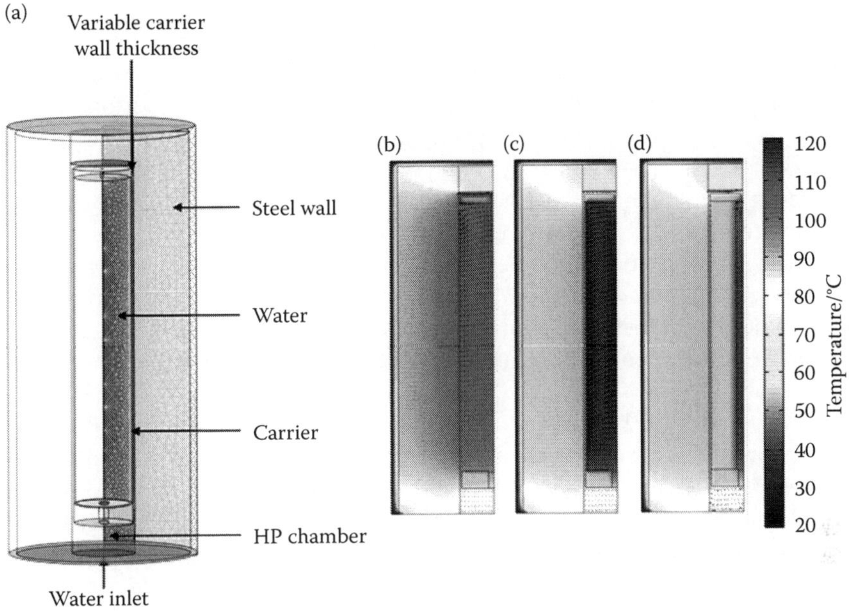


**FIGURE 14.6** The depiction of modified model geometry (a) and the predicted temperature distributions at a wall thickness of 0 mm (b), 5 mm (c), and 70 mm (d).

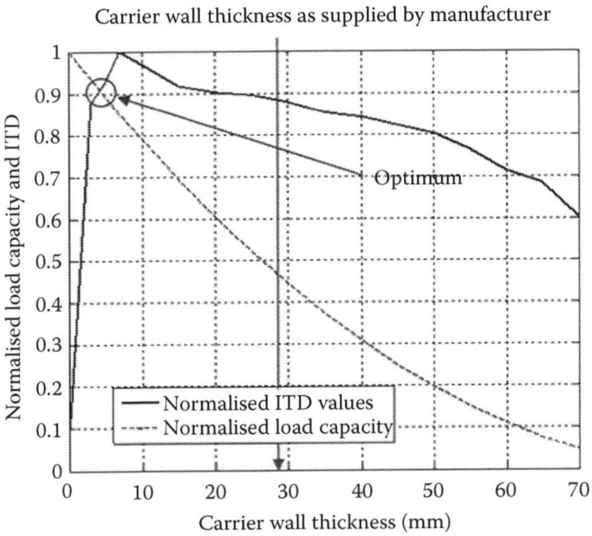


**FIGURE 14.7** The determination of optimum carrier wall thickness by evaluating temperature performance (integrated temperature distributor) and load capacity of the carriers with varying wall thickness.

centimeter for a very short time, leading to cell poration and cell death (Heinz et al., 2003). Overall treatment times are in the order of microseconds. Other potential applications of this technology are for enhanced extraction processes or softening of fruit and vegetable tissue, for example, for improving cutting performance and reducing cutting losses. Also, this technology can improve the quality attributes of foods compared to conventional thermal processing, such as flavor, color, and nutrients, among others. Being a continuous process, high throughputs are possible.

Published studies on numerical modeling of PEF processing include the utilization of such models for equipment characterization with respect to electric field, temperature and flow distribution (Buckow et al., 2010, 2011), the prediction of microbial or enzyme inactivation (Buckow et al., 2012; Knoerzer et al., 2011a) and equipment optimization to ensure uniform and effective processing (Knoerzer et al., 2012) (Figure 14.8).

Buckow et al. (2010) reported on the development of a three-dimensional (3D) model for a pilot-scale PEF system (Diversified Technologies Inc., Bedford, MA, USA) to predict electric field strength, flow and temperature distributions (Figure 14.9), and an extensive validation of the model predictions through temperature measurements within the constrained space of the treatment chamber's active zone and the second ground electrode (Figure 14.10). The authors investigated the treatment of salt solutions with different conductivities and whole milk, processed at two flow rates, and five different inlet temperatures. Pulses were applied at two different voltage settings, four pulse repetition rates, and three pulse widths. They were able to

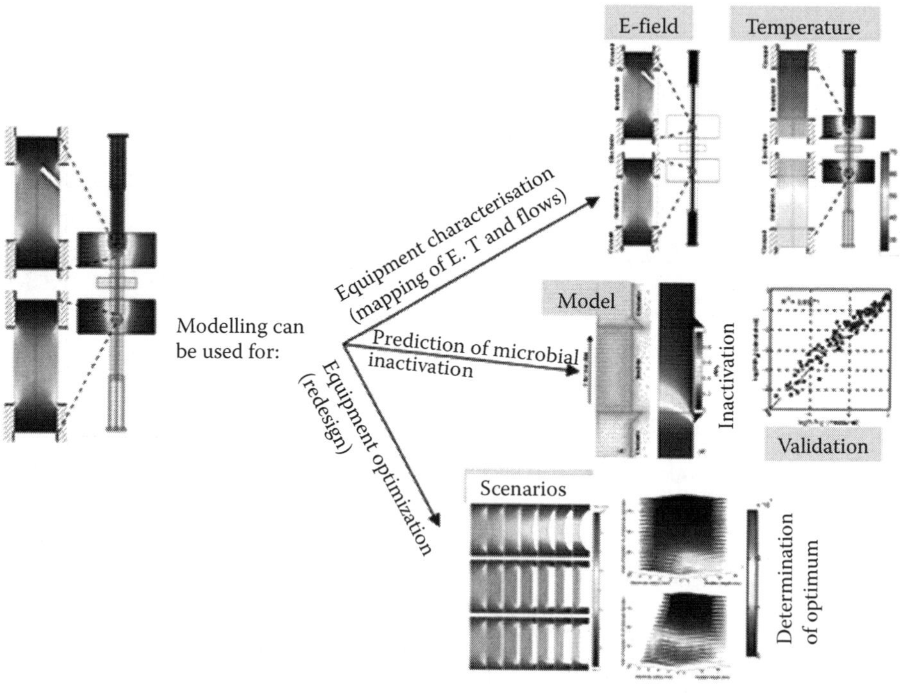


**FIGURE 14.8** Uses of numerical models describing pulsed electric processing.

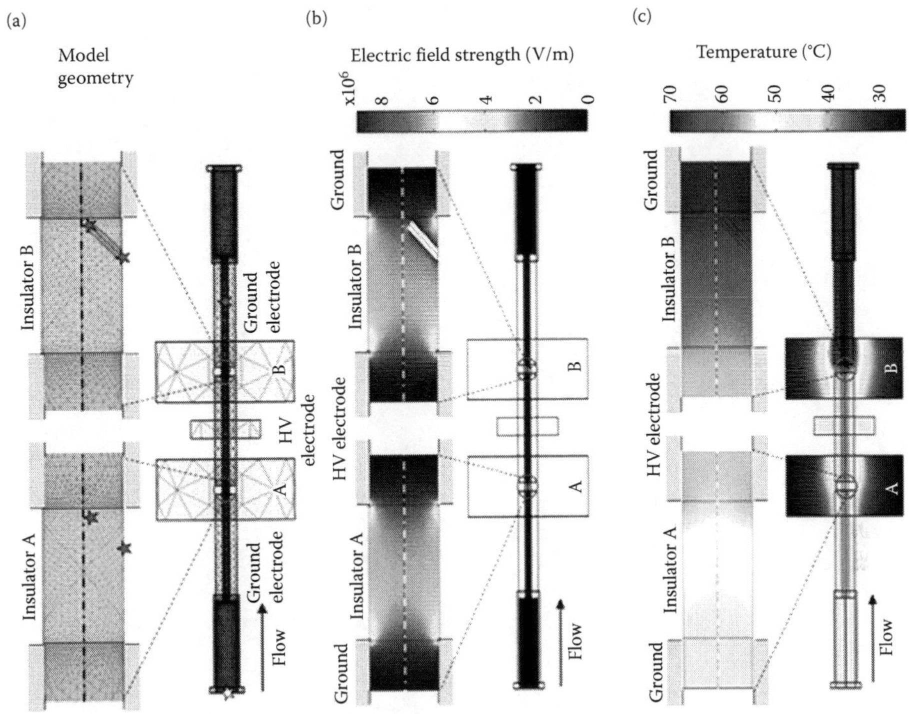


**FIGURE 14.9** A representation of the pilot-scale treatment chamber including magnification of the treatment zones (model geometry (a)), electric field strength distribution in a salt solution at 4 mS/cm, flow rate of 4 L/min, inlet temperature of 45°C, voltage of ~22 kV, pulse width of 5 us, and frequency of 600 Hz (b), and temperature distribution at these conditions (c).

utilize this model to characterize and evaluate the performance of the system as supplied by the manufacturer with respect to electric field strength, temperature, and flow distribution.

Buckow et al. (2011) applied this model further to derive simplified equations to estimate accurate electric field strengths and specific energy inputs from treatment variables such as voltage, pulse frequency, and duration, among others. Also evaluated were the effects of changing treatment chamber geometry and configuration on these process variables. A common approach for estimating the electric field strength is relating the applied voltage to the electrode gap. While this will give accurate predictions for parallel plate systems, it was found that for configurations used in continuously operating systems, such as cofield or colinear design, this approach always overpredicts the actual electric fields. This is similar for specific energy input, commonly estimated by multiplying voltage, current, pulse width, pulse repetition rate, and mass flow. Figure 14.11 shows (a) the correlation of relative electric field strength (i.e., actual electric field strength/estimated electric field strength for parallel plate configuration) and (b) the correlation of the relative specific energy input (i.e., the actual specific energy input/estimated specific energy input for parallel plate configuration) with the ratio of the electrode radius and electrode gap for different chamber configurations.

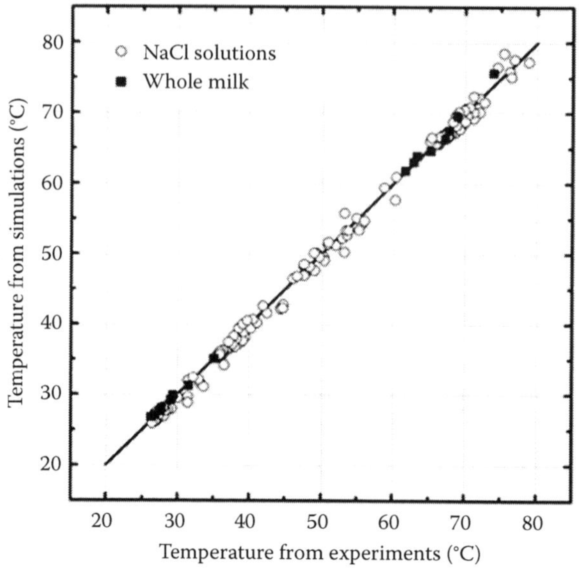


**FIGURE 14.10** The validation of the model for different fluids and ~50 process conditions (over 400 data points).

These configurations were "no inset" (where the insulator bore diameter is equal to the inner electrode diameter), "rectangular inset" (where the insulator bore diameter is smaller than the inner electrode radius), "chamfer edge inset" (which is identical to the rectangular inset with rounded edges of the insulator bore), and "elliptical inset" (where the insulator bore has in inward concave shape); see also Figure 14.13.

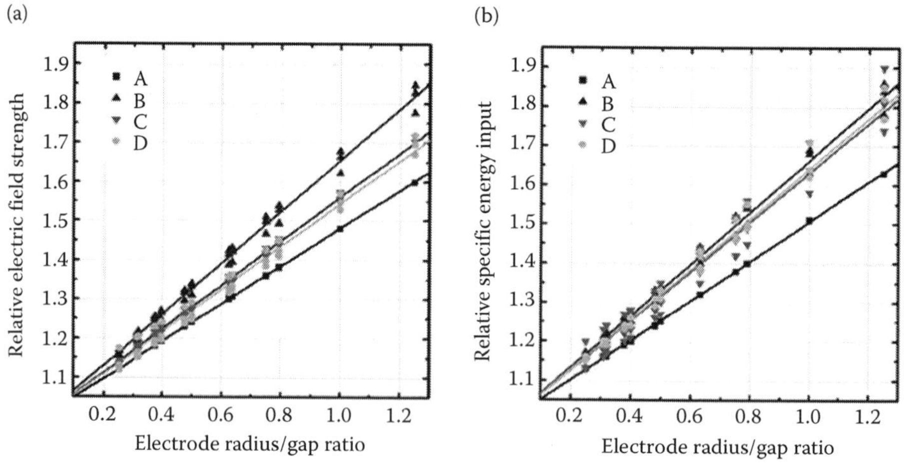


**FIGURE 14.11** Correlation of (a) the relative electric field strength and (b) the relative specific energy input with the ratio of electrode radius and gap for "no inset" (A), "rectangular inset" (B), "chamfer edge" (C), and "elliptical inset" (D) chamber configurations.

Buckow et al. (2012) then developed and validated a model for a laboratory-scale PEF system (Figure 14.12a) and evaluated the effect of the electric field on lactoperoxidase (LPO) degradation (an indicator for pasteurization) by coupling the predicted temperature distributions (e.g., Figure 14.12c) to predictive LPO degradation models accounting for the thermal component only.

The study indicated that the major effect for LPO inactivation comes from the elevated process temperatures as the predictions (thermal only degradation) were close to the measured degradation (combined thermal and PEF) for a number of process conditions; however, they found that there was also some additional inactivation caused by the electric field of up to 12%, potentially caused by the high-intensity electric pulses and induced electrochemical reactions.

Last, Knoerzer et al. (2012) developed an iterative algorithm that was capable of automatically changing the treatment chamber configuration and dimensions in the Multiphysics models and to identify, out of more than 100,000 scenarios, the one that showed the highest degree of electric field uniformity, together with sufficient throughput and lowest pressure drop, among other evaluation characteristics. The evaluation of the performance of the models was based on a parameter, referred to as dimensionless performance parameter (DPP), calculated by an equation derived by the authors (Equation 14.2), accounting for the treatment volume, pressure drop estimations, and electric field magnitude related to that achievable in parallel plate systems, electric field uniformity, and peaks of the electric field strength.

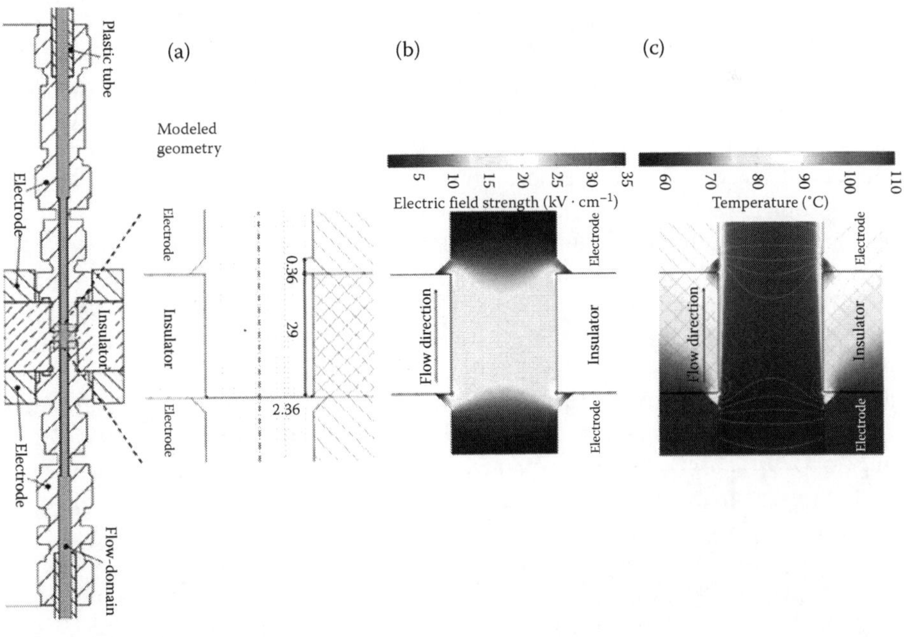


**FIGURE 14.12** A representation of the modeled geometry of the laboratory-scale PEF system (a), predicted electric field distribution (b), and predicted temperature distribution (c).

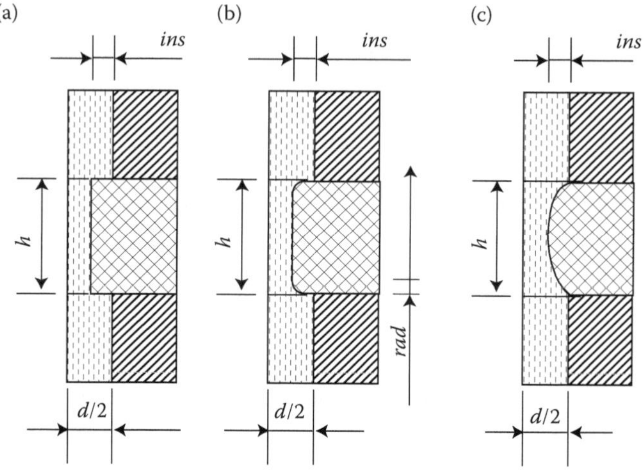


**FIGURE 14.13** Investigated chamber configurations, indicating the geometrical parameters varied in the model: (a) rectangular inset, (b) rectangular chamfered edge inset, and (c) elliptical inset scenario.

$$DPP = \left(\frac{V_{zone}}{V_{max}}\right)^{a_1} \cdot \left(\frac{(d-ins)^4}{d^4}\right)^{a_2} \cdot \left(\frac{E_{av}}{\frac{V_0}{h_{min}}}\right)^{a_3} \cdot \left(\frac{n_{av\pm10\%}}{n_{total}}\right)^{a_4} \cdot \left(\frac{E_{av}}{E_{max}}\right)^{a_5} \tag{14.2}$$

where $V_{zone}$ is the volume of the treatment zone (insulator region) of the respective scenario, $V_{max}$ the volume of the largest treatment zone considered, $E_{av}$ the average electric field strength of the insulator region, $V_0$ the applied potential, $h_{min}$ the minimum electrode distance (gap) of all scenarios investigated, $n_{av}\pm10\%$ the number of elements in the treatment zone with electric field strengths within 10% of the average electric field strength, $n_{total}$ the total number of elements in the treatment zone, and $E_{max}$ the maximum electric field strength in the respective scenario; $a_1$–$a_5$ are adjustable weighing parameters depending on the importance of the respective performance variable.

Three different chamber configurations (Figure 14.13) were studied and for each of these four different geometry parameters were varied: the internal diameter *d* of the electrodes ranging from 2 to 20 mm, the height *h* of the electrode gap ranging from 1 to 30 mm, a total inset *ins* (i.e., the internal diameter of the insulator) in a range of 0–90% of the electrode diameter *d*, and for the "rectangular rounded edge inset" models, also the chamfer radii *rad* ranging from 0% to 40% of the diameter reduction *ins* (Figure 14.13).

The algorithm first generated the models, then solved them, applied the performance evaluation by utilizing the DPP equation, and then identified the scenario that yielded the highest DPP value, which was found for configuration (b). The authors then set up a full 3D model of this configuration, built the new chamber and performed validation studies of the model for a salt solution, apple juice, and for various

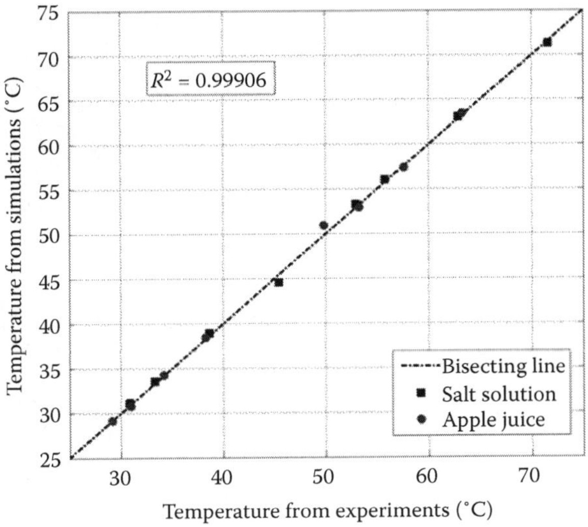


**FIGURE 14.14** A Parity plot of predicted and experimentally determined temperature values in the new treatment chamber.

process conditions. They found that the new design could be predicted well with respect to the temperatures generated in the treatment chamber (Figure 14.14).

### 14.2.3 Ultrasonics and Megasonics Processing

Ultrasound processing spans over a wide range of acoustic frequencies, starting as low as 18 kHz, up to several MHz. Applications are as diverse as the frequency spectrum is wide. At the lower frequency (18–200 kHz) end (also known as ultrasonics), the effects are caused mainly by instable cavitation. Traditional applications such as emulsification, cleaning, extraction (Gogate and Kabadi, 2009), and more novel applications used for improved drying (Sabarez et al., 2012) and beverage defoaming (Rodriguez et al., 2010) in airborne ultrasound systems can be listed. When using higher frequencies (>0.2 MHz, also known as megasonics), the effects can be either mechanical through standing pressure waves and microstreaming, and/or sonochemical (radical driven) or biochemical (stress response in living tissue). Novel high-frequency applications include the separation of particles in standing wave systems (Juliano et al., 2013), and texture improvement of processed fruits and vegetables, through produce-internal stress responses (Day et al., 2012).

Published studies on the numerical modeling of ultrasonics and megasonics processing include the utilization of the multiphysics models for equipment characterization with respect to acoustic pressure, temperature, and flow distribution (Trujillo and Knoerzer, 2009; Trujillo and Knoerzer, 2011), equipment optimization (Trujillo and Knoerzer, 2009), and for predicting particle separation in megasonics standing wave applications (Trujillo et al., 2013) (Figure 14.15).

Trujillo and Knoerzer (2011) reported on the development of a multiphysics model capable of simulating the formation of a sound pressure jet produced by a sonotrode

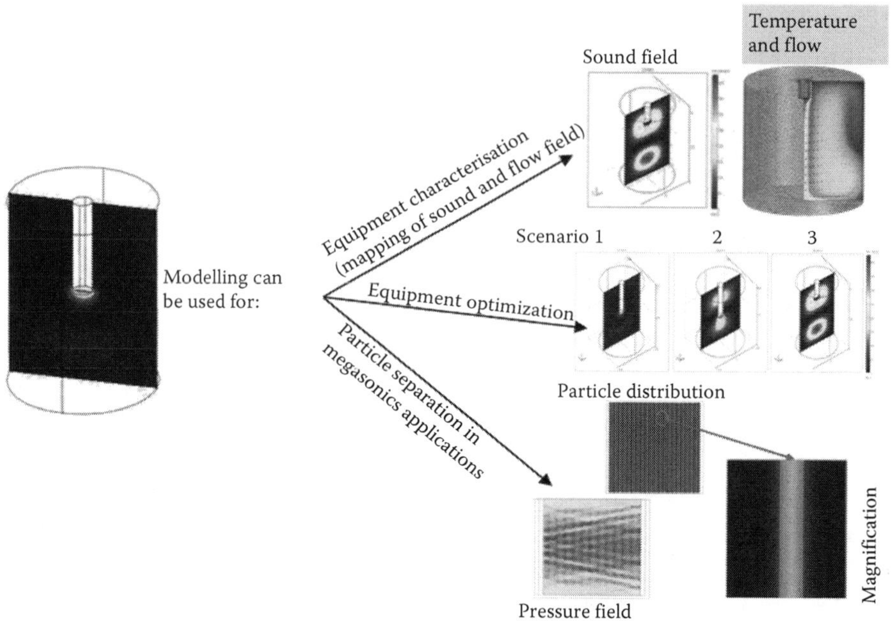


**FIGURE 14.15** Uses of numerical models describing low- and high-frequency ultrasound processing.

placed in water in a low-frequency (20 kHz) high-power ultrasound application. The acoustic power was dissipated within close proximity to the horn and the acoustic energy was completely converted into kinetic and thermal energy leading to a jet being formed and directed away from the sonotrode while the temperature was increasing in the bulk of the treated fluid. The model was validated by utilizing published data (Kumar et al., 2006) where fluid movement was measured by laser Doppler anemometry.

Figure 14.16 shows the computational representation of the system investigated by Kumar et al. (2006) and Trujillo and Knoerzer (2011) in 3D and two dimension (2D). Full 3D and axis-symmetric 2D models predicted almost identical values; therefore, the fact that the computational demand of the 3D model was very high, all further models for comparison with the Laser Doppler anemometry (LDA) data were solved in 2D only. Figure 14.17 shows a direct comparison of the velocity profile predicted by the model and the one measured by LDA for a specific power input of 35 kW/m$^3$. Both prediction and measurement show the jet being formed underneath the horn tip with much lower velocities throughout the rest of the reactor.

Apart from visual comparisons, the authors also performed a quantitative validation of the model by comparing the predicted values of the axial velocity at a number of heights and radii (Figure 14.18) and found good agreement.

Apart from low-frequency ultrasound applications, Trujillo et al. (2013) also published on the development of a multiphysics model for a high-frequency ultrasound application for separation of particles out of a continuous water phase. The simulated separation reactor is shown in Figure 14.19a. The model included solving for the

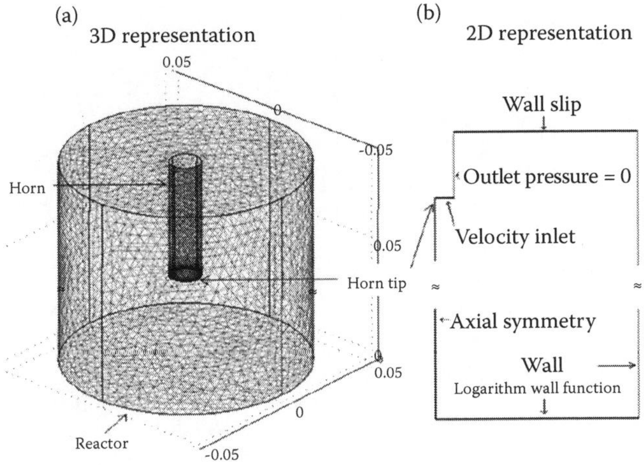


**FIGURE 14.16** The depiction of the geometry of the investigated system: (a) 3D representation and (b) 2D axis-symmetric representation, including the boundary conditions of the model.

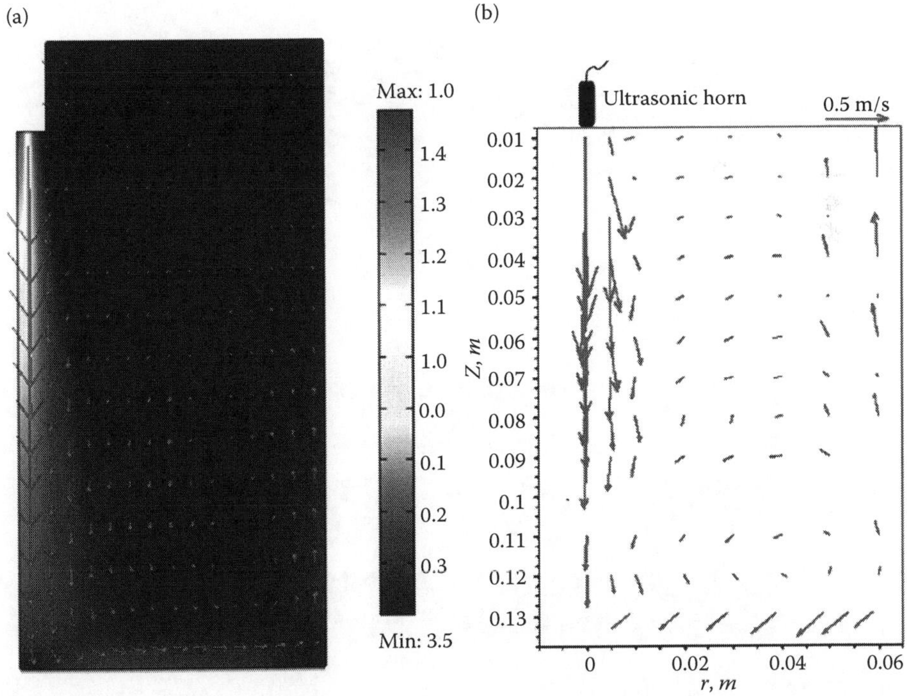


**FIGURE 14.17** A visual comparison of the predicted (a) and measured by LDA (b) flow profiles in the investigated system at a specific power input of 35 kW/m$^3$.

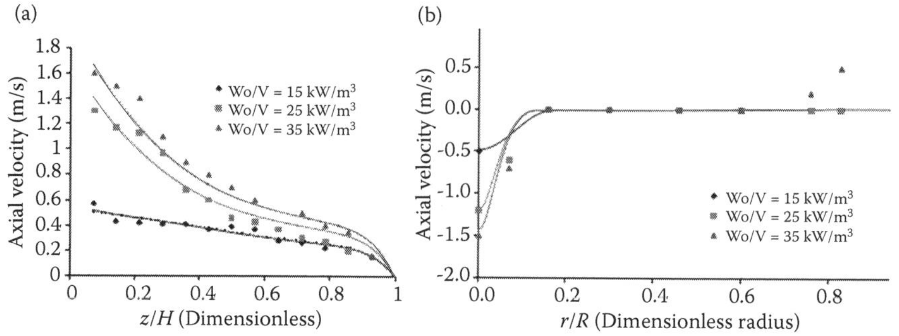


**FIGURE 14.18** The quantitative comparison of model predictions and experimentally determined values of (a) the axial velocity at different height levels under the horn tip and (b) different radii at height level of 13% of the total reactor height.

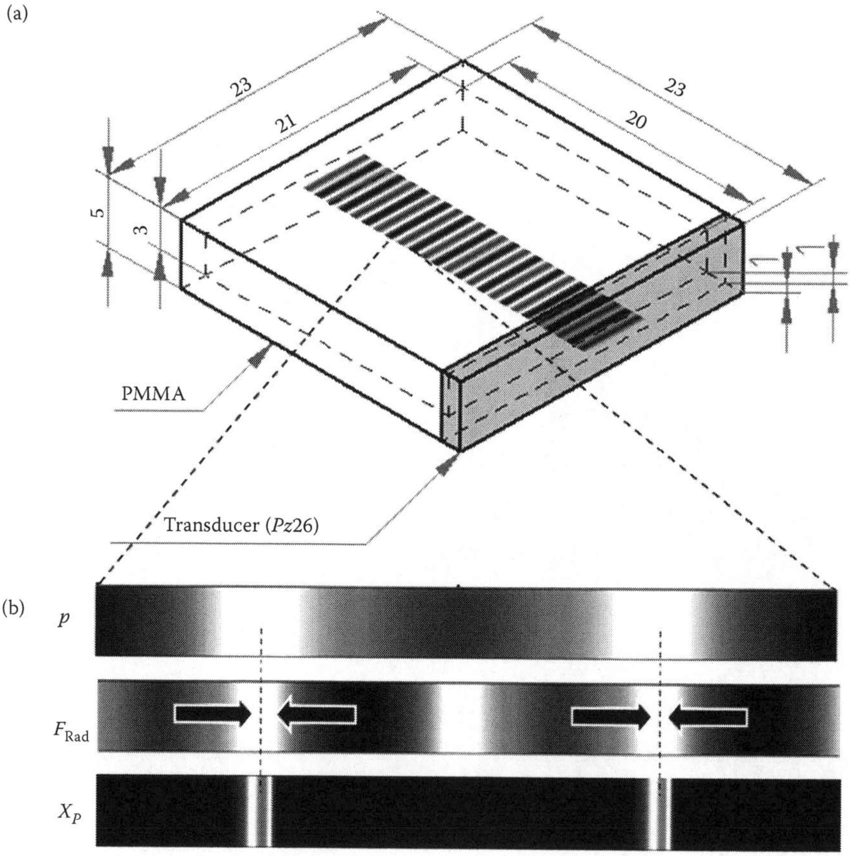


**FIGURE 14.19** A schematic representation of the investigated treatment chamber with pressure distribution for a fixed frequency of 1.54 MHz (a), magnification of one wavelength of the pressure distribution (b; $p$), the resulting acoustic force (b; $F_{\text{Rad}}$), and the particles concentrated at the nodes of the pressure wave at the fixed frequency (b; $X_p$).

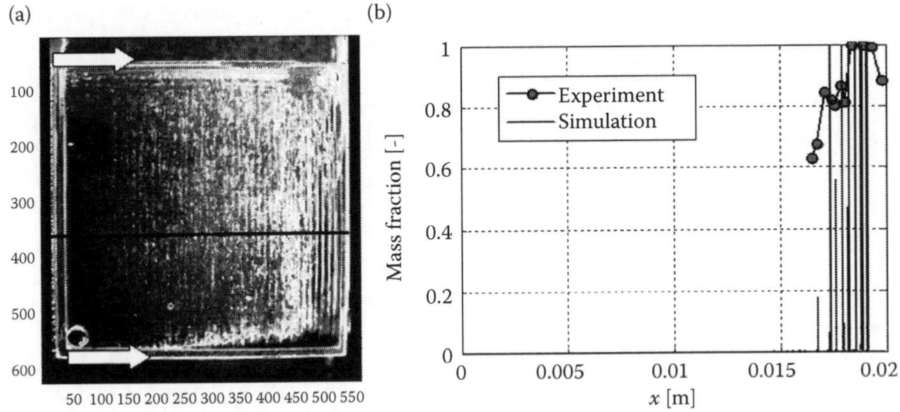


**FIGURE 14.20** (a) Digitized image of actual process at a discrete time of 120 s (false color representation; the black line indicating the area for comparison with the model prediction); (b) comparison of the mass fraction of the separated particles measured and predicted.

mechanical displacement of the reactor walls, leading to the formation of an acoustic pressure field (indicated in Figure 14.19a for a fixed frequency of 1.54 MHz as a thin band in the reactor and a magnified view in Figure 14.19b, $p$), followed by predicting the acoustic radiation force acting on suspended particles (Figure 14.19b, $F_{\text{Rad}}$). Finally, this (transient) force was utilized to solve for the movement of the particle phase to the nodes of the ultrasonic standing wave (Figure 14.19b, $X_P$) and frequency ramping, leading to active separation of the particles away from the transducer plate toward the reflector.

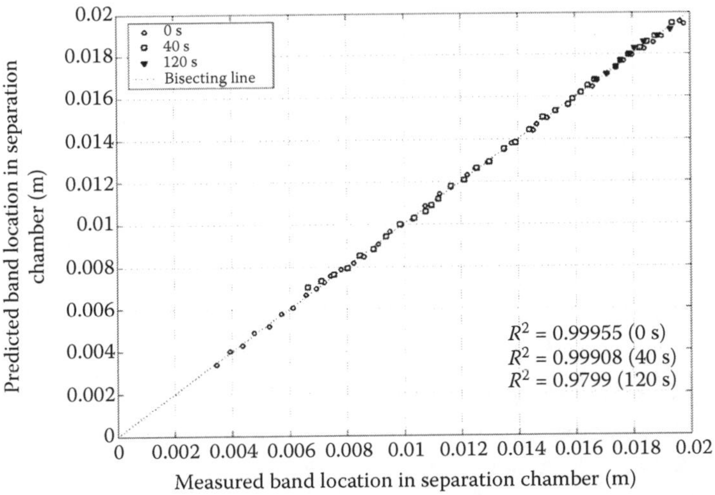


**FIGURE 14.21** A Parity plot of the measured and predicted band locations at discrete timesteps of 0, 40, and 120 s.

They then compared digitized images of the actual process at discrete times of 0, 40, and 120 s (Figure 14.20a) with the predicted particle band formation and transient band movement. As shown in Figure 14.20b, at a discrete time of 120 s, measurement and predictions agreed well. Figure 14.21 shows a parity plot of the measured and predicted band locations for all three timesteps; as can be seen, a very good agreement was found.

## 14.3 SIMULATION OF INDUSTRIAL FOOD DRYING OPERATIONS

Drying is one of the most important processing operations in the food manufacturing industry. It is an energy-intensive process, which constitutes a significant portion of the total production costs of dried-based products depending on the dryer design, operating conditions, properties of food materials, and so on. In addition, the process of drying usually affects the quality attributes of the product due to exposure at high temperatures or long drying times. The main challenge is to develop and optimize a drying process in terms of increased production throughput and reduced energy consumption without compromising the quality of the dried products.

In an industrial drying operation, tunnel dehydrators are the most widely used method for drying fruits (e.g., plums and grapes) and vegetables. The tunnel dehydrator basically consists of a tunnel (as a drying chamber) containing trays of product that are placed on mobile trolleys (also referred to as trucks) moving along the tunnel, a fan to circulate the heated air, and a heating unit to preheat the air before it is blown across the food product and then vented to the exhaust (Figure 14.22). The circulated air is directly heated by a gas burner and the heated air is forced into the tunnel by a fan. In tunnel drying, the raw materials are loaded onto the trolleys of trays and these trolleys are then fed into the drying tunnel at one end at regular intervals on a continuous basis and the dried product is unloaded at the other end of the tunnel.

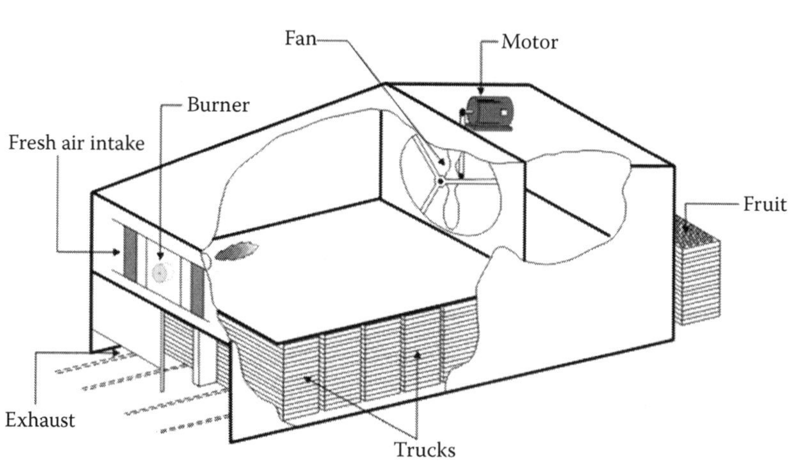


**FIGURE 14.22** Schematic diagram of a typical tunnel dryer for industrial drying of fruits (e.g., plums).

To date, this method of drying continues to be used at an industrial scale in fruit drying because it is much cheaper and easier to expand the capacity, in addition to its lower capital cost. However, the tunnel-drying system is inherently less efficient than other drying systems (e.g., continuous-belt dryer). There is a significant interest in the food and other processing industries to improve the efficiency in tunnel drying to maintain sustainability due to the prospect of a continuously increasing trend in fuel costs and the need for ecofriendly processes to mitigate environmental impact coupled with the rising consumer demand for high-quality products.

This section presents a characteristic case study of the industrial drying of plums in a tunnel dryer to illustrate the application of the modeling approach to determine the optimal design and operating conditions at an industrial scale.

A 2D axis-symmetric model was developed to describe the simultaneous transfer of momentum (air only), heat, and mass (air and food) occurring in convective air drying of fruits (e.g., plums). The governing PDEs describing the simultaneous transfer of heat, mass, and momentum in two distinct subdomains (air and food) during drying of plums were presented in previous studies (Sabarez, 2010, 2012). The nonisothermal turbulent flow of air in the drying chamber is described according to the standard k-ε model (COMSOL Multiphysics™, 2007).

The resulting systems of highly coupled nonlinear PDEs in the space–time domain together with the set of initial and boundary conditions were numerically solved by the finite element method coupled to the arbitrary Lagrange–Eulerian procedure to account for the shrinkage phenomenon using the commercial software package (COMSOL Multiphysics, Comsol AB, Stockholm, Sweden). The details of the numerical solution are presented in previous studies (Sabarez, 2010, 2012). Also, the solution of the governing PDEs requires knowledge of the thermophysical and transport properties of the product and air. The model parameters used in this work are given in previous studies (Sabarez, 2012, 2013).

Figure 14.23 shows a typical example of the comparison between the drying curves predicted by the model and the experimental drying tests performed at different air velocity levels (Sabarez, 2012). Generally, the measured and predicted data at different timesteps of the process banded closely around the straight line in the parity plot, which indicates the suitability of the model in describing the drying behavior of the material under the drying conditions tested. Also, Sabarez (2012) presented further validations to verify the predictive capability of the model over a range of conditions. The results confirm the validity of the model and demonstrate that the parameters used in the model are reasonable, indicating the suitability of the model to describe the drying process of plums under various conditions.

In general, the significant differences in drying kinetics between levels of the drying conditions investigated demonstrate the importance of optimizing the drying conditions during industrial drying of fruits. The advantage of the proposed numerical model is that the temperature and moisture distributions across the solid food domain as well as the changes of the condition of the drying air with location can be established at any time during drying (Figure 14.24). It should be noted that the modeling of the drying system comprises both material and equipment models, in which the material model describes the drying kinetics and the equipment model determines the changes of the condition of the drying medium with time and space

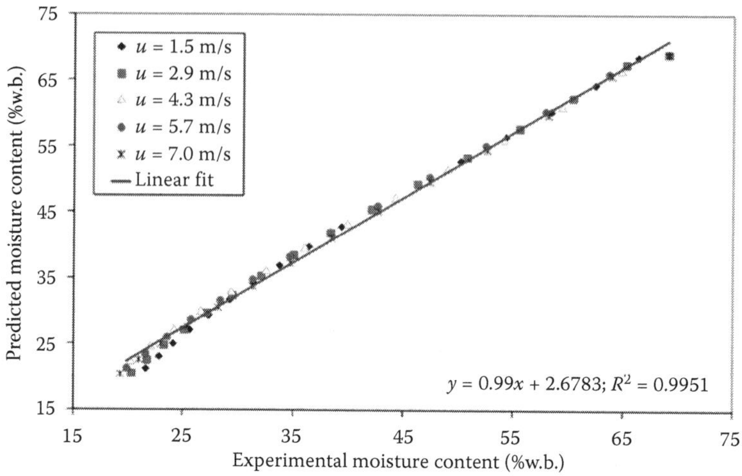


**FIGURE 14.23** Predicted versus experimental moisture content at various timesteps during the process and different airflow rates ($T = 80°C$; RH = 15%).

during drying. Together, these models constitute a complete modeling tool capable of predicting the dynamic behavior of the drying system. Thus, the prediction of the drying air stream conditions flowing across the product surface that would affect the drying behavior of the solid product at any time and position in the dryer is of particular importance in simulating the drying process of industrial drying systems where a systematic dynamic variation in drying conditions is typical (Sabarez, 2012).

In industrial tunnel dryers, the drying operation involves complex conditions interconnected with many factors associated in the design features and operational practices. These dryers are commonly operated in either a counterflow or parallel-flow mode of operation. In a counterflow configuration, the drying air is introduced into one end of the tunnel while the trolleys of food products enter at the other end and each moves in opposite directions. This configuration is characterized by having conditions most conducive to intense heating at the end of the drying cycle when the product is nearly dry and less heating in the early stages. The operation of the parallel-flow tunnel is opposite to that of the counterflow tunnel. The trolleys and drying air enter at the same end of the tunnel and progress through the tunnel in the same direction. This configuration is characterized by intense heating in the early stages where the product to be dried is still very wet followed by slow drying as the product approaches the cooler end. The results presented by Sabarez (2012) on the simulation of industrial drying of plums both in counterflow and parallel-flow modes of operation have shown that the parallel-flow operation would apparently result in shorter overall drying time to reach the desired final moisture content (20%) compared with the counterflow operation. The predicted overall drying times were also found to be in close agreement with those obtained in the industrial drying trials undertaken under similar drying conditions (Sabarez, 2010).

The conditions of the drying air (i.e., airflow, temperature, and relative humidity) are considered to be the main factors influencing the drying performance in

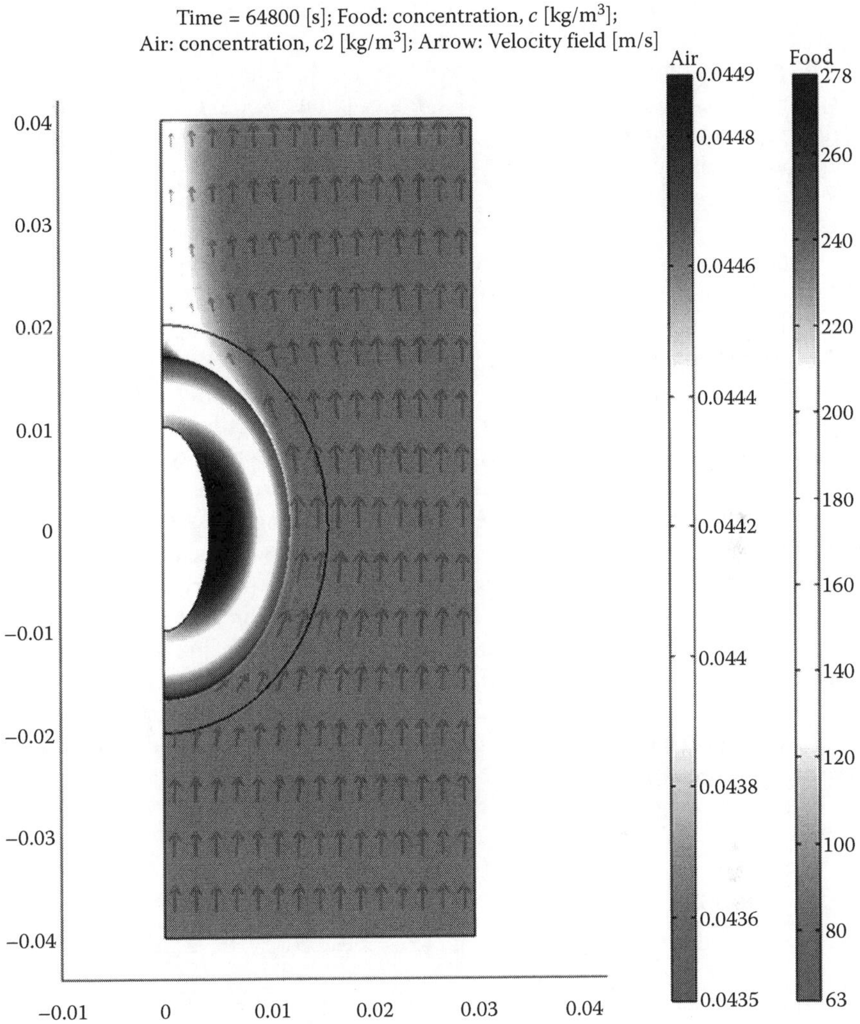


**FIGURE 14.24** Predicted product moisture concentration, moisture concentration, and velocity profiles of the drying air during drying of prunes ($T = 80°C$; RH = 15%; $u = 5.7$ m/s).

tunnel dryers. In this section, the effect of different air velocity levels was taken as an example to demonstrate the impact of this parameter on the drying kinetics and energy consumption. It is well known that the air velocity field greatly influences the heat and mass transfer rates at food/air interfaces. Therefore, the temperature and concentration of moisture in the product and the drying air are basically controlled by the level of air velocity and its distribution.

A number of simulation runs were undertaken to mimic the industrial tunnel drying of plums specifically in a parallel-flow mode of operation at various inlet air velocity conditions. The selected conditions are representative for the current commercial tunnel drying operation for plums. Figure 14.25 depicts the simulated effect of different levels of air velocity on both drying time and energy consumption. Clearly, it shows that as the air velocity increases, the energy consumption appears to increase. This is obvious as increases in air volume would result in an increased energy requirement to heat the large volume of air to the desired temperature level. On the other hand, it appears that the drying time significantly decreases as the air velocity increases, but only to a certain point, and then beyond this level the air velocity plays a proportionally decreasing role in reducing the drying time.

As can be observed from this plot, there appears to be an optimum level of air velocity to achieve better drying performance, which can be found at the intersection of the plots. Under these conditions, it can be seen that the optimum air velocity level is around 4–5 m/s. Hayashi (2007) suggested an air velocity of about 5 m/s as a sufficient level in drying most products. The result indicates that a further increase in air velocity beyond this level would significantly increase the energy consumption with minimal increase in throughput (i.e., slight reduction in drying time). For example, increasing the air velocity from 5 to 7 m/s would increase the energy consumption by as much as 26% without gaining any significant reduction in drying time. Consequently, there is little to be gained by using a very high air velocity. If the air velocity is too high, the costs of energy required to heat the excessive air would tend to offset the benefits of slight reduction in drying time. On the other hand, it shows that large savings in energy could be achieved if operating at lower air velocities, however, with the expense of longer drying times. Apart from reduced throughput, operating at longer drying times would also increase the labor costs associated in drying and possibly affect the product quality due to prolonged exposure times.

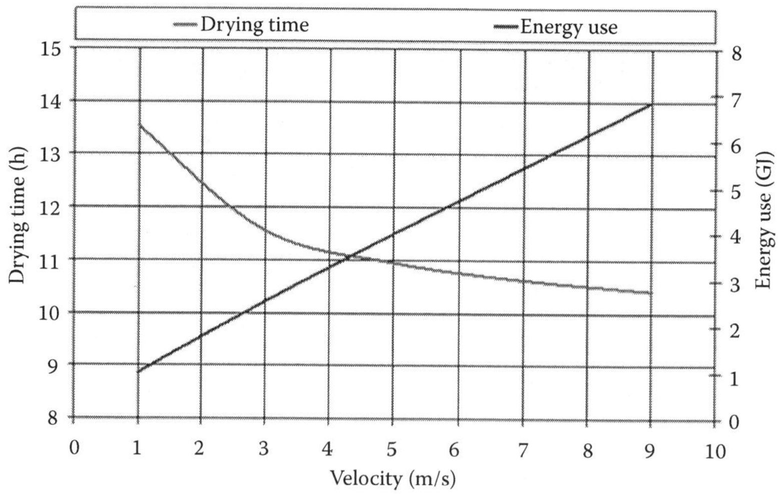


**FIGURE 14.25** The effect of air velocity on drying time and energy consumption (inlet: $T = 85°C$, RH = 15%; recycle: $T = 70°C$, RH = 30%, ratio = 90%; ambient: $T = 25°C$, RH = 65%).

## 14.4 SUMMARY AND OUTLOOK

It is widely established that innovative technologies are the means to meet a need and capture an opportunity, particularly around the manufacture of attractive, new, high-quality products with fresh-like quality attributes, ensured safety, a long shelf-life, and more sustainable manufacturing. The main incentive for applying these new technologies should focus on inducing disruptive innovations in the food manufacturing industry rather than providing merely incremental improvements of existing processes. Also, the redesign and optimization of traditional technologies will continue to meet their needs for improvements in order to be more sustainable.

Validated multiphysics models have been used to characterize, evaluate, and optimize existing equipment for innovative and traditional food processing technologies and such modeling strategies will be able to assist in further developing these technologies for effective and efficient implementation in the food manufacturing industry. Without such modeling capabilities, relying on the traditional approach of trial and error, development will be slow and in some instances, a sufficient performance justifying utilization in industry may never happen.

## REFERENCES


Barbosa-Canovas, G.V., Albaali, A., Juliano, P., and Knoerzer, K. 2011. Introduction to innovative food processing technologies: background, advantages, issues, and need for Multiphysics modeling. In: *Innovative Food Processing Technologies: Advances in Multiphysics Simulation* (Knoerzer, K., Juliano, P., Roupas, P., Versteeg, C. (eds.), Wiley Blackwell, Ames, USA, 3–22.

Buckow, R., Baumann, P., Schroeder, S., and Knoerzer, K. 2011. Effect of dimensions and geometry of co-field and co-linear pulsed electric field treatment chambers on electric field strength and energy utilisation. *Journal of Food Engineering*, 105(3), 545–556.

Buckow, R., Schroeder, S., Berres, P., Baumann, P., and Knoerzer, K. 2010. Simulation and evaluation of pilot-scale pulsed electric field (PEF) processing. *Journal of Food Engineering*, 101(1), 67–77.

Buckow, R., Semrau, J., Sui, Q., Wan, J., and Knoerzer, K. 2012. Numerical evaluation of lactoperoxidase inactivation during continuous pulsed electric field processing. *Biotechnology Progress*, 28(5), 1363–1375.

COMSOL Multiphysics. 2007. *Chemical Engineering Module Model Library*. COMSOL AB, Stockholm, Sweden.

Day, L., Xu, M., Oiseth, S.K., and Mawson, R. 2012. Improved mechanical properties of retorted carrots by ultrasonic pre-treatments. *Ultrasonics Sonochemistry*, 19(3), 427–434.

Gogate, P.R. and Kabadi, A.M. 2009. A review of applications of cavitation in biochemical engineering/biotechnology. *Biochemical Engineering Journal*, 44(1), 60–72.

Hayashi, E.M. 2007. Afghanistan fruit and vegetable dehydrator desk study. *Report for the United States Agency for International Development (USAID) Accelerating Sustainable Agricultural Program (ASAP)*.

Heinz, V., Toepfl, S., and Knorr, D. 2003. Impact of temperature on lethality and energy efficiency of apple juice pasteurization by pulsed electric fields treatment. *Innovative Food Science & Emerging Technologies*, 4(2), 167–175.

Juliano, P., Knoerzer, K., Fryer, P., and Versteeg, C. 2009. *C. botulinum* inactivation kinetics implemented in a computational model of a high pressure sterilization process. *Biotechnology Progress*, 25(1), 163–175.

Juliano, P., Temmel, S., Rout, M., Swiergon, P., Mawson, R., and Knoerzer, K. 2013. Creaming enhancement in a liter scale ultrasonic reactor at selected transducer configurations and frequencies. *Ultrasonics Sonochemistry*, 20(1), 52–62.

Knoerzer, K., Arnold, M., and Buckow, R. (2011a). Utilising Multiphysics modelling to predict microbial inactivation induced by pulsed electric field processing. *ICEF11—11th International Congress on Engineering and Food*, Athens, Greece, 22–26 May 2011.

Knoerzer, K., Baumann, P., and Buckow, R. 2012. An iterative modelling approach for improving the performance of a pulsed electric field (PEF) treatment chamber. *Computers & Chemical Engineering*, 37, 48–63.

Knoerzer, K., Buckow, R., Juliano, P., Chapman, B., and Versteeg, C. 2010. Carrier optimization in a pilot-scale high pressure sterilisation plant—An iterative CFD approach. *Journal of Food Engineering*, 97(2), 199–207.

Knoerzer, K. and Chapman, B. 2011. Effect of material properties and processing conditions on the prediction accuracy of a CFD model for simulating high pressure thermal (HPT) processing. *Journal of Food Engineering*, 104(3), 404–413.

Knoerzer, K., Juliano, P., Gladman, S., Versteeg, C., and Fryer, P. 2007. A computational model for temperature and sterility distributions in a pilot-scale high-pressure high-temperature process. *AIChE Journal*, 53(11), 2996–3010.

Knoerzer, K., Juliano, P., Roupas, P., and Versteeg, C. (2011b). *Innovative Food Processing Technologies: Advances in Multiphysics Simulation*. Wiley Blackwell, Ames, USA.

Kumar, A., Kumaresan, T., Pandit, A.B., and Joshi, J.B. 2006. Characterization of flow phenomena induced by ultrasonic horn. *Chemical Engineering Science*, 61(22), 7410–7420.

Olivier, S., Bull, M., Stone, G., Diepenbeek, R., Kormelink, F., Jacops, L., and Chapman, B. 2011. Strong and consistently synergistic inactivation of spores of spoilage-associated *Bacillus* and *Geobacillus* spp. by high pressure and heat compared with inactivation by heat alone. *Applied and Environmental Microbiology*, 77(7), 2317–2324.

Rodriguez, G., Riera, E., Gallego-Juarez, J.A., Acosta, V.M., Pinto, A., Martinez, I., and Blanco, A. 2010. Experimental study of defoaming by air-borne power ultrasonic technology. *International Congress on Ultrasonics Physics Procedia* (1), 135–139.

Sabarez, H.T. 2010. Improving prune dehydration cost efficiency. *Unpublished report prepared by CSIRO Animal, Food & Health Sciences for Horticulture Australia Limited (HAL)*, Australia.

Sabarez, H.T. 2012. Computational modelling of the transport phenomena occurring during convective drying of prunes. *Journal of Food Engineering*, 111(2), 279–288.

Sabarez, H.T. 2013. Mathematical modelling of the coupled transport phenomena and colour development: Finish drying of trellis-dried sultanas. *Drying Technology*, 32, 578–589.

Sabarez, H., Gallego-Juarez, J., and Riera, E. 2012. Ultrasonic-assisted convective drying of apple slices. *Drying Technology*, 30(9), 989–997.

Trujillo, F.J., Eberhardt, S., Moeller, D., Dual, J., and Knoerzer, K. 2013. Multiphysics modelling of the separation of suspended particles via frequency ramping of ultrasonic standing waves. *Ultrasonics Sonochemistry*, 20(2), 655–666.

Trujillo, F.J. and Knoerzer, K. 2009. An approach to model the acoustic streaming induced by an ultrasonic horn in a sonoreactor. *Institute of Food Technologists (IFT) Annual Meeting and Food Expo*, Anaheim, CA, USA, June 2009.

Trujillo, F.J. and Knoerzer, K. 2011. A computational modeling approach of the jet-like acoustic streaming and heat generation induced by low frequency high power ultrasonic horn reactors. *Ultrasonics Sonochemistry*, 18(6), 1263–1273.

# 15 New/Innovative Technologies

*George I. Katsaros and Petros S. Taoukis*

**CONTENTS**



## 15.1 INTRODUCTION

In a fast-growing world population facing changes at a historically unprecedented rate, research and innovations in the field of food process engineering and food packaging can significantly contribute in assuring sufficient quantity of high-quality food. Consumers demand safe food products of high nutritional value and bio-functional properties, superior sensory attributes, long shelf life and convenience in use, and yet fresh like, minimally processed and with "clean label," sustainably produced in an environmentally and energy-efficient way. These requirements pose extraordinary challenges for the continuous improvement of conventional processes as well as the development of novel products and processes. Conventional thermal treatment is applied for the production of safe products with a shelf-life increase when compared to untreated ones. Although thermal processes have been extensively studied and process optimization has been applied in most of the cases, there is a significant quality degradation of processed products. Thermal treatment may cause protein denaturation, carbohydrates solubility alteration, nonenzymatic browning through the Maillard reaction, and so on. Moreover, thermo-sensitive organoleptic characteristics such as flavor and taste are negatively affected. Nonthermal technologies are investigated, targeting the minimization of thermal treatment quality degradation of the final product.

The term nonthermal processing is often used to designate technologies that are effective at ambient or sub-lethal temperatures or when the main stress factor for microbial inactivation is not heat. An increase in consumer demand for fresher foods while at the same time providing a high degree of safety, has driven the growing interest in nonthermal preservation techniques because of their capability to inactivate microorganisms and enzymes in foods. Their main application is considered to be the thermal pasteurization replacement, but their applications in other food production steps aiming at increased yields, textural modifications of the treated raw materials or final products or high pressure (HP) biotechnology are being explored (Aertsen et al. 2009). Understanding the impact and potential of such technologies on food systems at the cellular level enables the design of tailor-made foods and the establishment of process–structure–function relationships. Based on this knowledge, new process design as well as the incorporation of novel technologies in traditional processes is possible. Consequently, the use of such processes for maintenance or even improvement of product quality via processing to fulfill the consumer preference and acceptance as well as industry needs, is a significant concept within the food industry. Improved mathematical tools have to be employed to systematically evaluate and optimally design processing, and shelf-life control of food products produced using novel technologies.

High hydrostatic pressure (HP), pulsed electric fields (PEFs), ultrasound (US), irradiation (IR), UV light (UV), high dense phase carbon dioxide (DPCD), ozonation (OZ), and cold plasma (CP) are used to exemplify scalable and flexible food manufacturing techniques, discussing the state of the art regarding the research and application of these emerging technologies and demonstrating the potential of establishing new routes of process and product development by interfacing food science and food manufacturing.

### 15.1.1 Nonthermal Technologies Characterization

The nonthermal technologies may be split into three major categories:

1. The technologies that have existed for a long time still have not been applied on an industrial level, nevertheless they are still subject of a significant

number of studies cited in the literature describing a solution for the development, design, and processing of food products. In this category ultrasonication may be included.

2. The more recent technologies have not been applied industrially either because there is either a lack of sufficient knowledge on the effects of their application on food constituents and products, or because of a lack of industrial scale equipment for their application. In this category, the cold plasma and high dense phase carbon dioxide technologies may be included.
3. The technologies that have already been applied in industries over the last years or could be applied; however, there is hesitation from industries in adapting these technologies into their production process. In this category, the high hydrostatic pressure, ozonation, and the pulsed electric fields, UV light, and IR technologies may be included.

Below, the state of the art for each of these technologies is presented.

#### 15.1.1.1 High Pressure

HP technology is among the most important non-thermal technologies and probably the most promising considering the results cited in the literature and the growing interest for its application in the food industry.

##### *15.1.1.1.1 State of the Art*

HP technology is being mostly applied as a pasteurization process to extend the shelf life and control safety risk in a wide range of food products, without the use of high temperatures potentially detrimental to their sensitive quality traits. HP treatments at ambient temperature are capable of inactivating undesirable microorganisms and enzymes. In many cases the combination of HP treatment with mildly elevated temperature is needed to achieve inactivation of the more pressure and/or temperature resistant microorganisms and enzymes. HP processing has been extensively studied for selected pre-packaged foods and its application assures safety, shelf life extension and nutrient preservation (Hayman et al. 2004). HP pasteurization is environmentally friendly and can retain the fresh-like characteristics of foods better than commonly used thermal treatments.

HP treatments, in general, are effective in inactivating most vegetative pathogenic and spoilage microorganisms at pressures above 200 MPa at chilled or process temperatures less than 45°C, but the rate of inactivation is strongly influenced by the peak pressure and the process time (Raoult-Wack 1994). Sterilization, that is, inactivation of spores such as *Clostridium botulinum* and *Bacillus* could be achieved through synergies of elevated heat and pressure (Ahn et al. 2007). The extent of inactivation depends on the type of microorganism, food composition, pH and water activity. Gram-positive bacteria are more resistant and significant variations in pressure resistance can be seen among strains of the same species. When validating a HP process, sufficient time after treatment should be allowed to confirm that sublethally injured organisms do not recover, and all viable organisms of concern during post-treatment storage are controlled. Yeasts are inactivated by short exposure to 300–400 MPa at 25°C but yeast ascospores require higher pressures. Bacterial spores require significantly higher process pressures, temperatures, and holding times than vegetative cells. *Bacillus amyloliquefaciens* spores have been used as an index for HP-heat resistance (Ahn et al. 2007). More research is needed to characterize and model the combined pressure-thermal resistance of pathogenic and spoilage microorganisms as a function of the food matrix, pH, and water activity. The extent and mechanisms of bacterial injury during HP pasteurization, and sterilization merit further investigation (Balasubramaniam et al. 2008).

HP can also modify protein structure (Galazka et al. 1996) and thus enzyme activity (Gomes and Ledward 1996). The HP mechanism for enzyme denaturation is governed by the *Le Chatelier* principle, which predicts that an application of pressure shifts the equilibrium to a state that occupies the smallest volume, so any reaction accompanied by volume decrease, is accelerated by elevated pressures (Cano et al. 1997). The effect of HP on the enzymes can be attributed to the fact that HP affects hydrogen bonds and alters the three-dimensional configuration of the molecules. The effect of HP on enzyme inactivation has been shown to be strongly dependent on the type of enzyme, pH, nature of the medium in which the enzyme is dispersed, temperature, and treatment time (Cheftel 1992; Nguyen et al. 2002). The effect of pressure on the activity of enzymes is important to food quality such as phenolases, pectinases, and peroxidases have been studied and reported in several publications (Alexandrakis et al. 2013; Boulekou et al. 2010; Guiavarc'h et al. 2005; Katsaros et al. 2009a, b, c). This issue has been addressed by several researchers who pointed out that pressure and heat denaturation represent processes underlying distinct mechanisms.

Apart from HP application as a nonthermal pasteurization technique, there are other potential applications. HP processing ranging from 200 to 350 MPa may denature proteins from the adductor muscle of mollusks such as oysters and clams. The treated muscle, which is responsible for closing the shell, will not be able to contract and the oyster will open. This exposes the meat for easy extraction, resulting in a significant yield increase (He et al. 2002). Another potential application is the cheese maturation enhancement (Malone et al. 2003), which is done by increasing the aminopeptidases activity responsible for the maturation process (Katsaros et al. 2009a). The gelatinization of starches under pressure is significantly different from that induced by heat, and hence they offer unique functional properties, like, for example, a formation of weak gels which could be used as fat replacer in dietary foods (Zhang et al. 2008). Pressure-induced protein gels opens up a possible generation of new textures as they retain their original flavor and color accompanied by a glossy appearance. Such gels can be applied for the manufacturing of milk products to improve yoghurt texture (Johnston et al. 1993) or increase cheese yield (López-Fandino et al. 1996).

Pressure treatment can be used to process both liquid and high-moisture-content solid foods. HP has small effect on chemical compounds and food quality (Grant et al. 2000). Pressure treatment can also be used to alter the functional and sensory properties of various food components, especially proteins, allowing these to be beneficially altered (Talens et al. 2002). HP has very little effect on low-molecular-weight compounds such as flavor compounds, vitamins, and pigments compared to thermal processes. For the HP treatment, the packaged food is deposed in a carrier

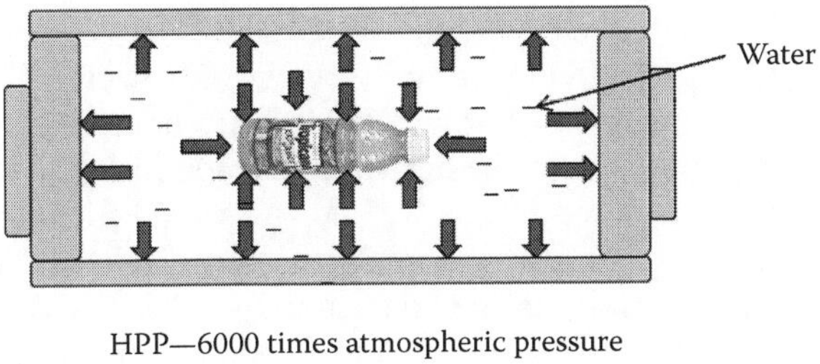


**FIGURE 15.1** HP treatment of a freshly squeezed orange juice for in-pack cold pasteurization.

and automatically loaded into the HP vessel and the vessel plugs are closed. The pressure transmitting media, usually water, is pumped into the vessel from one or both sides (Figure 15.1). After reaching the desired maximum pressure the pumping is stopped and in ideal cases no further energy input is needed to hold the pressure during dwell time. In contrast to thermal process where temperature gradients occur, all molecules in the HP vessel are subjected to the same amount of pressure at exactly the same time, due to the isostatic principle of pressure transmission (Heinz et al. 2009).

#### *15.1.1.1.2 Extrinsic Process Parameters*

In HP treatments, the process conditions are a function of the following extrinsic parameters:

- Applied pressure (MPa)
- Temperature when using Hurdle technology (thermal treatment) (C)
- Holding time (min)
- Pressure build-up time (min)
- Pressure release time (min)
- Number of pulses and time interval between pulses when pulsed pressures are used

Adiabatic heating should be considered when applying high pressures as evidenced from Figure 15.2 where a process parameters recording is depicted.

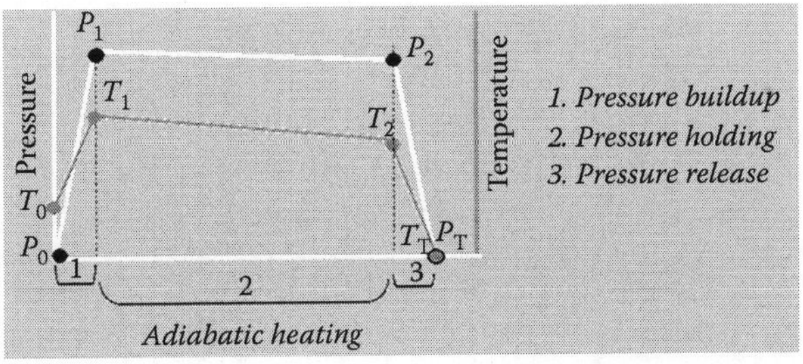


**FIGURE 15.2** Process parameters for the design of a HP treatment.

#### *15.1.1.1.3 Mathematical Modeling*

The systematic study of the technical parameters (that would allow for the use of the scientific achievements in a controlled and effective industrial process) is based on the kinetic approach of the destructive reactions of several factors that lead to spoilage or degradation of quality and functional properties of the food, during the HP treatments (Gould, 2002). This approach is detailed by Stoforos and Taoukis (2001) in analogy to the conventional thermal processing and constitutes the basis for the design, the treatment and the modeling of the procedure that would offer the possibility of a safe and effective industrial use. The degree of deterioration of an index sensitive to the HP process is a function of its resistance at various pressure–temperature conditions, depending also on the duration of the treatment. Assuming that the deterioration is described by a first-order reaction:

$$-\frac{dC}{dt} = kC, \quad k = f(T,P,...) \tag{15.1}$$

where $C$ is the concentration (or activation or population) of the HP sensitive index (e.g., number of microorganisms/mL, g/L, etc.), k is the reaction rate constant at constant conditions of the procedure (min$^{-1}$), $t$ the time of the treatment (min), $T$ the temperature during the HP treatment (K), and $P$ the pressure (MPa). Integrating Equation 15.1 leads to

$$\ln(C) - \ln(C_o) = \int_{t_a}^{t_b} -k\,dt \tag{15.2}$$

where $C_o$ is the initial concentration and subscripts $a$ and $b$ refer to the initial and the final condition, respectively (in this case, the beginning and the end of the procedure).

If the effect of pressure and temperature on the reaction rate is known, then the integral of Equation 15.2 can be calculated, using the appropriate Arrhenius (Equation 15.3) and Eyring (Equation 15.4) expressions:

$$k = k_{T\text{ref}} \exp\left[-\frac{E_a}{R}\left(\frac{1}{T} - \frac{1}{T_{\text{ref}}}\right)\right] \tag{15.3}$$

$$k = k_{P\text{ref}} \exp\left[-\frac{V_a}{R}\frac{(P - P_{\text{ref}})}{T}\right] \tag{15.4}$$

where $T_{\text{ref}}$ and $P_{\text{ref}}$ represent a reference temperature and a reference pressure, respectively, $E_a$ (J/mol) and $V_a$ (mL/mol) the activation energy and volume, respectively, and R the universal gas constant (8.314 J/(mol • K).

Using Equations 15.3 and 15.4 in Equation 15.2, assuming that the activation energy and volume depend on the pressure and the temperature, respectively, Equation 15.2 is finally expressed as follows:

$$\ln\left[\frac{C}{C_o}\right] = \int_0^t \left[-k_{\text{ref}} \exp\left[-\frac{E_a(P)}{R}\left(\frac{1}{T} - \frac{1}{T_{\text{ref}}}\right) - \frac{V_a(T)}{R}\frac{(P - P_{\text{ref}})}{T}\right]\right] dt \quad (15.5)$$

This equation forms the basis for the modeling of the HP treatment (e.g., for the determination of an equivalent time of treatment, e.g., the calculation of the F-value), in a similar way as this is done for conventional thermal treatments. The quantification of the effect of the treatment conditions and subsequent storage of products allows for the optimization of the overall production design. Products that have been manufactured under "optimized" conditions must be tested for their acceptability by representative consumer groups and appropriate sensory methodology.

#### *15.1.1.1.4 Energy Efficiency and Water Savings*

HP pasteurization, apart from retaining the fresh-like characteristics of foods better than commonly used thermal treatments, is environmentally friendly. An HP process requires power to increase pressure, and part of the power consumed is converted to heat for the temperature increase due to pressurization. Theoretically, the compression work and energy required for the temperature increase due to pressurization are about 52 and 70 kJ/kg upon compression of pure water up to 600 MPa, respectively. In addition, it should be noted that HP sterilization/pasteurization does not require a cooling process and decompression will decrease the temperature of the product. The theoretical total energy input into a HP process at 600 MPa for processing pure water is about 122 kJ/kg. During HP processing, water is used as a medium for pressure build up. This water is normally recycled since it is not in contact with the food (HP processing of packaged products), allowing for minimum water consumption. Although HP equipment is considered to be generally more expensive than conventional processing/packaging systems, significant energy cost savings may be accumulated over time through the use of HP rather than high temperature.

#### *15.1.1.1.5 Legal Approval*

In Europe, HP was classified as a novel technology since it was not used to a significant degree in the European food industry before May 15, 1997. High-pressured foods may be recognized as novel foods and consequently may fall under the Novel Foods Regulation, when a significant change occurs in these food products. In 2000, high-pressured fruit-based preparations were approved under the NFR by the European Commission. Currently, many different high-pressured food products are available on the EU market although they have not been approved under the NFR. The competent authorities of the member states agreed in July 2001 that the national authorities should decide on the legal status of high-pressured food products. The European Commission has concluded that HP was no longer considered to be a novel process. However, some member states were concerned that the HP foods should still be assessed for their safety and argued that data required for an assessment should be determined on a case-by-case basis. The approach concerning HP foods in the European Union may differ significantly between the member states and it may have an impact on the HP application. There may be room for different interpretations of the NFR since the definitions from the regulation—"significant degree" and "significant change" seem to be vague and unclear. In 2008, the European Commission published a Proposal for a New Novel Foods Regulation. However, according to the majority of experts, the Proposal did not change the situation of HP application in the EU.

As far as it concerns the United States, USDA has approved high hydrostatic pressure as an intervention method for *Listeria* contaminated pre-packed ready-to-eat (RTE) meat products, while U.S Food and Drug Administration (FDA) has accepted the commercial use of pressure-assisted thermal sterilization (PATS) processes for application in the production of low-acid foods (LAF) (February 2009).

#### 15.1.1.2 Pulsed Electric Fields

##### *15.1.1.2.1 State of the Art*

PEF technology offers the ability to inactivate microorganisms with minimal effects on the nutritional, flavor, and functional characteristics of food products due to the absence of heat, as in the case of HP. HP processing and PEF treatment are both volumetric treatments that inactivate microorganisms at any spot in the product. PEF technology is based on the application of pulses of high voltage (typically 20–80 kV/cm) delivered to the product placed between a set pair of electrodes that confine the treatment gap of the PEF chamber (Figure 15.3) (Raso and Heinz, 2006). The large field intensities are achieved through storing a large amount of energy in a capacitor bank (a series of capacitors) from a direct current power supply, which is then discharged in the form of high-voltage pulses (Zhang et al. 1995). The pulse caused by the discharge of electrical energy from the capacitor is allowed to flow through the food material for an extremely short period of time (1–100 μs) and can be conducted at moderate temperatures for less than 1 s (Deeth et al. 2007). When food is subjected to the electrical high-intensity pulses several events, such as resistance heating (Sastry and Barach 1995), electrolysis (Hülsheger and Niemann 1980), and disruption of cell membranes (Sitzmann 1995), can occur contributing to the inactivation of microorganisms. In fact, several theories exist for the destruction of bacterial cells by PEF, but they commonly describe damage on the cell membrane (electroporation), which affects its functioning and may lead to cell death (Deeth et al. 2007). PEF technology is mainly intended for preservation of pumpable fluid or semi-fluid foods. In particular it could be used to improve the shelf life of milk (Craven et al. 2008), green pea soups (Vega-Mercado et al. 1996), liquid whole eggs (Barbosa-Cánovas 2001) as well as fruit juices (Heinz et al. 2003). A combination of mild heat and PEF might also be helpful to achieve sufficient enzyme inactivation. When operating at high treatment temperatures and making use of synergetic heat effects, PEF energy input might be reduced close to the amount required for conventional thermal pasteurization, assuming 95% of heat recovery (Toepfl et al. 2006).

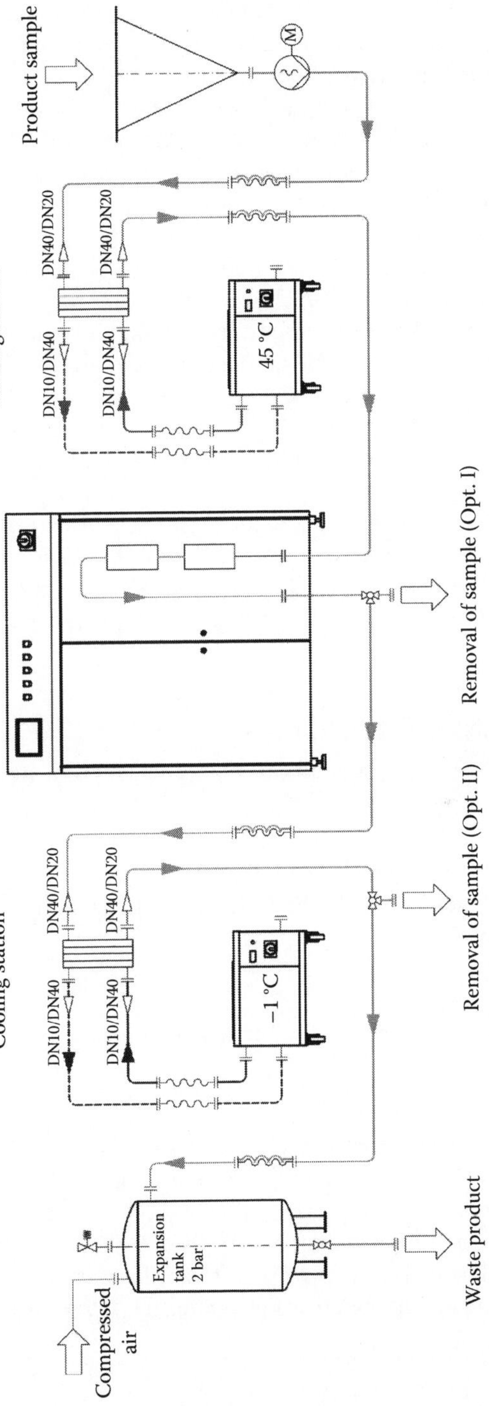


**FIGURE 15.3** Typical PEF process unit combined with thermal treatment for the production of food products.

Although the study of PEF technology has been focused on its ability to inactivate microorganisms in liquid food products at low temperatures, some other applications in the food industry have been explored as well, such as the enhancement of drying efficiency, decontamination of liquid waste, (Barbosa-Cánovas and Sepúlveda 2005) and the increase of the yield in fruit and vegetables juice extraction (Grimi et al. 2011). Among the most promising applications of PEF is the increase in the yield of oil extraction from olives, significantly impacting olive oil production. The current application in olive oil industries for yield increase is the malaxation process in which the olives are thermally treated at mild conditions (lower than 60°C) for a duration not higher than 30 min. The application of this technique results in an approximately 5% yield increase which is a significant percentage for the industries. Nevertheless, the temperature increase (even at a range from 50°C to 60°C) during the malaxation process has a negative effect on the extracted oil. The same aspect is of concern for industries producing grapes and tomato products (wineries, fruit, and tomato industries). The aim is to increase the yield of the extracted juices (from fruits and vegetables) applying technologies that minimally affect the quality of the extracted product.

The lack of understanding the factors that can affect the efficacy of PEF-treatment, such as proper processing conditions and design of equipment, type of microorganism and enzyme, suitability and properties of the treated food (Deeth et al. 2007) are currently limiting the validation of this technology for commercial application.

##### *15.1.1.2.2 Extrinsic Process Parameters*

In PEF treatments, the process conditions are a function of the following extrinsic parameters:

- Field strength (kV/cm)
- Temperature when using Hurdle technology (thermal treatment) (°C)
- Treatment time (μs)
- Pulse width (μs)
- Pulse shape
- Energy applied per liter or kilo of processed food product (J/L or J/kg)

##### *15.1.1.2.3 Mathematical Modeling*

The same approach as described in HP treatments can be adapted for the mathematical modeling of PEF parameters on the quality and safety indices of food products. Hülsheger and Niemann (1980) were the first to propose a mathematical model for inactivation of microorganisms with PEF. Their model was based on the dependence of the survival ratio $S$ ($N/N_o$ or the ratio of living cell count before and after PEF treatment) on the electric field intensity $E$ according to the following expression:

$$\ln(S) = -b_E(E - E_c) \tag{15.6}$$

where $b_E$ is the regression coefficient, $E$ is the applied electric field, and $E_c$ is the critical electric field obtained by the extrapolated value of $E$ for 100% survival. The critical electric field ($E_c$) has been found to be a function of the cell size (much lower for large cells) and pulse width. Hülsheger and Niemann (1980) proposed an inactivation kinetic model that relates microbial survival fraction ($S$) with PEF treatment time ($t$) in the form of

$$\text{ls}S = -b_t \ln(t/t_c) \tag{15.7}$$

where $b_t$ is the regression coefficient, $t$ is the treatment time, and $t_c$ is the extrapolate value of $t$ for 100% survival, or critical treatment time. The model can also be expressed as

$$S = \left(\frac{t}{t_c}\right)^{\frac{-(E - E_c)}{K}} \tag{15.8}$$

where $t$ is treatment time, $t_c$ is critical treatment time, $E_c$ is critical electric field intensity, and $K$ is the kinetic constant. A small value for the kinetic constant ($K$) indicates a wide span in the inactivation rate curve and lower sensitivity to PEF, whereas a large value implies a steep decline or higher susceptibility to PEF. Lower $E_c$ values would indicate less resistance to the PEF treatment.

#### *15.1.1.2.4 Energy Efficiency and Water Savings*

PEF treatment is usually considered to have a higher energy input than a thermal treatment. Increasing treatment temperature can improve treatment efficiency and reduce the energy consumption of PEF treatment. When using synergistic effects on microbial inactivation at an elevated treatment temperature of 35–65°C, the energy consumption by the PEF could be reduced from above 100 kJ/kg to less than 40 kJ/kg for a reduction of 6 log cycles (Heinz et al. 2003). When operating at elevated treatment temperatures and making use of synergetic heat effects, the energy input of PEF treatment might be reduced closer to the 20 kJ/kg required for conventional thermal pasteurization with 95% of heat recovery capability (Toepfl et al. 2006). In relation to reducing the energy consumption of PEF treatments, a number of options exist: (a) optimization of the design of pulse generation circuits, (b) optimization of the operation of PEF treatment, (c) use of a regenerative heat exchanger to minimize the cooling requirement, and (d) use of synergistic effects of elevated treatment temperature and PEF treatment. Moreover, no water is used during PEF processing.

#### *15.1.1.2.5 Legal Approval*

There is no special European legislation on products treated with PEF processing technologies in the food industry except for the general EC Regulation 258/97 concerns, which entered into force on May 15, 1997 concerning novel foods and novel food ingredients. Products treated with PEF processing are considered as novel and should follow the principles mentioned in the EC Regulation 258/97.

#### 15.1.1.3 Ozonation

##### *15.1.1.3.1 State of the Art*

Ozonation (OZ) is the technique in which ozone is applied as a strong oxidant and potent disinfecting agent. Ultraviolet radiation (188 nm wavelength) and corona discharge methods can be used to generate ozone. The bactericidal effects of ozone have been documented on a wide variety of organisms, including Gram positive and Gram negative bacteria as well as spores and vegetative cells. There are numerous potential application areas of ozone in the food industry such as food surface hygiene, sanitation of food plant equipment, and treatment of fruits and vegetables, targeting for an increase in the shelf-life of products. Notably, when ozone is applied to food, it leaves no residue since it decomposes quickly.

Ozone destroys microorganisms by the progressive oxidation of vital cellular components. The bacterial cell surface has been suggested as the primary target of ozonation. Two major mechanisms have been identified in ozone destruction of the target organisms: first mechanism is that ozone oxidizes sulfhydryl groups and amino acids of enzymes, peptides, and proteins to shorter peptides. The second mechanism is that ozone oxidizes polyunsaturated fatty acids to acid peroxides (Victorin 1992). Ozone degradation of the cell envelope unsaturated lipids results in cell disruption and subsequent leakage of cellular contents. Double bonds of unsaturated lipids are particularly vulnerable to ozone attack. In Gram negative bacteria, the lipoprotein and lipopolysaccharide layers are the first sites of destruction resulting in increases in cell permeability and eventually cell lysis (Kim et al. 1999). Chlorine selectively destroys certain intracellular enzyme systems; ozone will cause widespread oxidation of internal cellular proteins causing rapid cell death. Cellular death can also occur due to the potent destruction and damage of nucleic acids. Thymine is more sensitive to ozone than cytosine or uracil. Ozone also destroys viral RNA and alters polypeptide chains in viral protein coats.

Ozonation of fruit juices is under investigation from researchers. Ozonation of liquid phases is most frequently accomplished by injecting ozone gas (mixtures of air/ozone or oxygen/ozone) through a sparger into a liquid. Usually, the studies on ozone absorption in the aqueous systems are carried out in stirred-tank reactors or bubble columns (Cullen et al. 2009). Subsequent approval of ozone as a direct additive led to the application of ozone for processing various fruit juices such as apple cider (Steenstrup and Floros 2004), orange juice (Patil et al. 2009; Tiwari et al. 2008), blackberry juice (Tiwari et al. 2009a), strawberry juice (Tiwari et al. 2009b), and so on. A typical bubble column reactor is a cylindrical vessel with a gas diffuser to sparge ozone in a gaseous state into either a liquid phase or a liquid–solid dispersion. Many empirical correlations have been proposed to estimate design parameters for a bubble column based on the physical and chemical properties of the material under investigation including experimental conditions. In liquid treatment applications, ozonation is limited by the mass-transfer process governing the overall performance of ozone contactors. The design parameters for a bubble column are: gas–liquid specific interfacial area, individual mass-transfer coefficient, flow behavior, bubble size and distribution, and coalescence of bubbles (Zhao et al. 2004).

#### 15.1.1.3.2 Extrinsic Process Parameters

In OZ treatments, the process conditions are a function of the following extrinsic parameters:

- Flow rate (L/min)
- Ozone concentration (μg/mL)
- Treatment time (min)

#### 15.1.1.3.3 Mathematical Modeling

The degree of deterioration of an index sensitive to the OZ process is a function of its resistance at various process conditions (flow rate and ozone concentration), depending also on the duration of the treatment. Assuming that the deterioration is described by a first-order reaction:

$$-\frac{dC}{dt} = kC, \quad k = f(F, O_c \ldots) \tag{15.9}$$

where $C$ is the concentration (or activation or population) of the OZ sensitive index (e.g., number of microorganisms/mL, g/L, etc.), $k$ is the reaction rate constant at constant conditions of the procedure ($min^{-1}$), t the time of the treatment (min), $F$ the flow rate during the OZ treatment (L/min), and $O_c$ the ozone concentration (μg/mL).

#### 15.1.1.3.4 Energy Efficiency and Water Savings

Ozone is formed from oxygen zapped at high voltages. The gas is unstable and therefore quickly reverts to oxygen. Ozone generators were maintenance intensive and required a lot of energy, but over the past years the equipment developed for ozone treatments has become more efficient with lower energy demands. Water is needed for the ozone treatment but it can be recycled and reused for a significant number of processes. Nevertheless, ozone has environmental advantages over the traditional chlorination of water, since chlorination produces dangerous by-products that pollute.

#### 15.1.1.3.5 Legal Approval

In response to a Food Additive Petition submitted during August 2000, the U.S. Food and Drug Administration formally approved the use of ozone as an antimicrobial agent for the treatment, storage, and processing of foods in gas and aqueous phases. The approval was published on June 26, 2001 (FDA, 2001). According to the approval, ozone (CAS Reg. No. 10028-15-6) may be safely used in the treatment, storage, and processing of foods, including meat and poultry, in accordance with the following prescribed conditions:

1. The additive is an unstable, colorless gas with a pungent, characteristic odor, which occurs freely in nature. It is produced commercially by passing electrical discharges or ionizing radiation through air or oxygen.
2. The additive is used as an Antimicrobial agent.

3. The additive meets the specifications for ozone in the Food Chemicals Codex.
4. The additive is used in contact with food, including meat and poultry in the gaseous or aqueous phase in accordance with current industry standards of good manufacturing practice.
5. When used on raw agricultural commodities, the use is consistent with section of pesticides.

#### 15.1.1.4 Ultrasound (US)

##### *15.1.1.4.1 State of the Art*

Ultrasound (US) has been used for many years to homogenize foods and in-plant cleaning. Over the past years this technology was combined with heat treatments under pressure to reduce the amount of heating needed for microbial destruction or enzyme inactivation. In general, US could be applied for the pasteurization of a number of fruit and vegetable juices (fresh or concentrated) or smoothies. The main advantages of its application are the reduced energy consumption, the minimal flavor loss in liquid foods (for example juices), the increase in homogeneity, and the breakdown of agglomerates of bacteria (Awad et al. 2012; Mason et al. 1996; Piyasena et al. 2003; Vercet et al. 2001). One of the positive effects of US application is the removal of occluded oxygen from treated juices (Knorr et al. 2004) resulting in ascorbic acid and vitamins retention during the shelf-life (Solomon et al., 1995). Tiwari et al. (2009a, b) reported a maximum degradation of only 5% in the ascorbic acid content of orange juice when sonicated at high acoustic energy densities (0.81 W/mL) and treatment times (10 min).

Apart from pasteurization purposes, US could be used for the assisted freezing and drying of foods. In ultrasound assisted freezing, higher heat transfer coefficients can be obtained when applying ultrasonic power, thus the freezing rate can be greatly enhanced (reduced freezing time). Application of US on a solid–liquid system results in the appearance of a phenomenon referred to as micro-jetting (Bhaskaracharya et al. 2009), where an inrush of fluid from the bubble toward the surface is observed. This effect leads to rapid heat and mass transfer at the solid surface as the boundary layer is disrupted. Research related to the use of ultrasound to assist immersion freezing in particular, showed that the freezing rate represented by the characteristic freezing time can be significantly improved from 8% and up to 24% as compared to that of conventional immersion freezing, which makes the immersion chilling and freezing (ICF) process assisted by ultrasound very attractive from the economical point of view (Delgado et al. 2009; Li and Sun 2002; Li and Sun 2003).

US-assisted air drying has been studied the last years (Chemat et al. 2011). During this technique, a series of rapid and successive compressions and rarefactions in the food material ("sponge effect") induced by US take place. Water is expelled toward the surface of the product, and is evaporated by the hot gas stream. The deformation of the material caused by ultrasound is also responsible for the creation of microscopic channels that reduces the diffusion boundary layer and increases the convective mass transfer in the food product (Fernandes et al. 2008). Studies have shown that US might be more effective at low air velocities and low temperatures (Mulet et al. 2011).

#### 15.1.1.4.2 *Extrinsic Process Parameters*

When applying ultrasound to a process, it is necessary to know and measure the main parameters that control the process. The way the ultrasound energy is applied is unique for every application and therefore, should be specifically determined. The main extrinsic process parameters for this technology are:

- Amplitude of the process ($\mu m^{-1}$)
- Exposure time (min)
- Combined temperature (°C)
- Frequency (kHz)

#### 15.1.1.4.3 *Mathematical Modeling*

The degree of deterioration of an index sensitive to the US process is a function of its resistance at various process conditions (amplitude, frequency, temperature), depending also on the duration of the treatment. Assuming that the deterioration is described by a first-order reaction:

$$-\frac{dC}{dt} = kC, \quad k = f(A,F,T) \tag{15.10}$$

where $C$ is the concentration (or activation or population) of the US sensitive index (e.g., number of microorganisms/mL, g/L, etc.), $k$ is the reaction rate constant at constant conditions of the procedure ($min^{-1}$), $t$ the time of the treatment (min), $F$ the frequency (kHz), $A$ the amplitude ($\mu m^{-1}$), and $T$ the treatment temperature ($K$).

#### 15.1.1.4.4 *Energy Efficiency and Water Savings*

The current US systems have an energy efficiency around 85% which simply means that most of the power sent to the transducer is transferred into the medium (Patist and Bates 2008). According to Herceg et al. (2012), US-treated milk at amplitude levels 90 μm, temperature 60°C and treatment time 9 min reduced *S. aureus* significantly, while the energy requirements for this operation were 38 J for a product volume equal to 200 mL. The respective energy requirements for conventional thermal pasteurization of the same liquid food (juice in this case), with heat regeneration or recovery is about 7 kJ (Kozempel et al. 1998). The application of US does not require the use of water since it is implemented directly to the product. US could also be used to reduce the energy consumption during freezing and drying by reducing the process time, thus assisting these two widely used processes in the food industry.

#### 15.1.1.4.5 *Legal Approval*

The process could be considered as novel, but, due to its dominant physical effect no significant changes in composition, nutrient value, and content of undesired substances is expected for the products. The general EC Regulation 258/97 concerns apply. The formation of radicals by cavitation might result in undesired changes of product quality.

### 15.1.1.5 Dense Phase Carbon Dioxide

#### *15.1.1.5.1 State of the Art*

Dense phase carbon dioxide (DPCD), also known as pressurized carbon dioxide ($CO_2$), has been investigated as an alternative nonthermal pasteurization technique for food production. This technique makes use of $CO_2$ in or near its supercritical state (temperature of 32°C and pressure of 74 bar). The antimicrobial effect of pressurized $CO_2$ is attributed in the following:

- Dissolution of $CO_2$ into the liquid phase of food products,
- Modification of the cell membrane,
- Reduction in intracellular pH and disruption of the electrolyte balance,
- Inhibition of enzyme reactions and cellular metabolism,
- Extraction of compounds from cells and cell membranes.

Given that the properties of $CO_2$ are temperature and pressure dependent, the optimal processing conditions—and thus the influence of the above-mentioned interactions—can vary for differing food products to be sterilized. DPCD has been shown to be very effective on vegetative bacteria, with total inactivation realized in many cases. DPCD sterilization has been shown to be highly effective for liquid products, although there is limited data on the organoleptic and nutritional properties of the products after treatment. For solid products, the microbial inactivation is strongly dependent on the physical characteristics of the product, and in particular how the matrix influences the $CO_2$ penetration into the product. Apart from DPCD pasteurization and sterilization techniques, supercritical fluid extraction is a potential alternative to conventional extraction methods using organic solvents for extracting biologically active components from plants (Hamburger et al. 2004; Andras et al. 2005). Supercritical fluid extraction can prevent the oxidation of lipids. Supercritical $CO_2$ extraction can achieve a higher yield and quality of essential oils, flavors, and natural aromas than conventional steam distillation. The mean percentage yields of cedar wood oil for supercritical $CO_2$ extraction and steam distillation were 4.4% and 1.3%, respectively (Eller and King 2000). Supercritical fluids have also been used to remove moisture from foods. Conventional air drying is the most commonly used drying process in the food industry. However, the high air temperature, which is typical 65–85°C, may cause damage to the microstructure of food materials and have a negative effect on the color, texture, taste, aroma, and nutritional value of the dried products. The supercritical fluid can retain the microstructure of dried foods much better than that of air-dried foods (Brown et al. 2008).

#### *15.1.1.5.2 Extrinsic Process Parameters*

When applying DPCD, it is necessary to know and measure the main parameters that control the process. The main extrinsic process parameters for this technology are:

- Pressure (MPa)
- Exposure time (min)
- Combined temperature of treatment (C)
- Depressurization rate (MPa/min)

#### 15.1.1.5.3 *Mathematical Modeling*

The degree of deterioration of an index sensitive to DPCD process is a function of its resistance at various process conditions (pressure, temperature), depending also on the duration of the treatment. Assuming that the deterioration is described by a first-order reaction:

$$-\frac{dC}{dt} = kC, \quad k = f(P,T) \tag{15.11}$$

where $C$ is the concentration (or activation or population) of the DPCD sensitive index (e.g., number of microorganisms/mL, g/L, etc.), $k$ is the reaction rate constant at constant conditions of the procedure ($min^{-1}$), $t$ the time of the treatment (min), $P$ the pressure (MPa), and $T$ the treatment temperature (K).

#### 15.1.1.5.4 *Energy Efficiency and Water Savings*

DPCD sterilization eliminates thermal cycling (as used in UHT), thus resulting in a considerable reduction in the energy that is associated with extreme temperature swings. It also minimizes energy losses associated with heat. For liquids, the process can be continuous, thus only requiring energy to start up and maintain the process (pumps, heaters). No water is used in the DPCD sterilization process, in contrast to methods such as chemical sterilization that often makes use of a dilute solution of bleach or hydrogen peroxide followed by a washing step.

#### 15.1.1.5.5 *Legal Approval*

The US Food and Drug Administration issued a "Guidance for Industry: Juice HACCP Hazards and Control Guidance" in March 2004. The section that involves DPCD is under section V, Process Validation, subsection 5.34 "Dense Phase $CO_2$ Processing Systems," mentioning that DPCD is a technology that could be used in juice production since it is effective in reducing vegetative pathogens. It mentions that pressure and residence time may be used to optimize the bactericidal effects of $CO_2$. For these processes, $CO_2$ concentration is critical and that the process is performed under ambient conditions and temperature is not monitored as a critical factor. For the European Community, DPCD is considered as novel technology so DPCD processed foods to be sold in the EC market should be in accordance to the commission recommendation 97/618 EC concerning the scientific information and the safety assessment report required.

### 15.1.1.6 Cold Atmospheric Plasma

#### 15.1.1.6.1 *State of the Art*

Plasma is a high-energy gas that is created when an electrical current is passed through a gas. A well-known application of this principle is fluorescent lighting. Until recently, plasmas could only be created at relatively high temperatures in a vacuum and the use of plasma in the agro-sector and the packaging industry was therefore impractical. In the last few years, technological breakthroughs have made it possible to produce cold plasma under atmospheric conditions. Thus, it is now possible to apply the plasma technology in the food industry. Although plasma is a surface treatment, it can reach shadow places since it is a gas and in this way, it differs from light treatments. Several gases can be used for creation of plasma. Noble gases are used when radical formation is unwanted while oxygen is used for the production of ozone in the plasma. Depending on the gases used, temperature, activity, and stability of the plasma can be varied. No commercial equipment is available at the moment. CP treatment has been proposed for fruit juice processing. Plasma, a mix of ionized gas molecules and free electrons, is often referred to as the fourth state of matter. Plasma is a hot ionized gas consisting of approximately equal numbers of positively charged ions and negatively charged electrons. The characteristics of plasmas are significantly different from those of ordinary neutral gases. For example, because plasmas are made up of electrically charged particles, they are strongly influenced by electric and magnetic fields while neutral gases are not. Plasma sterilization, a new application of low-pressure plasma, is a promising technique and an alternative to other conventional sterilization methods like high-temperature sterilization, ethylene oxide sterilization and sterilization by radiation, especially for treatment of heat-sensitive materials. It promotes an efficient inactivation of the microorganisms and minimizes damage to the materials. It has been reported that the atmospheric cold plasma generated by this method has antibacterial activity against various bacterial strains studied which is in agreement with other studies done by the use of plasma generated by other methods as reported by Kayes et al. (2007). CP has proven sterilization (kill) capability against both Gram-positive and Gram-negative bacteria in different extents depending on special strain characteristics (Moman and Najmaldeen 2010).

#### *15.1.1.6.2 Extrinsic Process Parameters*

The main process parameters for cold plasma application are:

- Voltage level (kV)
- Gas mixture
- Mode of plasma exposure (direct or indirect)
- Process time (min)

#### *15.1.1.6.3 Mathematical Modeling*

The degree of deterioration of an index sensitive to CP process is a function of its resistance at various process conditions (voltage gases concentrations), depending also on the duration of the treatment. Assuming that the deterioration is described by a first-order reaction:

$$-\frac{dC}{dt} = kC, \quad k = f(V,Gc) \tag{15.12}$$

where $C$ is the concentration (or activation or population) of the US sensitive index (e.g., number of microorganisms/mL, g/L, etc.), k is the reaction rate constant at constant conditions of the procedure ($min^{-1}$), t the time of the treatment (min), V the voltage level (kV), and $T$ the treatment temperature (K).

##### *15.1.1.6.4 Energy Efficiency and Water Savings*

The nonthermal plasma technology is both energy efficient (lower energy demands than 100 W for the plasma generator) and a dry technology. It is a low-cost solution with zero residue. This approach offers a significant potential reduction in water usage as no water is required.

##### *15.1.1.6.5 Legal Approval*

From a European perspective, foods that are produced by novel means are subject to the Novel Food Regulation 258/97. However, if a treatment is considered rather than a new ingredient or functional compound, for example, it is sufficient to demonstrate substantial equivalence of the product under consideration. The minimum requirements for packaging materials that are in contact with foods and act as a barrier to protect foods are regulated by guideline EC 1935/2004. Treatment by plasma gas may not alter the barrier or functional properties of the material to pose a risk. Similar requirements are used by the FDA for the introduction of plasma-treated (food) products to the US market.

## 15.1.2 Ultraviolet Light

### 15.1.2.1 State of the Art

In the group of technologies related to light technology is ultraviolet light and pulsed white light. UV light processing involves the use of radiation from the ultraviolet region, ranging from 100 to 400 nm. The germicidal range, which inactivates bacteria and viruses, is located between 100 and 280 nm. Pulsed white light treatment involves the use of intense and short duration pulses of broad spectrum "white light." The spectrum for pulsed light treatments includes UV and near infrared (80% of the electromagnetic energy is within the range of 170–260 nm) (Ohlsson and Bengtsson 2002). Both light treatments result in considerable inactivation of microorganisms on surfaces without effects on the product quality. Since light technologies have the disadvantage of casting shadows, applications for rough surfaces can be difficult. Equipment for these technologies is now available. The germicidal properties of ultraviolet IR are due to the DNA absorption of the UV light, causing crosslinking between neighboring pyrimidine nucleoside bases (thymine and cytosine) in the same DNA strand (Miller and others 1999). Due to the mutated base, formation of the hydrogen bonds to the purine bases on the opposite strand is impaired. DNA transcription and replication is thereby blocked, compromising cellular functions and eventually leading to cell death. The amount of crosslinking is proportional to the amount of UV exposure.

### 15.1.2.2 Extrinsic Process Parameters

The main extrinsic process parameters for UV light include:

- Wavelength (nm)
- Fluence (mJ/cm$^2$)
- Time (ms)

#### 15.1.2.3 Mathematical Modeling

The degree of deterioration of an index sensitive to UV light process is a function of its resistance at various process conditions (Fluence, Wavelength), depending also on the duration of the treatment. Assuming that the deterioration is described by a first-order reaction:

$$-\frac{dC}{dt} = kC, \quad k = f(F,W) \tag{15.13}$$

where $C$ is the concentration (or activation or population) of the UV sensitive index (e.g., number of microorganisms/mL, g/L, etc.), $k$ is the reaction rate constant at constant conditions of the procedure ($min^{-1}$), $t$ the time of the treatment (min), $F$ the fluence ($mJ/cm^2$), and $W$ the wavelength (nm).

#### 15.1.2.4 Energy Efficiency and Water Savings

UV light technology is used for the disinfection of fruits, vegetables, nuts, as well as for drinking water. The energy spent to inactivate microorganisms in these cases is less compared to the energy required for alternative technologies such as boiling (for water disinfection) or chlorine usage for fruits and vegetables (energy and significant amount of water needed for this application). No water is needed.

#### 15.1.2.5 Legal Approval

The Code 21CFR179.41, issued by the FDA, Department of Health and Human Services, approves the use of pulsed UV light in the production, processing, and handling of food. In the Section 179.41 of Chapter I of the Food and Drugs Guide it is mentioned that pulsed light may be safely used for the treatment of foods under the following conditions:

1. The radiation sources consist of xenon flashlamps designed to emit broadband radiation consisting of wavelengths covering the range of 200–1000 nm, and operated so that the pulse duration is no longer than 2 ms;
2. The treatment is used for surface microorganism control;
3. Foods treated with pulsed light shall receive the minimum treatment reasonably required to accomplish the intended technical effect; and
4. The total cumulative treatment shall not exceed 12.0 joules/square centimeter ($J/cm^2$)

Correspondingly to the US FDA, the European Union considers UV light as an IR. Regulations for the use of the IR process in the EU are not harmonized and still the member staff discuss on the foodstuffs that should be allowed to be treated by ionizing radiation. Food IR may be authorized only if:

1. There is a reasonable technological need.
2. It presents no health hazard and is carried out under the conditions proposed.
3. It is of benefit to the consumers,
4. It is not used as a substitute for hygiene and health practice or for good manufacturing or agricultural practice.

## 15.1.3 Irradiation

### 15.1.3.1 State of the Art

IR is the deliberate process of exposing an item to certain types of radiation energy to bring about desirable changes. Ionizing radiation is radiant energy that has the ability to break chemical bonds. There are three types of ionizing radiation that can potentially be used in food IR:

> Electron beams (machine generated), x-rays—(machine generated), and gamma rays (occur naturally from radioactive decay of Cesium 137 or Cobalt 60). Cobalt-60 is most commonly used for food IR, though the electron beam is finding increasing application. Currently, there are a number of nonfood-related products being irradiated (cosmetics, wine corks, hospital supplies, medical products, packaging materials) mostly to achieve nonthermal sterilization. The radiation dose refers to the amount of these gamma rays absorbed by the product and is measured in Grays (Gy). 1 Gy = 1 Joule of absorbed energy/kg of product. Most treatment levels are on the order of 1–10 kGy (1 kGy = 1000 Gy). Depending on the dose, the process may be characterized as food pasteurization, food product quality improvement, or food sterilization. This technology has been applied over the past decades in more than 40 countries worldwide. In Table 15.1, the FDA approvals for food IR are presented. For consumer awareness the food-irradiated products are labeled with the label depicted in Figure 15.4.

**TABLE 15.1**
**Approved Food Products and Corresponding IR Doses Based on FDA**

**FDA Approvals for Food IR**

| Foods | Uses | Maximum Dose |
|---|---|---|
| All foods | Control insects | 1 kGy |
| Fresh foods | Delays maturation | 1 kGy |
| Poultry | Controls pathogen microorganisms | 3 kGy |
| Fresh iceberg lettuce and fresh spinach | Control of pathogen and extension of shelf-life | 4.0 kGy |
| Red meat (beef, lamb, and pork) | Controls spoilage and pathogen microorganisms | 4.5 kGy (fresh)<br>7 kGy (frozen) |
| Molluscan shellfish (oysters, mussels, and clams), fresh/frozen | For control of Vibrio bacteria and other foodborne microorganisms | 5.5 kGy |
| Dehydrated enzyme | Controls insects and microorganisms | 10 kGy |
| Spices and dry vegetable seasoning | Decontaminates and controls insects and microorganisms | 30 kGy |

*Source:* Adapted from FDA, Terms of the 1958 Food Additive Amendment to the Food, Drug and Cosmetic Act.

**FIGURE 15.4** Mandatory label used for IR food products for consumer awareness.

#### 15.1.3.2 Extrinsic Process Parameters

The main extrinsic process parameters for IR are:

- The dose (kGy)
- The temperature (°C)
- The treatment time (min)

#### 15.1.3.3 Mathematical Modeling

The degree of deterioration of an index sensitive to the IR process is a function of its resistance at various process conditions (dose, temperature), depending also on the duration of the treatment. Assuming that the deterioration is described by a first-order reaction:

$$-\frac{dC}{dt} = kC, \quad k = f(D,T) \tag{15.14}$$

where $C$ is the concentration (or activation or population) of the IR sensitive index (e.g., number of microorganisms/mL, g/L, etc.), $k$ is the reaction rate constant at constant conditions of the procedure ($\text{min}^{-1}$), $t$ the time of the treatment (min), D the IR dose (kGy), and $T$ the temperature (K).

#### 15.1.3.4 Energy Efficiency and Water Savings

The IR may be applied in various food products. The energy spent to control insects and inactivate microorganisms depends on the applied energy required and could be significantly low or similar to other conventional technologies with the advantage that IR application is the only one with such promising results (especially in the case of spices pasteurization). No water is needed.

#### 15.1.3.5 Legal Approval

It is the same as presented in UV light technology.

### 15.1.4 Nonthermal Technologies Applications

Most nonthermal technologies applications are referring to the replacement of conventional thermal treatment. Nevertheless, there are other applications for each of the above-mentioned technologies as presented below:

**US**. US technology apart from enzyme denaturation, microbial cell membrane lysis could be used for the crystallization and deaeration of liquid products. US could be also used for the homogenization of liquids.

**PEF**. PEF could be used for the continuous flow pasteurization of liquid foods such as juices, milks, and so on, as well as for batch pasteurization of nonliquid foods. One of the most important potential applications of PEF technology is the yield increase in juice industries and olive oils from olives, as well as the extraction of antioxidant and aromatic compounds from herbs and plants.

**UV light and IR**. These technologies are mainly used for the pasteurization of meats, fisheries, herbs and spices, nuts, and so forth. For UV its main application is the disinfection of drinking water especially for reverse osmosis equipments as a pasteurization technique to inactivate viruses and microorganisms. For IR, the spices and dried herbs and vegetables decontamination is the main application.

**DPCD**. DPCD is targeting in the pasteurization/sterilization of liquid foods only such as milk and fruit juices. The effect of the process parameters on the quality characteristics of the treated products is still under investigation from researchers.

**CP**. Although plasma is a surface treatment it can reach shadow places since it is a gas and in this way it differs from light treatments. It is effective in surface decontamination of fruits and vegetables as well as spices. The quality characteristics deterioration for the treated products are still under investigation.

**OZ**. Ozonation is mainly used for the decontamination of surfaces that food products are produced. It could be used for the surface decontamination of fruits and vegetables, avoiding the chemicals used in the corresponding production processes.

**HP**. High-pressure technology is among the nonthermal technologies the one with the higher potential for industrial applications. Several food products are nowadays produced using this technology for shelf-life extension. It is of high importance that the process is an in-pack pasteurization technique not allowing for cross-contamination. It is independent of the shape and size of treated products since pressure is considered to be applied uniformly in the whole food mass. Apart from pasteurization purposes, HP could be used for opening oysters resulting in an increased yield of meat extraction, reduce the maturation time of cheese as well as help in the formation of weak gels to be used as fat replacer in dietary foods, and so on.

In Table 15.2, the advantages, limitations (disadvantages), potential products to be treated, and products already available in the worldwide market are presented for each of these nonthermal technologies.

### 15.1.5 SWOT Analysis for the Nonthermal Technologies

SWOT analysis is used to evaluate the strengths, weaknesses, opportunities, and threats involved in nonthermal technologies. The SWOT analysis involves

**TABLE 15.2**

**Advantages, Limitations (Disadvantages), Potential Products to be Treated and Products Already Available in the Worldwide Market for Each of the Nonthermal Technologies**

| Technology | Advantages of Technology Application | Disadvantages of Technology Application | Potential Products for Processing Using Specific Technology | Products Already Available in the World Market Processed Using This Specific Technology |
|---|---|---|---|---|
| Ultrasonication | • Inactivation of microorganisms and enzymes<br>• Food homogenization | • Temperature increase<br>• Denaturation of proteins and vitamins<br>• Application only to liquid foods prior packaging | Most liquid foods such as milk and fruit and vegetables juices for homogenization | Homogenization of *Aloe vera* juice_*Aloe vera* gel break down for juice extraction |
| Ultraviolet light and IR | Microorganisms inactivation | • Nutritional and organoleptic characteristics deterioration<br>• Severe food legislation for its application<br>• Inactivation of microorganisms only at points that food may be exposed on IR<br>(successful application dependent on food shape) | May be applied for the surface inactivation of microorganisms of foods such as red meat, poultries, sea-foods, spices, and nuts.<br>Pasteurization and disinfection of drinking water. | Due to severe legislation of application of this technology, its sole application is the pasteurization of packaged spices.<br>Units established in reverse osmosis systems for the disinfection of drinking water. |
| Ozonation | Inactivation of microorganisms | • Vitamins oxidation<br>• Organoleptic characteristics deterioration | May be applied for the surface inactivation of microorganisms in fruits and vegetables (for the replacement of chlorine) | Not available industrial applications |
| Dense phase carbon dioxide | Inactivation of microorganisms | • Nutritional and organoleptic characteristics deterioration<br>• Application only to liquid foods | Application in milk for microorganisms inactivation.<br>Application in fruit juices for yeasts and molds inactivation | Not available industrial applications |

| | | | | |
|---|---|---|---|---|
| Cold plasma | Inactivation of microorganisms<br>Application in nonliquid foods independently of their shape | • Nutritional and organoleptic characteristics deterioration<br>• Promising technology for the surface inactivation of microorganisms on meats, vegetables, fruits, and seafoods | Potential application for the surface pasteurization of meats, vegetables, fruits, and sea-foods | Not available industrial applications |
| Pulsed electric fields | • Inactivation of microorganisms and enzymes<br>• Liquid foods homogenization<br>• Pasteurization of mostly liquid foods or nonliquid foods processed into a chamber fulfilled with a liquid medium<br>• Major PEF application is the yield increase of juice extraction from fruits and vegetables and olive oil from olives | Mostly applied to liquid foods._Appropriate equipment (vessel fulfilled with appropriate liquid medium) could be used for the processing of nonliquid foods | • Fruit and vegetable juices for yield increase and pasteurization<br>• Milk pasteurization<br>• Wine production for yield increase during vinification | Although it seems there was one industry producing orange juice pasteurized using PEF, this unit does not produce this product anymore |
| High hydrostatic pressure | • Inactivation of microorganisms and enzymes, while simultaneously retaining organoleptic, nutritional and textural characteristics<br>• Instant and uniform pressure transfer to the food<br>• The pressure transfer is independent of the size and shape of the processed product | • Different optimal process conditions for different food products<br>• High capital cost<br>• Batch-type processing (semi-continuous processing could be achieved if more than one HP units were used in parallel) | • Pasteurization of fruits, vegetables, and their juices<br>• Pasteurization of milk and milk products<br>• Pasteurization of meat and meat products (the most promising application of this technology)<br>• Ripening time reduction and pasteurization of cheeses | • In the worldwide market lots of HP-treated products may be found such as fruit juices, dairy and meat products, fruit preparations, wet salads, crustaceans, avocado based products, etc. |

**TABLE 15.2 (contiued)**

**Advantages, Limitations (Disadvantages), Potential Products to be Treated and Products Already Available in the Worldwide Market for Each of the Nonthermal Technologies**

| Technology | Advantages of Technology Application | Disadvantages of Technology Application | Potential Products for Processing Using Specific Technology | Products Already Available in the World Market Processed Using This Specific Technology |
|---|---|---|---|---|
| | • Food products high in water concentration do not deform due to pressure applied.<br>• In-pack cold pasteurization of food products increasing significantly the shelf-life of foods.<br>• May be applied to liquid, semi-liquid, and not liquid foods.<br>• This technology may be applied to various food industries since significant number of studies have been conducted and cited in the literature demonstrating the potentials for its application<br>• Brined cheeses ripening time reduction<br>• Selective increase of specific enzymes activity<br>• Production of food products with novel rheological characteristics<br>• Energy efficient and environmental friendly technology | | • Pasteurization of fish and fish products<br>• Crustacean meat extraction yield increase<br>• Pasteurization of fruit preparations and jams_retention of fresh-like characteristics<br>• Wines and beers clarification and pasteurization<br>• Pasteurization of deli salads, dips, and ready meals | |

specifying the objective of the new technologies implications identifying the internal and external factors that are favorable and unfavorable to achieving that objective. Identification of SWOTs is important because they can inform later steps in planning process to achieve the objective. This would inform the potential technology users of the importance of the application.

- Strengths: characteristics of the business or project that give it an advantage over others
- Weaknesses: characteristics that place the team at a disadvantage relative to others
- Opportunities: elements that the project could exploit to its advantage
- Threats: elements in the environment that could cause trouble for the business or project

In all new developments of minimal processing, it is not only the improved eating quality of products that is important. The process must be capable of operating in a factory environment and not just in a laboratory with highly qualified staff; it should ensure a financial benefit to the manufacturer; and it should be sufficiently flexible to accommodate a wide range of products, often having short production runs and brief product life cycles (Manvell 1996). In Table 15.3, the SWOT analysis for nonthermal technologies compared to conventional thermal treatment is presented.

### 15.1.6 Technologies Readiness Levels

The interest in food industries is growing over the past few years for the application of novel nonthermal technologies for the production of novel products or superior products compared to conventional ones (longer shelf-life and better quality and organoleptic characteristics). Despite their interest, it is of significant importance for their decision to know the technology readiness level (TRL) of each of these technologies. The TRL is dependent on the research published in the cited literature, of the maturation level of the equipment producing industries for industrial scale equipment and the advantages of the technology application when compared to conventional technologies. The primary purpose of using TRL is to help management in making decisions concerning the development and transitioning of technology. It should be viewed as one of several tools that are needed to manage the progress of research and development activity within an organization. The TRL is a 9-scale scoring of technologies and provides a common understanding of technology status and risk management with regards the technology application; it may be used to make decisions concerning technology funding and transition of technology. Levels 1–4 relate to creative and innovative technologies before or during the lab phase. Levels 5–9 relate to existing technologies and to industrial applicability phase. Based on all the above described, the TRL scoring for each of the presented technologies is depicted in Table 15.4.

**TABLE 15.3**
**SWOT Analysis for Nonthermal Technologies Compared to Conventional Thermal Treatment**

| Strengths | Weaknesses |
|---|---|
| • Significant increase of the shelf-life of nonthermally pasteurized food products<br>• Nutritional and organoleptic characteristics similar to untreated products<br>• Potential of in-pack cold pasteurization avoiding cross-contaminations<br>• Availability of food products at longer distance markets<br>• Production of safe products minimizing the risk for foodborne illnesses<br>• Energy efficient and environmental-friendly technologies | • Different optimal process conditions for treated products<br>• Significantly high capital cost<br>• Training of personnel to handle the equipment of the novel technologies |
| **Opportunities** | **Threats** |
| • The increased shelf-life of the novel technologies treated products resulting in the reduction of food waste can counterbalance the increased capital cost<br>• Opportunities for spin-off industries development as well as for novel technologies equipment production industries. | The effectiveness of the application and adaptation of the novel technologies is a function of:<br>• The investment cost for a food industry<br>• The added cost for processing compared to a conventional product<br>• The consumer acceptance for the new products. |

### 15.1.7 Personnel Skills and Expertise

Over the last decades there is a boost on studies conducted to prove technology feasibility, and enhance development and demonstration as cited in the literature. Despite the significant number of studies published, some critical issues have to be addressed and solved before the industrial application of these nonthermal technologies. One of these issues is the expertise and the skills of the personnel involved in the design, demonstration, and application-adaptation of these technologies in the flow chart of the production lines of food industries. The major issue is the determination of the process conditions optimization taking into account the data cited in the literature, aiming at novel products design and production that will have a long shelf-life and their quality characteristics will be superior to the conventional products (if conventional products exist).

Considering the work that has been carried out over the past years from the nonthermal equipment design and production industries, based on knowledge and experience from the growing application (referring to industrial and lab level), the control and operation of the equipment does not demand any specialized personnel, since all the procedures are mostly automated.

## TABLE 15.4
## TRL Levels for Each of the Nonthermal Technologies

<table>
<tr><th>Technology Readiness level</th><th>Score</th><th>Technology</th></tr>
<tr><td rowspan="6">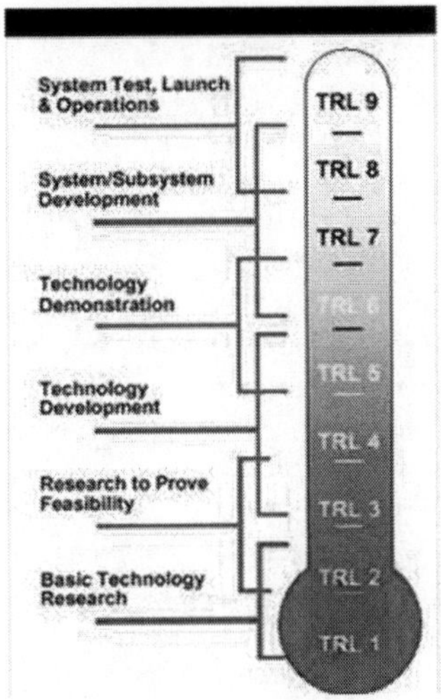

<ul>
<li>TRL 1—basic principles observed</li>
<li>TRL 2—technology concept formulated</li>
<li>TRL 3—experimental proof of concept</li>
<li>TRL 4—technology validated in lab</li>
<li>TRL 5—technology validated in relevant environment (industrially relevant environment in the case of key enabling technologies)</li>
<li>TRL 6—technology demonstrated in relevant environment (industrially relevant environment in the case of key enabling technologies)</li>
<li>TRL 7—system prototype demonstration in operational environment</li>
<li>TRL 8—system complete and qualified</li>
</ul></td><td>9</td><td>High hydrostatic pressure technology:<br>This technology has already been applied industrially for the production of various food products with prolonged shelf-life worldwide. Many research teams study the potentials of its application and present their works by a significant number of published papers cited in the literature. Some of the results may be directly applied in the food industry. Combining the above with the high maturation level of the appropriate HP equipment production industries, the TRL may be scored as TRL = 9.</td></tr>
<tr><td>8</td><td>Pulsed electric fields technology:<br>This technology has been extensively studied over the last decades and the results are promising for its application in an industrial level but it does not seem to have the same success as the HP technology. In the market, the number of companies that can provide the specific industrial equipment for PEF technology is limited (mainly focused in the manufacturing of lab scale equipment). Hence, PEF technology may be scored as TRL = 8.</td></tr>
<tr><td>8</td><td>Ultrasonication technology:<br>As in the case of PEF technology, ultrasonication technology can be applied satisfactory in a pilot scale, but its implementation in industrial level is still at an early stage (TRL = 8).</td></tr>
<tr><td>8</td><td>Ultraviolet IR technology:<br>This technology has been already applied in individual food industries abroad (Greek legislative frame is prohibitory). Although the know-how and the suitable equipment are available, the persuasion of the companies and the consumers for the safety and the advantages of this technology is still pending (TRL = 8).</td></tr>
<tr><td>8</td><td>IR technology:<br>The same applies to this technology as in the ultraviolet IR technology (TRL = 8).</td></tr>
<tr><td>5</td><td>High dense phase carbon dioxide technology:<br>This technology is still under research study. The effect of high dense phase carbon dioxide technology on food quality and safety is being studied, taking into consideration the process parameters. The appropriate equipment is available for industrial applications. This technology may be scored as TRL = 5.</td></tr>
</table>

*continued*

**TABLE 15.4 (continued)**
**TRL Levels for Each of the Nonthermal Technologies**

| Technology Readiness level | Score | Technology |
|---|---|---|
| • TRL 9—actual system proven in operational environment (competitive manufacturing in the case of key enabling technologies; or in space)<br>*Source*: http://ec.europa.eu/research | 5 | *Cold plasma technology:*<br>This technology is still under research. Although the results of the cold plasma technology application on food ingredients are promising, it has not been studied yet on the whole food. It is only available for pilot-scale experiments. This technology may be scored as TRL = 5. |
| | 8 | *Ozonation technology:*<br>This technology has been extensively studied over the last decades and the results are promising for its application in an industrial level but it does not seem to have the same success as other nonthermal technologies (HP technology). In the market, the number of companies that can provide the specific industrial equipment for ozonation technology is limited (mainly focused in the manufacturing of lab-scale equipment). Hence, this technology may be scored as TRL = 8. |

## REFERENCES


Aertsen A., Meersman F., Hendrickx M., Vogel R. and Michiels C.W. 2009. Biotechnology under high pressure: Applications and implications. *Trends in Biotechnology* 27, 434–441.

Ahn J., Balasubramaniam V.M. and Yousef A.E. 2007. Inactivation kinetics of selected heat resistant aerobic and anaerobic bacterial surrogate spores by pressure-assisted thermal processing. *International Journal of Food Microbiology* 113(3), 321–329.

Alexandrakis Z., Katsaros G., Stavros P., Katapodis P., Nounesis G. and Taoukis P. 2014. Comparative structural changes and inactivation kinetics of pectin methylesterases from different orange cultivars processed by high pressure. *Food and Bioprocess Technology* 7(3), 853-867.

Andras C.D., Simandi B., Orsi F., Lambrou C., Missopolinou-Tatala D., Panayiotou C., et al. 2005. Supercritical carbon dioxide extraction of okra (*Hibiscus esculentus* L.) seeds. *Journal of the Science of Food and Agriculture* 85, 1415–1419.

Awad T.S., Moharram H.A., Shaltout O.E., Asker D. and Youssef M.M. 2012. Applications of ultrasound in analysis, processing and quality control of food: A review. *Food Research International* 48(2012), 410–427.

Balasubramaniam V.M., Farkas D. and Turek E.J. 2008. Preserving foods through high-pressure processing. *Food Technology* 62(11), 32–38.

Barbosa-Cánovas G.V. 2001. Developments in pulsed electric fields—USA research and consortium activities. *Proceedings of Electric Field Processing the potential to make a difference*. Campden and Chorleywood Food Research Association. Chipping Campden.

Barbosa-Cánovas G.V. and Sepúlveda D. 2005. Present status and the future of PEF technology. In: Gustavo V. Barbosa-Cánovas, María S. Tapia, M. Pilar Cano, Olga Martín-Belloso and Antonio Martínez, Editors, *Novel Food Processing Technologies*, CRC Press, Boca Raton, USA.

Bhaskaracharya R.K., Kentish S. and Ashokkumar M. 2009. Selected applications of ultrasonics in food processing. *Food Engineering Reviews* 1, 31–49.

Boulekou S.S., Katsaros G.I. and Taoukis P.S. 2010. Inactivation kinetics of peach pulp pectin methylesterase as a function of high hydrostatic pressure and temperature process conditions. *Food and Bioprocess Tech* 3(5), 699–706.

Brown Z.K., Fryer P.J., Norton I.T., Bakalis S. and Bridson R.H. 2008. Drying of foods using supercritical carbon dioxide—Investigations with carrot. *Innovative Food Science and Emerging Technologies* 9(3), 280–289.

Cano M.P., Hernandez A. and De Ancos B. 1997. High pressure and temperature effects on enzymes inactivation in strawberry and orange products. *Food Science* 62, 85–88.

Cheftel J.C. 1992. Effect of high hydrostatic pressure on food constituents: An overview. In: C. Balny, R. Hayashi, K. Heremans and P. Masson, Editors *High Pressure and Biotechnology,* pp. 195–209. John Libbey and Co. Ltd., London.

Chemat F., Huma Z.-e. and Kamran Khan, M. 2011. Applications of ultrasound in food processing: Processing, preservation and extraction. *Ultrasound Sonochemistry* 18, 813–835.

Craven H., Swiergon P., Ng S., Midgely J., Versteeg C. and Coventry M. 2008. Evaluation of pulsed electric field and minimal heat treatments for inactivation of pseudomonads and enhancement of milk shelf-life. *Innovative Food Science and Emerging Technologies* 9, 211–216.

Cullen P.J., Tiwari B.K., O'Donnell C.P. and Muthukumarappan K. 2009. Modelling approaches to ozone processing of liquid foods. *Trends Food Sci. Technol.* 20(3–4), 125–136.

Deeth H., Datta N., Ross A. and Dam X. 2007. Pulsed electric field technology: Effect on milk and fruit juices. In: G. Tewari and V. Juneja, Editors, *Advances in Thermal and Non-Thermal Food Preservation*, Blackwell Publishing, Oxford, UK.

Delgado A.D., Zheng L. and Sun D.-W. 2009. Influence of ultrasound on freezing rate of immersion-frozen apples. *Food and Bioprocess Technology: An International Journal* 2, 263–270.

Eller F.J. and King J. W. 2000. Supercritical carbon dioxide extraction of cedarwood oil: A study of extraction parameters and oil characteristics. *Phytochemistry Analysis* 11(4), 226–231.

FDA. Terms of the 1958 Food Additive Amendment to the Food, Drug and Cosmetic Act.

FDA. 2001. Ozone in Food Processing Applications. Rules and Regulations, Vol. 66, No 123.

Fernandes F.A.N., Oliveira F.I.P. and Rodrigues S. 2008. Use of ultrasound for dehydration of papayas. *Food and Bioprocess Technology: An International Journal* 1(4), 339–345.

Galazka V.B., Dickinson E. and Ledward D.A. 1996. Effect of high pressure on the emulsifying behaviour of b-lactoglobulin. *Food Hydrocolloids* 10, 213–219.

Gomes M.R.A. and Ledward D.A. 1996. Effect of high pressure treatment on the activity of some polyphenoloxidases. *Food Chemistry* 56, 1–5.

Gould G. 2002. The evolution of High Pressure Processing of Foods. In: M. Hendrickx and D. Knorr, Editors *Ultra High Pressure Treatments of Foods*, Food Engineering Series.

Grant S., Patterson M. and Ledward D., 2000. Food processing gets freshly squeezed. *Chemistry and Industry* 24, 55–58.

Grimi N., Mamouni F., Lebovka N., Vorobiev E. and Vaxelaire J. 2011. Impact of apple processing modes on extracted juice quality: Pressing assisted by pulsed electric fields. *J. Food Eng.* 103, 52–61.

Guiavarc'h Y., Segovia O., Hendrickx M. and Van Loey A. 2005. Purification, characterization, thermal and high pressure inactivation of a pectin methylesterase from white grapefruit (*Citrus paradisi*). *Innovative Food Science and Emerging Technologies* 6, 363–371.

Hamburger M., Baumann D., and Adler S. 2004. Supercritical carbon dioxide extraction of selected medicinal plants—Effects of high pressure and added ethanol on yield of extracted substances. *Phytochemical Analysis* J5(1), 46–54.

Hayman M.M., Baxter I., O'Riordan P.J. and Stewart C.M. 2004. Effects of high-pressure processing on the safety, quality, and shelf life of ready-to-eat meats. *Journal of Food Protection* 67(8), 1709–1718.

He H., Adams R.M., Farkas D.F. and Morrissey M.T. 2002. Use of high-pressure processing for oyster shucking and shelf-life extension. *Journal of Food Science* 67(2), 640–645.

Heinz V., Knoch A. and Lickert T. 2009. Product innovation by high pressure processing. *New Food* (2), 43–44.

Heinz V., Toepfl S. and Knorr D. 2003. Impact of temperature on lethality and energy efficiency of apple juice pasteurization by pulsed electric fields treatment. *Innovative Food Science and Emerging Technologies* 4, 167–175.

Herceg Z., Jambrak A.R., Lelas V. and Thagard S.M. 2012. Inactivation of *S. aureus* and *E. coli* in milk by ultrasound. *Food Technology and Biotechnology* 50(1), 46–52.

Hülsheger H. and Niemann E. 1980. Lethal effects of high-voltage pulses on *E. coli* k12. *Radiation and Environmental Biophysics* 18, 281–288.

Johnston D.E., Austin B.A. and Murphy R.J. 1993. Properties of acid-set gels prepared from high-pressure treated skim milk. *Milchwissenschaft* 48, 206–209.

Katsaros G., Tsevdou M., Panagiotou T. and Taoukis P. 2010. Kinetic study of high pressure microbial and enzyme inactivation and selection of pasteurization conditions for Valencia Orange Juice. *International Journal of Food Science and Technology* 45(6), 1119–1129.

Katsaros G.I., Giannoglou M.N. and Taoukis P.S 2009a. Kinetic study of the combined effect of high hydrostatic pressure and temperature on the activity of *L. bulgaricus* aminopeptidases. *Journal of Food Science* 74 (5), E219–E225.

Katsaros G.I., Katapodis P. and Taoukis P.S 2009b. High hydrostatic pressure inactivation kinetics of the plant proteases ficin and papain. *Journal of Food Engineering* 91(1), 42–48.

Katsaros G.I., Katapodis P. and Taoukis P.S. 2009c. Modeling the effect of temperature and high hydrostatic pressure on the proteolytic activity of kiwi fruit juice. *Journal of Food Engineering* 94(1), 40–45.

Kayes M.M., Critzer F.J., Wintenberg K.K., Roth J.R., Thomas C.M. and Golden D.A. 2007. Inactivation of food born pathogens using a one atmosphere uniform glow discharge plasma. *Food Borne Pathogens and Disease* 4(1), 50–56.

Kim J.G., Yousef A.E. and Dave S. 1999. Application of ozone for enhancing the microbiological safety and quality of foods: A review. *Journal of Food Protection* 62(9), 1071–1087.

Knorr D., Zenker M., Heinz V., Lee D.U. 2004. Applications and ultrasonics in food potential of processing. *Trends in Food Science & Technology* 15, 261–266.

Kozempel M., Annous B.A., Cook R., Scullen O.J. and Whiting R. 1998. Inactivation of microorganisms with microwaves at reduced temperature. *Journal of Food Protection* 61(5), 582–585.

Li B. and Sun D.-W. 2002. Novel methods for rapid freezing and thawing of foods—A review. *Journal of Food Engineering* 54, 175–182.

Li B. and Sun D.-W. 2003. Microstructural change of potato tissues frozen by ultrasound-assisted immersion freezing. *Journal of Food Engineering* 57(14), 337–345.

López-Fandino R., Carrascosa A.V. and Olano A. 1996. The effects of high pressure on whey protein denaturation and cheese making properties of raw milk. *Journal of Dairy Science* 79, 929–36.

Malone A.S., Wick C., Shellhammer T.H. and Courtney P.D. 2003. High pressure effects on proteolytic and glycolytic enzymes involved in cheese manufacturing. *Journal of Dairy Science* 86(4), 1139–46.

Manvell C. 1996. Opportunities and problems of minimal processing and minimally processed foods. EFFOST/GDL Congress on Minimal Processing of Foods, November 6–8, 1996.

Mason T.J., Paniwnyk L. and Lorimer J.P. 1996. The uses of ultrasound in food technology. *Ultrasonics Sonochemistry* 3, 253–260.

Miller R., Jeffrey W., Mitchell D. and Elasri M. 1999. Bacterial responses to ultraviolet light. *American Society for Microbiology* 65(8), 535–541.

Mulet A., Cárcel J.A., García-Pérez and Riera E. 2011. Ultrasound-assisted hot air drying of foods. H. Feng, G.V. Barbosa-Cánovas and J. Weiss, Editors *Ultrasound Technologies for Food and Bioprocessing*, pp. 511–534. Springer Science + Business Media, LLC, New York, USA.

Nguyen B.L., Van Loey A., Fachin D., Verlent I. and Hendrickx I.M. 2002. Purification, characterization, thermal, and high pressure inactivation of pectin methylesterase from bananas (cv Cavendish). *Biotechnology and Bioengineering* 78(6), 683–691.

Ohlsson T. and Bengtsson N. 2002. *Minimal Processing Technologies in the Food Industry*. Woodhead Publishing, Cambridge, England, 13–14.

Patil S., Bourke P., Frias J., Tiwari B., and Cullen P.J. 2009. Inactivation of Escherichia coli in orange juice using ozone. *Innovative Food Science and Emerging Technologies* 10(4), 551–557. http://dx.doi.org/10.1016/j.ifset.2009.05.011

Patist A. and Bates D. 2008. Ultrasonic innovations in the food industry: From the laboratory to commercial production. *Innovative Food Science and Emerging Technologies* 9, 147–154.

Piyasena P., Mohareb E. and McKellar R.C. 2003. Inactivation of microbes using ultrasound: A review. *International Journal of Food Microbiology* 87, 207–216. KBBE.2013.2.5-02.

Raja M. Moman and Hmeda Najmaldeen. 2010. The bactericidal efficacy of cold atmospheric plasma technology on some bacterial strains. *Egypt. Acad. J. Biol. Sci.* 2(2), 43–47.

Raoult-Wack A.L. 1994. Recent advances in the osmotic dehydration of foods. *Trends Food Sci. Technol* 5, 255–260.

Raso J. and Heinz V. Editors 2006. Pulsed electric fields technology for the food industry. *Food Engineering Series*. Springer-Verlag, Heidelberg.

Sastry S. and Barach J. 1995. Ohmic and inductive heating. *Journal of Food Science* 65(Suppl), 42–46.

Sitzmann W. 1995. High-voltage techniques for food preservation. In: W. Gould, Editor, *New Methods of Food Preservation*. Blackie Academic and Professional, London, UK.

Solomon O., Svanberg U. and Sahlstrom A. 1995. Effect of oxygen and fluorescent light on the quality of orange juice during storage at 8°C. *Food Chemistry* 53(4), 363–368.

Steenstrup L.D. and Floros J.D. 2004. Inactivation of *E. coli* 0157:H7 in apple cider by ozone at various temperatures and concentrations. *J. Food Proc. Preserv.* 28:103–116.

Stoforos N.G and Taoukis P.S. 2001. High pressure process evaluation through kinetic modeling. In: Welti-Chanes J, Barbosa-Canovas V, Aguilera J.M., Lopez-Leal L.C., Wecshe-Ebeling P., Lopez-Malo A., Palou –Garcia E. Editors. *Proceedings of the Eighth International Congress of Engineer*ing and Food, ICEF8. April 9–13, 2000, Puebla, Mexico. Technomic Publishing Co. Inc., Lancaster, Basel. Vol. II, pp. 1437–1441.

Talens P., Martínez-Navarrete N., Fito P. and Chiralt A. 2002. Changes in optical and mechanical properties during osmodehydrofeezing of kiwi fruit. *Innovative Food Science and Emerging Technologies* 3, 191–199.

Tiwari B.K., O'Donnell C.P., Muthukumarappan K. and Cullen P.J. 2009a. Anthocyanin and colour degradation in ozone treated blackberry juice. *Innov. Food Sci. Emerg* Technol. 10(1), 70–75.

Tiwari B.K., O'Donnell C.P., Patras A., Brunton N. and Cullen P.J. 2009b. Effect of ozone processing on anthocyanins and ascorbic acid degradation of strawberry juice. *Food Chem* 113(4), 1119–1126.

Tiwari B.K., Muthukumarappan K., O'Donnell C.P. and Cullen P.J. 2008. Kinetics of freshly squeezed orange juice quality changes during ozone processing. *J. Agric. Food Chem.* 56(15), 6416–6422.

Toepfl A., Mathys Heinz V. and Knorr D. 2006. Review: Potential of high hydrostatic pressure and pulsed electric fields for energy efficient and environmentally friendly food processing. *Food Reviews International* 22, 405–423.

Vega-Mercado H., Martin-Belloso O., Chang F.-J., Barbosa-Cánovas G. and Swanson B. 1996. Inactivation of *Escherichia coli* and *Bacillus subtilis* suspended in pea soup using pulsed electric fields. *Journal of Food Processing and Preservation* 20 (6), 501–510.

Vercet A., Burgos J., Crelier S. and Lopez-Buesa, P. 2001. Inactivation of proteases and lipases by ultrasound. *Innovative Food Science & Emerging Technologies* 2, 139–150.

Victorin K. 1992. Review of the genotoxicity of ozone. *Mutation Research* 277, 221–238.

Zhang G., Sofyan M. and Hamaker B.R. 2008. Slowly digestible state of starch: mechanism of slow digestion property of gelatinized maize starch. *Journal of Agricultural and Food Chemistry* 56, 4695–702.

Zhang Q., Barbosa-Cánovas G. and Swanson B. 1995. Engineering aspects of pulsed electric field pasteurization. *Journal of Food Engineering* 25(2), 261–281.

Zhao W., Shi H. and Wang D. 2004. Modeling of mass transfer characteristics of bubble column reactor with surfactant present. *Journal of Zhejiang University Science* 5(6), 714–720.

# Index

## D

## G

## H

## M

## O



## P

## R



## S

## Y



## Z